# DATA SOURCEBOOK FOR FOOD SCIENTISTS AND TECHNOLOGISTS

# DATA SOURCEBOOK FOR FOOD SCIENTISTS AND TECHNOLOGISTS

*Edited by*

Y. H. Hui

VCH

Y. H. Hui
President
American Food and Nutrition Center
Cutten, California 95534-0034, USA

**Library of Congress Cataloging-in-Publication Data**

Data sourcebook for food scientists and technologist / editor, Y. H. Hui.

p. cm.
Includes bibliographical references and index.
ISBN 1-56081-009-2
1. Food—Composition. I. Hui, Y. H. (Yiu H.)
TX531.D37 1991
664—dc20 91-32078
CIP



**British Library Cataloguing-in-Publication Data**

Data sourcebook for food scientists and technologists
I. Hui, Y. H. (Yiu H.), *1940-*
016.6411

ISBN 1-56081-0092





Printed in the United States of America

ISBN 1-56081-009-2 VCH Publishers
ISBN 3-527-28044-8 VCH Verlagsgesellschaft

Printing History:
10 9 8 7 6 5 4 3 2 1

Published jointly by:

VCH Publishers, Inc.
220 East 23rd Street
Suite 909
New York, NY 10010

VCH Verlagsgesellschaft mbH
P. O. Box 10 11 61
D-6940 Weinheim
Federal Republic of Germany

VCH Publishers (UK) Ltd.
8 Wellington Court
Cambridge CB1 1 HZ
United Kingdom

# PREFACE

This handbook provides domestic and foreign food scientists and technologists easy access to certain useful scientific, technical, and legal information most of which is normally distributed over many documents in the United States. As a reference book, it covers the following groups of information:

1. Functional and economical chemicals used in processing foods and on raw agricultural products respectively.
2. Classification and profiles of pathogens ingested in foods, beverages, and drinking water.
3. The application of biotechnology in food processing and related areas.
4. Product categories: fruits and vegetables, meat and poultry, bakery products, and alcoholic beverages.
5. U.S. government's recommended nomenclature for food fishes.
6. Safety of workers in food processing plants.
7. Sanitation in food processing.
8. Laws and regulations governing the safety and economic attributes of domestic and imported food products.

Because this handbook covers a variety of information, it can be used by professionals in government, industry, and academia. Further, students and instructors in food science, technology, engineering and related fields will benefit from this book when they do library research for term papers, tests, and research projects.

I personally believe that this handbook will serve well all domestic and foreign libraries associated with:

1. Academic food science, technology, and engineering programs.
2. Food companies, ingredient and equipment suppliers, and commercial laboratories.
3. Government research laboratories.

In general, this book will be useful if one needs answers to those questions listed under different areas in the table presented at the end of this preface.

Those questions are not meant to be exhaustive. The data in this book will provide answers to many other scientific, technical, and legal questions in food processing and related areas. Interested persons will find the complete and detailed table of contents (in the front matter) useful in determining how this book can serve them. However, the above questions serve to provide the reader with a general idea of coverage in this handbook.

Perhaps, you may agree with me on the unique qualities of this handbook if we consider the following perspectives:

1. Information management: data from literature reprocessed for easy access and understanding.
2. Interdisciplinary approach: covering many disciplines within food science and technology.
3. Areas covered: chemistry, biology, nutrition, manufacturing processes, formulations, selected products, laws and regulations, definitions, etc.
4. Emphasis: processes, practices, and chemical substances legally permitted and recommended for usage and application in the manufacturing of foods and beverages.

Most of the information, though available in the literature, has not been assembled or appeared together in one source or in any current reference book in the market.

All legal language has been reworded or removed to facilitate easy understanding. Directions for obtaining original materials are provided. Since some of the information has been extracted from very costly legal documents, its easy access here has eliminated the need of purchasing original publications that contain much unneeded data.

Let me explain that, although I determined what data appear in this handbook, I am not the author of the information. Rather, I have served as an editor and a manager of information. After having determined what information to include, I identified, extracted, collated, and edited it. Therefore, I would like to emphasize the following:

1. Most of the materials have been abstracted from documents that are either in public domain or from copyrighted materials. For all of them, proper reference and credit are given. Where applicable, permission to do so is indicated.
2. This book contains a fair amount of technical and legal information, most of which has been paraphrased or reorganized from original documents, and the interpretation or editing are entirely the editor's. Every reasonable effort has been made to assure accuracy in the process. Because of the voluminous amount of materials involved, this editor does not assume any liability for errors or omissions. Persons using this source should always consult the original document referenced to obtain further details and should use their own interpretation when necessary.
3. My role in handling certain legal information is management and I assume no legal expertise. The legal information contained in this book cannot and should not be used as *prima facie* evidence in a court of law. If such is needed, consult the original documents or an attorney-at-law for professional assistance.

Y. H. Hui
November 1991
Cutten, California

**A Table Listing Examples of Questions That Will Be Answered by Data in This Book**

Food Chemicals

1. What are the functional and chemical characteristics of common chemicals used in food processing?
2. How may I ascertain if one or more of a group of chemicals is used as a pesticide (or related formulation) on raw agricultural products? Where is the information located?
3. Has a tolerance been established for this pesticide (or a related chemical) in a specific processed food? Where is the information located?

Biotechnology

1. What are the U.S. government agencies responsible for regulating biotechnology products and proposals on research and development?
2. What are some examples of applying biotechnology to food processing and related areas?
3. What are the definitions for some terms commonly encountered in biotechnology?
4. How are the fermentative productions of amino acids related to such factors as fundamental considerations; enzyme processes; application of gene technology; auxotrophic mutants, regulatory mutants and auxotrophic-regulatory mutants; and biosynthetic precursors?
5. What types of fermented foods are available in industrial societies and third world countries? What are the microorganisms used? What are the substrates? What are the products?
6. What are the fundamentals of manufacturing fermented dairy products?
7. What are the classifications and sources of enzymes? What are the representative industrial enzymes of the six major classes? What are the profiles of enzymes with industrial-scale applications? What are the major classes of enzymes of commercial interest and some of their sources?
8. What are the biology and properties of some major enzymes used in food processing and related areas?
9. What are some examples of applying gene technology to food processing enzymes?
10. When using fermentation to produce enzymes, what are the procedures? Microorganisms used? Techniques employed? Culture and media conditions? Requirements for fermentation?

11. When using enzymes in food processing, what enzymes are matched with what processing steps? What are the profiles of major enzymes used in food processing? What are the typical trade names and manufacturers for enzymes used in food processing?
12. How are yeasts applied in food processing?
13. What are the enzymes used in diagnostics and food analysis? What are the enzymes used in the processing of various food products?
14. What are the differences and similarities in the production of the fermented products–beer, whiskey, and bread?
15. When using immobilized enzymes, what are the considerations in relation to such factors as fundamental requirements, advantages, classification of methods, characteristics of an enzyme carrier, classification of enzyme matrixes, classes of immobilization techniques, Michaelis constants, optimum pH, and activation energies?
16. When applying the immobilized cell systems, what are the considerations in relation to such factors as fundamental requirements; nature of the catalytic reaction; organic and inorganic supports; available methods; examples of cells entrapped in gels, polyacrylamide, photo-crosslinking resin prepolymers, urethane polymers, collagen, gelatin, agar or agarose, carrageenan, alginate, or membrane reactors?
17. What are the industrial applications of immobilized cells?
18. What are some examples of immobilized cell systems activated *in situ*? Immobilized microbial cells utilized in analysis? Immobilized plant cells?
19. What are the iron complex products obtained from microorganisms?
20. What are the lipid contents of some microorganisms?
21. In relation to microbial products, what are the considerations in relation to such factors as classification, metabolic functions, and sources of coenzymes?
22. What are the various microbial $\alpha$-glucosidase inhibitors?

Foodborne Diseases

1. If I have a description of the symptoms of a suspected food poisoning, what type of *tentative* diagnosis can I make about the causative agent, sampling, prevention, and so on?
2. What are the characteristic features of water- and foodborne pathogenic bacteria and parasites?
3. What are some well-known unsafe herbs? Why are they unsafe?
4. What are the definitions for various terms associated with food poisoning?

**Table Listing Questions (continued)**

Fruits and Vegetables

1. What is the water content for each of the major fruits and vegetables?
2. What is the pH for each of the major fruits and vegetables?
3. What are some characteristic features of those organisms that can spoil canned fruits or vegetables?
4. What are the storage requirements for fresh fruits and vegetables in relation to temperature, humidity, time, water content, and specific heat?
5. What are some characteristic features of using controlled atmospheres to store fruits and vegetables?
6. Which fresh and processed fruit or vegetable has a standard for grades promulgated by the U.S. Department of Agriculture?
7. How is a common fruit or vegetable processed?

Meat and Poultry Products

1. What are the general considerations for processing meats?
2. What raw ingredients are used to prepare processed meats?
3. Which types of meats are used in comminuted processed meats?
4. What are the different categories of perishable cooked and cured meat products?
5. What are the basic considerations in curing and smoking meat products?
6. What are the basic considerations for canning meats?
7. What are the formulas for curing some meat and poultry products?
8. What are the formulas for canning some meat and poultry products?
9. What are the commercial and legal classifications for sausages?
10. What different types of animal casings are used for making sausages?
11. What factors must be considered in meat fermentation?
12. What are the microorganisms used in preparing starter cultures for curing meat products?
13. What are the formulas for different types of sausages: ground, summer, dry, emulsion, specialty, and so on?
14. What are the defect criteria established by the federal government for meat from animal carcasses?
15. What are the definitions for special terms used in meat and poultry labeling?
16. What food additives are used in meat processing? What are the federal requirements for their usage?
17. For which meat products has the federal government issued definitions and standards? What are the contents of definitions and standards for

individual processed meat products? What is the required percentage of meat in processed meat products sold in the United States?

18. What food additives are used in poultry processing? What are the federal requirements for their usage?
19. For which poultry products has the federal government issued definitions and standards? What are the contents of definitions and standards for individual processed poultry products? What is the minimal or maximal percentage requirement of poultry meat in processed poultry products?

Bakery Products

1. What are the fundamental considerations in the manufacture of baked goods?
2. What are the pH values of various bakery products?
3. What are the baking times and temperatures for various baked goods?
4. What are the factors causing external and internal bread faults?
5. What are the formulas for some rolls, cakes, cookies, and variety baked goods?
6. What are the United States enrichment standards for cereals and flours?
7. What are the different processes for making sour doughs?

Fish List

1. What is the market name for a fish if one knows its scientific name?
2. What is the scientific name for a fish if one knows the market or regional name?

Alcoholic Beverages

1. What are the fundamental considerations in manufacturing beer?
2. What are the classical beer types throughout the world?
3. What are the legal classes and types of beer in the United States?
4. What are the legal definitions for major terms used in the manufacture of beer in the United States?
5. What are the flows of processing materials in malting, brewing, fermentation, and finishing?
6. What are the components of a quality assurance program in a brewery?
7. What are the fundamental considerations in the manufacture of wine and brandy?
8. What are the legal classes and types of wine manufactured in the United States?

**Table Listing Questions (continued)**

9. What are the legal definitions for major terms used in the manufacture of wine in the United States?
10. What chemical substances are authorized for the treatment of wine and juice in the United States? What are the limitations?
11. What materials are authorized for treatment of distilling material in the United States and what are the limitations?
12. What processes are authorized for the treatment of wine, juice, and distilling material in the United States and what are the limitations?
13. What factors must be considered in the manufacturing of distilled beverage spirits?
14. What are the legal definitions for some major terms used in the manufacture of distilled beverage spirits in the United States?
15. What are the legal classifications of liquor in the United States?
16. What natural flavoring substances are permitted in alcoholic beverages? What are the limitations?
17. What legal natural flavoring substances and natural substances are used in conjunction with flavors in the manufacture of alcoholic beverages?

Workers Safety And Sanitation in Food Processing Plants

1. What are the safety hazards in a food processing plant? What are the safety controls?
2. What are the occupational hazards associated with various jobs in food processing and related areas, for example, agricultural migrant workers, meat processing, eateries?
3. What are the biological and chemical agents commonly associated with injuries to and diseases of workers in food processing plants?
4. What are the exposure standards for chemicals associated with the food processing industry?
5. What is the hazard analysis critical control point system in food processing? What are the seven principles? What are hazards? What are risk assessments?
6. What cleaning and sanitizing operations are used in a food service or food processing establishment?
7. When using chemical and physical agents in a cleansing operation, what are the factors to be considered in relation to efficiency, properties, etc.?
8. What are the FDA food defect action levels? What are the defects? What are the foods?
9. What are the FDA food action levels for poisonous or deleterious substances? What are the poisonous substances? What are the foods?

10. What legal food color additives are exempt from certification?
11. What legal food color additives are subject to certification?
12. What are the legal uses and limitations of food preservatives?
13. What are the synthetic flavoring substances and adjuvants permitted in food?

Laws and Regulations

1. What are the responsibilities of the U.S. Food and Drug Administration (FDA) in relation to the relevant federal statutes?
2. What are the responsibilities of the FDA in safeguarding health and economic integrity of the food system in the United States?
3. What types of regulations does the FDA promulgate?
4. What types of enforcement actions does the FDA use?
5. What administrative and legal terms are used by the FDA?
6. What are the food additive regulations promulgated by the FDA in relation to listing, classification, limitations, categories (e.g., generally recognized as safe), and other factors?
7. What are the food standards promulgated by the FDA?
8. What does the FDA Inspection Operations Manual cover?
9. What does the FDA Regulatory Procedures Manual cover?

# ACKNOWLEDGMENTS

I wish to express my sincere appreciation to all my colleagues and friends for the generous help given during the preparation of this book.

Professor P. Stanfield is a professional colleague who has contributed much to the completion of this book and she knows that I am grateful.

Part of the contents of this text have been obtained from government documents, most of which are unauthored. All materials used are credited with the specific sources at the beginning or end of each section or chapter. Where applicable, the source is stated in the footnote of a table. Although all such writings are in public domain and can be reproduced where the need arises, I would like to thank all those government experts who have painstakingly composed the voluminous amount of laws and regulations. As a "manager" of information, I would not have been able to develop and complete the book without their anonymous contribution.

The chapter on biotechnology has been completely derived from publications of VCH Publishers, Inc. and I appreciate their permission to do so.

The continuous support of Dr. Stanley Kudzin, my editor at VCH, Inc., has been most gratifying. The readers are the best judge of the professionalism and excellent work of the production staff of VCH, Publishers, Inc.

Finally, I am deeply grateful to my family for their indispensable and long-enduring patience and support.

To all who have assisted in the preparation of this book, I express gratitude and recognize that factual errors which may appear in the text are solely my responsibility.

# CONTENTS

**Chapter Headings**

## Chapter Headings and Subheadings

### Chapter 1 Functional and Economic Food Chemicals



### Chapter 2 Biotechnology

Part II Product Categories Data

Chapter 4: Fruits and vegetables

Unit II Poultry Products

## Part C Safety, Sanitation, and Regulations

### Chapter 9 Safety and Sanitation

## Chapter 10 Laws and Regulations

# PART A
# CHEMICAL AND BIOLOGICAL DATA

# CHAPTER 1
# FUNCTIONAL AND ECONOMIC FOOD CHEMICALS

In the United States, the use of chemicals in the processing of foods and beverages and the protection of raw agricultural products is regulated by U.S. federal agencies including the U.S. Environmental Protection Agency (USEPA), the U.S. Food and Drug Administration (USFDA), and the U.S. Department of Agriculture (USDA).

This chapter provides two main groups of information:

1. The major properties of the most common functional food chemicals permitted to be used in the processing of foods and beverages.
2. A listing of three groups of pesticide chemicals. The USEPA has established tolerances for two of them: raw agricultural products; and processed foods and beverages. No tolerance for residues has been established for the third group which includes inert ingredients used in those chemical formulations that are applied to raw agricultural products.

The citation for U.S. legal documents follows the standard format. For example "40 CFR 180" means Volume 40 of the U.S. Code of Federal Regulations, Part 180. All legal federal documents are for sale by the U.S. Government Printing Office, Washington, D.C.

## 1.1. Functional Food Chemicals

**Table 1.1.1. Properties of Major Chemicals Used in Food Processing**

Usual or common names
1. Usage
2. Description
3. General formula

Acacia (Gum arabic)
1. Emulsifier, stabilizer, thickener
2. Dried, gummy substance
3. A polysaccharide

Acetaldehyde (Ethanol)
1. Flavoring agent
2. Liquid, colorless; flammable; characteristic odor
3. $CH_3CHO$

Acetanisole (*p*-Methoxyacetophenone)
1. Flavoring agent
2. Solid, pale yellow; with hawthorn-like odor
3. $C_9H_{10}O_2$

Acetic Acid
1. Glacial acidifier, flavoring agent
2. Liquid, clear, colorless; pungent and acrid odor; acidic flavor when diluted
3. $CH_3COOH$

Acetoin (Acetyl methyl carbinol; Dimethyl ketol; 2-Hydrox-2-butanone)
1. Flavoring agent
2. Monomer (colorless, pale liquid; characteristic buttery odor), dimer (white crystalline powder)
3. $CH_3CH(OH)COCH_3$

Acetone (2-Propanone; Dimethyl ketone)
1. Extraction solvent
2. Liquid; clear, colorless; volatile; flammable; characteristic odor
3. $CH_3COCH_3$

Acetone Peroxides
1. Bleaching, maturation, dough conditioning agents
2. Usually solid containing an edible carrier
3. Although mainly 2,2-hydroperoxypropane, usually a mixture of chemicals

Acetophenone (Methyl phenyl ketone)
1. Flavoring substance
2. Liquid; colorless; sweet and pungent odor
3. $C_6H_5COCH_3$

**Table 1.1.1. (continued)**

Usual or common names
1. Usage
2. Description
3. General formula

Acetylated Monoglycerides
1. Emulsifier, lubricant, solvent, coating agent, texture-modifying agent
2. Liquid (thin, clear) to solid (no color to pale yellow); odor similar to acetic acid; bland taste
3. Glycerin esters (complete and incomplete), acetic acid, and fatty acids (edible and capable of forming triglycerides)

Adipic Acid (Hexanedioc acid; 1,4-Butanedicarboxylic acid)
1. Neutral pH, buffering and neutralizing agent
2. Solid, crystalline, nonhygroscopic, white
3. $HOOC(CH_2)_4COOCH$

Agar
1. Stabilizer, thickener, emulsifier
2. Solid, agglutinated strips to powder; pasty taste; slight, characteristic odor
3. Dried hydrophyllic polygalactoside, sulfuric acid ester of a linear galactan

DL-Alanine (DL-2-Aminopropanoic acid)
1. Amino acid (nonessential), as nutrient and dietary supplement
2. Powder; white, odorless, crystalline; sweet taste
3. $Ch_3CH(NH_2)COOH$

L-Alanine (L-2-Aminopropanioc acid)
1. Nonessential amino acid, a nutrient, dietary supplement
2. Powder; white, odorless, crystalline; sweet taste
3. $CH_3CH(NH_2)COOH$

Alginic Acid
1. Emulsifier, stabilizer, thickener
2. Fibrous powder; white to white-yellow; no odor; no taste
3. $(C_6H_8O_6)n$, a carbohydrate in pyransoe ring form, hydrophilic and colloidal

Allyl Cyclohexanepropionate (Allyl-3-cyclohexanepropionate)
1. Flavor
2. Liquid; colorless; pineapple odor
3. $C_{12}H_{20}O_2$

Allyl Hexanoate (Allyl caproate)
1. A flavor
2. Liquid; pale yellow; strong pineapple odor
3. $CH_3(CH_2)COOCH_2CH{=}CH_2$

Allyl Ionone
1. Flavor

2. Liquid; yellow; strong pineapple-like odor
3. $C_{16}H_{24}O$

Allyl Isothiocynate (Volatile mustard oil)
1. Flavor
2. Liquid; pale yellow or no color; pungent odor; acrid taste; strongly refractive index
3. $CH_2=CH–CH_2–N=C=S$

Almond Oil (Bitter almond oil free from prussic acid)
1. Flavor
2. Liquid; pale yellow or colorless; odor similar to almond; mild, slightly astringent taste
3. Triglycerides and fatty acids

Aluminum Ammonium Sulfate (Ammonium alum)
1. Buffering agent
2. Solid, crystals, granules, or powder; colorless; odorless; sweet, astringent taste; solutions acidic to litmus
3. $AlNH_4(SO_4)_2 \cdot 12H_2O$

Aluminum Potassium Sulfate (Potassium alum)
1. Buffering and firming agent
2. Crystals or powder; no color; no odor; sweet, astringent taste; acidic to litmus in solution
3. $AlK(SO_4)_2 \cdot 12H_2O$

Aluminum Sodium Sulfate (Soda alum; Sodium alum)
1. Buffering and firming
2. Crystals, granules, or powder; no color; no odor; saline, astringent taste; may be hydrated
3. $AlNa(SO_4)_2$, sometimes $AlNa(SO_4)_2 \cdot 12H_2O$

Aluminum Sulfate
1. Firming agent
2. Powder or crystalline; white; no odor; sweet taste, mildly astringent; available as a 50% water solution of $Al_2(SO_4)2 \cdot 18H_2O$, sometimes slightly colored and turbid; acid to litmus
3. Hydrated form freely soluble in water; anhydrous form slowly soluble in water; $Al_2(SO_4)_2 \cdot xH_2O$

Ambrette Seed Oil
1. Flavor
2. Liquid; yellow to amber and clear; musky odor
3. Triglycerides and fatty acids

Ammonium Alginate (Algin)
1. Emulsifier, stabilizer, thickener
2. Powder, fibrous or granular; white to yellow
3. $(C_6H_7O_6NH_6)_n$, ammonium alginate

**Table 1.1.1. (continued)**

| Usual or common names<br>1. Usage<br>2. Description<br>3. General formula |
|---|
| Ammonium Bicarbonate Alkali<br>1. Leavening agent<br>2. Crystals or powder; white<br>3. $NH_4HCO_3$ |
| Ammonium Carbonate (A mixture of ammonium bicarbonate and ammonium carbamate)<br>1. Buffering agent<br>2. Powder or translucent mass; white; undergoing changes upon exposure to air (opaqueness followed by conversion to ammonium bicarbonate and the release of carbon dioxide and ammonia)<br>3. $NH_4HCO_3$ and $NH_2COONH_4$ (ratios vary) |
| Ammonium Chloride (Sal ammoniac)<br>1. Baking aid (dough conditioner, yeast food)<br>2. Crystalline white powder; mildly hygroscopic; saline, sodium chloride taste<br>3. $NH_4Cl$ |
| Ammonium Hydroxide (Ammonia water, Ammonia solution)<br>1. Alkali<br>2. Solution of ammonia ($NH_3$) in water; clear and colorless; ammonia vaporizes rapidly when exposed to air.<br>3. Aqueous; $NH_4OH$ |
| Ammonium Phosphate, Dibasic (Diammonium phosphate)<br>1. Buffering agent, dough conditioner, leavening agent, yeast food<br>2. Crystalline powder or granules; white; no odor; taste similar to sodium chloride (saline)<br>3. $(NH_4)_2HPO_4$ |
| Ammonium Phosphate, Monobasic (Monoammonium phosphate)<br>1. Buffering agent, dough conditioner, leavening agent, yeast food<br>2. Crystalline powder or granules; white; no odor<br>3. $NH_4HPO_4$ |
| Ammonium Saccharin (1,2-Benziisothiazolin-3-one 1,1-dioxide ammonium salt)<br>1. Artificial sweetener<br>2. Crystalline or powder; white; extremely sweet<br>3. $C_7H_8N_2O_3S$ |
| Ammonium Sulfate<br>1. Baking yeast food, dough conditioner<br>2. Crystalline or powder; white<br>3. $(NH_4)_2SO_4$ |

α-Amylcinnamaldehyde (Amylcinnamaldehyde)
1. Flavoring
2. Liquid; yellow; odor similar to jasmine
3. $C_{14}H_{18}O$

Amyris Oil (West Indian sandalwood oil)
1. Flavoring
2. Liquid,viscous; clear, pale yellow; odor resembling sandalwood
3. $C_{10}H_{12}O$

Angelica Root Oil
1. Flavoring
2. Liquid; pale yellow to deep amber; pungent odor, bittersweet taste
3. Triglycerides and fatty acids

Angelica Seed Oil
1. Flavoring
2. Liquid; pale yellow; taste sweeter and aroma more delicate than angelica root oil.
3. Triglycerides and fatty acids

Anise Oil
1. Flavoring
2. Liquid; no color to pale yellow; highly refractive; anise odor and taste.
3. Triglycerides and fatty acids

Anisole (Methylphenyl ether)
1. Flavoring
2. Liquid; no color; phenolic and anise-like odor
3. $C_7H_8O$

Anisyl Acetate (*p*-Methoxybenzyl acetate)
1. Flavoring
2. Liquid; colorless to pale yellow; flowery and fruity odor
3. $C_{10}H_{12}O_3$

Anisyl Alcohol
1. Flavoring agent
2. Liquid; no color or pale yellow; flowery odor
3. $C_8H_{10}O_2$

L-Arginine (L-1-Amino-4-guanidovaleric acid)
1. Amino acid, may be essential nutrient, diet supplement
2. Solid, crystalline or powder; white; alkaline; capable of removing carbon dioxide from air
3. $C_6H_{14}N_4O_2$

L-Arginine Monohydrochloride (L-1-Amino-4-guanidovaleric acid monohydrochloride)
1. Amino acid, may be essential nutrient, diet supplement
2. Solid, crystalline or powder; white; odorless

**Table 1.1.1. (continued)**

Usual or common names
1. Usage
2. Description
3. General formula

3. $C_6H_{14}N_4O_2 \cdot HCI$

Ascorbic Acid (Vitamin C; L-Ascorbic acid)
1. Nutrient, diet supplement, antioxidant, preservative
2. Solid, crystalline, powder; white, slightly yellow
3. $C_6H_8O_6$

Ascorbyl Palmitate (Pammitoyl L-ascorbic acid)
1. Antioxidant
2. Powder; white to pale yellow; odor similar to citrus
3. $C_{22}H_{38}O_7$

L-Asparagine (L-$\alpha$-Aminosuccinamic acid)
1. Amino acid, nutrient, dietary supplement
2. Crystalline (powder); white; MP ~234°C (453°F); acid to litmus in solution
3. $H_2NOCCH_2CH(NH_2)COOH$

Aspartame (*N*-L-$\alpha$-Aspartyl-L-phenylalanine 1-methyl ester)
1. Nonnutritive sweetener
2. Crystalline powder; white; no odor; taste sweet
3. $C_{14}H_{18}N_2O_5$

DL-Aspartic Acid (DL-Amino succinic acid)
1. Amino acid, nutrient, dietary supplement
2. Crystalline; white or colorless; no odor; acidic taste , MP and decomposition point 280°C (536°F)
3. $HOOCCH_2CH(NH_2)COOH$

L-Aspartic Acid (L-Aminosuccinic acid)
1. Amino acid, nutrient, dietary supplement
2. Crystalline or powder; white; no odor; slightly acidic taste; MP ~270°C (518°F)
3. $HOOCCH_2CH(NH_2)COOH$

Azodicarbonamide
1. Flour-maturing agent
2. Crystalline powder; yellow, orange, red; melting and decomposition point >180°C (356°F)
3. $C_2H_4N_4O_2$

Balsam Peru Oil
1. Flavoring
2. Liquid, viscous; yellow, pale green; sweet, balsamic odor
3. Triglycerides and fatty acids

Basil Oil
1. Flavoring
2. Liquid; light yellow; spicy odor
3. Triglycerides and fatty acids

Bay Oil (Myrcia oil)
1. Flavoring
2. Liquid; yellow to brownish yellow; pleasant aromatic odor; spicy and pungent taste; acid to litmus in alcohol solutions
3. Triglycerides and fatty acids

Beeswax, White (White wax)
1. Polish and glaze for candy, flavoring
2. Solid; white, yellowish; faint and unique odor
3. Complex formula

Beeswax, Yellow (Yellow wax)
1. Flavoring, polish and glaze for candy
2. Solid; yellow, gray, brown; odor pleasant and resembling honey; brittle in cold, pliable at (>35°C (95°F)
3. Complex formula

Benzaldehyde (Benzoid aldehyde)
1. Flavoring
2. Liquid; no color; almond-like odor; hot (burning) taste; oxidized to benzoic acid when exposed to air and deteriorates under light
3. $C_7H_6O$

Benzoid Acid
1. Preservative, antimicrobial
2. Crystalline scales or needles; white; odor resembling benzaldehyde; volatile in steam, sublimes at ~100°C (212°F)
3. $C_6H_5COOH$

Benzophenone (Diphenyl ketone benzoylbenzene)
1. Flavoring
2. Solid, crystalline or flaky; white; odor pleasant, delicate, resembling a rose
3. $C_{13}H_{10}O$

Benzoyl Peroxide (Benzoyl superoxide)
1. Bleaching agent
2. Crystal; no color; reactive when dry, may explode spontaneously. MP and decomposition points 103-106°C (217-223°F).
3. $C_{14}H_{10}O_4$

Benzyl Acetate
1. Flavoring
2. Liquid, no color, flowery odor
3. $C_9H_{10}O_2$

Benzyl Alcohol (Phenyl carbinol)

**Table 1.1.1. (continued)**

Usual or common names
1. Usage
2. Description
3. General formula

1. Flavoring
2. Liquid; no color; odor slightly aromatic, resembling benzaldehyde; MP and decomposition points 206°C (403°F); soluble; taste hot, burning, sharp; neutral to litmus
3. $C_7H_8O$

Benzyl Benzoate
1. Flavoring
2. Liquid, oil; no color; aromatic odor; hot, burning taste
3. $C_{14}H_{12}O_2$

Benzyl Butyrate (Benzyl *n*-butyrate)
1. Flavoring
2. Liquid; no color, fruity odor; resembling plum
3. $C_{11}H_{14}O_2$

Benzyl Cinnamate
1. Flavoring
2. Solid; white, pale yellow; sweet and balsamic odor
3. $C_{16}H_{14}O_2$

Benzyl Phenylacetate
1. Flavoring
2. Liquid; no color; sweet, flowery odor, resembling honey essence
3. $C_{15}H_{14}O_2$

Benzyl Propionate
1. Flavoring
2. Liquid; no color; sweet, fruity odor
3. $C_{10}H_{12}O_2$

Benzyl Salicylate
1. Flavoring
2. Liquid; no color; sweet, delicate odor
3. $C_{14}H_{12}O_3$

Bergamot Oil, Expressed
1. Flavoring
2. Liquid; green, yellow-green, yellow-brown; sweet, fruity odor
3. Triglycerides and fatty acids

BHA (Butylated hydroxyanisole)
1. Antioxidant, preservative
2. Solid, waxy; white or slightly yellow; faint and characteristic odor; MP 48 and 63°C (118 and 145°F).

3. $C_{11}H_6O_2$, mainly 3-BHA (3-*tert*-butyl-4-hydroxyanisole) and variable amounts of the minor 2-BHA (2-*tert* butyl-4-hydroxyanisole)

BHT (Butylated hydroxytoluene; 2,6-Di-*tert*-butyl-*p*-cresol)
1. Antioxidant, preservative
2. Crystals; white; delicate and characteristic odor
3. $C_{15}H_{24}O$

Biotin [*cis*-Hexahydro-2-oxo-*1H*-thieno(3,4) imidazole-4-baleric acid]
1. Vitamin, essential nutrient, dietary supplement
2. Crystalline powder; white; stable to heat and resistant to oxidation
3. $C_{10}H_{16}N_2O_3S$

Birch Tar Oil, Rectified
1. Flavoring
2. Liquid; clear, brown, dark; leathery odor
3. Triglycerides and fatty acids

Bois de rose Oil
1. Flavoring
2. Liquid, oily; no color, pale yellow; flowery odor; camphoraceous essence

Bornyl Acetate (*levo*-Bornyl acetate)
1. Flavoring
2. May be liquid (no color), semicrystalline mass, or crystalline (white solid); pine odor
3. $C_{12}H_{20}O_2$

Brominated Vegetable Oil
1. Stabilizer (for beverages), flavoring aid
2. Liquid, oily, viscous; pale yellow to dark brown; bland fruity odor; bland taste
3. Triglycerides and fatty acids

2-Butanone (Methyl ethyl ketone)
1. Flavoring
2. Liquid; clear, no color; sharp and penetrating odor
3. $CH_3COCH_2CH_3$

Butyl Acetate (*n*-Butyl acetate)
1. Flavoring
2. Liquid; clear, no color; strong, fruity odor
3. $CH_3COOCH_2CH_2CH_2CH_3$

Butyl Alcohol (1-Butanol)
1. Flavoring
2. Liquid; clear, no color; penetrating odor resembling wine
3. $CH_3(CH_2)_2CH_2OH$

Butylated Hydroxymethylphenol (4-Hydroxymethyl-2,6-di-*tert*-butylphenol)
1. Antioxidant
2. Solid, crystal; white; faint and unique odor

**Table 1.1.1. (continued)**

Usual or common names
1. Usage
2. Description
3. General formula

3. $C_{15}H_{24}O_2$

Butyl Butyrate (*n*-Butyl, *n*-Butyrate)
1. Flavoring
2. Liquid; no color; fruity odor, resembling pineapple in dilute solution.
3. $CH_3CH_2CH_2COOC_4H_9$

Butyl Butyryllactate (Butyrllactic acid; Butyl ester lactic acid; Butyl ester; Butyrate)
1. Flavoring
2. Liquid, no color, mild odor, resembling butter
3. $C_{11}H_{20}O_4$

1,3-Butyl Glycol
1. Solvent for flavorings
2. Liquid, viscous and hygroscopic; no color, clear; slightly characteristic taste; nearly odorless
3. $CH_2OHCH_2CHOHCH_3$

*tert*-Butylhydroquinone (Mono-*tert*-butylhydroquinone; TBHQ)
1. Antioxidant, preservative
2. Crystal; white; unique odor
3. $C_{10}H_{14}O_2$

Butyraldehyde (Butyl aldehyde)
1. Flavoring
2. Liquid; clear, colorless; sharp, pungent odor
3. $CH_3(CH_2)_2CHO$

Butyric Acid
1. Flavoring
2. Liquid; clear, colorless; sharp and penetrating odor, resembling rancid butter
3. $CH_3(CH_2)_2COOH$

Caffeine (1,3,7-Trimethylxanthine)
1. A stimulant in beverages
2. Powder (white solid), needles (glistening); no odor; bitter taste; solutions neutral to litmus, anhydrous or hydrated (1 $H_2O$); hydrate is efflorescent.
3. $C_8H_{10}N_4O_2$

Calcium Acetate
1. Sequestrant
2. Solid, bulky, finely divided; white; no odor
3. $Ca(C_2H_3O_2)_2$

Calcium Alginate (Algin)
1. Emulsifier, stabilizer, thickener
2. Fibrous or granular powder; white or yellow; no odor; no taste
3. $[(C_6H_7O_6)_2Ca]_n$

Calcium Ascorbate
1. Antioxidant, preservative
2. Crystal, powder; white, pale yellow
3. $C_{12}H_{14}CaO_{12}\bullet 2H_2O$

Calcium Bromate
1. Dough conditioner, maturing agent
2. Crystal, powder; white
3. $Ca(BrO_3)_2\bullet H_2O$

Calcium carbonate
1. Dietary supplement, nutrient, alkali, anticaking/firming agent, dough conditioner, yeast food
2. Fine, microcrystalline powder; white.
3 $CaCO_3$

Calcium Chloride, Anhydrous
1. Firming agent, sequestrant
2. Solid, hard, fragments, granules; white; deliquescent; no odor
3. $CaCl_2$

Calcium Chloride Solution
1. Firming agent, sequestrant
2. Solution; no color to slightly colored, may be clear or slightly turbid; available in 35-45% solutions
3. $CaCl_2$

Calcium Citrate
1. Buffer, firming agent, sequestrant
2. Solid, fine powder; white; no odor
3. $Ca_3(C_6H_5O_7)_2\bullet 4H_2O)$

Calcium Disodium EDTA [Calcium disodium ethylenediamine tetracetate; Calcium disodium (ethylenedinitrilo) tetracetate; Calcium disodium edetate]
1. Preservative, sequestrant
2. Crystalline granules, powder; white or off-white; no odor; saline taste; hygroscopic, stable in air
3. $C_{10}H_{12}CaN_2O_8\bullet 2H_2O$

Calcium Gluconate
1. Buffer, firming agent
2. Crystalline granules, powder; white; no odor; no taste; stable in air
3. $[CH_2OH(CHOH)_4COO]_2Ca$

Calcium Glycerophosphate
1. Nutrient, diet supplement

**Table 1.1.1. (continued)**

Usual or common names
1. Usage
2. Description
3. General formula

2. Fine powder; white; no taste; hygroscopic.
3. $C_3H_7CaO_6P$

Calcium Hydroxide (Slaked lime)
1. Buffer, firming agent
2. Powder; white; bitter taste, resembling alkaline
3. $Ca(OH)_2$

Calcium Iodate
1. Dough conditioner, maturing agent
2. Powder; white; slight odor
3. $Ca(IO_3) \cdot 2H_2O$

Calcium Lactate
1. Buffer, dough conditioner, yeast food
2. Crystal, powder, granules; white or cream colored; containing 1-5 molecules of water of crystallization (pentahydrate efflorescent); anhydrous at 120°C (248°F)
3. $[Ca_3CH(OHCOO]_2Ca \cdot xH_2O$

Calcium Lactobionate [Calcium 4-(β-D-Galactosido)-D-gluconate]
1. Firming agent (e.g., dry pudding mixes)
2. Powder, free flowing; white; no odor; bland taste; forming double salts with chloride, bromide, gluconate, etc.; decomposition point ~120°C(248°F)
3. $C_{24}H_{42}CaO_{24}$

Calcium Oxide (Lime)
1. Nutrient, alkali, dietary supplement, yeast food
2. Masses, granules, powder; white or gray
3. CaO

Calcium Pantothenate (*dextro*-Calcium pantothenate)
1. Vitamin
2. Powder; white; slightly hygroscopic; no odor; bitter taste; stable in air
3. $[HOCH_2C(CH_3)_2CH(OH)CONH(CH_2)_2COO]_2Ca$

Calcium Peroxide (Calcium dioxide; Calcium superoxide)
1. Dough conditioner, oxidizing agent
2. Powder or granules; white to yellowish; no odor; no taste; unstable in moist air; aqueous slurry alkaline
3. $CaO_2$

Calcium Phosphate (Dibasic dicalcium phosphate)
1. Dough conditioner, nutrient supplement, yeast food
2. Powder; white; no odor; no taste; stable in air; anhydrous or dihydrated

3. $CaHPO_4 \cdot 2H_2O$

Calcium Phosphate, Monobasic (Monocalcium phosphate; Calcium biophosphate; Acid calcium phosphate)

1. Buffer, dietary supplement, dough conditioner, firming agent, leavening agent, sequestrant, yeast food
2. Crystals or granules; white; anhydrous or monohydrated
3. $Ca(H_2PO_4)_2$

Calcium Phosphate, Tribasic (Tricalcium phosphate; Precipitate calcium phosphate)

1. Anticaking agent, buffer, dietary supplement, nutrient
2. Powder; white; no odor; no taste; stable in air
3. $\sim 10CaO \cdot 3P_2O_5 \cdot H_2O$

Calcium Propionate

1. Mold and rope inhibitor, preservative
2. Crystal; white; odor resembling propionic acid; water solution slightly alkaline
3. $[CH_3CH_2COO]_2Ca$

Calcium Pyrophosphate

1. Buffer, dietary supplement, neutralizing agent, nutrient
2. Powder, fine; white; no odor; no taste
3. $Ca_2P_2O_7$

Calcium Saccharin (1,2-Benzisothiazolin-3-one 1,1-Dioxide calcium salt)

1. Nonnutritive sweetener
2. Crystals or powder; white; faintly aromatic odor; intensely sweet taste
3. $C_{14}H_8CaN_2O_6S_2 \cdot 3\frac{1}{2}H_2O$

Calcium Silicate

1. Anticaking agent
2. Powder, free-flowing; white; high absorption power; powdery in a large amount of water or liquid
3. CaO and $SiO_2$ in different proportions

Calcium Stearate

1. Anticaking agent, binder, emulsifier
2. Powder, fine bulky; white; yellow
3. Calcium stearate and calcium palmitate mixed in varying proportions

Calcium Stearoyl 2-Lactylate

1. Dough conditioner, stabilizer, whipping agent
2. Powder; cream colored; odor resembling caramel
3. Occuring as a mixuture of calcium salts

Calcium Sulfate

1. Dough conditioner, firming agent, nutrient, sequestrant, yeast food
2. Powder, fine; white, yellow; no odor; anhydrous or dihydrated
3. $CaSO_4 \cdot xH_2O$

**Table 1.1.1. (continued)**

Usual or common names
1. Usage
2. Description
3. General formula

Cananga Oil
1. Flavoring
2. Liquid; yellow; harsh, flowery odor
3. Triglycerides and fatty acids

Candelilla Wax
1. Masticatory substance in chewing gum base, surface-finishing agent
2. Wax, hard; opaque to translucent, yellow, brown
3. Complex formula

Caramel (Caramel color)
1. A natural colorant
2. Liquid; brown, black, resembling burnt sugar; tasteless when diluted
3. A complex formula

Caraway Oil
1. Flavoring
2. Liquid; no color, pale yellow; odor and taste characteristic of caraway spice
3. Triglycerides and fatty acids

Carbon, Activated (Activated charcoal; Decolorizing carbon; Active carbon)
1. Substance used to remove odor, color, taste; purification agent
2. Solid, porous, coarse granules, fine powder; black, no taste
3. Carbon

Cardamon Oil
1. Flavoring agent
2. Liquid; no color, pale yellow; camphoraceous and penetrating odor; decomposes under light
3. Triglycerides and fatty acids

Carmine (Carminic acid)
1. A food colorant
2. Crystals; red; darkens at >130°C (266°F), decomposes at 250°C (482°F)
3. $C_{22}H_{20}O_{13}$

Carnauba Wax
1. Candy glaze and polish
2. Solid, hard, brittle, with a resinous fracture; light brown, pale yellow
3. Complex formula

$\beta$-Carotene
1. Nutrient, food color, dietary supplement
2. Crystals, powder; red
3. $C_{40}H_{56}$

Carrageenan
1. Emulsifier, gelling agent, stabilizer, thickener
2. Fine powder; yellow, tan, white; no odor; pasty taste
3. A polysaccharide

Carrot Seed Oil
1. Flavoring
2. Liquid; yellow, amber; aromatic and pleasant odor
3. Triglycerides and fatty acids

Carvacrol
1. Flavoring
2. Liquid; no color, pale yellow; spicy and pungent odor, resembling thymol
3. $C_{10}H_{14}O$, a mixture of the isomeric carvacrols (isopropyl *o*-creols)

*d*-Carvone (*dextro*-Carbone, *d*-1-Methyl-4-isopropenyl-6-cyclohexen-2-one)
1. Flavoring
2. Liquid; no color, light yellow; odor resembling caraway
3. $C_{10}H_{14}O$

*l*-Carvone (*Levo*-Carvone; *l*-1-Methyl-4-isopropenyl-6-cyclohexen-2-one)
1. Flavoring
2. Liquid; no color, pale yellow, straw; odor resembling spearmint
3. $C_{10}H_{14}O$

$\beta$-Caryophyllene
1. Flavoring
2. Liquid oily, no color, faintly yellow; odor resembling cloves
3. $C_{15}H_{24}$, mixture of sesquitepenes as in several essential oils

Casarilla Oil (Sweetwood bark oil)
1. Flavoring
2. Liquid; yellow, brown, amber; spicy and pleasant odor
3. Triglycerides and fatty acids

Cassia Oil (Cinnamon oil)
1. Flavoring
2. Liquid; yellow, brown; odor and taste resembling cassia cinnamon; aging and light can darken and thicken the oil.
3. Triglycerides and fatty acids

Castor Oil
1. Component in protective coatings, release agent, anticaking agent
2. Liquid, viscous; no color, yellow; faint, unique odor; bland, unique taste
3. Triglycerides and fatty acids

Cedar Leaf Oil (Thuja oil; White cedar leaf oil)
1. Flavoring
2. Liquid; no color or yellow; strong, camphoraceous odor resembling sage
3. Triglycerides and fatty acids

Celery Seed Oil

**Table 1.1.1. (continued)**

Usual or common names
1. Usage
2. Description
3. General formula

1. Flavoring
2. Liquid; yellow, green, brown, odor; aromatic, pleasant odor
3. Triglycerides and fatty acids

Cellulose, Microcrystalline (Cellulose gel or powder)
1. Agent in anticaking, binding, disintegrating, dispersing, and making tablets
2. Powder, fine crystalline, free-flowing (nonfibrous); white; no odor; compressed agent in tablets decomposes in the presence of light.
3. $C_6H_{10}O_5$

Chamomile Oil, English
1. Flavoring
2. Liquid; pale blue, light greenish blue; strong, aromatic, flowery odor; color darkens on aging.
3. Triglycerides and fatty acids

Chamomile Oil, German (Hungarian chamomile oil)
1. Flavoring
2. Liquid, viscous when cold; blue, green, browning under light; strong, unique, aromatic odor; bitter taste
3. Triglycerides and fatty acids

Chlorine
1. Antimicrobial, oxidation/bleaching agent
2. Gas; green, yellow; packaged under pressure in cylinders; a respiratory irritant and can cause burn (skin, eye) and death (suffocation).
3. $Cl_2$

Cholic Acid (Cholalic acid; 3,7,12-Trihydroxycholanic acid)
1. Emulsifier
2. Plates, powder; no color, white; bitter taste and sweet aftertaste
3. $C_{24}H_{40}O_5$

Choline Bitartrate [(2-Hydroxyethyl)trimethylammonium bitartrate]
1. Dietary supplement
2. Powder, hygroscopic, crystalline; white; no odor; acidic taste
3. $[HOCH_2CH_2N(CH_3)]_3HC_4H_4O_6$

Choline Chloride [(2-Hydroxyethyl) trimethylammonium chloride]
1. Dietary supplement
2. Crystals or crystalline powder; white; odor slightly resembling trimethylamine; hygroscopic
3. $[HOCH_2CH_2N(CH_3)_3]^+Cl^-$

Cinnamaldehyde (Cinnamic aldehyde; Cinnamal)

1. Flavoring
2. Liquid; yellow, strongly refractive; odor characteristic of cinnamon oil; burning and aromatic taste; decomposes under light.
3. $C_9H_8O$

Cinnamic Acid (3-Phenylpropenoic acid)
1. Flavoring
2. Crystalline scales; white; odor resembling honey and flowers
3. $C_9H_8O_2$

Cinnamon Bark Oil, Ceylon
1. Flavoring
2. Liquid; yellow; odor resembling cinnamon; spicy and burning taste
3. Triglycerides and fatty acids

Cinnamon Leaf Oil
1. Flavoring
2. Liquid; brown; spicy odor
3. Triglycerides and fatty acids

Cinnamyl Acetate
1. Flavoring
2. Liquid; no color and faint yellow; sweet balsamic and flowery odor
3. $C_{11}H_{12}O_2$

Cinnamyl Alcohol
1. Flavoring agent
2. Solid; white, yellow; characteristic, balsamic odor
3. $C_9H_{10}O$

Cinnamyl Anthranilate
1. Flavoring
2. Powder; red, yellow; characteristic, fruity, balsamic odor, resembling anthranilates
3. $C_{16}H_{15}NO_2$

Cinnamyl Formate
1. Flavoring agent
2. Liquid; no color or yellow; balsamic odor, uniquely cinnamon
3. $C_{10}H_{10}O_2$

Cinnamyl Isovalerate
1. Flavoring
2. Liquid; no color, pale yellow; mixed odor, spicy-fruity, fruity- flowery
3. $C_{14}H_{18}O_2$

Cinnamyl Propionate
1. Flavoring
2. Liquid; no color, pale yellow; fruity, flowery odor
3. $C_{12}H_{14}O_2$

Citral (Gerania, Neral)

**Table 1.1.1. (continued)**

Usual or common names
1. Usage
2. Description
3. General formula

1. Flavoring
2. Liquid; pale yellow; odor resembling lemon; containing *trans* and *cis* isomers
3. $C_{10}H_{16}O$, *trans*- (Gerania) or *cis*- (Neral) 3,7-Dimethyl-2,6-octadien-1-al

Citric Acid
1. Acidifier, flavor, sequestrant, dispersing agent
2. Powder, crystalline; translucent, colorless or white; odorless; strong and acidic taste; anhydrous or monohydrated (efflorescent in dry air)
3. $C_6H_8O_7$

Citronellal (3,7-Dimethyl-6-octen-1-al)
1. Flavor
2. Liquid; faintly yellow; intense odor resembling lemon, citronella, and rose
3. $C_{10}H_{18}O$

Citronellol (3,7-Dimethyl-6-octen-1-ol)
1. Flavoring
2. Liquid, oily; no color; odor resembling rose
3. $C_{10}H_{20}O$

Citronellyl Acetate (3,7-Dimethyl-6-octen-1-yl acetate)
1. Flavoring
2. Liquid; no color; odor resembling fruit
3. $C_{12}H_{22}O_2$

Citronellyl Butyrate (3,7-Dimethyl-6-octen-1-yl butyrate)
1. Flavoring
2. Liquid; no color; strong odor, resembling fruits and roses
3. $C_{14}H_{26}O_2$

Citronellyl Formate (3,7-Dimethyl-6-octen-1-yl formate)
1. Flavoring
2. Liquid; no color; strong, fruity odor resembling flowers
3. $C_{11}H_{20}O_2$

Citronellyl Isobutyrate (3,7-Dimethyl-6-octen-1-yl isobutyrate)
1. Flavoring
2. Liquid; no color; odor resembling fruits and roses
3. $C_{14}H_{26}O_2$

Citronellyl Propionate (3,7-Dimethyl-6-octen-1-yl propionate)
1. Flavoring
2. Liquid; colorless; odor resembling roses or fruits
3. $C_{13}H_{24}O_2$

Clary Oil (Clary sage oil; Oil of muscatel)
1. Flavoring
2. Liquid; pale yellow or yellow; herbaceous odor and winelike bouquet
3. Triglycerides and fatty acids

Clove Leaf Oil
1. Flavoring
2. Liquid; pale yellow
3. Triglycerides and fatty acids

Clove Oil
1. Flavoring
2. Liquid; colorless or slightly yellow; odor and taste characteristic of cloves; aging and air exposure darkens and makes oil more viscous.
3. Triglycerides and fatty acids

Clove Stem Oil
1. Flavoring
2. Liquid; yellow to light brown; odor and taste characteristic of clove
3. Triglycerides and fatty acids

Cognac Oil, Green (Wine yeast oil)
1. Flavoring
2. Liquid; green; odor resembling cognac
3. Triglycerides and fatty acids

Copaiba Oil
1. Flavoring
2. Liquid; colorless or pale yellow; aromatic odor, characteristic of copaiba balsam; bitter, pungent taste
3. Triglycerides and fatty acids

Copper Gluconate
1. Nutrient, dietary supplement
2. Powder, fine solid; blue
3. $[CH_2OH(CHOH)_4COO]_2Cu$

Coriander Oil
1. Flavoring
2. Liquid; colorless or slightly yellow; odor and taste characteristic of coriander
3. Triglycerides and fatty acids

Costus Root Oil
1. Flavoring
2. Liquid, viscous; yellow, brown; odor resembling a combination of violet, orris, and vetivert
3. Triglycerides and fatty acids

Cresyl Acetate (*p*-Tolyl acetate)
1. Flavoring

**Table 1.1.1. (continued)**

Usual or common names
1. Usage
2. Description
3. General formula

2. Liquid; clear, colorless; strong and flowery odor
3. $C_9H_{10}O_2$

Cubeb Oil
1. Flavoring
2. Liquid; colorless to green; spicy odor; acrid taste
3. Triglycerides and fatty acids

Cuminic Aldehyde (*p*-Cuminic aldehyde, Cumaldehyde, *p*-Isopropylbenzaldehyde, Cuminal)
1. Flavoring
2. Liquid; colorless to yellow; strong, pungent odor resembling cumin oil
3. $C_{10}H_{12}O$

Cumin Oil
1. Flavoring
2. Liquid; yellow to brown; strong, unpleasing odor
3. Triglycerides and fatty acids

Cyclamen Aldehyde [2-Methyl-3-(*p*-isopropylphenyl)propionaldehyde]
1. Flavoring
2. Liquid; no color to pale yellow; flowery odor
3. $C_{13}H_{18}O$

L-Cysteine Monohydrochloride (L-2-Amino-3-mercaptopropanoic acid monohydrochloride)
1. Nutrient, supplement
2. Crystalline powder; white; no odor; acidic taste; anhydrous form MP and decomposition point ~175°C (347°C)
3. $C_3H_7NO_2S \cdot HCI \cdot H_2O$

L-Cystine [3,3′-Dithiobis-(2-amino)-propanoic acid]
1. Nutrient, supplement
2. Crystals, white; no odor
3. $HOOCCH(NH_2)CH_2SSCH_2–CH(NH_2)COOH$

Decalactone
1. Flavoring
2. Liquid; no color; odor resembling coconut and fruit at high concentration and butter essence in low concentration
3. $C_{10}H_{18}O_2$

Decanal (Aldehyde C-10; Capraldehyde)
1. Flavoring
2. Liquid; colorless, pale yellow; odor fatty and flowery when diluted

3. $C_{10}H_{20}O$

Decanoic Acid (Capric acid)

1. Byproduct of the production of food additives, defoaming agent
2. Crystals; white; unpleasant and rancid odor
3. $CH_3(CH_2)_8COOH$

1-Decanol, Natural (Decyl alcohol; Alcohol C-10; Nonylcarbionol)

1. Flavoring agent
2. Liquid; no color; flowery odor similar to orange blossoms
3. $CH_3(CH_2)_8CH_2OH$

Dehydroacetic Acid (3-Acetyl-6-methyl-1,2-pyran-2,4(3*H*)-dione; Methylaceto-pyrone; DHA)

1. Preservative
2. Crystalline powder; white; no odor, faint and acidic taste
3. $C_3H_8O_4$

Desoxycholic Acid (Deoxycholic acid; 13$\alpha$, 12$\alpha$-Dihydroxycholanic acid)

1. Emulsifying agent
2. Crystalline powder; white
3. $C_{24}H_{40}O_4$

Dexpanthenol [D(+)-Pentothenyl alcohol; Panthenol]

1. Nutrient, dietary supplement
2. Liquid, viscous, hygroscopic; clear; faint, characteristic odor; solutions alkaline to litmus
3. $HOCH_2C(CH_3)_2CH(OH)CO–NH(CH_2)_2CH_2OH$

Diacetyl (2,3-Butanedione; Dimethyl diketone; Dimethylglyoxal)

1. Flavoring
2. Liquid; clear, yellow-green; strong, pungent odor; diluted solution buttery in odor and flavor
3. $C_4H_6O_2$

Diacetyl Tartaric Acid Esters of Mono- and Diglycerides

1. Emulsifiers
2. Can be liquid (viscous, sticky), fatty substance or solid (waxy), depending on iodine value of oils or fats used in the production process; ester forms have a faint acidic odor.
3. Triglycerides and fatty acids

Diatomaceous Silica

1. Filtering aid in food processing
2. Powder; white, gray, buff colored
3. Formula complex

Diethyl Malone (Ethyl malonate; Malonic ester)

1. Flavoring
2. Liquid; colorless; slightly fruity odor

**Table 1.1.1. (continued)**

Usual or common names
1. Usage
2. Description
3. General formula

3. $C_7H_{12}O_4$

Diethyl Sebacate (Ethyl sebacate)
1. Flavoring
2. Liquid; colorless, pale yellow; slight odor
3. $C_2H_5OOC(CH_2)_8COOC_2H_5$

Diethyl Succinate (Ethyl succinate)
1. Flavoring
2. Liquid; no color, pleasant odor
3. $C_2H_5OOCCH_2CH_2COOC_2H_5$

Dilauryl Thiodipropionate
1. Preservative, antioxidant
2. Crystalline flakes; white; sweet odor resembling ester.
3. $(C_{12}H_{25}OOCCH_2CH_2)_2S$

Dill Seed Oil, European
1. Flavoring
2. Liquid; yellow; odor and taste resembling caraway
3. Triglycerides and fatty acids

Dill Seed Oil, Indian (Dill seed oil, East Indian; Dill oil, Indian)
1. Flavoring
2. Liquid; light yellow to light brown; odor and taste caraway-like and harsh
3. Triglycerides and fatty acids

Dillweed oil, American (Dill oil; Dill herb oil)
1. Flavoring
2. Liquid; yellow; characteristic odor and taste
3. Triglycerides and fatty acids

Dimethyl Anthranilate (Methyl-*N*-methyl anthranilate)
1. Flavoring
2. Liquid; yellow, bluish fluorescence; odor resembling grape
3. $C_9H_{11}NO_2$, containing small amounts of methyl anthranilate

Dimethyl Benzyl Carbinol ($\alpha, \alpha$,-Dimethylphenethyl alcohol)
1. Flavoring
2. Crystals, white; can exist in a liquid supercooled form (colorless to pale yellow); flowery odor
3. $C_{10}H_{14}O$

Dimethyl Benzyl Carbinyl Acetate ($\alpha, \alpha$,-Dimethylphenethyl acetate)
1. Flavoring

2. Liquid, solid at room temperature; no color; flowery and fruity odor
3. $C_{12}H_{16}O_2$

3,7-Dimethyl-1-Octanol (Dimethyl octanol; Tetrahydrogeraniol)
1. Flavoring
2. Liquid; no color; odor resembling a sweet rose
3. $C_{10}H_{22}O$

Dimethylpolysiloxane (Dimethyl silicone)
1. Defoaming
2. Liquid, viscous; clear, colorless
3. A mixture of fully methylated linear siloxane polymers containing repeating units of the formula $[(CH_3)_2SiO]$ and stabilized with trimethylsiloxy end-blocking units of the formula $[(CH_3)_3SiO^-]$

Dioctyl Sodium Sulfosuccinate (DSS)
1. Emulsifier, wetting agent
2. Solid; white; odor characteristic of octyl alcohol
3. $C_{20}H_{37}NaO_7S$

Disodium EDTA [Disodium ethylene diamine tetracetate; Disodium (ethylenedinitrilo) tetraacetate; Disodium edetate]
1. Preservative, sequestrant, stabilizer
2. Crystalline powder; white
3. $C_{10}H_{14}N_2Na_2O_8 \cdot 2H_2O$

Disodium Guanylate (Sodium 5′-guanylate; Disodium guanosine-5′-monophosphate)
1. Flavor enhancer
2. Crystalline powder; colorless or white; characteristic taste
3. $C_{10}H_{12}N_5Na_2O_8P \cdot xH_2O$ (where $x$ is usually 7.)

Disodium Inosinate (Sodium 5′-inosinate; Disodium inosine 5′-monophosphate)
1. Flavor enhancer
2. Crystals or powder; white; characteristic taste
3. $C_{10}H_{11}N_4Na_2O_8P \cdot xH_2O$ (where $x$ is usually 7.5).

Dodecalactone
1. Flavoring
2. Liquid; no color; odor resembling fruit and coconut and butter (in dilute concentrations)
3. $CH_3(CH_2)_6CHCH_2CH_2CH_2$

Erythorbic Acid (D-Araboascorbic acid)
1. Antioxidant, preservative
2. Crystals, powder; white, yellow, discoloration (darkens) under light; air stable when dry; MP (decomposition point) = 164°C to 171°C (327°F to 340°F).

**Table 1.1.1. (continued)**

Usual or common names
1. Usage
2. Description
3. General formula

3. $C_6H_8O_6$

Estragole (*p*-Allylanisole)
1. Flavoring agent
2. Liquid; colorless or light yellow; odor resembling anise
3. $C_{10}H_{12}O$

Ethyoxylated Mono- and Diglycerides (Polyoxethylene 20); Mono- and Diglycerides of Fatty Acids (Polyglycerate 60)
1. Dough conditioner, emulsifier
2. Liquid, oily; yellow; faint, characteristic odor; mildly bitter taste
3. Mixture of stearate, palmitate, and myristate partial esters of glycerin condensed with 20 mol ethylene oxide per mole of alphamonoglyceride

Ethoxyquin (6-Ethoxy-1,2-dihydro-2,2,3-trimethylquinolin)
1. Antioxidant (pears and apples)
2. Liquid; clear; the decrease of antioxidant activity with aging may darken compound.
3. $C_{14}H_{19}NO$

Ethyl Acetate
1. Flavoring agent
2. Liquid; no color; fragrant odor, faintly resembling acetone; burning taste; inflammable; volatile at low temperature
3. $CH_3COOC_2H_5$

Ethyl Acetoacetate (Acetoacetic ester; Ethyl 3-oxobutanoate)
1. Flavoring
2. Liquid, mobile; clear, no color or light yellow; pleasant odor
3. $C_6H_{10}O_3$, equilibrium mixture of keto- and enol- ethyl acetoacetate

Ethyl Acrylate
1. Flavoring
2. Liquid, mobile; clear, colorless; fruity, harsh, penetrating odor; lachrymatous (tears)
3. $CH_2{=}CHCOOC_2H_5$

Ethyl Alcohol (Alcohol; Ethanol)
1. Extraction solvent, vehicle
2. Liquid, mobile; clear, colorless; faint, characteristic odor; burning taste; inflammable; BP = 78°C (172°F); may be denatured or undenatured
3. $C_2H_5OH$

Ethyl *p*-Anisate (Ethyl *p*-methoxybenzoate)

1. Flavoring
2. Liquid; no color, yellow; fruity, light odor resembling anise
3. $C_{10}H_{12}O_3$

Ethyl Anthranilate (Ethyl *o*-aminobenzoate)
1. Flavoring
2. Liquid; clear, colorless to amber; flowery, orange-blossom odor.
3. Complex chemical structure

Ethyl Benzoate
1. Flavoring
2. Liquid; no color; fruity odor
3. $C_9H_{10}O_2$

2-Ethylenebutraldehyde
1. Flavoring
2. Liquid, mobile; clear, colorless
3. $C_6H_{12}O$

Ethyl Butyrate
1. Flavoring agent
2. Liquid; colorless; fruity odor
3. $CH_3CH_2CH_2COOC_2H_5$

2-Ethylbutyric Acid
1. Flavoring
2. Liquid; clear, no color; rancid odor
3. $C_6H_{12}O_2$

Ethylcellulose
1. Protective coating for vitamin and mineral tablets, binder and filler for vitamin preparations
2. Solid; white or light tan; heat sensitive with loss of unique properties, e.g., discoloration at $>240^oC$ ($464^oF$)
3. Polysaccharide

Ethyl Cinnamate (Ethyl 3-phenylpropenoate)
1. Flavoring
2. Liquid, oily; no color; odor slighly resembling cinnamon and sometimes characterized by a sweet, balsamic honey scent.
3. $C_{11}H_{12}O_2$

Ethylene Dichloride
1. Dichloro-extraction solvent
2. Liquid, oily; colorless, clear; odor resembling chloroform; sweet taste; inflammable
3. $C_2H_4Cl_2$

Ethyl Formate
1. Flavoring

**Table 1.1.1. (continued)**

| Usual or common names<br>1. Usage<br>2. Description<br>3. General formula |
|---|
| 2. Liquid; no color; characteristic odor, inflammable. |
| 3. $HCOOC_2H_5$ |
| Ethyl Heptanoate (Ethyl heptoate) |
| 1. Flavoring |
| 2. Liquid; no color; fruity odor resembling wine; similar taste but harsh aftertaste |
| 3. $CH_3(CH_2)_5COOC_2H_5$ |
| Ethyl Hexanoate (Ethyl caproate; Ethyl capronate) |
| 1. Flavoring |
| 2. Liquid; no color; fruity odor |
| 3. $CH_3(CH_2)_4COOC_2H_5$ |
| Ethyl Isovalerate (Ethyl-2-methyl butyrate) |
| 1. Flavoring |
| 2. Liquid; clear, colorless; fruity odor, resembling apple essence in diluted form |
| 3. $(CH_3)_2CHCH_2COOC_2H_5$ |
| Ethyl lactate (Ethyl 2-hydroxypropionate) |
| 1. Flavoring |
| 2. Liquid; no color; unique odor |
| 3. $CH_3CHOHCOOC_2H_5$ |
| Ethyl Laureate (Ethyl dodecanoate) |
| 1. Flavoring |
| 2. Liquid, oily; no color; odor lightly fruity and flowery |
| 3. $CH_3(CH_2)_{10}COOC_2H_5$ |
| Ethyl Maltol (3-Hydroxy-2-ethyl-4-pyrone) |
| 1. Flavoring |
| 2. Powder, crystalline; white; unique odor; sweet taste resembling fruit in dilute solution; MP = 90°C (194°F) |
| 3. $C_7H_8O_3$ |
| Ethyl Methyl Phenylglycidate (Aldehyde C-16; Strawberry aldehyde) |
| 1. Flavoring |
| 2. Liquid; no color and pale yellow; strongly fruity odor resembling strawberry |
| 3. $C_{12}H_{14}O_3$. |
| Ethyl Nonanoate (Ethyl pelargonate) |
| 1. Flavoring |

2. Liquid; clear, no color; odor resembling fat, fruit, and cognac
3. $CH_3(CH_2)_7COOC_2H_5$

Ethyl Octanoate (Ethyl caprylate; Ethyl Octoate)
1. Flavoring
2. Liquid; no color; odor fruity and flowery
3. $CH_3(CH_2)_6COOC_2H_5$

Ethyl Oxyhydrate (Rum ether)
1. Flavoring
2. Liquid; clear, colorless; sharp odor resembling rum
3. Complex chemical structure

Ethyl Phenylacetate
1. Flavoring
2. Liquid; no color, sweet, pleasant odor resembling honey
3. $C_{10}H_{12}O_2$

Ethyl Phenylglycidate
1. Flavoring
2. Liquid; no color, light yellow; odor and taste resembling strawberry
3. $C_{11}H_{12}O_3$

Ethyl Propionate
1. Flavoring
2. Liquid; transparent, no color; odor resembling rum
3. $CH_3CH_2COOC_2H_5$

Ethyl Salicylate
1. Flavoring
2. Liquid; no color; wintergreen odor
3. $C_9H_{10}O_3$

Ethyl Vanillin (3-Ethoxy-4-hydroxybenzaldehyde)
1. Flavoring
2. Crystals; white, yellow; odor and taste resembling vanilla; solutions acidic to litmus; light decomposes product.
3. $C_9H_{10}O_3$

Eucalyptol (1:8 Cineol; Anhydride of menthane 1:8 diole; 1:8 Oxido-*p*-methane; 1:8 Epoxy-p-menthane)
1. Flavoring
2. Liquid; no color; unique odor; cool, pungent taste
3. $C_{10}H_{18}O$

Eucalyptus Oil
1. Flavoring
2. Liquid; colorless, pale yellow; aromatic, camphoraceous odor; pungent taste, spicy and cool
3. Triglycerides and fatty acids

**Table 1.1.1. (continued)**

Usual or common names
1. Usage
2. Description
3. General formula

Eugenol (4-Allyl-2-methoxyphenol; Eugenic acid; 4-Allylguiacol)
1. Flavoring
2. Liquid; colorless, pale yellow; strong, aromatic odor resembling clove; spicy, pungent taste; discoloration and thickening in air
3. $C_{10}H_{12}O_2$

Eugenyl Acetate (4-Allyl-2-methoxyphenyl acetate; Eugenol acetate; Acetyl eugenol; Acetugenol)
1. Flavoring
2. Liquid form or solid form at room temperature; odor resembling clove
3. $C_{12}H_{14}O_3$

Fennel Oil
1. Flavoring
2. Liquid; colorless or yellow; odor and taste characteristic of fennel
3. Triglycerides and fatty acids

Ferric Ammonium Citrate, Brown (Iron ammonium citrate)
1. Nutrient, supplement
2. Scales or granules (transparent brown, red) or powder (brown, yellow); no odor to slightly ammoniacal; slightly metallic taste; deliquescent in air; decomposition in light; solutions slightly acidic to neutral
3. $(NH_4)_xFe(C_6H_5O_7)_y$ (exact structure undefined)

Ferric Ammonium Citrate, Green (Iron ammonium citrate)
1. Anticaking agent (e.g., for sodium chloride), supplement, nutrient
2. Scales, granules, powder (transparent green), crystals (green); no odor; slightly metallic taste; solutions acid to litmus; decomposing in air
3. $(NH_4)_xFe(C_6H_5O_7)_y$ (exact structure undefined)

Ferric Phosphate (Iron phosphate; Ferric orthophosphate)
1. Nutrient, supplement
2. Powder; white, buff, yellow; odorless
3. $FePO_4 \bullet H_2O$ (2-4 molecules of water of hydration)

Ferric Pyrophosphate (Iron pyrophosphate)
1. A nutrient, supplement
2. Powder; tan, yellow
3. $Fe_4(P_2O_7)_3 \bullet xH_2O$

Ferrous Fumarate
1. A supplement, nutrient
2. Powder; red, orange, brown; no odor

3. $C_4H_2FeO_4$

Ferrous Gluconate

1. Coloring adjunct, nutrient, supplement
2. Powder or granules; yellow, gray, green; odor resembling burnt sugar; soluble; acidic to litmus
3. $C_12H_{22}FeO_{14}\cdot 2H_2O$

Ferrous Sulfate

1. Nutrient, supplement
2. Crystals or granules; blue, green; no odor; saline taste; oxidized by moist air to basic ferric sulfate (brown, yellow)
3. $FeSO_4\cdot 7H_2O$

Ferrous Sulfate, Dried

1. Nutrient, supplement
2. Powder; gray, white, buff
3. $FeSO_4\cdot xH_2O$ (mainly $FeSO_4\cdot H_2O$)

Fir Needle Oil, Canadian (Balsam Fir oil)

1. Flavoring
2. Liquid; no color or faintly yellow; odor resembling balsam
3. Triglycerides and fatty acids

Fir Needle Oil, Siberian (Pine Needle oil)

1. Flavoring
2. Liquid; colorless or faintly yellow; odor resembling balsam
3. Triglycerides and fatty acids

Food Starch, Modified (Modified food starch)

1. Binder, colloidal stabilizer, thickener
2. Powder or granules (white), flakes, amorphous powders, or coarse particles (pregelatinized)
3. Polysaccharide

Folic Acid (Pteroylmonoglutamic acid; Folacin)

1. Vitamin supplement, nutrient
2. Crystals; yellow; chars at 482°F (250°C); salts can be barium, lead, and sodium; sensitive to heat, acids, oxygen, and ultraviolet radiation (in solution only).
3. $C_{19}H_{19}N_7O_6$

Formic Acid

1. Flavoring
2. Liquid; no color; pungent odor; corrosive
3. HCOOH

Fumeric Acid (*trans*-Butenedioic acid; *trans*-1,2-Ethylenedicarboxylic acid)

1. Acidifier, flavoring
2. Granule or powder; white or crystalline; odorless

**Table 1.1.1. (continued)**

Usual or common names
1. Usage
2. Description
3. General formula

3. $C_4H_4O_4$

Furfural (2-Furaldehyde; Pyromucic aldehyde)
1. Flavoring
2. Liquid, oily; no color or yellow (prolonged storage, brown, red); odor characteristic of the cyclic aldehydes (almond)
3. $C_5H_4O_2$

Garlic Oil
1. Flavoring
2. Liquid; clear, orange, red; strong, pungent odor characteristic of garlic
3. Triglycerides and fatty acids

Geraniol (*trans*-3,7-Dimethyl-2,6-octadien-1-ol)
1. Flavoring
2. Liquid; colorless; odor resembling rose; may contain other isomeric and closely related terpenic alcohols.
3. $C_{10}H_{18}O$

Geranium Oil, Algerian (Rose geranium oil, Algerian)
1. Flavoring
2. Liquid; yellow; odor resembling roses and geraniol
3. Triglycerides and fatty acids

Geranyl Acetate (3,7-Dimethyl-2,6-octadien-1-yl acetate)
1. Flavoring
2. Liquid; colorless; pleasant, flowery odor; may contain other isomeric and closely related terpenic esters.
3. Complicated formula

Geranyl Benzoate (3,7-Dimethyl-2,6-octadien-1-yl benzoate)
1. Flavoring
2. Liquid; slightly yellow; flowery odor, resembling ylang ylang oil; may contain other isomeric and closely related terpenic esters.
3. $C_{17}H_{22}O_2$

Geranyl Butyrate (3,7-Dimethyl-2,6-octadien-1-yl butyrate)
1. Flavoring
2. Liquid; no color, slightly yellow; fruity and rosy odor; may contain other isomeric and closely-related terpenic esters.
3. $C_{14}H_{24}O_2$

Geranyl Formate (3,7-Dimethyl-2-6-octadien-1-yl formate)
1. Flavoring

2. Liquid; pale yellow; fresh odor, resembling leafy roses; may contain other isomeric and closely related terpenic esters.
3. $C_{11}H_{18}O_2$

Geranyl Phenylacetate (3,7-Dimethyl-2,6-octadien-1-yl phenylacetate)

1. Flavoring agent
2. Liquid; yellow; odor resembling honey and roses; may contain other isomeric and closely related terpenic esters.
3. $C_{18}H_{24}O_2$

Geranyl Propionate (3,7-Dimethyl-2,6-octadien-1-yl propionate)

1. Flavoring
2. Liquid; colorless; fruity and flowery odor; may contain other isomeric and closely related terpenic esters.
3. $C_{13}H_{22}O_2$

Gibberellic Acid

1. Enzyme activator, plant growth hormones
2. Crystalline powder; white, yellow to no color; MP = 234°C (453 °F)
3. $C_{19}H_{22}O_6$

Ginger Oil

1. Flavoring
2. Liquid; yellow, aromatic odor characteristic of ginger
3. Triglycerides and fatty acids

Glucono Delta-Lactone

1. Acidifier, leavening agent, sequestrant
2. Crystalline powder; white; no odor; decomposition point = ~153°C (307°F)
3. $C_6H_{10}O_6$

L-Glutamic Acid (Glutamic acid; L-2-Aminopentanedoic acid)

1. Salt substitute, supplement, nutrient
2. Crystalline powder; white; no odor; saturated solution acidic
3. $HOOCCH_2CH_2CH(NH_2)COOH$

Glutamic Acid Hydrochloride (2-Aminopentanedioic acid hydrochloride)

1. Flavoring, salt substitute
2. Crystalline powder; white; solutions acidic to litmus
3. $C_5H_9NO_4{\cdot}HCl$

L-Glutamine (L-2-Aminoglutarimic acid)

1. Nutrient, supplement
2. Crystals or powder; white; odorless; slightly sweet taste; solutions acidic to litmus; MP and decompostion point = 185°C (365°C)
3. $H_2NOCCH_2CH_2CH(NH_2)COOH$

Glycerin (Glycerol)

1. Bodying agent, humectant, plasticizer

**Table 1.1.1. (continued)**

Usual or common names
1. Usage
2. Description
3. General formula

2. Syrupy liquid; clear and colorless; odor slightly characteristic; hygroscopic; solutions neutral
3. $CH_2OH \bullet CHOH \bullet CH_2OH$

Glycerol Ester of Partially Dimerized Rosin
1. Masticatory substance in chewing gum base
2. Resin, hard; amber colored
3. Complicated chemical structure

Glycerol Ester of Partially Hydrogenated Wood Rosin
1. Masticatory substance in chewing gum base
2. Resin, hard; pale amber.
3. Complicated chemical structure

Glycerol Ester of Tall Oil Rosin
1. Masticatory substance in chewing gum base
2. Resin, hard; pale-amber
3. Complicated chemical structure

Glycine (Aminoacetic acid; Glycocoll)
1. Supplement, nutrient
2. Crystalline Powder; white; no odor; sweet taste; solutions acidic to litmus
3. $H_2NCH_2COOH$

Grapefruit Oil, Expressed (Grapefruit oil, coldpressed; Oil of Shaddock)
1. Flavoring
2. Liquid; yellow, red; may contain a flocculent separation of waxy substance
3. Triglycerides and fatty acids

Guar Gum
1. Emulsifier, stabilizer, thickener
2. Powder; white, yellowish; no odor
3. Polysaccharide

Gum Guaiac (Guaiac resin)
1. Antioxidant, preservative
2. Powder; brown, yellow, discoloration in air; balsamic odor; slightly acidic taste
3. A polysaccharide

Heptanal (Aldehyde C-7; Heptaldehyde)
1. Flavoring
2. Liquid; colorless to faintly yellow; penetrating odor
3. $CH_3(CH_2)_5CHO$

2-Heptanone (Methyl amyl ketone)
1. Flavoring agent
2. Liquid, mobile; clear, colorless; odor resembling ketone
3. $CH_3CO(CH_2)_4CH_3$

3-Heptanone (Ethyl butyl ketone)
1. Flavoring
2. Liquid, mobile; clear, colorless; odor resembling ketone
3. $CH_3(CH_2)_3COCH_2CH_3$

Heptyl Alcohol (Enanthic alcohol)
1. Flavoring
2. Liquid; colorless; odor resembling fat and citrus
3. $CH_3(CH_2)_5CH_2OH$

Heptylparaben (Heptyl-*p*-hydroxybenzene)
1. Antimicrobial, preservative
2. Crystals (colorless) or crystalline powder (white); faint odor; burning taste
3. $C_{14}H_{20}O_3$

Hexanoic Acid (Caproic acid)
1. Flavoring
2. Liquid, oily; colorless, pale yellow; odor resembling cheese
3. $CH_3(CH_2)_4COOH$

Hexyl Alcohol, Natural (1-Hexanol; Alcohol C-6)
1. Flavoring
2. Liquid, mobile; clear, colorless; mild, sweet, penetrating odor
3. $CH_3(CH_2)_4CH_2OH$

$\alpha$-Hexylcinnamaldehyde
1. Flavoring
2. Liquid; pale yellow; odor resembling jasmine
3. $C_{15}H_{20}O$

L-Histidine [L-$\alpha$-Amino-4 (or 5)-imidazolepropionic acid]
1. Supplement, nutrient
2. Crystals or crystalline powder; odorless; slightly bitter taste; MP = 277-288$^{o}$C (531-550 $^{o}$F)
3. $C_6H_9N_3O_2$

L-Histidine Monohydrochloride (L-$\alpha$-Amino-4 (or 5)-imidazolepropionic acid monohydrochloride)
1. Supplement, nutrient
2. Crystals or crystalline powder; no odor; slightly acidic taste; MP = 250$^{o}$C (482$^{o}$F), also decomposition point
3. $C_6H_9N_3O_2 \cdot HCl \cdot H_2O$

Hops Oil
1. Flavoring

**Table 1.1.1. (continued)**

Usual or common names
1. Usage
2. Description
3. General formula

2. Liquid; yellow to green; aromatic and unique odor; discoloration with increased viscosity during storage

Hydrochloric Acid
1. Acidifier
2. Liquid; clear, colorless to slightly yellow; corrosive and pungent odor; care in handling
3. HCl

Hydrogen Peroxide
1. Bleach and oxidizing agent, preservative, starch modifier
2. Liquid; clear and colorless; odor resembling ozone; safe and stable when properly handled; when handled improperly, it decomposes and releases heat in the presence of dirt and other foreign substances.
3. $H_2O_2$

Hydroxycitronella (7-Hydroxy-3,7-dimethyl octanal)
1. Flavoring
2. Liquid; no color; sweet and flowery odor, resembling lily
3. $C_{10}H_{20}O_2$

Hydroxycitonellal Dimethyl Acetal (7-Hyroxy-3,7-dimethyl octanal; Acetal)
1. Flavoring
2. Liquid; no color; light and flowery odor
3. $C_{12}H_{26}O_3$

Hydroxylated Lecithin
1. Clouding agent, emulsifier
2. Fluid to plastic, depending on the level of free fatty acid and soybean oil, with or without diluents; light yellow in color, with a "bleached" characteristic

Hydroxypropyl Cellulose
1. Emulsifier, film former, protective colloid, stabilizer, suspending agent, thickener
2. Powder white
3. Polysaccharide

Hydroxypropyl Methylcellulose (Propylenetclycol ether of methylcellulose)
1. Emulsifier, stabilizer, thickener
2. Powder or granules (off-white, fibrous); commercial products vary in characteristics depending on the ingredients present; aqueous solutions

are surface active, forming films during drying; sol when hot and gel when cold (interchangeable).
3. Polysaccharide

Indole
1. Flavoring
2. Crystalline, flaky; white; unpleasant odor when concentrated, flowery when diluted
3. $C_8H_7N$

Inositol (1,2,3,5/4,6-cyclohexanehexol; *i*-Inositol; *meso*-Inositol)
1. Supplement, nutrient
2. Crystals or crystalline powder; white; no odor; sweet taste; air stable; solutions neutral to litmus
3. $C_6H_{12}O_6$

$\alpha$-Ionone (2,6,6-Trimethyl-2-cyclohexenyl-3-butene-2-one)
1. Flavoring
2. Liquid; no color or pale yellow; woody-violet odor
3. $C_{13}H_{20}O$

$\beta$-Ionone [4(2,6,6-Trimethyl-1-cyclohexenyl)-3-butene-2-one]
1. Flavoring
2. Liquid; pale yellow; fruity, woody odor
3. $C_{13}H_{20}O$

Iron, Electrolytic
1. Supplement, nutrient
2. Powder, amorphous; lusterless, gray, black; stable in dry air
3. Fe

Iron, Reduced
1. Supplement, nutrient
2. Powder; gray, black, no luster; passing through a 100-mesh sieve, particles crystalline under microscope; stable in dry air
3. Fe

Isoamyl Acetate (Amyl acetate; *c*-Methyl butyl acetate)
1. Flavoring
2. Liquid; colorless; fruity odor, resembling a pear
3. $CH_3COOC_5H_{11}$ (acetates of mixed isomeric amyl alcohols, 2-methyl-1-butanol and 3-methyl-1-butanol predominating)

Isoamyl Butyrate (Amyl butyrate)
1. Flavoring
2. Liquid; colorless; strong, characteristic odor
3. $CH_3(CH_2)_2COOC_5H_{11}$ (butyrates of mixed isomeric amyl alcohols, 2-methyl-1-butanol and 3-methyl-1-butanol predominating)

Isoamyl Formate (Amyl formate)

**Table 1.1.1. (continued)**

Usual or common names
1. Usage
2. Description
3. General formula

1. Flavoring
2. Liquid; no color; odor resembling plums
3. $HCOOC_5H_{11}$ (formates of mixed isomeric amyl alcohols, 2-methyl-1-butanol and 3-methyl-1-butanol predominating)

Isoamyl Hexanoate (Amyl hexanoate; Isoamyl caproate; Pentyl hexanoate)
1. Flavoring
2. Liquid; no color; fruity odor
3. $CH_3(CH_2)_4COOC_5H_{11}$ (hexanoates of mixed isomeric amyl alcohols, 2-methyl-1-butanol and 3-methyl-1-butanol predominating)

Isoamyl Isovalerate (Amyl valerate; Amyl isovalerianate)
1. Flavoring
2. Liquid; clear, colorless; fruity odor, resembling apple
3. $(CH_3)_2CHCH_2COOCH_2CH_2CH(CH_3)_2$ (iovalerates of mixed isomeric amyl alcohols, 2-methyl-1-butanol and 3-methyl-1-butanol predominating)

Isoamyl Salicylate (Amyl salicylate)
1. Flavoring
2. Liquid; no color; pleasant, characteristic odor
3. $C_{12}H_{16}O_3$ (salicylates of mixed isomeric amyl alcohols, 2-methyl-1-butanol and 3-methyl-1-butanol predominating)

Isobornyl Acetate
1. Flavoring
2. Liquid; clear, colorless, discoloration with age; agreeable odor, resembleing camphor with a touch of pine needles or hemlock
3. $C_{12}H_{20}O_2$

Isobutyl Acetate
1. Flavoring
2. Liquid; clear and no color; fruity odor resembling banana when diluted
3. $CH_3COOCH_2CH(CH_3)_2$

Isobutyl Alcohol (Isobutanol)
1. Flavoring
2. Liquid, mobile; clear, colorless; characteristic, penetrating odor resembling wine
3. $(CH_3)_2CHCH_2OH$

Isobutyl Cinnamate
1. Flavoring
2. Liquid; colorless; sweet, fruity odor resembling balsamic

3. $C_{13}H_{16}O_2$

Isobutylene-Isoprene Copolymer (Butyl rubber)

1. Masticatory substance in chewing gum base
2. Dried coagulum
3. A synthetic copolymer containing form 0.5 to 2.0 molar percent of isoprene, with the rest consisting of isobutylene

Isobutyl Phenylacetate

1. Flavoring
2. Liquid; no color; odor of roses and honey
3. $C_{12}H_{16}O_2$

Isobutyl Salicylate

1. Flavoring
2. Liquid; no color; odor resembling orchid.
3. $C_{11}H_{14}O_2$

Isobutyraldehyde

1. Flavoring
2. Liquid, mobile; clear, colorless; pungent odor
3. $(CH_3)_2CHCHO$

Isobutyric Acid (2-Methyl propanoic acid; Isopropyl formic acid)

1. Flavoring
2. Liquid; no color; strong and penetrating odor resembling butter
3. $C_4H_8O_2$

Isoeugenol (2-Methoxy-4-propenylphenol)

1. Flavoring
2. Liquid, viscous; pale yellow; flowery odor, suggestive of carnation
3. $C_{10}H_{12}O_2$

Isoeugenyl Acetate (2-Methoxy-4-propenyl phenyl acetate)

1. Flavoring
2. Crystals; white; spicy odor resembling clove
3. $C_{12}H_{14}O_3$

DL-Isoleucine (DL-2-Amino-3-methylvaleric acid)

1. Supplement, nutrient
2. Crystalline powder; white; slightly bitter taste; MP (also decomposition point) = 292°C (558°F)
3. $C_6H_{13}NO_2$

L-Isoleucine (L-2-Amino-3-methylvaleric acid)

1. Supplement, nutrient
2. Crystalline leaflets or crystalline powder (white); bitter taste; sublimate at 168-170°C (334-338°F)
3. $C_6H_{13}NO_2$

Isopropyl Acetate

**Table 1.1.1. (continued)**

Usual or common names
1. Usage
2. Description
3. General formula

1. Flavoring
2. Liquid, mobile; clear, colorless; characteristic odor
3. $C_5H_{10}O_2$

Isopropyl Alcohol (2-Propanol; Ispropanol)
1. Solvent for extraction
2. Liquid; clear, colorless; characteristic odor; slightly bitter taste; flammable
3. $CH_3CHOHCH_3$

Isopulegol (*p*-Menth-8-en-3-ol)
1. Flavoring
2. Liquid; no color; harsh camphoraceous odor, combining mint, rose-leaf, and geranium
3. $C_{10}H_{18}O$

Isovaleric Acid (Isopropylacetic acid)
1. Flavoring
2. Liquid; colorless; acidic taste; unpleasant odor
3. $(CH_3)CHCH_2COOH$

Juniper Berries Oil
1. Flavoring
2. Liquid; colorless, yellow, greenish; characteristic odor; aromatic, bitter taste; storage accompanied by polymerization
3. Triglycerides and fatty acids

Kaolin
1. Anticaking
2. Powder, fine; white yellow, gray; earthy taste; when wet, odor resembling clay, accompanied by discoloration
3. Complex chemical structure

Karaya Gum (Sterculia gum)
1. Emulsifier, stabilizer, thickener
2. Broken and irregular pieces (suggesting crystalline); yellow, pink, brown, horny (occasionally mixed with fragments and/or pieces of bark); slightly acetous odor; pasty and slightly acetous taste
3. Polysaccharide

Kelp
1. A supplement for iodine
2. Chopped coarse particles or finely ground powder; dark green, olive brown; salty, characteristic taste

3. Polysaccharide

Labdanum Oil

1. Flavoring
2. Liquid viscous; yellow; strong, balsamic odor, odor of diluted oil resembling ambergris; discoloration upon standing
3. Triglycerides and fatty acids

Lactated Mono- and Diglycerides

1. Emulsifier, stabilizer
2. Soft to hard and waxy solid
3. Mixture of partial lactic and fatty acid esters of glycerin

Lactic Acid (2-Mhydroxypropionic acid)

1. Acidifier
2. Liquid, syrupy; colorless or yellow; hydroscopic
3. $CH_3CHOHCOOH$, a mixture of lactic acid and lactic acid lactate

Lactylated Fatty Acid Esters of Glycerol and Propylene Glycol (Propylene glycol lactostearate)

1. Emulsifier, plasticizer, stabilizer, surface-active agent, whipping agent
2. Soft to hard waxy solids, may be in liquid form.
3. A mixture of fatty acid esters of lactic acids and its polymers

Lactylic Esters of Fatty Acids

1. Emulsifier, surface-active agent
2. May be liquid or waxy solid.
3. A mixture of fatty acid esters of lactic acid and its polymers

Lanolin, Anhydrous (Wool fat)

1. Masticatory substance in chewing gum base
2. Semisolid, fatlike substance; yellowish white when purified
3. Triglycerides and fatty acids

Laurel Leaf Oil (Bay leaf oil)

1. Flavoring
2. Liquid, yellow, spicy and aromatic odor
3. Triglycerides and fatty acids

Lauric Acid (Dodecanoic acid)

1. Defoaming agent, used for manufacturing other food-grade additives
2. Crystalline powder; white, yellow, glossy in appearance
3. $CH_3(CH_2)_{10}COOH$

Lauryl Alcohol, Natural (Alcohol C-12; 1-Dodecanol)

1. Flavoring agent
2. Liquid, no color >21$^oC$ (69.8$^oF$) odor characteristically fatty
3. $CH_3(CH_2)_{10}CH_2OH$

Lauryl Aldehyde (Aldehyde C-12; Dodecanal)

1. Flavoring

**Table 1.1.1. (continued)**

| Usual or common names<br>1. Usage<br>2. Description<br>3. General formula |
|---|
| 2. Liquid, solidifies at low temperatures; no color or yellow; characteristically fatty odor |
| 3. $CH_3(CH_2)_{10}CHO$ |
| Lavender Oil |
| 1. Flavoring |
| 2. Liquid; colorless, yellow; unique odor; taste resembling lavender flowers |
| 3. Triglycerides and fatty acids |
| Lecithin |
| 1. Emulsifier, Stabilizer, and Thickener |
| 2. Plastic to fluid, changing consistency depending on free fatty acid and soybean content, with or without dilutents; yellow, brown (bleached or unbleached); nut-like odor; bland taste |
| 3. Complex mixture of acetone-insoluble phosphatides' consisting mainly of phosphatidyl choline, phosphatidyl ethanolamine, phosphatidyl serine, and phosphatidyl inositol, combining with varying quantities of other substances, such as triglycerides, fatty acids, and carbohydrates |
| Lemongrass Oil |
| 1. Flavoring |
| 2. Liquid; East Indian type, also known as Cochin, Native, and British Indian lemongrass oil, yellow, brown; strong lemon-like odor; West Indian type, also known as Madagascar, Guatemala lemongrass oil, lighter with lemon odor |
| 3. Triglycerides and fatty acids |
| Lemon Oil |
| 1. Flavoring |
| 2. Liquid; green, yellow; characteristic odor; taste resembling lemon peel |
| 3. Triglycerides and fatty acids |
| Lemon Oil, Distilled |
| 1. Flavoring agent |
| 2. Liquid; colorless, pale yellow; odor charactieristic of fresh lemon peel |
| 3. Triglycerides and fatty acids |
| DL-Leucine (DL-2-Amino-4-methylvaleric acid) |
| 1. Supplement, nutrient |
| 2. Small crystals (white) or crystalline powder; no odor; somewhat bitter taste; MP (also decomposition point) = 290°C (554°F); solutions slightly acid to neutral |

3. $C_6H_{13}NO_2$

L-Leucine (L-2-Amino-4-methylvaleric acid)
1. Supplement, nutrient
2. Small plates (white, lustrous), or crystalline powder (white); sublimate 150°C (302°F)
3. $C_6H_{13}NO_2$

Lime Oil, Distilled
1. Flavoring
2. Liquid; colorless, green, yellow

Limestone, Ground
1. Texturizing and release agent, modifier for chewing gum base and chewing gum
2. Microcrystalline powder, fine; white or off-white; no odor; no taste; stable in air
3. $CaCO_3$

*d*-Limonine (*d-p*-Mentha-1,8-diene; Cinene)
1. Flavoring
2. Liquid; no color; pleasant odor resembling mild citrus
3. $C_{10}H_{15}$

Linaloe Wood Oil
1. Flavoring
2. Liquid; no color or yellow; pleasing and flowery odor
3. Triglycerides and fatty acids

Linalool (3,7-Dimethyl-1-1,6-octadien-3-ol)
1. Flavoring
2. Liquid; colorless; pleasant, flowery odor
3. $C_{10}H_{18}O$

Linalyl Acetate (3,7-Dimethyl-1,6-octadien-3-yl acetate)
1. Flavoring
2. Liquid; colorless or slightly yellow, pleasant and flowery odor
3. $C_{12}H_{20}O_2$

Linalyl Benzoate (3,7-Dimethyl-1,6-octadien-3-yl benzoate)
1. Flavoring
2. Liquid; yellow, brown; odor similar to tuberose
3. $C_{17}H_{22}O_2$

Linalyl Isobutyrate (3,7-Dimethyl-2,6-octadien-3-yl isobutyrate)
1. Flavoring
2. Liquid; colorless, slightly yellow; fruity odor
3. $C_{14}H_{24}O_2$

Linalyl Propionate (3,7-Dimethyl-2,6-octadien-3-yl-propionate)
1. Flavoring

**Table 1.1.1. (continued)**

Usual or common names
1. Usage
2. Description
3. General formula

2. Liquid; no color; sweet, flowery odor
3. $C_{13}H_{22}O_2$

Locust Bean Gum (Carob bean gum)
1. Emulsifier, stabilizer, thickener
2. Powder; white, yellow; no odor
3. Made up of high molecular weight hydrocolloidal polysaccharide (galactan and mannan units linked by glycosidic linkages, chemically described as galactommannan)

Lovage Oil
1. Flavoring
2. Liquid; yellow, green, brown; odor and taste both strong and characteristic; increased darkening and viscosity when exposed to air and light
3. Triglycerides and fatty acids

L-Lysine Monohydrochloride (2,6-Diaminohexanoic acid hydrochloride)
1. Supplement, nutrient
2. Crystalline powder, free-flowing; white; no odor; MP (also decomposition point) = 260°C (500°F)
3. $NH_2(CH_2)_4CH(NH_2)COOH \cdot HCl$

Mace Oil
1. Flavoring
2. Liquid; colorless or pale yellow; characteristic odor; taste resembling nutmeg; commercial products may be East Indian or West Indian, both similar in physical and chemical properties.
3. Triglycerides and fatty acids

Magnesium Carbonate
1. Alkalizing agent, anticaking agent, carrier, color-retention agent, drying agent
2. Friable masses or a bulky powder; light white; no odor; air stable
3 $MgCO_3$, may be hydrated.

Magnesium Chloride
1. Color-retention agent, firming agent
2. Flakes or crystals very deliquescent; colorless; odorless
3. $MgCl_2 \cdot 6H_2O$

Magnesium Hydroxide
1. Alkalizing agent, color-retention agent, drying agent
2. Bulky powder; white

3. $Mg(OH)_2$

Magnesium Oxide
1. Alkalizing agent, neutralizer
2. Bulky (white) powder (referred to as light magnesium oxide); may be a dense (white) powder (referred to as heavy magnesium oxide).
3. MgO

Magnesium Phosphate, Dibasic (Dimagnesium phosphate)
1. A dietary supplement, nutrient
2. Crystalline powder; white; no odor
3. $MgHPO_4 \cdot 3H_2O$

Magnesium Phosphate, Tribasic (Trimagnesium phosphate)
1. A dietary supplement, nutrient
2. Crystalline powder; white; no odor; no taste
3. $Mg_3(PO_4)_2 \cdot xH_2O$, hydration with 4, 5, or 8 water molecules

Magnesium Silicate
1. Anticaking agent
2. Powder, nongritty, fine; white; no odor; no taste; pH of 6.3-9.5

Magnesium Stearate
1. Anticaking agent, binder, emulsifier
2. Powder, fine, bulky, unctuous, greaselike, and free of grit; white
3. Magnesium with varying proportions of stearic and palmitic acids

Magnesium Sulfate (Epsom salt)
1. A supplement, nutrient
2. Crystals, small, needlelike; colorless; cool, saline, and bitter taste; efflorescent in warm, dry air; solutions neutral
3. $MgSO_4 \cdot 7H_2O$

Malic Acid (DL-Malic acid; Hydroxysuccinic acid)
1. Acidifying agent, flavoring
2. Crystalline powder or granules; white or nearly white; strong, acidic taste; solutions optically active
3. $C_4H_6O_5$

Maltol (3-Hydroxy-2-methyl-1-4-pyrone)
1. Flavoring
2. Crystalline powder; white; odor resembling caramel, butterscotch, suggestive of a fruity-strawberry aroma when dilute
3. $C_6H_6O_3$

Mandarin Oil, Expressed
1. Flavoring
2. Liquid; clear, orange, yellow, brown, fluorescent (bluish) in diffused light, oil obtained from unripe fruit is green; pleasant odor resembling oranges
3. Triglycerides and fatty acids

**Table 1.1.1. (continued)**

Usual or common names
1. Usage
2. Description
3. General formula

Manganese Chloride
1. Dietary supplement, nutrient
2. Crystals, large, irregular; pink, translucent
3. $MnCl_2 \cdot 1.4H_2O$

Manganese Gluconate
1. Dietary supplement, nutrient
2. Powder; pink
3. $[CH_2OH(CHOH)_4COO]_2Mn \cdot 3H_2O$

Manganese Glycerophosphate
1. Dietary supplement, nutrient
2. Powder; white or pink; odorless; tasteless
3 $C_3H_7MnO_6 \cdot xH_2O$

Manganese Hypophosphite
1. Dietary supplement, nutrient
2. Powder, granular or crystalline; pink; no odor; no taste; stable in air, explosion when mixed with oxidiaing agents such as nitrates, chlorates, etc.
3. $Mn(PH_2O_2)_2 \cdot xH_2O$

Manganese Sulfate
1. Dietary supplement, nutrient
2. Powder, granular; pink; odorless
3. $MnSO_4 \cdot H_2O$

Mannitol (D-Mannitol; Mannite; 1,2,3,4,5,6-Hexanehexol)
1. Dietary supplement, nutrient, texturizing agent
2. Crystalline solid; white; odorless; sweet taste
3. $C_6H_{14}O_6$, made up of D-Mannitol and a small quantity of sorbitol

Marjoram Oil, Spanish
1. Flavoring
2. Liquid; faintly yellow; camphoraceous odor
3. Triglycerides and fatty acids

Marjoram Oil, Sweet
1. Flavoring
2. Liquid, oily; yellow, green; spicy or cardamon essence
3. Triglycerides and fatty acids

Masticatory Substances
1. Natural masticatory substance in chewing gum base
2. Essentially coagulated or concentrated latices of vegetable origin

3. Polysaccharides

*Mentha Arvensis* Oil, Dementholized (Cornmint oil, dementholized)
1. Flavoring
2. Liquid; colorless or yellow; unique, minty odor
3. Triglycerides and fatty acids

Menthol (3-*p*-Menthanol)
1. Flavoring
2. Crystals (colorless), hexagonal (often needlelike), or fused masses, or as crystalline powder; pleasant, peppermint odor
3. $C_{10}H_{20}O$

DL-Methionine [DL-2-Amino-4-(methylthio)butyric acid]
1. Dietary supplement, nutrient
2. Platelets or powder; white; characteristic odor; pH of dilute solutions 5.6-6.1
3. $C_5H_{11}NO_2S$

L-Methionine [L-2-Amino-4-(methylthio)butyric acid]
1. Dietary supplement, nutrient
2. Plates (lustrous, white) or powder (white, crystalline); faint and characteristic odor
3. $C_5H_{11}NO_2S$

*p*-Methoxybenzaldehyde (Anisic aldehyde; *p*-Anisaldehyde)
1. Flavoring
2. Liquid; colorless or faintly yellow; characteristic and hawthorn-like odor
3. $C_8H_8O_2$

Methyl Alcohol (Methanol)
1. Extraction solvent
2. Liquid; clear; odorless; inflammable
3. $CH_3OH$

*p*-Methyl Anisole (*p*-Cresylmethyl ether; Methyl *p*-Cresol)
1. Flavoring agent
2. Liquid; colorless; penetrating odor of ylang ylang
3. $C_8H_{10}O$

Methyl Anthranilate (Methyl 2-aminobenzoate)
1. Flavoring
2. Liquid; colorless, faint yellow, fluorescent (bluish); odor resembling grape
3. $C_8H_9NO_2$

Methyl Benzoate
1. Flavoring
2. Liquid; colorless; fruity odor, resembling cananga oil
3. $C_8H_8O_2$

**Table 1.1.1. (continued)**

Usual or common names
1. Usage
2. Description
3. General formula

α-Methylbenzyl Acetate (Methyl phenylcarbinyl acetate; α-Phenylethyl acetate)
1. Flavoring
2. Liquid; colorless; odor suggestive of gardenia
3. $C_{10}H_{12}O_2$

α-Methylbenzyl Alcohol (Methyl phenylcarbinol; α-Phenethyl alcohol)
1. Flavoring
2. Liquid; no color; odor mildly hyacinth-gardenia; congeals below room temperature
3. $C_8H_{10}O$

Methylcellulose
1. Binder, bodying agent, bulking agent, emulsifier, film former, stabilizer, thickener
2. Powder or granules (white, fibrous); water solutions of methylcellulose surface-active, forming films upon drying; sol (hot) to gel (cool) and reversible
3. Polysaccharide

α-Methylcinnamaldehyde
1. Flavoring
2. Liquid; yellow; odor resembleing cinnamon
3. $C_{10}H_{10}O$

Methyl Cinnamate
1. Flavoring
2. Solid; white, yellow; fruity odor resembling balsam
3. $C_{10}H_{10}O_2$

Methyl Cyclopentenolone (3-Methyl cyclopentane-1,2-dione)
1. Flavoring
2. Crystalline powder; white; nutty odor, suggesting maple-licorice when diluted
3. $C_6H_8O_2$

Methylene Chloride (Dichloromethane)
1. Extraction solvent
2. Liquid; clear, colorless; odor suggestive of chloroform; nonflammable
3. $CH_2Cl_2$

Methyl Ester of Rosin, Partially Hydrogenated
1. Masticatory substance in chewing gum base

2. Liquid, resin; amber
3. A complex chemical formula

Methylethylcellulose
1. Emulsifier, foaming agent, stabilizer
2. Fibrous solid or powder; white, cream color; odorless
3. Polysaccharide

Methyl Eugenol (Eugenyl methyl ether; 1,2-Dimethoxy-4-allylbenzene)
1. Flavoring
2. Liquid; colorless, faint yellow; odor resembling delicate clove carnation
3. $C_{11}H_{14}O_2$

Methyl Formate
1. Fumigant
2. Liquid; colorless; pleasant odor; inflammable, handle with caution; BP = ~32°C (896°F)
3. $HCOOCH_3$

6-Methyl-5-hepten-2-one (Methyl heptenone)
1. Flavoring
2. Liquid; faintly yellow; odor resembling citrus-lemongrass
3. $C_8H_{14}O$

Methyl Isoeugenol (4-Allyl-1,2-dimethoxy benzene; Isoeugenyl methyl ether; 4-Propenyl veratrole)
1. Flavoring
2. Liquid; colorless or pale yellow, delicate odor resembling clove and carnation
3. $C_{11}H_{14}O_2$

Methyl *β*-Naphthyl Ketone (2′-Acetonaphthone)
1. Flavoring
2. Crystalline solid; white or nearly white; odor resembling orange blossom
3. $C_{12}H_{10}O$

Methyl-2-Octynoate (Methyl heptine carbonate)
1. Flavoring
2. Liquid; colorless to faint yellow; strong and unpleasant odor, resembling violet when diluted
3. $CH_3(CH_2)_4{=}CCOOCH_3$

Methylparaben (Methyl *p*-hydroxybenzoate)
1. Antimicrobial agent, preservative
2. Crystals or powder, small; colorless, white; odorless; slighly burning taste
3. $C_8H_8O_3$

4-Methyl-2-Pentanone (Methyl isobutyl ketone)
1. Flavoring
2. Liquid mobile; clear, colorless; odor resembling ketone

**Table 1.1.1. (continued)**

Usual or common names
1. Usage
2. Description
3. General formula

3. $CH_3COCH_2CH(CH_3)_2$

Methyl Phenylacetate
1. Flavoring
2. Liquid; colorless; odor resembling honey
3. $C_9H_{10}O_2$

Methyl Salicylate (Wintergreen oil; Gaultheria oil)
1. Flavoring
2. Liquid; colorless, yellow, or reddish; odor and taste resembling wintergreen; BP and decomposition point = 219-224°C (426-435°F)
3. $C_8H_8O_3$

2-Methylundecanal (Aldehyde C-12 MNA; Methyl *n*-nonyl acetaldehyde)
1. Flavoring
2. Liquid; colorless or pale yellow; odor resembling fatty acids, diluted product suggestive of a floral essence
3. $C_{12}H_{24}O$

Mineral Oil, White (Liquid petrolatum)
1. Binder, defoaming agent, fermentation aid, lubricant, protective coating, release agent
2. Liquid, oily; colorless, transparent, no fluorescence; no odor; no taste when cold; hot oil smells like petroleum
3. Triglycerides and fatty acids

Mono- and Diglycerides
1. Emulsifier, stabilizer
2. Liquid (yellow), plastic (ivory colored), and hard solids (ivory colored); bland taste and odor
3. Mixtures of glycerol mono- and diesters, with small amounts of triesters, edible fats or oils, etc.

Monoammonium L-Glutamate (Monoammonium glutamate; Ammonium glutamate)
1. Flavor enhancer, salt substitute
2. Crystalline powder, free-flowing; white; no odor
3. $C_5H_{12}N_2O_4 \cdot H_2O$

Monoglyceride Citrate
1. Syngergist and solubilizer for antioxidants
2. Waxy, like lard, soft; white, ivory; bland odor and taste
3. Mixture of glyceryl monooleate and its citric acid monoester

Monopotassium L-Glutamate (Monopotassium glutamate; Potassium glutamate; MPG)
1. Flavor enhancer, salt substitute
2. Crystalline powder, free flowing; white; no odor; hygroscopic
3. $C_5H_8KNO_4 \cdot H_2O$

Monosodium L-Glutamate (Monosodium glutamate; Sodium glutamate; MSG)
1. Flavor enhancer
2. Crystals or crystalline powder, free flowing; white; taste slightly sweet or salty
3. $C_5H_8NNaO_4 \cdot H_2O$

Myristic Acid (Tetradecanoic acid)
1. Basic ingredient for manufacturing food additives, defoaming agent
2. Crystalline substance, hard (white, pale yellow, glossy), may be a powder (white, yellow)
3. $CH_3(CH_2)_{12}COOH$

Myrrh Oil
1. Flavoring
2. Liquid; light brown or green; unique odor; discoloration and increased viscosity in the presence of air and light
3. Triglycerides and fatty acids

Nerol (*cis*-3,7-Dimethyl-2,6-octadien-1-Ol)
1. Flavoring
2. Liquid; colorless; odor resembling fresh, sweet roses
3. $C_{10}H_8O$, containing geranoils and other terpenic alcohols

Nerolidol (3,7,11-Trimethyl-1,6,10-dodecatrien-3-ol)
1. Flavoring
2. Liquid; colorless or straw colored; flowery odor, resembling rose and apple
3. $C_{15}H_{26}O$

Niacin (Nicotinic acid; 3-Pryridine carboxylic acid)
1. Supplement, nutrient
2. Crystals or crystalline powder; white, light yellow; no or slight odor
3. $C_6H_5NO_2$

Niacinamide (Nicotinamide)
1. Supplement, nutrient
2. Crystalline powder; white; no or slight odor; solutions neutral to litmus
3. $C_6H_6N_2O$

$\gamma$-Nonalactone (Aldehyde C-18)
1. Flavoring
2. Liquid; colorless or yellow; strong odor, resembling coconut
3. $C_9H_{16}O_2$

Nonanal (Aldehyde C-9; Pelargonic aldehyde)

**Table 1.1.1. (continued)**

Usual or common names
1. Usage
2. Description
3. General formula

1. Flavoring
2. Liquid; colorless or light yellow; strong odor resembling fat, diluted product resembling an essence of orange and rose
3. $CH_3(CH_2)_7CHO$

Nonyl Acetate
1. Flavoring
2. Liquid; no color; fruity, flowery odor
3. $CH_3COO(CH_2)_8CH_3$

Nonyl Alcohol (1-Nonanol; Alcohol C-9)
1. Flavoring
2. Liquid; colorless; odor resembling rose, citrus
3. $CH_3(CH_2)_7CH_2OH$

Nutmeg Oil (Myristica oil)
1. Flavoring
2. Liquid; colorless or light yellow; odor and taste resembling nutmeg; soluble in alcohol
3. Triglycerides and fatty acids

Octanal (Aldehyde C-8; Caprylic aldehyde)
1. Flavoring
2. Liquid; colorless or pale yellow; pungent odor, resembling fat, and fruity
3. $CH_3(CH_2)_6CHO$

Octanoic Acid (Caprylic acid)
1. Ingredient for making food additives, defoaming agent
2. Liquid, oily; colorless; unpleasant odor; rancid, burning taste
3. $CH_3(CH_2)_6COOH$

1-Octanol, Natural (Alcohol C-8; Octyl alcohol; Capryl alcohol)
1. Flavoring
2. Liquid; colorless; penetrating odor, resembling fat
3. $CH_3(CH_2)_6CH_2OH$

Octyl Acetate
1. Flavoring
2. Liquid; colorless; fruity odor resembling orange and jasmine
3. $CH_3COO(CH_2)_7CH_3$

Octyl Formate
1. Flavoring
2. Liquid; colorless; fruity odor

3. $C_8H_{17}OOCH$

Oleic Acid (*cis*-9-Octadecenoic acid)

1. An ingredient used in making food additives, defoaming agent, binder, lubricant
2. Liquid, oily when freshly prepared; colorless or pale yellow, the presence of air darkens the product; odor and taste resembling lard; heating in air decomposes to form vapors.
3. $CH_3(CH_2)_7CH{=}CH(CH_2)_7COOH$

Olibanum Oil (Oil of frankincense)

1. Flavoring
2. Liquid; slightly yellow; balsamic odor, with a slight lemon scent
3. Triglycerides and fatty acids

Onion Oil

1. Flavoring
2 Liquid; clear, yellow, amber, orange; pungent, penetrating, odor; onion-like taste
3. Triglycerides and fatty acids

Orange Oil (Sweet orange oil)

1. Flavoring
2. Liquid; yellow, orange; unique odor; taste resembling the outer portion of fresh, sweet orange peel
3. Triglycerides and fatty acids

Orange Oil, Bitter

1. Flavoring
2. Liquid; yellow, brown; aromatic odor, resembling Seville orange; aromatic and bitter taste; substance degraded by light; alcohol solutions neutral to litmus
3. Triglycerides and fatty acids

Orange Oil, Distilled

1. Flavoring
2. Liquid; colorless or faintly yellow; odor resembling fresh orange peel
3. Triglycerides and fatty acids

Origanum Oil, Spanish

1. Flavoring
2. Liquid; yellow, brown, red; pungent, spicy odor, resembling thyme oil
3. Triglycerides and fatty acids

Orris Root Oil

1. Flavoring
2. Solid at room temperature (yellow or brown), MP = 38-50°C (100.4-122°F); liquid (yellow or brown)
3. Triglycerides and fatty acids

**Table 1.1.1. (continued)**

Usual or common names
1. Usage
2. Description
3. General formula

Oxystearin
1. Crystallization inhibitor (salad and cooking oils), defoaming agent, sequestrant
2. Waxlike; tan, light brown; bland taste
3. Mixture of glycerides of partially oxidized stearic acid and other fatty acids

Palmitic Acid (Hexadecanoic acid)
1. An ingredient used in manufacturing food additives, defoaming agent
2. Crystalline, hard (white, yellow, glossy), powder (white, yellow); odor and taste fairly characteristic
3. $C_{16}H_{32}O_2$

DL-Panthenol (DL-Panthenyl alcohol; Racemic pantothenyl alcohol)
1. A dietary supplement, nutrient
2. Crystalline powder; white, cream color; faint, characteristic, odor; solutions neutral or alkaline to litmus
3. $HOCH_2C(CH_3)_2CH(OH)CONH(CH_2)_2CH_2OH$

Papain
1. Proteolytic enzyme
2. Amorphous powder; white, light tan
3. Complicated structure

Paraffin, Synthetic (Fischer-Tropsch paraffin)
1. Masticatory substance in chewing gum base
2. Waxy, hard; white; no odor and no taste
3. A mixture of paraffin hydrocarbons

Parsley Herb Oil
1. Flavoring
2 Liquid; yellow, brown; odor resembling parsley herb
3. Triglycerides and fatty acids

Pectin
1. Emulsifier, gelling agent, stabilizer, thickener
2. Powder, coarse to fine; yellow, white; no odor; pasty taste
3. A polysaccharide

Pennyroyal Oil (Pennyroyal oil, imported)
1. Flavoring
2. Liquid; yellow; aromatic mint odor
3. Triglycerides and fatty acids

Pentaerythritol Ester of Partially Hydrogenated Wood Rosin

1. Maticatory substance in chewing gum base
2. Resin, hard; amber
3. Complex structure

Pentaerythritol Ester of Wood Rosin
1. Masticatory substance in chewing gum base
2. Resin, hard; amber
3. Complex chemical structure

2,2-Pentanedione (Acetyl propionyl)
1. Flavoring
2. Liquid; clear, yellow, green; strong, pungent odor, when diluted resembling an essence of butter
3. $C_5H_8O_2$

2-Pentanone (Methyl propyl ketone)
1. Flavoring
2. Liquid,mobile; clear, colorless; flowery odor
3. $CH_3COCH_2CH_2CH_3$

Peppermint Oil
1. Flavoring
2. Liquid; colorless or slightly yellow; heavy, penetrating, odor, resembling peppermint; taste first pungent, then a cooling sensation, which is emphasized when the mouth is used for breathing.
3. Triglycerides and fatty acids

Pepper Oil, Black
1. Flavoring
2. Liquid; colorless, green; peppery odor; mild taste
3. Triglycerides and fatty acids

Pepsin
1. Proteolytic enzyme
2. Scales (lustrous, transparent, translucent), masses (granular, spongy, yellow, brown), powder (fine, white, yellow), inoffensive odor; slightly acidic or salty taste; slightly hygroscopic; no degradation under 100°C (212°F), activity lost in alkali solution at >70°C (158°F), solution (2% pepsin) acid to litmus
3. Complex chemical structure

Perlite (Expanded perlite)
1. Filter aid for processing
2. Volcanic rock, dense, glassy structure; gray to brown
3. Made up of fused sodium-potassium-aluminum silicate with water (3% to 5%)

Petitgrain Oil, Paraguay
1. Flavoring

**Table 1.1.1. (continued)**

Usual or common names
1. Usage
2. Description
3. General formula

2. Liquid; yellow, brown; harsh, bitter-sweet, flowery odor
3. Triglycerides and fatty acids

Petrolatum (White petrolatum; Yellow petrolatum)
1. Defoaming agent, lubricant, protective coating, release agent
2. Mass, greasy, solid, semisolid; white, yellow, amber, transparent in thin layers, slightly fluoresent; no odor; no taste
3. Made of a mixture of semisolid and purified hydrocarbons

Petroleum Wax (Refined paraffin wax; Microcrystalline wax)
1. Defoaming agent, masticatory substance in chewing gum base, protective coating
2. Wax, translucent; MP = 48-93°C (120-200°F)
3. Containing solid hydrocarbons paraffinic

Petroleum Wax (Synthetic wax; Ethylene Polymer)
1. Defoaming agent, masticatory substance in chewing gum base, protective coating
2. Wax; amber, white, MP = 85-116°C (185-240°F)
3. Solid hydrocarbons (paraffinic)

$\alpha$-Phellandrene (*p*-Mentha-1,5-diene)
1. Flavoring
2. Liquid; colorless, pale-yellow; herbaceous odor with a mild mintlike essence
3. $C_{10}H_{16}$

Phenethyl Acetate (2-Phenylethyl acetate)
1. Flavoring
2. Liquid; no color; flowery odor
3. $C_{10}H_{12}O_2$

Phenethyl Alcohol (2-Phenylethyl alcohol)
1. Flavoring
2. Liquid; no color; odor resembling roses
3. $C_8H_{10}O$

Phenethyl Isobutyrate
1. Flavoring
2. Liquid; colorless, pale yellow; fruity, roselike odor
3. $C_{12}H_{16}O_2$

Phenethyl Isovalerate
1. Flavoring

2. Liquid; colorless or pale yellow; fruity and roselike odor
3. $C_{13}H_{18}O_2$

Phenethyl Phenylacetate
1. Flavoring
2. Liquid, becoming solid at <26°C (78.8°F); colorless or pale yellow; odor resembling roses and hyacinth
3. $C_{16}H_{16}O_2$

Phenethyl Salicylate
1. Flavoring
2. Crystals; white; odor resembling balsam and roses
3. $C_{15}H_{14}O_3$

Phenoxyethyl Isobutyrate
1. Flavoring
2. Liquid; clear, colorless; odor resembling honey and roses
3. $C_{12}H_{16}O_3$

Phenylacetaldehyde ($\alpha$-Toluic aldehyde)
1. Flavoring
2. Liquid, increasing in viscosity with storage; oily; colorless, pale yellow; penetrating odor, resembling hyacinth when diluted
3. $C_8H_8O$

Phenylacetaldehyde Dimethyl Acetal
1. Flavoring
2. Liquid; colorless; heavy and characteristic odor
3. $C_{10}H_{14}O_2$

Phenylacetic Acid ($\alpha$-Toluic acid)
1. Flavoring
2. Crystals; white, glistening; unpleasant, persisting odor, resembling geranium leaf and rose when diluted
3. $C_8H_8O_2$

DL-Phenylalanine (DL-$\alpha$-Amino-$\beta$-phenylpropionic acid)
1. Dietary supplement, nutrient
2. Platelets, crystalline; white; no odor; optically inactive
3. $C_6H_5CH_2CH(NH_2)COOH$

L-Phenylalanine (L-$\alpha$-Amino-$\beta$-phenylpropionic acid)
1. Supplement, nutrient
2. Platelets (colorless, white), powder (crystalline); slightly characteristic odor; faintly bitter taste; MP (decomposition point) = 283°C (541°F)
3. $C_6H_5CH_2CH(NH_2)COOH$

3-Phenyl-1-Propanol (Phenylpropyl alcohol; Hydrocinnamyl alcohol)
1. Flavoring agent

**Table 1.1.1. (continued)**

Usual or common names
1. Usage
2. Description
3. General formula

2. Liquid, viscosity slight; colorless; uniquely sweet odor, resembling hyacinth-mignonette
3. $C_9H_{12}O$

2-Phenylpropionaldehyde (Hydratropic aldehyde; α-Methylphenyl acetaldehyde)
1. Flavoring
2. Liquid; yellow, white; characteristically flowery odor
3. $C_9H_{10}O$

3-Phenylpropionaldehyde (Hydrocinnamaldehyde; Phenylpropyl-aldehyde)
1. Flavoring
2. Liquid; colorless, pale yellow; pungent, flowery odor, resembling hyacinth
3. $C_9H_{10}O$

2-Phenylpropionaldehyde Dimethyl Acetal (Hydratropic aldehyde dimethyl acetal)
1. Flavoring
2. Liquid; colorless, pale yellow; odor resembling mushroom
3. $C_{11}H_{16}O_2$

3-Phenylpropyl Acetate
1. Flavoring
2. Liquid; colorless; characteristically spicy and flowery odor
3. $C_{11}H_{14}O_2$

Phosphoric Acid (Orthophosphoric acid)
1. Acidifier, sequestrant
2. Soluble; no color; available products may contain different concentrations, e.g., 75%, 85%, etc.
3. $H_3PO_4$

Pimenta Leaf Oil (Pimento leaf oil)
1. Flavoring
2. Liquid; yellow, light brown, darkens during storage; color changes to blue, then brown upon contact with iron; spicy odor
3. Triglycerides and fatty acids

Pimenta Oil (Pimento oil; Allspice oil)
1. Flavoring
2 Liquid; colorless, yellow, red, darkening during storage; odor and taste resembling allspice; decomposed by light
3. Triglycerides and fatty acids

Pine Needle Oil, Dwarf (Pine needle oil)
1. Flavoring
2. Liquid; colorless or yellow; pleasant and aromatic odor; bitter and pungent taste
3. Triglycerides and fatty acids

Pine Needle Oil, Scotch
1. Flavoring
2. Liquid; colorless or yellow; odor resembling turpentine
3. Triglycerides and fatty acids

Piperonal (3,4-Methylenediosybenzaldehyde; Heliotropine; Piperonyl aldehyde)
1. Flavoring
2. Crystalline solid; white; sweet, flowery odor, resembling heliotrope, byodor of safrole absent
3. $C_8H_6O_3$

Poloxamer 331 [$\alpha$-Hydro-$w$-hydroxsy poly(oxyethylene) poly(oxypropylene) (51-57 mol) poly(oxyethylene) block copolymer]
1. Solubilizing and stabilizing agent in flavor concentrates
2. Liquid; no color
3. A polymer (from ethylene oxide and propylene oxide reacting), average MW = 3800

Poloxamer 407 [$\alpha$-Hydro-$w$-hydroxypoly(oxyethylene) poly(oxypropylene) (63-71 mols) poly(oxethylene) block copolymenr]
1. Solubilizing and stabilizing agent in flavor concentrates
2. Solid; white; MP = 52 - 56$^{o}$C (125.6 - 132.8$^{o}$F)
3. A copolymer of ethylene oxide and propylene oxide, ~MW = 12,500

Polyethylene
1. Masticatory substance in chewing gum base
2. Partially crystalline and amorphous resin; white, translucent; products vary in molecular weight, molecular weight distribution, degree of chain branching, and extent of crystallinity
3. A polymerized ethylene, MW = 2000-21,000

Polyethylene Glycols (PEG)
1. Binders, coating agents, dispersing agents, flavoring adjuvants, plasticizing agents
2. Liquid, viscous, hygroscopic; clear, colorless; solid, resembling paraffin (white, waxy, or flakes); pH = 4.0-7.5 (11.20 concentration); products identified by number for MW (<600 for liquids, >1000 for solids); higher MW produces higher solubility in water, an increase in vapor pressure, hygroscopicity, but with a reduced solubility in organic solvents; a higher

**Table 1.1.1. (continued)**

Usual or common names
1. Usage
2. Description
3. General formula

molecular weight is also accompanied by an increased solidification point, specific gravity, flash point, and viscosity.
3. Addition polymers of ethylene oxide and water; MW = 200-9500

Polyglyceral Esters of Fatty Acids
1. Emulsifiers
2. A mixture of fatty chemicals, liquid (yellow, amber), oily (viscous), solid (tan, brown), plastic, waxy substances (soft, hard); hydrophilic to lipophilic; dispersible in water
3. A diverse mixture of chemical substances

Polyisobutylene
1. Masticatory substance for chewing gum base
2. Soft and gummy for low molecular weights, tough and elastic for high molecular weights; color light; no odor; no taste
3. A polymer

Polysorbate 20 [Polyoxyethylene (20) sorbitan monolaurate]
1. Emulsifier, stabilizer
2. Liquid; lemon, yellow amber; slight odor; bitter taste
3. A mixture of esters of sorbitol and sorbitol anhydrides condensed with ethylene oxide

Polysorbate 60 [Polyoxyethylene (60) sorbitan monostearate]
1. Emulsifier, stabilizer
2. Liquid, oily, semigel; lemon, orange color; slight, characteristic odor; bitter taste
3. A mixture of stearate and palmitate partial esters of sorbitol and sorbitol anhydrides

Polysorbate 65 [Polyoxyethylene (20) sorbitan tristearate]
1. Emulsifier, stabilizer
2. Waxy solid; tan; slightly characteristic odor; bitter taste
3. A mixture of stearate and palimitate partial esters of sorbitol and its anhydrides

Polysorbate 80 [Polyoxyethylene (20) sorbitan monoleate]
1. Emulsifier, stabilizer
2. Liquid, oily; yellow, orange; odor slightly characteristic; bitter taste
3. A mixture of oleate partial esters of sorbitol and sorbitol anhydrides

Polyvinyl Acetate
1. Masticatory substance in chewing gum base

2. Resin; white, pale yellow
3. Polymerized resin of vinyl acetate

Potassium Acid Tartrate (Potassium bitartrate; Cream of tartar)
1. Acidifier, buffer
2. Crystals (colorless, slightly opaque), powder (crystalline); pleasant, acidic taste; solution (saturated) acidic to litmus
3. KOOCCH(OH)CH(OH)COOH

Potassium Alginate (Algin)
1. Emulsifier, stabilizer, thickener
2. Powder, fibrous, granular; white, yellow; no odor; no taste
3. $(C_6H_7O_6K)_n$ (potassium salt of alginic acid)

Potassium Bicarbonate
1. Alkali, leavening agent
2. Crystals (colorless, transparent), powder (white, granular); odorless; stable in air; solutions are neutral or alkaline to phenolphthalein.
3 $KHCO_3$

Potassium Bromate
1. Dough conditioner, maturing agent
2. Crystals or granular powder; white; pH = 5-9 for 5% (by weight) solutions
3. $KBrO_3$

Potassium Carbonate
1. Alkali
2. Powder (anhydrous salt, white, granular), crystals or granules (hydrated, small, white, translucent); no odor; strong alkaline, taste; deliquescent; solutions alkaline and slightly turbid; potassium bicarbonate forms upon exposure to carbon dioxide.
3. $K_2CO_3$

Potassium Chloride
1. Supplement, gelling agent, nutrient, salt substitute, yeast food
2. Crystals or powder, granular; white; no odor; saline taste; stable in air; solutions neutral to litmus
3. KCl

Potassium Citrate
1. Buffer, sequestrant, and numerous applications in food processing
2. Crystals or powder, granular; transparent, white; odorless; saline taste; deliquescent in presence of moist air
3. $C_6H_5K_3O_7 \cdot H_2O$

Potassium Gibberellate
1. Enzyme activator
2. Crystalline powder; white; odorless and tasteless; deliquescent, pH = 6.0 for a 5% (by weight) solution

**Table 1.1.1. (continued)**

Usual or common names
1. Usage
2. Description
3. General formula

3. $C_{19}H_{21}KO_6$

Potassium Glycerophosphate
1. Supplement, nutrient
2. Liquid, syrupy; slightly yellow; solutions alkaline to litmus
3. $C_3H_7K_2O_6P \cdot 3H_2O$

Potassium Hydroxide
1. Alkali
2. Many forms (pellets, flakes, sticks, fused masses); white in color; deliquescent; absorbing carbon dioxide in the presence of air
3. KOH

Potassium Iodate
1. Dough conditioner, maturing agent
2. Crystalline powder; white; no odor; pH = 5-8 for a 5% (by weight) solution
3. $KIO_3$

Potassium Iodide
1. Dietary supplement, nutrient
2. Crystals (transparent), granular powder (white); stable in dry air, hygroscopic in moist air; pH = 6-10 for a 5% (by weight) solution
3. KI

Potassium Metabisulfite (Potassium pyrosulfite)
1. Antioxidant, preservative
2. Crystals, freeflowing, powder or granules; white, colorless; solutions acid to litmus
3. $K_2S_2O_5$

Potassium Nitrate
1. Color fixative in meat and meat products, preservative
2. Granules or cylindrical sticks; white, yellow; deliquescent; solution (10% by weight) alkaline to litmus
3. $KNO_3$

Potassium Phosphate, Dibasic (Dipotassium monophosphate; Dipotassium phosphate)
1. Buffer, sequestrant, yeast food
2. Solid, granules; colorless or white; deliquescent in moist air; a 1% (wt) solution has a pH of about 9.
3. $K_2HPO_4$

Potassium Phosphate, Monobasic (Potassium biphosphate; Monobasic potassium biphosphate; Potassium dihydrogen phosphate; Monopotassium phosphate)
1. Buffer, sequestrant, yeast food
2. Crystals (colorless), crystalline or granular powder (white); odorless; stable in air; pH 4.2-4.7 for a 1% (by weight) solution
3. $KH_2PO_4$

Potassium Phosphate, Tribasic (Tripotassium phosphate)
1. Emulsifier
2. Crystals or granules; white; odorless; hygroscopic; pH = 11.5 for a 1% (by weight) solution
3. $K_3PO_4$

Potassium Polymetaphosphate (Potassium Kurrol's salt)
1. Emulsifier (fat), moisture-retaining agent
2. Powder; white
3. $(KPO_3)_x$

Potassium Pyrophosphate (Tetrapotassium pyrophosphate)
1. Emulsifer, texturizer
2. Crystals (colorless) or granular substance (white); hygroscopic; pH = 10.5 for a 1% (by weight) solution
3. $K_4P_2O_7$

Potassium Sorbate (2,4-Hexadienoic acid; Potassium salt)
1. Preservative
2. Crystals, crystalline powder or pellets; white; MP (decomposition point) = 270°C (518°F)
3. $C_6H_7KO_2$

Potassium Sulfate
1. Regulating water and numerous applications in food processing
2. Crystals, crystalline powder; colorless or white; bitter, saline, taste; pH = 5.5-8.5 for a 5% (by weight) solution
3. $K_2SO_4$

Potassium Sulfite
1. Antioxidant, preservative
2. Granular powder; white; odorless; oxidation in the presence of air
3. $K_2SO_3$

Potassium Tripolyphosphate (Pentapotassium triphosphate; Potassium triphosphate)
1. Texturizer
2. Granules or powder; white; hygroscopic; pH = 9.2-10.1 for a 1% (by weight) solution
3. $K_5P_3O_{10}$

**Table 1.1.1. (continued)**

Usual or common names
1. Usage
2. Description
3. General formula

L-Proline (L-2-Pyrrolidine carboxylic acid)
1. Dietary supplement, nutrient
2. Crystals or crystalline powder; white; odorless; faintly sweet taste
3. $C_5H_9NO_2$

Propenylguathol (1-Ethoxy-2-hydroxy-4-propenylbenzene)
1. Flavoring
2. Crystalline powder; white; odor resembling vanilla
3. $C_{11}H_{15}O_2$

Propionaldehyde
1. Flavoring
2. Liquid, mobile; clear, colorless; penetrating odor.
3. $CH_3CH_2CHO$

Propionic Acid
1. Mold and rope inhibitor, preservative
2. Liquid, oily; odor faintly pungent and rancid
3 $CH_3CH_2COOH$

*p*-Propyl Anisole (Dihydroanethole)
1. Flavoring
2. Liquid; colorless, pale yellow; odor resembling anise and sassafra essence
3. $C_{10}H_{14}O$

Propylene Glycol (1,2-Propanediol; 1,2-Dihydroxypropane; Methyl glycol)
1. Humectant, solvent, wetting agent
2. Liquid, viscous; clear, no color; faint taste; odorless; absorbing moisture in the presence of moist air
3. $CH_3 \bullet CH(OH) \bullet CH_2OH$

Propylene Glycol Alginate (Hydroxypropyl alginate; Algin derivative)
1. Emulsifier, stabilizer, thickener
2. Fibrous or granular powder; white, yellow; odorless and tasteless
3. $(C_9H_{14}O_7)_n$

Propylene Glycol Monostearate
1. Emulsifier, stabilizer
2. Flakes or beads; white; bland odor and taste
3. A mixture of propylene glycol mono- and diesters of stearic acid and palmitic acids

Propyl Gallate (Gallic acid; Propyl ester)
1. Antioxidant

2. Powder; white or off-white; odorless; slightly bitter taste
3. $C_{10}H_{12}O_5$

Propylparaben (Propyl *p*-Hydroxybenzoate)
1. Antimicrobial agent, preservative
2. Crystals (small, colorless), powder (white)
3. $C_{10}H_{12}O_3$

PVP (Polyvinylpyrrolidone; Providone; Poly[1-(2-oxo-1-pyrrolidinyl)ethylene]
1. Bodying agent, classifying agent, clarifying agent, dispersant, stabilizer, tableting aid
2. Powder; white, tan; slight odor; pH = 3-7 for a 5% (by weight) solution
3. Polymer of 1-vinyl-2-pyrrolidone, obtained from chemical synthesis

Pyridoxine Hydrochloride (5-Hydroxy-6-methyl-3,4-pyridinedimethanol hydrochloride; Vitamin $B_6$ hydrochloride)
1. Dietary supplement, nutrient
2. Crystals, powder; white; stable in air, decomposition in sunlight; MP = 206°C (403°F); solutions acidic to litmus
3. $C_8H_{11}NO_3 \cdot HCl$

Quinine Hydrochloride
1. Flavoring
2. Crystals, needle-like; white, glistening; no odor; bitter taste; effloresces in warm air, solutions neutral or alkaline to litmus.
3. $C_{20}H_{24}N_2O_2 \cdot HCl \cdot 2H_2O$

Quinine Sulfate
1. Flavoring
2. Crystals (needle-like); white, discoloration in light; no odor; persistent and bitter taste; solution (saturated) neutral or alkaline to litmus
3. $(C_{20}H_{24}N_2O_2)_2 \cdot H_2SO_4 \cdot 2H_2O$

Rhodinol
1. Flavoring
2. Liquid; colorless; odor resembling rose
3. A mixture of terpenic alcohols

Rhodinyl Acetate
1. Flavoring
2. Liquid; colorless, pale yellow; odor resembling fresh roses
3. A mixture of acetates of terpene alcohols

Rhodinyl Formate
1. Flavoring
2. Liquid; colorless or pale yellow; leafy odor, resembling rose
3. A mixture of formates of terpene alcohols

Riboflavin (Vitamin $B_2$)
1. Dietary supplement, nutrient

**Table 1.1.1. (continued)**

Usual or common names
1. Usage
2. Description
3. General formula

2. Crystalline powder; yellow, orange; slight odor; MP = 280°C (536°F); solution decomposes in light
3. $C_{17}H_{20}N_4O_6$

Riboflavin 5′-Phosphate Soldium (Riboflavin 5′-phosphate ester monosodium salt)
1. Dietary supplement, nutrient
2. Crystalline powder, fine; orange yellow; slight odor; solution decomposes in light, hygroscopic.
3. $C_{17}H_{20}N_4NaO_9P \cdot 2H_2O$

Rice Bran Wax
1. Coating agent, masticatory substance in chewing gum base
2. Wax, hard, crystalline, refined; tan, light brown
3. Complex chemical structure

Rose Oil
1. Flavoring
2. Liquid, viscous at 25°C (77°F), cooling gradually to a translucent, crystalline mass; colorless or yellow; odor and taste resembling roses
3. Triglycerides and fatty acids

Rosemary Oil
1. Flavoring
2. Liquid; colorless, yellow; odor resembling rosemary; warm and camphoraceous taste
3. Triglycerides and fatty acids

Rue Oil
1. Flavoring
2. Liquid; yellow, amber; odor resembling fat
3. Triglycerides and fatty acids

Saccharin (*o*-Benzosulfimide; Gluside; 1,2-Benzisothiazolin-3-one 1,1-dioxide)
1. Nonnutritive sweetener
2. Crystals (white), powder (white); faintly aromatic odor; intensely sweet taste
3. $C_7H_5NO_3S$

Sage Oil, Dalmatian
1. Flavoring
2. Liquid; yellow, green; odor and flavor resembling thujone, camphoraceous essence

3. Triglycerides and fatty acids

Sage Oil, Spanish

1. Flavoring
2. Liquid, oily; colorless, pale yellow; camphoraceous odor with an essence of cineole
3. Triglycerides and fatty acids

Sandalwood Oil, East Indian

1. Flavoring agent
2. Liquid, relatively viscous; pale yellow; heavy and persistent odor
3. Triglycerides and fatty acids

Santalol

1. Flavoring
2. Liquid, viscous; colorless or pale yellow; odor resembling sandalwood
3, $C_{15}H_{24}O$

Santalyl Acetate

1. Flavoring
2. Liquid; colorless or pale yellow; odor resembling sandalwood oil
3. $C_{17}H_{26}O_2$

Savory Oil, Summer Variety (Summer savory oil)

1. Flavoring
2. Liquid; yellow, brown; spicy odor resembing thyme or origanum
3. Triglycerides and fatty acids

DL-Serine (DL-2-Amino-3-hydroxypropanoic acid)

1. Dietary supplement, nutrient
2. Crystals or powder; MP (decomposition) = 246°C (475°F); optically inactive
3. $C_3H_7NO_3$

L-Serine

1. Supplement, nutrient
2. Powder, crystalline; white; odorless; sweet taste; MP (decomposition point) = 228°C (442°F)
3. $C_3H_7NO_3$

Silicon Dioxide (Synthetic amorphous silica)

1. Anticaking agent, carrier, conditioning agent, defoaming agent
2. Solid (amorphous and noncrystalline), powder (white, nongritty, light, fluffy), beads or granules (white micocellular); hygroscopic
3. $SiO_2$

Sodium Acetate

1. Buffer
2. Crystals (colorless, transparent), powder (colorless, transparent, granular); no odor; efflorescent in warm air

**Table 1.1.1. (continued)**

Usual or common names
1. Usage
2. Description
3. General formula

3 $C_2H_3NaO_2 \cdot 3H_2O$

Sodium Acetate, Anhydrous
1. Buffer
2. Powder, granular; white; no odor; hygroscopic
3. $C_2H_3NaO_2$

Sodium Acid Pyrophosphate (Disodium pyrophosphate; Disodium dihydrogen pyrophosphate)
1. Buffer, leavening agent, sequestrant
2. Powder, free flowing (white), masses (fused); pH = 4 for a 1% (by weight) solution
3. $Na_2H_2P_2O_7$

Sodium Alginate (Algin)
1. Emulsifier, stabilizer, thickener
2. Powder, fibrous, granular; white, yellow
3. $(C_6H_7O_6Na)_n$

Sodium Aluminum Phosphate, Acidic
1. Leavening agent
2. Powder; white; no odor
3. $NaAl_3H_{14}(PO_4)_8 \cdot 4H_2O$ or $Na_3Al_2H_{15}(PO_4)_8$

Sodium Aluminum Phosphate, Basic (Kasal)
1. Emulsifier
2. Powder; white; odorless
3. A mixture of alkaline sodium aluminum phosphate with approximately 30% dibasic sodium phosphate

Sodium Ascorbate (Vitamin C sodium; Sodium L-ascorbate)
1. Antioxidant, dietary supplement, nutrient
2. Crystals; white; pH = 7.5 for a 10% (by weight) solution
3. $C_6H_7NaO_6$

Sodium Benzoate
1. Antimicrobial agent, preservative
2. Powder or flakes, crystalline or granular; white
3. $C_7H_5NaO_2$

Sodium Bicarbonate (Baking soda)
1. Alkali, leavening agent

2. Powder, crystalline; white; stable in dry air, degrading in the presence of moist air; cold, unagitated solution alkaline to litmus with increasing alkalinity under heat and agitation
3. $NaHCO_3$

Sodium Bisulfate (Sodium acid sulfate; Nitre cake)
1. Acidifier
2. Crystals or granules; white; decomposing into sodium sulfate and free sulfuric acid in alcohol; aqueous solutions acidic
3. $NaHSO_4$

Sodium Bisulfite
1. Preservative
2. Crystals or granular powder; white or yellow; odor resembling sulfur dioxide; unstable in air
3. $NaHSO_3$ and $Na_2S_2O_5$ (different ratios)

Sodium Carbonate
1. Alkali
2. Crystals or granular, crystalline powder; colorless, white; hydration extent varies; anhydrous salt hygroscopic; MP (dehydration point) = 32°C (89.6°F); efflorescent hydrates; aqueous solutions alkaline to litmus
3. $Na_2CO_3 \cdot xH_2O$

Sodium Carboxymethylcellulose (CMC1; Cellulose gum)
1. Stabilizer, thickener
2. Powder or granules; white or cream colored; powder hygroscopic; pH = 6.5-8.5 for a 1% (by weight) solution
3. Polysaccharide

Sodium Chloride (Salt)
1. Flavoring and intensifier
2. Crystals; white; can be transparent to opaque; at low RH (<75%), dry in the presence of moist air); at higher RH, salt deliquiescent
3. NaCl

Sodium Citrate
1. Buffer, nutrient for cultured buttermilk, sequestrant
2. Crystals or powder; colorless, white; may be anhydrous or hydrated with two molecules of water of crystallization.
3. $C_6H_5NaO_7 \cdot 2H_2O$

Sodium Dehydroacetate [Sodium 3-(1-Hydroxyethylidene)-6-methyl-1,2-pyran-2,4(3*H*)-dione]
1. Preservative
2. Powder; white; odorless; slight taste
3. $C_8H_7NaO_4 \cdot H_2O$

Sodium Diacetate (Sodium hydrogen diacetate)

**Table 1.1.1. (continued)**

Usual or common names
1. Usage
2. Description
3. General formula

1. Mold and rope inhibitor, preservative, sequestrant
2. Crystals; hygroscopic; odor resembling acetic acid; pH = 4.5-5.0 for a 10% (by weight) solution
3. $CH_3COONa{\bullet}CH_3COOH\ {\bullet}xH_2O$

Sodium Erythorbate
1. Antioxidant, preservative
2. Powder or granules, crystalline; white; odorless; stable in dry air, solutions decomposing in the presence of air, heat, light, and trace metals
3. $C_6H_7NaO_6{\bullet}H_2O$

Sodium Ferric Pyrophosphate (Sodium iron pyrophosphate)
1. Supplement, nutrient
2. Powder; white, brown; odorless
3. $Na_8Fe_4(P_2O_7)_5{\bullet}xH_2O$

Sodium Ferrocyanide (Yellow Prussiate of soda)
1. Anticaking agent (for sodium chloride)
2. Crystals (yellow) or crystalline powder
3. $Na_4Fe(CN)_6{\bullet}10H_2O$

Sodium Gluconate
1. Supplement, nutrient, sequestrant
2. Powder, granular, crystalline; white, light brown
3. $CH_2OH(CHOH)_4COONa$

Sodium Hydroxide (Caustic soda)
1. Alkali
2. Masses (flakes, sticks, pellets, or fused); white; absorption of carbon dioxide and moisture in the presence of air
3. $NaOH$

Sodium Lauryl Sulfate
1. Surfactant
2. Crystals, small; white, pale yellow; slight odor
3. A mixture of sodium alkylsulfates, mainly of sodium lauryl sulfate, $[CH_3(CH_2)_{10}CH_2OSO_3Na]$

Sodium Metabisulfite (Sodium pyrosulfate)
1. Antioxidant, preservative
2. Crystals (colorless), powder (white, yellow); odor resembling sulfur dioxide; solutions acidic to litmus
3. $Na_2S_2O_5$

Sodium Metaphosphate (Sodium polyphosphate; Glassy sodium phosphate; Sodium hexametaphosphate; Sodium tetraphosphate; Graham's salt; Kurrol's salt; Sodium trimetaphosphate; Sodium tetrametaphosphate; Insoluble sodium metaphosphate)
1. Emulsifier, sequestrant, texturizer
2. Crystals to amorphous solids; colorless; hygroscopic
3. A group of phosphates ranging from crystalline to amorphous solids, e.g., $P_2 0_5$

Sodium Methylate (Sodium methoxide)
1. Catalyst used in the transesterification of fats
2. Powder, amorphous, free-flowing; white; hygroscopic; reactive with oxygen and carbon dioxide; decomposing in water; inflammable; dangerous and harmful (contact must be avoided)
3. $CH_3ONa$

Sodium Nitrate
1. Color fixative in meat products, preservative
2. Crystals (colorless), granules (crystalline); odorless; deliquescent in the presence of moist air
3. $NaNO_3$

Sodium Nitrite
1. Color fixative in meat, preservative
2. Powder, granular; white, yellow; fused masses or sticks (white or opaque); mild, saline taste; solutions alkaline to litmus
3. $NaNO_2$

Sodium Phosphate, Dibasic (Disodium monohydrogen phosphate; Disodium phosphate)
1. Buffer, dietary supplement, emulsifier, nutrient, texturizer
2. Powder, anhydrous; white; hygroscopic;, water of hydration usually two
3. $Na_2HPO_4$

Sodium Phosphate, Monobasic (Monosodium phosphate; Sodium biphosphate; Monosodium dihydrogen phosphate)
1. Buffer, dietary supplement, emulsifier, nutrient
2. Powder, anhydrous; hygroscopic; odorless; pH = 4.1-4.7 for a 1% (by weight) solution; water of hydration usually one or two
3. $NaH_2PO_4$

Sodium Phosphate, Tribasic (Trisodium phosphate)
1. Buffer, dietary supplement, emulsifier, nutrient
2. Crystals, granules, or powder; pH = 11.5-12.0 for a 1% (by weight) solution; water of hydration varies from 1 to 12
3. $Na_3PO_4$

Sodium Potassium Tartrate (Rochelle salt)

**Table 1.1.1. (continued)**

Usual or common names
1. Usage
2. Description
3. General formula

1. Buffer, sequestrant
2. Crystals (colorless), powder (white); saline and cooling taste; efflorescent in warm dry air
3. $KOOCCH(OH)\cdot(CH(OH)\cdot COONa\cdot 4H_2O$

Sodium Propionate
1. Mold and rope inhibitor, preservative
2. Crystals (colorless, transparent) powder (granular, crystalline); odorless, with a faint aceticbutyric acid scent; deliquescent in the presence of moist air; pH = 8.0-10.5 for a 10% (by weight) solution
3. $CH_3CH_2COONa$

Sodium Pyrophosphate (Tetrasodium diphosphate; Tetrasodium pyrophosphate)
1. Buffer, dietary supplement, emulsifier, nutrient
2. Powder, crystalline, granular; white; may be anhydrous or water of hydration 10 (efflorescent in dry air); pH = 10 for a 1% (by weight) solution
3. $Na_4P_2O_7$

Sodium Saccharin (1,2-Benzisothiazolin-3-one 1,1-Dioxide sodium salt; Sodium-*o*-benzosulfimide; Soluble saccharin)
1. Nonnutritive sweetener
2. Crystals (white), powder, crystalline (white); odorless to odor faintly aromatic; intensely sweet taste; powdered form efflorescent
3. $C_7H_4NNaO_3S\cdot 2H_2O$

Sodium Sesquicarbonate
1. Alkalizer, neutralizer in certain dairy products
2. Flakes or powder (crystalline); solutions alkaline to litmus
3. $Na_2CO_3NaHCO_3\cdot 2H_2O$

Sodium Stearoyl-2-Lactylate
1. Emulsifier, dough conditioner, stabilizer, whipping agent
2. Powder; cream colored, slightly hygroscopic; mildly caramel odor
3. A mixture of sodium salts of stearoyl lactylic acids, with small amounts of other sodium salts of related acids

Sodium Stearyl Fumarate
1. Dough conditioner
2. Powder, fine; white
3. $C_{22}H_{39}NaO_4$

Sodium Sulfate
1. An ingredient for preparing caramel
2. Crystals or fine powder; colorless, white; may be anhydrous or a water of crystallization of 10; hydrate product efflorescent; pH = 8-10 for a 5% (by weight) solution
3. $Na_2SO_3$

Sodium Sulfite (Exsiccated sodium sulfite)
1. Antioxidant, preservative
2. Powder; white, pink, brown; odorless; cool, saline, sulfurous taste; oxidation in the presence of air; solutions alkaline to litmus and phenolphthalein
3. $C_4H_4Na_2O_6 \cdot 2H_2O$

Sodium Thiosulfate (Sodium hyposulfite)
1. Antioxidant, sequestrant
2. Crystals, large (colorless), coarser crystalline powder; deliquescent in moist air, efflorescent in dry air [>33°C (91.4°F)]; solutions neutral or slightly alkaline to litmus
3. $Na_2S_2O_3 \cdot 5H_2O$

Sodium Tripolyphosphate (Pentasodium triphosphate; Triphosphate; Sodium triphosphate)
1. Texturizer
2. Granules, powder; white; slightly hygroscopic; may be anhydrous or a water of hydration of six; pH = 9.5 for a 1% (by weight) solution
3. $Na_5P_3O_{10}$

Sorbic Acid (2,4-Hexadienoic acid)
1. Preservative
2. Powder, free-flowing; white; characteristic odor
3. $CH_3CH{=}CHCH{=}CHCOOH$

Sorbitan Monostearate
1. Defoaming agent, emulsifier, stabilizer
2. Waxy substance, hard; cream, tan colored; bland odor and taste
3. A mixture of partial stearic and palmitic acid esters of sorbitol and its mono- and dianhydrides.

Sorbitol (D-Sorbitol; D-Glucitol; D-Sorbite; D-Sorbol; 1,2,3,4,5,6-hexanehexol)
1. Humectant, sequestrant, texturizing agent
2. Flakes or granules; white; hygroscopic; sweet taste; MP = 89 - 101°C (192.2 - 213°F); neutral to litmus
3. $C_6H_{14}O_6$

Spearmint Oil
1. Flavoring
2. Liquid; colorless or yellow; odor and taste resembling spearmint
3. Triglycerides and fatty acids

**Table 1.1.1. (continued)**

Usual or common names
1. Usage
2. Description
3. General formula

Spike Lavender Oil
1. Flavoring
2. Liquid; yellow; camphoraceous odor resembling lavender
3. Triglycerides and fatty acids

Stannous Chloride
1. Antioxidant, reducing agent
2. Crystals; white or colorless; odorless
3. $SnCl_2 \cdot 2H_2O$

Stearic Acid (Octadecanoic acid)
1. Defoaming agent, lubricant, an ingredient used in producing food additives
2. Crystalline substance, hard (white, yellow, glossy), powder (white, yellow); slight odor; taste resembling tallow
3. $C_{18}H_{36}O_2$

Stearyl Monoglyceridyl Citrate
1. Emulsion stabilizer
2. Waxy substance, soft, lardy; white, brown
3. Complex chemical strucutre

Succinic Acid (Butanedioic acid)
1. Buffer, neutralizing agent, and other applications in in food processing
2. Crystals; colorless, white; acidic taste; odorless
3. $HOOCCH_2CH_2COOH$

Succinylated Monoglycerides
1. Dough conditioner, emulsifier
2. Waxy substance; white; bland taste; MP = 60°C (140°F)
3. A mixture of succinic acid esters of mono- and diglycerides

Sulfuric Acid
1. Acidifier
2. Liquid, oily, corrosive; clear, colorless; extreme caution is needed when used under all applications.
3. $H_2SO_4$

Tangerine Oil, Expressed
1. Flavoring
2. Liquid, oily; red, brown, orange; pleasant aroma, resembling orange; oils obtained from unripe fruit may be green.
3. Triglycerides and fatty acids

Tannic Acid

1. Clarifier
2. Amorphous powder, glistening scales, and spongy masses; yellow, white, brown; usually odorless, with an occasional faint odor; astringent taste
3. Complex chemical structure

Tarragon Oil (Estragon oil)
1. Flavoring
2. Liquid; yellow, amber; mildly spicy odor resembling licorice and sweet basil
3. Triglycerides and fatty acids

Tartaric Acid [L(+)-Tartaric acid]
1. Acidifier, sequestrant
2. Crystals (colorless or translucent), fine granular, crystalline powder; white; no odor; acidic taste; stable in air; solutions dextrorotatory
3. $C_4H_6O_6$

Terpene Resin, Natural
1. Masticatory substance in chewing gum base
2. Resinous powder
3. Complex chemical structure

Terpene Resin, Synthetic
1. Masticatory substance in chewing gum base
2. Resinous powder
3. Complex chemical structure

Terpineol (Menthen-1-ol-8)
1. Flavoring
2. Liquid; clear, colorless; odor resembling lilac
3. $C_{10}H_{13}O$

Terpinyl Propionate (Menthen-1-yl-propionate)
1. Flavoring
2. Liquid; colorless or pale yellow; flowery odor resembling lavender
3. $C_{13}H_{22}O_2$ (a mixture of alpha-, beta-, and gamma-isomers)

Thiamine Hydrochloride (Aneurine hydrochloride; Thiamine chloride; Vitamin $B_1$; Vitamin $B_1$ hydrochloride)
1. Dietary supplement, nutrient
2. Crystals, small white, crystalline powder; faint odor; anhydrous product absorbing 1-4% water in the presence of air; MP (decomposition point) = 248°C (478°F)
3. $C_{12}H_{17}ClN_4OS \cdot HCl$

Thiamine Mononitrate (Thiamine nitrate; Vitamin $B_1$ mononitrate)
1. Dietary supplement, nutrient
2. Crystals (white), crystalline powder; faint odor
3. $C_{12}H_{17}N_5O_4S$

L-Threonine (L-2-Amino-3-hydroxybutyric acid)

**Table 1.1.1. (continued)**

Usual or common names
1. Usage
2. Description
3. General formula

1. Dietary supplement, nutrient
2. Powder, crystalline; white; odorless; faintly sweet taste; MP (decomposition point) = 256°C (493°F)
3. $C_4H_9NO_3$

Thyme Oil
1. Flavoring agent
2. Liquid; colorless or yellow or red; pleasing odor; pungent, persisitent taste; degraded by light
3. Triglycerides and fatty acids

Titanium Dioxide
1. Colorant
2. Powder, amorphous; white; no odor; no taste
3. $TiO_2$

*dl-α*-Tocopherol
1. Antioxidant, a supplement, nutrient, a vitamin
2. Liquid, viscous; yellow, amber, clear; discoloration and oxidation in the presence of air and light; odorless
3. $C_{29}H_{50}O_2$

Tocopherols Concentrate, Mixed
1. Antioxidant, dietary supplement, nutrient, vitamin E source
2. Liquid, viscous; clear, brown, red; odorless; oxidizes and darkens slowly
3. Complex chemical structure

*d-α*-Tocopheryl Acetate (*d-α*-tocopherol acetate)
1. Dietary supplement, nutrient, a source of vitamin E
2. Liquid, oily and viscous; yellow, clear; odorless; decomposed by light; unstable when alkali.
3. $C_{31}H_{52}O_3$

*d-α*-Tocopheryl Acid Succinate (*d-α*-tocopherol acid succinate)
1. A dietary supplement, nutrient, crystalline form of vitamin E
2. Powder, crystalline; white; odorless and tasteless; stable in air, unstable to heat and alkali
3. $C_{33}H_{54}O_5$

*p*-Tolyl Isobutyrate (*p*-Cresyl isobutyrate)
1. Flavoring
2. Liquid; colorless; odorless
3. $C_{11}H_{14}O_2$

Tragacanth
1. Emulsifier, stabilizer, thickener
2. Pieces; white, yellow; ~2.5 mm in thickness, translucent, horny in texture, with a short fracture; odorless; pasty taste
3. Polysaccharide

Triacetin (Gylceryl triacetate)
1. Humectant, solvent
2. Liquid, oily; colorless; odor resembling fat; bitter taste; distillation point = 258-270°C (496-518°F)
3. $C_9H_{14}O_6$

Tributyrin (Glyceryl tributyrate; Butyrin)
1. Flavoring
2. Liquid, oily; colorless; unique odor; bitter taste; distillation point = 305-310°C (581-590°F)
3. $C_{15}H_{26}O_6$

Trichlorethylene (Ethylene trichloride)
1. Extraction solvent
2. Liquid, mobile; clear, colorless; odor resembling chloroform
3. Chlorinated hydrocarbon

Triethyl Citrate (Ethyl citrate)
1. Sequestrant
2. Liquid, oily; colorless; odorless
3. $C_{12}H_{20}O_7$

DL-Tryptophan (DL-$\alpha$-Amino-3-indolpropionic acid)
1. Dietary supplement, nutrient
2. Powder, crystals or crystalline; white; odorless; optically inactive
3. $C_{11}H_{12}N_2O_2$

L-Tryptophan (L-$\alpha$-Amino-3-indolpropionic acid)
1. Dietary supplement, nutrient
2. Crystals (white, yellow), crystalline powder; odorless; slightly bitter taste
3. $C_{11}H_{12}N_2O_2$

L-Tyrosine [L-$\beta$-(*p*-Hydroxyphenyl)alanine]
1. Dietary supplement, nutrient
2. Powder, needles or crystalline; white
3. $C_9H_{11}NO_3$

$\gamma$-Undecalactone (Aldehyde C-14; Peach aldehyde)
1. Flavoring
2. Liquid; colorless or yellow; strong, fruity odor, resembling peach when diluted
3. $C_{11}H_{20}O_2$

Undecanal (Aldehyde C-11 undecyclic; *n*-Undecyl aldehyde)

**Table 1.1.1. (continued)**

Usual or common names
1. Usage
2. Description
3. General formula

1. Flavoring
2. Liquid; colorless, pale yellow; sweet odor
3. $CH_3(CH_2)_9CHO$

10-Undecenal (Aldehyde C-11 undecylenic; Undecene-10-al)
1. Flavoring
2. Liquid; colorless, pale yellow; odor resembling fat
3. $CH_2{+}CH(CH_2)_8CHO$

Valeric Acid (Pentanoic acid)
1. Flavoring
2. Liquid, mobile; clear, colorless; unpleasant, rancid, penetrating odor
3. $CH_3(CH_2)_3COOH$

L-Valine (L-2-Amino-3-methylbutyric acid)
1. Dietary supplement, nutrient
2. Powder, crystalline; white; odorless; unique taste; pH = 5.5-7.0 for a 5% (by weight) solution; MP = 315°C (599°F)
3. $C_5H_{11}NO_2$

Vanillin (4-Hydroxy-3-methoxy benzaldehyde)
1. Flavoring
2. Crystals, needlelike; white, pale yellow; smell and taste like vanilla; decomposing in light
3. $C_8H_8O_3$

Vitamin A
1. Dietary supplement, nutrient
2. Liquid, oily; yellow, red; solidifies when cold; solid form available commercially; odorless to fish like aroma; bland taste; unstable in light and air
3. Derivative of $C_{20}H_{30}O$; retinol or esters, may contain preservatives and antioxidants

Vitamin $D_2$ (Calciferol; Ergocalciferol)
1. Dietary supplement, nutrient
2. Crystals; white; odorless; decomposes in air and light.
3. $C_{28}H_{44}O$

Vitamin $D_3$; Activated 7-dehydrocholesterol; Cholecalciferol
1. Dietary supplement, nutrient
2. Crystals; white; odorless; decomposition in the presence of air and light
3. $C_{27}H_{44}O$

Xanthan Gum
1. Bodying agent, emulsifier, foam enhancer, stabilizer, suspending agent, stabilizer, thickener
2. Powder; cream-colored; solutions neutral
3. Polysaccharide

Zinc Sulfate
1. Dietary supplement, nutrient
2. Prisms or small needles (transparent, colorless), granular, crystalline powder; no odor; efflorescent in dry air; solution (5% by weight) has a pH of 4.4-6.0.
3. $ZnSo_4 \cdot 7H_2O$

---

Major sources of data are: 1) 21 CFR 170-199, 1991; 2) U.S. Food and Drug Administration, Inspections Operations Manual, Washington, D.C., 1991.

Minor reference materials include: 1) Food Chemical Codex, 3rd ed., National Academy of Sciences, National Research Council, Washington, DC, 1981; 2) U. S. Pharmacopeia, 11th ed., U. S. Pharmacopeial Convention, Rockville, Maryland, 1991; 3) Merck Index, 11th ed., Merck & Co., Rahway, New Jersey, 1989.

## 1.2. Economic Food Chemicals

**Table 1.2.1. Alphabetical Listing of Pesticide Chemicals for Which the Environmental Protection Agency Has Established Tolerances for Their Residues in Raw Agricultural Products**

---

Acephate
Acetaldehyde
Acetic acid
Alachlor
Aldicarb
Allethrin (allyl homolog of cinerin I)
Aluminum isopropoxide and aluminum secondary butoxide
Aluminum phosphite
Aluminum tris(*o*-ethylphosphonate)
Ametryn
2-Amino-4,5-dihydro-6-methyl-4-propyl-*s*-triazolo-(1,5-$\alpha$)-pyrimidin-5-one
4-Amino-6-(1,1-dimethylethyl)-3-(methylthio)-1,2,4-triazin-5(4*H*)-one
4-Aminopyridine
Amitraz
Ammonia

**Table 1.2.1. (continued)**

Ammoniates of [ethylenebis(dithiocarbamato)]zinc and ethylenebis[dithiocarbamic acid] bimolecular and trimolecular cyclic anhydrosulfides and disulfides; interim tolerance
Ammonium nitrate
Ammonium sulfamate
Aqueous extract of seaweed meal
Asulam
Atrazine
Avermectin $B^1$ and its $\delta$-8,9-isomer
*Bacillus popilliae*, viable spores
*Bacillus thuringiensis berliner*, viable spores
Barban
Benomyl
Bensulfuron methyl ester
Bentazon
6-Benzyladenine
Bifenox
Bifenthrin
Biphenyl
1,1-Bis(*p*-chlorophenyl)-2,2,2-trichloroethanol
Boron
Bromacil
Bromoxynil
Bufencarb
Butanoic anhydride
Butralin
*sec*-Butylamine
Butyl benzyl phthalate
*N*-Butyl-*N*-ethyl-$\alpha$-$\alpha$-$\alpha$-trifluoro-2,6-dinitro-*p*-toluidine
Cacodylic acid
Calcium arsenate
Calcium cyanide
Calcium hypochlorite
Captafol
Captan
Carbaryl
Carbaryl (1-naphthyl *N*-methylcarbamate and its metabolite 1-naphthol, calculated as carbaryl; interim tolerance
Carbon dioxide
Carbon tetrachloride

Carbofuran
Carbophenothion
Carboxin
Chloramben
Chlordimeform
Chlorimuron ethyl
2-Chloroallyl diethyldithiocarbamate
2-Chloro-*N,N*-diallylacetamide
2-Chloro-1-(2,4,dichlorophenyl)vinyldiethyl phosphate
2-{[4-Chloro-6-(ethylamino)-*S*-triazin-2-yl] amino}-2-methylpropionitrile
2-Chloro-*N*-isopropylacetanilide
Chloroneb
*p*-Chlorophenoxyacetic acid
1-(4-Chlorophenoxy)-3,3-dimethyl(1*H*-1,2,4-triazol-1-yl)-2-butanone
2-(*M*-Chlorophenoxy)propionic acid
2-(2-Chlorophenyl)methyl-4,4-dimethyl 3-isoxazolidinone
Chloropicrin
Chlorosulfamic acid
Chlorothalonil
Chlorotoluene
2-Chloro-1-(2,4,5-trichlorophenyl)vinyldimethyl phosphate
Chloroxuron
Chlorpyrifos
Chlorpyrifos-methyl
Chlorsulfuron
Chlorthiophos
CIPC
Clofentezine
Clopyralid
*Colletrotorichum gloeosporioides* f.sp. *aeschynomene*
Combustion product gas
Coordination product of zinc ion and maneb
Coordination product of zinc ion and maneb; interim tolerance
Copper
Copper carbonate, basic
Coumaphos
Cross-linked nylon-type encapsulating polymer
Cross-linked polyurea-type encapsulating polymer
Cross-linked polyurea-type encapsulating polymer (alachlor)
Crufomate
Cyano(3-phenoxyphenyl)methyl-4-chloro-$\alpha$-(methylethyl) benzenacetate

**Table 1.2.1. (continued)**

[1 α-(S),3 α-(Z)]-(±)-Cyano(3-phenoxyphenyl) methyl 3-(2-chloro-3,3,3-trifluoro-1-propenyl)-2,2-dimethylcyclopropane-carboxylate.
Cycloheximide
Cyfluthrin
Cyhexatin
Cypermethrin
Cyprazine
Cyromazine
2,4-D
Dalapon
Daminozide
Demeton
Desmedipham
Dialifor
*N,N*-Diallyl dichloroacetamide
Diatomaceous earth
Diazinon
Dicamba
Dichlobenil
Dichlone
*S*-2,3-Dichloroallyl diisopropylthiocarbamate
1,1-Dichloro-2,2-bis(*p*-ethylphenyl) ethane
2,4-Dichloro-6-*O*-chloroanilino-*s*-triazine
3,5-Dichloro-*N*-(1,1-dimethyl-2-propynyl)benzamine
2,2-Dichloro-*N*-(1,3-dioxolan
2,2-Dichloro-*N*-(1,3-dioxolan-2-ylmethyl)-*N*-2-prophenylacetamide
2,6-Dichloro-4-nitroaniline
4-(2,4-Dichlorophenoxy) butyric acid
3-(3,5-Dichorophenyl)-5-ethenyl-5-methyl-2,4-oxazolidinedione
2,4-Dichlorophenyl *p*-nitrophenyl ether
2,2-Dichlorovinyldimethyl phosphate
Diclofop-methyl
Diethatyl-ethyl
*O,O*-Diethyl-*O*-(2-diethylamino-6-methyl-4-pyrimidinyl) phosphorothioate and its oxygen analog
*O,O*-Diethyl-*S*-[2-(ethylthio)ehtyl] phosphorodithioate
*O,O*-Diethyl-*O*-(*p*-methylsultinyl) phospherothioate
*N,N*-Diethyl-2-(1-naphthalenyloxy)propionamide
*O,O*,Diethyl-*O*-phenylphosphorothioate
Difenzoquat

Diflubenzuron
2-[4,5-Dihydro-4-methyl-4(1-methylethyl)-5-oxo-1H-imidazol-2-yl]-3-quinoline carboxylic acid
*S*-(*O*,*O*-Diisopropyl phosphorodithioate) of *N*-(2-mercaptoethyl) benzenesulfonamide
Dimethipin
Dimethoate including its oxygen analog
*O*,*O*-Dimethyl *O*-*p*-(dimethylsulfamoyl)-phenyl phosphorothioate including its oxygen analog
3,5-Dimethyl-4-(methylthio)phenyl methylcarbamate
*O*,*O*-Dimethyl *S*-[4-oxo-1,2,3-benzotriazin-3(4*H*)-ylmethyl] phosphorodithioate
*O*,*O*-Dimethyl-*S*-[4-oxo-1,2,3-benzotriazin-3(4*H*)-ylmethyl] phosphorodithioate residues and/or its metabolites in milk
Dimethyl phosphate of 3-hydroxy-*N*,*N*-dimethyl-*cis*-crotonamide
Dimethyl phosphate of 3-hydroxy-*N*-methyl-*cis*-crotonamide
Dimethyl phosphate of $\alpha$-methylbenzyl-3-hydroxy-*cis*-crotonate
Dimethyl tetrachloroterephthalate
Dimethylformamide
*N* *N*-Dimethylpiperidium chloride
Dimethyl sulfoxide
Dimethyl-(2,2,2-trichloro-1-hydroxyethyl) phosphonate
2,6-Dimethyl-4-trideclymorpholine
Dinitramine
4,6-Dinitro-*O*-cresol and its sodium salt
2,4-Dinitro-6-octylphenyl crotonate and 2,6-dinitro-4-octylphenyl crotonate
Dinoseb
Dioxathion
Diphenamid
Diphenylamine
Dipropetryn
Dipropyl isocinchomeronate
Diquat
Diuron
Dodine
Egg solids (whole)
Endosulfan
Endothall
Endothall [7-oxabicyclo-(2,2,1) heptane 2,3-dicarboxylic acid]; interim tolerance
Endrin
EPN

**Table 1.2.1. (continued)**

| |
|---|
| Ethalfluralin |
| Ethephon |
| Ethion |
| Ethoprop |
| Ethofumesate |
| Ethoxyquin |
| 5-Ethoxy-3-trichloromethyl-1,2,4-thiadiazole |
| 2-[1-(Ethoxyimino)butyl]-5-[2-(ethylthio)propyl]-3-hydroxy-2-cyclohexane-1-one |
| *S*-Ethylcyclohexylethyl thiocarbamate |
| Ethyl-4,4′-dichlorobenzilate |
| *S*-Ethyldiisobutyl thiocarbamate |
| *S*-Ethyl dipropylthiocarbamate |
| *S*-Ethyl hexahydro-1*H*-azepine 1-carbothioate |
| Ethyl-3-methyl-4-(methylthio)phenyl(1-methylethyl) phosphoramidate |
| *O*-Ethyl-*O*-[4-(methylthio)phenyl] *S*-propyl phosphorodithioate |
| *O*-Ethyl *S*-phenyl ethylphosphonodithioate |
| Ethylene |
| Ethylene dibromide |
| Ethylene glycol |
| Ethylene oxide |
| *S*-[2-(Ethylsulfinyl)ethyl]-*O*,*O*-dimethylphosphorothioate |
| FD & C Blue No. 1 |
| Fenarimol |
| Fenoxaprop-ethyl |
| Fenridazon |
| Fenthion |
| Ferbam |
| Fluazifop-butyl |
| Fluchloralin |
| Flucythrinate |
| Fluometuron |
| Fluridone |
| Fluorine compounds |
| ($\alpha$-RS,2R)-Fluvalinate[(RS)-$\alpha$-cyano-3-phenoxybenzyl (R)-2-[2-chloro-4-(trifuoro-methyl)anilino]3-methylbutanoate] |
| Folpet |
| Formaldehyde |
| Formetanate hydrochloride |
| Gibberellins |

Glyodin
Glyphosate
Gossyplure
Heptachlor and heptachlor epoxide
Heptachlor; interim tolerance
Hexachlorophene
(*Z*)-11-Hexadecenal
Hexakis(2-methyl-2-phenylpropyl)distannoxane
Hexazinone
Hexythiazox
*Hirsutella thompsonii*
Hydrogen cyanide
Hydrogenated castor oil
Imazlil
Imazethapyr, ammonium salt
Inorganic bromide residues resulting from soil treatment with ethylene dibromide
Inorganic bromide residues in peanut hay and peanut hulls
Inorganic bromide residues resulting from fumigation with methyl bromide
Inorganic bromides resulting from soil treatment with 1,2-dibromo-3-chloropropane
Inorganic bromides resulting from soil treatment with combinations of chloropicrin, methylbromide, and proparagyl bromide
Interim tolerances
Iodene-detergent complex
Iprodione
Isobutyric acid
Isomate-M
Isopropalin
Isopropyl M-chlorocarbanilate (CIPC); interim tolerance
Isopropyl carbanilate (IPC); interim tolerance
Kontrol H.V.
Lactic acid
Lactofen
Lead arsenate
Lindane
Linuron
Linseed oil, boiled
Magnesium phosphide
Malathion
Maleic hydrazide

**Table 1.2.1. (continued)**

| |
|---|
| Maneb |
| Manganous dimethyldithiocarbamate |
| Mefluidide |
| Menthol |
| Mercaptobenzothiazole |
| *N*-(Mercaptomethyl)phthalimide-*S*-(*O*,*O*- dimethyl phosphorodithioate) and its oxygen analog |
| Metalaxyl |
| Methamidophos |
| Methanearsonic acid |
| Methazole |
| Methidathion |
| Methomyl |
| Methoprene |
| 1-(8-Methoxy-4,8-dimethylnonyl)-4-(methylethyl)benzene |
| Methoxychlor |
| Methyl $\alpha$-eleosterate |
| Methylene chloride |
| 2-Methyl-4-chlorophenoxyacetic acid |
| 1-Methyl 2{ethoxy[1-methyl(ethyl)amino] phosphinothioyl} benzoate |
| 4-(2-Methyl-4-chlorophenoxy) butyric acid |
| Methyl 3-[(dimethoxyphosphenyl)oxy] butenoate, $\alpha$ and $\beta$ isomers |
| 6-Methyl-1,3-dithiolo(4,5-b) quinoxalin-2-one |
| Methyl eugenol/malathion combination |
| Methyl 2-(4-isopropyl-4-methyl-5-oxo-2-imidazolin-2-yl)-P-Toluate and methyl 6-(4-isopropyl-4-methyl-5-oxo-2-imidazolin-2-yl)-M-toluate |
| Methyl-3-[[[[(4-methoxy-6-methyl-1,3,5-triazin-2-yl)amino]carbonyl]amino]sulfonyl]-2-thiophenecarboxylate |
| 4-(Methylsulfonyl)-2,6-dinitro-*N*,*N*-dipropylaniline |
| Metobromuron |
| Metolachlor |
| metsulfuron methyl |
| Mineral oil |
| Monocarbamide dihydrogen sulfate |
| Myclobutanil |
| Naled |
| $\alpha$-Naphthaleneacetamide |
| 1-Naphthaleneacetic acid |
| $\beta$-Naphthoxyacetic acid |
| *N*-1-Naphthyl phthalamic acid |

Nicotine-containing compounds
Nicotine
Nitrapyrin
Nitrogen
Norea
Norflurazon
*Nosema locustae*
Nuclear polyhedrosis virus of *Heliothis zea*
*N*-Octylbicycloheptenedicarboximide
2-*N*-Octyl-4-isothiazolin-3-one
(*E*,*Z*)-3,13-Octadecadien-1-ol-acetate and (*Z*,*Z*)-3,13-octadecadien-1-01 acetate
Orthoarsenic acid
Oryzalin
Oxadiazon
Oxamyl
Oxyfluorfen
Oxytetracycline hydrochloride
Paraformaldehyde
Paraquat
Parathion or its methyl homolog
Parathion (*O*,*O*, diethyl-*o*-*p*-nitrophenylthiophosphate) or its methyl homolog; interim tolerance
Pendimethalin
Pentachloronitrobenzene
Pentachloronitrobenzene; interim tolerance
Pentane
Perfluidone
Permethrin
Phenmedipham
Phenothiazine; interim tolerance
*O*-Phenylphenol and its sodium salt
Phorate
Phosalone
Phosphamidon
Phytophthora palmivora
Picloram
Pine oil
Piperonyl butoxide
Pirimiphos-ethyl
Pirimiphos-methyl
Plant volatiles/pheromone

**Table 1.2.1. (continued)**

| |
|---|
| Poly-*N*-acetyl-*D*-glucosamine |
| Polyamide polymer derived from sebacic acid |
| Polybutenes |
| Poly-D-Glucosamine |
| Poly(oxy-1,2-ethanediyl), $\alpha$-isooctadyl-$\omega$-hydroxy |
| Polyoxymethylene copolymer |
| Polyvinyl chloride |
| Potassium oleate and related $C_{12}$-$C_{18}$ fatty acid potassium salts |
| Potassium ricinoleate and related $C_{12}$-$C_{18}$ fatty acid potassium salts |
| Profenofos |
| Profluralin |
| Prometryn |
| Propanil |
| Propargite |
| Propazine |
| Propionic acid |
| *S*-Propyl butyethylthiocarbamate |
| *S*-Propyl diprophylthiocarbamate |
| Pseudomonas Fluorescens Eg-1053 |
| Pyrazon |
| Pyrethrins |
| Quizaolofop-ethyl |
| Ronnel |
| Sesone |
| Silvex |
| Silvex (2-(2,4,5-Trichlorophenoxy)-Propionic acid); interim tolerance |
| Simazine (2-chloro-4,6-bisethylamino)-*S*-triazine) |
| Simazine (2-chloro-4,6-bisethylamino)-*S*-triazine) and metabolites |
| Sodium arsenite |
| Sodium arsenite (calculated as $AS_2O_3$): interim tolerance |
| Soldium chlorate |
| Sodium chlorite |
| Sodium diacetate |
| Sodium dehydroacetate |
| Sodium dimethyldithiocarbamate |
| Sodium propionate |
| Sodium salt of acifluorfen |
| Sodium salt of fomesafen |
| Sodium trichloroacetate |
| Sulfur dioxide |

Sulfuric acid
Tartar emetic
Tebuthiuron
Tefluthrin
Temephos
Terbacil
Terbufos
Terbuthylazine
Terbutryn
1,2,4,5-Tetrachloro-3-nitrobenzene
Tetradifon
Tetraethyl pyrophosphate
Tetrahydro-5,5-dimethyl-2(1*H*)-pyrimidinone[3-(4-trifluoromethyl)phenyl]-1-[2-4-(trifluoromethyl)phenyl]ethenyl(2-propenylidene)hydrazone
Tetraiodoethylene
Thiabendazole
Thidiazuron
Thiobencarb
2-(Thiocyanomethylthio)benzothiazole
Thiodicarb
Thiophanate-methyl
Thiram
Toxaphene
Toxaphene (chlorinate camphene containing 67-69% chlorine); interim tolerance
Tralomethrin
1-Triacontanol
*S,S,S*-Tributyl phosphorotrithioate
Tributylphosphorotrithioite
*S*-2,3,3-Trichloroallyl disopropyl thiocarbamate
Triclopyr
Trichlorobenzyl chloride
1,1,1-Trichloroethane
2,3,6-Trichlorophenylacetic acid
(*E*)-4-Tridecen-1-yl acetate and (*Z*)-4-tridecen-1-yl acetate
Trifluralin
Triforine
2,3,5-Triiodobenzoic acid
2,2,5-Trimethyl-3-dichloroacetyl-1,3-oxazolidine
3,7,11-Trimethyl-1,6,10-dodecatrine-1-ol and 3,7,11-trimethyl-2,6,10-dodecatriene-3-01

**Table 1.2.1. (continued)**

3,4,5-Trimethylphenyl methylcarbamate and 2,3,5-trimethylphenyl methylcarbamate
Triphenyltin hydroxide
Xylene
Zinc phosphide
Zinc sulfate, basic
Zineb
Zineb (Zinc Ethylenebisdithiocarbamate); interim tolerance
Ziram

Source: Adapted from 40 CFR 180, 1991.

**Table 1.2.2. Alphabetical Listing of Inert Ingredients That Are Used in Pesticide Chemicals and for Which There Are No Tolerance Requirements in Raw Agricultural Products**

| Inert Ingredient(s): Use(s) |
|---|
| Acetic acid: catalyst |
| Acetic anhydride: solvent cosolvent |
| Acetone: solvent, cosolvent |
| Alkanoic and alkenoic acids, mono- and diesters of $\alpha$-hydro-$\omega$-hydroxypoly(oxyethylene): emulsifiers |
| Alkyl ($C_8$-$C_{34}$) benzenesulfonic acid and its ammonium, calcium, magnesium, potassium, sodium, and zinc salts: surfactants, related adjuvants of surfactants |
| $\alpha$-Alkyl ($C_9$-$C_{18}$) $\omega$-hydroxy poly(oxyethylene): surfactant, related adjuvant of surfactants |
| $\alpha$-Alkyl ($C_{12}$-$C_{15}$) $\omega$-hydroxypoly(oxyethylene) sulfate, ammonium, calcium, magnesium, potassium, sodium, and zinc salts: surfactants, related adjuvants of surfactants |
| $\alpha$-Alkyl ($C_8$-$C_{14}$) $\omega$-hydroxypoly(oxypropylene) block copolymer with polyoxyethylene: surfactants, related adjuvants of surfactants |
| $\alpha$-($p$-Akylphenyl)-$\omega$-hydroxy poly(oxyethylene): surfactant, related adjuvant of surfactants |
| Alkyl ($C_8$-$C_{18}$) sulfate and its ammonium, calcium, isopropylamine, magnesium, potassium, sodium, and zinc salts: surfactants |
| Almond shells: solid diluent and carrier |
| $\alpha$-($p$-Nonylphenyl)-$\omega$-hydroxy poly(oxyethylene) mixture of dihydrogen phospahte and monohydrogen phosphate esters and the corresponding ammon- |

ium, calcium, magnesium, monoethanolamine, potassium, sodium, and zinc salts of the phosphate esters: surfactants, related adjuvants of surfactants

α-[*p*-(1,1,3,3-Tetramethylbutyl)phenyl]-ω-hydroxypoly (oxyethylene): surfactant, related adjuvant of surfactants

Aluminum hydroxide: diluent, carrier

Aluminum oxide: diluent

Aluminum stearates: surfactants

Ammonium bicarbonate: surfactant, suspending agent, dispersing agents

Ammonium carbamate: synergist in aluminum phosphide formulators

Ammonium chloride: intensifier when used with ammonium nitrate as a dessicant or defoliant. Fire suppressant in aluminum phosphide and magnesium phosphide formulations

Ammonium hydroxide: solvent, cosolvent, neutralizer, solubilizing agent

Ammonium stearate: surfactant

Ammonium sulfate: solid diluent, carrier

Ammonium thiosulfate: intensifier when used with ammonium nitrate as desiccant or defoliant

Amyl acetate: solvent, cosolvent, attractant

Animal glue: surfactant, adhesive

Apple pomace: surfactant, adhesive

Ascorbyl palmitate: preservative

Attapulgite-type clay: solid diluent, carrier, thickener

*Bacillus thuringiensis* fermentation solids and/or solubles: diluent, carrier

Beeswax: coating agent

Bentonite: solid diluent, carrier

Benzoic acid: preservative for formulation

Butane: propellant

n-Butanol

α-Butyl-ω-hydroxy poly(oxypropylene) block polymer with poly(oxyethylene): surfactants, related adjuvants of surfactants

α-(*p-tert*-Butylphenyl)-ω-hydroxy poly(oxyethylene) mixture of dihydrogen phosphate and monohydrogen phosphate esters and the corresponding ammonium, calcium, magnesium, monoethanolamine, potassium, sodium, and zinc salts of the phosphate esters: surfactants, related adjuvants of surfactants

Butylated hydroxyanisole: antioxidant

Butylated hydroxytoluene: antioxidant

Calcareous shale: solid diluent carrier

Calcite: solid diluent carrier

Calcium carbonate: solid diluent carrier

Calcium chloride: stabilizer

**Table 1.2.2. (continued)**

| Inert Ingredient(s): Use(s) |
|---|
| Calcium citrate: solid diluent, carrier |
| Calcium phosphate: solid diluent, carrier |
| Calcium hydroxide: solid diluent, carrier |
| Calcium hypochlorite: sanitizing and bleaching agent |
| Calcium oxide: solid diluent, carrier |
| Calcium salt of partially dimerized rosin, conforming to 21 CFR 172.210: coating agent |
| Calcium silicate: solid diluent, carrier |
| Calcium stearate: solid diluent, carrier |
| Carnauba wax: coating agent |
| Carrageenan, conforming to 21 CFR 172.620: thickener |
| Casein: surfactant, emulsifier, wetting agent |
| Castor oil, polyoxyethylated: surfactant, related adjuvant of surfactants |
| $\alpha$-Cellulose: solid diluent, carrier |
| Charcoal, activated: carrier |
| Citric acid: sequestrant |
| Citrus meal: solid diluent, carrier |
| Cocoa shells: solid diluent, carrier |
| Coconut oil, surfactant, emulsifier, wetting agent |
| Coconut shells: solid diluent and carrier |
| Cod liver oil: solvent, cosolvent |
| Coffee grounds: solid diluent, carrier |
| Corn cobs: solid diluent, carrier |
| Corn dextrin: solid diluent, carrier |
| Corn meal: solid diluent, carrier |
| Corn oil: solvent, cosolvent |
| Cornstarch: solid diluent, carrier |
| Corn syrup: rehydrating agent |
| Cottonseed oil: safener |
| Coumarone-indene resin, conforming to 21 CFR 172.215: component of coating agent |
| Croscarmellose sodium: disintegent, solid diluent, carrier, and thickener |
| Dextrin: surfactant, suspending agent, dispersing agent |
| Dextrose: solid diluent, carrier |
| Diacetyl tartaric acid esters of mono- and diglycerides of edible fatty acids: emulsifier |
| Dialkyl ($C_8$-$C_{18}$) dimethylammonium chloride: flocculating agents in the manufacture of silica, hydrated silicas for use as solid diluents, carriers |
| Diatomite (diatomaceous earth): solid diluent carrier |

Dichlorodifluoromethane: propellant
Dichlorotetrafluoroethane: propellant
Diethylene glycol abietate: surfactant, related adjuvant of surfactants
1,2-Dihydro-6-ethoxy-2,2,4-trimethylquinolene: antioxidant
3,6-Dimethyl-4-octyn-3,6-diol: surfactant, related adjuvant of surfactants
Dimethylpolysiloxane (as defined in 21 CFR 172.340): defoaming agent
$\alpha$-(*o,p*-Dinonylphenyl)-alpha-hydroxy poly (oxyethylene): surfactant, related adjuvant of surfactants
$\alpha$-(*o,p*-Dinonylphenyl)-omega-hydroxy poly(oxyethylene) mixture of dihydrogen phosphate and monohydrogen phosphate esters and the corresponding ammonium, calcium, magnesium, monoethanolamine, potassium, sodium, and zinc salts of the phosphate esters: surfactants, related adjuvants of surfactants
Dipropylene glycol: solvent, cosolvent
Disodium phosphate: anticaking agent, conditioning agent
Disodium zinc ethylenediamine tetraacetate dihydride: sequestrant
$\alpha$-(*p*-Dodecylphenyl)-$\omega$-hydroxy poly (oxyethylene): surfactant, related adjuvant of surfactants
Dodecylbenzene sulfonic acid, amine salts: surfactants, related adjuvants of the surfactants
Dolomite: solid diluent, carrier
Expoxidized linseed oil: surfactant, related adjuvant of surfactants
Epoxidized soybean oil: surfactant, related adjuvant of surfactants
Ethoxylated lignosulfonic acid, sodium salt: surfactant
Ethyl acetate: solvent, cosolvent
Ethyl alcohol: solvent, cosolvent
2-Ethyl-1-hexanol: solvent, adjuvant of surfactants
Ethylene methylphenylglycidate: synthetic flavoring
Ethylene oxide adducts of 2,4,7,9-tetramethyl-5-decynediol: surfactants, related adjuvants of surfactants
Ethylenediamine tetraacetic acid: sequestrant
Ethylenediamine tetraacetic acid, tetrasodium salt: sequestrant
Fatty acids, conforming to 21 CFR 172.860: binder, defoaming agents, lubricant
FD & C Blue No. 1: dye
Ferric sulfate: solid diluent, carrier
Fish meal: solid diluent, carrier
Fish oil: solvent, cosolvent
Furcelleran: thickener
Glycerol: thickener
Glyceryl monostearate: emulsifier
Glycerol mono-, di-, and triacetate: solvent, cosolvent

**Table 1.2.2. (continued)**

| Inert Ingredient(s): Use(s) |
|---|
| Granite: solvent, cosolvent |
| Graphite: solid diluent, carrier |
| Grape pomace, dried: solid diluent, carrier |
| Guar gum: surfactant, related adjuvant of surfactants |
| Gum arabic (acacia): surfactant, suspending agent, dispersing agent |
| Gypsum: solid diluent, carrier |
| Hexamethylene tetraamine: preservative |
| Hexane: solvent |
| *n*-Hexylalcohol: solvent, cosolvent |
| Hydrochloric acid: solvent neutralizer |
| $\alpha$-Hydro-$\omega$-hydroxy poly(oxypropylene): solvent neutralizer |
| Hydroxyethylcellulose: solvent, neutralizer |
| Hydroxypropylmethylcellulose: thickener |
| Iron oxide: solid diluent, carrier |
| Isopropyl alcohol: solvent, cosolvent, stabilizer, inhibitor |
| Kaolinite-type clay: solid diluent, carrier |
| Lactic acid: solvent |
| Lactose: solid diluent, carrier |
| Lard: solid diluent, carrier |
| Lauryl alcohol: surfactant |
| $\alpha$-Lauryl-$\omega$-hydroxy poly (oxyethylene) sulfate, sodium salt: surfactants, related adjuvants of surfactants |
| $\alpha$-Lauryl-$\omega$-hydroxypoly(oxyethylene): emulsifier |
| Licorice root |
| Lignosulfonate, ammonium, calcium, magnesium, potassium, sodium, and zinc salts: surfactants, related adjuvants of surfactants |
| Lecithin: emulsifier |
| Magnesium carbonate: anticaking agent, conditioning agent |
| Magnesium chloride: safener |
| Magnesium oxide: solid diluent, carrier |
| Magnesium lime: solid diluent, carrier |
| Magnesium silicate: solid diluent, carrier |
| Magnesium stearate: surfactant |
| Magnesium sulfate: solid diluent, carrier, safener |
| Magnesium oxide: solid diluent, carrier |
| Methyl alcohol: solvent |
| Methyl isobutyl ketone: solvent |
| Methylated silicones: antifoaming agent |
| Methylcellulose: thickener |

Methyl chloride: propellant

Methyl *n*-amyl ketone: solvent, cosolvent

Methyl esters of fatty acids derived from edible fats and oils: solvents, cosolvents

Methyl esters of higher fatty acids conforming to 21 CFR 573.640: antidusting agents, surfactants

Methyl ester of rosin, partially hydrogenated (as defined in 21 CFR 172.615): surfactant, related adjuvant of surfactants

Mica: solid diluent, carrier

Mineral oil (USP): diluent, solvent

Modified polyester resin derived from ethylene glycol, fumaric acid, and rosin: resinous coating

Molasses: attractant

Monoammonium phosphate: postharvest fumigation in formulation with aluminum phosphide

Mono- and diglycerides of $C_8$-$C_{18}$ fatty acids: surfactants, related adjuvants of surfactants

Montmorillonite-type clay: solid diluent, carrier

Montmorillonite-type clay treated with polytetrafluorethylene

$\alpha$-(*p*-Nonylphenyl-$\omega$-hydroxypoly (oxypropylene) block polymer with poly (oxyethylene): surfactants, related adjuvants of surfactants

$\alpha$-(*p*-Nonylphenyl)-$\omega$-hydroxypoly (oxyethylene): surfacants, related adjuvants of surfactants

$\alpha$-(*p*-Nonylphenyl)-$\omega$-hydroxypoly (oxyethylene) mixture of dihydrogen phosphate and monohydrogen phosphate esters and the corresponding ammonium, calcium, magnesium, monoethanolamine, potassium, sodium, and zinc salts of the phosphate esters: surfactants, related adjuvants of surfactants

$\alpha$-(*p*-Nonylphenyl)-$\omega$-hydroxypoly (oxethylene) sulfate, ammonium, calcium, magnesium, potassium, sodium, and zinc salts: surfactants, related adjuvants of surfactants

$\alpha$-(*p*-Nonylphenyl) poly (oxypropylene) block polymer with poly (oxethylene): surfactants, related adjuvants of surfactants

Oatmeal: solid diluent, carrier

Oats: solid diluent, carrier

$\alpha$-*cis*-9-Octadecenyl-$\omega$-hydroxy poly (oxyethylene): surfactant, related adjuvant of surfactants

Octyl and decyl glucosides mixture with a mixture of octyl and decyloligosaccharides and related reaction products: surfactants, related adjuvants of surfactants

Oleic acid: diluent

**Table 1.2.2. (continued)**

| Inert Ingredient(s): Use(s) |
|---|
| Oleic acid diester of alpha-hydro-omega-hydroxy poly(oxyethylene): surfactant, related adjuvant of surfactants |
| $\alpha$-Oleoyl-$\omega$-hydroxypoly (oxyethylene): emulsifier |
| Orange pomace: solid diluent, carrier |
| Oxalic acid: caclium chelating hard water inhibitor |
| Palmitic acid: diluent |
| Peanut shells: solid diluent, carrier |
| Pentaerythritol ester of maleic anhydride modified wood rosin: plasticizer |
| Pentaerythritol ester of modified resin: plasticizer |
| Perchloroethylene: solvent, cosolvent |
| Petrolatum, conforming to 21 CFR 172.880: coating agent |
| Petroleum hydrocarbons, light odorless conforming to 21 CFR 172.884: solvent, diluent |
| Petroleum hydrocarbons, synthetic isoparafinic, conforming to 21 CFR 172.882: solvent, diluent |
| Petroleum wax, conforming to 21 CFR 172.886(d): coating agent |
| Phosphoric acid: buffer |
| Phosphorus oxychloride: catalyst |
| Pine lignin: adsorbent |
| $\beta$-Pinene polymers: surfactants, related adjuvants of surfactants |
| Polyethylene, conforming to 21 CFR 177.1520(c): binder, carrier, and coating agent |
| Polyethylene, glycol [$\alpha$-, hydro-$\omega$-hydroxypoly (oxyethylene)], conforming to 21 CFR 178.3750: surfactants, related adjuvants of surfactants |
| Polyethylene, oxidized conforming to 21 CFR 172.260: coating agent |
| Polyglycerol esters of fatty acids conforming to 21 CFR 172.854: surfactants, related adjuvants of surfactants |
| Polyglyceryl phthalate esters of coconut oil fatty acids: surfactants, related adjuvants of surfactants |
| Poly(methylene-*p*-*tert*-butylphenoxy)-poly(oxyethylene) ethanol: coating agent |
| Poly(methylene-*p*-nonylphenoxy)-poly(oxyethylene) ethanol: coating agent |
| Poly(oxy-1,2-ethanediyl)-$\alpha$-(carboxymethyl)-$\omega$-(nonylphenoxy): surfactant |
| Polyoxyethylene(20)sorbitan monostearate: surfactant, related adjuvant of surfactants |
| Poly(oxypropylene) block polymer with poly(oxyethylene): surfactants, related adjuvants of surfactants |
| Poly(vinyl pyrrolidone): surfactant, related adjuvant of surfactants |
| Polymerized sodium methacrylate: pH control |

Polymers derived from the monomers including acrylic acid, sodium form, butyl acrylate, ethyl acrylate, methacrylic acid and its ammonium and potassium salts, and methyl methacrylate: surfactant, related adjuvant of surfactant
Polysorbate 65, conforming to 21 CFR 172.838: emulsifier
Potassium aluminum silicate: solid diluent, carrier
Potassium chloride: solid diluent, carrier
Potassium hydroxide: neutralizer
Potassium phosphate: buffer
Potassium sulfate: solid diluent
Propane: propellant
*n*-Propanol: solvent, cosolvent
Propionic acid: catalyst
Propyl gallate: antioxidant
Propylene glycol: solvent, cosolvent
Propylene glycol alginate (as defined in 21 CFR 172.858): defoaming agent
Propylene oxide: stabilizer
Propyl-*p*-hydroxybenzoate: preservative for formulations
Pyrophyllite: solid diluent, carrier
Rhodamine B: dye
Rice bran: solid diluent, carrier
Rosin, partially dimerized (as defined in 21 CFR 172.615): surfactant, related adjuvant of surfactants
Rosin, partially hydrogenated (as defined in 21 CFR 172.615): surfactant, related adjuvant of surfactants
Rosin, wood: surfactant, related adjuvant of surfactants
Salts of fatty acids, conforming to 21 CFR 172.863: binders, emulsifiers, anticaking agents
Sand: solid diluent, carrier
Secondary alkyl ($C_{11}$-$C_{15}$) poly(oxyethylene) acetate, sodium salt: surfactant
Shellac, bleached, refined, food grade, arsenic- and rosin-free: coating agent
Silica, hydrated: solid diluent, carrier
Silicon dioxide, fumed, amorphous: flow control, anticaking and carrier agent
Soap (sodium or potassium salts of fatty acids): surfactant, emulsifier, wetting agent
Soapstone: solid diluent
Sodium acetate: buffer
Sodium acid pyrophosphate: surfactant, suspending agent, dispersing agent, buffer
Sodium alginate: stabilizer
Sodium $\alpha$-olefin sulfoante (sodium $C_{14}$-$C_{16}$) (olefin sulfonate): surfactants, related adjuvants of surfactants

**Table 1.2.2. (continued)**

| Inert Ingredient(s): Use(s) |
|---|
| Sodium aluminum silicate: solid diluent, carrier |
| Sodium benzoate: anticaking agent |
| Sodium bicarbonate: neutralizer |
| Sodium carboxymethylcellulose: surfactant, related adjuvant of surfactants |
| Sodium chloride: solid diluent, carrier |
| Sodium diisobutyl naphthalene sulfonate: surfactant, related adjuvant of surfactants |
| Sodium dioctylsulfosucciante: surfactant, related adjuvant of surfactants |
| Sodium dodecylphenoxybenzene disulfonate: surfactants, related adjuvants of surfactants |
| Sodium hexametaphosphate: surfactant, emulsifier, wetting agent, suspending agent, dispersing agent, buffer |
| Sodium hydroxide: neutralizer |
| Sodium isopropylisohexyl naphthalene sulfonate: surfactant, related adjuvant of surfactants |
| Sodium *N*-lauroyl-*N*-methyltaurine: surfactant, related adjuvant of surfactants |
| Sodium laurylglyceryl ether sulfonate: surfactant, related adjuvant of surfact-ants |
| Sodium metasilicate: surfactant, emulsifier, buffer, wetting agent, dispersing agent |
| Sodium mono-, di-, and triisopropyl naphthalene sulfonate: surfactants, related adjuvants of surfactants |
| Sodium mono- and dimethyl naphthalene sulfonates: surfactants, related adjuvants of surfactants |
| Sodium monoalkyl and dialykyl ($C_8$-$C_{16}$) phenoxybenzene disulfonate mixtures containing not less than 70% of the monoalkylated product: surfactants, related adjuvants of surfactants |
| Sodium *N*-oleoyl-*N*-methyltaurine: surfactant, related adjuvant of surfactants |
| Sodium oleyl sulfate: surfactant, related adjuvant of surfactants |
| Sodium *N*-palmitoyl-*N*-methyltaurine: surfactant, related adjuvant of surfactants |
| Sodium mono-, di-, and tributyl naphthalene sulfonates: surfactants, related adjuvants of surfactants |
| Sodium propionate: preservative for formulation |
| Sodium salt of sulfated oleic acid: surfactant, related adjuvant of surfactants |
| Sodium silicate: surfactant, emulsifier, wetting agent, stabilizer, inhibitor |
| Sodium sulfate: solid diluent, carrier |
| Sodium sulfite: stabilizer |

Sodium tripolyphosphate: buffer, surfactant, suspending agent, dispersing agent, anticaking agent, conditioning agent
Sorbitan fatty acid esters ($C_{12}$, $C_{14}$, $C_{16}$, $C_{18}$): surfactants, related adjuvants of surfactants
Sorbic acid (and potassium salt): preservatives for formulations
Sorbitol: antidusting agent
Soy protein, isolated: adhesive
Soybean flour: surfactant
Soybean oil: solvent, cosolvent
Soybean oil-derived fatty acids: solvents, cosolvents
Sperm oil conforming to 21 CFR 172.210: coating agent
Starch (potato, tapioca, and wheat): solid diluent, carrier
Stearic acid: diluent
$\alpha$-Stearoyl-$\omega$-hydroxy poly (oxyethylene): emulsifier, surfactant, related adjuvant of surfactants
Sucrose: solid diluent carrier, safener
Sucrose octaacetate: adhesive
Sulfuric acid: pH control agent
Sulfurous acid: preservative
Synthetic paraffin, conforming to 21 CFR 172.275: carrier, binder, and carrying agent
Synthetic petroleum wax, conforming to 21 CFR 172.888: binder, carrier, and coating agent
Talc: solid diluent, carriers
Tall oil with fatty acids under 58%, rosin acids under 44%, and unsaponifiables under 8%: surfactants, related adjuvants of surfactants
Tartrazine: dye
$\alpha$-[*p*-(1,1,3,3-Tetramethylbutyl)phenyl]-$\omega$-hydroxy poly(oxyethylene): surfactant, related adjuvant of surfactants
Tetrahydrofurfuryl alcohol: solvent, cosolvent
2,4,7,9-Tetramethyl-5-decyn-4,7-diol: surfactant, related adjuvant of surfactants
Tetrasodium pyrophosphate: anticaking agent, conditioning agent
Tricalcium phosphate: Surfactant, suspending agent, dispersing agent, anticaking agent, conditioning agent
1,1,1-Trichloroethane: solvent, cosolvent
Trichlorofluoromethane: propellant
Tridecyl poly(oxyethylene)acetate, sodium salt: surfactants, related adjuvants of surfactants
Trisodium phosphate: surfactant, emulsifier, wetting agent
Urea: stabilizer, inhibitor
Vermiculite: solid diluent, carrier

**Table 1.2.2. (continued)**

| Inert Ingredient(s): Use(s) |
|---|
| Walnut shells: solid diluent, carrier |
| Wheat bran: solid diluent, carrier |
| Wintergreen oil: attractant |
| Wood flour: solid diluent and carrier |
| Xanthan gum: thickener |
| Xylene conforming to 21 CFR 172.884(b)(4): solvent, cosolvent |
| Zeolite (hydrated alkali aluminum silicate): solid diluent, carrier |
| Zinc oxide: coating agent |
| Zinc sulfate (basic and monohydrate): coating agent |
| Zinc sulfate (basic and monohydrate): solid diluent, carrier |

Source: Adapted from 40 CFR 180, 1991.

**Table 1.2.3. Alphabetical Listing of Pesticides for Which the Environmental Protection Agency Has Established Tolerances for Their Residues in Processed Foods**

Acephate
Aldicarb
Aluminum phosphide
4-Amino-6-(1,1-dimethylethyl)-3-(methylthio)-1,2,4-triazin-5(4*H*)-one
Avermectin $B_1$ and its delta 8,9-geometric isomer
Benomyl
1,1-Bis(*p*-chlorophenyl)-2,2,2-trichloroethanol
Bromide ion and residual bromine
Captan
Carbofuran; tolerances for residues
Carbon dioxide
Carbophenothion
Chlordimeform
1-(4-Chlorophenoxy)-3,3-dimethyl-1-[1H-1,2,4-triazol-1-yl]-2-butanone
Chlorpyrifos
Chloropyrifosmethyl
Clopyralid
Combustion product gas
Copper
Cyfluthrin
Cyano(3-phenoxyphenyl)methyl-4-chloro-$\alpha$-(1-methylethyl)benzene acetate

Cyfluthrin
Cyhexatin
2,4-D
Dalapon
Dialfor
Diatomaceous earth
Diazinon
Dicamba
3-(3,5-Dichlorophenyl)-5-ethenyl-5-methyl-2,4-oxazolidinedione
2,2-Dichlorovinyl dimethyl phosphate
*O,O*-Dimethyl-*S*[4-oxo-1,2,3-benzotriazin-3(4*H*)-yl-methyl] phosphorodithioate
Dimethyl phosphate of 3-hydroxy-*N*-methyl-*cis*-crotonamide
2,2-Dimethyl-1,3-benzodioxol-4-ol methylcarbamate
*O,O*-Dimethyl-*O*(4-nitro-*m*-tolyl) phosphorothioate
Diquat
Endosulfan
Endothall
Ethephon
Ethion
2-[1-(Ethoxyimino)butyl)]-5-[2-(ethylthio)propyl)]-3-hydroxy-2-cyclohexene-1-one
Ethylene oxide
Ethyl formate
Ethyl-3-methyl-4-(methylthio)phenyl(1-methylethyl) phosphoramidate
*O*-Ethyl O-[4-(methylthio)phenyl]-*S*-propyl phosphorodithioate
Fenarimol
Fluazifop-butyl
Flucythrinate
Formetanate hydrochloride
Fumigants for grain-mill machinery
Fumigants for processed grains used in production of fermented malt beverages
Glyphosate
Hexakis
Hydrogen cyanide
Imazalil
Inorganic bromide
Iprodione
Magnesium phosphide
Malathion
Maleic hydrazide

**Table 1.2.3. (continued)**

| |
|---|
| *N*-(Mercaptomethyl)phthalimide-*S*-(*O*,*O*-dimethyl phosphorodithioate) and its oxygen analog |
| Metalaxyl |
| Metaldehyde |
| Methomyl |
| Methoprene |
| 1-Methoxycarbonyl-1-propen-2-yl dimethylphosphate and its $\beta$-isomer |
| Methyl chloride |
| Methyl formate |
| *N*-Octylbicycloheptene dicarboximide |
| Nitrogen |
| Norflurazon |
| Oryzalin |
| Oxyfluorfen |
| Paraformaldehyde |
| Paraquat |
| Phosalone |
| Picloram |
| Piperonyl butoxide |
| Pirimiphosmethyl |
| Propargite |
| Propetamphos |
| Propylene oxide |
| Pyrethrins |
| Quizalofop-ethyl |
| Resmethrin |
| Simazine |
| Tralomethrin |
| Tetradifon |
| Thiabendazole |
| Trifluralin |
| Toxaphene |
| Triforine |
| Zinc ion and maneb coordination product |

Source: Adapted from 40 CFR 185, 1991. Tolerance has been established for malathion which may result in a processed food from contact with containers or equipment.

# CHAPTER 2
# BIOTECHNOLOGY

In the last decade, the term biotechnology has assumed increasing significance in food production and related applications. VCH Publishers, Inc. have been publishing reference books on biotechnology since the early 1980's. Currently, its collection of titles in this discipline covers topics ranging from amino acids (production) to yeasts (molecular genetics). This chapter has made extensive use of the information in these books.

After a careful study of the books has been made, many of the useful data were selected, rearranged, and systematically presented in tables in this chapter, some of which include a small amount of data from other sources. Subject coverage in this chapter includes:

1. U.S. government agencies regulating the discipline of biotechnology.
2. Application of genetic engineering in food and food-related areas.
3. Examples of hybridization and gene technology.
4. A glossary of terms most commonly used in biotechnology.
5. Amino acid production.
6. Food fermentation.
7. Enzymes: classification, sources, properties, gene technology, fermentation and production, and applications in food processing.
8. Immobilized enzymes.
9. Immobilized cell systems.
10. Microbial products: vitamins, coenzymes, iron, nucleotides, and nucleosides.

The citation of federal legal documents follows the standard format. For example, "51 FR 23339" refers to the U.S. Federal Register, Volume 51, Page 23339. All federal legal documents are available from the U.S. Government Printing Office, Washington, D.C.

## 2.1. U. S. Government Agencies Regulating Biotechnology Products and Proposals

**Table 2.1.1. U. S. Agencies Responsible for Approval of Commercial Biotechnology Products**

| Biotechnology Products | Responsible Agencies |
|---|---|
| Foods and food additives | FDA[a], FSIS |
| Human drugs, medical devices, and biologics | FDA |
| Animal drugs | FDA |
| Animal biologics | APHIS |
| Other contained uses | EPA |
| Plants and animals | APHIS, FSIS, FDA |
| Pesticide microorganisms released in the environment | EPA, APHIS |
| Other uses (microorganisms): | |
| Intergeneric combination | EPA, APHIS |
| Intrageneric combination: | |
| Pathogenic source organism | |
| Agricultural use | APHIS |
| Nonagricultural use | EPA, APHIS |
| Nonpathogenic source organisms | EPA Report |
| Nonengineered pathogens: | |
| Agricultural use | APHIS |
| Nonagricultural use | EPA, APHIS |
| Nonengineered nonpathogens | EPA Report |

Source: 51 FR 23339.

[a]Designates lead agency where jurisdictions may overlap. Abbreviations:

FSIS: Food Safety and Inspection Service, under the Assistant Secretary of Agriculture for Marketing and Inspection Services, is responsible for food use.

FDA: Food and Drug Administration is involved in relation to a food use.

APHIS, Animal and Plant Health Inspection Service, is involved when the microorganism is plant pest, animal pathogen, or regulated article requiring a permit.

EPA: Environmental Protection Agency requirement will only apply to environmental release under a "significant new use rule" that EPA intends to propose.

**Table 2.1.2. Jurisdiction for Review of Planned Introduction of Genetically Engineered Organisms**

| Proposed Research | Responsible Agencies |
|---|---|
| Contained research, no release in environment: | |
| Federally funded | Funding agency[a] |
| Nonfederally funded | NIH or S&E voluntary review, APHIS[b] |
| Foods/food additives, human drugs, medical devices, biologics, animal drugs: | |
| Federally funded | FDA[c], NIH guidelines and review |
| Nonfederally funded | FDA[c], NIH voluntary review |
| Plants, animals and animal biologics: | |
| Federally funded | Funding agency[a], APHIS[b] |
| Nonfederally funded | APHIS[b], S&E voluntary review |
| Pesticide microorganisms: | |
| Genetically engineered: | |
| Intergeneric | EPA[d], APHIS[b], S&E voluntary review |
| Pathogenic intrageneric | EPA[d], APHIS[b], S&E voluntary review |
| Intrageneric nonpathogen | EPA[d], S&E voluntary review |
| Nonengineered: | |
| Nonindigenous pathogens | EPA[d], APHIS |
| Indigenous pathogens | EPA[d], APHIS |
| Nonindigenous nonpathogen | EPA[d] |
| Other uses (microorganisms) released in the environment: | |
| Genetically engineered: | |
| Intergeneric organisms | |
| Federally funded | Funding agency[a], APHIS, EPA[d] |
| Commercially funded | EPA, APHIS, S&E voluntary review |
| Intrageneric organisms | |
| Pathogenic source organisms | |
| Federally funded | Funding agency[a], APHIS[b], EPA[d] |
| Commercially funded | APHIS[b], EPA[d] (if nonagricultural use) |

**Table 2.1.2. (continued)**

| Proposed Research | Responsible Agencies |
|---|---|
| Intrageneric combination | |
| Nonpathogenic source organisms | EPA Report |
| Nonengineered | EPA Report[e], APHIS[b] |

Source: 51 FR 23305.

[a]Review and approval of research protocols conducted by NIH, S&E, or NSF.

[b]EPA: jurisdiction for research on a plot greater than 10 acres.

[c]APHIS: issues permits for the importation and domestic shipment of certain plants and animals, plant pests and animal pathogens, and for the shipment or release in the environment of regulated articles.

[d]EPA reviews federally funded environmental research only when it is for commercial purposes.

[e]Designates lead agency where jurisdictions may overlap.

Abbreviations: NIH = National Institutes of Health; NSF = National Science Foundation; S&E = United States Department of Agriculture Science and Education; APHIS = Animal and Plant Health Inspection Service; EPA = Environmental Protection Agency.

## 2.2. Genetic Engineering and Its Applications in Food and Food-Related Areas

**Table 2.2.1. Some Plants Propagated Through Tissue Culture for Production or Breeding**

| Vegetable crops | Fruit and nut trees | Fruit and berries |
|---|---|---|
| Asparagus | Almond | Blackberry |
| Beet | Apple | Grape |
| Brussels sprouts | Banana | Pineapple |
| Cauliflower | Coffee | Strawberry |
| Eggplant | Date | |
| Onion | Grapefruit | |
| Spinach | Lemon | |
| Sweet potato | Olive | |
| Tomato | Orange | |
| | Peach | |

Source: Adapted from U. S. Congress, 1981.

**Table 2.2.2. Representative List of Tissue Culture Programs of Commercial Significance in the United States**

| Industry | Application | Economic Benefits |
|---|---|---|
| Asparagus industry | Rapid multiplication of seed stock | Improved productivity, earliness, and spear quality |
| Chemical and pharmaceutical | Biosynthesis of chemicals | Reduced production costs |
| | Propagation of medicinal plants | High volumes of plants for planting |
| Citrus industry | Virus elimination | Improved quality, high productivity |
| Coffee industry | Disease resistance breeding | Disease resistance |
| Pineapple industry | Mass propagation | Improved quality in higher volumes |
| Strawberry industry | Mass propagation | Rapid introduction of new strains |

Source: Adapted from U. S. Congress, 1981.

**Table 2.2.3. Some Representative Pending and Potential Environmental Applications of Genetically Engineered Organisms**

Microorganisms

Bacteria as pesticides. "Ice-minus" bacteria to reduce frost damage to agricultural crops.

Bacteria carrying *Bacillus thuringiensis* toxin to reduce loss of corn crops to black cutworm.

Mycorrhizal fungi to increase plant growth rates by improving efficiency of root uptake of nutrients.

Plant symbionts. Nitrogen-fixing bacteria to increase nitrogen available to plants, and decrease need for fertilizers.

Toxic waste disposal. Bacteria engineered to enhance their existing abilities to degrade compounds found in sludge in waste treatment plants.

Bacteria engineered to enhance their abilities to degrade compounds in landfills, dumps, runoff deposits, and contaminated soils.

Heavy metal recovery. Engineered enhancements possible to several species of bacteria now used to recover metals from low-grade ores (e.g., copper and cobalt).

**Table 2.2.3. (continued)**

Pollution control. Possible increased utility of bacteria in purifying water supplies of phosphorus, ammonia, and other compounds.

Viruses as pesticides. Insect viruses with narrowed host specificity or increased virulence against specific agricultural insect pests, including cabbage looper, pine beauty moth, cutworms, and other pests.

<u>Plants</u>

Herbicide resistance or tolerance to:
- Glyphosate
    - Atrazine
    - Sulfonylurea (chlorosulfuron and sulfometuron)
    - Imidazolinone
    - Bromoxynil
    - Phosphinotricin

Disease resistance to:
- Crown gall disease (tobacco)
- Tobacco mosaic virus (and related viruses)
- Potato leaf roll virus

Pest resistance
- BT-toxin-protected crops, including tobacco (principally as research tool) and tomato.
- Seeds with enhanced anti-feedant content to reduce losses to insects while in storage.

Enhanced tolerance to environmental factors, including:
- Salt
- Drought
- Temperature
- Heavy metals

Nitrogen-fixation enhancements
- Nonlegumes enhanced to fix nitrogen, independent of association with symbiotic bacteria.

Engineered marine algae
- Algae enhanced to increase production of such compounds as $\beta$-carotene and agar, or to enhance ability to sequester heavy metals (e.g., gold and cobalt) from seawater.

<u>Animals</u>

Livestock and poultry
- Livestock species engineered to enhance weight gain or growth rates, reproductive performance, disease resistance, or coat characteristics.
- Livestock animals engineered to function as producers for pharmaceutical drugs, especially of mammalian compounds that require post-synthesis modifications in the cell.

Fish
- Triploid salmon produced by heat shock for use as game fish in lakes and streams.
- Fish with enhanced growth rates, cold tolerance, or disease resistance for use in aquaculture
- Triploid grass carp for use as aquatic weed control agents.

Source: Adapted from U. S. Congress, 1988.

**Table 2.2.4. Biotechnology Products for Food Processing: Enzymes, Amino Acids and Antibiotics**

<u>Enzymes</u>

Amylases
Amyloglucosidase
Anticyanase
L-asparaginase
Catalase
Cellulase
Dextranase
'Diagnostic enzymes'
Esterase-lipase
Glucanase
Glucose dehydrogenase
Glucose isomerase
Glucose oxidase
Glutamic decarboxylase
Hemi-cellulase
Hespiriginase
Invertase
Lactase
Lipase
Microbial rennet
Naringinase
Pectinase
Pentosanase
Proteases
Streptokinase-streptodornase
Uricase

<u>Amino Acids</u>

L-alanine
L-arginine
L-aspartic acid
L-citrulline
L-glutamic acid
L-glutamine
L-glutathione
L-histidine
L-homoserine
L-isoleucine
L-leucine
L-lysine
L-methionine
L-ornithine
L-phenylalanine
L-proline
L-serine
L-threonine
L-tryptophan
L-tyrosine
L-valine

<u>Antibiotics</u>

Nisin
Pimaricin

Source: Adapted from U. S. Congress, 1981.

**Table 2.2.5. Biotechnology Products for Food Processing: Organic Acids, Solvents, Vitamins, Growth Factors, Nucleosides and Nucleotides, Flavors, Processing Aids, and Others**

| Organic Acids | Solvents |
|---|---|
| Acetic acid | Ethanol |
| Citric acid | 2,3-butanediol |
| Comenic acid | |
| Erythorbic acid | Miscellaneous Products and Processeing Aids |
| Fumaric acid | |
| Gluconic acid | |
| Itaconic acid | Acetoin |
| 2-keto-D-gluconic acid | Acyloin |
| $\alpha$-ketoglutaric acid | Alginate |
| Lactic acid | Anka-pigment (red) |
| Malic acid | Blue cheese flavor |
| Oxalic acid | Curdlan |
| Succinic acid | Desferrioxamine |
| Tartaric acid | Dihydroxyacetone |
| Urocanic acid | Dextran |
| | Diacetyl (from acetoin) |
| Vitamins and Growth Factors | Guanylic acid |
| | Inosinic acid |
| Gibberellins | Insecticide (*Bacillus thuringiensis*) |
| Riboflavin | Pectin |
| Vitamin $B_{12}$ | Picibanil |
| Zeearalanol | Pullulan |
| | Ribose |
| Nucleosides and Nucleotides | Scleroglucan |
| | Sorbose (from sorbitol) |
| 5-ribonucleotides and nucleosides | Starter cultures |
| Orotsic acid | Sterol oxidations |
| Ara-A-(9-$\beta$-D-arabino-furanosyl) | Steroid oxidations |
| 6-azauridine | Xanthan |

Source: Adapted from U. S. Congress, 1981.

## 2.3. Hybridization and Gene Technology

**Table 2.3.1. The Use of Hybridization to Obtain A Yeast Strain for the Production of Low-Carbohydrate Beer**

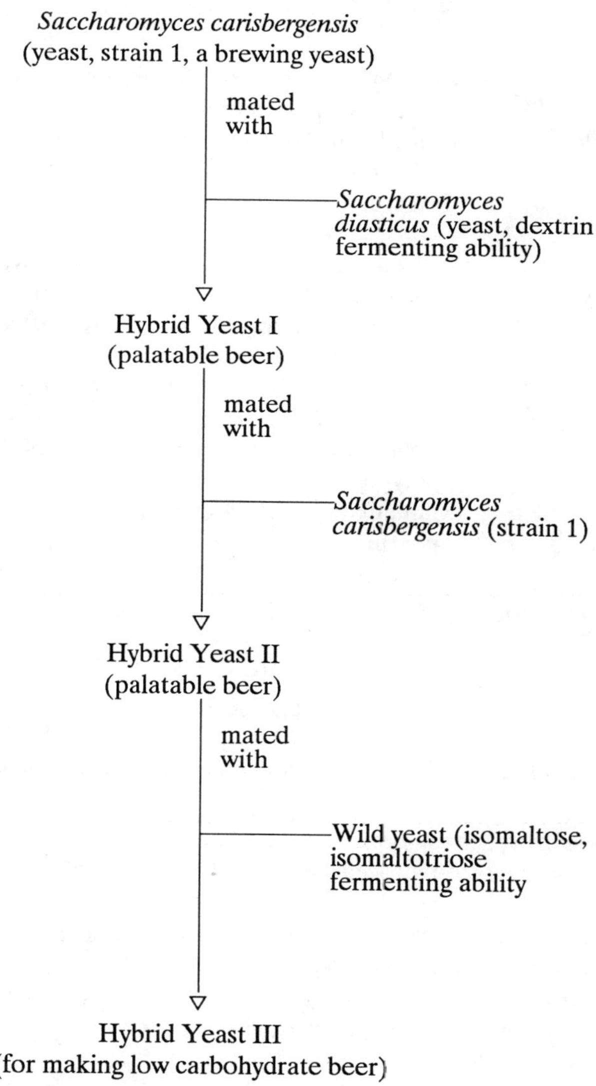


Source: Adapted from U. S. Congress, 1981.

**Table 2.3.2. Examples of the Application of Gene Technology to Improve the Manufacture of Biotechnology Products**

| Product Class | Application of Gene Technology |
|---|---|
| Amino Acids<br>L-glutamic acid<br>and others | Gene amplification increases amino acid production |
| Nucleotides<br>5'-GMP | Cloning of gene for bottleneck enzyme elevates guanosine production |
| Polysaccharides<br>xanthan | Cloned genes permit growth on new carbon source and production of altered biopolymer |
| Vitamins<br>ascorbic acid | Construction of novel pathway |
| Alcohols<br>ethanol | Cloned glucoamylase gene permits growth of yeast on cheaper carbon source |
| Enzymes<br>glucoamylase | Increased gene copy number causes increased enzyme production |
| Peptide sweeteners<br>aspartame | *Escherichia coli* clones contain gene encoding up to 150 repeats of aspartylphenylalanine, which can be obtained by enzymatic degradation of the polypeptide |
| Therapeutic<br>polypeptides | Manipulation of host genome and cloned genes permits expression and secretion of heterologous proteins |

Source: Adapted from Bigelis, 1989. Used with permission.

## 2.4 A Glossary for Biotechnology

**Table 2.4.1. Common Terms Used in Biotechnology**

Aerobic: growing only in the presence of oxygen

Algorithm: a step-by-step procedure for solving a problem or modeling a process

Alkaloids: a group of nitrogen-containing organic substances found in plants; many are pharmacologically active, e.g., nicotine, caffeine, and cocaine.

Allele: alternate forms of the same gene. For example, the genes responsible for eye color (blue, brown, green, etc) are alleles.

Amino acids: the building blocks of proteins. There are 20 common amino acids; they are joined together in a strictly ordered "string" which determines the character of each protein.

Anaerobic: growing only in the absence of oxygen

Antibody: a protein component of the immune system in mammals' found in the blood

Anticodon: a triplet of nucleotides on transfer RNA molecules that binds to the complementary condon on messenger RNA during the polypeptide-producing (translation) process. The amino acid carried by the transfer RNA is inserted into the growing polypeptide chain.

Antigen: a large molecule, usually a protein or carbohydrate, which, when introduced in the body, stimulates the production of an antibody that will react specifically with the antigen

Aromatic chemical: an organic compound containing one or more six-membered rings

Aromatic polymer: large molecules consisting of repeated structural units of aromatic chemicals

Artificial insemination: the manual placement of sperm into the uterus or oviduct

Bacteriophage (or phage): a virus that multiplies in bacteria. *Bacteriophage lambda* is commonly used as a vector in recombinant DNA experiments.

Baculovirus: a virus of the family Bacloviridae found only in invertebrates, and presently being pursued as a potential biocontrol agent

Bioassay: determination of the relative strength of a substance (such as a drug) by comparing its effect on a test organism with that of a standard preparation

Biomass: plant and animal material

Biome: a community of living organisms in a major ecological region

Biosynthesis: The production of a chemical compound by a living organism

Biotechnology: the collection of industrial processes that involve the use of biological systems. For some of these industries, these processes involve the use of genetically engineered microorganisms.

Blastocyst: an early developmental stage of the embryo; the fertilized egg undergoes several cell divisions and forms a hollow ball of cells called the blastocyst.

Callus: the cluster of plant cells that results from tissue culturing a single plant cell

Carbohydrates: the family of organic molecules consisting of simpler sugars such as glucose and sucrose, and sugar chains (polysaccharides) such as starch and cellulose

**Table 2.4.1. (continued)**

Catalyst: a substance that enables a chemical reaction to take place under milder than normal conditions (e.g. lower temperatures); biological catalysts are enzymes; nonbiological catalysts include metallic complexes.

Cell fusion: the fusing together of two or more cells to become a sing cell

Cell lysis: Disruption of the cell membrane allowing the breakdown of the cell and exposure of its contents to the environment

Cellulase: an enzyme that degrades cellulose to glucose

Cellulose: a polysaccharide composed entirely of several glucose units linked end to end; it constitutes the major part of cell walls in plants.

Chimera: an individual composed of a mixture of genetically different cells

Chloroplast(s): those structure within plant cells where photosynthesis occurs; contain small circular DNA molecules that replicate independently.

Chromosomes(s): the physical structures within a cell's nucleus, composed of DNA-protein complex, and containing the hereditary material, i.e., genes; in bacteria, the DNA molecule in a single, closed circle (no associated protein) comprising a cell's genome

Clone: a group of genetically identical cells or organisms asexually descended from a common ancestor. All cells in the clone have the same genetic material and are exact copies of the original

Codon: a series of three adjacent nucleotides in a DNA molecule that directs ("codes for") the insertion of a specific amino acid into a growing protein chain

Conjugation: the reproductive process by which DNA is transferred between bacteria during cell-to-cell contact

Conjugation tube: the bridge-like structure by which cell-to-cell contact is maintained during conjugation

Conjugative plasmid: a plasmid capable of initiating and directing the process of conjugation (compare *nonconjugative plasmid*).

Convergent evolution: the evolution of similar structures or similar life strategies in unrelated species because of adaptation in response to similar selection pressures (e.g., in similar ecological habitats)

Crossing-over: a genetic event that can occur during cellular replication, which involves the breakage and reunion of DNA molecules.

Cryptic plasmid: a plasmid of undetermined function

Cultivar: often used to refer to plant strains. (See *strain*.)

Cytogenetics: a branch of biology that deals with the study of heredity and variation by the methods of both cytology (the study of cells) and genetics

Cytoplasm: the protoplasm of a cell, external to the cell's nuclear membrane

Cytoplasm(ic): the substance within a cell external to the nuclear membrane; pertaining to or contained in the cytoplasm.

Diploid: a cell with double the basic chromosome number

Disarmed transposon: a transposon altered in a manner rendering it incapable of movement, most often by deletion of the transposase gene, which encodes the enzyme necessary for excision. (Compare *transposon.*)

DNA vector: a vehicle for transferring DNA from one cell to another

DNA (deoxyribonucleic acid): the genetic material found in all living organisms; every inherited characteristic has its origin somewhere in the code of each individual's complement of DNA.

Dominant gene: a characteristic whose expression prevails over alternative characteristics for a given trait

Embryo transfer: implantation of an embryo into the oviduct or uterus

Endotoxin: poison produced by some gram-negative bacteria, present in the cellular membrane, and released only upon cell rupture; composed of a complex lipopolysaccharide (fat-like molecule plus sugar molecule) and more heat-stable than protein exotoxins.(Compare *exotoxin.*)

Endogenous: developing or originating within the organism, or arising from causes within the organism

Enzyme: a functional protein that catalyzes a chemical reaction. Enzymes control the rate of metabolic processes in an organism; they are the active agents in the fermentation process

Episome: a DNA molecule that may exist either as an integrated part of a chromosomal DNA molecule of the host or as an independently replicating DNA molecule (plasmid) free of the host chromosome

Epistasis: the interaction of genes at different loci resulting in the masking of a character

Epizootic: a disease affecting many animals of one kind at the same time; analogous to the term epidemic in human populations

*Escherichia coli*: a bacterium that commonly inhabits the human intestine; it is a favorite organism for many microbiological experiments.

Estrogens: Female sex hormones.

Estrus ("heat"): the period in which the female will allow the male to mate her

Eukaryote: a higher, compartmentalized cell characterized by its extensive internal structure and the presence of a nucleus containing the DNA. All multicellular organisms are eukaryotic. The simpler cells, the prokaryotes, have much less compartmentalization and internal structure; bacteria are prokaryotes

Exotic: Describing a species not originating in the place where it is found; a nonnative, introduced species

Exotoxin: a poison excreted by some gram-negative or gram-positive organisms; composed of protein

**Table 2.4.1. (continued)**

Extrachromosomal DNA: DNA not associated with the chromosomes [e.g., plasmid DNA or organelle (mitochondria or chloroplast) DNA].

F factor: see *fertility factor*

Feral: pertaining to organisms in the wild

Fermentation: the biochemical process of converting raw material such as glucose into a product such as ethanol

Fertility factor: an episome capable of transferring a copy of itself from its host bacterial cell (an F plus cell) to a bacterial cell not harboring an F factor (an F-cell). When the F factor is integrated into the host chromosome (the resulting cell is called an Hfr cell), the factor is capable of mobilizing transfer of the bacterial chromosome to an F- cell.

Fibroblast: a cell that gives rise to connective tissues

Gamete: a mature reproductive cell (haploid set of chromosomes) capable of fusing with a similar cell of opposite sex to yield a zygote; also called a sex cell

Gene: the hereditary unit; a segment of DNA coding for a specific protein

Gene expression: the manifestation of the genetic material of an organism as specific traits

Gene family: a group of related genes exhibiting a high degree of homology in function and nucleotide base sequence

Gene mapping: Determining the relative locations of different genes on a given chromosome

Gene pool: the sum total of genes in a breeding population

Gene probe: a molecule of known structure and/or function used to locate and identify a specific region or nucleotide sequence of a genome; usually a piece of complementary DNA that has been labeled with a tracer substance such as a dye or radioactive label

Genetic code: the biochemical basis of heredity consisting of codons (base triplets along the DNA sequence) that determine the specific amino acid sequence in proteins and that are the same for all forms of life studied so far

Genetic drift: Changes of gene frequency in small population due to chance preservation or extinction of particular genes

Genetic engineering: a technology used at the laboratory level to alter the hereditary apparatus of a living cell so that the cell can produce more or different chemicals, or perform completely new functions; these altered cells are then used in industrial production.

Genetic variance: the fraction of the phenotypic variance due to differences in the genetic constitution of individuals in a population. (Compare *phenotypic variance*)

Genome: the basic chromosome set of an organism-the sum total of its genes

Genotype: the genetic constitution of an individual or group

Germ cell: the sex cell of an organism (sperm or egg, pollen or ovum); it differs from other cells in that it contains only half the usual number of chromosomes; germ cells fuse during fertilization.

Germ line: the earliest, primitive state of development; pertaining to tissues or cell lineages production gametes. (Compare *somatic*).

Germplasm: the total genetic variability available to an organism, represented by the pool of germ cells or seed

Glycopeptides: chains of amino acids with attached carbohydrates

Glycoprotein: a conjugated protein in which the nonprotein group is a carbohydrate.

Gram-negative/positive: a classification of bacteria based on differential staining utilizing the Gram-Wiegert procedure; primarily as a result of an organism's cell membrane structure, gram-negative organisms stain red and gram-positive organisms stain purple.

Haploid: a cell with only one set (half of the usual number) of chromosomes

Heterozygous: When the two genes controlling a particular trait are different, the organism is heterozygous for that trait.

High copy number plasmid: a plasmid present in multiple copies within a single host bacterium; copy number (single, low, and high) is dependent on both plasmid and host cell factors.

Homologous sequence: nucleic acid segments having an identical or nearly identical linear order of nucleotide base pairs

Homology: degree of relatedness in appearance, function, or structure

Homozygous: When the two genes controlling a particular trait are identical for a pair of chromosomes the organism is said to be homozygous for that trait.

Horizontal transfer: the passage of genetic material from one organism to another via nonsexual mechanisms

Hormones: the "messenger" molecules of the body that help coordinate the actions of various tissues; they produce a specific effect on the activity of cells remote from their point of origin.

Host: in the context of recombinant DNA technology, the organism used for growth and reproduction of virus, plasmid, or other foreign DNA source

Hybrid: a new variety of plant or animal that results from cross-breeding two different existing varieties

Hybridization: a procedure in which single-stranded nucleic acid segments are allowed to bind to homologous sequences, forming hybrid double-stranded helices; also, a process (e.g., in plants, producing offspring resulting from a cross between two genetically unlike entities); the formation of new cells resulting from fusion of whole cells or cell parts of different genotypes

**Table 2.4.1. (continued)**

Hydrocarbon: all organic compounds that are composed only of carbon and hydrogen

Ice minus (ice-): a bacterium lacking a functional gene coding for a protein that promotes the formation of ice crystals by providing a physical nucleus around which ice crystallizes; the gene has been deleted from strains of *Pseudomonas syringae Pseudomanas fluorescens*, and *Erwinia herbicola*, the organisms around which debate recently has focused.

Ice plus (ice+): a bacterium with an intact, functional ice-nucleating gene.

Immunoproteins: all the proteins that are part of the immune system (including antibodies, interferon, and cytokines).

*In vitro*: outside the living organism and in an artificial environment.

*In vivo*: within the living organism.

Introgression: the entry or introduction of a gene or genes from one population into another (most often in nature via sexual reproduction, or hybridization).

IS (insertion sequence): one of a class of different nucleotide sequences found in bacteria that are capable of spontaneous movement from one chromosomal location to another. Chromosomal material may be mobilized during IS movement; movement may result in mutation at the original and/or new sites of insertion. (Compare *transposable element*.)

Leukocytes: the white cells of blood

Line: See *strain* and *cultivar*.

Lipids: water insoluble biomolecules, such as cellular fats and oils

Lipopolysaccharides: complex substances composed of lipids and polysaccharides

Locus (pl. loci): the physical location on a chromosome occupied by a particular gene or its alleles

LTR (long terminal repeat): one of a class of nucleotide sequences (300-1200 base pairs in length) associated with tumor viruses and cellular oncogenes that promote gene activity and are similar to transposons. (Compare *transposable element*.)

Lymphoblastoid: referring to malignant white blood cells

Lymphokines: the biologically active soluble factor produced by white blood cells

Maleic anhydride: an important organic chemical used in the manufacture of synthetic resins, in fungicides, in the dyeing of cotton textiles, and to prevent rancidity and the oxidation of fats and oils during storage and rancidity

Meiosis: the process by which chromosomes are duplicated then divided during the formation of haploid cell (gametes).

Mendelian genetics: classical method of observing inheritance of a trait(s) in the offspring of crosses between individuals differing in that trait; results in accordance with Mendel's laws

Mendelian: referring to a trait that is controlled by a single gene, and that shows a simple dominant/recessive pattern of inheritance

Messenger RNA: ribonucleic acid molecules that serve as a guide for protein synthesis

Metabolism: the sum of the physical and chemical processes involved in the maintenance of life and by which energy is made available

Mitochondria: structures in higher cells that serve as the "powerhouse" for the cell, producing chemical energy

Monoclonal antibodies: antibodies derived from a single source or clone of cells which recognize only one kind of antigen

Mutants: organisms whose visible properties with respect to some trait differ from the norm of the population due to mutations in its DNA.

Mutation: any change that alters the sequence of bases along the DNA, changing the genetic material.

Myeloma: a malignant disease in which tumor cells of the antibody producing system synthesize excessive amounts of specific proteins.

*n*-alkanes: straight chain hydrocarbons–main constituents of petroleum.

Natural selection: the process of differential reproductive success by which genes in a population increase or decrease in frequency with the passage of generations, depending on their contribution to the survival of offspring in which they are carried; arguably the most important of the several mechanisms by which evolution takes place, discovered by Darwin and first described in 1858-59.

Nif genes: the genes for nitrogen fixation present in certain bacteria

Nonconjugative plasmid: a plasmid incapable of initiating or directing the process of conjugation. (Compare *conjugative plasmid.*)

Nontransferable plasmid: See *nonconjugative plasmid*.

Nucleic acid: a macromolecule composed of sequences of nucleotide bases; DNA or RNA

Nucleotide (base): the unit of nucleic acids; the molecules consist of one of four bases–adenine, guanine, cytosine, or thymine/uracil (DNA/RNA) attached to a phosphate-sugar group; the sugar group is deoxyribose in DNA: in RNA it is ribose.

Nucleus: the membrane-enclosed structure in eukaryotes that contains the chromosomes

Organelle: a structure in the cytoplasm of a cell that is specialized in its ultrastructure and biochemical composition to serve a particular function (e.g., mitochondria, endoplasmic reticulum, chloroplast).

**Table 2.4.1. (continued)**

Organic compounds: chemical compounds based on carbon chains or rings, which contain hydrogen, and also may contain oxygen, nitrogen, and various other elements

Parthenogenesis: reproduction in animals without male fertilization of the egg

Pathogen: a specific causative agent of disease

Pathogenic: able to cause disease; often utilized to express inactivation or lethality

P element: a transposable DNA sequence present in the fruit fly, *Drosophila*. (Compare *transposon*).

Peptide: short chain of amino acids

pH: a measure of the acidity or basicity of a solution; on a scale of 0 (acidic) to 14 (basic): for example, lemon juice has a pH of 2.2 (acidic), water has a pH of 7.0 (neutral), and a solution of baking soda has a pH of 8.5 (basic).

Phage: See *Bacteriophage*.

Phenotype: the observable characteristics of an organism produced by the interaction of the genotype and the environment. (Compare *genotype*).

Phenotypic variance: the variation among genetically identical individual organisms in external appearance caused by the interaction of environment with the genotype during development. (Compare *genetic variance*).

Phylogenetic: referring to genetic similarities between different organisms as a result of descent from a common ancestor in evolutionary history

Plasmid: an extrachromosomal, circular piece of DNA found in the cytoplasm and capable of replicating and segregating independently of the host chromosome. (Compare *conjugative*; *cryptic*; *high copy number*; *nonconjugative*; *nontransferable plasmid*.)

Plastid: any specialized organ of the plant cell other than the nucleus, such as the chloroplast

Pleiotropy: the production of diverse phenotypic effects produced by a mutation in a single gene

Ploidy: describes the number of sets of chromosomes present in the organism; for example, human are diploid, having two homologous sets of 23 chromosomes (one set from each parent) for a total of 46 chromosomes; many plants are haploid, having only one copy of each chromosome.

Polymer: a long-chain molecule formed from smaller repeating structural units

Polymerase: an enzyme that assembles a number of similar or identical subunits into a macromolecule (e.g. DNA ploymerase and RNA polymerase)

Polysaccharide: a long chain carbohydrate containing at least three molecules of simple sugars linked together; examples would include cellulose and starch.

Primary producer: an organism (plant or microorganism) that uses photosynthesis to convert solar energy into chemical energy that can be used by nonphotosynthetic organisms (e.g., humans)

Progestogens: hormones involved with ovulation

Prokaryote: an organism lacking organelles and whose DNA is not enclosed within membrane-bound, structurally discrete nuclei; bacteria, viruses, and blue-green algae are prokaryotes. (Compare *eukaryote*.)

Prostaglandin: refers to a group of naturally occurring, chemically related long-chain fatty acids that have certain physiological effects (stimulate contraction of uterine and other smooth muscles, lower blood pressure, affect action of certain hormones)

Protein: a linear polymer of amino acids; proteins are products of gene expression and the functional and structural components of cells.

Protoplast fusion: a means of achieving genetic transformation by joining two protoplasts or joining a protoplast with any of the components of another cell

Protoplast: a cell without a wall

Recessive gene: any gene whose expression is dependent on the absence of a dominant gene. (Compare *dominant*.)

Recombinant DNA: the hybrid DNA produced by joining pieces of DNA from different sources

Recombination: the formation of a new association of genetic material; usually applied to the process of meiosis, during a stage of which the genetic material packaged into gametes is mixed and reconstituted in any of an enormous number of possible combinations

Replicon: a genetic element possessing sequences specifying the initiation and control of the process by which DNA is precisely duplicated

Restriction enzyme: an enzyme within a bacterium that recognizes and degrades DNA from foreign organisms, thereby preserving the genetic integrity of the bacterium; in recombinant DNA experiments, restriction enzymes are used as tiny biological scissors to cut up foreign DNA before it is recombined with a vector.

Retrovirus: a family of viruses whose genetic material is RNA and is further characterized by the presence of reverse transcriptase in virion; also called tumor virus

Reverse transcriptase: an enzyme capable of directing the production of a single-strand DNA copy from an RNA template

Rhizosphere: a region of the soil closely surrounding plant roots that is affected by root excretions

**Table 2.4.1. (continued)**

RNA (ribonucleic acid): a molecule existing in three forms (messenger RNA, transfer RNA, and ribosomal RNA) responsible for translating the genetic information encoded in DNA into a protein product; hereditary material of some viruses

Selective advantage: an organism's increased probability of reproduction and producing offspring, conferred by its genetic characteristics

Selective pressure: the influence of factor extrinsic to an organism (i.e., environmental factors) on its ability to compete with other organisms for reproductive success.

Sibling species: independent, reproductive populations that are genetically distinct from one another yet very closely related, and often difficult or impossible to distinguish by morphological or other criteria.

Somatic cell: one of the cells composing parts of the body (e.g. tissues, organs) other than a germ cell

Species: taxonomic category subordinate to a genus composed of individuals with common characteristics that distinguish them from other groups of the same taxonomic level; in sexually reproducing organisms, a group of interbreeding natural populations that are genetically distinct from other such groups

Strain: a pure culture of organisms within a species, characterized by one or more particular physical or genetic properties. See *cultivar.*

Tissue culture: an in vitro method of propagating healthy cells from tissues, such as fibroblasts from skin

Toxin: See *endotoxin* and *exotoxin*.

Transduction: the process by which foreign DNA becomes incorporated into the genetic complement of the host cell

Transformation: the transfer of genetic information by DNA separated from the cell

Transposable element: a class of DNA sequences capable of insertion into genome at numerous positions, and of moving from one area of a genome to another area or another genome

Transposase: enzyme that assists movement of DNA during the process of transposition

Transposon: a type of transposable element incapable of autonomous existence, often shuttling genetic material back and forth between cell chromosomes, between smaller replicons, and between chromosomes and replicons

Triplet: Three consecutive bases along a nucleic acid chain. See *codon, anticodon.*

Vector: A transmission agent; a DNA vector is a self-replicating DNA molecule that transfers a piece of DNA from one host to another.

Vertical transfer: The passage of genetic material from one organism to another through the germ line, i.e., sexual mechanisms; in bacteria, through genome replication and cell division.

Virus: An infectious agent that requires a host cell in order for it to replicate. It is composed of either RNA or DNA wrapped in a protein coat.

Zygote: A cell formed by the union of two mature reproductive cells.

Source: Adapted from U. S. Congress, 1981, 1988.

## 2.5. Amino Acids Production

**Table 2.5.1. Fundamentals of Amino Aicds Production**

- Background information
- Microbial production of amino acids
  - Production of amino acids by wild strains
  - Production of amino acids by auxotrophic mutants
  - Production of amino acids by regulatory mutants
  - Production of amino acids from biosynthetic Precursors
- Enzymatic synthesis of amino acids
  - Hydrolytic enzymes
    - L-$\alpha$-Amino-$\epsilon$-caprolactam hydrolase
    - 2-Amino-$\Delta^2$-thiazoline-4-carboxylate hydrolase
    - Hydantoinase
  - Ammonia lyases
    - Aspartase
    - Phenylalanine ammonia lyase
  - Arginine Deiminase
  - Pyridoxal 5'-Phosphate Enzymes
    - Aspartate $\beta$-decarboxylase
    - $\beta$-Tyrosinase
    - Tryptophanase
    - Cysteine desulfhydrase
    - Tryptophan synthase
    - $\beta$-Chloro-D-alanine hydrogenchloride lyase
    - L-Methionine $\gamma$-lyase
    - Serine hydroxymethyltransferase

**Table 2.5.1. (continued)**

- L-Threonine aldolase
- Other Enzymes
  - Amino acid dehydrogenases
  - Glutamine synthetase
- Enzymatic Resolution of Racemic Amino Acids
  - Introduction
  - Enzymatic Methods
    - Resolution by enzymatic assymetric derivatization
    - Resolution by assymetric hydrolysis
      - Esterase method
      - Amidase method
      - Aminoacylase method
- Use of Amino Acids
- Production Figures and Economic Aspects

Source: Adapted from Soda, Tanaka, and Esaki, 1983. Used with permission.

**Table 2.5.2. Fermentative Processes and Amino acids Production**

| Amino Acid | Organisms | Yields (g/L)[a] | Note On Organisms |
|---|---|---|---|
| L-Arginine | *Corynebacterium glutamicum* | 25 | Amino acid analog -resistant |
| | *Brevibacterium flavum* | 35 | |
| | *Bacillus subtilis* | 28 | |
| | *Serratia marcescens* | 100 | |
| L-Glutamic acid | *Corynebacterium* spp. | 100 | Biotin-requiring, poor TCA cycle activity, altered cell permeability |
| | *Brevibacterium* spp. | 80 | |
| | *Arthrobacter paraffineus* | 62 | |
| L-Glutamine | *Corynebacterium glutamicum* | 37 | Specific excretion of glutamine |
| | *Flavobacterium ringense* | 25 | |

| | | | |
|---|---|---|---|
| L-Histidine | *Corynebacterium glutamicum* | 15 | Amino acid analog -resistant |
| | *Brevibacterium flavum* | 10 | |
| | *Serratia marcescens* | 23 | |
| L-Isoleucine | *Corynebacterium glutamicum* | 10 | Amino acid analog -resistant |
| | *Brevibacterium flavum* | 30 | |
| | *Serratia marcescens* | 25 | |
| L-Leucine[b] | *Brevibacterium lactofermentum* | 30 | Amino acid analog -resistant |
| | *Serratia marcescens* | 15 | |
| L-Lysine | *Corynebacterium glutamicum* | 42 | Amino acid analog -resistant |
| | *Brevibacterium flavum* | 34 | |
| | *Brevibacterium lactofermentum* | 70 | |
| L-Pheyl-alanine[c] | *Corynebacterium glutamicum* | 9.5 | Amino acid analog -resistant |
| | *Brevibacterum lactofermentum* | 25 | |
| | *Bacillus subtilis* | 6 | |
| L-Proline | *Corynebacterium glutamicum* | 15 | Wild type, analog-resistant or auxotrophic |
| | *Brevibacterium flavum* | 35 | |
| | *Serratia marcescens* | 22 | |
| L-Serine[d] | *Corynebacterium glycinophilum* | 14 | Precursor conversion; high serine hydorxy-methyltransferase activity permits production from glycine |
| | *Pseudomonas* sp. | 12 | |
| | *Hyphomicrobium methylovorum* | 24 | |
| L-Threonine | *Corynebacterium glutamicum* | 14 | Amino acid analog -resistant |
| | *Brevibacterium flavum* | 17.5 | |
| | *Serratia marcescens* | 25 | |
| | *Escherichia coli* | 6 | |

**Table 2.5.2. (continued)**

| Amino Acid | Organisms | Yields (g/L)[a] | Note On Organisms |
|---|---|---|---|
| L-Tryptophan | *Corynebacterium glutamicum* | 12 | Amino acid analog -resistant |
| | *Brevibacterium flavum* | 20 | |
| L-Valine[c] | *Corynebacterium glutamicum* | 30 | Amino acid analog -resistant |
| | *Brevibacterium lactofermentum* | 31 | |
| | *Serratia marcescens* | 10 | |

Source: Adapted from Bigelis, 1989. Used with permission.
[a]Maximum reported. [b]Also produced by extraction after protein hydrolysis. [c]Also produced by enzymatic synthesis. [d]Also produced by chemical synthesis.

**Table 2.5.3. Enzymatic Processes and Amino acids Production**

| Amino Acid | Enzymes | Processes |
|---|---|---|
| L-Alanine[a] | *E. coli* aspartase and *Pseudomonas dacunhae* L-aspartate $\beta$-decarboxylase | Converson of ammonium fumarate to L-aspartic acid and then to L-alanine sequentially by immobilized cells |
| L-Aspartic acid | *E. coli* aspartase | Converson of ammonium fumarate to L-aspartic acid by immobilized cells |
| L-Lysine[b] | *Cryptococcus laurentii* L-aminocaprolactam hydrolase and *Achromobacter obae* aminocaprolactamracemase | Conversion of synthetic DL-$\alpha$-amino-$\epsilon$-caprolactam to L-lysine by two enzymes |
| L-Methionine[c] | *Aspergillus oryzae* aminoacylase | Converson of syntetic DL-methionine to *N*-acetyl derivative and then asymmetric hydrolysis by immobilized aminoacylase |
| L-Phenylalanine[d] | *Rhodotorula rubra* phenylalanine ammonia lyase | Conversion of *trans*-cinnamic acid to L-phenylalanine by immobilized cells in presence of 7.85 mol/L ammonia |

Source: Adapted from Bigelis, 1989. Used with permission.
[a]DL-Alanine is also produced by chemical synthesis.
[b]L-Lysine is also produced by bacterial fermentation.
[c]DL-Methionine is also synthesized by chemical means; aminoacylase has also been used to synthesize L-valine, L-phenylalanine, and other L-amino acids.
[d]Most L-phenylalanine is made by bacterial fermentation.

**Table 2.5.4. Application of Gene Technology to Amino Acid Production**

| Amino Acid | Enzymes | Processes |
|---|---|---|
| L-Aspartic acid | *Escherichia coli* | Recombinant strain with cloned aspartase gene has 5-30-fold greater enzyme activity |
| | *Serratia marcescens* | Recombinant strains with cloned aspartase gene have 39-fold more enzyme activity |
| L-Glutamic acid | *Corynebacterium glutamicum* | Introduction of high copy number plasmid isolated from *C. glutamicum* increases L-glutamic acid production |
| L-Histidine | *Corynebacterium glutamicum* | Introduction of cloned *his*G gene for ATP phosphoribosyltransferase doubles L-histidine production by recipient and induces L-histidine overproduction by a wild-type strain |
| | *Brevibacterium* sp. | Introduction of recombinant DNA with histidine gene induces L-histidine overproduction |
| L-Phenyl-alanine | *Corynebacterium glutamicum* | Introduction of cloned chorismate mutase and prephenate dehydrogenase genes results in 50% increase in produciton of L-phenylalanine to 19 g/L |
| | *Corynebacterium glutamicum* | Introduction of cloned prephenate dehydrogenase gene increases this enzyme's specific activity 6-fold |

**Table 2.5.4. (continued)**

| Amino Acid | Enzymes | Processes |
|---|---|---|
| l-Proline | *Brevibacterium lactofermentum* | Amplification of cloned phosphoenolpyruvate gene improves L-proline production 1.8-fold |
| | *Serratia marcescens* | Amplification of cloned *pro*A and *pro*B genes improves L-proline production 1.5-fold; plasmids with cloned proline *dpr*-1 gene promote excretion of 75 g/L L-proline |
| L-Threonine | *Corynebacterium glutamicum* | Expression of cloned E. coli threonine operon increases L-threonine production 4-fold |
| | *Brevibacteirum lactofermentum* | Amplification of clone homoserine dehydrogenase (HD) gene doubles HD activity and elevates L-threonine production 1.4-fold to 25 g/L; coexistence of HD and homoserine kinase genes in a threonine producing strain raises produciton 30% to 33 g/L and reduces by-products |
| | *Escherichia coli* | Introduction of plasmid-borne threonine operon increases production of L-threonine under optimal conditions |
| L-Tryptophan | *Brevibacterium lactofermentum* | Cloning of tryptophan gene cluster |
| L-Tyrosine | *Corynebacterium glutamicum, Brevibacterium* sp. | Recombinant strains of both species produce 5.3 g/L L-tyrosine |

Source: Adapted from Bigelis, 1989. Used with permission.

**Table 2.5.5. Amino Acid Production by Auxotrophic Mutants**

| Amino Acid | Microorganism | Genetic Markers[a] | Product Yield (g/L) |
|---|---|---|---|
| L-Aspartic acid | *Brevibacterium flavum* | Citrate synthase$^-$ | 10.6 |
| L-Citrulline | *Bacillus subtilis* K | Arg$^-$ | 16.5 |
| | *Corynebacterium glutamicum* | Arg$^-$ | 10.7 |
| L-Leucine | *Corynebacterium glutamicum* | Phe$^-$, His$^-$, Ile$^-$ | 16 |
| L-Lysine | *Corynebacterium glutamicum* | Homoser$^-$ | 13 |
| | *Brevibacterium flavum* | Thr$^-$, Met$^-$ | 34 |
| | *Brevibacterium lactofermentum* | Homoser$^-$ | 20.2 |
| L-Ornithine | *Corynebacterium glutamicum* | Cit$^-$ | 26 |
| | *Corynebacterium hydrocarbolactum* | Cit$^-$ | 9 |
| | *Brevibacterium lactofermentum* | Cit$^-$ | 40% molar yield |
| | *Arthrobacter paraffineus* | Cit$^-$ | 8 |
| L-Proline | *Brevibacterium* sp. | His$^-$ | 23 |
| | *Corynebacterium glutamicum* | Ile$^-$ | 14.8 |
| | *Brevibacterium flavum* | Ile, SG$^r$ | 35 |
| L-Threonine | *Arthrobacter paraffineus* | Ile$^-$ | 9 |
| | *Escherichia coli* | DAP$^-$, Met$^-$, Ile-revertant | 13 |
| | *Candida guillermondii* | Ile$^-$, Met$^-$, Trp$^-$ | 4 |
| | *Escherichia coli* | Met$^-$, Val$^-$ | 10.5 |
| L-Valine | *Corynebacterium glutamicum* | Ile$^-$, Leu$^-$ | 30 |
| | *Arthrobacter paraffineus* | Ile$^-$ | 9 |

Source: Soda, Tanaka, and Esaki, 1983. Used with permission.
[a]SG$^r$ = Sulfaguanidine resistance. DAP = $\alpha,\epsilon$-Diaminopimelate.

**Table 2.5.6. Amino Acid Production by Regulatory Mutants and Auxotrophic-Regulatory Mutants**

| Amino Acid | Microorganism | Genetic Markers | Product Yield (g/L) |
|---|---|---|---|
| L-Arginine | *Bacillus subtilis* | ArgHX$^r$, 6AU$^r$ | 28 |
| | *Corynebacterium glutamicum* | Ile revertant, D-Ser$^s$, D-Arg$^r$, ArgHX$^r$, 2TA$^r$ | 25 |
| | *Brevibacterium flavum* | 2TA$^r$, guanine$^-$ | 34.8 |
| | *Serratia marcescens* | Arg(Fr), Arg(Rr), Arg(D) | 25 |
| | *Bacillus subtilis* | ArgHX$^r$, 5HUR$^r$, TRA$^r$, 6FT$^r$, 6AU$^r$ | 17 |
| L-Citrulline | *Bacillus subtilis* | ArgHX$^r$, 6AU$^r$, Arg$^-$ | 26.2 |
| L-Histidine | *Corynebacterium glutamicum* | TRA$^r$ | 7 |
| | *Corynebacterium glutamicum* | TRA$^r$, Leu$^-$ | 11 |
| | *Corynebacterium glutamicum* | TRA$^r$, 6MG$^r$, 8AG$^r$, 4TU$^r$, 6MP$^r$, 5MT$^r$ | 15 |
| | *Serratia marcescens* | Histidase$^-$, TRA$^r$, 2MH$^r$ | 13 |
| | *Streptomyces coelicolor* | His-revertant | 13.5 |
| L-Isoleucine | *Serratia marcescens* | IleHX$^r$, ABA$^r$ | 12 |
| | *Brevibacterium flavum* | AHV$^r$, OMT$^r$ | 14.5 |
| | *Corynebacterium glutamicum* | Met$^-$, AHV$^r$, TIL$^r$, AEC$^r$, ETH$^r$, AZL$^r$, ABA$^r$ | 9.7 |
| L-Leucine | *Serratia marcescens* | ABA$^r$, Ile-revertant | 13.5 |
| | *Brevibacterium lactofermentum* | TA$^r$, Met$^-$, Ile$^-$ | 28 |
| L-Lysine | *Brevibacterium flavum* | Thr$^s$, Met$^s$ | 25 |
| | *Brevibacterium flavum* | AEC$^r$ | 32 |
| | *Brevibacterium lactofermentum* | AEC$^r$, Ala$^-$, CCL$^r$, ML$^r$ | 60 |

| | | | |
|---|---|---|---|
| | *Brevibacterium lactofermentum* | AEC$^{r}$, AHV$^{r}$ | 29 |
| | *Candida pelliculosa* | AEC$^{r}$ | 3.2 |
| | *Corynebacterium glutamicum* | AEC$^{r}$, Homoser$^{-}$, Leu$^{-}$, Pant$^{-}$ | 42 |
| L-Methionine | *Corynebacterium glutamicum* | Thr$^{-}$, ETH$^{r}$, SME$^{r}$, MetHX$^{r}$ | 2 |
| L-Phenyl-alanine | *Brevibacterium flavum* | MFP$^{r}$ | 2.2 |
| | *Bacillus subtilis* | 5FT$^{r}$ | 6 |
| | *Corynebacterium glutamicum* | PFP$^{r}$, PAP$^{r}$, Tyr$^{-}$ | 9.5 |
| | *Brevibacterium lactofermentum* | 5MT$^{r}$, PFP$^{r}$, Dec$^{s}$, Tyr$^{-}$, Met$^{-}$ | 24.8 |
| L-Serine | *Corynebacterium glutamicum* | OMS$^{r}$, MSE$^{r}$, ISE$^{r}$ | 3.8 |
| L-Threonine | *Escherichia coli* | AHV$^{r}$, Met$^{-}$, Ile$^{-}$ | 6.1 |
| | *Brevibacterium flavum* | AHV$^{r}$, Met$^{-}$ | 18 |
| | *Cornybacterium glutamicum* | AHV$^{r}$, AEC$^{r}$, Met$^{-}$ | 14 |
| | *Brevibacterium flavum* | AEC$^{r}$, AHV$^{r}$ | 14.8 |
| | *Serratia marcescens* | Ile$^{-}$, DAP$^{-}$, AHV$^{r}$ | 12.7 |
| | *Serratia marcescens* | Ile$^{-}$, DAP$^{-}$, AHV$^{r}$, AEC$^{r}$, Thr deaminase$^{-}$, Thr dehydrogenase$^{-}$ | 25 |
| L-Trypto-phan | *Brevibacterium flavum* | 5FT$^{r}$, MFP$^{r}$, Phe$^{-}$ | 6.2 |
| | *Corynebacterium glutamicum* | 5MT$^{r}$, 4MT$^{r}$, 6FT$^{r}$, TrpHX$^{r}$, PFP$^{r}$, PAP$^{r}$, TyrHX$^{r}$, PheHX$^{r}$, Phe$^{-}$, Tyr$^{-}$ | 12 |
| | *Brevibacterium flavum* | 5FT$^{r}$, AZS$^{r}$, PFP$^{r}$, Tyr$^{-}$ | 10.5 |
| L-Tyrosine | *Brevibacterium flavum* | MFP$^{r}$ | 1.9 |

**Table 2.5.6. (continued)**

| Amino Acid | Microorganism | Genetic Markers | Product Yield (g/L) |
|---|---|---|---|
| L-Tyrosine (continued) | *Corynebacterium glutamicum* | 3AT$^r$, PAP$^r$, PFP$^r$, TyrHX$^r$, Phe$^-$ | 17.6 |
| L-Valine | *Brevibacterium flavum* | TA$^r$ | 31 |

Source: Soda, Tanaka, and Esaki, 1983. Used with permission.
Abbreviations that do not represent amino acids are as follows: ABA $\alpha$-aminobutyrate; AEC $S$-($\beta$-aminoethyl)-cysteine; AG azaguanine; AHV $\alpha$-amino-$\beta$-hydroxy-valerate; AT amino-tyrosine; AU azauracil; AZL azaleucine; AZS azaserine; CCL $\alpha$-chlorocaprolactam; Dec decoynine; DAP $\alpha,\epsilon$-diaminopimelate; ETH ethionine; FT fluorotryptophan; HUR hydroxyuridine; HX hydroxamate; ISE isoserine; MFP $m$-fluorophenylalanine; MG mercaptoguanine; MH methylhistidine; ML $\gamma$-methyllysine; MP mercaptopurine; MSE $\gamma$-methylserine; MT methyltryptophan; OMS $O$-methylserine; OMT $O$-methylthreonine; PAP $p$-aminophenylalanine; PFP $p$-fluorophenylalanine; SME selenomethionine; TA thiazole alanine; TIL thioisoleucine; TRA 1,2,4-triazole alanine; TU thiouracil. Arg(Fr) lack of feedback inhibition by arginine; Arg(Rr) lack of repression of arginine operon; Arg(D) deficiency of arginine-degrading enzymes.

**Table 2.5.7. Amino Acid Production from Biosynthetic Precursors**

| Amino Acid | Microorganism | Precursor |
|---|---|---|
| L-Histidine | *Brevibacterium flavum* | Histidinol |
| L-Isoleucine | *Serratia marcescens* | DL-$\alpha$-Aminobutyrate |
| | *Bacillus subtilis* | DL-$\alpha$-Aminobutyrate |
| | *Corynebacterium* sp. | DL-$\alpha$-Aminobutyrate |
| | *Serratia marcescens* | D-Threonine |
| | *Bacillus, Brevibacterium, Micrococcus, Aerobacter* | DL-$\alpha$-Hydroxybutyrate |
| | *Brevibacterium, Micrococcus, Corynebacterium* | DL-$\alpha$-Bromobutyrate |
| L-Methionine | *Pseudomonas denitrificans* | 2-Hydroxy-4-methyl-thiobutyrate |
| L-Phenylalanine | *Pseudomonas denitrificans* | 2-Hydroxy-3-phenyl-propionate |

| | | |
|---|---|---|
| L-Serine | *Corynebacterium glycinophilum* | Glycine |
| | *Pseudomonas* sp. | Glycine |
| | *Arthrobacter globiformis* | Glycine |
| | *Sarcina albida* | Glycine |
| | *Nocardia butanica* | Glycine |
| | *Pseudomonas* sp. | Glycine |
| L-Threonine | *Bacillus subtilis* | L-Homoserine |
| | *Proteus rettgeri* | L-Homoserine |
| | *Enterobacter* | DL-Homoserine |
| L-Tryptophan | *Claviceps purpurea* | Indole |
| | *Hansenula anomala* | Anthranilic acid |

Source: Soda, Tanaka, and Esaki, 1983. Used with permission.

## 2.6. Fermented Foods

**Table 2.6.1. Fermented Foods in North America and Europe**

| Products | Raw Materials | Microorganisms |
|---|---|---|
| Anchovy of Northern type | Herring, sardines | *Vibrio costicola* |
| Baked goods | Wheat flour | *Saccharomyces cerevisiae* |
| Beer | Malt | *Saccharomyces cerevisiae, Brettanomyces bruxellensis, Lactobacillus brevis* |
| Brandy | Wine | *Saccharomyces cerevisiae* |
| Cocoa | Cocoa beans | Yeasts, lactic acid bacteria, acetic acid bacteria, bacilli |
| Coffee | Coffee beans | Enterobacteria, lactic acid bacteria, yeasts |
| Fruit brandy | Cherry, pears | *Saccharomyces cerevisiae* |
| Fruit wine | Fruit juices | *Saccharomyces cerevisiae* |
| Ham | Pork | *Vibrio costicola, Staphylococcus* spp. |
| Hard cheese | Milk | *Streptococcus salivarius* ssp. *thermophilus, Lactobacillus helveticus, Propionibacterium freudenreichii* |

**Table 2.6.1. (continued)**

| Products | Raw Materials | Microorganisms |
|---|---|---|
| Kefir | Milk | *Candida kefir, Lactobacillus kefir, L. acidophilus, Streptococcus lactis* |
| Kom, whysky | Cereals | *Saccharomyces cerevisiae* |
| Kwass | Malt, bread | Yeasts, lactic acid bacteria |
| Mold cheese | Milk | Starter, *Penicillium caseicolum* or *camembertii, Penicillium roquefortii* |
| Olives | Olives | *Lactobacillus* spp., *Leuconostoc* spp., *Pediococcus* spp. |
| Quark | Milk | Starter or *Streptococcus salivarius* ssp., *thermophilus, Lactobacillus delbrueckii* ssp. *bulgaricus* |
| Raw sausage | Beef, pork, mutton | *Lactobacillus* spp., *Staphylococcus* spp., *Micrococcus varians* |
| Sauerkraut | White cabbage | *Leuconostoc mesenteroides, Lactobacillus plantarum, L. brevis* |
| Schneidebohnen, lactic acid fermented | French beans | Lactic acid bacteria |
| Smear coat cheese | Milk | Starter, *Brevibacterium linens* |
| Sour cream butter | Cream | Starter, containing: *Streptococcus lactis,* ssp. *lactis, S. lactis* ssp. *cremoris, S. lactis* ssp. *diacetylactis, Leuconostoc mesenteroides* ssp. *cremoris* |
| Sour dough | Rye and wheat flour | *Lactobacillus sanfranzisko, Lactobacillus brevis, Lactobacillus plantarum, Lactobacillus fructivorams, Lactobacillus fermentum, Torulopsis holmii, Saccharomyces cerevisiae, Pichia saitoi, Candida krusei* |
| Sour milk | Milk | *Streptococcus lactis* ssp. *cremoris, Streptococcus lactis* ssp. *diacetylactis, Leuconostoc mesenteroides* ssp. *cremoris* |

| | | |
|---|---|---|
| Soy sauce | Rice, wheat, soybeans | *Aspergillus oryzae, Lactobacillus* spp., *Pediococcus* spp., *Zygosaccharomyces rouxii, Torulopsis* spp. |
| Sparkling wine | Wine | *Saccharomyces cerevisiae* |
| Tea | Leaves of the tea plant | (endogenous enzymes) |
| Topinambur brandy | Jerusalem artichoke | *Saccharomyces cerevisiae Kluyveromyces marxianus* |
| Vegetable juices | Beets, carrots white cabbage | *Streptococcus lactis, Lactobacillus casei, Lactobacillus bavaricus* |
| Vinegar | Wine, malt, ethanol | *Acetobacter aceti, A. pasteurianus, A. hansenii, Gluconobacter oxydans* |
| Wine | Grape juice | *Saccharomyces cerevisiae, Leuconostoc oenos* |
| Vodka | Potato, cereals | *Saccharomyces cerevisiae* |
| Yoghurt | Milk | *Streptococcus salivarius* ssp. *thermophilus, Lactobacillus delbrueckii* ssp. *bulgaricus* |

Source: Bottazzi, 1983; Gilliland, 1985; King and Cheetham, 1987; Rehm and Reed, 1983, 1986, 1988, 1987, 1989; Tsao, 1985; Vogel, 1983; Wood, 1985.

**Table 2.6.2. Food Fermentation Fungi**

| Fungus | Product | Fungus | Product |
|---|---|---|---|
| *Actinomucor elegans* | Tempeh | *M. subtilissimus* | Sofu |
| *Aspergillus flavus-columnaris* | Koji | *Neurospora sitophila* | Ontjom |
| | | *Penicillium album* | Cheese |
| | | *P. camemberti* | Cheese |
| *A. oryzae* | Koji | *P. caseicolum* | Cheese |
| *A. soyae* | Koji | *P. roqueforti* | Cheese |
| *Monascus purpureus* | (Ang-kak) | *Rhizopus arrhizus* | Tempeh |
| | | *R. oligosporus* | Tempeh |
| *Mucor hiemalis* | Sofu | *R. oryzae* | Tempeh |
| *M. silvaticus* | Sofu | | |

Source: Jacobson and Jolly, 1987. Used with permission.

**Table 2.6.3. Major Regional Fermented Foods**

- Oriental fermented foods
  - Soy sauce
    - Preparation of soybeans
    - Preparation of wheat
    - Koji process
    - Mash (moromi) stage
    - Pasteurization
  - Miso
    - Preparation of koji
    - Preparation of soybeans
    - Fermentation and aging
    - Modified indigenous procedure
  - Fermented whole soybeans (natto products)
  - Sufu
  - Meitauza
  - Lao-chao
  - Ang-kak
  - Puto
  - Ragi
  - Tapé
  - Tempeh
    - Preparation of soybeans and fermentation
    - Biochemical changes
    - Nutritional characteristics
  - Oncom
    - Preparation of peanuts and fermentation
    - Biochemical changes
  - Fish products
  - Kimchi
- Fermented foods of India
  - Idli
  - Waries
  - Papadam
  - Dhokla
  - Khaman
  - Kenima
  - Jalebies
  - Kanji
- Fermented foods of Africa
  - Dawadawa
  - Gari
  - Banku
  - Ogi
  - Injera
  - Kaffir beer
  - Merissa
- Other fermented products
  - Milk and grain products
    - Kushik
    - Tarhana
    - Kishk
  - Kaanga-kopuwai
  - Poi
  - Chicha
  - Pozol
  - Legume-based milk prdoucts
- Nutrition and health

Source: Adapted from Beuchat, 1983. Used with permission.

**Table 2.6.4. Properties of Fermented Foods of Non-European Origins**

Ang-kak (anka, red rice): available in China, Southeast, Asia, Syria
1. Description (usage): dry red powder (colorant)
2. Microorganism (substrate): *Monascus purpureus* (rice)

Bagoong: available in Philippines
1. Description (usage): paste (seasoning agent)
2. Microorganism (substrate): unknown (fish)

Bagni: available in Caucasus
1. Description (usage): liquid (drink)
2. Microorganism (substrate): unknown (millet)

Banku: available in Ghana
1. Description (usage): dough (staple)
2. Microorganism (substrate): lactic acid bacteria, yeasts (maize, cassava)

Bonkrek: available in Central Java (Indonesia)
1. Description (usage): solid (roasted or fried in oil, used as a meat substitute)
2. Microorganism (substrate): *Rhizopus oligosporus* (coconut press cake)

Bouza: available in Egypt
1. Description (usage): liquid (thick acidic)
2. Microorganism (substrate): unknown (wheat)

Braga: available in Romania
1. Description (usage): liquid (drink)
2. Microorganism (substrate): unknown (millet)

Burukutu: available in Savannah regions of Nigeria
1. Description (usage): liquid (creamy drink with suspended solids)
2. Microorganism (substrate): lactic acid bacteria, *Candida* spp., *Saccharomyces cerevisiae* (sorghum and cassava)

Busa: available in Tartars of Krim, Turkestan, Egypt
1. Description (usage): liquid (drink)
2. Microorganism (substrate): *Lactobacillus* and *Saccharomyces* (rice or millet, sugar)

Chee-fan: available in China
1. Description (usage): solid (eaten fresh, cheeselike)
2. Microorganism (substrate): *Mucor* sp., *Aspergillus glaucus* (soybean whey curd)

Chicha: available in Peru
1. Description (usage): spongy (eaten with vegetables)
2. Microorganism (substrate): *Aspergillus, Penicillium* spp., yeasts, bacteria (maize)

Chickwangue: available in Congo

**Table 2.6.4. (continued)**

1. Description (usage): paste (staple)
2. Microorganism (substrate): bacteria (cassava roots)

Chinese yeast: available in China
1. Description (usage): solid (eaten fresh or canned, used as side dish with rice)
2. Microorganism (substrate): Mucoraceous molds and yeasts (soybeans)

Darassum: available in Mongolia
1. Description (usage): liquid (drink)
2. Microorganism (substrate): unknown (millet)

Dawadawa (daddowa, uri, kpalugu, kinda, neteton): available in West Africa, Nigeria
1. Description (usage): solid, sun-dried (eaten fresh, supplement to soups, stews)
2. Microorganism (substrate): lactic acid bacteria, yeasts (African locust bean)

Dhokla: available in India
1. Description (usage): spongy (condiment)
2. Microorganism (substrate): unknown (Bengal gram and wheat)

Dosai (doa): available in India
1. Description (usage): spongy, pancakelike (breakfast food)
2. Microorganism (substrate): Yeasts, *Leuconostoc mesenteroides* (black gram and rice)

Fish sauce (nuoc-mam, patis, mampla, ngam-pyaye): available in Southeast Asia
1. Description (usage): liquid (seasoning agent)
2. Microorganism (substrate): bacteria (fish)

Gari: available in West Africa
1. Description (usage): Wet paste (eaten fresh as staple with stews, vegetables)
2. Microorganism (substrate): *Corynebacterium manihot*, *Geotrichum candidum* (cassava root)

Hamanatto: available in Japan
1. Description (usage): beans retain individual form, raisin-like, soft (flavoring agent for meat and fish, eaten as snack)
2. Microorganism (substrate): *Aspergillus oryzae, Streptococcus, Pediococcus* (whole soybeans, wheat flour)

Idli: available in Southern India
1. Description (usage): spongy, moist (bread substitute)
2. Microorganism (substrate): l*euconostoc mesenteroides*, *Torulopsis candida and Trichosporon pulluians* (rice and black gram)

Injera: available in Ethiopia

1. Description (usage): bread-like, moist (bread substitute)
2. Microorganism (substrate): *Candida guilliermondii* (teff, or maize wheat, barley, sorghum)

Jalebies: available in India, Nepal, Pakistan
1. Description (usage): pretzel-like, sirup filled (Confection)
2. Microorganism (substrate): *Saccharomyces bayanus* (wheat flour)

Jamin-bang: available in Brazil
1. Description (usage): bread or cake-like (bread substitute)
2. Microorganism (substrate): Yeasts and bacteria (maize)

Kaanga-kopuwai: available in New Zealand
1. Description (usage): soft, slimy (eaten as vegetable)
2. Microorganism (substrate): bacteria and yeasts (maize)

Kanji: available in India
1. Description (usage): liquid (sour, added to vegetables)
2. Microorganism (substrate): *Hansenula anomala* (rice and carrots)

Katsuobushi: available in Japan
1. Description (usage): solid, dry (seasoning agent)
2. Microorganism (substrate): *Aspergillus glaucus* (whole fish)

Kecap: available in Indonesia and vicinity
1. Description (usage): liquid (condiment, seasoning agent)
2. Microorganism (substrate): *Aspergillus oryzae, Lactobacillus, Hansenula, Saccharomyces* (soybeans, wheat)

Kenima: available in Nepal, Sikkim, Darjeeling district of India
1. Description (usage): solid (snack)
2. Microorganism (substrate): unknown (soybeans)

Kenkey: available in Ghana
1. Description (usage): Mush (steamed, eaten with vegetables)
2. Microorganism (substrate): unknown (maize)

Ketjap: available in Indonesia
1. Description (usage): sirup (seasoning agent)
2. Microorganism (substrate): *Aspergillus oryzae* (black soybeans)

Khaman: available in India
1. Description (usage): solid, cake-like (breakfast food)
2. Microorganism (substrate): unknown (Bengal gram)

Kimchi (kim-chee): available in Korea
1. Description (usage): solid and liquid (condiment)
2. Microorganism (substrate): lactic acid bacteria (vegetables, sometimes seafoods, nuts)

Kishk (kushuk, kushik): available in Egypt, Syria, Arab world
1. Description (usage): solid (dried balls dispersed rapidly in water)

**Table 2.6.4. (continued)**

2. Microorganism (substrate): lactic acid bacteria, *Bacillus* spp. (wheat, milk)

Lafun: available in West Africa, Nigeria
1. Description (usage): paste (staple food)
2. Microorganism (substrate): bacteria (cassava root)

Lao-chao: available in China, Indonesia
1. Description (usage): soft, juicy, glutinous (Eaten as such as dessert or combined with eggs, seafood)
2. Microorganism (substrate): *Rhizopus oryzae, R. chinensis, Chlamydomucor oryzae, Saccharomycopsis* sp. (rice)

Mahewu (Magou): available in South Africa
1. Description (usage): liquid (drink, sour and non-alcoholic)
2. Microorganism (substrate): lactic acid bacteria (*Lactobacillus delbruckii*) (maize)

Meitauza: available in China, Taiwan
1. Description (usage): solid (fried in oil or cooked with vegetables)
2. Microorganism (substrate): *Actinomucor elegans* (soybean cake)

Meju: available in Korea
1. Description (usage): paste (seasoning agent)
2. Microorganism (substrate): *Aspergillus oryzae, Rhizopus* spp. (soybeans)

Merissa: available in Sudan
1. Description (usage): liquid (Drink)
2. Microorganism (substrate): *Saccharomyces* sp. (sorghum)

Minchin: available in China
1. Description (usage): solid (condiment)
2. Microorganism (substrate): *Paecilomyces, Aspergillus, Cladosporium, Fusarium, Syncephalastum, Penicillium, Trichothecium* spp. (wheat gluten)

Miso (chiang, jang, doenjang, tauco, tao chieo): available in Japan, China
1. Description (usage): paste (soup base, seasoning)
2. Microorganism (substrate): *Aspergillus oryzae, Torulopsis etchellsii, Lactobacillus* bacteria, *Saccharomyces rouxii* (rice and soybeans or rice and other cereals such as barley)

Munkoyo: available in Africa
1. Description (usage): liquid (drink)
2. Microorganism (substrate): unknown (millet, maize or kaffir corn plus roots of munkoyo)

Nan (khab-z): available in India, Pakistan, Afghanistan, Iran
1. Description (usage): solid (snack)
2. Microorganism (substrate): unknown (unbleached wheat flour)

Natto: available in Northern Japan

1. Description (usage): solid (cake, as a meat substitute)
2. Microorganism (substrate): *Bacillus natto* (soybeans)

Ogi: available in Nigeria, West Africa

1. Description (usage): paste (staple, eaten for breakfast, weaning babies)
2. Microorganism (substrate): lactic bacteria, *Cephalosporium, Fusarium, Aspergillus, Penicillium* spp., *Saccharomyces cerevisiae, Candida mycoderma* (*C. valida* or *C. vini*) (Maize)

Oncom (ontjom, lontjom): available in Indonesia

1. Description (usage): solid (roasted or fried in oil, used as meat substitute)
2. Microorganism (substrate): *Neurospora intermedia*, less often *Rhizopus oligosporus* (peanut press cake)

Papadam: available in India

1. Description (usage): solid, crisp (condiment)
2. Microorganism (substrate): *Saccharomyces* spp. (black gram)

Peujeum: available in Java

1. Description (usage): solid (eaten fresh or fried)
2. Microorganism (substrate): unknown (banana, plantain)

Pito: available in Nigeria

1. Description (usage): liquid (drink)
2. Microorganism (substrate): unknown (guineacorn or maize or both)

Poi: available in Hawaii

1. Description (usage): semi solid (side dish with fish, meat)
2. Microorganism (substrate): *lactobacillus* spp., *Candida vini* (*Mycoderma vini*), *Geotrichum candidum* (taro corms)

Pozol: available in Southeastern Mexico

1. Description (usage): dough, spongy (diluted with water, drunk as basic food)
2. Microorganism (substrate): Molds, yeasts, bacteria (maize)

Prahoc: available in Cambodia

1. Description (usage): paste (seasoning agent
2. Microorganism (substrate): unknown (fish)

Puto: available in Philippines

1. Description (usage): solid (snack)
2. Microorganism (substrate): lactic acid bacteria, *Saccharomyces cerevisiae* (rice)

Rabdi: available in India

1. Description (usage): semi-solid (mush, eaten with vegetables)
2. Microorganism (substrate): unknown (maize and buttermilk)

Sierra rice: available in Ecuador

1. Description (usage): solid (brownish-yellow, seasoning)

**Table 2.6.4. (continued)**

2. Microorganism (substrate): *Aspergillus flavus, A. candidus, Bacillus subtilis* (unhusked rice)

Sorghum beer (Ibantu beer, kaffir beer, leting, joala, utshivala, mqomboti, igwelel): available in South Africa
1. Description (usage): liquid (drink, acidic and weakly alcoholic)
2. Microorganism (substrate): lactic acid bacteria, yeasts (sorghum, maize)

Soybean milk: available in China, Japan
1. Description (usage): liquid (drink)
2. Microorganism (substrate): lactic acid bacteria (soybeans)

Soy sauce (Chaing-yu, shoyu, toyo, kanjang, kecap, seeieu): available in Japan, China, Philippines, other parts of Orient
1. Description (usage): liquid (seasoning for meat, fish, cereals, vegetables)
2. Microorganism (substrate): *Aspergillus oryzae* or *A. soyae*, *Lactobacillus* bacteria, *Saccharomyces rouxii* (*Zygosaccharomyces* spp.) (soybeans and wheat)

Sufu (tahuri, taokaoan, tao-hu-yi): available in China, Taiwan
1. Description (usage): solid (soybean cheese, condiment)
2. Microorganism (substrate): *Actinomucor elegans, Mucor hiemalis, M. silvaticus, M. subtilissimus* (soybean whey curd)

Tao-si: available in Philippines
1. Description (usage): semi solid (seasoning agent)
2. Microorganism (substrate): *Aspergillus oryzae* (soybeans plus wheat flour

Taotjo: available in East Indies
1. Description (usage): semi solid (condiment)
2. Microorganism (substrate): *Aspergillus oryzae* (soybeans plus roasted wheat meal or glutinous rice)

Tapé: available in Indonesia and vicinity
1. Description (usage): soft solid (eaten fresh as staple)
2. Microorganism (substrate): *Saccharomyces cerevisiae, Hansenula anomala, Rhizopus oryzae, Chlamydomucor oryzae, Mucor sp., Endomycopsis fibuliger* (*Saccharomycopsis* sp.) (cassava or rice)

Tarhana: available in Turkey
1. Description (usage): solid powder (dried seasoning for soups)
2. Microorganism (substrate): lactic acid bacteria (parboiled wheat meal and yoghurt (2:1))

Tauco: available in West Java (Indonesia)
1. Description (usage): liquid (drink)
2. Microorganism (substrate): *Rhizopus oligosporus, Aspergillus oryzae* (soybeans, cereals)

Tempeh (tempe kedeke): available in Indonesia and vicinity, Surinam

1. Description (usage): solid (fried in oil, roasted, or used as meat substitute in soup)
2. Microorganism (substrate): *Rhizopus* spp., principally *R. oligosporus* (soybeans)

Thumba (bojah): available in West Bengal

1. Description (usage): liquid (drink, mildly alcoholic)
2. Microorganism (substrate): *Endomycopsis fibuliger* (millet)

Torani: available in India

1. Description (usage): liquid (seasoning for vegetables)
2. Microorganism (substrate): *Hansenula anomala, Candida guilliermondii, C. tropicalis, Geotrichum candidum* (rice)

Waries: India

1. Description (usage): spongy (spicy condiment eaten with vegetables, legumes, rice)
2. Microorganism (substrate): *Candida* spp., *Saccharomyces* spp. (black gram flour)

---

Source: Adapted from Beuchat, 1983. Used with permission.

**Table 2.6.5. Fundamentals of Manufacturing Fermented Dairy Products**

---

Description of the characteristics of the principal products

Taxonomy of lactic acid bacteria present in fermented products

- Morphological and physiological characteristics
- Chemical composition of the cell wall
- Serological reaction
- Genetic characteristics

Microbial composition of fermented dairy products

- Acid fermented milks: yoghurt, leben, acidophilus milk
- Slightly acid fermented milks; cultured milk, cultured buttermilk, cultured cream
- Slightly acid, thick fermented milks: long milk and viili
- Acid-alcoholic fermented milks: kefir, kumis
- Fermented milk products with extended shelf life: labneh, quark, laktofil

Description of the technology for each of the nine stages in manufacturing the major types of fermented products

1. Quality of the milk
2. Standardization
3. Temperature treatment
4. Homogenization
5. Length of fermentation
6. Cooling
7. Contamination
8. Production techniques
9. Packaging

**Table 2.6.5. (continued)**

Starter cultures for fermented dairy products
- Biological aspects of the association between *Streptococcus thermophilus* and *Lactobacillus bulgaricus*
- Technological aspects of the association *streptococcus thermophilus* - *Lactobacillus bulgaricus* and between Mesophilic streptococci
- Factors that interfere with the development op the cultures
- Yoghurt with microflora integrated with other microorganisms
- Preparation of special cultures

Biochemical changes in fermented milk
- Fermentation of lactose and production of lactic acid
- Production of aromatic compounds
- Hydrolysis of proteins and lipids
- Vitamin B content

Probiotic role of lactic acid bacteria and fermented milk products

Source: Adapted from Bottazzi, 1987. Used with permission.

**Table 2.6.6. Principal Fermented Milk Products**

| Product | Country of Origin | Type of Milk |
|---|---|---|
| Yoghurt | Turkey, Caucasus | Bovine |
| Leben raib | Egypt | Bovine, buffalo |
| Leben | Iraq | Ewe, goat |
| Mazum | Armenia | Bovine, ewe, buffalo |
| Koumis | Russia | Horse |
| Kefir | Russia | Bovine, ewe, goat |
| Dahi | India | Bovine, buffalo |
| Abdoogh | Afghanistan, Iran | Bovine, buffalo |
| Gioddu | Italy | Ewe |
| Viili | Finland | Bovine |
| Long milk | Scandinavia | Bovine |
| Yakult | Japan | Bovine |
| Acidophilus milk | USA | Bovine |
| Sour cream | USA, Russia | Bovine |
| Cultured milk | USA | Bovine |
| Cheese | USA | Bovine |
| Laktofil | Sweden | Bovine |
| Labneh | Israel | Bovine |
| Quark | Europe | Bovine |

Source: Adapted from Bottazzi, 1987. Used with permission

## 2.7. Enzymes: Classifications and Sources

**Table 2.7.1. Representative Industrial Enzymes of the Six Major Classes**

| Enzymes | Formed by Microorganisms |
|---|---|
| Oxidoreductases | |
| Glucose oxidase | *Aspergillus niger* |
| Catalase | *Aspergillus niger, Micrococcus lysodeicticus* |
| Pyranose-2-oxidase | *Polyporus obtusus* |
| Transferases | Not applicable |
| Hydrolases | |
| Bacterial α-amylase | *Bacillus amyloliquefaciens, B. licheniformis* |
| Fungal α-amylase | *Aspergillus oryzae, A. niger* |
| Glucoamylase | *Aspergillus niger, A. awamori, Endomycopsis* sp., *Rhizopus* sp. |
| β-Glucanase (licheninase) | *Bacillus subtilis, B. amyloliquefaciens* |
| Cellulase complex, containing endo-β-1,4-glucanase, cellobiohydrolase, β-glucosidase | *Trichoderma reesei (viride), Aspergillus niger* |
| α-1,3-Glucanase | *Trichoderma harzianum* |
| Dextranase | *Penicillium funiculosum* |
| Hemicellulase | *Bacillus subtilis, Aspergillus niger* |
| Pullulanase | *Bacillus* sp. |
| α-Galactosidase | *Aspergillus niger, Mortierella vinacea* |
| Invertase | Yeast |
| Lactase | *Kluayveromyces fragilis* |
| Bacterial neutral protease | *Bacillus amyloliquefaciens, B. subtilis* |
| Bacterial alkaline protease | *Bacillus licheniformis, B. subtilis, B. amyloliquefaciens* |
| Fungal protease | *Aspergillus niger* |
| Rennet | *Mucor miehei, M. pusillus, Endothia parasitica* |
| Streptokinase | *Streptococcus* sp. |
| L-Asparaginase | *Erwinia carotovora* |
| Uricase | *Bacillus fastidiosus* |

**Table 2.7.1. (continued)**

| Enzymes | Formed by Microorganisms |
|---|---|
| Fungal pectinase complex, containing polymethyl-galacturomases with endo-PMG and exo-PMG activities and poly-galacturonases with endo-PG and exo-PG activities | *Aspergillus niger, Trichoderma reesei* |
| Lyases | |
| Fungal pectinase complex, containing polymethyl-galacturonate lyase and polygalacturonate lyase | *Aspergillus niger, Trichoderma reesei* |
| Isomerases | |
| Glucose isomerase | *Bacillus coagulans, Actinoplanes missouriensis, Streptomyces* sp. |
| Ligases | Not applicable |

Source: Adapted from Borriss, 1987. Used with permission.
Note: PMG = polymethylgalacturonase; PG = polygalacturonase

**Table 2.7.2. Profiles of Enzymes With Industrial-Scale Applications**

| Groups Types | Commercial Applications | Types of Fermentation | Types of Products |
|---|---|---|---|
| Proteinases | | | |
| Bacterial alkaline | Food processing | Deep (submerged) fermentation, extracellular (enzyme in culture supernatant) | Formulated liquid, dried solid |
| Bacterical neutral | Brewing Food | Deep (submerged) fermentation, extracellular (enzyme in culture supernatant) | Formulated liquid, dried solid |

| | | | |
|---|---|---|---|
| Fungal | Baking | Deep (submerged) fermentation, extracellular (enzyme in culture supernatant) | Dried solid |
| | Cheese | Deep (submerged) fermentation, extracellular (enzyme in culture supernatant) | Formulated liquid |
| | Cheese | Solid-substrate fermentation | Dried solid |
| Amylolytic enzymes | | | |
| Bacterial α-amylase | Starch, brewing | Deep (submerged) fermentation, extracellular (enzyme in culture supernatant) | Formulated liquid, dried solid |
| Fungal α-amylase | Starch, baking | Deep (submerged) fermentation, extracellular (enzyme in culture supernatant) | Formulated liquid, dried solid |
| Amylo-glucosidase | Starch Brewing | Deep (submerged) fermentation, extracellular (enzyme in culture supernatant) | Formulated liquid Dried solid |
| Pullulanase neutral | Starch | Deep (submerged) fermentation, cellular (enzyme in harvested biomass) and extracellular (enzyme in culture supernatant) | Formulated liquid |

**Table 2.7.2. (continued)**

| Groups<br>Types | Commercial Applications | Types of Fermentation | Types of Products |
|---|---|---|---|
| Other Glycosidases | | | |
| Lactase ($\beta$-galactosidase) | Dairy | Deep (submerged) fermentation, cellular (enzyme in harvested biomass), solid-substrate fermentation | Formulated liquid, dried solid |
| Invertase | Confectionery | Deep (submerged) fermentation, cellular (enzyme in harvested biomass) | Formulated liquid, dried solid |
| Raffinase ($\alpha$-galactosidase) | Sugar refining | Deep (submerged) fermentation, cellular (enzyme in harvested biomass) | Immobilized |
| Cellulase | Food, fruit juices and soft drinks | Deep (submerged) fermentation, extracellular (enzyme in culture supernatant) | Formulated liquid, dried solid |
| $\beta$-1,3(4)-glucanase | Brewing | Deep (submerged) fermentation, extracellular (enzyme in culture supernatant) | Formulated liquid |
| Other Hydrolases | | | |
| Lipase | Food | Deep (submerged) fermentation, extracellular (enzyme in culture supernatant) | Dried solid |

| | | | |
|---|---|---|---|
| Pectinase[a] | Fruit juices and soft drinks, food | Deep (submerged) fermentation, extracellular (enzyme in culture supernatant), solid-substrate fermentation | Formulated liquid, dried solid |
| Other Enzymes[b] | | | |
| Glucose isomerase | Starch | Deep (submerged) fermentation, cellular (enzyme in harvested biomass) | Immobilized |
| Glucose oxidase | Food, fruit juices and soft drinksy | Deep (submerged) fermentation, cellular (enzyme in harvested biomass) | Formulated liquid |
| Catalase | Food, fruit juices and soft drinks | Deep (submerged) fermentation, cellular (enzyme in harvested biomass) | Formulated liquid |

Source: Adapted from Frost and Moss, 1987.

[a]Pectinase preparation usually contain a methylesterase, polygalacturonases (glycosidases), and also pectin lyases.

[b]Glucose oxidase and catalase are usually made together in a single fermentation. For use in food and drinks the enzymes are not separated during purification. For diagnostic use the glucose oxidase is normally separated from the catalase which may be sold separately for food use.

**Table 2.7.3 Major Classes of Enzymes of Commercial Interest and Some of Their Sources**

| Enzyme | Sources |
|---|---|
| α-Amylase | *Aspergillus niger, A. oryzae, Bacillus licheniformis, B. subtilis,* barley and wheat malt, bovine and porcine pancreas |
| β-Amylase | Barley and wheat malt, *Bacillus cereus, Bacillus* sp. |
| Catalase | *Aspergillus niger, Penicillium vitale, Micrococcus lysodeikticus,* bovine liver |
| Cellulase | *Aspergillus niger, Trichoderma reesei, Trichoderma* sp., *Penicillium funiculosum* |
| Dextranase | *Penicillium lilacium, P. funiculosum, Chaetomium gracile* |
| α-Galactosidase | *Mortierella vinaceae* |
| β-Galactosidase | *Aspergillus niger, A. oryzae, Candida pseudotropicalis, Kluyveromyces fragilis, K. marxianus var. lactis, Bacillus* sp. |
| β-Glucanase | *Aspergillus niger, Aspergillus* sp., *Bacillus subtilis, Penicillium emersonii* |
| Glucoamylase | *Aspergillus niger, A. oryzae, A. awamori, Rhizopus oryzae, R. niveus* |
| Glucose isomerase[a] | *Actinoplanes missouriensis, Arthrobacter globiformis, Bacillus coagulans, Flavobacterium (Microbacterium) arborescens, Streptomyces olivaceus, S. olivochromogenes, S. rubiginosus, S. albus* |
| Glucose oxidase | *Aspergillus niger, Penicillium amagasakiense, P. vitale* |
| β-Glucosidase | *Aspergillus niger, A. oryzae, Trichoderma reesei, Saccharomyces* sp., *Bacillus* sp., *Clostridium thermocellu, Prunus amygalus* |
| Hemicellulase | *Aspergillus niger* |
| Hesperiginase | *Aspergillus niger* |
| Inulinase | *Candida* sp., *Aspergillus* sp. |
| Invertase | *Saccharomyces cerevisiae, S. diastaticus* |
| Lipase | *Aspergillus niger, A. oryzae, Candida cylindracea, Rhizomucor miehei, Rhizomucor* sp., *Rhizopus arrhizus* var. *delemar, Rhizopus* sp., *Micrococcus* sp., *Lactobacillus* sp., *Penicillium roqueforti,* pancreatic tissue, edible forestomach of calves, lambs, and kids |

| | |
|---|---|
| Lipase-esterase | *Rhizomucor miehei*; calf, kid, and lamb glands |
| Lysozyme | Egg white |
| Naringinase | *Aspergillus niger, A. usamii* |
| Pectinase | *Aspergillus niger, A. alliaceus, Aspergillus* sp., *Rhizopus oryzae, Rhizopus* sp., *Penicillium* sp., *Botrytis* sp., *Sclerotinia libertiana, Coniothyrium diplodiella* |
| Phospholipase A2 | Porcine pancreas |
| Protease | *Aspergillus niger, A. oryzae, A. melleus, Bacillus licheniformis, B. subtilis, B. thermoproteolyticus, Bacillus* sp. (alkalophilic), *Rhizopus* sp., *Ananas comosus* (bromelain), *Ficus glabrata* (ficin), *Carica papaya* (papain), bovine and porcine stomachs (pepsin), pancreas (trypsin) |
| Pullulanase | *Bacillus acidopullulyticus, Klebsiella aerogenes* |
| Rennet | Fourth stomach of ruminant animals, *Cryphonectria parasitica*[b], *Bacillus cereus*[b], *Rhizomucor miehei*[b] *R. pusillus*[b]. |
| Tannase | *Asperghillus niger* |
| Xylanase | *Trichoderma reesei* |

Source: Adapted from Bigelis, 1989.
[a]Immobilized enzymes or cells.
[b]Produces substitute for rennin which resembles the action of rennet extract.

**Table 2.7.4. Industrial Applications of Proteolytic Enzymes**

| Industry | Enzyme | Application |
|---|---|---|
| Detergents | Alkaline proteases | Extensive use in laundry detergents for protein stain removal |
| Dairy | Calf rennet | Coagulation of milk protein (cheese production) |
| | Fungal proteases | Replacement of calf rennet |
| | Chymosin | Active component of calf rennet; Production by genetically engineered microorganisms is being developed |
| | Proteases | Production of enzyme-modified cheese; Processing of whey protein |

**Table 2.7.4. (continued)**

| Industry | Enzyme | Application |
|---|---|---|
| Leather | Trypsin | Bating of leather |
| | Other proteases | Dehairing and dewooling of skins |
| Beverages | Papain | Removal of turbidity in beverages |
| Baking | Neutral protease | Dough conditioner |
| Meat and fish | Papain | Meat tenderizing |
| | Several proteases | Recovery of protein from bones and waste fish |
| Food processing | Several proteases | Modification of protein-rich material, i.e., soy protein or wheat gluten |
| Sweeteners | Thermolysin | Reverse hydrolysis in aspartame synthesis |
| Medicine | Trypsin | Removal of dead tissue and dissolution of blood clots |
| | Chymopapain | Treatment of certain types of hernia |
| Insulin | Carboxypeptidase<br>Trypsin | Conversion of hog insulin into human insulin |

Source: Adapted from Poldermans, 1990. Used with permission.

## 2.8. Enzymes: Biology and Properties

**Table 2.8.1. Biology of Enzymes**

- Biological role of enzymes
  - Background information
  - Intracellular enzymes
  - Extracellular enzymes
  - Evolution of enzymes
    - Evolution in prokaryotes
    - Serine protease family
    - Evolution of new enzymatic properties
    - Role of introns in evolution
- Sources of enzymes
  - Animal and plant enzymes
  - Microbiel enzymes
    - Microorganisms important for production of technical enzymes

- Proteases
- Amylases
- Cellulases
- Isomerases and oxidoreductases

Biosynthesis of enzymes

- Secretion
  - Signal hypothesis
  - Signal sequences of prokaryotic secretory proteins
  - Signal peptidases
- Extracellular enzymes from prokaryotic and eukaryotic organisms
  - α-Amylase of *Bacillus amyloliquefaciens*
  - Glucoamylase of *Aspergillus*
- Regulation of enzyme biosynthesis
  - Control during transcription
  - Post-transcriptional control

Genetics of enzyme formation

- Genes controlling formation of extracellular enzymes
- Improvement of enzyme formation by traditional genetic methods
- Application of recombinant DNA techniques

Source: Adapted from Borriss, 1987. Used with permission.

**Table 2.8.2. Some Properties of Endomylases**

| Enzyme | Source | pH Range; pH Optimum | °C Optimum; °C Inactivation |
|---|---|---|---|
| Pancreatic amylase | Porcine pancreas | 5.5-8.5; 6.0-7.0 | 40-45; 75 |
| Bacterial α-amylase | *Bacillus subtilis*<br>*Bacillus amyloliquefaciens* | 4.5-9.0; 6.5-7.5 | 70-85; 95 |
| Thermostable bacterial α-amylase | *Bacillus licheniformis* | 5.8-8.0; 7.0 | 90-105; 120 |
| Fungal α-amylase | *Aspergillus oryzae*<br>*Aspergillus niger* | 4.0-7.0; 5.0-6.0 | 55-60; 80 |
| Malt α-amylase | Barley malt<br>Wheat malt | 3.5-7.5; 4.5-7.0 | 60-70; 85 |

Source: Adapted from Uhlig, 1990. Used with permission.

**Table 2.8.3. Some Properties of Glucamylases**

| Source | pH: Range; Optimum | °C: Optimum; Inactivation |
|---|---|---|
| *Aspergillus niger* (different types) | 3.0-6.0; 4.5-5.0 | 55-60; 90 |
| *Aspergillus awamori* | 2.0-7.0; 3.0-5.0 | 55-65; 85 |
| *Rhizopus niveus* | 3.0-6.0; 3.0-5.5 | 55-60; 70-80 |

Source: Adapted from Uhlig, 1990. Used with permission.

**Table 2.8.4. Profiles of Microbial Amylases and Related Enzymes**

| Enzyme (pH-Optimum) | Action | Source | MW |
|---|---|---|---|
| | Exo-Acting Amylases | | |
| Glucoamylase (4.5) | Hydrolysis of $\alpha$-D-(1-6)-, $\alpha$-D-(1-4)-, $\alpha$-D-(1-3)- linkages. Procudes glucose from starch and related carbohydrates | Fungi | 60000-100000 |
| $\beta$-Amylase (6-7)[a] | Hydrolysis of $\alpha$-D-(1-4)- bonds. Produces maltose from amylose | Bacteria | 35000-58000 |
| | Endo-Acting Amylases | | |
| $\alpha$-Amylase (5-7)[b] | Hydrolysis of $\alpha$-D-(1-4)-linkages | Bacteria Fungi | 50000 |
| | Debranching Enzyames | | |
| Pullulanase (6)[c] | Hydrolysis of pullulan to maltotriose | Bacteria | 100000 |
| Isoamylase (5-6.4) | Restricted to $\alpha$-1,6-bonds | Yeasts Bacteria | |

Source: Adapted from Borriss, 1987. Used with permission.
[a] SH-groups essential for enzyme activity. [b] Ca-metallo proteins. [c] Unable to degrade pullulan

**Table 2.8.5. Comparison of Properties of Various Lipase Preparations**

| Source | pH: Range; Optimum | °C | pH for Stability |
|---|---|---|---|
| Porcine pancreas | 6.5-9.5; 7.5-8.5 | 40-45 | 5.5-7.5 |
| *Rhizopus* species | 6.0-7.5; 7.0 | 35-40 | 4.0-8.0 |
| *Mucor javanicus* | 5.5-8.0; 7.0 | 40-45 | 4.5-6.5 |
| *Aspergillus niger* | | | |
| Short-chain fatty acids | 3.0-7.0; no data | 40-50 | 5.0-7.0 |
| Fats and oils | 7.5-9.0; no data | 40-45 | 5.0-7.0 |
| *Pseudomonas* | | | |
| One isoenzyme | No data; 4.0-5.0 | 50-60 | 4.5-10.0 |
| One isoenzyme | No data; 7.0-8.5 | 50-60 | 4.5-10.0 |
| *Candida cylindracea* | 5.0-7.5; no data | 40-45 | 4.5-8.5 |

Source: Uhlig, 1990. Used with permission.

**Table 2.8.6. Properties of Alkaline and Neutral Bacterial Proteases**

| Property | Alkaline | Neutral |
|---|---|---|
| pH range | 7.0-11.0 | 6.0-9.0 |
| pH for optimal stability | 7.5-9.5 | 6.0-8.0 |
| Inhibitors | DIFP, PMSF, potato + soy inhibitor | EDTA, sodium dodecyl sulfate |
| Specific substrates | Different esters, amides, and peptides | Few peptides and esters |

Source: Uhlig, 1990. Used with permission.
Note: DIFP = diisopropyl fluorophosphate. PMSF = phenylmethanesulfonyl fluoride. EDTA = ethylenediaminetetraacetic acid.

**Table 2.8.7. Classification and Applications of Commercial Proteinases**

| Class | Source (Specific Enzyme) | Use |
|---|---|---|
| Serine proteinases (alkaline) | Animal pancreas (trypsin, chymotrypsin) | Pharmaceutical |
| | *Bacillus amyloliquefaciens* (subtilisins) | Detergents<br>Food |
| | *Bacillus licheniformis* (subtilisins) | |
| | *Bacillus subtilis* (subtilisins) | |
| | Alkalophilic bacilli (subtilisins) | |

**Table 2.8.7. (continued)**

| Class | Source (Specific Enzyme) | Use |
|---|---|---|
| Thiol proteinases | Papaya (papain) | Meat tenderizing |
| | Pineapple (bromelain) | Beer haze prevention |
| Carboxyl proteinases (acid) | *Aspergillus* spp. | Baking, other food uses |
| | *Mucor* spp. | |
| | Calf stomach (rennin) | Milk-clotting in cheese manufacture |
| | Pig stomach (pepsin) | |
| Metallo-proteinases (neutral) | *Bacillus thermoproteolyticus* (thermolysin) | Aspartame manufacture |
| | *Bacillus amyloliquefaciens* | Food, brewing |
| | *Bacillus subtilis* | Food, brewing |
| | *Aspergillus* spp. | Food, brewing |

Source: Frost and Moss, 1987. Used with permission.

**Table 2.8.8. Properties of Microbial Extracellular Proteases**

| Properties | Acid Proteases (Carboxy-peptidases) | Neutral Proteases (Metallo-proteases) | Alkaline Proteases |
|---|---|---|---|
| Source | Fungi | Bacteria, fungi | Bacteria, fungi |
| Properties | Pepsin-like, rennin-like (milk-clotting) | Zinc containing, stabilized by $Ca^{2+}$ | Trypsin-like, serine residue at the active center |
| pH Range | | | |
| Maximum activity | 2-5 | 7-8 | 9-11 (12) |
| Maximum stability | 2-6 | 7-9 | 5-10 (12) |
| MW~ | 3500 | 4500 | 27500 (17000) |
| Inhibitors | DAN | EDTA | PMSF, DFP |

Source: Borriss, 1987. Used with permission.
Note: DAN = diazoacetyl norleucine methylester; EDTA = ethylenediamine tetraacetic acid; DFP = diisopropylphosphofluoridate; PMSF = phenyl-methansulfonyl-fluoride.

**Table 2.8.9. Properties of Proteases from Molds**

| Organism | pH Range for Hemoglobin | pH Range for Casein | Optimal pH for Stability |
|---|---|---|---|
| *Aspergillus saitoi*[a] | 3,0-4.5 | 2.5-3.0 | 2.0-5.0 |
| *Aspergillus oryzae* (acidic protease) | 3.0-4.0 | 2.5-3.0 | 5.0 |
| *Aspergillus oryzae*[b] (neutral protease) | 5.5-7.5 | No data | 7.0 |
| *Aspergillus oryzae*[c] (alkaline protease) | 6.0-9.5 | 6.5-10.0 | 7.0-8.0 |
| *Paecilomyces varioti* | 3.5-5.5 | 3.0 | 3.0-5.0 |
| *Rhizopus chinensis* | 5.0 | 2.9-3.3 | 3.8-6.5 |
| *Mucor pusillus* | 3.5-4.5 | 5.6 | 3.0-6.0 |

Source: Uhlig, 1990. Used with permission.

[a]Inhibitor = *N*-bromosuccinimide; [b]Inhibitor = ethylenediaminetetraacetic acid; [c]Inhibitor = diisopropyl fluorophosphate.

**Table 2.8.10. Optimal pH Values for Some Pectin Depolymerases**

| Source | Type | Optimal pH |
|---|---|---|
| *Aspergillus niger* | Endo-polymethylgalacturonase | 5.2 |
| *Aspergillus niger* | Endo-polymethylgalacturonase | 5.5 |
| *Falvobacterium pectinovorum* | Endo-polymethylgalacturonase | 7.4-8.2 |
| *Aspergillus niger* | Endo-polygalacturonase | 4.7-4.8 |
| *Aspergillus species* | Endo-polygalacturonase | 3.7-4.2 |
| *Coniothyrium diplodiella* | Endo-polygalacturonase | 4.4 |
| *Aspergillus sojae* | Endo-pectin transeliminase | 5.5 |
| *Bacterium polymyxa* | Endo-pectin transeliminase | 8.9-9.1 |

Source: Uhlig, 1990. Used with permission.

**Table 2.8.11. Properties of Lactases**

| Property | From Yeast | From *Aspergillus oryzae* |
|---|---|---|
| pH required for activity | 6.5-7.0 | 4.5-6.5 |
| Optimal temperature | 30-35°C | 40-50°C |
| Optimal stability | pH 6.5-7.0 at 30°C | pH 4.5 |

Source: Adapted from Uhlig, 1990. Used with permission.

**Table 2.8.12. Properties of Some β-1,4-galactosidases**

| Source | Optimal pH | Optimal °C |
|---|---|---|
| *Aspergillus niger* | 3.0-4.0 | 55 |
| *Aspergillus oryzae* | 4.8 | 46 |
| *Escherichia coli* | 6.9-7.5 | 45 |
| *Saccharomyces fragilis* | 6.5 | 50 |

Source: Adapted from Reimerdes, 1990. Used with permission.

**Table 2.8.13. Uses of Mammalian NAD/NADP Dehydrogenases and Alternative Microbial Sources**

| Enzyme | Applications | Occurrence in Microorganisms |
|---|---|---|
| L-Lactate dehydrogenase | Determination of: ADP, alanine aminotransferase, lactate, lecithin, creatine, pyruvate, pyruvate kinase, triglycerides | *Leuconostoc* sp<br>*Thermus caldophilus* |
| L-Malate dehydrogenase | Determination of: Aspartate aminotransferase, $CO_2$ | Yeast<br>*Bacillus stearothermophilus* |
| L-Glutamate dehydrogenase | Determination of: Ammonia, 5' nucleotidase, urea | Various[a] |

| | | |
|---|---|---|
| Diaphorase | Determination of: Triglycerides and coupling with other dehydrogenases | *Clostridium kluyveri* |

Source: Frost and Moss, 1987. Used with permission.
[a]Microbial enzymes are cofactor specific.

**Table 2.8.14. Pyridine Nucleotide Dehydrogenases Obtained from Bakers' Yeast**

| Enzyme | Application | Cofactor |
|---|---|---|
| Alcohol dehydrogenase | Alcohol determination | NAD |
| Isocitrate dehydrogenase | Isocitrate determination | NADP |
| 6-Phospogluconate dehydrogenase | Gluconate determination | NADP |
| Glucose-6-phosphate dehydrogenase | Glucose determination | NADP (yeast) NAD, NADP (*Leuconostoc*) |
| Aldehyde dehydrogenase | Uric acid determination | NADP |
| Glycerol-3-phosphate dehydrogenase | ATP determination | NAD |

Source: Frost and Moss, 1987. Used with permission.

**Table 2.8.15. Microbial NAD/NADP Dehydrogenases from Special Strains**

| Enzyme | Application | Source | Cofactor |
|---|---|---|---|
| Galactose dehydrogenase | Determination of galactose | *Pseudomonas fluorescens* | NAD |
| 12 $\alpha$-Hydroxy steroid dehydrogenase | Production of 12-ketochenode-oxycholic acid | *Clostridium* sp. | NADP |
| Formate dehyrogenase | Determination of oxalate | *Pseudomonas oxalacticus* *Candida boidinii* | NAD |
| Formate dehyrogenase | Producation of L-alanine from pyruvate | *Candida boidinii* | NAD |
| Alanine dehydrogenase | Producation of L-alanine from pyruvate | *Bacillus subtilis* | NAD |

Source: Frost and Moss, 1987. Used with permission.

**Table 2.8.16. Inducers for Some Commercial Enzymes**

| Enzyme | Inducer |
|---|---|
| Amyloglucosidase | Starch or maltose |
| $\alpha$-Amylase | Starch or maltodextrins |
| Catalase | Hydrogen peroxide |
| Cellulase | Cellulose, cellobiose |
| Glucose oxidase | Glucose |
| Invertase | Sucrose |
| Lactase | Lactose |
| Lipase | Lipids or fatty acids |
| Pectinases | Pectin |
| Pullulanase | Pullulan or maltose |
| Xylose (glucose) isomerase | Xylose or xylan |

Source: Frost and Moss, 1987. Used with permission.

**Table 2.8.17. Examples of Toxicity Studies Performed with Enzymes**

| Enzyme | Type of Study (Only Oral) | Result |
|---|---|---|
| | Microbial | |
| Cellulase<br>Amyloglu-cosidase<br>Lactase<br>Lipase<br>Protease<br>Rennet | Acute<br>500-40000 mg/kg body weight | non-toxic<br>no $LD_{50}$ established |
| Cellulase | 30 days mouse<br>0.9-2.1 g/kg body weight | non-toxic |
| Amyloglu-cosidase | 90 days rats<br>3.5-7 g/kg body weight | non-toxic |
| Lactase | 90 days rats<br>0.5-1.0% in the diet | non-toxic |
| Lipase | 30 days rats<br>0.5-5.0% in the diet | non-toxic |
| Protease | 90 days rats<br>5-10% in the diet | non-toxic |
| Rennet | 1 year rat and dog<br>8-40 mg/kg body weight | non-toxic |
| | 1 year rat and dog<br>0.1-2.5% in the diet | non-toxic |
| | Animal/Plant | |
| Chymo-trypsin | Acute<br>0-22.5 g/kg body weight | non-toxic<br>no $LD_{50}$ established |
| | 5 weeks rabbits<br>20 mg/kg body weight | non-toxic |
| | 12 days dogs<br>100 mg/kg body weight | non-toxic |
| Rennet | 7-9 weeks poultry<br>0.1-70.0 mg/pound feed | non-toxic |

Source: Adapted from Noordervliet and Toet, 1987. Used with permission.

## 2.9. Enzymes: Gene Technology

**Table 2.9.1. Enzymes, Cloned Genes, and Recombinant Hosts[a]**

| Enzyme[a] | Source of Cloned Gene | Recombinant Host |
|---|---|---|
| α-Amylase | Human salivary gland | *Saccharomyces cerevisiae* |
| | Mouse salivary gland | *S. cerevisiae* |
| | Mouse pancreas | *S. cerevisiae* |
| | *Saccharomycopsis fibuligera* | *S. cerevisiae* |
| | Wheat | *S. cerevisiae* |
| Cellulase | | |
| Cellobiohydrolase | *Trichoderma reesei* | *S. cerevisiae* |
| Endo-1,4-β-glucanase | *T. reesei* | *S. cerevisiae* |
| β-Glucosidase | *Aspergillus niger* | *S. cerevisiae* |
| | *Kluyveromyces fragilis* | *S. cerevisiae* |
| | *Candida pelliculosa* | *S. cerevisiae* |
| Bacterial cellulase | *Cellulomonas fimi* | *S. cerevisiae* |
| β-Galactosidase | *Kluyveromyces marxianus* var. *lactis* | *S. cerevisiae* |
| β-Glucanase | *Bacillus subtilis* | *S. cerevisiae* |
| Glucoamylase | *Aspergillus awamori* | *S. cerevisiae* |
| | *Rhizopus* sp. | *S. cerevisiae* |
| | *Saccharomycopsis fibuligera* | *S. cerevisiae* |
| | *Saccharomyces diataticus* | *S. cerevisiae* |
| | *Aspergillus niger* | *A. niger, A. awamori* |
| Invertase | *Saccharomyces cerevisiae* | *S. cerevisiae* |
| Protease | Calf (chymosin, rennin) | *S. cerevisiae* |
| | *Mucor miehei*[b] | *Aspergillus nidulans* |
| Lysozyme | Human | *S. cerevisiae* |

Source: Adapted from Bigelis, 1989.

[a]All of the enzymes shown are secreted by the recombinant host except β-glucanase and β-galactosidase. [b]Produces substitute for rennin.

**Table 2.9.2. Donor, Host, Vector for Modifying Proteinase Genes of *Bacillus* spp.**

| Donor Organism | Gene | Vector | Host Organism |
|---|---|---|---|
| *Bacillus stearothermophilus* | Thermostable neutral proteinase | Plasmid pTB90 | *Bacillus stearothermophilus* *B. subtilis* |
| *B. amyloliquefaciens* | Subtilisin | Plasmid pBS42 | *B. subtilis* I-168 |
| *B. amyloliquefaciens* ATCC 23350 | Neutral proteinase | Plasmid pUB110 | *B. subtilis* |
| *B. licheniformis* ATCC 14580 | Proteinase regulatory gene | Plasmid pUB110 | *B. subtilis* |
| *B. subtilis* I-168 | Subtilisin and partial deletion mutant | Plasmid pBS42 | *B. subtilis* I-168 |
| *B. amyloliquefaciens* ATCC 23844 | Alkaline proteinase neutral proteinase | Plasmids | *B. subtilis* |
| *B. subtilis* | Neutral proteinase deletion mutant | Plasmids | *B. subtilis* |
| *B. amyloliquefaciens* | Subtilisin directed *in vitro* mutants | Plasmids | *B. subtilis* proteinase deletion mutant |
| *B. amyloliquefaciens* | Neutral proteinase plus promotor and signal sequence | Plasmid pUB110 | *B. subtilis* |
| *B. subtilis* | Proteinases | No data | No data |
| *B. amyloliquefaciens* | Amylase promoter | Plasmids | *B. subtilis* |
| *B. amyloliquefaciens* ATCC 23844 | Proteinases plus promoter and regulators | Plasmids pBD64 pUB110 | *B. subtilis* |
| *B. subtilis* | Proteinase | Plasmid pLP33 | *B. licheniformis* |

Source: Adapted from Frost and Moss, 1987. Used with permission.

**Table 2.9.3. Cloned Microbial Genes Coding for Extracellular Enzymes with Industrial Importance**

| Gene Encoding | Donor |
|---|---|
| α-Amylase | *Bacillus subtilis* |
| | *B. amyloliquefaciens* |
| | *B. megaterium* |
| | *B. coagulans* |
| | *B. licheniformis* |
| | *B. stearothermophilus* |
| Neutral protease | *B. subtilis* |
| | *B. amyloliquefaciens* |
| | *B. stearothermophilus* |
| Alkaline protease | *B. subtilis* |
| | *B. amyloliquefaciens* |
| Endo-β-1,4-glucanase | *Clostridium thermocellum* |
| | *Cellulomonas finii* |
| | Alkalophilic *Bacillus* |
| Endo-β-1,3-1,4-glucanase | *B. subtilis* |
| | *B. amyloliquefaciens* |
| Xylanase | *B. pumilus* |
| Glucoamylase | *Aspergillus niger* |
| | *A. awamori* |
| | *Saccharomyces diastaticus* |
| Cellobiohydrolase | *Trichoderma reesei* |
| β-Glucosidase | *Aspergillus niger* |

Source: Adapted from Borriss, 1987. Used with permission.

**Table 2.9.4. Some Commercial Enzyme Genes which Have been Manipulated *in vitro***

| Gene | Donor | Host |
|---|---|---|
| α-Amylase | *Bacillus* sp. | *Bacillus subtilis* |
| | *Bacillus subtilis* | *bacillus subtilis* |
| α-Amylase (thermostable) | *Bacillus stearothermophilus* | *Bacillus subtilis*<br>*Escherichia coli* |
| | *Bacillus amyloliquefaciens* | *Bacillus subtilis* |
| | *bacillus licheniformis* | *Bacillus subtilis*<br>*Escherichia coli* |
| β-Amylase | *Bacillus stearothermophilus* | *Bacillus* sp. |
| Cellulases | *Bacillus* sp. | *Escherichia coli* |
| Cellulase (endoglucanase A) | *Clostridium thermocellum* | *Escherichia coli* |
| β-Glucanase | *Bacillus amyloliquefaciens* | *Escherichia coli* |
| Lactase (thermostable) | *Bacillus stearothermophilus* | *Bacillus subtilis* |
| Pectinases | *Erwinia carotovora* | *Escherichia coli* |
| Penicillin V amidase | *Bacillus sphaericus* | *Escherichia coli*<br>*Bacillus subtilis* |
| Peptidase | *Pseudomonas* sp. | *Saccharomyces cerevisiae* |
| Proteinase | *Bacillus licheniformis* | *Bacillus subtilis* |
| Xylanase | *Bacillus pumilus* | *Escherichia coli*<br>*Bacillus subtilis* |
| | *Aeromonas* sp. | *Escherichia coli* |
| Xylose isomerase | *Escherichia coli* | *Schizosaccharomyces pombe* |

Source: Adpated from Frost and Moss, 1987. Used with permission.

## 2.10. Enzymes: Fermentation, Production, and Microorganisms

**Table 2.10.1. Production of Enzymes by Fermentation**

- Background information
- Enzymes made by fermentation
  - Advantages of production by fermentation
  - Fields of use
  - Intracellular industrial enzymes
  - Extracellular industrial enzymes
  - Diagnostic and small volume enzymes
- Fermentation methods
  - Sources of enzymes
  - Strain selection and improvement
    - Strain selection
    - Strain improvement
      - "Classical" mutation and selection
      - Hybridization
      - Recombinant and technology
      - Increases in enzyme production by strain improvement
  - Medium formulation and preparation
  - Fermentation equipment and processes
    - Solid substrate fermentation
    - Submerged fermentation
      - Equipment
      - Operational methods
  - The influence of culture conditions on enzyme production in submerged culture
    - Temperature
    - pH
    - Dissolved oxygen tension
  - Economics of enzyme production
    - Variable costs
    - Semi-variable costs
    - Fixed costs
    - Factors affecting unit production costs
- Oxidoreductases
  - Pyridine nucleotide dehydrogenases
    - Types of source and uses
    - Fermentation-derived alternatives to mammalian NAD/NADP dehydrogenases

        - Fermentation medium and conditions
    - β-1,3(4)-Glucanases
        - Dextranases
        - Pectolytic enzymes
        - Microbial cell-lytic enzymes
    - α-Galactosidase
    - β-Galactosidases (lactase)
        - Nature of enzymes from different strains
        - Yeast lactases
        - Mold lactases
        - Bacterial lactases
    - β-Fructosidase (invertase)
  - Proteolytic enzymes
    - Types of commercially useful proteinases
    - *Bacillus* spp. proteinases
        - Background information
        - Biological control of enzyme production
        - Media
        - Types of fermentation process and equipment
        - Use of recombinant DNA techniques
    - Fungal proteinases
        - Microbial rennins
        - Other fungal proteinases
    - Proteinases from higher organisms and their genes
  - Other hydrolytic enzymes
    - Penicillin amidase
    - Aminoacylase
- Other enzymes
  - Transferases
    - Glycosyltransferases
    - Kinases
  - Xylose (glucose) isomerase
    - Commercial uses and history
    - Source organisms
    - Media and conditions of fermentation
    - Cloning of glucose isomerase genes
- Future developments in enzyme fermentation Technology
  - Developments associated with the organism
  - Developments associated with the process operation

---

Source: Adapted from Frost and Moss, 1987. Used with permission.

**Table 2.10.2. Examples of Enzymes Produced by Solid-Substrate Fermentation**

| Enzyme | Organism |
|---|---|
| Cellulase | *Trichoderma viride* |
| | *Trichoderma reesei* |
| | *Trichoderma harzianum* |
| Amyloglucosidase | *Aspergillus niger* |
| β-Glucosidase | *Aspergillus phoenicis* |
| α-Galactosidase | *Aspergillus awamori* |
| Invertase | *Aspergillus awamori* |
| Lactase | *Scopulariopsis* sp. |
| | *Aspergillus niger* |
| | *Fusarium moniliforme* |

Source: Frost and Moss, 1987. Used with permission.

**Table 2.10.3. Cultivation and Medium for the Production of Lactase from Molds**

| Species | Type of Cultivation | Medium |
|---|---|---|
| *Aspergillus niger* | Solid-substrate | Wheat bran plus trace metals |
| *Aspergillus oryzae* | Solid-substrate | Wheat bran |
| *Alternaria tenuis* | Submerged | Soy flour |
| *Fusarium moniliforme* | Submerged | Whey plus ammonium sulfate |
| | Solid-substrate | Wheat bran |
| *Trichoderma reesei* | Submerged | Lactose, defined |

Source: Frost and Moss, 1987. Used with permission.

**Table 2.10.4. Culture Temperature ($^{o}C$) for Thermostable Bacterial Lactase Producers**

| Organism | Culture Temperature ($^{o}C$) |
|---|---|
| *Bacillus circulans* | 45 |
| *Bacillus coagulans* | 45 |
| *Bacillus stearothermophilus* | 65 |
| *Bacillus stearothermophilus* | 50 |
| Bacterial strain NCIB 11259 | 45-55 |
| *Caldariella acidophila* | 87 |
| *Streptomyces coelicolor* | 40 |
| *Thermus* sp. | 75 |

Source: Frost and Moss, 1987. Used with permission.

**Table 2.10.5. Production Conditions for α-Galactosidase**

*Absidia griseola*

1. Maximum yield culture (EU/mL): 23.0
2. Assay substrate: Melibiose
3. Assay temperature ($^{o}C$): 40
4. Assay pH: 5.2

*Absidia* spp.

1. Maximum yield culture (EU/mL): 3.8
2. Assay substrate: Melibiose
3. Assay temperature ($^{o}C$): 40
4. Assay pH: 5.2

*Bacillus stearothermophilus*

1. Maximum yield culture (EU/mL): 0.4
2. Assay substrate: *p*-Nitrophenyl-$\alpha$-D-glalactoside
3. Assay temperature ($^{o}C$): 25
4. Assay pH: 7.0

*Circinella muscae*

1. Maximum yield culture (EU/mL): 8.2
2. Assay substrate: Melibiose
3. Assay temperature ($^{o}C$): 40
4. Assay pH: 5.2

*Monascus* sp.

1. Maximum yield culture (EU/mL): 5.8
2. Assay substrate: *p*-Nitrophenyl-$\alpha$-D-glalactoside

3. Assay temperature ($^{o}C$): 50
4. Assay pH: 4.5

*Mortierella vinacea*
1. Maximum yield culture (EU/mL): 1.7
2. Assay substrate: Melibiose
3. Assay temperature ($^{o}C$): 40
4. Assay pH: 5.2

*Saccharomyces carlsbergensis*
1. Maximum yield culture (EU/mL): 1.0
2. Assay substrate: *p*-Nitrophenyl-$\alpha$-D-glalactoside
3. Assay temperature ($^{o}C$): 30
4. Assay pH: 4.5

*Saccharomyces cerevisiae*
1. Maximum yield culture (EU/mL): 4.4
2. Assay substrate: Melibiose
3. Assay temperature ($^{o}C$): 40
4. Assay pH: 5.6

Source: Frost and Moss, 1987. Used with permission.

Note: Microorganisms such as *Aspergillus awamori*, *Apergillus niger*, and *Penicillium dupontii* also produce $\alpha$-galactosidase, but data such as those in the table are not available.

**Table 2.10.6. Strain and Medium for Microorganisms Producing Glucose Isomerase**

*Arthrobacter sp.*
1. Strain: NRRL B3724, 3725, 3726, 3727, 3728
2. Medium: Enzume produced constitutively, higher yield with xylose present; no metal ion required
3. Comment: None

*Actinoplanes missouriensis*
1. Strain: NRRL B3324
2. Medium: Enzyme produced constitutively; no metal ion required
3. Comment: None

*Bacillus coagulans*
1. Strain: HN68
2. Medium: Xylose required; metal ion required, not necessarily cobalt
3. Comment: None

*Bacillus coagulans*
1. Strain: No data

**Table 2.10.6. (continued)**

2. Medium: Enzyme produced constitutively; metal ion required, not necessarily cobalt
3. Comment: Constitutive mutant

*Bacillus licheniformis*
1. Strain: ATCC 31604
2. Medium: Xylose required; no metal ion required
3. Comment: Wild-type

*Bacillus licheniformis*
1. Strain: ATCC 31667
2. Medium: Enzyme produced constitutively; no metal ion required
3. Comment: Constitutive mutant

*Bacillus stearothermophilus*
1. Strain: 429
2. Medium: Xylose used; cobalt used
3. Comment: None

*Curtobacterium helvolum*
1. Strain: NCIB 10352
2. Medium: Xylose used; cobalt used
3. Comment: None

*Flavobacterium arborescens*
1. Strain: NRRL B11022
2. Medium: Enzyme produced constitutively; no metal ion required
3. Comment: Constitutive mutant

*Streptomyces acidodurans*
1. Strain: NRRL 11489
2. Medium: Xylose required; no metal ion required
3. Comment: Wild-type

*Streptomyces acidodurans*
1. Strain: NRRL 11495
2. Medium: Enzyme produced constitutively; no metal ion required
3. Comment: Constitutive mutant

*Streptomyces albus*
1. Strain: No data
2. Medium: Xylan used; cobalt used
3. Comment: None

*Streptomyces coelicolor*
1. Strain: NRRL 15398
2. Medium: Enzyme produced constitutively; no metal ion required
3. Comment: Constitutive mutant of ATCC 21666

*Streptomyces flavogriseus*

1. Strain: No data
2. Medium: Xylan used; metal ion required, not necessarily cobalt
3. Comment: None

*Streptomyces glaucescens*
1. Strain: NRRL B2900
2. Medium: Xylose used; cobalt used
3. Comment: None

*Streptomyces kanamyceticus*
1. Strain: KCC S-0433
2. Medium: Xylose used; no metal ion required
3. Comment: Thermostable enzyme

*Streptomyces olivaceus*
1. Strain: NRRL 3583, NRRL 3916
2. Medium: Xylose used; no metal ion required
3. Comment: Possible production strains

*Streptomyces olivochromogenes*
1. Strain: ATCC 21114
2. Medium: Xylose required; cobalt used
3. Comment: None

*Streptomyces olivochromogenes*
1. Strain: ATCC 21713
2. Medium: Enzume produced constitutively, higher yield with xylose present; no metal ion required
3. Comment: Constituent mutant

*Streptomyces thermoviolaceus*
1. Strain: ATCC 19283
2. Medium: Xylose required; no metal ion required
3. Comment: Thermostable enzyme, wild-type

*Streptomyces thermoviolaceus*
1. Strain: NRRL 15615
2. Medium: Enzyme produced constitutively; no metal ion required
3. Comment: Thermostable enzyme, constitutive mutant

*Streptomyces violaceus-ruber*
1. Strain: No data
2. Medium: Xylose required; no metal ion required
3. Comment: None

*Streptomyces wedmorensis*
1. Strain: ATCC 21175
2. Medium: Xylose required; cobalt used
3. Comment: None

*Streptomyces* sp.

**Table 2.10.6. (continued)**

1. Strain: 756
2. Medium: Xylose used; no metal ion required
3. Comment: None

*Zymomonas mobilis*

1. Strain: No data
2. Medium: Enzyme produced constitutively; no metal ion required
3. Comment: None

Source: Frost and Moss, 1987. Used with permission.

**Table 2.10.7. Some Production Requirements for Microbial Sources of Lipase[a]**

Bacteria

*Staphylococcus aureus*

1. Fermentation: Shake-flask
2. Activity (EU/mL): 4.5
3. Assay conditions: Olive oil 37$^{o}$C, pH 8.0

*Staphylococcus aureus*

1. Fermentation: Shake-Flask
2. Activity (EU/mL): 17.0
3. Assay conditions: Tributyrin, 37$^{o}$C pH 8.0

*Chromobacterium viscosium* var. *paralipolyticum*

1. Fermentation: 30 L-fermenter
2. Activity (EU/mL): 10.8
3. Assay conditions: Lard oil, 40$^{o}$C, pH 7.0

*Micrococcus caseolyticus*

1. Fermentation: Shake-flask
2. Activity (EU/mL): 15.0
3. Assay conditions: Olive oil, 40$^{o}$C, pH 9.5

*Bacillus licheniformis*

1. Fermentation: Shake-flask
2. Activity (EU/mL): 6.0
3. Assay conditions: Olive oil, 45$^{o}$C, pH 8.5

*Pseudomonas nitroreducens*

1. Fermentation: 20 L-fermenter
2. Activity (EU/mL): 500.0
3. Assay conditions: Olive oil, 37$^{o}$C, pH 9.0

*Pseudomonas fragi*

1. Fermentation: 20 L-fermenter
2. Activity (EU/mL): 75.0[b]
3. Assay conditions: Olive oil, 37°C, pH 9.0

*Alcaligenes* sp
1. Fermentation: 15 L-fermenter
2. Activity (EU/mL): 800.0
3. Assay conditions: Olive oil, 37°C, pH 8.7

*Bacillus stearothermophilus*
1. Fermentation: Shake-flask
2. Activity (EU/mL): Not available
3. Assay conditions: Tributyrin, 55°C, pH 8.8

Yeasts

*Candida cylindracea*
1. Fermentation: Shake-flask
2. Activity (EU/mL): None available
3. Assay conditions: Olive oil, 30°C, pH 7.0

*Candida cylindracea*
1. Fermentation: Shake-flask
2. Activity (EU/mL): None available
3. Assay conditions: Olive oil, 37°C, pH 7.0

*Torulopsis ernobii*
1. Fermentation: 30 L-fermenter
2. Activity (EU/mL): 110.0
3. Assay conditions: Olive oil, 37°C, pH 6.5

*Saccharomycopsis lipolytica*
1. Fermentation: Shake-flask
2. Activity (EU/mL): 40.0
3. Assay conditions: Olive oil, 30°C, pH 9.5

*Candiada lipolytica*
1. Fermentation: 2 L-fermenter
2. Activity (EU/mL): 1.0
3. Assay conditions: Olive oil, 37°C, pH 8.2

Molds

*Aspergillus luchuensis*
1. Fermentation: Solid-substrate
2. Activity (EU/mL): None available
3. Assay conditions: Olive oil, 40°C, pH 7.0

**Table 2.10.7. (continued)**

*Aspergillus niger*
1. Fermentation: Shake-flask
2. Activity (EU/mL): 4.0
3. Assay conditions: Olive oil, 37°C,pH 6.5

*Rhizopus delemar*
1. Fermentation: Submerged
2. Activity (EU/mL): 60.0
3. Assay conditions: Olive oil, 30°C, pH 5.6

*Rhizopus arrhizus*
1. Fermentation: 100 L-fermenter
2. Activity (EU/mL): 350.0
3. Assay conditions: Olive oil, 37°C, pH 7.5

*Rhizopus delemar*
1. Fermentation: Tower fermenter and 3 L-aerated stirred fermenter
2. Activity (EU/mL): 460.0
3. Assay conditions: Olive oil, 30°C, pH 8.5

*Geotrichum candidum*
1. Fermentation: Static liquid culture
2. Activity (EU/mL): 6.0
3. Assay conditions: Olive oil, 32°C, pH 8.5

*Geotrichum candidum*
1. Fermentation: Shake-flask
2. Activity (EU/mL): 170.0
3. Assay conditions: Olive oil, 30°C, pH 5.6

*Mucor lipolyticus*
1. Fermentation: Shake-flask
2. Activity (EU/mL): 90.0
3. Assay conditions: Olive oil, 37°C, pH 8.0

*Humicola lanuginosa*
1. Fermentation: 600 L-fermenter
2. Activity (EU/mL): 55.0
3. Assay conditions: Olive oil, 45°C, pH 8.0

*Myriococcum* sp.
1. Fermentation: 20 L-fermenter
2. Activity (EU/mL): 4.0
3. Assay conditions: Olive oil, 55°C, pH 8.0

*Phycomyces* spp.
1. Fermentation: Fermenter
2. Activity (EU/mL): 160.0
3. Assay conditions: Olive oil, 37°, pH 7.0

*Pencillium funiculosum*

1. Fermentation: Shake-flask
2. Activity (EU/mL): Not available
3. Assay conditions: Olive oil, 37°C, pH 8.0

*Byssochlamys fulva*

1. Fermentation: Shake-flask
2. Activity (EU/mL): Not avialable
3. Assay conditions: Olive oil, 37°C, pH 8.0

*Fusarium solani*

1. Fermentation: Shake-flask
2. Activity (EU/mL): Not available
3. Assay conditions: Olive oil, 37°C, pH 8.0

*Fusarium oxysporum*

1. Fermentation: Shake-flask to 2500 L
2. Activity (EU/mL): 220.0
3. Assay conditions: Tributyrin, 30°C, pH 7.0

---

Source: Adapted from: Frost and Moss, 1987. Used with permission.

[a]1 EU = the quantity of enzyme which releases 1 μmol of fatty acid per minute under the conditions stated. Where other units are specifed in the original papers, they have been, if possible, recalculated as EU

[b]45 EU/mL in cells, 30 EU/mL in culture fluid.

**Table 2.10.8. Grouping of Microorganisms Used for Enzyme Production**

---

Microorganisms That Have Traditionally Been Used in Food Or in Food Processing

*Bacillus subtilis* (including strains known under the names *B. mesentericus, B. natto, and B. amyloliquefaciens*)

*Aspergillus niger* (including strains known under the names *A. aculeatus, A. awamori, A. ficum, A. foetidus, A. japonicus, A. phoenicis, A. saitoi, and A. usamii*)

*Aspergillus oryzae* (including strains known under the names *A. sojae* and *A. effusus*)

| | |
|---|---|
| *Mucor javanicus* | *Saccharomyces cerevisiae* |
| *Rhizopus arrhizus* | *Kluyveromyces fragilis* |
| *Rhizopus oligosporus* | *Kluyveromyces lactis* |
| *Rhizopus oryzae* | *Leuconostoc oenos* |

**Table 2.10.8. (continued)**

| Microorganisms Accepted As Harmless Contaminants Present In Food | |
|---|---|
| *Bacillus stearothermophilus* | *Bacillus megaterium* |
| *Bacillus licheniformis* | *Bacillus circulans* |
| *Bacillus coagulans* | *Klebsiella aerogenes* |
| **Other Microorganisms** | |
| *Bacillus acidopullulyticus* | *Bacillus cereus* |
| *Mucor miehei* | *Trichoderma reesei* |
| *Mucor pusillus* | *Trichoderma harzianum* |
| *Endothia parasitica* | *Penicillium lilacinum* |
| *Actinoplanes missouriensis* | *Penicillium emersonii* |
| *Streptomyces albus* | *Sporotrichum dimorphosporum* |

Source: Adapted from Noordervliet and Toet, 1987. Used with permission.

**Table 2.10.9. Enzymes Produced by Microorganisms**

| Bacteria | Principal Enzymatic Activity |
|---|---|
| *Actinoplanes missouriensis* | Glucose isomerase |
| *Arthrobacter* sp. | Glucose isomerase |
| *Bacillus acidopullulaticus* | Pullulanase |
| *Bacillus cereus* | Protease |
| | $\beta$-amylase |
| | Iso-amylase |
| *Bacillus circulans* | Endo-$\beta$-glucanase |
| *Bacillus coagulans* | Glucose isomerase |
| *Bacillus licheniformis* | $\alpha$-amylase |
| | Protease |
| *Bacillus megaterium* | $\beta$-amylase |
| *Bacillus subtilis* | $\alpha$-amylase |
| (including strains known | $\beta$-amylase |
| under the name | Endo-$\beta$-glucanase |
| *B. mesentericus and* | Hemicellulase |
| *B. amyloliquefaciens*) | Protease |
| *Klebsiella aerogenes* | Dextranase |
| | Pullulanase |

| | |
|---|---|
| *Leuconostoc oenos* | Malic acid decarboxylase |
| *Micrococcus lysodeicticus* | Catalase |
| *Streptomyces albus* | Glucose isomerase |
| *Streptomyces olivaceus* | Glucose isomerase |
| *Streptomyces olivochromogenes* | Glucose isomerase |
| *Streptomyces rubiginosus* | Glucose isomerase |
| *Streptomyces* sp. | Xylanase |
| Fungi | Principal Enzymatic Activity |
| *Aspergillus melleus* | Protease |
| *Aspergillus niger* | $\alpha$-amylase |
| | Arabinofuranosidase |
| | Glucoamylase or amyloglucosidase |
| | Catalase |
| | Cellulase |
| | Endo-$\beta$-glucanase |
| | Glucose oxidase |
| | Hemicellulase |
| | Lipase |
| | Maltase or $\alpha$-glucosidase |
| | Pectinase |
| | Pectinesterase |
| | Pectinlyase |
| | Polygalacturonase |
| | Protease |
| | Cellobiase |
| | Tannase |
| | Xylanase |
| | Lactase |
| | $\alpha$-galactosidase |
| | Invertase |
| *Aspergillus oryzae* | $\alpha$-amylase |
| | Glucoamylase or amyloglucosidase |
| | Cellulase |
| | Endo-$\beta$-glucanase |
| | Hemicellulase |
| | Lipase |
| | Maltase or $\alpha$-glucosidase |
| | Pectinase |

**Table 2.10.9 (continued)**

| Fungi | Principal Enzymatic Activity |
|---|---|
| *Aspergillus oryzae* (continued) | Protease |
| | Tannase |
| | Lactase |
| *Candida lipolytica* | Lipase |
| *Endothia parasitica* | Protease |
| *Mortierella vinacea* sp. | $\alpha$-galactosidase |
| *Mucor javanicus* | Lipase |
| *Mucor miehei* | Lipase |
| | Esterase |
| | Protease |
| *Mucor pusillus* | Protease |
| | Lipase |
| *Penicillium emersonii* | Endo-$\beta$-glucanase |
| *Penicillium funicolosum* | Dextranase |
| *Penicillium lilacinum* | Dextranase |
| *Penicillium simplicissimum* | Pectinase |
| *Rhizopus arrhizus* | Lipase |
| | Glucoamylase or amyloglucosidase |
| *Rhizopus delemar* | $\alpha$-amylase |
| | Glucoamylase or amyloglucosidase |
| | Cellulase |
| | Endo-$\beta$-glucanase |
| | Hemicellulase |
| *Rhizopus niveus* | Lipase |
| | Glucoamylase or amyloglucosidase |
| *Rhizopus oryzae* | $\alpha$-amylase |
| | Glucoamylase or amyloglucosidase |
| | Cellulase |
| | Endo-$\beta$-glucanase |
| | Hemicellulase |
| | Maltase or $\alpha$-glucosidase |
| | Pectinase |
| *Sporotrichum dimorphosporum* | Cellulase |
| | Hemicellulase |
| | Xylanase |

| | |
|---|---|
| *Thielavia terrestris* | Cellulase |
| *Trichoderma harzianum* | $\beta$-glucosidase |
| *Trichoderma reesei* | Glucoamylase or amyloglucosidase |
| | Cellulase |
| | Hemicellulase |
| | Cellobiase or $\beta$-glucosidase |
| | Pectinase |
| Yeasts | Principal Enzymatic Activity |
| *Kluyveromyces fragilis* | Lactase |
| | Inulinase |
| | Invertase |
| *Kluyveromyces lactis* | Lactase |
| | Invertase |
| | $\alpha$-galactosidase |
| *Saccharomyces cerevisiae* | Invertase |

Source: Noordervliet and Toet, 1987. Used with permission.

**Table 2.10.10. Source Organisms of Some Commercially Important Enzymes**

| Bacterial Enzymes | Produced By | Bacterial Enzymes | Produced By |
|---|---|---|---|
| $\alpha$-Amylase | *Bacillus* spp. | Glucose isomerase | *Actinoplanes missouriensis* |
| $\beta$-Amylase | *Bacillus cereus* | | *Streptomyces* spp. |
| Dextran-sucrase | *Leuconostoc mesenteroides* | Penicillin acylase | *Bacillus megaterium* |
| $\beta$-1,3-Glucanase | *Bacillus subtilis* | Proteinase | *Escherichia coli* |
| Glucose isomerase | *Arthrobacter globiformis* | | *Bacillus subtilis* |
| | *Bacillus coagulans* | Pullulanase | *Klebsiella aerogenes* |

**Table 2.10.10. (continued)**

| Fungal Enzymes | Produced By | Fungal Enzymes | Produced By |
|---|---|---|---|
| α-Amylase | *Aspergillus oryzae* | Invertase | *Saccharomyces* spp. |
| Amyloglu-cosidase | *Aspergillus niger* | Lactase | *Kluyveromyces fragilis* |
| Catalase | *Aspergillus niger* | | *Aspergillus niger* |
| Cellulases | *Trichoderma reesei* | Lipase | *Candida lipolytica* |
| | *Penicillium funiculosum* | | *Rhizopus delemar* |
| | | Microbial rennin | *Mucor miehei* |
| Dextranase | *Penicillium* spp. | Pectinases | *Aspergillus niger* |
| Glucose oxidase | *Aspergillus niger* | Proteinases | *Aspergillus oryzae* |
| | | Raffinase | *Mortierella vinacea* |

Source: Adapted from Frost and Moss, 1987. Used with permission.

**Table 2.10.11. Enzymes from Sources Other Than Microorganisms**

| Animal Enzymes | Isolated from |
|---|---|
| α-Amylase | Pancreatic gland (bovine, porcine) |
| Trypsin | Pancreatic gland (bovine, porcine) |
| Chymotrypsin | Pancreatic gland (bovine, porcine) |
| Pepsin (acid protease) | Fundus portion of hog stomach |
| Rennet (milk clotting enzyme) | Fourth stomach of calf |
| Esterase | Pancreatic gland |
| Lipase | Pancreatic gland |
| Glucose oxidase | Liver |
| Catalase | Liver |
| Plant Enzymes | Isolated from |
| α-Amylase | Germinated barley (malt) |
| β-Amylase | Barley, wheat, rye, sweet potatoes, soy beans |
| R-Enzyme (debranching enzyme) | Beans, potatoes |
| Endo-β-glucanases | Germinated barley (malt) |
| Papain (SH-protease) | Latex of *Carica papaya* |
| Bromelain (SH-protease) | Stems of pineapple plants |
| Ficin (SH-protease) | Latex of *Ficus carica* |
| Lipoxygenase | Beans, potatoes |

Source: Adapted from Borriss, 1987. Used with permission.

**Table 2.10.12. Microbial Species Producing Amyloglucosidase**

| Type | Species |
|---|---|
| Bacteria | *Halobacterium sodomense* |
| | *Clostridium thermoamylolyticum* |
| Yeasts | *Endomycopsis fibuligera* |
| | *Schwanniomyces occidentalis* |
| | *Pichia burtonii* |
| | *Schwanniomyces castelli* |
| | *Schwanniomyces diastaticus* |
| | *Schwanniomyces alluvius* |
| | *Filobasidium capsuligenum* |
| Molds | *Talaromyces* sp. |
| | *Neurospora* sp. |

Source: Adapted from Frost and Moss, 1987. Used with permission.

**Table 2.10.13. A Partial List of Microorganisms Producing Cellulase**

| | |
|---|---|
| *Alternaria alternata* | *Pencillium purpurogenum* |
| *Aspergillus aculeatus* | *Scytalidium lignicola* |
| *Aspergillus fumigatus* | *Sporotrichum cellulophilum* |
| *Aspergillus ustus* | *Thermoascus aurantiacus* |
| *Cellulomonas uda* | *Thielavia terrestris* |
| *Clostridium thermocellum* | *Trichosporon cutaneum* |
| *Dichomitus squalens* | *Trichoderma harzianum* |
| *Pencillium janthinellum* | |

Source: Adapted from Frost and Moss, 1987. Used with permission.

**Table 2.10.14. Sources of Cholesterol Oxidase**

| Origin | Comments |
|---|---|
| *Nocardia Cholesterolicum* | Growth on glycerol, cholesterol inducer |
| *Nocardia rhodochrous* | Growth on glycerol, cholesterol inducer |
| *Schizophyllum commune* | Oleic acid affect production |
| *Gloeocystidium chrysocreas* | Soy oil used; may act as inducer |
| *Streptomyees* spp | None |
| *Arthrobacter simplex* | Constitutive mutant |

Source: Adapted from Frost and Moss, 1987. Used with permission.

## 2.11. Enzymes: Applications in Food Processing

**Table 2.11.1. Commercial Food Uses of Enzymes**

| Process | Enzymatic Action | Enzyme |
|---|---|---|
| | Starch Industry | |
| Glucose syrups<br>Maltodextrins | Hydrolysis of glycosidic links | α-amylase<br>β-amylase<br>Pullulanase |
| High fructose syrups | Isomerization of glucose | Glucose isomerase |
| Fructose syrups | Hydrolysis of fructans | Inulase |
| | Dairy Industry | |
| Cheese | Hydrolysis of kappa-casein | Rennet |
| Cheese flavor | Degradation of fat | Esterase/lipase |
| Milk preservation | Removal of $H_2O_2$ | Catalase |
| Low lactose milk<br>Sweet whey | Hydrolysis of lactose to glucose and galactose | Lactase |
| | Distillery Industry | |
| Fermentable sugars | Hydrolysis of glucosidic links and glucans | α-amylase<br>Amyloglucosidase<br>β-glucanase |
| | Wine and Fruit Juice Industry | |
| Clarification and filtration of juice | Hydrolysis of pectins and cellulose | Pectinase-group<br>Cellulase |
| Juice extraction | Hydrolysis of starch | Hemicellulase |
| Starch removal | Oxygen removal | Amylase |
| Antioxidant | | Glucose oxidase |
| | Brewing Industry | |
| Mashing | Hydrolysis of starch | α-amylase |
| Filterability | Hydrolysis of β-glucan | β-glucanase |
| Conditioning | Hydrolysis of complex forming proteins | Protease |

| | | |
|---|---|---|
| | Baking Industry | |
| Fermentability<br>Bread volume<br>Crumb structure<br>Crust color<br>Gluten regulation | Hydrolysis of glycosidic links<br>Hydrolysis of glycosidic links<br>Hydrolysis of proteins | α-amylase<br>Amyloglucosidase<br>α-amylase<br>Amyloglucosidase<br>Protease |
| | Products Requiring Protein Modification | |
| Solubility<br>Taste | Hydrolysis of proteins to functional peptides and amino acids | Protease |
| | Meat Industries | |
| Tenderization | Hydrolysis of proteins, myofibrillar breakdown | Protease |
| | Other Industries | |
| Soft-centered candies | Saccharose hydrolysis to glucose and fructose | Invertase |
| Fat modification | Hydrolysis of fat | Lipase/esterase |
| Fat antioxidant (oxygen removal) | Glucose to gluconic acid and $H_2O_2$ | Glucose oxidase |
| | $H_2O_2$ to $H_2O$ and $O_2$ | Catalase |
| Eggs anti-browning (glucose removal) | Glucose to gluconic acid and $H_2O_2$ | Glucose oxidase |
| | $H_2O_2$ to $H_2O$ and $O_2$ | Catalase |

Source: Adapted from Rehm and Reed, 1983, 1987; Gerhartz, 1990.

**Table 2.11.2. Profiles of Major Enzymes Used in Food Processing**

---

$\alpha$-Amylase
1. Sources:
   Barley malt
   Bacteria: *Bacillus subtilis, B. licheniformis*
   Fungi: *Aspergillus oryzae, A. niger, Rhizopus oryzae*
2. Applications: Starch processing, grain mashes, baking

$\beta$-Amylase
1. Source: Barley malt
2. Applications: Starch processing, grain mashes, baking

Bromelain
1. Source: Pineapples
2. Applications: Meat tenderizing

Catalase
1. Sources:
   Bovine liver
   Bacteria: *Micrococcus lysodeikticus*
   Fungi: *A. niger*
2. Applications: Peroxide-treated milk

Cellulase
1. Sources: Fungi (*A. niger, Trichoderma reesei*)
2. Applications: Plant macerates, feeds, digestive aids

$\beta$-Glucanase
1. Sources:
   Bacteria (*B. subtilis*)
   Fungi; (*A. niger*)
2. Applications: Brewing, facilitation of juice filtration

Ficin
1. Source: Figs
2. Applications: Brewing, meat, feeds, seafoods

Glucoamylase
1. Sources: Fungi (*Aspergillus oryzae, A. niger, Rhizopus oryzae*)
2. Applications: Starch processing, brewers' and distillers' mashes

Glucose oxidase
1. Sources: Fungi (*A. niger, Penicillium amagaskinense, P. vitale*)
2. Applications: Beverages, eggs, fruit juices, wine

Glucose isomerase
1. Sources:
   Bacteria: *Bacillus coagulans*

Fungi: *Streptomyces olivaceus, S. olivochromogenes, S. rubiginosus, Actinoplanes missouriensis*

2. Applications: Glucose conversion to fructose; dietetic foods, reduction of glucose content

Invertase

1. Sources: Yeast (*Saccharomyces cerevisiae*)
2. Applications: Confectionary, baking

Lactase

1. Sources:
   Yeast (*Saccharomyces lactis*)
   Fungi (*Aspergillus oryzae, A. niger, Rhizopus oryzae*)
2. Applications: Milk, whey

Lipoxygenase

1. Source: Soybeans
2. Application: Whitener

Lysozyme

1. Source: Egg white
2. Application: Milk

Lipase/esterase

1. Sources:
   Oral glands: kid, calf, lamb
   Fungi (*Aspergillus oryzae, A. niger, Rhizopus oryzae*)
2. Applications: Dairy products, flavor

Milk coagulants

1. Sources:
   Calf abomasum, bovine abomasum
   Fungi (*Endothia parasitica, Mucor miehei, M. pusillus*)
2. Applications: Cheesemaking

Papain

1. Source: *Carica papaya*
2. Applications: Baking, brewing, meat

Pectinase

1. Sources: Fungi (*Aspergillus oryzae, A. niger, Rhizopus oryzae*)
2. Applications: Fruit juices, wine

Pepsin

1. Sources: Porcine stomach, bovine abomasum
2. Applications: Cheesemaking

Protease

1. Sources:
   Bacteria: *Bacillus subtilis, B. licheniformis*
   Fungi: *Aspergillus oryzae, A. niger, Rhizopus oryzae*

2. Applications: Meat, cheese flavoring

Rennet extract

1. Sources: Calf, cow abomasa
2. Applications: Cheesemaking

Source: Adapted from Peppler and Reed, 1987. Used with permission.

**Table 2.11.3. Typical Trade Names and Manufacturers for Enzymes Used in Food Processing**

| Enzyme | Trademark | Manufacturer |
|---|---|---|
| Bacterial α-amylase | Tenase | Miles Laboratories |
| | Optiamyl | Miles Kali-Chemie |
| | BAN | Novo-Nordisk |
| | Rapidase | Gist-Brocades |
| Bacterial thermostable α-amylase | Termamyl | Novo-Nordisk |
| | Taka-therm II | Miles Laboratories |
| | Taka-lite | Miles Laboratories |
| | Opti-therm | Miles Kali-Chemie |
| | Maxamyl | Gist-Brocades |
| Amyloglucosidase | Diazyme | Miles Laboratories |
| | Optidex | Miles Kali-Chemie |
| | AMG | Novo-Nordisk |
| | Spezyme | Finn Biochemical |
| | Amigase | Gist-brocades |
| Debranching Enzyme | Dextrozyme (contains AMG) | Novo-Nordisk |
| | Promozyme | Novo-Nordisk |
| β-Amylase | β-Amylase | Amano |
| | Spezyme BBA | Finn Biochemicals |
| Fungal amylase | Biozyme | Amano |
| | MKC fungal amylase | Miles Kali-Chemie |
| | Clarase | Miles Laboratories |
| | Fungamyl | Novo-Nordisk |
| Fungal protease | Amano "A" | Amano |
| | Fungal protease | Miles Laboratories |
| Neutral protease | HT Proteolytic | Miles Laboratories |
| | Neutrase | Novo-Nordisk |

Source: Hagen and Pedersen, 1990. Used with permission.

**Table 2.11.4. Applications of Yeasts in Food Processing**

| Yeast | Synonym | Application |
|---|---|---|
| *Saccharomyces bayanus* *S. Cerevisiae* | *Sasccharomyces cerevisiae* | Wine fermentation, bread leavening, ale fermentation, wine fermentation, ethanol fermentation |
| *S. carlsberginsis* | *S. uvarum, S. cerevisiae* | Lager beer fermentation |
| *S. exiguus* | *Torulopsis holmii* | Bread leavening (sour dough) |
| *S. rosei* | *Torulospora delbrueckii* | Dough leavening (sweet dough) |
| *S. rouxii* | *Zygosaccharomyces rouxii* | Shoyus, miso |
| *Candida utilis* | *Torula utilis, Torulopsis utilis* | Fodder, single cell protein |
| *C. shehatae* | | Xylose fermentation |
| *C. pseudotropicalis* | *Candida kefyr* | Lactose fermentation |
| *Kluyveromyces lactis* | *K. marxianus* var. *lactis* | Milk, lactose fermentation |
| *K. fragilis* | *K. marxianus* var. *marxianus* | Lactose fermentation |
| *Pachysolen tannophilus* | None | Xylose fermentation |
| *Phaffia rhodozyma* | None | Fish/poultry feeds |
| *Pichia stipitis* | None | Xylose fermentation |

Source: Jacobson and Jolly, 1987. Used with permission.

**Table 2.11.5. Enzymes Used In Diagnostics And Food Analysis**

Acetyl-CoA synthetase (acetate-CoA ligase)
1. Source: yeast
2. $M_r$, $10^3$ dalton enzyme (subunit): 151 (78)
3. $K_m$, mol/L (substrate): 2.8 x $10^{-4}$ (acetate); 3.5 x $10^{-5}$ (CoA)
4. pH optimum: 7.6
5. Isoelectric point: no data
6. Substrate measured: acetate

Adenylate kinase (myokinase)
1. Source: porcine muscle
2. $M_r$, $10^3$ dalton enzyme (subunit): 21
3. $K_m$, mol/L (substrate): 5 x $10^{-4}$ (AMP); 3 x $10^{-4}$ (ATP); 1.58 x $10^{-3}$ (ADP)
4. pH optimum: ca. 7.5
5. Isoelectric point: 6.1
6. Substrate measured: AMP

Alcohol dehydrogenase
1. Source: yeast
2. $M_r$, $10^3$ dalton enzyme (subunit): 141 (35)
3. $K_m$, mol/L (substrate): 1.3 x $10^{-2}$ (ethanol); 7.8 x $10^{-4}$ (acetaldehyde); 7.4 x $10^{-5}$ ($NAD^+$); 1.08 x $10^{-5}$ (NADH)
4. pH optimum: 9.0
5. Isoelectric point: 5.4-5.8
6. Substrate measured: ethanol, other alcohols, aldehydes

Aldehyde dehydrogenase
1. Source: yeast
2. $M_r$, $10^3$ dalton enzyme (subunit): 207 (57)
3. $K_m$, mol/L (substrate): 9 x $10^{-6}$ (acetaldehyde); 1.3 x $10^{-4}$ ($NAD^+$)
4. pH optimum: 9.3
5. Isoelectric point: no data
6. Substrate measured: acetaldehyde, ethanol

D-Amino acid oxidase
1. Source: porcine kidney
2. $M_r$, $10^3$ dalton enzyme (subunit): 100 (50 or 38-39)
3. $K_m$, mol/L (substrate): 1.8 x $10^{-3}$ (D-alanine); 1.8 x $10^{-4}$ (oxygen)
4. pH optimum: 9.5
5. Isoelectric point: no data
6. Substrate measured: penicillin, hormones

Amyloglucosidase (glucamylase, glucan 1,4-$\alpha$-glucosidase)
1. Source: *Aspergillus niger*

2. $M_r$, $10^3$ dalton enzyme (subunit): 97 (48)
3. $K_m$, mol/L (substrate): 2.2 x $10^{-7}$ (amylopectin); 3.2 x $10^{-5}$ (amylose); 1.85 x $10^{-2}$ (maltose)
4. pH optimum: ca. 5.0
5. Isoelectric point: 4.2
6. Substrate measured: starch, glycogen

Ascorbate oxidase
1. Source: *Cucurbita* species
2. $M_r$, $10^3$ dalton enzyme (subunit): 140 (65)
3. $K_m$, mol/L (substrate): 2.4 x $10^{-4}$ (L-ascorbate)
4. pH optimum: 5.6-7.0
5. Isoelectric point: 5.0-5.5
6. Substrate measured: ascorbate; (removal of ascorbate interference)

Batroxobin (*Baothrops atrox* serine proteinase, reptilase)
1. Source: *Bothrops atrox* venom (different zoological varieties)
2. $M_r$, $10^3$ dalton enzyme (subunit): 32-43
3. $K_m$, mol/L (substrate): 1.6-2.9 x $10^{-4}$ (benzoyl-Phe-Val-Arg-*p*-nitroanilide
4. pH optimum: 7.4-8.2
5. Isoelectric point: 6.6
6. Substrate measured: fibrinogen

Catalase
1. Source: bovine liver
2. $M_r$, $10^3$ dalton enzyme (subunit): 232 (57.5)
3. $K_m$, mol/L (substrate): See footnote 'a.'
4. pH optimum: 6.8-7.0
5. Isoelectric point: 5.4-5.8
6. Substrate measured: uric acid, cholesterol (indicator enzyme)

Cholesterol esterase
1. Source: *Pseudomonas fluorescens*
2. $M_r$, $10^3$ dalton enzyme (subunit): ca. 129 (27)
3. $K_m$, mol/L (substrate): 7 x $10^{-5}$ (cholesterol oleate) 5.5 x $10^{-5}$ (cholesterol linoleate)
4. pH optimum: 7.3-7.6
5. Isoelectric point: 4.5
6. Substrate measured: cholesterol esters

Cholesterol oxidase
1. Source: *Norcardia erythropolis*
2. $M_r$, $10^3$ dalton enzyme (subunit): 59
3. $K_m$, mol/L (substrate): 1 x $10^{-6}$ (cholesterol)
4. pH optimum: 7.5

**Table 2.11.5. (continued)**

5. Isoelectric point: 4.85
6. Substrate measured: cholesterol

Cholesterol oxidase
1. Source: *Streptomyces* species
2. $M_r$, $10^3$ dalton enzyme (subunit): 55
3. $K_m$, mol/L (substrate): 6 x $10^{-4}$ (cholesterol)
4. pH optimum: 7.0-7.5
5. Isoelectric point: 4.5
6. Substrate measured: cholesterol

Choline kinase
1. Source: yeast
2. $M_r$, $10^3$ dalton enzyme (subunit): 67-68
3. $K_m$, mol/L (substrate): 1.5 x $10^{-5}$ (choline); 1.4 x $10^{-4}$ (ATP)
4. pH optimum: 8.0-9.5
5. Isoelectric point: ca. 6
6. Substrate measured: phosphatidylcholine

Choline oxidase
1. Source: *Arthrobacter globiformis*
2. $M_r$, $10^3$ dalton enzyme (subunit): 84 or 71[b]
3. $K_m$, mol/L (substrate): 1.2 x $10^{-3}$ (choline); 8.7 x $10^{-2}$ (betaine aldehyde)
4. pH optimum: 7.5
5. Isoelectric point: 4.5
6. Substrate measured: phospholipids

Citrate lyase [citrate(*pro-3S*)-lyase]
1. Source: *Aerobacter aerogenes*
2. $M_r$, $10^3$ dalton enzyme (subunit): 575 (73.8)
3. $K_m$, mol/L (substrate): 2.1 x $10^{-4}$ (citrate)
4. pH optimum: 8.0-9.0
5. Isoelectric point: 8.0
6. Substrate measured: citrate

Citrate synthase
1. Source: porcine heart
2. $M_r$, $10^3$ dalton enzyme (subunit): 100 (50)
3. $K_m$, mol/L (substrate): 2.5 x $10^{-4}$ (citrate) 2.8 x $10^{-5}$ (CoA)
4. pH optimum: 8.0
5. Isoelectric point: 5.05
6. Substrate measured: acetate

Creatinase (creatine amidinohydrolase)
1. Source: *Pseudomonas* species
2. $M_r$, $10^3$ dalton enzyme (subunit): 94 (47)

3. $K_m$, mol/L (substrate): 1 x $10^{-2}$ (creatine)
4. pH optimum: 8.0
5. Isoelectric point: 4.8
6. Substrate measured: creatine, creatinine

Creatininase (creatinine amidohydrolase)

1. Source: *Pseudomonas* species
2. $M_r$, $10^3$ dalton enzyme (subunit): 175 (23)
3. $K_m$, mol/L (substrate): 3 x $10^{-2}$ (creatinine) 6 x $10^{-2}$ (creatine)
4. pH optimum: 7.8
5. Isoelectric point: 4.7
6. Substrate measured: creatinine

Creatinine deiminase

1. Source: *Corynebacterium lilium*
2. $M_r$, $10^3$ dalton enzyme (subunit): 200
3. $K_m$, mol/L (substrate): 6 x $10^{-3}$ (creatine)
4. pH optimum: 7.5-9.0
5. Isoelectric point: 4.2
6. Substrate measured: creatine

Dihydrolipoamide dehydrogenase (diaphorase)

1. Source: porcine heart
2. $M_r$, $10^3$ dalton enzyme (subunit): 114 (57)
3. $K_m$, mol/L (substrate): 5 x $10^{-3}$ (lipoamide); 2 x $10^{-3}$ (lipoate); 2 x $10^{-4}$ ($NAD^+$); 2.7 x $10^{-4}$ (ferricyanide)
4. pH optimum: 4.8 (diaphorase) 5.6-6.5 (lipoate reduction)
5. Isoelectric point: 5.9-7.2[c]
6. Substrate measured: NADH (indicator enzyme)

Esterase (carboxylesterase)

1. Source: porcine liver
2. $M_r$, $10^3$ dalton enzyme (subunit): 168 (42)
3. $K_m$, mol/L (substrate): 5 x $10^{-4}$ (ethyl *n*-butyrate) 4.4 x $10^{-4}$ (methyl *n*-butyrate) 1.5 x $10^{-4}$ (phenyl *n*-butyrate)
4. pH optimum: 8.6-8.8
5. Isoelectric point: 5.0
6. Substrate measured: triglycerides

Factor Xa (thrombokinase)

1. Source: bovine plasma
2. $M_r$, $10^3$ dalton enzyme (subunit): 47.2 (Xa$\alpha$); 44.2 (Xa$\beta$)
3. $K_m$, mol/L (substrate): 5 x $10^{-4}$ (benzoyl-Ile-Glu-Gly-Arg-*p*-nitroanilide)
4. pH optimum: 8.3
5. Isoelectric point: no data
6. Substrate measured: prothrombin

**Table 2.11.5. (continued)**

Formaldehyde dehydrogenase
1. Source: *Pseudomonas* species
2. $M_r$, $10^3$ dalton enzyme (subunit): 150 (75)
3. $K_m$, mol/L (substrate): 6.7 x $10^{-5}$ (formaldehyde) 1.2 x $10^{-4}$ ($NAD^+$)
4. pH optimum: 7.8
5. Isoelectric point: 5.25
6. Substrate measured: formaldehyde

Formate dehydrogenase
1. Source: *Candida boidinii*
2. $M_r$, $10^3$ dalton enzyme (subunit): 74 (36)
3. $K_m$, mol/L (substrate): 1.3 x $10^{-2}$ (formate); 9 x $10^{-5}$ ($NAD^+$)
4. pH optimum: 7.5-8.5
5. Isoelectric point: no data
6. Substrate measured: formate, oxalate

$\beta$-Fructosidase (invertase, saccharase)
1. Source: yeast
2. $M_r$, $10^3$ dalton enzyme (subunit): 270 (135)
3. $K_m$, mol/L (substrate): 9.1 x $10^{-3}$ (sucrose) 2.4 x $10^{-1}$ (raffinose)
4. pH optimum: 3.4-4.0
5. Isoelectric point: 4.02, 4.24[c]
6. Substrate measured: sucrose

$\beta$-D-Galactose dehydrogenase
1. Source: *Pseudomonas fluorescens*
2. $M_r$, $10^3$ dalton enzyme (subunit): 64 (32)
3. $K_m$, mol/L (substrate): 7 x $10^{-4}$ (D-galactose) 2.4 x $10^{-4}$ ($NAD^+$) 2.3 x $10^{-3}$ ($NADP^+$)
4. pH optimum: 9.1-9.5
5. Isoelectric point: 5.13
6. Substrate measured: galactose, raffinose

$\alpha$-Galactosidase
1. Source: *Escherichia coli*
2. $M_r$, $10^3$ dalton enzyme (subunit): 329 (82)
3. $K_m$, mol/L (substrate): 2 x $10^{-2}$ (galactose)
4. pH optimum: 6.0-7.0
5. Isoelectric point: 5.1
6. Substrate measured: raffinose

$\beta$-Galactosidase
1. Source: *Escherichia coli*
2. $M_r$, $10^3$ dalton enzyme (subunit): 465 (116)

3. $K_m$, mol/L (substrate): 3.85 x $10^{-3}$ (lactose); 9.5 x $10^{-4}$ (2-nitrophenyl-$\beta$-D-galactoside); 4.45 x $10^{-4}$ (4-nitrophenyl-$\beta$-D-galactoside)
4. pH optimum: 8.0
5. Isoelectric point: 4.61
6. Substrate measured: lactose; (marker enzyme)

Gluconokinase

1. Source: *Escherichia coli*
2. $M_r$, $10^3$ dalton enzyme (subunit): 67 (34)
3. $K_m$, mol/L (substrate): 7 x $10^{-5}$ (D-gluconate); 5 x $10^{-4}$ (ATP)
4. pH optimum: 8.0
5. Isoelectric point: no data
6. Substrate measured: D-gluconate

Glucose dehydrogenase

1. Source: *Bacillus megaterium*
2. $M_r$, $10^3$ dalton enzyme (subunit): 118 (28)
3. $K_m$, mol/L (substrate): 4.75 x $10^{-2}$ (glucose); 4.5 x $10^{-3}$ ($NAD^+$)
4. pH optimum: 8.0 (tris-HCl); 9.0 (acetate)
5. Isoelectric point: no data
6. Substrate measured: glucose

Glucose oxidase

1. Source: *Aspergillus niger*
2. $M_r$, $10^3$ dalton enzyme (subunit): 160 (79)
3. $K_m$, mol/L (substrate): 3.3 x $10^{-2}$ (glucose); 2 x $10^{-4}$ (oxygen)
4. pH optimum: 5.5-6.5
5. Isoelectric point: 4.2
6. Substrate measured: glucose

Glucose 6-phosphate dehydrogenase

1. Source: *Leuconostoc mesenteroides*
2. $M_r$, $10^3$ dalton enzyme (subunit): 103.7 (54.8)
3. $K_m$, mol/L (substrate): 6.4 x $10^{-5}$ (glucose-6-phosphate); 1.15 x $10^{-4}$ ($NAD^+$); 7 x $10^{-6}$ ($NADP^+$)
4. pH optimum: 7.8
5. Isoelectric point: 4.6
6. Substrate measured: glucose-6-phosphate, sucrose, ATP; (marker enzyme)

Glucose 6-phosphate isomerase

1. Source: yeast
2. $M_r$, $10^3$ dalton enzyme (subunit): 120 (60)
3. $K_m$, mol/L (substrate): 0.7-1.5 x $10^{-3}$ (glucose 6-phosphate); 2.3 x $10^{-4}$ (fructose 6-phosphate)
4. pH optimum: 7.6

**Table 2.11.5. (continued)**

5. Isoelectric point: 5.0-5.4[c]
6. Substrate measured: fructose

$\alpha$-Glucosidase (maltase)
1. Source: yeast
2. $M_r$, $10^3$ dalton enzyme (subunit): 68
3. $K_m$, mol/L (substrate): $1.8\text{-}2.8 \times 10^{-4}$ (4-nitrophenyl-$\alpha$-D-glucoside)
4. pH optimum: 7.5-8.0
5. Isoelectric point: 5.6-5.9[c]
6. Substrate measured: $\alpha$-amylase

$\beta$-Glucosidase
1. Source: *Amygdalae dulces*
2. $M_r$, $10^3$ dalton enzyme (subunit): 135 (65)[c] + 90
3. $K_m$, mol/L (substrate): $6 \times 10^{-3}$ (2-nitrophenyl-$\beta$-D-glucoside)
4. pH optimum: 4.4-6.0[c]
5. Isoelectric point: 7.3[c]
6. Substrate measured: $\alpha$-amylase, amygdalin

Glutamate dehydrogenase
1. Source: bovine liver
2. $M_r$, $10^3$ dalton enzyme (subunit): oligomers of 332, e.g., 2200 (55)
3. $K_m$, mol/L (substrate): $1.8 \times 10^{-3}$ (L-glutamate); $7 \times 10^{-4}$ (2-oxoglutarate)
4. pH optimum: 8.5-9.0 (glutamate oxidation)
5. Isoelectric point: 4.5
6. Substrate measured: glutamate, 2-oxoglutarate, ammonia

Glutamate dehydrogenase
1. Source: bovine liver
2. $M_r$, $10^3$ dalton enzyme (subunit): oligomers of 332, e.g., 2200 (55)
3. $K_m$, mol/L (substrate): $3.2 \times 10^{-3}$ (ammonia); $7.0 \times 10^{-4}$ ($NAD^+$); $4.7 \times 10^{-5}$ ($NADP^+$); $2.5 \times 10^{-5}$ (NADH); $2.4 \times 10^{-5}$ (NADPH)
4. pH optimum: 7.8 (reductive amination)
5. Isoelectric point: 4.5
6. Substrate measured: glutamate, 2-oxoglutarate, ammonia

Glutamic-oxaloacetic transaminase (aspartate aminotransferase)
1. Source: porcine heart
2. $M_r$, $10^3$ dalton enzyme (subunit): 94 (47)
3. $K_m$, mol/L (substrate): $8.9 \times 10^{-3}$ (L-glutamate); $8.8 \times 10^{-5}$ (oxaloacetate); $3.9 \times 10^{-3}$ (L-aspartate); $4.3 \times 10^{-4}$ (2-oxoglutarate)
4. pH optimum: 8.0-8.5
5. Isoelectric point: 5.0
6. Substrate measured: malate

Glutamic-pyruvic transaminase (alanine aminotransferase)

1. Source: porcine heart
2. $M_r$, $10^3$ dalton enzyme (subunit): 115
3. $K_m$, mol/L (substrate): 2.5 x $10^{-2}$ (L-glutamate); 3 x $10^{-4}$ (pyruvate); 2.8 x $10^{-2}$ (L-alanine); 4 x $10^{-4}$ (2-oxoglutarate)
4. pH optimum: 8.0-8.5
5. Isoelectric point: no data
6. Substrate measured: lactate

Glycerol dehydrogenase

1. Source: *Enterobacter aerogenes*
2. $M_r$, $10^3$ dalton enzyme (subunit): 340 (56)
3. $K_m$, mol/L (substrate): 1.7 x $10^{-2}$ (glycerol); 1.3 x $10^{-3}$ (dihydroxy-acetone); 1.5 x $10^{-4}$ ($NAD^+$); 1.4 x $10^{-5}$ (NADH)
4. pH optimum: 9.0
5. Isoelectric point: no data
6. Substrate measured: glycerol, triglycerides

Glycerol kinase

1. Source: *Bacillus stearothermophilus*
2. $M_r$, $10^3$ dalton enzyme (subunit): 230 (58)
3. $K_m$, mol/L (substrate):4.4 x $10^{-5}$ (glycerol); 6 x $10^{-5}$ (ATP)
4. pH optimum:10.0-10.5
5. Isoelectric point: no data
6. Substrate measured: glycerol, triglycerides, lipase

Glycerol 3-phosphate dehydrogenase

1. Source: rabbit muscle
2. $M_r$, $10^3$ dalton enzyme (subunit): 78 (37.5)
3. $K_m$, mol/L (substrate): 1.1 x $10^{-4}$ (glycerol 3-phosphate); 3.8 x $10^{-4}$ ($NAD^+$)
4. pH optimum: 7.5-8.6
5. Isoelectric point: 6.45
6. Substrate measured: glycerol, triglycerides

Glycerol 3-phosphate oxidase

1. Source: *Aerococcus viridans*
2. $M_r$, $10^3$ dalton enzyme (subunit): 75
3. $K_m$, mol/L (substrate): 3.2 x $10^{-3}$ (L-glycerol 3-phosphate)
4. pH optimum: 7.5-8.5
5. Isoelectric point: 4.2
6. Substrate measured: triglycerides

Hexokinase

1. Source: yeast
2. $M_r$, $10^3$ dalton enzyme (subunit): 104 (52)

**Table 2.11.5. (continued)**

3. $K_m$, mol/L (substrate): 1.0 x $10^{-4}$ (D-glucose); 7.0 x $10^{-4}$ (D-fructose); 2.0 x $10^{-4}$ (ATP); 5.0 x $10^{-5}$ (D-mannose)
4. pH optimum: 8.0-9.0
5. Isoelectric point: 4.7
6. Substrate measured: glucose, fructose, mannose, ATP; creatine kinase

3-Hydroxybutyrate dehydrogenase

1. Source: *Rhodopseudomonas spheroides*
2. $M_r$, $10^3$ dalton enzyme (subunit): 85
3. $K_m$, mol/L (substrate): 4.1 x $10^{-4}$ (D-3-hydroxybutyrate); 8.0 x $10^{-5}$ ($NAD^+$)
4. pH optimum: 6.2-6.9 (reduction); 7-9 (oxidation)
5. Isoelectric point: no data
6. Substrate measured: 3-hydroxybutyrate

Isocitrate dehydrogenase

1. Source: porcine heart
2. $M_r$, $10^3$ dalton enzyme (subunit): ca. 60
3. $K_m$, mol/L (substrate): 2.6 x $10^{-6}$ (isocitrate); 9.2 x $10^{-6}$ (NADPH); 1.3 x $10^{-4}$ (2-oxoglutarate); 1 x $10^{-7}$ ($NADP^+$)
4. pH optimum: 7.0-7.5
5. Isoelectric point: 7.4
6. Substrate measured: isocitrate

D-Lactate dehydrogenase

1. Source: *Lactobacillus leichmanii*
2. $M_r$, $10^3$ dalton enzyme (subunit): 68 or 80[d]
3. $K_m$, mol/L (substrate): 7 x $10^{-2}$ (D-lactate); 1.2 x $10^{-3}$ (pyruvate); 7.1 x $10^{-5}$ (NADH)
4. pH optimum: 7.0
5. Isoelectric point: no data
6. Substrate measured: D-lactate; glutamic-oxaloacetic and glutamic-pyruvic transaminases

L-Lactate dehydrogenase

1. Source: porcine heart
2. $M_r$, $10^3$ dalton enzyme (subunit): 140 (35)
3. $K_m$, mol/L (substrate): 1.5 x $10^{-4}$ (pyruvate); 3.3 x $10^{-3}$ (L-lactate); 1.1 x $10^{-5}$ (NADH); 6.7 x $10^{-5}$ ($NAD^+$)
4. pH optimum: 7.0
5. Isoelectric point: see footnote 'c.'
6. Substrate measured: L-lactate, pyruvate, citrate, ADP

Lipase (triacylglycerol acylhydrolase)

1. Source: *Pseudomonas* species

2. $M_r$, $10^3$ dalton enzyme (subunit): 32
3. $K_m$, mol/L (substrate): 9 x $10^{-4}$ (tributyrin)
4. pH optimum: 7.0-8.0
5. Isoelectric point: 4.3
6. Substrate measured: triglycerides

Luciferase, bacterial (alkanal monooxygenase, FMN-linked)

1. Source: *Photobacterium fischeri*
2. $M_r$, $10^3$ dalton enzyme (subunit): 80 (40 + 40)
3. $K_m$, mol/L (substrate): 4-8 x $10^{-7}$ ($FMNH_2$)
4. pH optimum: 6.8
5. Isoelectric point: no data
6. Substrate measured: NADH

Luciferase, firefly (*Photinus*-luciferin 4-monooxygenase)

1. Source: *Photinus pyralis*
2. $M_r$, $10^3$ dalton enzyme (subunit): 100 (50)
3. $K_m$, mol/L (substrate): 5 x $10^{-5}$ (ATP)
4. pH optimum: 7.5-7.8
5. Isoelectric point: 6.2-6.3
6. Substrate measured: ATP

Malate dehydrogenase

1. Source: porcine heart mitochondria
2. $M_r$, $10^3$ dalton enzyme (subunit): 67 (34)
3. $K_m$, mol/L (substrate): 4 x $10^{-4}$ (L-malate); 3.3 x $10^{-5}$ (oxaloacetate)
4. pH optimum: 7.4-7.5
5. Isoelectric point: 6.1-6.4
6. Substrate measured: malate, oxaloacetate, acetate, citrate; (marker enzyme)

Mannose 6-phosphate isomerase

1. Source: yeast
2. $M_r$, $10^3$ dalto n enzyme (subunit): 45
3. $K_m$, mol/L (substrate): 0.8-1.35 x $10^{-3}$ (mannose 6-phosphate)
4. pH optimum: 7.0-7.2
5. Isoelectric point: no data
6. Substrate measured: mannose

NADH peroxidase

1. Source: *Streptococcus faecalis*
2. $M_r$, $10^3$ dalton enzyme (subunit): 120 (59)
3. $K_m$, mol/L (substrate): 1.7 x $10^{-5}$ (NADH); 2.8 x $10^{-5}$ ($H_2O_2$)
4. pH optimum: 6.0
5. Isoelectric point: no data
6. Substrate measured: $H_2O_2$ (indicator enzyme)

**Table 2.11.5. (continued)**

Nitrate reductase
1. Source: *Aspergillus* species
2. $M_r$, $10^3$ dalton enzyme (subunit): 200 (97)
3. $K_m$, mol/L (substrate): $3.2 \times 10^{-4}$ (nitrate)
4. pH optimum: 7.5
5. Isoelectric point: no data
6. Substrate measured: nitrate

Peroxidase
1. Source: horseradish
2. $M_r$, $10^3$ dalton enzyme (subunit): ca. 40
3. $K_m$, mol/L (substrate): see footnote 'a.'
4. pH optimum: 6.0-7.0
5. Isoelectric point: see footnote 'c.'
6. Substrate measured: $H_2O_2$ (indicator enzyme and marker enzyme)

Phosphatase, alkaline
1. Source: calf intestine
2. $M_r$, $10^3$ dalton enzyme (subunit): 140 (69)
3. $K_m$, mol/L (substrate): see footnote 'a.'
4. pH optimum: 9.8
5. Isoelectric point: 5.7
6. Substrate measured: (marker enzyme)

6-Phosphogluconate dehydrogenase
1. Source: yeast
2. $M_r$, $10^3$ dalton enzyme (subunit): 100
3. $K_m$, mol/L (substrate): $1.6 \times 10^{-4}$ (6-phosphogluconate); $2.6 \cdot 10^{-5}$ ($NADP^+$)
4. pH optimum: 8.0
5. Isoelectric point: no data
6. Substrate measured: gluconate

Phospholipase C
1. Source: *Bacillus cereus*
2. $M_r$, $10^3$ dalton enzyme (subunit): 23
3. $K_m$, mol/L (substrate): $2 \times 10^{-2}$ (phosphatidylcholine)
4. pH optimum: 8.0
5. Isoelectric point: 7.0
6. Substrate measured: phosphatidylcholine

Phospholipase D
1. Source: *Streptomyces chromofuscus*
2. $M_r$, $10^3$ dalton enzyme (subunit): 57 or 50[b]

3. $K_m$, mol/L (substrate):1.43 x $10^{-3}$ (phosphatidylcholine); 5.6 x $10^{-2}$ (sphingomyelin)
4. pH optimum: 8.0
5. Isoelectric point: 5.1
6. Substrate measured: trigylcerides, choline phospholipids

Plasmin
1. Source: human or bovine plasma
2. $M_r$, $10^3$ dalton enzyme (subunit): 87-91
3. $K_m$, mol/L (substrate): 1.7 x $10^{-4}$ (tosyl-Gly-Pro-Lys-*p*-nitroanilide)
4. pH optimum: 7.4
5. Isoelectric point: 6.1-8.4[c]
6. Substrate measured: $\alpha_2$-antiplasmin

Pyruvate kinase
1. Source: rabbit muscle
2. $M_r$, $10^3$ dalton enzyme (subunit): 237 (57)
3. $K_m$, mol/L (substrate): 7.0 x $10^{-5}$ (phosphoenolpyruvate); 1.0 x $10^{-2}$ (pyruvate); 3.0 x $10^{-4}$ (ADP); 8.6 x $10^{-4}$ (ATP)
4. pH optimum: 7.0-7.8
5. Isoelectric point: 5.98
6. Substrate measured: ADP, IDP, creatine kinase, phosphoenol-pyruvate

Pyruvate oxidase
1. Source: *Pediococcus* species
2. $M_r$, $10^3$ dalton enzyme (subunit): 150 (38)
3. $K_m$, mol/L (substrate): 1.7 x $10^{-3}$ (pyruvate); 5 x $10^{-4}$ (phosphate)
4. pH optimum: 6.5-7.5
5. Isoelectric point: 4.0
6. Substrate measured: pyruvate, ADP, glutamic-oxaloacetic and glutamic-pyruvic transminases

Sarcosine oxidase
1. Source: *Pseudomonas*
2. $M_r$, $10^3$ dalton enzyme (subunit): 174 (110, 44, 21, 10)
3. $K_m$, mol/L (substrate): 4 x $10^{-3}$ (sarcosine); 1.2 x $10^{-2}$ (*N*-methylalanine); 1.3 x $10^{-4}$ (oxygen)
4. pH optimum: 8.0
5. Isoelectric point: no data
6. Substrate measured: sarcosine, creatinine

Sorbitol dehydrogenase (L-iditol dehydrogenase)
1. Source: sheep liver
2. $M_r$, $10^3$ dalton enzyme (subunit): 115
3. $K_m$, mol/L (substrate): 0.7-1.1 x $10^{-3}$ (sorbitol); 1.8 x $10^{-4}$ (xylitol)
4. pH optimum: 7.9-8.1

**Table 2.11.5. (continued)**

5. Isoelectric point: no data
6. Substrate measured: sorbitol, xylitol

Succinyl-CoA synthetase (succinate-CoA ligase, GDP-forming)

1. Source: porcine heart
2. $M_r$, $10^3$ dalton enzyme (subunit): 75 (42.5 + 34.5)
3. $K_m$, mol/L (substrate): 4-8 x $10^{-4}$ (succinate); 5-10 x $10^{-6}$ (GTP); 5-20 x $10^{-6}$ (CoA)
4. pH optimum: 8.3
5. Isoelectric point: 5.8-6.4[c]
6. Substrate measured: succinate

Sulfite oxidase

1. Source: chicken liver
2. $M_r$, $10^3$ dalton enzyme (subunit): 110 (55)
3. $K_m$, mol/L (substrate): 2.5 x $10^{-5}$ (sulfite)
4. pH optimum: 8.5
5. Isoelectric point: no data
6. Substrate measured: sulfite

Thrombin

1. Source: bovine plasma
2. $M_r$, $10^3$ dalton enzyme (subunit): 39 ($\alpha$), 28 ($\beta$)
3. $K_m$, mol/L (substrate): 1.3 x $10^{-3}$ (benzoyl-Arg-*p*-Nitroanilide); 5.9 x $10^{-6}$ (tosy;-Gly-Pro-Arg-*p*-nitroanilide
4. pH optimum: 9.0
5. Isoelectric point: 5.3-5.75
6. Substrate measured: antithrombin III, fibrinogen, heparin, coagulation status

Trypsin

1. Source: bovine pancreas
2. $M_r$, $10^3$ dalton enzyme (subunit): 23.3
3. $K_m$, mol/L (substrate): 9.4 x $10^{-4}$ (benzoyl-Arg-*p*-nitroanilide); 4.3 x $10^{-6}$ (benzoyl-Arg-ethyl ester)
4. pH optimum: 8.0
5. Isoelectric point: 10.5-10.8
6. Substrate measured: $\alpha_1$-proteinase inhibitor, $\alpha_2$-macroglobulin

Urate oxidase (uricase)

1. Source: *Arthrobacter protophormiae*
2. $M_r$, $10^3$ dalton enzyme (subunit): 170 (40)
3. $K_m$, mol/L (substrate): 6.6 x $10^{-5}$ (urate)
4. pH optimum: 9.0
5. Isoelectric point: no data

6. Substrate measured: urate

Urease

1. Source: jack beans
2. $M_r$, $10^3$ dalton enzyme (subunit): 480 (75-83)
3. $K_m$, mol/L (substrate): 1.05 x $10^{-2}$ (urea)
4. pH optimum: 7.0
5. Isoelectric point: 5.0-5.1
6. Substrate measured: urea; (marker enzyme)

Xanthine oxidase

1. Source: cow milk
2. $M_r$, $10^3$ dalton enzyme (subunit): 283 (150)
3. $K_m$, mol/L (substrate): 1.7 x $10^{-6}$ (xanthine); 2.4 x $10^{-5}$ (oxygen)
4. pH optimum: 8.5-9.0
5. Isoelectric point: 6.2
6. Substrate measured:xanthine, hypoxanthine, phosphate

Source: Kresse, 1990. Used with permission.

[a]Enzyme reaction does not obey Michaelis-Menten kinetics.

[b]Depending on method (gel filtration or polyacrylamide gel electrophoresis in the presence of sodium dodecylsulfate).

[c]Isoenzymes or multiple species.

[d]Depending on method (ultracentrifugation or gel filtration).

**Table 2.11.6. Enzymes Used in the Processing of Dairy Products**

| Enzymes | Sources |
|---|---|
| Rennet Enzymes | |
| Chymosin[a] (rennin, calf rennet) | Calf, lamb, kid |
| Pepsin[a] | Calf, hog, chicken |
| Rennet Substitutes | |
| Rennet Substitutes | *Mucor miehei, Mucor pusillus, Endothia parasitica, Bacillus subtilis* |
| Other Rennet Substitutes | |
| Other Rennet Substitutes | *Mucor racemosus, Bacillus cereus, Bacillus mesenterius, Bacillus polymyxa, Bacillus licheniformis, Bacillus megaterium, Irpex lacteus, Absidia racemosa, Rhizopus* species, *Aspergillus oryzae, Aspergillus nidulans, Candida* species |

**Table 2.11.6 (continued)**

| | |
|---|---|
| b-1,4-Galactosidases | |
| $\beta$-1,4-Galactosidases[a] | *Kluyveromyces lactis, Kluyveromyces fragilis, Saccharomyces lactis, Aspergillus niger, Aspergillus oryzae, Bacillus stearothermophilus, Bacillus subtilis* |
| Other Enzymes | |
| Glucose isomerase | *Bacillus* species |
| Lysozymes | Chicken egg, microbial |
| Peroxidase | Bovine milk |
| Catalase[a] | Bovine liver |
| Superoxide dismutase | Bovine milk, erythrocytes |
| Thiol oxidase[a] | Bovine milk |
| Acid phosphatases[a] | Potato |
| Alkaline phophatases[a] | Calf intestine |
| Lipases | Pancreas, *Streptococcus* species, *Penicillium* species, *Aspergillus* species |
| Glucose oxidase[a] | *Aspergillus niger, Penicillium* species |
| Other Proteinases | |
| Papain[a] | *Carica papaya* |
| Ficin[a] | *Ficus carica* |
| Trypsin[a] | bovine, hog |
| Chymotrypsin[a] | bovine, hog |

Source: Reimerdes, 1990. Used with permission.
[a]Enzyme is also used in immobilized form.

**Table 2.11.7. Major Milk-Clotting Enzymes Used in Cheesemaking**

| Coagulant | Source | Active Enzyme(s) |
|---|---|---|
| Calf rennet | Calf's abomasum | Chymosin |
| Bovine rennet | Cow's abomasum | Chymosin/pepsin |
| Pepsin | Porcine stomach | Pepsin |
| Microbial rennet | Fungal culture filtrates[a] | Proteases |
| Rennet blends | 1:1 mixtures of 1 + 3; 2 + 3; or 3 + 4 | As above |

Source: Adapted from Peppler and Reed, 1987. Used with permission.
[a]*Endothia parasitica, Mucor miehei, M. pusillus*

**Table 2.11.8. Representative Conditions in the Conversion of Starch to High-fructose Corn Syrup**

| Process Step and Related Enzyme | Level, % | DE, % | °C | pH | $Ca^{2+}$, mg/L |
|---|---|---|---|---|---|
| Starch conversion | | | | | |
| Primary liquefaction | | | | | |
| α-Amylase | 0.01-0.02 | 2-5 | 105-115 | 5.5-6.5 | 50-100 |
| Secondary liquefaction | | | | | |
| α-Amylase | 0.1-0.2 | 15-20 | 90-95 | 5.5-6.5 | 50-100 |
| Saccharification | | | | | |
| Amyloglu-cosidase | 0.1 | 98-99 | 55-65 | 4-4.5 | 50-100 |
| Isomerization | | | | | |
| Glucose isomerase | No data | 42% (fruc-tose) | 55-60 | 7-8 | 2 |
| Enrichment | No data | 55, 95% fructose | No data | No data | No data |

Source: Goldstein, 1990. Used with permission.
Note: "% DE" refers to conversion attained, using dextrose equivalent (DE) as end point. During the isomerization stage glucsoe isomerase requires 25-100 mg $Mg^{2+}$/L.

**Table 2.11.9. Some Properties of Enzymes Used in Starch Processing**

Fungal $\alpha$-amylase: obtained from *Aspergillus niger, Aspergillus oryzae, and Aspergillus awamori*; used in limited hydrolysis of starch

1. pH range; °C range: 4-6; 50-70
2. Typical pH; oC: 4.5; 65
3. Activators: none

Bacterial $\alpha$-amylase: obtained from mesophilic bacteria (*Bacillus subtilis, Bacillus amyloliquefaciens*) and used in limited hydrolysis of starch

1. pH range; °C range: 6-7; 40-80
2. Typical pH; °C: 4.5; 6.5
3. Activators: none

Bacterial $\alpha$-amylase: obtained from thermophilic bacteria (*Bacillus licheniformis*) and used in high-temperature liquefaction of starch

1. pH range; °C range: 5.5-6.5; 90-120
2. Typical pH; °C: 5.8; 110
3. Activators: calcium ions

Bacterial $\alpha$-amylase: obtained from thermophilic bacteria (*Bacillus stearothermophilus*) and used in high-temperature liquefaction of starch

1. pH range; °C range: 4.5-6.5; 90-120
2. Typical pH; °C: 6; 115
3. Activators: calcium ions

$\beta$-Amylase: obtained from *Bacillus polymyxa,* wheat, and soybean, and used in formation of low-dextrose equivalent syrups and in baking

1. pH range; °C range: 4-7; 55-75
2. Typical pH; °C: 5; 60
3. Activators: none

Pullulanase: obtained from *Bacillus* species, and *Aerobacter aerogenes*, and used in debranching, 1,6-bond degradation

1. pH range; °C range: 3.5-4.5; 55-65
2. Typical pH; °C: 4; 60
3. Activators: calcium ions

Amyloglucosidase: obtained from *Aspergillus niger* and used in saccharification, primarily 1,4-bond degradation

1. pH range; °C range: 3.5-4.5; 55-65
2. Typical pH; °C: 4; 60
3. Activators: none

Source: Goldstein, 1990. Used with permission

**Table 2.11.10. Enzymes Used by the Starch and Sweetener Industries**

| Enzyme | Action | Source | Optimal pH | Optimal °C |
|---|---|---|---|---|
| $\alpha$-Amylase | Random, endo-hydrolysis of $\alpha$-1,4-glucosidic linkages | *Aspergillus oryzae* | 5 | 55 |
| | | *A. niger* | 5 | 55 |
| | | *Bacillus subtilis* | 6-7 | 80 |
| | | *B. licheniformis* | 7-8 | 95 |
| | | Cereal | 5-6 | 55 |
| $\beta$-Amylase | Exo-enzyme; splits maltose from non-reducing end | Cereal | 5.5 | 55 |
| Glucoamylase | Exo-enzyme; splits glucose from non-reducing end, also splits $\alpha$-1,6-bonds at branch points but at slower rate | *Aspergillus oryzae* | 4-5 | 60 |
| | | *A. niger* | 4-5 | 60 |
| | | *Rhizopus oryzae* | 3-5 | 55 |
| Pullulanase | Splits $\alpha$-1,6-glucosidic linkages in pullulan and amylopectin | *Klebsiella aerogenes* | 5 | 50 |
| Glucose isomerase | Ketol isomerase, converts glucose to fructose (also xylose isomerase) | *Bacillus coagulans* | 8 | 60 |
| | | *Streptomyces albus* | 7 | 65 |

Source: Adapted from Peppler and Reed, 1987. Used with permission.

**Table 2.11.11. Use of Pectinases in Fruit and Vegetable Processing**

| Process | Enzyme | Objective |
|---|---|---|
| Clarification | Pectinases (PTE, PG, PE) | Clear juices, improved filtration |
| Depectinization | Pectinases, amylases | Juice concentrates, improved stability |
| Precipitation of pectin | Pectinesterase (PE) | Cider production |
| Digestion of mash | Pectinases, cullulases, hemicellulases | Improved juice yield and pressing ability |

**Table 2.11.11. (continued)**

| Process | Enzyme | Objective |
|---|---|---|
| Maceration | Pectin glycosidase (PG) | Stable vegetable pulp |
| Liquifaction | Pectinases, cellulases, hemicellulases, amylases | Entire fruit, pulp concentrates, dry products |

Source: Adapted from Uhlig, 1990. Used with permission.
Note: PTE = pectin transeliminase; PG = polygalacturonase; PE = pectin methylesterase

**Table 2.11.12. Enzymes in the Hydrolysis of Proteins**

| Substrate | Enzyme | % Hydrolysis | Application |
|---|---|---|---|
| Fish waste | Papain | 10-15 | Extraction of fish oil, feeds |
| Fish | Pancreatic proteinase | ca. 10 | Non-bitter-tasting hydrolysates, dietetics |
| Bone meat | Neutral bacteral protease | 2-5 | Soups |
| Whey protein | Nneutral bacterial protease, pancreatic proteinase | 2-8 | Improvement of foaming power |
| Casein | Pancreatic proteinase | 2-2.5 | Soluble, nonbitter casein peptides, improved foam stability |
| Casein | Pancreatic proteinase | 37 | Medical infusion solutions, microbiological nutrient media |
| Gelatin | Pancreatic proteinase, neutral bacterial proteinase | up to 25 | Cosmetics |

Source: Adapted from Uhlig, 1990. Used with permission.

**Table 2.11.13. Comparison of Enzyme Conversion in the Production of Beer, Whiskey, and Bread**

<table>
<tr><th>Criteria</th><th>Brewing</th><th>Distilled Beverages (Whiskey)</th><th>Baking (Lean Dough)</th></tr>
<tr><td>Substrate</td><td>Barley malt</td><td>Corn plus 10% malt</td><td>Flour plus 0.2% malt</td></tr>
<tr><td>Consistency</td><td>Thick mask</td><td>Thick mash</td><td>Dough</td></tr>
<tr><td>Stages</td><td><u>Stage 1</u><br>Enzyme conversion to solubilize starch and protein. Partial saccharification to fermentable sugars (50-65°C)<br>Extraction of soluble portion and boiling of resulting wort to inactivate enzymes<br><br><u>Stage 2</u><br>Yeast fermentation of all fermentable sugars (10-20°C). No further conversion by malt enzymes</td><td><u>Stage 1</u><br>Boil mash with 1-2% malt to solubilize starch. Lower temp. to 60°C and add 8% conversion malt for partial conversion to fermentable sugars<br><br><u>Stage 2</u><br>Yeast fermentation of preformed sugars and sugars formed by continuing enzyme action (30°C)</td><td><u>One Stage</u><br>Formation of fermentable sugars and yeast fermentation occur simultaneously<br>Temp. 30°C</td></tr>
<tr><td>Limiting factor in rate of yeast fermentation in <u>Stage 2</u></td><td>Yeast cell counts and yeast activity</td><td>Yeast cell counts and yeast activity in the beginning; enzyme conversion at the end of the fermentation</td><td>Yeast cell counts and enzyme activity</td></tr>
</table>

Source: Adapted from Peppler and Reed, 1987. Used with permission.

**Table 2.11.14. Enzymes in the Production of Wine**

| Process | Problems | Enzymes Used |
|---|---|---|
| Grapes | None | None |
| Mash | Pressability, e.g., type of grape | Pectinases |
| | *Botrytis* purified | β-glucanases |
| | Low free running juice | |
| Pressing | None | None |
| Must | Microbiological contamination | Pectinases |
| (heating) | Removal of mucous substance | Proteinases |
| Fermentation | Formation of foam | Proteinases |
| Young wine | Clarification | Proteinases |
| | Protein stabilization | Pectinases |

Source: Adapted from Uhlig, 1990. Used with permission.

## 2.12. Enzymes: Immobilization

**Table 2.12.1. Fundamentals of Enzyme Immobilization**

Background information
Classification of immobilized enzymes
Matrixes for enzyme immobilization
  Morphological classification
    Non-porous matrixes
    Porous matrixes
  Chemical classification
    Organic carriers
      Natural polymers
      Proteins
      Carbon materials
      Synthetic polymers
    Inorganic carriers
Methods of enzyme immobilization
  Crosslinking method
  Support-binding method
    Physical-adsorption method
    Ionic-binding method
    Metal-linking method
    Covalent-binding method
      Functional groups of enzymes for covalent binding
      Coupling reactions for covalent binding
  Entrapment method
    Gel-entrapment method
    Fiber-entrapment method
    Microencapsulation
      Phase-separation method
      Interfacial-polymerization method
      Liquid-drying method
      Liquid-surfactant-membrane method
  Immobilized-soluble-enzyme method
    Soluble enzymes without derivatization
    Derivatized soluble enzymes
Comparison of immobilization techniques

| | |
|---|---|
| Comparison of immobilization techniques | pH-activity profile |
| Effect of immobilization methods on the kinetics and properties of enzymes | Temperature-activity profile |
| Kinetic parameters | Optimum temperature |
| Michaelis constant | Activation energy |
| Maximum reaction velocity | Stability |
| | pH-stability |
| | Heat and storage stability |
| | Operational stability |

Source: Adapted from Kennedy and Cabral, 1987. Used with permission.

**Table 2.12.2. Advantages Of Immobilized Enzymes Over Soluble Enzymes**

1. Enzymes can be reused.
2. Processes can be operated continuously and can be readily controlled.
3. Products are easily separated.
4. Effluent problems and materials handling are minimized.
5. In some cases, enzyme properties (activity and stability) can be altered favorably by immobilization.

Source: Adapted from Kennedy and Cabral, 1987. Used with permission.

**Table 2.12.3. Classification of Immobilization Methods for Enzymes**

| Methods for Insoluble Enzymes | | Methods for Soluble Enzymes |
|---|---|---|
| Binding | Meteal binding | Ultrafiltration membranes |
| Cross-linking | Covalent binding | Hollow fiber devices |
| Carrier binding | Entrapping | |
| Physical adsorption | Gel entrapping | |
| Ionic binding | Fiber entrapping | |
| | Microencapsulation | |

Source: Adapted from Kennedy and Cabral, 1987. Used with permission.

**Table 2.12.4. Characteristics of An Enzyme Carrier**

| | | |
|---|---|---|
| Large surface area | Chemical, mechanical, and thermal stability | particle size |
| Permeability | High rigidity | Resistance to microbial attack |
| Hydrophilic character | Suitable shape and | Regenerability |
| Insolubility | | |

Source: Adapted from Kennedy and Cabral, 1987. Used with permission.

**Table 2.12.5. Comparison of Enzyme Carriers**

| Morphological Classification | Advantages | Disadvantages |
|---|---|---|
| Non-porous | Low diffusional limitations | Low surface area<br>Low enzyme loading<br>Use of fine particles:<br>Difficult to remove<br>Difficult to use continuously<br>High pressure drops<br>Limited flow rates |
| Porous | High surface area<br>High enzyme loading<br>Protection from external environment | High internal surfact area<br>High diffusional limitations<br>High cost (controlled pore)<br>High pressure drops (gels)<br>Low surface area available (broad pore) |

Source: Adapted from Kennedy and Cabral, 1987. Used with permission.

**Table 2.12.6. Chemical Classification of Enzyme Matrixes**

| Organic | | Inorganic |
|---|---|---|
| <u>Natural Polymers</u> | | <u>Minerals</u> |
| Polysaccharides | Cellulose | Attapulgite clays |
| | Starch | Bentonite |
| | Dextran | Kieselgur |
| | Agar and agarose | Pumice stone |
| | Alginate | |
| | Carrageenan | |
| | Chitin and chitosan | |
| Proteins | Collagen | |
| | Gelatin | |
| | Albumin | |
| | Silk | |
| Carbon materials | Many | |

| Synthetic polymers | | Fabricated materials |
|---|---|---|
| Polystyrene | Many | Non-porous glass |
| Polyacrylate types | Polyacrylates and polymethacrylates | Controlled pore glass |
| | Polyacrylamide | Controlled pore metal oxides |
| | Hydroxyalkyl methacrylates | Metals |
| | Glycidyl methacrylates | |
| | Maleic and anhydride polymers | |
| | Vinyl and allyl polymers | |
| | Polyamides | |

Source: Adapted from Kennedy and Cabral, 1987. Used with permission.

**Table 2.12.7. Adsorbents with Affinity Groups For Enzyme Immobilization**

| Matrix | Derivative | Matrix | Derivative |
|---|---|---|---|
| Agarose | Concanavalin-A- | Cellulose | Phenoxyacetyl- |
| | Lecithin- | | Palmitoyl- |
| | Heptyl- | | Tannin- |
| | Octyl- | | Tannin-TEAE aminohexyl- |
| | Decyl- | | |
| | Amminohexadecyl- | Silica | Octadecyl- |

Source: Adapted from Kennedy and Cabral, 1987. Used with permission.

**Table 2.12.8. Ion Exchange for Ionic Binding of Enzymes**

| Anion exchanger | | Cation exchanger |
|---|---|---|
| DEAE-cellulose | Amberlite$^{R}$ IRA 94 | CM-cellulose |
| AE-cellulose | Amberlite$^{R}$ IRA 910 | Cellulose phosphate |
| TEAE-cellulose | Amberlite$^{R}$ IRA 938 | CM-Sephadex$^{R}$ |
| DEAE-Sephadex$^{R}$ | | SP-Sephadex$^{R}$ |
| QAE-Sephadex$^{R}$ | | Dextran sulfate |
| DEAE-Bio-Gel$^{R}$ A | | Amberlite$^{R}$ IRC-50 |
| Amberlite$^{R}$ IRA 93 | | Amberlite$^{R}$ IRC-200 |

Source: Adapted from Kennedy and Cabral, 1987. Used with permission.

**Table 2.12.9. Comparison of the Attributes of Three Classes of Immobilization Techniques: Crosslinking, Physical Adsorption, Entrapping**

| Characteristic | Crosslinking | Physical Adsorption | Entrapping |
|---|---|---|---|
| Preparation | Intermediate | Simple | Difficult |
| Binding force | Strong | Weak | Intermediate |
| Enzyme activity | Low | Intermediate | Low |
| Regeneration of carrier | Impossible | Possible | Impossible |
| Cost of immobilization | Intermediate | Low | Intermediate |
| Stability | High | Low | High |
| General applicability | No | Yes | Yes |
| Protection of enzyme from microbial attack | Somewhat | No | Yes |

Source: Adapted from Kennedy and Cabral, 1987. Used with permission.

**Table 2.12.10. Comparison of the Attributes of Three Classes of Immobilization Techniques: Ionic Binding, Metal Binding, Covalent Binding**

| Characteristic | Ionic Binding | Metal Binding | Covalent Binding |
|---|---|---|---|
| Preparation | Simple | Simple | Difficult |
| Binding force | Intermediate | Intermediate | Strong |
| Enzyme activity | High | High | High |
| Regeneration of carrier | Possible | Possible | Rare |
| Cost of immobilization | Low | Intermediate | High |
| Stability | Intermediate | Intermediate | High |
| General applicability | Yes | Yes | No |
| Protection of enzyme from microbial attack | No | No | No |

Source: Adapted from Kennedy and Cabral, 1987. Used with permission.

**Table 2.12.11. Michaelis Constant of Immobilized Enzymes**

| Enzyme (Substrate) | Immobilization method | $K_m$ (mM) for Enzyme: Free (Immobilized) |
|---|---|---|
| Asparaginase (L-asparagine) | Entrapping (polyacrylamide gel) | $1.8 \times 10^{-2}$ (3.3) |
| Urease (urea) | Entrapping (liquid membrane capsules) | 3.4 ($1.8 \times 10^{2}$) |
| Ficin (BAEE) | Peptide binding (CM-cellulose azide) | 20 (2.0) |
| Aminoacylase (Ac-DL-methionine) | Ionic binding (DEAE-cellulose) | 5.7 (3.5) |
| | Alkylation (iodoacetyl cellulose) | 5.7 (6.7) |
| Trypsin (TAME) | Ionic binding (DEAE-sephadex) | 0.12 (2.5) |
| | Peptide binding (CNBr-sepharose) | 0.12 (6.2) |

Source:

**Table 2.12.12. Optimum pH of Immobilized Enzymes**

| Enzyme (Substrate) | Immobilization method | Optimum pH for Enzyme: Free (Immbilized) |
|---|---|---|
| Trypsin (TAME) | Ionic binding (DEAE-Sephadex) | 8.5 (7.0) |
| Trypsin (TAME) | Peptide binding (CNBr-Sepharose) | 8.5 (9.5) |
| Invertase (Sucrose) | Ionic binding (DEAE-Cellulose) | 5.4 (3.4) |
| | Entrapping (Polyacrylamide gel) | 5.0 (5.5) |
| | Physical adsorption(Bentonite) | 5.2 (6.3) |
| NADH dehydrogenase (NADH) | Crosslinking (Glutaraldehyde) | 6.0 (5.5) |

Source: Adapted from Kennedy and Cabral, 1987. Used with permission.

**Table 2.12.13. Activation Energies of Immobilized Enzymes**

| Enzyme (Substrate) | Immobilization Method | Enzyme Activation Energy (kJ/mol): Free (Imobilized) |
|---|---|---|
| Aminoacylase (Ac-DL-methionine) | Entrapping (polyacrylamide gel) | 28.140 (23.940) |
| Glucoamylase (maltodextrin) | Schiff's base formation (titanium activated porous silica) | 57.120 (47.460) |
| Glucoamylase (soluble starch) | (titanium activated celite) | 44.520 (37.800) |
| Glucoamylase (soluble starch) | (titanium activated controlled pore glass) | 44.520 (35.280) |

Source: Adapted from Kennedy and Cabral, 1987. Used with permission.

## 2.13. Immobilized Cell Systems

**Table 2.13.1. Fundamentals of Immobilized Cell Systems**

- Background information
- Whole cell immobilization versus enzyme immobilization
- Immobilized cell fermentations versus conventional fermentations
- Cell supports
  - Organic supports for cell immobilization
  - Inorganic supports for cell immobilization
  - General support criteria
- Immobilization techniques for cells
  - Immobilization of cells without a support
  - Cross-linking
  - Copolymerization
  - Covalent binding
- Adsorption
  - Organic adsorption carriers
    - Ion-exchange resin adsorption carriers
    - Cellulose adsorption carriers
    - Lectin-cell adsorption
  - Inorganic adsorption carriers
- Entrapment
  - Synthetic polymers
    - Polyacrylamide
    - Methacrylates
    - Photo-crosslinkable resin prepolymer
    - Polyurethane
    - Epoxy resin
  - Natural polymers
    - Thermal gelation

            - Ionotropic gelation
            - Precipitation techniques
        - Membrane reactors
    - Liquid membrane encapsulation
- Interactions between cells and carriers
    - Carrier surface area and cell loading
    - Microenvironmental effects
        - Charge of the matrix
        - Hydrophobicity of the matrix
        - Porosity of the matrix
        - Chelating and adsorbing properties of the matrix
        - Protective properties of the matrix
        - Proximity effects
- Reactor types for immobilized cells
    - Batch reactors
    - Continuous reactors
- Activity of immobilized cells
    - Mass transport
    - Effectiveness factor
    - Catalytic stability
- Applications of immobilized cells
    - Production of biochemicals
    - Analysis
- Immobilized plant cells
- Immobilized animal cells

Source: Adapted from Brodelius and Vandamme, 1987. Used with permission.

**Table 2.13.2. Complexity of Immobilized Cell Systems in Terms of the Nature of the Catalytic Reaction**

1. Reactions involving single enzymes (bioconversions)
2. Reactions involving multi-enzyme systems, without specific cofactor requirement
3. Cofactor requiring systems (single or multi-step reactions)
4. Reactions involving a complete metabolic pathway (fermentations yielding primary or secondary metabolites, i.e., *de novo* synthesis)

Source: Adapted from Brodelius and Vandamme, 1987. Used with permission.

**Table 2.13.3. Methods to Reduce Limitations of Using Immobilized Whole Cells**

1. Use of surface bound cells; the overall capacity of a reactor will, however, be decreased considerably.
2. Decrease of particle size; results in increased effective transfer area but the optimal particle size is highly dependent on reactor design and performance.

**Table 2.13.3. (continued)**

3. Increase of substrate concentration in the bulk phase; may result in problems with product isolation, substrate recirculation, and increased viscosity of the substrate solution.
4. Increase of the pore size of the carrier; results in an increased diffusion rate; can readily be achieved with large cells.
5. Decrease of cell loading; the overall capacity of a reactor will be decreased.

Source: Adapted from Brodelius and Vandamme, 1987. Used with permission.

**Table 2.13.4. Organic Supports for Cell Immobilization**

| Polysaccharides | Synthetic Polymers |
|---|---|
| Cellulose | Polyacrylamide |
| Agar/agarose | Methacrylate |
| Chitosan | Polyurethane |
| Dextran | Epoxy resin |
| Carrageenan | Polystyrene |
| Alginate | Polyester |
| Pectate | Polypropylene |
| Xanthan gum | Polyphenylene oxide |
| Proteins | Polyvinyl alcohol |
| Collagen | Polyvinyl chloride |
| Gelatin | |
| Albumin | |
| Fibrin | |

Source: Adapted from Brodelius and Vandamme, 1987. Used with permission.

**Table 2.13.5. Inorganic Supports for Cell Immobilization**

Ungrafted: alumina, ceramics, ferromagnet, glass, magnesia, sand, silica, titania, zirconia

Grafted: supports grafted with various coupling agents

Source: Adapted from Brodelius and Vandamme, 1987. Used with permission.

**Table 2.13.6. Some Examples of Microbial Cells Covalently Linked to Various Supports**

| Cell Species | Support | Coupling Agent | Product |
|---|---|---|---|
| *Acetobacter* sp. | Metal hydroxides | No data | Acetic acid |
| *Aspergillus niger* | Glycidyl methacrylate | Glutaraldehyde | Gluconic acid |
| *Micrococcus luteus* | CM-cellulose | Carbodiimide | Urocanic acid |
| *Saccharomyces carlsbergensis* | Aminopropyl silica | Glutaraldehyde | Ethanol |
| *S. cerevisiae* | Hydroxyalkyl methacrylate | 1,6-Diaminohexane + glutaraldehyde | Killer toxin |
| *S. cerevisiae* | Cellulose | Cyanuric chloride | Ethanol |
| *Zygosaccharomyces lactis* | Hydroxyalkyl methacrylate | 1,6-Diaminohexane + glutaraldehyde | β-Galactosidase |

Source: Adapted from Brodelius and Vandamme, 1987. Used with permission.

**Table 2.13.7. Some Examples of Microbial Cells Adsorbed to Various Supports**

| Cell Species | Support | Enzyme | Product |
|---|---|---|---|
| *Azotobacter aceti* | Ceramic | Multi-enzyme | Acetic acid |
| *Aspergillus niger* | Glass | Multi-enzyme | Citric acid |
| *Aspergillus* sp. | ECTEOLA-cellulose | Invertase | Glucose + fructose |
| *Escherichia coli* | Dowex 1 | No data | No data |
| *Kluyveromyces lactis* | Polyphenylene oxide | β-Galactosidase | Glucose + galactose |
| *Nocardia erythropolis* | DEAE-cellulose | Cholesterol oxidase | Cholest-4-ene-3-one |
| *Pichia polymorpha* | Wood chips | Inulinase | Fructose |
| *Pseudomonas aeruginosa* | Polypropylene | No data | Denitrified water |
| *P. aeruginosa* | Polyvinyl chloride | No data | Denitrified water |

**Table 2.13.7. (continued)**

| Cell Species | Support | Enzyme | Product |
|---|---|---|---|
| *Pseudomonas* sp. | Anthracite | Multi-enzyme | Phenol degradation |
| *Saccharomyces cerevisiae* | Amberlite | Multi-enzyme | Ethanol |
| *S. cerevisiae* | Brick | Multi-enzyme | Ethanol |
| *S. cerevisiae* | ECTEOLA-cellulose | Multi-enzyme | Ethanol |
| *S. cerevisiae* | Wood chips | Multi-enzyme | Ethanol |

Source: Adapted from Brodelius and Vandamme, 1987. Used with permission.

**Table 2.13.8. Methods Available for the Immobilization of Whole Cells by Entrapment in Gels**

| Gel Formation Mechanism | Polymer |
|---|---|
| Polymerization | Polyacrylamide polymethacrylate |
| Cross-linking | Various prepolymers, proteins |
| Polycondensation | Polyurethane, epoxy resin |
| Thermal gelation | Collagen, gelatin, agar/agarose, $\chi$-carrageenan |
| Inotropic gelation | Alginate, chitosan |
| Precipitation | Cellulose, cellulose triacetate |

Source: Adapted from Brodelius and Vandamme, 1987. Used with permission.

**Table 2.13.9. Some Examples of Microbial Cells Entrapped in Polyacrylamide**

| Cell Species | Enzyme | Substrate | Product |
|---|---|---|---|
| *Acetobacter suboxydans* | Multi-enzyme | Glycerol | Dihydroxy acetone |
| *A. xylinum* | Multi-enzyme | Glycerol | Dihydroxy acetone |
| *Achromobacter aceris* | NAD-kinase | $NAD^+$ + ATP | $NADP^+$ + ADP |

| | | | |
|---|---|---|---|
| *A. butyri* | Polyphosphate glucokinase | Glucose + $(P_i)_n$ | Glucose-6-phosphate + $(P_i)_{n-1}$ |
| *A. liquidum* | L-Histidine ammonia lyase | L-Histidine | Urocanic acid |
| *Arthrobacter globiformis* | 3-Ketosteroid-$\Delta^1$-dehydrogenase | Hydrocortisone | Prednisolone |
| *A. globiformis* | 20 $\beta$-Dehydrogenase | 20-Oxohydrocortisone | Hydrocortisone |
| *A. simplex* | 3-Ketosteroid-$\Delta^1$-dehydrogenase | Hydrocortisone | Prednisolone |
| *Arthrobacter* sp. | Multi-enzyme | Phenanthrene benzpyrene | Degradation products |
| *Aspergillus niger* | Multi-enzyme | Glucose | Gluconic acid |
| *A. terreus* | Multi-enzyme | Glucose | Itaconic acid |
| *Bacillus megaterium* | 20 $\alpha$- and 20 $\beta$-Hydroxysteriod-dehydrogenase | Cortisol | 20-Hydroxy derivatives |
| *B. subtilis* | Multi-enzyme | Malt + yeast extracts | $\alpha$-Amylase |
| *Bacillus* sp | Multi-enzyme | Peptone | Bacitracin |
| *Bacillus* sp | Multi-enzyme | Hydantoins | *N*-Carbamoyl-D-amino acids |
| *Brevibacterium ammoniagenes* | Multi-enzyme | Panthotenate + cysteine + ATP | Coenzyme A |
| *B. ammoniagenes* | Polyphosphate NAD kinase | $NAD^+$ + $(P_i)_n$ | $NADP^+$ + $(P_i)_{n-1}$ |
| *B. ammoniagenes* | Fumarase | Fumarate | L-Malic acid |
| *Caldariella acidophila* | $\beta$-Galactosidase | Lactose | Glucose + galactose |
| *Candida lipolytica* | Multi-enzyme | Medium | Citric acid |
| *C. rugosa* | Fumarase | Fumarate | L-Malic acid |
| *C. tropicalis* | Multi-enzyme | Phenol + oxygen | Carbon dioxide + water |

**Table 2.13.9. (continued)**

| Cell Species | Enzyme | Substrate | Product |
|---|---|---|---|
| *Chlorella regularis* | No data | Sea water | Enriched uranium |
| *Clostridium butyricum* | Multi-enzyme | Alcohol factory wastes | Hydrogen |
| *Coryne-bacterium dismutans* | Multi-enzyme | $^{14}C$-Glucose | $^{14}C$-Alanine |
| *C. glutamicum* | Multi-enzyme | Glucose | Glutamic acid |
| *Curvularia lunata* | 11 $\beta$-Hydroxylase | Reichstein's compound S | Cortisol |
| *Escherichia alcalescens* | $\beta$-Galactosidase | Lactose | Glucose + galactose |
| *E. alcalescens* | Aspartase | Ammonium fumarate | L-Aspartic acid |
| *E. coli* | Penicillin amidase | Penicillin G | 6-APA |
| *E. coli* | $\beta$-Lactamase | $\beta$-Lactam antibiotics | Hydrolyzed lactam ring |
| *E. coli* | Aspartase | Ammonium fumarate | L-Aspartic acid |
| *E. coli* | $\beta$-Galactosidase | Lactose | Glucose + galactose |
| *E. coli* | Tryptophan synthase | Indole + serine | L-Tryptophan |
| *E. coli* | Multi-enzyme | Indole + pyruvate + $NH_3$ | L-Tryptophan |
| *E. coli* | Multi-enzyme | L-Glutamate + L-cysteine + glycine | Glutathione |
| *Gluconobacter melanogenes* | Multi-enzyme | L-Sorbose | L-Sorbosone |
| *G. melanogenes + Pseudo-monas syrina* | Multi-enzyme | L-Sorbose | 2-Keto-L-gulonic acid |
| *G. oxydans* | Multi-enzyme | Glycerol | Dihydroxy-acetone |
| *Hansenula polymorpha* | Methanol oxidase | Methanol | Formalde-hyde |

| | | | |
|---|---|---|---|
| *Kluyvera citrophila* | Multi-enzyme | Medium | Penicillin acylase |
| *Kluyveromyces lactis* | β-Galactosidase | Lactose | Glucose + galactose |
| *Lactobacillus bulgaricus* | β-Galactosidase | Lactose | Glucose + galactose |
| *L. casei* | Multi-enzyme | L-Malic acid | L-Lactic acid |
| *Mycobacterium phlei* | Multi-enzyme | Cholesterol derivatives | Androstedione derivatives |
| *Nocardia corallina* | Multi-enzyme | Propylene | Propylene oxide |
| *Nocardia* sp. | Multi-enzyme | Androst-4-ene-dione | Methylperhydroindanone propionic acid |
| *Penicillium chrysogenum* | Multi-enzyme | Glucose + $NH_3$ + phenylacetate | Penicillin G |
| *Propionibacterium shermanii* | Multi-enzyme | Glucose | Nucleosides and nucleotides |
| *Propionibacterium* sp. | Multi-enzyme | Glucose/starch | Acetic and propionic acid |
| *Proteus rettgeri* | Penicillin amidase | Penicillin G | 6-APA |
| *Pseudomonas putida* | Multi-enzyme | Benzene | Degradation products |
| *P. putida* | L-Arginine deiminase | L-Arginine | L-Citrulline |
| *P. testosteroni* | 11 β-Hydroxylase | Reichsteins's compound S | Cortisol |
| *Rhizopus nigricans* | 11 α-Hydroxylase | Progesterone | 11 α-Hydroxyprogesterone |
| *Saccharomyces cerevisiae* | 17 β-Hydroxysteroid dehydrogenase | 3-Methoxy-Δ-1,3,5,10,9,11,8,14-secoestratetraene-14,17-dione | 3-Methoxy-Δ-1,3,5,10,9,11,8,14-secoestratetraene-17 β-ol-14-one |

**Table 2.13.9. (continued)**

| Cell Species | Enzyme | Substrate | Product |
|---|---|---|---|
| *S. cerevisiae* | Glutathione synthetase | L-Glutamic acid + L-cysteine + glycine | Glutathione |
| *S. cerevisiae* | Alcohol dehydrogenase | Ethanol + $NAD^+$ | Acetaldehyde + NADH |
| *S. cerevisiae* | Multi-enzyme | Glucose | Ethanol |
| *S. cerevisiae* | Invertase | Sucrose | Glucose + fructose |
| *S. cerevisiae* | Multi-enzyme | Sucrose | Glucose + gluconic acid |
| *S. cerevisiae* | Catalase | Hydrogen peroxide | Water + oxygen |
| *Streptococcus faecalis* | Arginine deaminase | L-Arginine | Ornithine/ putrescine |
| *S. lactis* | Multi-enzyme | Medium | Nisin |
| *Streptomyces clavuligerus* | Multi-enzyme | Glucose | Cephalosporins |
| *S. fradiae* | Multi-enzyme | Starch, meat, and yeast extracts | Protease |
| *S. viridochromogenes* | No data | Sea water | Enriched uranium |
| *Tieghemella orchedis* | 11 $\alpha$- and 11 $\beta$-Hydroxylase | Steroids | 11-Hydroxy steroids |
| *Umbilicaria pustulata* | Decarboxylase/ esterase | Orcellinic acid/depside | Orcinol + $CO_2$ |
| *Xanthomonas citri* | Multi-enzyme | Medium | Cephalexin |

Source: Adapted from Brodelius and Vandamme, 1987. Used with permission.

**Table 2.13.10. Some Examples o Microbial Cells Entrapped in Photo-crosslinking Resin Prepolymers**

| Cell Species | Enzyme | Substrate | Product |
|---|---|---|---|
| *Arthrobacter simplex* | 3-Ketosteroid-Δ-dehydrogenase | Hydrocortisone | Prednisolone |
| *Brevibacterium ammonia genes* | Multi-enzyme | Panthotenate + cysteine + ATP | Coenzyme A |
| *Coryne bacterium* sp. | 9 $\alpha$-Hydroxylase | 4-Androstene-3, 17-dione | 9 $\alpha$-Hydroxy-4-androstene-3,17-dione |
| *Curvularia lunata* | 11 $\beta$-Hydroxylase | Reichstein's compound S | Cortisol |
| *Enterobacter aerogenes* | Trans-glycosylase | Uracil arabinoside + adenine | Adenine arabinoside |
| *Hansenula jadinii* | Multi-enzyme | CMP + choline + glucose + $P_i$ | CDP-choline |
| *Nocardia rhodocrous* | Steroid-$\Delta^1$-dehydrogenase | 4-Androstene-3, 17-dione | 1,4-Androstadiene-3,17-dione |
| *N. rhodocrous* | Steroid-$\Delta^1$-dehydrogenase | 3 $\beta$-Hydroxy-$\Delta^5$-steroids | 3-Keto-$\Delta^4$-steroids |
| *Propioni bacterium* sp. | Multi-enzyme | Medium | Vitamin $B_{12}$ |
| *Rhizopus stolonifer* | 11 $\alpha$-Hydroxylase | Progesterone | 11 $\alpha$-Hydroxyprogesterone |
| *Rhodotorula minuta* | Stereospecific hydroxylase | ($\pm$)-Menthyl succinate | (−)-Menthol |
| *Saccharomyces cerevisiae* | Multi-enzyme | Adenosine + glucose + $P_i$ | ATP |
| *Saccharomyces* sp. | Multi-enzyme | Glucose | Ethanol |

Source: Adapted from Brodelius and Vandamme, 1987. Used with permission.

**Table 2.13.11. Some Examples of Microbial Cells Entrapped in Urethane Polymers**

| Cell Species | Enzyme | Substrate | Product |
|---|---|---|---|
| *Arthrobacter simplex* | 3-Ketosteroid-$\Delta^1$-dehydrogenase | Hydrocortisone | Prednisolone |
| *Aspergillus phoenicis* | 11 $\beta$-Hydroxylase | Progesterone | 11 $\beta$-Hydroxy-progesterone |
| *Escherichia coli* | Penicillin G acylase | Penicillin G | 6-APA |
| *Enterobacter aerogenes* | Transglycosylase | Uracil arabinoside | Adenine arabinoside |
| *Nocardia rhodocrous* | Steroid-$\Delta^1$-dehydrogenase | 3 $\beta$-Hydroxy-$\Delta^5$-steroids | 3-Keto-$\Delta^4$-steroids |
| *N. rhodocrous* | Steroid-$\Delta^1$-dehydrogenase | Testosterone | $\Delta^1$-Dehydro-testosterone |
| *Porphyridium cruentam* | Multi-enzyme | Light + $CO_2$ | Pollysac-charides |
| *Propionibacterium* sp. | Multi-enzyme | Medium | Vitamin $B_{12}$ |
| *Rhizopus stolonifer* | 11 $\alpha$-Hydroxylase | Progesterone | 11 $\alpha$-Hydroxyprogesterone |
| *Rhodotorula minuta* | Stereospecific hydroxylase | ($\pm$)-Menthyl succinate | (−)-Menthol |

Source: Adapted from Brodelius and Vandamme, 1987. Used with permission.

**Table 2.13.12. Some Examples of Microbial Cells Entrapped in Collagen**

| Cell Species | Enzyme | Substrate | Product |
|---|---|---|---|
| *Acetobacter* sp. | Multi-enzyme | Ethanol | Acetic acid |
| *Anacystis nidulans* | Multi-enzyme | Water | Oxygen |
| *A. nidulans* | Multi-enzyme | Nitrite | Ammonia |
| *Arthrobacter simplex* | 3-Ketosteroid-$\Delta^1$-dehydrogenase | Hydrocortisone | Prednisolone |
| *Aspergillus niger* | Multi-enzyme | Sucrose | Citric acid |

| | | | |
|---|---|---|---|
| *Brevibacterium flavum* | Multi-enzyme | Glucose | Glutamic acid |
| *Clostridium butyricum* | Multi-enzyme | Alcohol factory wastes | Hydrogen |
| *Corynebacterium lilium* | Multi-enzyme | Glucose | Glutamic acid |
| *Erwinia herbicola* | β-Tyrosinase | Phenol + pyruvate + $NH_3$ | L-Tyrosine |
| *Klebsiella pneumonia* | Multi-enzyme | Glucose + nitrogen | Ammonia |
| *Pseudomonas aeruginosa* | No data | Waste water | Plutonium-free water |
| *Saccharomyces cerevisiae* | Invertase | Sucrose | Glucose + fructose |
| *Serratia marcescens* | Multi-enzyme | Glucose | 2-Keto gluconic acid |
| *Streptomyces griseus* | Multi-enzyme | Glucose | Candicidin |

Source: Adapted from Brodelius and Vandamme, 1987. Used with permission.

**Table 2.13.13. Some Examples of Microbial Cells Entrapped in Gelatin**

| Cell Species | Enzyme | Substrate | Product |
|---|---|---|---|
| *Actinoplanes missouriensis* | Glucose isomerase | Glucose | Fructose |
| *Arthrobacter* X-4 | Xanthine oxidase | 1-Methylxanthine | 1-Methyluric acid |
| *Aspergillus phoenicis* | 11β-Hydroxylase | Progesterone | 11β-Hydroxyprogesterone |
| *Saccharomyces cerevisiae* | Invertase | Sucrose | Glucose + fructose |
| *S. uvarum* | Multi-enzyme | Molasses | Ethanol |
| *Streptomyces aureofaciens* | Multi-enzyme | Daunomycinone | 13-Dihydroxydaunomycinone |
| *Streptomyces* sp. | Glucose isomerase | Glucose | Fructose |

Source: Adapted from Brodelius and Vandamme, 1987. Used with permission.

**Table 2.13.14. Some Examples of Microbial Cells Entrapped in Agar or Agarose**

| Cell Species | Enzyme | Substrate | Product |
|---|---|---|---|
| *Arthrobacter globiformis* | 3-Ketosteroid-$\Delta^1$-dehydrogenase | Hydrocortisone | Prednisolone |
| *Azotobacter chroococcum* | Nitrogenase + multi-enzyme | Medium | Nitrogen compounds |
| *Bacillus subtilis* | Multi-enzyme | Medium | Proinsulin |
| *Chlorella vulgaris* | Multi-enzyme | Carbon dioxide + light | Oxygen |
| *Clostridium butyricum* | Multi-enzyme | Alcohol factory wastes | Hydrogen |
| *C. butyricum* | Multi-enzyme | Glucose | Hydrogen |
| *Escherichia coli* | $\beta$-Galactosidase | Lactose | Glucose + galactose |
| *E. coli* | Thymidine phosphorylase | Thymidine | Thymine |
| *E. coli* | Penicillin acylase | Phenylacety-ADCA | 7-ADCA |
| *E. coli* | Aspartase | Ammonium fumarate | L-Aspartic acid |
| *Leuconostoc mesenteroides* | Multi-enzyme | L-Malate | L-Lactic acid |
| *Prototheca zopfii* | No data | Kepone | Kepone-free effluent |
| *Propionibacterium* sp. | Multi-enzyme | Medium | Vitamin $B_{12}$ |
| *Providencia* sp. PCM 1298 | L-Amino acid oxidase | L-Amino acids | 2-Oxo acids |
| *Pseudomonas putida* | Multi-enzyme | Caffeine | Decaffeinated solution |
| *Rhizopus nigricans* | 11 $\alpha$-Hydroxylase | Progesterone | 11 $\alpha$-Hydroxy-progesterone |
| *Rhodospirillum rubrum* | Multi-enzyme | Malate | Hydrogen |
| *Saccharomyces pastorianus* | Invertase | Sucrose | Glucose + fructose |

| | | | |
|---|---|---|---|
| *Schizosaccharo myces pombe* | Multi-enzyme | L-Malic acid | L-Lactic acid |
| *Streptomyces* T59-235 | Multi-enzyme | Medium | Tylosin |
| *S. tendae* | Multi-enzyme | Medium | Nikkomycin |
| *S. virido-chromogenes* | No data | Sea water | Enriched uranium |
| *Thiobacillus denitrificans* | Multi-enzyme | Nitrate | Nitrite |

Source: Adapted from Brodelius and Vandamme, 1987. Used with permission.

**Table 2.13.15. Some Examples of Microbial Cells Entrapped in Carrageenan**

| Cell Species | Enzyme | Substrate | Product |
|---|---|---|---|
| *Acetobacter suboxydans* | Multi-enzyme | Medium | L-Sorbose |
| *A. xylinum* | Multi-enzyme | Glycerol | Dihydroxy acetone |
| *Achromobacter liquidum* | L-Histidine ammonia lyase | L-Histidine | Urocanic acid |
| *Aspergillus oryzae* | Aminoacylase | *N*-Acyl-D,L-amino acids | L-Amino acids |
| *A. phoenicis* | 11 $\beta$-Hydroxylase | Progesterone | 11 $\beta$-Hydroxy-progester-one |
| *Bacillus amylolique-faciens* | Multi-enzyme | Medium | $\alpha$-Amylase |
| *B. subtilis* | Multi-enzyme | Glucose | Arginine |
| *Brevibacter-ium flavum* | Fumarase | Fumarate | L-Malic acid |
| *B. fuscum* | Multi-enzyme | Dehydrocholic acid | 12-Ketoch-enodeoxy-cholic acid |
| *Candida tropicalis* | Multi-enzyme | $n$-Alkanes/$n$-alcohols/$n$-Monoacids/ $\alpha, \omega$-diols | $\alpha, \omega$-Dodec-ane- + $\alpha, \omega$-tridecane-dioic acids |

**Table 2.13.15. (continued)**

| Cell Species | Enzyme | Substrate | Product |
|---|---|---|---|
| *Corynebacterium glutamicum* | Multi-enzyme | Glucose | L-Glutamate |
| *Enterobacter aerogenes* | Multi-enzyme | Glucose | 2,3-Butanediol |
| *Escherichia coli* | Glutathione synthase + acetate kinase | L-Glutamic acid + L-cysteine + glycine + acetyl-phosphate | Glutathione |
| *E. coli* | Aspartase | Fumarate | L-Aspartic acid |
| *E. coli* | Penicillin amidase | Penicillin | 6-APA |
| *Gluconobacter suboxydans* | D-Arabitol dehydrogenase | D-Arabitol | D-Xylulose |
| *Myxococcus xanthus* | Multi-enzyme | Medium | Hydrolases |
| *Penicillium urticae* | Multi-enzyme | Medium | Patulin |
| *Propionibacterium* sp. | Multi-enzyme | Medium | Vitamin $B_{12}$ |
| *Pseudomonas dacunhae* | L-Aspartate decarboxylase | L-Aspartic acid | L-Alanine |
| *P. putidum* | L-Arginine deiminase | L-Arginine | L-Citrulline |
| *Saccharomyces bayanus* | Multi-enzyme | Glucose | Ethanol |
| *S. cerevisiae* | Multi-enzyme | Glucose | Ethanol |
| *S. cerevisiae* | Multi-enzyme | Glucose | Glycerol |
| *S. cerevisiae* | Glutathione synthase | L-Glutamic acid + L-cysteine + glycine | Glutathione |
| *Sarcina lutea* | Urease | Urea | $NH_3 + CO_2$ |
| *Serratia marcescens* | Multi-enzyme | Medium | L-Isoleucine |
| *S. marcescens* | Multi-enzyme | Medium | L-Arginine |
| *Streptomyces griseus* | Glucose isomerase | Glucose | Fructose |

| | | | |
|---|---|---|---|
| *S. phaeo-chromo-genes* | Glucose isomerase | Glucose | Fructose |
| *Streptomyces* T52-235 | Multi-enzyme | Medium | Tylosin |
| *S. tendae* | Multi-enzyme | Medium | Nikkomycin |
| *Trichoderma reesei* | Multi-enzyme | Medium | Cellulase |
| *Xanthomonas citri* | Multi-enzyme | Medium | Cephalexin |
| *Zymomonas mobilis* | Multi-enzyme | Glucose | Ethanol |

Source: Adapted from Brodelius and Vandamme, 1987. Used with permission.

**Table 2.13.16. Some Examples of Microbial Cells Entrapped in Alginate**

| Cell Species | Enzyme | Substrate | Product |
|---|---|---|---|
| *Anabaena sp.* | Multi-enzyme | Light + air + water | Ammonia |
| *Arthrobacter simplex* | Steroid-$\Delta^1$-hydrogenase | Cortisol | Prednisolone |
| *Aspergillus phoenicus* | 11 $\beta$-Hydroxylase | Progesterone | 11 $\beta$-Hydroxy-progester-one |
| *Chlorella pyrenoides* | Multi-enzyme | Carbon dioxide + light | Oxygen |
| *Claviceps purpurea* | Multi-enzyme | Medium | Claviceps alkaloids |
| *Clostridium acetobuty licum* | Multi-enzyme | Glucose | Acetone/ butanol |
| *C. beijerinckii* | Multi-enzyme | Glucose | Isopropyl alcohol/ butanol/ ethanol |
| *Coriolus decolorize* | Multi-enzyme | Waste water | Decolorized water |
| *Curvularia lunata* | 11 $\beta$-Hydroxylase | Reichstein's compound S | Cortisone |
| *Fusarium lini* | Multi-enzyme | D-Xylose | Ethanol |

**Table 2.13.16. (continued)**

| Cell Species | Enzyme | Substrate | Product |
|---|---|---|---|
| *Gluconobacter oxydans* | Multi-enzyme | Glycerol | Dihydroxy acetone |
| *G. oxydans* | Glucose oxidase | Glucose | Gluconic acid |
| *G. suboxydans* | Glucose oxidase | Glucose | Gluconic acid |
| *Kluyveromyces fragilis* | Multi-enzyme | Sugars | Ethanol |
| *K. marxianus* | Multi-enzyme | Sugars | Ethanol |
| *Lactobacillus bulgaricus* | Multi-enzyme | Sugars | Lactic acid |
| *L. delbrueckii* | Multi-enzyme | Sugars | Lactic acid |
| *L. vaccino-stereus* | Multi-enzyme | Xylose | Lactic acid |
| *Leuconostoc mesenteroides* | Multi-enzyme | L-Malate | L-Lactate |
| *L. oenos* | Multi-enzyme | L-Malate (wine) | L-Lactate |
| *Methanosar-cina barkeri* | Multi-enzyme | Methanol | Methane |
| *Mucor sp.* | Multi-enzyme | D-Xylose | Ethanol |
| *Mycobacterium phlei* | Multi-enzyme | Sterols | $C_{19}$-Steroids |
| *M. Pyl* | Monooxygenase | Propylene/ ethylene | Propylene oxide/ethyl-ene oxide |
| *Nitrosomonas europaea* | Multi-enzyme | $[NH_4]^+$ | $[NO_2]^-$ |
| *Nocardia erythropolis* | Multi-enzyme | Sterols | $C_{22}$-Steroids |
| *Pachysolen tannophilus* | Multi-enzyme | D-Xylose | Ethanol |
| *Penicillium chrysogenum* | Multi-enzyme | Glucose | Penicillin G |
| *Photobacterium phosphoreum* | Multi-enzyme | Medium | Light |
| *Propionibac-terium* sp. | Multi-enzyme | Medium | Vitamin $B_{12}$ |
| *Providencia* sp. | L-Amino acid oxidase | L-Amino acids | 2-Oxoacids |
| *Pseudomonas dentrificans* | Multi-enzyme | Nitrate | Denitrified water |

| | | | |
|---|---|---|---|
| *Pseudomonas* sp.) | Phenylalanine | L-Phenylalanine | L-Tyrosine |
| *Rhizopus stolonifer*) | 11 $\alpha$-Hydroxylase | Progesterone | 11 $\alpha$-Hydroxyprogesterone |
| *Rhodopseudomonas capsulata* | Multi-enzyme | Malate | Hydrogen |
| *Saccharomyces cerevisiae* | Multi-enzyme | Glucose | Ethanol |
| *S. cerevisiae* | Multi-enzyme | D-Xylose | Ethanol |
| *Schizosaccharomyces pombe* | Multi-enzyme | L-Malate | L-Lactate |
| *Streptomyces* T54-235 | Multi-enzyme | Medium | Tylosin |
| *S. tandae* | Multi-enzyme | Medium | Nikkomycin |
| *S. viridochromogenes* | No data | Sea water | Enriched uranium |
| *Trichoderma* sp. | $\beta$-Glucosidase | Cellobiose/ salicin | Glucose |
| *Trigonopsis variabilis* | D-Amino acid oxidase | D-Amino acids | 2-Oxo acids |
| *Zymomonas mobilis* | Multi-enzyme | Glucose | Ethanol |

Adapted from Brodelius and Vandamme, 1987. Used with permission.

**Table 2.13.17. Some Examples of Microbial Cells Entrapped In Membrane Reactors**

| Cell Species | Enzyme | Substrate | Product |
|---|---|---|---|
| *Escherichia coli* | Multi-enzyme | Medium | $\beta$-Lactamase |
| *Lactobacillus delbrueckii* | Multi-enzyme | Glucose | Lactic acid |
| *Pseudomonas denithrificans* | Multi-enzyme | Nitrate | Denitrified effluent |
| *P. fluorescens* | L-Histidine ammonia lyase | L-Histidine | Urocanic acid |
| *Rhodotorula mucilaginosa* | Multi-enzyme | Benzaldehyde | Benzyl alcohol |

**Table 2.13.17. (continued)**

| Cell Species | Enzyme | Substrate | Product |
|---|---|---|---|
| *Saccharomyces cerevisiae* | Multi-enzyme | Glucose | Ethanol |

Source: Adapted from Brodelius and Vandamme, 1987. Used with permission.

**Table 2.13.18. Some Examples of Immobilized Cell Systems Activated *in situ***

| Cell Species Matrix | Reaction, Application | Incubation Time | Fold Increase In Activity or Cell Number |
|---|---|---|---|
| *Arthrobacter globiformis* Polyacrylamide | Hydrocortisone transformation | 24 h | 5.7 |
| *A. simplex* Polyacrylamide | Cortisol to prednisolone | 5 x 48 h | 6.5 |
| *A. simplex* Alginate | Cortisol to prednisolone | 3 x 24 h | 20 |
| *Azotobacter chroococcum* Agar | Nitrogen fixation | 4 days | 2 |
| *Bacillus subtilis* Polyacrylamide | Production of $\alpha$-amylase | 7 x 5 h | 8.5 |
| *Bacillus* sp. Polyacrylamide | Production of bacitracin | 7 x 5 h | 4.5 |
| *Candida tropicalis* Polymethacrylamide | Phenol degradation | 13 h | 14 |
| *Curvularia lunata* Polyacrylamide | Reichstein's compound S to cortisol | 16 h | 4 |

| | | | |
|---|---|---|---|
| *Escherichia coli* | | | |
| Polyacrylamide | β-Lactam degradation | overnight | 2 |
| *Leuconostoc mesenteroides* | | | |
| Alginate or agar | Degradation of malate in wine | 24-48 h | $10^3$-$6 \cdot 10^4$ |
| *Penicillium chrysogenum* | | | |
| Polyacrylamide | Production of penicillin G | 3 x 5 h | 1.75 |
| *Providencia* sp. | | | |
| Alginate | Amino acids to keto acids | 16 h | 2.5 |
| *Psuedomonas denitrificans* | | | |
| Alginate | Denitrification of water | 3 x 20 h | Restored to original activity |
| *P. putida* | | | |
| Polyacrylamide | Benzene degradation | Overnight | 3 |
| *Saccharomyces cerevisiae* | | | |
| Carrageenana | Ethanol production | 80 h | $10^3$ |
| *S. cerevisiae* | | | |
| Alginate | Ethanol production | 35 h | 7 |
| *S. carlsbergensis* | | | |
| Carrageenan | Ethanol production | 5 days | $10^3$ |
| *Serratia marcescens* | | | |
| Carrageenan | Production of L-isoleucine | 4 days | $10^3$ |

Source: Adapted from Brodelius and Vandamme, 1987. Used with permission.

**Table 2.13.19. Industrial Applications of Immobilized Cells**

| Cell Species | Immobilization Method | Utilized Enzyme | Application |
|---|---|---|---|
| *Actinoplanes missouriensis* | Entrapment in glutaraldehyde cross-linked gelatin | Glucose isomerase | Isomerization of glucose to fructose |
| *Bacillus coagulans* | Cross-linking with glutaraldehyde | Glucose isomerase | Isomerization of glucose to fructose |
| *Bacillus* sp. | Cross-linked with glutaraldehyde | $\beta$-Galactosidase | Hydrolysis of lactose |
| *Brevibacterium flavum* | Entrapment in carrageenan | Fumarase | Conversion of fumarate to L-malic acid |
| *Escherichia coli* | Entrapment in carrageenan | Aspartase | Conversion of ammonium fumarate to L-aspartic acid |
| *E. coli* | Entrapment in polyacrylamide | Penicillin acylase | Production of 6-APA from penicillins |
| *E. coli* | Entrapment in glutaraldehyde cross-linked gelatin | Penicillin acylase | Production of 6-APA from penicillins |
| *E. coli* + *Pseudomonas ducunhae* | Entrapment in carrageenan | Aspartase + L-aspartate $\beta$-decarboxylase | Production of L-alanine from fumarate |
| *Mortierella vinacea* | Cross-linking with glutaraldehyde | $\alpha$-Galactosidase | Hydrolysis of raffinose |
| *Proteus rettgeri* | Binding to glycidylmethacrylate with glutaraldehyde | Penicillin acylase | Production of 6-APA from penicillins |
| *Streptomyces albus* | Heat treatment + filter reactor | Glucose isomerase | Isomerization of glucose to fructose |
| *S. olivaceus* | Cross-linking with glutaraldehyde | Glucose isomerase | Isomerization of glucose to fructose |

Source: Adapted from Brodelius and Vandamme, 1987. Used with permission.

**Table 2.13.20. Some Examples of Immobilized Microbial Cells Utilized In Analysis**

| Cell Species<br>Support | Transducer | Substance | Sensitivity ($\mu$M) |
|---|---|---|---|
| *Azotobacter vinelandii*<br>Dialysis membrane | Ammonia gas electrode | Nitrate | 10 |
| *Bacterium cadaveris*<br>Dialysis membrane | Ammonia gas electrode | L-Aspartate | 50 |
| *Brevibacterium lactofermentum*<br>Cellophane membrane | Oxygen electrode | Sugars | 200 |
| *Citrobacter freudii*<br>Collagen | pH electrode | Cephalo-sporins | 60 ($\mu$g mL) |
| *Clostridium acidiurici*<br>Dialysis membrane | Ammonia gas electrode | L-Serine | 50 |
| *C. butyricum*<br>Polyacrylamide | Platinum electrode | BOD[a] | 10 (ppm) |
| *C. butyricum*<br>Acetylcellulose filter | Platinum electrode | Formic acid | 200 |
| *Escherichia coli*<br>Cellophane membrane | $CO_2$ electrode | Glutamic acid | $5 \cdot 10^4$ |
| *Hansenula anomala*<br>Dialysis membrane | Oxygen electrode | BOD[a] | 10 |
| *Lactobacillus arabilosus*<br>Agar | pH electrode | Nicotinic acid | 0.4 |

**Table 2.13.20. (continued)**

| Cell Species<br>Support | Transducer | Substance | Sensitivity ($\mu$M) |
|---|---|---|---|
| *Leuconostoc mesenteroides*<br>Acetylcellulose filter | Oxygen electrode | Phenyl-alanine | 0.6 |
| *Methylomonas flagellata*<br>Acetylcellulose filter | Oxygen electrode | Methane | 13 |
| *Nocardia opaca*<br>Polyacrylamide | Glassy carbon electrode | 3-Keto-4-ene steroid | 2.5 |
| *Proteus morganii*<br>Dialysis membrane | $H_2S$ electrode | L-Cysteine | 300 |
| *Pseudomonas fluorescens*<br>Collagen | Oxygen electrode | Glucose | 10 |
| *Pseudomonas* sp.<br>Cellulose acetate-nitrate filter | Ammonia gas electrode | L-Histidine | 100 |
| *Pseudomonas* sp.<br>Dialysis membrane | Ammonia gas electrode | Nitrilotri-acetic acid | 100 |
| *Saccharomyces cerevisiae*<br>Polyacrylamide | Thermistor | Arsenate | 2000 |
| *S. cerevisiae*<br>Polyacrylamide | Thermistor | 2,4-Dinitro-phenol | 1000 |
| *S. cerevisiae*<br>Polyacrylamide | Thermistor | Glucose | 1000 |

| | | | |
|---|---|---|---|
| *S. cerevisiae*<br>Collagen | Oxygen electrode | Nystatin | 0.5 (U/mL) |
| *S. cerevisiae*<br>Alginate | Oxygen electrode | Thiamine | 0.2 |
| *Sarcina flava*<br>Dialysis membrane | Ammonia gas electrode | L-Glutamine | 20 |
| *Streptococcus flecium*<br>Dialysis membrane | Ammonia gas electrode | L-Arginine | 50 |
| *Trichosporon brassicae*<br>Acetylcellulose filter | Oxygen electrode | Acetic acid | $8 \cdot 10^4$ |
| *T. brassicae*<br>Acetylcellulose filter | Oxygen electrode | Ethanol | 45 |

Source: Adapted from Brodelius and Vandamme, 1987. Used with permission.
[a]BOD = biological oxygen demand

**Table 2.13.21. Some Examples of Immobilized Plant Cells**

| Cell Species<br>Enzyme | Support | Substrate | Product |
|---|---|---|---|
| *Capsicum frutescens*<br>Multi-enzyme | Polyurethane foam | Sucrose | Capsaicin |
| *Catharanthus roseus*<br>Cathenamine reductase | Agarose | Cathenamine | Ajmalicine isomers |
| *C. roseus*<br>Multi-enzyme | Alginate/agarose/agar/carrageenan | Tryptamine + secologanin | Ajmalicine isomers |

**Table 2.13.21. (continued)**

| Cell Species<br>Enzyme | Support | Substrate | Product |
|---|---|---|---|
| *C. roseus* | | | |
| Multi-enzyme | Alginate/agarose/ agar/carrageenan | Sucrose | Ajmalicine |
| *C. roseus* | | | |
| Multi-enzyme | Alginate | Sucrose | Serpentine |
| *C. roseus* | | | |
| Multi-enzyme | Polyacrylamide | Sucrose | Amjalicine/ serpentine |
| *Daucus carota* | | | |
| 5 β-Hydroxy- lase | Alginate | Digitoxigenin | Periplogenin |
| *D. carota* | | | |
| 5 β-Hydroxy- lase | Alginate | Gitoxigenin | 5-Hydroxy- gitoxigenin |
| *Digitalis lanata* | | | |
| 12 β-Hydroxy- lase | Alginate | Digitoxin | Digoxin |
| *D. lanata* | | | |
| 12 β-Hydroxy- lase | Alginate | Methyldigitoxin | Methyldigi- toxin |
| *Glycine max* | | | |
| Multi-enzyme | Hollow fibers | Sucrose | Phenolics |
| *Mentha* sp. | | | |
| No data | Polyacrylamide | Menthone | Neomenthol |
| *Morinda citrifolia* | | | |
| Multi-enzyme | Alginate | Sucrose | Anthraqui- nones |
| *Petunia hybrida* | | | |
| Multi-enzyme | Hollow fibers | Sucrose | Phenolics |
| *Solanum aviculara* | | | |
| Multi-enzyme | Polyphenylene oxide | Sucrose | Steroid glycosides |

Source: Adapted from Brodelius and Vandamme, 1987. Used with permission.

## 2.14. Microbial Products

**Table 2.14.1. Characteristics of Primary and Secondary Metabolism**

| Primary metabolism | Secondary metabolism |
|---|---|
| General solution to biological problem | Specific solution to biological problem |
| Important for growth | Unnecessary for growth |
| Physiological role known | Physiological role difficult to asess |
| Present throughout life-cycle | Acting primarily during delimited part of developmental cycle |
| Present under various growth conditions | Operation heavily dependent on growth conditions |
| Ubiquitous | Non-ubiquitous |
| Single definite product formed | Families (mixtures) of related products formed |
| Often products with relatively simple chemical constitution | Frequently products with complex chemical constitution |

Source: Adapted from Hütter, 1986. Used with permission.

**Table 2.14.2. Fundamentals of Iron Complex Products**

- List of siderophores
- Function of siderophores
- Isolation, characterization, and biological activity of siderophores
- Bacterial sideropores
  - Enterobactin (enterochelin)
  - Agrobactin
  - Parabactin
  - Vibriobactin
  - Azotochelin
  - Pyochelin
  - Pseudobactin
  - Pyoverdines
  - Mycobactins
  - Ferrioxamines
  - Citrate-hydroxamates (schizokinen, arthrobactin, aerobactin)
- Fungal siderophores
  - Ferrichoromes
  - Coprogėns
  - Fusigens (fusarinines)
  - Rhodotorulin acid

Source: Adapted from Winkelmann, 1986. Used with permission.

**Table 2.14.3. List of Siderophores**

| Siderophores | Producing Organisms |
|---|---|
| Enterobactin (=Enterochelin) | *Escherichia coli, Salmonella typhimurium, Klebsiella aerogenes, Shigella dysenteriae, Enterobacter aerogenes* |
| Agrobactin | *Agrobacterium tumefaciens* |
| Parabactin | *Paracoccus denitrificans* |
| Vibriobactin | *Vibrio cholerae* |
| Azotochelin | *Azotobacter vinelandii* |
| Azotobactin | *A. vinelandii* |
| Pyochelin | *Pseudomonas aeruginosa* |
| Ferrithiocin | *Streptomyces antibioticus* |
| Ferribactin | *Pseudomonas fluorescens* ATCC 13525 |
| Pseudobactin | *P. fluorescens* B10 |
| Pseudobactin A | *P. fluorescens* B10 |
| Pyoverdine (Pf) | *P. fluorescens* |
| Pyoverdine (Pa) | *P. aeruginosa, P. fluorescens W, P. chlororaphis, P. putida* |
| Mycobactin P | *Mycobacterium phlei* |
| Mycobactin A | *M. aurum* |
| Mycobactin Av | *M. avium* |
| Mycobactin F | *M. fortuitum* |
| Mycobactin H | *M. thermoresistible* |
| Mycobactin K | *M. kansasii* |
| Mycobactin M | *M. marinum* |
| Mycobactin R | *M. terrae* |
| Mycobactin S | *M. smegmatis* |
| Mycobactin T | *M. tuberculosis* |
| Nocobactin NA | *Nocardia asteroides, N. sylvoderifera, N. paraffinae, N. uniformis* |
| Nocobactin NB | *N. brasiliensis* |
| Nocobactin NC | *N. caviae, N. phenotolerans* |

| | |
|---|---|
| Ferrioxamines ($A_1$, $A_2$, B, C, $D_1$, $D_2$, E, F, G, H) | *Streptomyces pilosus, S. griseoflavus, S. viridochromogenes, S. olivaceus, S. aureofaciens, S. galilaeus, S. lavendulae, S. polychromogenes, S. griseus* |
| Ferrioxamine E | *Pseudomonas stutzeri, Chromobacterium violaceum* |
| Ferrioxamine B | *Arthrobacter simplex* |
| Schizokinen | *Bacillus megaterium, Anabaena sp.* |
| Arthrobactin | *Arthrobacter pascens* |
| Aerobactin | *Enterobacter aerogenes, E. cloacae, Escherichia coli* ColV, *Shigella flexneri, S. boydii* |
| Ferrichrome | *Aspergillus niger, A. quadricinctyum, Penicillium parvum, P. adametzii, P. brefeldianum, P. duclauxii, P. resticulosum, Ustilago sphaerogena, U. longissima, U. maydis, U. puellaris U. scabiosae, U. segetum var. avenae, U. trichophora, Cintractia sorghi-vulgaris, Farysia olivacea, Moesziomyces penicillaris, Schizonella melanogramma, Sorosporium cenchri, Spacelotheca andropogonis, Sporisorium sorghi,* |

**Table 2.14.3 (continued)**

| Siderophores | Producing Organisms |
|---|---|
| Ferrichrome (continued) | *Graphiola phoenicis,*<br>*Protomyces macrosporus,*<br>*Tilletiaria anomala* |
| Tetraglycylferrichrome | *Neovossia indica* |
| Ferrichrome A | *Ustilago sphaerogena,*<br>*U. maydis,*<br>*U. scabiosae,*<br>*Schizonella melanogramma,*<br>*Sphacelotheca andropogonis* |
| Malonichrome | *Fusarium roseum* |
| Ferrichrome C | *Cryptococcus melibiosum,*<br>*Neurospora crassa,*<br>*Aspergillus oryzae,*<br>*A. duricaulis* |
| Ferricrocin | *A. aureolus,*<br>*A. auratus,*<br>*A. fischeri,*<br>*A. fumigatus*<br>*A. strameniu,*<br>*A. unilateralis,*<br>*A. viridi-nutans,*<br>*Neurospora crassa* |
| Ferrichrysin | *Aspergillus melleus, A. oryzae,*<br>*A. ochraceus, A. terreus* |
| Ferrirubin | *A. ochraceus,*<br>*Penicillium rugulosum,*<br>*P. variabile,*<br>*Paecilomyces varioti,*<br>*Spicaria sp.* |
| Ferrirhodin | *Aspergillus nidulans,*<br>*A. versicolor* |
| Asperchromes<br>(A, B1, B2, B3, C, D1,<br>D2, D3, E) | *A. ochraceus*<br>*Fusarium roseum* |
| Fusarinins (A, B) | *F. cubense,* |
| Fusigen (Fusarinin C) | *F. roseum,*<br>*Aspergillus fumigatus,*<br>*Giberella fujijuroi,* |

| | |
|---|---|
| | *Penicillium chrysogenum,* |
| | *Paecilomyces varioti* |
| Triacetylfusigen | *Aspergillus fumigatus,* |
| (Triacetylfusarinin) | *A. breviceps* |
| | *A. aureolus,* |
| | *Fusarium roseum,* |
| | *Aspergillus deflectus,* |
| | *Penicillium javanicum,* |
| | *P. purpurogenum* |
| Dimerum acid | *Fusarium dimerum* |
| Coprogen | *Neurospora crassa,* |
| | *Penicillium camemberti,* |
| | *P. chrysogenum,* |
| | *P. citrinum,* |
| | *P. notatum,* |
| | *P. patulum,* |
| | *P. urticae* |
| Triornicin | *Epicoccum purpurascens* |
| Isotriornicin | *E. purpurascens* |
| Neocoprogen I | *Curvularia subulata* |
| Neocoprogen II | *C. subulata* |
| Neurosporin | *Neurospora crassa* |
| Coprogen B | *Fusarium sp.,* |
| | *F. dimerum,* |
| | *Mydrothecium roridum,* |
| | *M. striatisporum,* |
| | *M. verrucaria,* |
| | *Nectria cinnabarina,* |
| | *Neurospora crassa* |
| Rhodotorulic acid | *Rhodotorula pilimanae,* |
| | *R. glutinis,* |
| | *R. gramminis,* |
| | *R. rubra,* |
| | *Leucosporidium scottii,* |
| | *Rhodosporidium toruloides,* |
| | *Sporidiobolus johnsonii,* |
| | *S. ruinenii,* |
| | *Sporobolomyces albo-rubescens,* |
| | *S. hispanicus,* |
| | *S. pararoseus,* |
| | *S. roseus* |

**Table 2.14.3 (continued)**

| Siderophores | Producing Organisms |
|---|---|
| Rhodotorulic acid (continued) | *Ustilago violacea*, *U. major* |

Source: Adapted from Winkelmann, 1986. Used with permission.

**Table 2.14.4. Lipid Content (% w/w) of Some Oleaginous Microorganisms**

| | |
|---|---|
| Bacterium | |
| *Arthrobacter* AK19 | |
| | 78 |
| Algae | |
| *Chlorella pyrenoidosa* | |
| *C. vulgaris* | 37 |
| *Dunaliella salina* | 40 |
| *Monalanthus salina* | 47 |
| *Neochloris oleoabundans* | 70 |
| *Ochromonas danica* | 35-54 |
| *Scenedesmus dimorphus* | 39-71 |
| *S. obliquus* | 16-40 |
| | 49 |
| Yeasts | |
| *Candida curvata* | 57-58 |
| *Candida* sp. 107 (NCYC 911) | 42 |
| *Cryptococcus albidus* | 55-63 |
| *C. terricolus* | 65 |
| *Endomycopsis vernalis* | 65 |
| *Geotrichum candidum* | 50 |
| *Lipomyces lipofer* | 64 |
| *L. starkeyi* | 63 |
| *Rhodosporidium toruloides*[a] | 66 |
| *Rhodotorula glutinis* | 71 |
| *R. graminis* | 41 |
| *Trichosporon cutaneum* | 45 |
| *T. pullulans* | 65 |
| Molds | |
| Phycomycetes | |
| *Blakeslea trispora* | |
| *Cunninghamella echinulata* | 37 |
| *C. elegans* | 45 |
| *Entomophthora coronata* | 56 |
| *Mortierella isabellina* | 47 |
| *M. vinacea* | 63-86 |
| *Mucor circinelloides* | 66 |
| *M. ramannianus* | 65 |
| *Pyhthium ultimum* | 56 |
| | 48 |
| Ascomycetes | |
| *Aspergillus nidulans* | 51 |
| *A. terreus* | 57 |
| *Fusarium bulbigenum* | 50 |
| *Gibberella fujikuroi* | 48 |
| *Penicillum lilacinum* | 56 |
| *P. spinulosum* | 64 |
| Hyphomycetes | |
| *Cladosporium herbarum* | 49 |
| Ustilaginaceae | |
| *Spacelotheca reiliana* | 41 |
| *Tolyposporium ehrenbergii* | 41 |
| *Ustilago zeae* | 59 |

Source: Adapted from Ratledge, 1986. Used with permission.
[a]Includes *Rhodotorula gracilis*

**Table 2.14.5. Fundamentals for Manufacturing Vitamins**

- General remarks
  - Definition
  - Historical details
  - Classification and Role
  - Sources
  - Production
- Vitamin $B_{12}$
  - Introduction
  - Structure
  - Properties
  - Sources
  - Biosynthesis
  - Production
    - Fermentation of *Pseudomonas dentrificans*
    - Fermentation of *Propionibacterium shermanii*
    - Fermentation of *Rhodopseudomonas protamicus*
    - Isolation
    - Assays
- Vitamin $B_2$
  - Introduction
  - Structure
  - Properties
  - Sources
  - Biosynthesis
  - Production
    - Semisynthetic process: production of D-ribose by fermentation
    - Biochemical process
      - Production of riboflavin by fermentation of *Ashbya gossypii*
      - Production of riboflavin by fermentation of a DNA recombinant strain of *Bacillus subtilis*
    - Isolation of riboflavin
    - Assays
- Vitamin C
  - Introduction
  - Structure
  - Properties
  - Sources
  - Biosynthesis
  - Production
    - Processes starting with reduction of D-glucose
    - Processes starting with oxidation of D-glucose
      - The L-idonic acid pathway
      - The 2,5-diketo-D-gluconic acid pathway
    - Assays
- Provitamin A ($\beta$-Carotene)
  - Introduction
  - Biosynthesis
  - Production
    - Fermentation of *Blakeslea trispora*
    - Isolation of B-carotene
    - Assays
- Biotin
  - Introduction
  - Structure
  - Properties
  - Sources
  - Biosynthesis
  - Production
    - Direct production by fermentation
    - Production by biotransformation

**Table 2.14.5. (continued)**

| | |
|---|---|
| of related products | Biosynthesis |
| Production by co-oxidation | Production |
| of a main substrate and | Production by yeasts |
| an analogue of biotin | Production by fermentation |
| Isolation | of fungi |
| Assays | Isolation |
| Provitamin $D_2$ (Ergosterol) | Assays |
| Introduction | Economic Importance |

Source: Adapted from J. Florent, 1986. Used with permission.

**Table 2.14.6. Basic Data on Coenzymes**

| | |
|---|---|
| Background information | Production of Lipoic Acid |
| Production of adenosine 5′-triphosphate | Production of *S*-adenosyl-methionine, *S*-adenosyl-homocysteine, and related nucleosides |
| Production of ATP | *S*-Adenosylmethionine |
| Application of ATP generating systems to the synthesis of useful compounds using ATP as an energy source | *S*-Adenosylhomocysteine and related nucleosides |
| Production of nicotinamide adenine dinucleotide and nicotinamide adenine dinucleotide phosphate | Production of Coenzyme A and 4′-phosphopantetheine |
| Production of NAD | Coenzyme A |
| Production of NADP | 4′-Phosphopantetheine and other intermediates in Coenzyme A biosynthesis |
| Reduction of NAD and NADP | Production of nucleoside diphosphate coenzymes |
| Production of flavin adenine dinucleotide | Production of pyridoxal 5′-phosphate |
| Production of pyrroloquino-line quinone | Biotin |
| Production of Coenzyme Q | 5′-Deoxyadenosylcobalmin and methylcobalamin |
| Production of Menaquinone-6 | |

Source: Adapted from Shimizu and Yamada, 1986. Used with permission.

**Table 2.14.7. Classification, Metabolic Functions, and Sources of Coenzyme**

| Coenzyme | Function | Vitamin source |
|---|---|---|
| Oxidoreduction enzyme | | |
| NAD | Hydrogen and electron transport | Nicotinic acid |
| NADP | Hydrogen and electron transport | Nicotinic acid |
| Flavin mononucleotide | Hydrogen and electron transport | Riboflavin |
| FAD | Hydrogen and electron transport | Riboflavin |
| Pyrroquinoline quinone | Hydrogen and electron transport | None |
| Coenzyme Q | Hydrogen and electron transport | None |
| Menaquinone-6 | Hydrogen and electron transport | None |
| Lipoic acid | Hydrogen and acyl transfer | None |
| Cytochromes | Electron transport | None |
| Ferredoxins | Electron transport and hydrogen activation | None |
| Thioredoxins | Hydrogen transport | None |
| Group transfer coenzymes | | |
| S-Adenosylmethionine | Transmethylation | None |
| Methylcobalamin | Transmethylation | Vitamin B12 |
| Coenzyme M | Methyl transfer | None |
| Tetrahydrofolic acid and conjugates | Transfer of formyl, hydroxymethyl and methyl groups | Folic acid |
| Coenzyme A | Transacylation, etc. | Pantothenic acid |
| 4′-Pnospnopantetheine (or acyl carrier protein) | Transfer of acyl and amino acid groups, etc. | Pantothenic acid |
| Nucleoside diphosphate coenzymes | Transfer of sugars and phosphorylcholine | None |
| Pyridoxal 5′-phosphate | Transamination decarboxylation, etc. | Vitamin $B_6$ |

**Table 2.14.7. (continued)**

| Coenzyme | Function | Vitamin source |
|---|---|---|
| Biotin (or $CO_2$-biotin enzymes) | Carboxylation, trans-carboxylation, etc. | Biotin |
| Thiamine pyrophosphate | $C_2$-group transfer | Thiamine |
| Adenosine 3′-phosphate-5′-phosphosulfate | Sulfate transfer | None |
| ATP | Phosphorylation, pyrophosphorylation and transfer of adenosyl group | None |
| Isomerization coenzymes | | |
| 5′-Deoxyadenosyl cobalamin | Carboxyl shifts | Vitamin $B_{12}$ |
| UDP-sugars, etc. | Sugar isomerization | None |

Source: Adapted from Shimizu and Yamada, 1986. Used with permission.

**Table 2.14.8. Production of Useful Compounds by Combination with ATP-Generating Systems**

| Product | ATP-Generating System | Microorganism and/or Enzyme |
|---|---|---|
| Nucleoside 5′-triphosphates | Glycolysis | Baker's yeast etc. |
| GTP | Ribotidation[a] | *Brevibacterium ammoniagenes* |
| ppGpp[b] and pppGpp[c] | Ribotidation | *B. ammoniagenes* |
| Nucleoside diphosphate sugars | Glycolysis | Baker's yeast etc. |
| CDP-choline | Glycolysis | Brewer's yeast etc. |
| NAD | Ribotidation | *B. ammoniagenes* |
| | Cell-free enzyme | Acetate kinase, adenylate kinase and NAD pyrophosphorylase |
| NADP | Glycolysis | Baker's yeast and *B. ammoniagenes* |

| | | |
|---|---|---|
| | Cell-free enzyme | Acetate kinase, adenylate kinase, NAD pyrophosphorylase and NAD kinase |
| FAD | Ribotidation | *B. ammoniagenes* or *Sarcina lutea* |
| | Glycolysis | Baker's yeast and *Arthrobacter globiformis* |
| Coenzyme A | Ribotidation | *B. ammoniagenes* or *S.lutea* |
| | Glycolysis | Baker's yeast and *B. ammoniagenes* |
| S-Adenosyl methionine | Cell-free enzyme | Pyruvate kinase, adenylate kinase, inorganic pyrophosphatase and methionine adenosyl-transferase |
| Glutamine | Glycolysis | Baker's yeast and glutamine synthetase |
| Glutathione | Glycolysis | Baker's yeast |
| | Cell-free enzyme | Acetate kinase and *Escherichia coli* |
| Gramicidin S | Cell-free enzyme | Acetate kinase and gramicidin S synthetase |
| Glucose 6-phosphate | Cell-free enzyme | Acetate kinase and hexokinase |

Source:
[a]This process involves direct ribotidation of bases or its modification.
[b]Guanosine 3′-diphosphate, 5′-diphosphate
[c]Guanosine 3′-diphosphate, 5′-triphosphate

**Table 2.14.9 Fundamentals of Obtaining Nucleic Acids, Nucleotides, and Related Compounds from Microorganisms**

Background information
- Nucleic acids
  - Ribonucleic acid (RNA)
    - Yeast
    - Bacteria
  - Deoxyribonucleic Acid (DNA)
  - Polyribonucleotides

**Table 2.14.9 (continued)**

- Nuclease $P_1$
  - History
  - Properties
  - Production of 5′-Nucleotides
  - Applications as a biochemical tool
- Purine nucleotides
  - Inosine monophosphate (IMP)
    - *Bacillus subtilis*
    - *Brevibacterium ammoniagenes*
  - Xanthosine monophosphate (XMP)
  - Guanosine monophosphate (GMP)
  - Cyclic adenosine 3′,5′-monophosphate
    - *Escherichia coli*
    - *Salmonella typhimurium*
    - *Brevibacterium liquefaciens*
    - *Corynebacterium murisepticum* and *microbacterium*
    - Hydrocarbon-utilizing bacteria
  - Adenosine triphosphate (ATP)
    - *Brevibacterium ammoniagenes* (adenine; glucose)
    - *Saccharmyces cerevisiae* (adenine; glucose)
    - *Saccharomyces cerevisiae* (adenosine; glucose)
    - *Erwinia carotovora* and *Saccharomyces cerevisiae* (adenine and inosine; glucose)
    - *Escherichia coli* (AMP; glucose-6-phosphate or glucose)
    - *Candida boidinii* (AMP, adenosine, or adenine; methanol)
    - Photosynthetic bacterial chromatophores (ADP; light)
    - Cyanobacteria (blue-green algae) (ADP; light)
    - Immobilized yeast mitochondria (ADP)
    - *Bacillus stearothermophilus* (ADP; acetylphosphate)
  - Deoxyribonucleoside polyphosphate
- Purine Nucleosides
  - Inosine
    - *Bacillus subtilis*
    - *Brevibacterium ammoniagenes*
  - Guanosine
  - Adenosine
  - Xanthosine
  - Purine Arabinosides
  - Purine 3′-deoxynucleosides
- Pyrimidine-related Compounds

Orotic acid and orotidine
Uridine monophosphate (UMP)
Cytosine
Pseudouridine
Purine-primidine-related Antibiotics

Source: Adapted from Kuninaka, 1986. Used with permission.

**Table 2.14.10. *De novo* Synthesis of Nucleotides and Related Compounds by *Brevibacterium ammoniagenes***

| Markers of strain | Product | Remark |
|---|---|---|
| $2FA^R$ | Adenine | None |
| $ade^-$ | IMP, hypoxanthine | None |
| $ade^-$ $gua^-$ | IMP, hypoxanthine | None |
| $ade^L$ | IMP, hypoxanthine | $Mn^{2+}$ insensitive |
| $ade^L$ | IMP, hypoxanthine | Permeable mutant |
| $gua^-$ | XMP | None |
| $gua^-$ $ade^L$ | XMP | |
| Wild | XMP | Addition of psicofuranine or deocyinine |
| $ade^L$ $6MG^R$ | Inosine | None |
| $ade^L$ $gua^-$ $6MG^R$ $6MTP^R$ | Inosine | None |
| Wild | UMP | Addition of 6-azauracil |
| Wild | Omp | Addition of 6-azauracil |
| $2FA^R$ | Cytosine | None |
| Wild | Ribose-5-phosphate | None |
| | PRPP | None |

Source: Adapted from Kuninaka, 1986. Used with permission.

**Table 2.14.11. Salvage Synthesis of Nucleotides and Related Compounds by *Brevibacterium ammoniagenes***

| Precursor | Product | Remark |
|---|---|---|
| Hypoxanthine | IMP | None |

**Table 2.14.11. (continued)**

| Precursor | Product | Remark |
|---|---|---|
| Guanine | GMP, GDP, GTP | None |
| Adenine | AMP, ADP, ATP | None |
| Uracil, orotic acid | UMP | None |
| Orotic acid + 6-azauracil | OMP | None |
| Purine analogue | Purine analogue nucleotide | None |
| Pyrimidine analogue | Pyrimidine analogue nucleotide | None |
| 6-Azauracil | 6-Azauridine | None |
| 1,2,4-Triazole-3-carboxamide | Ribavirin | None |
| Nicotinic acid + adenine | NAD | None |
| FMN + adenine | FAD | None |
| XMP (conversion) | GMP, GDP, GTP | Decoyinine resistant |
| XMP, GMP (conversion) | pppGpp, ppGpp, pGpp | Decoyinine resistant |

Source: Adapted from Kuninaka, 1986. Used with permission.

**Table 2.14.12. Production of Nucleoside Diphosphate Sugars by the Yeast Fermentation System**

| Product | Substrate | Yeast | Yield: $\mu$mol/mL; %[a] |
|---|---|---|---|
| UDP-glucose | UMP and glucose | Brewer's yeast | 13,6; 68 |
| | | *Torulopsis candida* | 60; 60 |
| UDP-galactose | UMP and galactose | *T. candida* | 69.8; 42 |
| | UMP and lactose | *T. candida* | 13; 65 |
| | UMP and glucose | *T. candida* | 17; 85 |
| UDP-*N*-acetyl-glucosamine | UMP, gluocse and glucosamine | Baker's yeast | 6.7; 40 |
| | | *Debaryomyces subglobosus* | 10; 40 |

| | | | |
|---|---|---|---|
| GDP-mannose | GMP and glucose | Baker's yeast | 9; 45 |
| | | *Hansenula jadinii* | 16; 80 |

Source: Adapted from Kuninaka, 1986. Used with permission.
[a]% molar yields based on the nucleotides used.

**Table 2.14.13. Properties of Microorganisms Required for the Production of Purine Nucleosides**

Production of Inosine

1. Adenylosuccinate [succinoadenosine 5′-monophosphate (SAMP)] synthetase and IMP-dehydrogenase should be absent or present at a very low level.
2. Nucleosidase and/or nucleoside phosphorylase should be absent or present at a very low level.
3. Regulation of enzymes involved in IMP formation, such as PRPP amidotransferase, by AMP and GMP should be released.

Production of Guanosine

1. Both SAMP synthetase and GMP reductase should be lacking or present at a very low level.
2. Nucleosidase and/or nucleoside phosphorylase should be absent or present at a very low level.
3. Regulation of enzymes participating directly in GMP synthesis, IMP dehydrogenase, and GMP synthetase, by GMP should be released.
4. Regulation of enzymes involved in de novo biosynthesis of IMP, such as PRPP amidotransferase, by AMP and GMP should be released.
5. Both IMP dehydrogenase and GMP synthetase should be more active than nucleotidase to avoid formation of inosine as by-product.

Production of Adenosine

1. Both IMP dehydrogenase and AMP deaminase should be absent or present at a very low level.
2. Adenosine degrading enzymes (nucleosidase and nucleoside phosphorylase) should be present at a very low level.
3. Regulation of SAMP synthetase by AMP should be released.
4. Regulation of enzymes involved in de novo biosynthesis of IMP, such as PRPP amidotransferase and SAMP lyase, by AMP and GMP should be released.

**Table 2.14.13. (continued)**

Production of Xanthosine

1. Both SAMP synthetase and GMP synthetase should be absent or present at a very low level.
2. Xanthosine degrading enzymes (nucleosidase and nucleoside phosphorylase) should be present at a very low level.
3. Feedback inhibition of IMP dehydrogenase by XMP should be weakened.
4. Microorganisms should be grown in an adenine and guanine limited medium. (Enzymes involved in de novo biosynthesis of IMP, such as PRPP amidotransferase and SAMP lyase, are not regulated by XMP but regulated by AMP and GMP.)

Source: Adapted from Kuninaka, 1986. Used with permission.

**Table 2.14.14. Representative Methods for Producing ATP**

| Biological Means | Starting Material | ATP Formed (g/L) | Molar Yield (%) |
|---|---|---|---|
| Brevibacterium ammoniagenes (resting cells) | Adenine, glucose | 10.9 | 83 |
| *Saccharomyces cerevisiae* (dried cells) | Adenosine, glucose | 102.5 | 50 |
| *Escherichia coli* (dried cells) | AMP, glucose | 14.0 | 92 |
| *Candida boidinii* (sorbitol-treated cells) | AMP, methanol | 30.0 | 60 |
| | Adenosine, methanol | 100.0 | 77 |
| | Adenine, methanol | 4.1 | 40 |
| Photosynthetic bacteria (immobilized chromatophores) | ADP, light | 0.4 | 80 |
| Cyanobacteria (intact cells) | ADP, light | 0.17 | 17 |
| *Candida tropicalis* (immobilized mitochondria) | ADP | 0.24 | 60 |
| *Bacillus stearothermophilus* (purified enzymes) | ADP, acetylphosphate | 10.0 | 100 |

Source: Adapted from Kuninaka, 1986. Used with permission.

**Table 2.14.15. Fundamentals of Microbial Glycosides Inhibitors**

| | |
|---|---|
| Background information | validamine and derivatives |
| Assay methods | Glycopeptide inhibitors |
| Inhibitors of α-glucosidases | Polypeptide inhibitors |
| Pseudo-oligosaccharide inhibitors | Inhibitors of other glycosidases |
| Acarbose and homologous inhibitors | Mechanism of action of α-glucosidase inhibitors |
| Other pseudo-oligosaccharide inhibitors | Applications of glycosidase inhibitors |
| Monosaccharide inhibitors | Scientific applications |
| Nojirimycin and 1-deoxynojirimycins | Medical applications of α-glucosidase inhibitors |
| Valienamine, valiolamine, | |

Source: Adapted from Müller, 1986. Used with permission.

**Table 2.14.16. Microbial α–glucosidase Inhibitors**

| Microorganism | Composition | Specificity |
|---|---|---|
| Actinomycetes of various genera | Proteins; pseudo-oligosaccharides (acarbose series) | Salivary, pancreatic α-amylase; intestinal glucoamylase and disaccharidases; microbial α-amylases |
| *Streptomyces flavochromogenes* | Pseudo-oligo-saccharides | Salivary, pancreatic α-amylases; intestinal sucrase; *Rhizopus* glucoamylase |
| *Streptomyces diastaticus* var. *amylostaticus* | Pseudo-oligo-saccharides (amylostatins) | α-amylase; microbial a-amylases; malt amylase; *Rhizopus* amylase |
| *Streptomyces fradiae* | Acidic polypeptide (MW 6500) | Mammalian α-amylases |

**Table 2.14.16. (continued)**

| Microorganism | Composition | Specificity |
|---|---|---|
| *Bacilli, Streptomyces lavendulae* | Nojirimycin, l-deoxy-nojirimycin | Intestinal glucoamylase and disaccharidases; *Rhizopus* glucoamylase; plant, microbial, fungal β-glucosidases |
| *Streptomyces* sp. | Oligosaccharide (?) (MW ~600) | Salivary α--amylase; glucoamylase |
| *Streptomyces calidus* | Glycopeptide (MW 10000-20000) | Pancreatic α-amylase; intestinal maltase and sucrase |
| *Streptomyces violaceoruber* | Polypeptide (MW 5000-10000) | Pancreatic α-amylase |
| *Streptomyces tendae* | Polypeptides (HOE 467a: MW = 7958) (tendamistat) | Mammalian α-amylases |
| *Streptomyces* sp. | Pseudo-oligo-saccharides ("amino sugars") | α-Amylases; intestinal sucrase and maltase |
| *Streptomyces calvus* | Pseudo-oligo-saccharides (adiposins) | Microbial and mammalian α-amylases; *Rhizopus* glucoamylase; intestinal disaccharidases |
| *Streptomyces dimorphogenes* | Pseudo-oligo-saccharides (trestatins) | Pancreatic, microbial α-amylases |
| *Streptomyces griseosporeus* | Polypeptides (MW ~8500)(Haim I and II) | Mammalian α-amylases |

| | | |
|---|---|---|
| *Streptomyces myxogenes* | Pseudo-oligo-saccharides (oligostatins) | Salivary, pancreatic α-amylase; *Rhizopus* glucoamylase |
| *Streptomyces viridosporus* | Neutral polypeptide (MW 8000) | Pancreatic α-amylase |
| *Aspergillus niger* | Glycopeptide (MW ~10500) | Aspergillus glucoamylase |
| *Streptomyces hygroscopicus* | Valiolamine | Intestinal maltase and sucrase |
| *Clado sporium herbarum* | Glycopeptide (MW 18000) | Salivary, pancreatic α-amylase |
| *Cladosporium cladosporoides* | Acidic glycopeptide (MW ~25000) (tomastachin) | Mammalian α-amylases |
| *Streptomyces griseofuscus* | Oligosaccharide (MW 2200-2500) | Pancreatic α-amylase |
| *Streptomyces flavochromogenes* | Pseudo-oligo-saccharides | Microbial amylase; *Rhizopus* glucoamylase |
| *Streptomyces corchorushii* | Polypeptides (MW = 4085, 4387) (Paim I and II) | Pancreatic α-amylase (pig, dog, cow, horse) |

Source: Adapted from Müller, 1986. Used with permission.

**Table 2.14.17. Other Microbial Glycosidase Inhibitors**

| Specificity | Microorganism | Composition |
|---|---|---|
| Plant and microbial β-amylase | *Streptomyces* sp. | Glucosyl 1-deoxynojirimycin |
| Fungal cellulase | *Aspergillus terreus* | MW <1200 |
| Dictyostelium cellulase | *Dictyostelium discoidem* | Protein (MW ~20000) |

**Table 2.14.17 (continued)**

| Specificity | Microorganism | Composition |
|---|---|---|
| Streptococcus dextranase | *Streptococcus mutans* | Proteins |
| Viral neuraminidase | *Streptomyces pseudoverticillus, Streptomyces rimosus* | 5-Alkylbenzene-1,3-disulfates ("panosialin") |
| Viral, bacterial, and mammalian neuraminidase | *Streptomyces verticillus* | "Siastatin A"; 2-acetamido-3,4-dihydropiperidine-5-carboxylic acid ("siastatin B") |
| Plant and microbial $\beta$-glucosidase | *Streptomyces lavendulae* | 5-Deoxy-5-amino-D-gluco-pyranose ("nojirimycin") |
| Plant and microbial $\beta$-glucosidase | *Streptomyces lavendulae, Bacillus* | 1-Deoxynojirimycin |
| Plant and microbial $\beta$-glucosidase | *Streptomyces lavendulae* | 5-Deoxy-5-amino-D-manno-pyranose ("nojirimycin B") |
| Bovine liver $\beta$-galactosidase | *Streptomyces alboverticillatus* | 1-(1,2-Dihydroxy-ethyl)-3-hydroxy-methyl-9-pyrido(3,4)-indole ("pyridindolol") |
| Bovine liver $\beta$-galactosidase | *Streptomyces xanthophaeus* | Isoflavone rhamnosides |
| Bovine liver $\beta$-galactosidase | *Streptomyces nigellus* | *p*-Hydroxyphenyl-acetaldioxime |
| Mammalian and plant $\alpha$-mannosidase | *Streptomyces lavendulae* | 5-Deoxy-5-amino-D-mannopyranose ("nojirimycin B") |
| Microbial $\beta$-mannosidase | *Streptomyces lavendulae* | 5-Deoxy-5-amino-D-mannopyranose ("nojirimycin B") |
| *Neuorspora trehalase* | *Neurospora crassa* | Protein |
| Microbial and mammalian trehalase | *Streptomyces lavendulae* | 1-Deoxynojirimycin |
| Bacterial $\beta$-glucuronidase | *Micromonospora* sp. | Glycoprotein (MW 8700) |
| Egg-white lysozyme | *Streptomycetes* | Aminoglycoside antibiotics |

Source: Adapted from Müller, 1986. Used with permission.

## 2.15. References


Beuchat, L. R. Indigenous fermented foods. In Biotechnology, Volume 5: *Food and Feed Production with Microorganisms*, G. Reed, ed., VCH, New York, 1983.

Borriss, R. Biology of enzymes. In Biotechnology, Volume 7a: *Enzyme Technology*, J. F. Kennedy, ed., VCH, New York, 1987.

Bottazzi, V. Other fermented dairy products. In Biotechnology, Volume 5: *Food and Feed Production with Microorganisms*, G. Reed, ed., VCH, New York, 1983.

Brodelius, P. and E. J. Vandamme. Immobilized cell systems. In Biotechnology, Volume 7a: *Enzyme Technology*, J. F. Kennedy, ed., VCH, New York, 1987.

Florent, J. Vitamins. In Biotechnology, Volume 4: *Microbial Products II*, Rehm, H. J. and G. Reed, eds., VCH, New York, 1986.

Frost, G. M. and D. A. Moss. Production of enzymes by fermentation. In Biotechnology, Volume 7a: *Enzyme Technology*, J. F. Kennedy, ed., VCH, New York, 1987.

Gerhartz, W., ed. *Enzymes in Industry: Production and Applications*, VCH, New York, 1990.

Gilliland, S. E., ed. *Bacterial Starter Cultures for Foods*, CRC Press, Boca Raton, FL, 1985

Goldstein, W. E. Enzymes in starch processing and baking. In *Enzymes in Industry: Production and Applications*, W. Gerhartz, ed., VCH, New York, 1990.

Goldstein, W. E. Safety and environmental aspects. In *Enzymes in Industry: Production and Applications*, W. Gerhartz, ed., VCH, New York, 1990.

Hagen, H. A. and S. Pedersen. Glucose isomerization. In *Enzymes in Industry: Production and Applications*, W. Gerhartz, ed., VCH, New York, 1990.

Hütter, R. Overproduction of microbial metabolites. In Biotechnology, Volume 4: *Microbial Products II*, Rehm, H. J. and G. Reed, eds., VCH, New York, 1986.

Jacobson, G. K. and S. O. Jolly. Yeasts, molds, and algae. In Biotechnology, Volume 7b: *Gene Technology*, Jacobson, G. K. and S. O. Jolly, eds., VCH, New York, 1989.

Kennedy, J. F. and J. M. S. Cabral. Enzyme immobilization. In Biotechnology, Volume 7a: *Enzyme Technology*, J. F. Kennedy, ed., VCH, New York, 1987.

Kessler, C. Enzymes in genetic engineering. In *Enzymes in Industry: Production and Applications*, ed., W. Gerhartz. VCH, New York, 1990.

King, R. D. and P. S. J. Cheetham, eds. *Food Biotechnology*, Elsevier, New York, 1987.

Kresse, G. B. Enzymes in analysis and medicine. In *Enzymes in Industry: Production And Applications*, W. Gerhartz, ed., VCH, New York, 1990.

Kuninaka, A. Nucleic acids, nucleotides, and related compounds. In Biotechnology, Volume 4: *Microbial Products II*, Rehm, H. J. and G. Reed, eds., VCH, New York, 1986.

Müller, L. Microbial glycosidase inhibitors. In Biotechnology, Volume 4: *Microbial Products II*, Rehm, H. J. and G. Reed, eds., VCH, New York, 1986.

Noordervliet, P. F. and D. A. Toet. Safety in enzyme technology. In Biotechnology, Volume 7a: *Enzyme Technology*, J. F. Kennedy, ed., VCH, New York, 1987.

Peppler, H. J. and G. Reed. Enzymes in food and feed processing. In Biotechnology, Volume 7a: *Enzyme Technology*, J. F. Kennedy, ed., VCH, New York, 1987.

Poldermans, B. Proteolytic enzymes. In *Enzymes in Industry: Production and Applications*, W. Gerhartz, ed., VCH, New York, 1990.

Ratledge, C. Lipids. In Biotechnology, Volume 4: *Microbial Products II*, Rehm, H. J. and G. Reed, eds., VCH, New York, 1986.

Rehm, H. J. and G. Reed, eds. Biotechnology, Volume 5: *Food And Feed Production With Microorganisms*, VCH, New York, 1983.

Rehm, H. J. and G. Reed, eds. Biotechnology, Volume 4: *Microbial Products II*, VCH, New York, 1986.

Rehm, H. J. and G. Reed, eds. Biotechnology, Volume 8: *Microbial Degradations*, VCH, New York, 1986.

Rehm, H. J. and G. Reed, eds. Biotechnology, Volume 6b: *Special Microbial Processes*, VCH, New York, 1988.

Rehm, H. J. and G. Reed, eds. Biotechnology, Volume 7b: *Gene Technology*, VCH, New York, 1989.

Rehm, H. J. and G. Reed, eds. Biotechnology, Volume 7a: *Enzyme Technology*, VCH, New York, 1987.

Reimerdes, E. H. Dairy products. In *Enzymes in Industry: Production And Applications*, W. Gerhartz, ed., VCH, New York, 1990.

Shimizu, S. and H. Yamada. Coenzymes. In Biotechnology, Volume 4: *Microbial Products II*, Rehm, H. J. and G. Reed, eds, VCH, New York, 1986.

Soda, K. Tanaka, H., and N. Esaki. Amino acids. In Biotechnology, Volume 3: *Biomass, Microorganisms for Special Applications, Microbial Products I, Energy from Renewable Resources*, H. Dellweg, ed., VCH, New York, 1983.

Tsao, G. T., ed. *Annual Reports On Fermentation Processes*, Academic Press, San Diego, CA, 1985.

U.S. Congress, Office of Technology Assessment. *New Developments in Biotechnology–Field-Testing Engineered Organisms: Genetic and Ecological Issues*. OTA-BA-350, U.S. Government Printing Office, Washington, D.C., 1988.

U.S. Congress, Office of Technology Assessment. *Impacts of Applied Genetics: Microorganisms, Plants, and Animals*. U.S. Government Printing Office, Washington, D.C., 1981.

Uhlig, H. Enzymes in meat processing. In *Enzymes in Industry: Production And Applications*, W. Gerhartz, ed., VCH, New York, 1990.

Uhlig, H. Fruit, vegetables, and wine processing. In *Enzymes in Industry: Production And Applications*, W. Gerhartz, ed., VCH, New York, 1990.

Uhlig, H. Hydrolysis of protein, fat, and cellulose. In *Enzymes in Industry: Production And Applications*, W. Gerhartz, ed., VCH, New York, 1990.

Uhlig, H. Survey of industrial enzymes. In *Enzymes in Industry: Production And Applications*, W. Gerhartz, ed. VCH, New York, 1990.

van Scheltinga, A. H. T. Fermentation. In *Enzymes in Industry: Production and Applications*, ed., W. Gerhartz. VCH, New York, 1990.

Vogel, H. C., ed. *Fermentation and Biochemical Engineering Handbook: Principles, Process Design, and Equipment*, Noyes, Park Ridge, NJ, 1983.

Winkelmann, G. Iron complex compounds (siderophores). In Biotechnology, Volume 4: *Microbial Products II*, Rehm, H. J. and G. Reed, eds., VCH, New York, 1986

Wood, B. J. B., ed. *Microbiology of Fermented Foods*, Elsevier, New York, 1985.

# CHAPTER 3
# FOODBORNE DISEASES AND ASSOCIATED TERMINOLOGY

This chapter presents some harmful biological agents occurring in foods and beverages. The following types of data are provided:

1. The classification of foodborne diseases according to symptoms.
2. Important properties of major pathogenic bacteria.
3. A description of major parasites.
4. A list of herbs that should not be used in foods, beverages, or drugs.
5. A glossary of terms commonly associated with foodborne diseases.

To illustrate the usefulness of the above data, two examples are provided here.

If a person exhibits upper gastrointestinal tract signs and symptoms (nausea, vomiting) initially or if such signs predominate, what type of foodborne disease can it be? There are several possibilities. One is mushroom poisoning and another antimony poisoning. By classifying foodborne diseases according to symptoms, it has narrowed the diagnosis to a few alternatives.

Food scientists, technologists, and engineers are all aware that there are many types of pathogens in food. These harmful organisms can cause food poisoning. Since many of these professionals are not microbiologists, they are unfamiliar with professional reference books in the area. Occasionally, they may wish to compile certain information about these harmful organisms. This chapter has summarized the major bacterial pathogens in foods and drinking water by providing the following data for each:

1. Description.
2. Disease transmitted.
3. Sources.
4. Oxygen requirement.
5. pH (for growth).
6. $^{o}$C (for growth).
7. Resistant to heat: temperature and time.

Although all data in this chapter are available in many professional reference books, they are rarely arranged in one document for easy access.

## 3.1. Classifying Foodborne Diseases

**Table 3.1.1 Classifying Foodborne Diseases According to Symptoms**

A. Upper gastrointestinal tract signs and symptoms (nausea, vomiting) occur first or predominate
   1. Incubation (latency) period usually less than 1 hour
      a. Fungal agents
      b. Chemical agents
   2. Incubation (latency) period 1 to 6 hours
      a. Bacterial agents
      b. Chemical agents
   3. Incubation (latency) period usually 7 to 12 hours
      a. Fungal agents

B. Sore throat and respiratory signs and symptoms occur
   1. Incubation (latency) period less than 1 hour
      a. Chemical agents
   2. Incubation (latency) period 12 to 72 hours
      a. Bacterial agents
   3. Incubation (latency) period usually 12 to 72 hours
      a. Bacterial agents
   4. Incubation (latency) period greater than 72 hours
      a. Viral agents
      b. Parasitic agents

C. Neurological signs and symptoms (visual disturbances, tingling, paralysis) occur
   1. Incubation (latency) period usually less than 1 hour
      a. Fungal agents
      b. Chemical agents
      c. Toxic animals
      d. Plant toxicants
   2. Incubation (latency) period 1 to 6 hours
      a. Chemical agents
      b. Toxic animals
   3. Incubation (latency) period usually 12 to 72 hours
      a. Bacterial agents
   4. Incubation (latency) period greater than 72 hours
      a. Chemical agents

D. Generalized infection signs and symptoms (fever, chills, malaise, aches) occur

1. Incubation (latency) period greater than 72 hours
    a. Bacterial agents
    b. Viral agents
    c. Parasitic agents

E. Allergic-type symptoms (facial flushing, itching) occur
    1. Incubation (latency) period less than 1 hour
        a. Bacterial (and animal) agents
        b. Chemical agents
    2. Incubation (latency) period 1 to 6 hours
        a. Toxic animals

Source: Adapted from Bryan, 1982.

**Table 3.1.2. Foodborne Diseases Causing Upper Gastrointestinal Tract Signs and Symptoms (Nausea, Vomiting) To Occur First or Predominate**

Incubation (Latency) Period Usually Less Than 1 Hour

Fungal Agents

Gastrointestinal irritating: group mushroom poisoning

1. Agent: possibly resin-like substances in some mushrooms
2. Incubation: 30 minutes to 2 hours
3. Symptoms: nausea, vomiting, retching, diarrhea, abdominal cramps
4. Foods: many varieties of wild mushrooms
5. Specimens: vomitus
6. Causative factors: eating unknown varieties of mushrooms; mistaking toxic mushrooms for edible varieties

Chemical Agents

Antimony poisoning

1. Agent: antimony in glaze
2. Incubation: a few minutes to 1 hour
3. Symptoms: vomiting, abdominal pain, diarrhea
4. Foods: high-acid foods and beverages
5. Specimens: vomitus, stools, urine
6. Causative factors: using antimony-containing utensils; storing high-acid foods in gray enamelware

Cadmium poisoning

1. Agent: cadmium in plated utensils
2. Incubation: 15 to 30 minutes
3. Symptoms: nausea, vomiting, abdominal cramps, diarrhea, shock
4. Foods: high-acid foods and beverages, candy love beads or cake decorations

**Table 3.1.2. (continued)**

5. Specimens: vomitus, stools, urine, blood
6. Causative factors: using cadmium-containing utensils, storing high-acid foods in cadmium-containing containers; ingesting cadmium-containing foods

Copper poisoning

1. Agent: copper in pipes and utensils
2. Incubation: few minutes to few hours
3. Symptoms: metallic taste, nausea, vomiting (green vomitus), abdominal pain, diarrhea
4. Foods: high-acid food and beverages
5. Specimens: vomitus, gastric washings, urine, blood
6. Causative factors: storing high-acid foods in copper utensils or using copper pipes for dispensing high-acid beverages; faulty backflow preventor valves in vending machines

Fluoride poisoning

1. Agent: sodium fluoride in insecticides
2. Incubation: few minutes to 2 hours
3. Symptoms: salty or soapy taste, numbness of mouth, vomiting, diarrhea, abdominal pain, pallor, cyanosis, dilated pupils, spasms, collapse, shock
4. Foods: any accidentally contaminated food, particularly dry foods, such as dry milk, flour, baking powder, and cake mixes
5. Specimens: vomitus, gastric washings
6. Causative factors: storing insecticides in same area as foods; mistaking pesticides for powdered foods

Lead poisoning

1. Agent: lead in earthenware vessels, pesticides, paint, plaster, putty
2. Incubation: 30 minutes or longer
3. Symptoms: metallic taste, burning of mouth, abdominal pain, milky vomitus, blood or black stools, foul breath, shock, blue gum line
4. Foods: high-acid foods and beverages stored in lead-containing vessels, any accidentally contaminated food
5. Specimens: vomitus, gastric washings, stools, blood, urine
6. Causative factors: using lead-containing vessels; storing high-acid foods in lead-containing vessels; storing pesticides in same area as food

Tin poisoning

1. Agent: tin in tinned cans
2. Incubation: 30 minutes to 2 hours

3. Symptoms: bloating, nausea, vomiting, abdominal cramps, diarrhea, headache
4. Foods: high-acid foods and beverages
5. Specimens: vomitus, stools, urine, blood
6. Causative factors: using uncoated tin containers for storing acid foods

Zinc poisoning

1. Agent: zinc in galvanized containers
2. Incubation: a few minutes to a few hours
3. Symptoms: pain in mouth and abdomen, nausea, vomiting, dizziness
4. Foods: high-acid foods and beverages
5. Specimens: vomitus, gastric washings, urine, blood, stools
6. Causative factors: storing high-acid foods in galvanized cans

Incubation (Latency) Peroiod 1 to 6 hours

Bacterial Agents

Staphylococcal intoxication

1. Agent: exoenterotoxins A, B, C, D, and E of *Staphylococcus aureus*; staphylococci from noses, skin, and lesions of infected humans and animals and from udders of cows
2. Incubation: 1 to 8 hours, mean 2 to 4 hours
3. Symptoms: nausea, vomiting, retching, abdominal pain, diarrhea, prostration
4. Foods: ham, meat, and poultry products, cream-filled pastry, food mixtures, leftover foods
5. Specimens: for affected victims, vomitus, stools, rectal swabs; for carriers, nasal swabs, swabs of lesions, analysis of swab
6. Causative factors: inadequate refrigeration, workers touching cooked food; preparing foods several hours before serving; workers with infections containing pus; holding foods at warm (bacterial incubating) temperatures; fermentation of abnormally low-acid foods

Chemical Agents

Nitrite poisoning

1. Agent: nitrites or nitrates used as meat-curing compounds or ground water from shallow wells
2. Incubation: 1 to 2 hours
3. Symptoms: Nausea, vomiting, cyanosis, headache, dizziness, weakness, loss of consciousness, chocolate-brown colored blood

**Table 3.1.2. (continued)**

4. Foods: cured meats, any accidentally contaminated food, exposed to excessive nitrification
5. Specimens: blood
6. Causative factors: using excessive amounts of nitrites or nitrates in foods for curing or for covering up spoilage, mistaking nitrites for common salt and other condiments, inadequate refrigeraion

Incubation (Latency) Period Usually 7 to 12 hours

Fungal Agents

Cyclopeptide and gyromitrin groups of mushroom poisoning

1. Agent: cyclopeptide and gyromitrin in some mushrooms
2. Incubation: 6 to 24 hours
3. Symptoms: abdominal pain, feeling of fullness, vomiting, protracted diarrhea, loss of strength, thirst, muscle cramps, feeble rapid pulse, collapse, jaundice, drowsiness, dilated pupils, coma, death
4. Foods: *Amanita phalloides, A. verna, Galerina antumnalis, Gyromitra esculenta* (false morels) and similar species of mushrooms
5. Specimens: urine, blood, vomitus
6. Causative factors: eating certain species of *Amanita, Galerina,* and *Gyrojitra* mushrooms; eating unknown varieties of mushrooms; mistaking toxic mushrooms for edible varieties

Source: Adapted from Bryan, 1982.

**Table 3.1.3. Foodborne Diseases Causing Sore Throat and Respiratory Signs and Symptoms**

Incubation (Latency) Period Less Than 1 Hour

Chemical agents

Sodium hydroxide poisoning

1. Agent: sodium hydroxide in bottle washing compounds, detergents, drain cleaners, or hair straighteners
2. Incubation: a few minutes
3. Symptoms: burning of lips, mouth and throat; vomiting, abdominal pain, diarrhea
4. Foods: bottled beverages
5. Specimens: vomitus

6. Causative factors: inadequate rinsing of bottles cleaned with caustic (sodium hydroxide)

Incubation (Latency) Period 12 to 72 Hours
Bacterial Agents
$\beta$-Hemolytic streptococcal infections
1. Agent: *Streptococcus pyogenes* from throat and lesions of infected humans
2. Incubation: 1 to 3 days
3. Symptoms: sore throat, fever, nausea, vomiting, rhinorrhea, sometimes a rash
4. Foods: raw milk, foods containg eggs
5. Specimens: throat swabs, vomitus
6. Causative factors: workers touching cooked foods; workers with infections containing pus; inadequate refrigeration; inadequate cooking or reheating; preparing foods several hours before serving

Source: Adapted from Bryan, 1982.

**Table 3.1.4. Foodborne Diseases Causing Lower Gastrointestinal Tract Signs and Symptoms (Abdominal Cramps, Diarrhea) To Occur First or Predominate**

Incubation (Latency) Period Usually 7 to 12 Hours
Bacterial Agents
*Bacillus cereus* gastroenteritis
1. Agent: Exoenterotoxic of *B. cereus* organism in soil
2. Incubation: 8 to 16 hours; rare reports of 2 to 4 hours
3. Symptoms: nausea, abdominal pain, diarrhea; some reports of vomiting
4. Foods: cereal products, rice, custards and sauces, meatloaf
5. Specimens: stools, vomitus
6. Causative factors: inadequate refrigeration; holding foods at warm (bacterial incubating) temperatures; preparing foods several hours before serving; inadequate reheating of leftovers

*Clostridium perfringens* gastroenteritis
1. Agent: endoenterotoxin formed during sporulation of *C. perfringens* in intestines; organism in feces of infected humans, other animals, and in soil
2. Incubation: 8 to 22 hours, mean 10 hours

**Table 3.1.4. (continued)**

3. Symptoms: abdominal pain, diarrhea
4. Foods: cooked meat, poultry, gravy, sauces, and soups
5. Specimens: stools
6. Causative factors: inadequate refrigeration; holding foods at warm (bacterial incubating) temperature; preparing foods seveal hours before serving; inadequae reheating of leftovers

Incubation (Latency) Period Usually 12 to 72 hours

Bacterial Agents

Cholera

1. Agent: endoenterotoxin of *Vibrio cholerae*, classical and El Tor biotypes, from feces of infected humans
2. Incubation: 1 to 3 days
3. Symptoms: Profuse, watery diarrhea (rice-water stools), vomiting, abdominal pain, dehydration, thirst, collapse, reduced skin turgor, wrinkled fingers, sunken eyes
4. Foods: raw fish and shellfish, foods washed or prepared with contaminated water
5. Specimens: stools
6. Causative factors: obtaining fish and shellfish from sewage-contaminated waters in endemic areas; poor personal hygiene; infected workers touching foods; inadequate cooking; using contamianted water to wash or freshen foods; inadequate sewage disposal; using night soil as fertilzer

Pathogenic *Escherichia coli* gastroenteritis

1. Agent: enterotoxigenic or invasive strains of *E. coli* from feces of infected humans and other animals
2. Incubation: 4 to 48 hours, mean 10 to 24 hours
3. Symptoms: abdominal pain, diarrhea, nausea, vomiting, fever, chills, headache, muscular pain
4. Foods: various foods, water
5. Specimens: stools, rectal swabs
6. Causative factors: infected workers touching foods; inadequate refrigeration; inadequate cooking; inadequate cleaning and disinfection of equipment

Salmonellosis

1. Agent: various serotypes of *Salmonella* from feces of infected humans and other naimals
2. Incubation: 6 to 72 hours, mean 18 to 36 hours

3. Symptoms: abdominal pain, diarrhea, chills, fever, nausea, vomiting, malaise
4. Foods: poultry and meat and their products, egg products, other foods contaminated by salmonellae
5. Specimens: stools, rectal swabs
6. Causative factors: inadequate refrigeration; holding foods at warm (bacterial incubating) temperatures; inadequate cooking and reheating; preparing foods several hours before serving; cross contamination; inadequate cleaning of equipment; infected workers touching cooked foods; obtaining foods from contaminated sources

Shigellosis

1. Agents: *Shigella flexneri*, *S. dysenteriae*, *S. sonnei*, and *S. boydii*
2. Incubation: 24 to 72 hours
3. Symptoms: abdominal pain, diarrhea, bloody and mucoid stools, fever
4. Foods: any contaminated foods, frequently salads, water
5. Specimens: stools or rectal swabs
6. Causative factors: infected workers touching foods; inadequate refrigeration; inadequate cooking and reheating

*Vibrio parahaemolyticus* gastroenteritis

1. Agent: *V. parahaemolyticus* from sea water or seafoods
2. Incubation: 2 to 48 hours, mean 12 hours
3. Symptoms: abdominal pain, diarrhea, nausea, vomiting, fever, chills, headache
4. Foods: raw seafoods, shellfish
5. Specimens: stools or rectal swabs
6. Causative factors: inadequate cooking; inadequate refrigeration; cross contamination; inadequate cleaning of equipment; using sea water in food preparation

Viral gastroenteritis

1. Agent: enteric viruses (ECHO viruses, Coxsackie viruses, reoviruses, adenoviruses)
2. Incubation: 3 to 5 days
3. Symptoms: dairrhea, fever, vomiting, abdominal pain, sometimes respiratory symptoms
4. Foods: variable
5. Specimens: stools
6. Causative factors: poor personal hygiene, infected workers touching foods, inadequae cooking and reheating

Parasitic Agents

**Table 3.1.4. (continued)**

Amoebic dysentery (amoebiasis)
1. Agent: *Entamoeba histolytica* from feces of infected humans
2. Incubation: 5 days to several months, mean 3 to 4 weeks
3. Symptoms: abdominal pain, constipation or diarrhea containing blood and mucus
4. Foods: raw vegetables and fruits
5. Specimens: stools
6. Causative factors: poor personal hygiene; infected workers touching foods; inadequate cooking

Beef tapeworm infection (taeniasis)
1. Agent: *Taenia saginata* from flesh of infested cattle
2. Incubation: 3 to 6 months
3. Symptoms: vague discomfort, hunger pain, loss of weight, abdominal pain
4. Foods: raw or insufficiently cooked beef
5. Specimens: stools
6. Causative factors: lack of meat inspection; inadequate cooking; inadequate sewage-disposal; sewage-contaminated pastures

Fish tapeworm infection (diphyllobothriasis)
1. Agent: *Diphyllobothrium latum* from flesh of infested fish
2. Incubation: 5 to 6 weeks
3. Symptoms: vague gastrointestinal discomfort; anemia may occur.
4. Foods: raw or insufficiently cooked freshwater fish
5. Specimens: stools
6. Causative factors: inadequate cooking; inadequate sewage disposal; sewage contaminated lakes

Giardiasis
1. Agent: *Giardia lamblia* from feces of infected humans
2. Incubation: 1 to 6 weeks
3. Symptoms: abdominal pain, mucoid diarrhea, fatty stools
4. Foods: raw vegetables and fruits, water
5. Specimens: stools
6. Causative factors: personal hygiene; infected workers touching foods; inadequate cooking; inadequate sewage disposal

Pork tapeworm infection (taeniasis)
1. Agent: *Taenia solium* from flesh of infested swine
2. Incubation: 3 to 6 weeks
3. Symptoms: vague discomfort, hunger pains, loss of weight
4. Foods: raw or insufficiently cooked pork
5. Specimens: stools

6. Causative factors: lack of meat inspection; inadequate cooking; inadequate sewage disposal; sewage-contaminated pastures

Source: Adapted from Bryan, 1982.

**Table 3.1.5. Foodborne Diseases Causing Neurological Signs and Symptoms (Visual Disturbances, Tingling, Paralysis)**

Incubation (Latency) Period Usually Less Than 1 Hour

Fungal Agents

Ibotenic acid group of mushroom poisoning

1. Agent: ibotenic acid and muscinol in some mushrooms
2. Incubation: 30 to 60 minutes
3. Symptoms: drowsiness and state of intoxication, confusion, muscular spasms, delirium, visual disturbances
4. Foods: *Amanita muscaria, A. pantherina*, and related species of mushrooms
5. Specimens: mushrooms, urine
6. Causative factors: eating *Amanita muscaria* and related species of mushrooms; eating unknown varieties of mushrooms; mistaking toxic mushrooms for edible varieties.

Muscarine group of mushroom poisoning

1. Agent: muscarine in some mushrooms
2. Incubation: 15 minutes to 2 hours
3. Symptoms: excessive salivation, perspiration, tearing, reduced blood pressure, irregular pulse, pupils constricted, blurred vision, asthmatic breathing
4. Foods: *Clitocybe dealbata, C. rivulosa* and many species of *Inocybe* and *Boletus* mushrooms
5. Specimens: vomitus
6. Causative factors: eating muscarine group of mushrooms; eating unknown varieties of mushrooms; mistaking toxic mushrooms for edible varieties

Chemical Agents

Organophosphorus poisoning

1. Agent: organic phosphorus insecticides, such as parathion, TEPP, diazinon, malathion
2. Incubation: a few minutes to a few hours
3. Symptoms: nausea, vomiting, abdominal cramps, diarrhea, headache, nervousness, blurred vision, chest pain, cyanosis, confusion, twitching, convulsions

**Table 3.1.5. (continued)**

4. Foods: any accidentally contaminated food
5. Specimens: blood urine, fat biopsy
6. Causative factors: spraying foods just before harvesting; storing insecticides in same area as foods; mistaking pesticides for powdered foods

Toxic Animals

Paralytic shellfish poisoning

1. Agent: saxitoxin and similar toxins from plankton, *Gonyaulax* species, which are consumed by shellfish
2. Incubation: a few minutes to 30 minutes
3. Symptoms: tingling, burning, numbness around lips and finger tips, giddiness, incoherent speech, respiratory paralysis
4. Foods: mussels and clams
5. Specimens: clams or mussels, sea water
6. Causative factors: harvesting shellfish from waters with high concentratoin of *Gonyaulax*

Tetraodon poisoning

1. Agent: Tetrodotoxin from intestines and gonads of puffer-type fish
2. Incubation: 10 minutes to 3 hours
3. Symptoms: tingling sensation of fingers and toes, dizziness, pallor, numbness of mouth and extremities, gastrointestinal symptoms, hemorrhage and desquamation of skin, eyes fixed, twitching, paralysis, cyanosis
4. Foods: puffer-type fish
5. Specimens: gonads, liver, skin, muscle of fish
6. Causative factors: eating puffer-type fish; failure to effectively remove intestines and gonads from puffer-type fish if they are to be eaten

Plant Toxicants

Jimson weed

1. Agent: tropane alkaloids in Jimson weed
2. Incubation: less than 1 hour
3. Symptoms: abnormal thirst, photophobia, distorted sight, difficulty in speaking, flushing, delirium coma, rapid heart beat
4. Foods: any part of plant, tomatoes grafted to Jimson weed stock
5. Specimens: urine
6. Causative factors: eating any part of Jimson weed or eating tomatoes from tomato plants grafted to Jimson weed stock

Water hemlock poisoning

1. Agent: resin and cicutoxin in hemlock root

2. Incubation: 15 to 60 minutes
3. Symptoms: excessive salivation, nausea, vomiting, stomach pain, frothing at mouth, irregular breathing, convulsions, respiratory paralysis
4. Foods: root of water hemlock *Cicuta virosa* and *C. masculate*
5. Specimens: urine
6. Causative factors: eating water hemlock; mistaking waterhemlock root for wild parsnip, sweetpotato, or carrot

Incubation (Latency) Period 1 to 6 Hours

Chemical Agents

Chlorinated hydrocarbon poisoning

1. Agent: chlorinated hydrocarbon insecticides such as aldrin, chlordane, DDT, dieldrin, endrin, lindane, and toxaphene
2. Incubation: 30 minutes to 6 hours
3. Symptoms: nausea, vomiting, paresthesia, dizziness, muscular weakness, anorexia, weight loss, confusion
4. Foods: any accidentally contaminated food
5. Specimens: blood, urine, stools, gastric washings
6. Causative factors: storing insecticides in same area as food; mistaking pesticides for powdered foods

*Ciguatera* Poisoning

1. Agent: ciguatoxin in intestines, roe, gonads, and flesh of tropical marine, fish
2. Incubation: 3 to 5 hours, sometimes longer
3. Symptoms: tingling and numbness about mouth, metallic taste, dry mouth, gastrointestinal symptoms, watery stools, muscular pain, dizziness, dilated eyes, blurred vision, prostration, paralysis
4. Foods: numerous varieties of tropical fish
5. Specimens: fish gonads, liver, intestines, and muscles
6. Causative factors: eating liver, intestines, roe, gonads, or flesh of tropical reef fishes; large reef fish are more commonly toxic

Incubation (Latency) Period Usually 12 to 72 Hours

Bacterial Agents

Botulism

1. Agent: exoneurotoxins A, B, E, and F of *Clostridium botulinum*; spores found in soil and animal intestines
2. Incubation: 2 hours to 8 days, mean 18 to 36 hours
3. Symptoms: vertigo; double or blurred vision; dryness of mouth; difficulty in swallowing; speaking and breathing; descending

**Table 3.1.5. (continued)**

muscular weakness; constipation; pupils dilated or fixed; respiratory paralysis; gastrointestinal symptoms may precede neurological symptoms; frequently fatal.
4. Foods: home canned low-acid foods, vacuum-packed fish; fermented fish eggs, fish, and marine mammals
5. Specimens: blood, stool
6. Causative factors: inadequate heat processing of canned foods and smoked fish; uncontrolled fermentations

Incubation (Latency) Period Greater Than 72 Hours

Chemical Agents

Mercury poisoning

1. Agent: methyl and ethyl mercury compounds from industrial waste and organic mercury in fungicides
2. sIncubation: 1 week or longer
3. Symptoms: numbness, weakness of legs, spastic paralysis, impairment of vision, blindness, coma
4. Foods: grains treated with mercury-containing fungicide; pork, fish, and shellfish exposed to mercury compounds
5. Specimens: urine, blood, hair
6. Causative factors: streams polluted with mercury compounds; feeding animals grains treated with mercury fungicides; eating mercury-treated grains or meat from animals fed such grains

Triorthocresyl phosphate poisoning

1. Agent: triorthocresyl phosphate used as extracts or as substitute cooking oil
2. Incubation: 5 to 21 days, mean 10 days
3. Symptoms: gastrointestinal symptoms, leg pain, ungainly high-stepping gait, foot and wrist drop
4. Foods: cooking oils, extracts, and other foods contaminated with triorthocresyl phosphate
5. Specimens: suspect foods; autopsy to ascertain myelin sheath and spinal cord lesions; bIopsy to study gastrocnemius muscle
6. Causative factors: using compound as food extractant or as a cooking or salad oil

Source: Adapted from Bryan, 1982.

**Table 3.1.6. Foodborne Diseases Causing Generalized Infection Signs and Symptoms (Fever, Chills, Malaise, Aches)**

---

Incubation (Latency) Period Greater Than 72 Hours

Bacterial Agents

Brucellosis

1. Agent: *Brucella abortus, B. melitensis*, and *B. suis* from tissues and milk of infected animals
2. Incubation: 7 to 21 days
3. Symptoms: fever, chills, sweats, weakness, malaise, headache, muscle and joint pain, loss of weight
4. Foods: raw milk, goat cheese
5. Specimens: blood
6. Causative factors: failure to pasteurize milk; livestock infested with brucellae

Q fever

1. Agent: *Coxiella burnetii* from tissue and milk of infected animals
2. Incubation: 14 to 26 days, mean 20 days
3. Symptoms: chills, headache, weakness, malaise, perspiration, fever, cough, chest pain
4. Foods: raw milk (rare)
5. Specimens: blood
6. Causative factors: ingestion of contaminated raw milk; failure to pasteurize milk (145°F; 30 minutes; 161°F, 15 seconds)

Typhoid fever

1. Agent: *Salmonella typhi* from feces of infected humans
2. Incubation: 7 to 28 days, mean 14 days
3. Symptoms: malaise, headache, fever, cough, nausea, vomiting, constipation, abdominal pain, chills, rose spots, bloody stools
4. Foods: shellfish, foods contaminated by workers, raw milk, cheese, watercress, water
5. Specimens: stools, rectal swabs, blood
6. Causative factors: infected workers touching foods; poor personal hygiene; inadequate cooking; inadequate refrigeration; inadequate sewage disposal; obtaining foods from unsafe sources; harvesting shellfish from sewage-contaminated waters

Viral Agents

Hepatitis A (infectious hepatitis)

1. Agent: hepatitis virus A from feces, urine, blood of infected humans and other primates

**Table 3.1.6. (continued)**

2. Incubation: 10 to 50 days, mean 24 days
3. Symptoms: fever, malaise, lassitude, anorexia, nausea, abdominal pain, jaundice
4. Foods: shellfish, any food contamianted by hepatitis viruses, water
5. Specimens: urine blood
6. Causative factors: infected workers touching foods; poor personal hygiene; inadequate cooking; harfesting (and subsequent sale or consumption of) shellfish from sewage contamianted waters; inadequate sewage disposal

Parasitic Agents

Angiostrongyliasis (eosinophillis meningoencephalitis)

1. Agent: *Angiostrongylus contonensis* (rat lung worm) from rodent feces and soil
2. Incubation: 14 to 16 days
3. Symptoms: gastroenteritis, headache, stiff neck and back, low-grade fever
4. Foods: raw crabs, prawns, slugs, shrimp, snails
5. Specimens: blood
6. Causative factors: inadequate cooking

Toxoplasmosis

1. Agent: *Toxoplasma gondii* from tissue and flesh of infected animal
2. Incubation: 10 to 13 days
3. Symptoms: fever, headache, myalgia, rash
4. Foods: raw or insufficiently cooked meat (rare)
5. Specimens: biopsy of lymph nodes, blood
6. Causative factors: inadequate cooking of meat of sheep, swine, and cattle

Trichinosis

1. Agent: *Trichinella spiralis* (roundworm) from flesh of infested swine, bear
2. Incubation: 4 to 28 days, mean 9 days
3. Symptoms: gastroenteritis, fever, edema about eyes, muscular pain, chills, prostration, labored breathing
4. Foods: pork, bear meat, walrus flesh
5. Specimens: muscle biopsy
6. Causative factors: eating raw or inadequately cooked pork or bear meat; inadequate cooking or heat processing; feeding uncooked or inadequately heat processed garbage to swine

Source: Adapted from Bryan, 1982.

**Table 3.1.7. Foodborne Disease Causing Allergic-Type Symptoms (Facial Flushing, Itching)**

---

Incubation (Latency) Period Less Than 1 Hour

Bacterial (and Animal) Agents

Scombroid poisoning

1. Agent: histamine-like substance produced by *Proteus* spp. or other bacteria from histidine in fish flesh
2. Incubation: a few minutes to 1 hour
3. Symptoms: headache, dizziness, nausea, vomiting, peppery taste, burning throat, facial swelling and flushing, stomach pain, itching of skin
4. Foods: tuna, mackerel, pacific dolphin
5. Specimens: vomitus
6. Causative factors: inadequate refrigeration of scombroid fish

Chemical Agents

Chinese Restaurant Syndrome

1. Agent: monosodium glutamate (MSG)
2. Incubation: a few minutes to 1 hour
3. Symptoms: burning sensation in back of neck, forearms, chest; feeling of tightness; tingling, flushing, dizziness, headache, nausea
4. Foods: Chinese foods
5. Specimens: foods
6. Causative factors: using excessive amounts of MSG as a flavor intensifier

Nicotinic acid (niacin) poisoning

1. Agent: sodium nicotinate used as color preservatives
2. Incubation: a few minutes to 1 hour
3. Symptoms: flushing, sensation of warmth, itching, abdominal pain, puffing of face and knees
4. Foods: meat or other food in which sodium nicotinate has been added
5. Specimens: foods
6. Causative factors: using sodium nicotinate as color preservative

Incubation (Latency) Period 1 to 6 Hours

Toxic Animals

Hypervitaminosis A

1. Agent: vitamin A
2. Incubation: 1 to 6 hours

**Table 3.1.7. (continued)**

3. Symptoms: headache, gastrointestinal symptoms, dizziness, collapse, convulsions, insomnia, desquamation of skin
4. Foods: liver and kidney of Arctic mammals
5. Specimens: blood
6. Causative factors: eating liver and kidney of animals from cold regions

Source: Adapted from Bryan, 1982.

## 3.2 Pathogenic Bacteria

**Table 3.2.1 Profiles of Major Pathogenic Bacteria**

*Bacillus anthracis*: Bacillaceae (family); *Bacillus* (genius)

1. Description: gram-positive, sporogenous, immotile rod
2. Disease transmitted: anthrax
3. Sources: soil, water, and infected animals
4. Oxygen requirement: facultative anaerobic
5. pH (for growth): range, none; optimum, 7.0-7.2
6. $^{o}$C (for growth): range, 12-43; optimum, 35-37
7. Resistant to 100$^{o}$C for 10-12 minutes

*Brucella abortus* (*B. melitensis*): family undefined; *Brucella* (genius)

1. Description: gram-negative, immotile rod
2. Disease transmitted: brucellosis
3. Sources: milk products of infected animals; feces
4. Oxygen requirement: aerobic
5. pH (for growth): range, 6.0-7.4; optimum, 7.0-7.2
6. $^{o}$C (for growth): range, 8-43; optimum, 37
7. Resistant to 60$^{o}$C for 10 minutes

*Brucella melitensis*: family undefined; *Brucella* (genius)

1. Description: gram-negative, immotile rod
2. Disease transmitted: brucellosis
3. Sources: milk products of infected animals; feces
4. Oxygen requirement: aerobic
5. pH (for growth): range, 6.0-7.4; optimum, 7.0-7.2
6. $^{o}$C (for growth): range, 8-43; optimum, 37
7. Resistant to 60$^{o}$C for 10 minutes

*Campylobacter jejuni*: Spirillaceae (family); *Campylobacter* (genius)

1. Description: gram-negative, slender, curved bacteria, motile by means of a single, polar flagellum

2. Disease transmitted: gastroenteritis
3. Sources: milk products of infected animals; poultry
4. Oxygen requirement: obligate microaerophile
5. pH (for growth): range, 5.5-8.0; optimum, 6.5-7.5
6. °C (for growth): range, 30-45; optimum, 42-45
7. Resistant to 48°C for 7.2-12.8 minutes

*Clostridium botulinum* (types A, B, and E): Bacillaceae (family); *Clostridium* (genius)

1. Description: gram-positive, motile rod
2. Disease transmitted: botulism
3. Sources: canned food; soil and water
4. Oxygen requirement: obligate anaerobic
5. pH (for growth): range, 5.4 or higher; optimum, 7.0
6. °C (for growth): range, 10-37; optimum, 20-30
7. Spores resistant to 100°C for 5 hours, 105°C for 2 hours, or 120°C for 10 minutes

*Clostridium perfringens*: Bacillaceae (family); *Clostridium* (genius)

1. Description: gram-positive, sporogenous, immotile rod
2. Disease transmitted: gas gangrene
3. Sources: cooked food left standing, soil, and carrier's feces
4. Oxygen requirement: obligate anaerobic
5. pH (for growth): range, 5.7 or higher; optimum, 7.2
6. °C (for growth): range, ~50; optimum, 44
7. Resistant to 90°C for 35 minutes or 100°C for 5 minutes. The pathogenic bacteria are resistant to 100°C for 1-4 hours.

*Corynebacterium diphtheriae*: Coryneform group of bacteria; *Corynebacterium* (genius)

1. Description: gram-positive, immotile, club-shaped rod; propionic fermentation
2. Disease transmitted: diphtheria
3. Sources: human respiratory tract, other mucus membranes, and skin
4. Oxygen requirement: aerobic
5. pH (for growth): range, none; optimum, 7.2-7.8
6. °C (for growth): range, 15-40; optimum, 34-37
7. Resistant to 58°C for 10 minutes

*Escherichia coli* (enteropathogenic): Enterobacteriaceae (family); *Escherichia* (genius)

1. Description: gram-negative, short, plump rod; mixed-acid fermentation; motile
2. Disease transmitted: infantile diarrhea, septicemia, pyelitis, pyelonephritis
3. Sources: in the intestine and feces

**Table 3.2.1. (continued)**

4. Oxygen requirement: facultative anaerobic
5. pH (for growth): range, 4.4-9.6; optimum, 6.0-7.6
6. $^{o}$C (for growth): range, 10-45; optimum, 37
7. Resistant to 55$^{o}$C for 1 hour or 60$^{o}$C for 15 minutes

*Escherichia freundii* (enteropathogenic): Enterobacteriaceae (family); *Escherichia* (genius)

1. Description: gram-negative, short rod; motile or immotile
2. Disease transmitted: food poisoning
3. Sources: free-living organism in the intestine and feces
4. Oxygen requirement: facultative anaerobic
5. pH (for growth): range, none; optimum, 6.0-7.6
6. $^{o}$C (for growth): range, 20-45; optimu, 37
7. Sensitive to heat

*Leptospira icterohaemorrhagiae*: Spirochaetaceae (family); *Leptospira* (genius)

1. Description: gram-negative, tightly coiled, thin, flexible organism; motility by axial filament
2. Disease transmitted: leptospirisis
3. Sources: water, food, and infected animals
4. Oxygen requirement: aerobic
5. pH (for growth): range, none; optimum, 7.2-7.4
6. $^{o}$C (for growth): range, 20-37; optimum, 25-30
7. Resistant to 50-55$^{o}$C for 30 minutes

*Listeria monocytogenes*: family undefined; *Listeria* (genius)

1. Description: small, short, nonsporing, gram-positive rods, flagellated, motile
2. Disease transmitted: food poisoning
3. Sources: beef and poultry carcasses, milk, cheeses
4. Oxygen requirement: aerobic
5. pH (for growth): range, 4.4-9.6; optimum, 7.0
6. $^{o}$C (for growth): range, 1.0-45; optimum, 30-37
7. Resistant to 62.8$^{o}$C for 30 minutes

*Mycobacterium tuberculosis*: Mycobacteriaceae (family); *Mycobacterium* (genius)

1. Description: gram-positive, pleomorphic, immotile rod; acid-fast
2. Disease transmitted: tuberculosis
3. Sources: infected cattle and milk
4. Oxygen requirement: aerobic
5. pH (for growth): range, 4.5-8.0; optimum, 6.8-7.2
6. $^{o}$C (for growth): range, 30-44; optimum, 37
7. Resistant to 60$^{o}$C for 20-30 minutes

*Paracolobactrum arizonae*: Enterobacteriaceae (family); *Paracolobactrum* (genius)

1. Description: gram-negative, motile rod
2. Disease transmitted: food poisoning
3. Sources: widely distributed in lizards, snakes, and warm-blooded animals
4. Oxygen requirement: facultative anaerobic
5. pH (for growth); range, 4.5-9.6; optimum, 6.0-7.5
6. $^{o}$C (for growth): range, 5-49; optimum, 35-37
7. Resistant to 55$^{o}$C for 40 minutes (ca)

*Proteus vulgaris* (*P. mirabilits*): Enterobacteriaceae (family); *Proteus* (genius)

1. Description: gram-negative, actively motile rod
2. Disease transmitted: gastroenteritis
3. Source: putrefying materials
4. Oxygen requirement: facultative anaerobic
5. pH (for growth): range, 4.4-9.2; optimum, 6.0-7.0
6. $^{o}$C (for growth): range, 10-43; optimum, 30-37
7. Resistant to 55$^{o}$C for <60 minutes

*Proteus morganii*: Enterobacteriaceae (family); *Proteus* (genius)

1. Description: gram-negative, motile rod
2. Disease transmitted: diarrhea
3. Sources: intestinal canals and in normal or diarrheal stools
4. Oxygen requirement: facultative anaerobic
5. pH (for growth): range, 4.0-9.0; optimum, 6-0-8.5
6. $^{o}$C (for growth): range 10-45; optimum, 30-37
7. Resistant to 55$^{o}$C for 40 minutes (ca)

*Pseudomonas aeruginosa*: Pseudomonadaceae (family); *Pseudomonas* (genius)

1. Description: gram-negative, small, slender rod; actively motile
2. Disease transmitted: Pseudomoniasis
3. Sources: soil, water, sewage, normal intestinal flora, and human skin
4. Oxygen requirement: aerobic
5. pH (for growth): range, 6.0-9.3; optimum, 6.6-8.4
6. $^{o}$C (for growth): range 5-42; optimum, 37
7. Resistant to 50$^{o}$C for 1 hour

*Salmonella enteritidis*: Enterobacteriaceae (family); *Salmonella* (genius)

1. Description: gram-negative, motile rod; mixed-acid fermentation
2. Disease transmitted: enteric fever, gastroenteritis, salmonella, septicemia
3. Sources: sewage water, foods, feces of infected animal
4. Oxygen requirement: facultative anaerobic
5. pH (for growth): range, 4.5-8; optimum, 7.2-7.3
6. $^{o}$C (for growth): range, 15-48.8; optimum, 37
7. Resitant to 53.3$^{o}$C for 6 hours

**Table 3.2.1. (continued)**

*Shigella dysenteriae* (*S. flexneri* and *S. sonnei*): Enterobacteriaceae (family); *Shigella* (genius)
1. Description: gram-negative, immotile rod
2. Disease transmitted: dysentery
3. Sources: tap water, milk products, intestinal tract and feces of infected men and monkeys
4. Oxygen requirement: facultative anaerobic
5. pH (for growth): range, 6-8; optimum, 7.2-7.4
6. $^{O}$C (for growth): range, 10-40; optimum, 37
7. Resistant to 60$^{O}$C for 30 minutes

*Staphylococcus aureus*: Micrococcaceae (family); *Staphylococcus* (genius)
1. Description: gram-positive, immotile coccus; grouped in grapelike clusters
2. Disease transmitted: food poisoning
3. Sources: human respiratory tract and skin; suppuration of wounds
4. Oxygen requirement: facultative anaerobic
5. pH (for growth): range, 4.5-9.8; optimum, 7.2-7.6
6. $^{O}$C (for growth): range, 6.7-45; optimum, 35-37
7. Resistant to 60$^{O}$C for 30-60 minutes or 48.8$^{O}$C for 12-18 hours

*Staphylococcus pyogenes*: Streptococcaceae (family); *Streptococcus* (genius)
1. Description: gram-positive coccus in chains; hemolytic; lactic acid homofermenter
2. Disease transmitted: infection, e.g., septicemias
3. Sources: normal flora of carrier's throat, nasal passage
4. Oxygen requirement: facultative anaerobic
5. pH (for growth): range, 4.0-9.6; optimum, 7.4-7.6
6. $^{O}$C (for growth): range, 20-40; optimum, 35-37
7. Resistant to 60$^{O}$C for 30 minutes

*Streptococcus faecalis*: Streptococcaceae (family); *Streptococcus* ((genius)
1. Description: gram-positive; occurring in chains
2. Disease transmitted: food poisoning
3. Sources: normal flora of intestinal tract and sometimes in food
4. Oxygen requirement: facultative anaerobic
5. pH (for growth): range, 4.0-9.6; optimum, 7.4-7.6
6. $^{O}$C (for growth): range, 10-45; optimum, 37
7. Resistant to 62.8$^{O}$C for 30 minutes

*Vibrio cholerae*: Vibrionaceae (family): *Vibrio* (genius)
1. Description: gram-negative, motile, curved rod; mixed-acid fermentation
2. Disease transmitted: cholera
3. Sources: contaminated water, soil, and feces
4. Oxygen requirement: aerobic
5. pH (for growth): range, 6.0-9.3; optimum, 6.6-8.4

6. °C (for growth): range, 23-37; optimum, 34
7. Resistant to 56°C for 15 minutes

*Vibrio parahaemolyticus*: Vibrionaceae (family): *Vibrio* (genius)

1. Description: gram-negative, polymorphous rod; halophilic; motile
2. Disease transmitted: food poisoning of infectious type resembling salmonellosis
3. Sources: seafood, water, and soil
4. Oxygen requirement: facultative anaerobic
5. pH (for growth): range, 4.8-9.4; optimum, 7.5-8.8
6. °C (for growth): range, 5-37; optimum, 37
7. Resistant to 80°C for 15 minutes

*Vibrio vulnificus*: Vibrionaceae (family); *Vibrio* (genius)

1. Description: gram-negative, polymorphous rod; halophilic; motile
2. Disease transmitted: food poisoning of infectious type, resembling salmonellosis
3. Sources: seafood, water, and soil
4. Oxygen requirement: facultative anaerobic
5. pH (for growth): range, 4.8-9.4; optimum, 7.5-8.8
6. °C (for growth): range, 5-37; optimum, 37
7. Resistant to 80°C for 15 minutes

*Yersinia enterocolitica:* family undefined; *Yersinia* (genius)

1. Description: gram-negative, non-spore-forming, motile, rods and cocci
2. Disease transmitted: food poisoning
3. Sources: chocolate milk, milk powder, chow mein, tofu, and pasteurized milk
4. Oxygen requirement: facultative anaerobic
5. pH (for growth): range, 6.8-9.0; optimum, 7.0-7.2
6. °C (for growth): range 20-40, optimum, 32-34
7. Resistant to 62.8°C for 0.24-0.96 minutes

---

Sources: Adapted from Banwart, 1989; Cliver, 1990; Doyle, 1989; Jay, 1986; Pierson, 1986.

3.3. Parasites

**Table 3.3.1 List of Food-Associated Parasites**

| Parasite | Pathogenicity | Stage Transmitted | Material Contaminated | Reservoir Hosts |
|---|---|---|---|---|
| Protozoa | | | | |
| Amoebas | | | | |
| *Dientamoeba fragilis* | Yes | Trophozoite | Food, water | Monkey |
| *Endolimax nana* | No | Cyst | Food, water | Monkey |
| *Entamoeba coli* | No | Cyst | Food, water | Monkey, dog |
| *Entamoeba histolytica* | Yes | Cyst | Food, water | Monkey, pig |
| *Iodamoeba butschlii* | No | Cyst | Food, water | Pig |
| Flagellates | | | | |
| *Chilomastix mesnili* | No | Cyst | Food, water | Monkey |
| *Enteromonas hervei* | No | Cyst | Food, water | Unknown |
| *Enteromonas hominis* | No | Cyst | Food, water | Unknown |
| *Giardia lamblia* | Yes | Cyst | Food, water | Rat, gerbil, guinea pig, beaver, dog, raccoon, sheep, pronghorn |
| *Retortamonas intestinalis* | No | Cyst | Food, water | Unknown |
| *Trichomonas hominis* | No | Trophozoite | Food, water | Primates |
| *Trichomonas tenax* | No | Trophozoite | Food, water | Unknown |
| Ciliates | | | | |
| *Balantidium coli* | Yes | Cyst | Food, water | Pig, monkey |

| | | | | |
|---|---|---|---|---|
| Sporozoa | | | | |
| *Isospora* spp. | Yes | | | |
| *Sarcocystis lindemanni* | Yes | Cyst | Food, water | Unknown primates |
| *Sarcocystis* spp. | Yes | Muscle cyst | Meat, poultry | Cow, dog |
| *Toxoplasma gondii* | Yes | Oocyst, trophozoite | Food, water, meat | Cat, mammals, birds |
| Platyhelminthes | | | | |
| Termatodes | | | | |
| *Centrocestus armatus* | Yes | Metacercaria | Fish | Dog, rabbit, rat, mouse |
| *Centrocestus formosanus* | Yes | Metacercaria | Fish | Rat, cat, dog, birds |
| *Clonorchis sinensis* | Yes | Metacercaria | Fish | Dog, cat, other carnivores |
| *Diorchitrema amplicaecale* | Yes | Metacercaria | Fish | Dog |
| *Diorchitrema formosanum* | Yes | Metacercaria | Fish | Rat |
| *Diorchitrema pseudocirratum* | Yes | Metacercaria | Fish | Dog, cat |
| *Echinochasmus perfoliatus* | No | Metacercaria | Fish | Dog, cat, pig, fox |
| *Echinostoma ilocanum* | Yes | Metacercaria | Mollusks | Rat, dog |
| *Echinostoma lindoense* | Yes | Metacercaria | Mollusks | Rat, mouse |
| *Fasciola gigantica* | Yes | Metacercaria | Vegetables | Cattle, camel, herbivorous mammals |
| *Fasciola hepatica* | Yes | Metacercaria; adult | Vegetables, liver | Sheep, herbivorous and ominvorous mammals |
| *Fasciolopsis buski* | Yes | Metacercaria | Vegetables | Pig |
| *Gastrodiscoides hominis* | Yes | Metacercaria | Vegetables | Pig, rat, deer (*Tragulus napu*), |

**Table 3.3.1. (continued)**

| Parasite | Pathogenicity | Stage Transmitted | Material Contaminated | Reservoir Hosts |
|---|---|---|---|---|
| | | | | rhesus monkey |
| *Haplorchis microrchia* | Yes | Metacercaria | Fish | birds |
| *Haplorchis pumilio* | Yes | Metacercaria | Fish, frogs | Birds, mammals |
| *Haplorchis taichui* | Yes | Metacercaria | Fish | Birds, mammals |
| *Haplorchis yokogawai* | Yes | Metacercaria | Fish, shrimp | Birds, cat, dog, water rat |
| *Heterophye heterophyes* | Yes | Metacercaria | Fish | Dog, cat, fox, fish-eating mammals |
| *Himasthla muehlensi* | Yes | Metacercaria | Clams | Birds |
| *Isoparorchis hypselobagri* | No | Metacercaria | Fish | Catfish, eel |
| *Metagonimus minutus* | Yes | Metacercaria | Fish | |
| *Metagonimus yokogawai* | Yes | Metacercaria | Fish | Fish-eating mammals, Pelicans |
| *Nanophyetus salmincola* | Yes | Metacercaria | Fish | Dog, fox, cat, bear, mink |
| *Opisthorchis felineus* | Yes | Metacercaria | Fish | Dog, fox, cat, seal, wolverine |
| *Paragonimus westermani* | Yes | Metacercaria | Crustaccea | Tiger, leopard |
| *Stellantchasmus falcatus* | Yes | Metacercaria | Fish | Dog, cat, shrew, birds |
| *Watsonius watsoni* | Yes | Metacercaria | Vegetables | Primates |
| Cestodes | | | | |
| *Braunia jasseyensis* | Yes | Pleurocercoid | Fish (probably) | Unknown |
| *Digramma brauni* | No | Pleurocercoid | Fish | Birds |

| | | | | |
|---|---|---|---|---|
| *Diphyllobothrium latum* | Yes | Pleurocercoid | Fish | Dog, bear |
| *Diplogonoporus grandis* | Yes | Pleurocercoid | Fish | Whales |
| *Echinococcus granulosus* | Yes | Egg | Food, water | Sheep, cattle, pig, other mammals |
| *Echinococcus multilocularis* | Yes | Egg | Food, water | Mammals |
| *Hymenolepis diminuta* | Yes | Cysticercus | Grain contaminated with beetles | Mouse, rat |
| *Hymenolepis nana* | Yes | Egg | Food | Mouse, rat |
| *Ligula intestinalis* | No | Pleurocercoid | Fish | Birds |
| *Multiceps multiceps* | Yes | Egg | Food, water | Sheep |
| *Multiceps serialis* | Yes | Egg | Food, water | Dog, other canines |
| *Spirometra* sp. | Yes | Procercoid, pleuroceroid | Water with cyclops, meat | Dog, cat |
| | | | | Dog, cat |
| *Taenia saginata* | Yes | Cysticercus, egg | Beef | Unknown |
| *Taenia solium* | Yes | Cysticercus | Pork | Gibbon |
| Nematodes | | | | |
| *Angiostrongylus cantonensis* | Yes | Larva | Vegetables, mollusks, prawns | Rat |
| *Antiostrongylus costaricensis* | Yes | Larva | Mollusks (slugs on vegetables) | Rat |
| *Anisakis* sp. | Yes | Larva | Fish | Whale |
| *Ascaris lumbricoides* | Yes | Egg | Vegetables | Pig |
| *Capillaria hepatica* | Yes | Egg | Food, water | Rodents, other mammals |
| *Capillaria philippinensis* | Yes | Larva | Fish | Rat |

**Table 3.3.1. (continued)**

| Parasite | Pathogenicity | Stage Transmitted | Material Contaminated | Reservoir Hosts |
|---|---|---|---|---|
| *Contracaecum* sp. | Yes | Larva | Fish | Birds, marine mammals |
| *Dioctophyma renale* | Yes | Larva | Fish | Mammals |
| *Echinocephalus* sp. | Yes | Larva | Oysters | Elasmobranchs |
| *Enterobius vermicularis* | Yes | Egg | Food, water | Chimpanzee |
| *Gnathostoma spinigerum* | Yes | Larva | Fish, frog | Felines, mink, raccoon, dog |
| *Gnathostoma* sp. | Yes | Larva | Fish, poultry | Pig |
| *Haemonchus contortus* | Yes | Larva | Vegetables | Herbivores |
| *Phocanema decipiens* | Yes | Larva | Fish | Seal |
| *Porrocaecum* sp. | Yes | Larva | Fish | Mammals, birds |
| *Rhabditis* sp. | No | Larva | Mollusks, water | Invertebrates |
| *Syngamus laryngeus* | No | Larva | Food, water | Bovines |
| *Syphacia obvelata* | No | Egg | Food, water | Rat, mouse |
| *Toxocara canis* | Yes | Egg | Food, water | Canines |
| *Toxocara cati* | Yes | Egg | Food, water | Felines |
| *Trichinella spiralis* | Yes | Larva | Meat | Mammals |
| *Trichostrongylus* sp. | Yes | Larva | Vegetables | Herbivores |
| *Trichuris trichiura* | Yes | Egg | Food, water | Monkey, pig |
| Acanthocephala | | | | |
| *Moniliformis moniliformis* | Yes | Larva | Grain contaminated with beetles | Rat, mouse |

Source: Adapted from Bier, 1981.

## 3.4 Unsafe Herbs

**Table 3.4.1 Some Unsafe Herbs that Should Not Be Used in Foods, Beverages or Drugs**

| Botanical Name | Common Names | Poisonous Nature |
|---|---|---|
| *Arnica montana* L. | Arnica, arnica flower, wolf's-bane, leopard's bane, mountain tobacco, flores arnicae | Aqueous and alcoholic extracts of the plant contain choline, plus two unidentified substances that affect the heart and vascular systems. Arnica, an active irritant, can produce violent toxic gastroenteritis, nervous distrubances, change in pulse rate, intense muscular weakness, collapse, and death. |
| *Atropa belladonna* L. | Belladonna, deadly nightshade | Poisonous plant that contains the toxic solanaceous alkaloids hyoscyamine, atropine, and hyoscine |
| *Solanum dulcamara* L. | Bittersweet twigs, dulcamara, bittersweet, woody nightshade, climbing nightshade | Poisonous. Contains the toxic glycoalkaloid solanine; also solanidine and dulcamarin. |
| *Sanguinaris canadensis* L. | Bloodroot, sanguinaria, red puccoon | Contains the poisonous alkaloid sanguinarine and other alkaloids. |
| *Cytisus scoparius* (L.) Link. | Broom-tops, scoparius, spartium, scotch broom, Irish broom, broom | Contains toxic sparteine, isosparteine and other alkaloids; also hydroxytyramine. |
| *Aesculus hippocaster anum* L. | Buckeyes, aesculus, horse chestnut | Contains a toxic coumarin glycoside, aesculin (esculin). A poisonous plant. |

**Table 3.4.1 (continued)**

| Botanical Name | Common Names | Poisonous Nature |
|---|---|---|
| *Acorus calamus* L. | Calamus, sweet flag, sweet root, sweet cane, sweet cinnamon | Oil of calamus, Jammu variety, is a carcinogen (causes cancer). FDA regulations prohibit marketing of calamus as a food or food additive. |
| *Heliotropium europaeum* L. | Heliotrope | A poisonous plant. It contains alkaloids that produce liver damage. Not to be confused with garden heliotrope (*Varleriana officinalis* L.) |
| *Conium maculatum* L. | Hemlock, conium, poison hemlock, spotted hemlock, spotted parsley, St. Bennet's herb, spotted cowbane, fool's parsley | Contains the poisonous alkaloid coniine and four other closely related alkaloids. Often confused with water hemlock (*Cicuta maculata* L.). Not to be confused with hemlock, hemlock spruce, etc. [*Tsuga canadensis* (L). Carr.]. |
| *Hyoscyamus niger* L. | Henbane, hyoscyamus, black henbane, hog's bean, poison tobacco, devil's eye | Contains the alkaloids hyoscyamine, hyoscein (scopolamine), and atropine. A poisonous plant. |
| *Exagonium purga* (Wenderoth) Betham; *Ipomoea jalapa* Nutt, and Coxe; *Ipomoea purga* (Wenderoth) Hayne; *Exagonium jalapa* (Wenderoth) Baillon | Jalap root, jalap, true jalap, jalapa, Vera Cruz jalap, high john root, (possibly also known as high John the conqueror, John conqueror, St. John the conqueror root, hi John conqueror) | A large twining vine of Mexico, this plant has undergone many name changes. The drug is a powerful, drastic cathartic. Purgative powers of jalap reside in its resin. In overdoeses, jalap may produce dangerous hypercatharsis. |

| | | |
|---|---|---|
| *Datura stramonium* L. | Jimson weed, datura, stramonium, apple of Peru, Jamestown weed, thornapple, tolguacha | Contains the alkaloids atropine, hyoscyamine, and scopolamine. Illegal drug for nonprescription use. A poisonous plant. |
| *Convallaria majalis* L. | Lily of the valley, convallaria, May lily | Contains the toxic cardiac glycosides convallatoxin, convallarin, and convallamarin. Poisonous plant. |
| *Lobelia inflata* L. | Lobelia, Indian tobacco, wild tobacco, asthma weed, emetic weed | A poisonous plant that contains the alkaloid lobeline plus a number of other pyridine alkloids. Overdoses of the plant or extracts of the leaves or fruits produce vomiting, sweating, pain, paralysis, depressed temperatures, rapid but feeble pulse, colapse, coma, and death. |
| *Mandragora officinarum* L. | Mandrake, mandragora, European mandrake | The plant is a poisonous narcotic similar in its properties to belladonna. Contains the alkaloids hyoscyamine, scopolamine, and mandragorine. |
| *Podophyllum peltatum* L. | Mandrake, May apple, podophyllum, American mandrake, devil's apple, umbrella plant, vegetable calomel, wild lemon, vegetable mercury | A poisonous plant, it contains podophyllotoxin, a complex polycyclic substance, and other constituents. |
| *Phoraden dron flavescens* (Pursh.) Nutt; *Viscum flavescens* (Pursh.) | Mistletoe, viscum, American mistletoe | Poisonous. Contains the toxic pressor amines $\beta$-phenylethylamine and tyramine. |

**Table 3.4.1 (continued)**

| Botanical Name | Common Names | Poisonous Nature |
|---|---|---|
| *Phoradendron juniperinum* Engelm. | Mistletoe, viscum, juniper mistletoe | May be poisonous. Little is known about its properties. |
| *Viscum album* L. | Mistletoe, viscum, European mistletoe | Poisonous. Contains the toxic pressor amines $\beta$-phenylethylamine and tyramine. |
| *Ipomoea purpurea* (L) Roth | Morning glory | Contains a purgative resin. In addition, morning glory seeds contain amides of lysergic acid but with a potency much less than that of LSD. |
| *Vinca major* L. and *Vinca minor* L. | Periwinkle, vinca, greater periwinkle, lesser periwinkle | Contains pharmacologically active, toxic alkaloids such as vinblastine and vincristine that have cytotoxic and neurological actions and can injure the liver and kidneys. |
| *Hypericum perforatum* L. | St. Johnswort, hypericum, klamath weed, goatweed | A primary photosensitizer for cattle, sheep, horses and goats. Contains hypericin, a fluorescent pigment, as a photosensitizing substance. |
| *Euonymus europaeus* L. | Spindle-tree | Violent purgative |
| *Dipteryx odorata* (Aubl.) Willd.; *Coumarouna odorata* (Aubl.) and *Dipteryx* | Tonka bean, tonco bean, tonquin bean | Active constituent of seed is coumarin. Dietary feeding of coumarin to rats and dogs causes extensive liver damage, growth retardation, and testicular atrophy. FDA regulations prohibit marketing of coumarin as a food or food additive. |

| | | |
|---|---|---|
| *oppositifolia* (Aubl.) Willd.; *Coumarouna oppositifolia* (Aubl.) | | |
| *Euonymus atropur-pureus* Jacq. | Wahoo bark, euonymus, burning bush, wahoo | The poisonous principle has not been completely identified. Laxative. |
| *Eupatorium rugosum* Houtt.; *E. ogeratoides* L.f. and *E. urticaefolium* | White snakeroot, (also called snakeroot, richweed) | Poisonous plant. Contains a toxic, unsaturate alcohol called tremetol combined with a resin acid. Causes "trembles" in catle and other livestock. Milk sickness is produced in humans by ingestion of milk, butter, and possibly meat from animals poisoned by this plant. |
| *Artemisia absinthium* Linne | Wormwood, absinthium, absinth, absinthe, madderwort, wermuth, mugwort, mingwort, warmot, magenkraut, herba absinthii | Contains a volatile oil (oil of wormwood) that is an active narcotic poison. Oil of wormwood is used to flavor absinthe, an alcoholic liqueur illegal in this country because its use can damage the nervous system and cause mental deterioration. |
| *Corynanthe yohimbi* Schum.; *Pausinystalia yohimbe* (Schum.) Pierre | Yohimbe, yohimbi | Contains the toxic alkaloid yohimbine (quebrachine) and other alkaloids. |

Source: Anonymous, 1983.

## 3.5. Glossary, Microbiology, and Pathogens

**Table 3.5.1. A Glossary of Terms Associated with Microbiology and Pathogens**

Abacterial: free from bacteria

Abbe condenser: an attachment to a microscope consisting of a mirror and a series of wide-angled achromatic lenses placed beneath the stage to give strong illumination

Abscess: a localized collection of pus in a cavity formed by the disintegration of tissues

Acetobacter: a genus of aerobic bacilli that secure their energy by the oxidation of alcohol to acetic acid

Acid-fast: the property of not being readily decolorized by acids or other means when stained

*Actinomyces*: a genus of parasitic fungi characterized by a radiating arrangement of the mycelium and small spores

Acute: having a short, severe course

Adenovirus: a virus causing mild infection of the upper respiratory tract

Aerobe: an organism that grows best in the presence of free oxygen

Aerogens: gas-producing bacteria

Afebrile: without symptoms of fever

Agar-agar: a gelatinous substance prepared from Ceylon moss and used as a solidifying agent in culture media. It dissolves in boiling water and solidifies at about 38°C

Agglutination: the phenomenon of clumping of the cells distributed in a fluid, observed when certain bacterial cultures are treated with serum containing an agglutinin specific for the particular organism

Agglutinin: an antibody found in an immune serum that, when added to a suspension of its homologous microorganism causes the organisms to adhere to one another, forming clumps

Agglutinogen: the agglutinable substance or antigen that stimulates the formation of agglutinins when injected into the animal body

Agglutinoid: heat-killed agglutinin, which has lost its property of causing agglutination, although it can still unite with its corresponding agglutinogen

Agranulocytosts: complete or nearly complete absence of granular leukocytes (granulocytes) from the bone marrow and the blood

Alastrim: a specific, contagious, virus-caused, eruptive fever, resembling a mild smallpox and probably related to that disease

Alexin: see Complement

Allergen: any substance capable of inducing or exciting a condition of allergy or hypersensitiveness

Allergy: a condition of unusual or exaggerated specific susceptibility to a substance that is harmless in similar amounts for the majority of member of the same species; examples: asthma, hay fever, hives

Alum: aluminum and potassium sulfate, $AlK(SO_4)_2 + 12\ H_2O$

Amboceptor: see Cytolysin

Amebiasis: the state of being infected with amoebae

Amino acid: an organic acid in which one or more hydrogen atoms have been replaced by the amino group NH. Amino acids are the building stones of the protein molecule

Amphitrichous: having one or a tuft of flagella at each end

Ampule: a small glass container capable of being sealed so as to preserve its contents in a sterile condition

Anabolism: the constructive process by which simple substances are converted by living cells into more complex compounds

Anaerobe: an organism that can grow without either air or free oxygen

Anaphylactic shock: a violent attack of symptoms produced by a second injection of a foreign protein in a sensitive animal

Anaphylactogen: a substance that is capable of causing anaphylaxis

Anaphylaxis: allergy in which the allergen is an antigenic protein

Anatoxin: see Toxoid

Andrade's indicator: a solution of acid fuchsin, 0.5 g in 100 mL of water, which is decolorized to a yellow color by sodium hydroxide

Anemia: a condition in which the blood is deficient either in quantity or quality; the latter may consist of a decrease in the amount of hemoglobin or in the number of red blood corpuscles

Anode: the positive pole of a galvanic battery or other electric source

Anorexia: lack or loss of appetite for food

Anthrax: a fatal infectious disease of cattle and sheep, transmissible to man, caused by *Bacillus anthracis*

Antianaphylaxis: a state of desensitization in which allergic phenomena do not occur, believed due to free antibodies in the blood preventing the antigen from reaching the sensitized tissues

Antibiotic: literally, growth inhibiting; usually denotes bacteriostatic substance produced by microorganisms

Antibody: the substance produced in the animal body against an antigen and exerting a specific antagonistic influence on the substance that has stimulated its formation

Anticomplementary: the property of reducing or destroying the power of complement

**Table 3.5.1 (continued)**

Antiformin: a strongly alkaline sodium hypochlorite solution, used as a disinfectant. Because it does not dissolve acid-fast organisms, it is used to isolate them

Antigen: a substance that prevents or inhibits the growth of microorganisms without necessarily destroying them

Antiserum: a serum containing specific antibodies, obtained from the blood of an animal that has received repeated doses of an antigen or that has recovered from an infection

Antitoxin: an antibody, found in the blood serum and in other body fluids, that has been incited by injections of a corresponding toxin and is specifically antagonistic to it

Arborescent: a branching treelike growth

Arnold sterilizer: an apparatus used for sterilizing by actively streaming steam

Arthritis: inflammation of a joint

Arthus phenomenon: a transient, swollen, edematous, local reaction occurring after repeated subcutaneous injections of horse serum into a rabbit, earlier injections having been without local effect. Still further injections produce a firm indurated area with local necrosis

Ascitic fluid: serous fluid that accumulates in the peritoneal cavity in cases of dropsy of that area; sometimes used to fortify artificial culture media

Asepsis: absence of septic matter or freedom from microorganisms

*Aspergillus*: a genus of fungi characterized by rounded conidiophores thickly set with chains of colored conidia

Asporogenous: not producing spores

Asthma: a condition marked by recurrent attacks of paroxysmal dyspnea, with wheezing, cough, and a sense of constriction, due to spasmodic contraction of the bronchi

Atabrine: an acridine preparation used for the treatment of malaria

Athlete's foot: ringworm of the feet and toes caused by infection with various members of the genus *Trichophyton* and other fungi

Atopy: a term devised by Coca to designate the inherited general tendency to become sensitized which may occur in human beings

Atrichous: without flagella

Attenuation: the process of reducing or weakening virulence of a microorganism

Autoclave: an apparatus for sterilizing by steam under pressure

Authohemagglutinins: specific substances in blood serum which agglutinate the cell of the same blood as that in which they are contained

Autolysin: a substance, present in an organism, that is capable of causing the disintegration of the cells or tissues of the same organism

Autolysis: disintegration of an organism by its own enzymes

Autopsy: post mortem examination of a body

Autosensitization: sensitization toward one's own tissues; can be effected in an animal by injecting its own lens of placenta protein

Autotrophic: capable of living on inorganic matter

Auxochrome: a chemical group that furnishes salt-forming properties and is responsible for transferring the color of a dye to a substance upon which it acts

*Azotobacter*: a genus of bacteria able to fix free nitrogen

Babes-Ernst bodies: metachromatic granules

*Bacillus*: a rod-shaped bacterium; genus of rod-shaped bacteria of the family Bacteriaceae

Bacteremia: a state in which microorganisms invade the bloodstream but do not multiply. This term has been used synonymously with septicemia

Bacteria: minute, one-celled, microscopic, plantlike organisms which multiply by fission and lack chlorophy II

Bacterial synergism: production of reactions by two or more species of bacteria growing together, that are not produced by either of the organisms growing alone

Bactericide: any agent that destroys bacteria

Bactericidin: a substance, present in the body fluids, that kills bacteria

Bacteriolysin: an antibody capable of destroying or dissolving specific bacterial cells that have stimulated its formation

Bacteriolysis: destruction or dissolution of bacteria inside or outside the animal body

Bacteriophage: an ultramicroscopic agent that produces a transmissible dissolution of specific bacterial cells and is regarded by some as a living agent and by other as an enzyme

Bacteriostasis: prevention of the growth of bacteria, such as results from the addition of certain dyes to culture media

Beef extract: a watery extract of the soluble constituents of beef, evaporated to semisolid consistency at low temperature in vacuum

Binary fission: a form of asexual reproduction involving simple cell division of the cytoplasm and nucleus, if present, into two equal parts

Biogenesis: the origin of living things from things already living

Bipolar: at both poles or ends of a cell

Bordet-Gengou phenomenon: see Complement fixation

Botulism: a type of food poisoning caused by the toxin of *Clostridium botulinum* in improperly canned or preserved foods

Breed count: a microscopic method for counting bacteria on a dried milk film

Bronchopneumonia: inflammation of the lungs that usually begins in the terminal bronchioles which become clogged with a mucopurulent exudate

**Table 3.5.1 (continued)**

Brownian movement: dancing motion of minute particles suspended in a liquid
Bubo: inflammatory swelling of a lymphatic gland, particularly in the axilla or groin
Buccal: pertaining to the cheek
Budding: a form of asexual reproduction in which the body divides into two unequal parts, the larger part being the parent and smaller one the bud
Buffer: any substance in a fluid which, upon addition of relatively large amounts of acid or alkali, tends to lessen the change in hydrogen ion concentration
Cachexia: a marked state of general ill health and malnutrition
Camp fever: a popular name for typhus fever
Cancer: a malignant tumor, made up chiefly of epithelial cells
Capsule: a broad, colorless, mucoid or gelatinous layer that surrounds some bacteria
*Carboxydomonas*: an autotrophic genus of bacteria that obtain their energy by the oxidation of carbon monoxide to carbon dioxide
Carcinoma: cancer
Cardiac: Pertaining to the heart
Carrier: an individual who harbors specific organisms of a disease in his body without manifesting symptoms but who serves as a means of conveying infection
Catabolism: the process of destruction or breakdown of tissues and cells from complex to simpler compounds
Catalase: an enzyme that decomposes hydrogen peroxide, liberating free oxygen
Catalyst: a substance that accelerates a chemical or physical reaction without itself being destroyed or changed
Catarrh: inflammation of the mucous membranes, especially of the nose and throat, with a discharge
Cathode: the negative electrode or pole of a galvanic cell
Cellulose: a complex carbohydrate forming the skeleton of most plant cells
Centrigrade thermometer: a thermometer with 100 degrees between the freezing point and the boiling point of water, the former being at zero and the latter at 100 degrees. To convert from Centigrade to Fahrenheit scale: (degrees C x 9/5) + 32 = degrees F.
Centrifuge: a machine for separating the more solid constituents of a fluid by rotation
Chancre: the primary lesion in syphilis
Chancroid: a soft, nonsyphilitic, venereal sore caused by *Hemophilus ducreyl*
Chemotaxis: phenomenon shown by living organisms of moving toward (positive) or away from (negative) certain other cells or substances that exert a chemical influence; also chemotropism

Chemotherapy: the prevention or treatment of diseases or inhibition of their parasitic causes by chemical disinfection

Chickenpox: varicella; a rather mild, highly contagious virus disease characterized by fever and the appearance of vesicles

Cholera: an acute infectious disease caused by *Vibrio comma* and characterized by its epidemicity, copious watery discharges, cramps, "rice-water" stool, prostration, and suppression of urine.

*Cholera infantum*: A common, and often fatal, noncontagious diarrhea of young children, prevailing in the summer months

*Cholera nostra*: Acute gastroenteritis, with diarrhea, cramps, and vomiting, occurring in summer and autumn and usually caused by improper food

Chromogenic: producing a pigment

Chromophore: any chemical group whose presence gives a specific color to a compound and that unites with certain other groups (auxochromes) to form dyes

Chronic: having a slow, comparatively mild course, continued for a long period

*Clostridium*: A genus of bacteria that are anaerobic or microaerophilic and that form spores

Coagulant: an agent that causes coagulation or clotting

Coccus: an organism that is round or spherical in shape

Cohesion: the force which unites the particles of a body

Colitis: inflammation of the colon

Colony: a group of bacteria, on a solid medium, usually derived from the multiplication of a single organism and visible to the naked eye

Colorimeter: an instrument for measuring color

Colostrum: the fluid secreted by the mammary gland a few days before and after parturition

Comma bacillus: *Vibrio comma*, the cause of Asiatic cholera

Commensalism: living together of two species, one of which is benefitted by the association, while the other is apparently neither benefitted nor harmed

Communicable: readily transferred from one person to another

Comparator block: a simple colorimeter consisting of a block of wood with holes in which to place the test tubes to be compared and transverse holes through which to view the colors

Complement: a thermolabile, normal blood constituent that reacts with sensitized cells to produce visible lysis; also called alexin

Complement fixation: When antigen unites with its specific antibody, complement, if present, is taken into the combination and becomes inactive or fixed. Its absence or presence as free active complement can be shown by adding sensitized blood cells, or blood cells and hemolytic

**Table 3.5.1 (continued)**

amboceptor, to the mixture. If free complement is present, hemolysis occurs; if not, no hemolysis is observed.

Complementoid: complement that has lost its activity by heating but still retains its binding property with amboceptors

Complication: a disease or diseases concurrent with another infection

Conjunctivitis: inflammation of the conjunctiva

Convalescence: the state of recovery following an attack of a disease

Convalescent carrier: a person who carries the infecting organism in his body during the period of convalescence from the infection

Convalescent serum: serum of a patient who has recovered from an infectious disease and containing specific antibodies against the cause of the disease

*Corynebacterium*: a genus of Corynebacteriaceae consisting of slender, gram-positive, nonmotile, rod-shaped organisms with a tendency to form club-shaped and pointed forms and to show uneven staining

Coryza: an acute catarrhal condition of the nasal mucous membrane, attended by a discharge from the nostrils

Cover slip: a thin glass plate that covers a mounted object to be studied microscopically

Cowpox: vaccinia; a virus disease related to and milder than smallpox, which confers immunity against smallpox. Vaccinia virus is used in vaccination against human smallpox

*Coxiella*: a genus of Rickettsieae; pleomorpohic rods or coccoid forms; the cause of Q fever, a typhyus-like, mild disease

Cosackie virus: several immunotypes of type A and one of type B have caused epidemics resembling polimyelitis in children.

Crateriform: depressed or hollowed; used to describe liquefaction of gelatin by certain bacteria

Culture: a growth of microorganisms

Culture medium: any substance or preparation suitable for and used for the growth and cultivation of microorganisms

Cyanosis: bluish discoloration of the skin, caused by insufficient oxygenation of the, blood

Cystitis: nflammation of the bladder

Cytochrome: a respiratory pigment, widely distributed in plant and animal cells, that is easily oxidized and reduced under suitable conditions

Cytolysin: an antibody that produces dissolution of cells; also called amboceptor or sensitizer

Cytolysis: the process of dissolution of cells

Danysz phenomenon: decrease of the neutralizing effect of an antitoxin when a toxin is added to it in divided portions instead of all at once

Decay: decomposition of organic compounds under aerobic conditions to form amino acids which are then further broken down

Denitrifying bacteria: bacteria capable of reducing nitrates to nitrites, nitrogen, or gaseous ammonia

Deoxyribonucleic acid (DNA): one of two types of nucleic acids; found only in nuclei, where its presence in genes indicates a central role in the genetic mechanism

Dermatitis: inflammation of the skin

Dermatosis: any skin disease

Dermomycosis: any skin disease caused by fungus

Dermotropic: having a selective affinity for the skin

Desensitization: the process of rendering a person insensitive, over a period of time, to a specific protein to which he is sensitive until he no longer reacts to it

Desiccator: a closed vessel for apparatus or chemicals that are to be dried and kept free from moisture, usually containing a dehydrating agent

Desquamation: the shedding of epithelial elements, chiefly of the skin, in scales or sheets

Deviation of complement: the Neisser-Wechsberg phenomenon of the failure of lysis to occur when considerable excess of amboceptor is added to normal serum (complement) and antigen. The complement unites with unbound amboceptor rather than with amboceptor which is united with antigen.

Dialysis: the separation of crystalloids from colloids by their different diffusibility through a semipermeable membrane

Dick test: a test for susceptibility to scarlet fever by the development of local redness following intradermal injection of toxin of the causative *Streptococcus*

Diphtheria: an acute infectious disease of the mucous membranes of the upper respiratory tract, characterized by patches of pseudomembrane and caused by *Corynebacterium diphtheriae*

Disease: a definite state of ill health having a characteristic train of symptoms, involving the whole body or any of its parts, and with the etiology, pathology, and prognosis either known or unknown

Droplet infection: infection by means of small droplets of sputum and nasal discharges that have been thrown into the air during talking, coughing, or sneezing, and that may remain viable for hours or days. Such diseases as measles, chickenpox, parotitis, vaccinia, and diphtheria are spread in this way.

Dye: a material used for staining or coloring, consisting of benzene rings with chromphore and auxochrome groups

**Table 3.5.1 (continued)**

Dysentery: a term given to a number of disorders marked by inflammation of the intestines, especially of the colon, and attended by pain in the abdomen and frequent stools containing blood and mucous
Dyspnea: difficult or labored breathing
ECHO virus: causes mild infections in the intestional tract, or may be present as harmless saprophytes; twenty distinct types
Eczema: an inflammatory skin disease with vesiculation, infiltration, watery discharge, and development of scales and crusts
Edema: the presence of abnormally large amounts of fluid in the intercellular tissue spaces of the body
Electrolysis: chemical decomposition produced by passing a direct current of electricity through the compound
Elementary bodies: minute intracellular bodies found in lesions produced by some viruses, e.g., in vaccinia, and which may be the infectious virus particles
Empyema: accumulation of pus in a cavity of the body, especially the chest
Encephalitis: inflammation of the brain
Endemic: said of a disease prevalent in the usual number in a given population
Endocarditis: inflammation of the endocardium or epithelial lining membrane of the heart
Endoenzyme: any intracellular enzyme retained in the cell and not excreted into the surrounding medium
Endotoxin: any toxin retained with the bacterial cell and liberated only after the death or disintegration of the organism
Enteric: pertaining to the intestines
Enteritis: inflammation of the intestine, chiefly of the small intestine
Enterotoxin: a toxin specific for the cells of intestinal mucosa, giving rise to symptoms of food poisoning
Enzyme: an organic catalyst produced by a living cell
Epidemic: a regional outbreak of an infectious disease that attacks many persons and spreads rapidly
Epidemiology: the study of disease outbreaks
Epizootic: any disease of animals that spreads rapidly and is widely diffused
Erysipelas: an acute, febrile, contagious disease caused by a strain of *Streptococcus pyogenes* and marked by chills, fever, and intense local redness of the skin and mucous membranes
Erythema: redness of the skin
Erythrocyte: a red blood corpuscle
Etiology: disease causation
Exanthem: any eruptive disease, e.g., variola, vaccina, scarlet fever

Exotoxin: a toxin that is excreted by the cell into the surrounding medium and may be separated by filtration

Exudate: any substance that is thrown out from the tissues and becomes deposited in or upon them

Facultative aerobe: a microorganism that is fundamentally an anaerobe but can grow in the presence of free oxygen

Fahrenheit thermometer: a thermometer in which the space between the freezing point and the boiling point of water is divided into 180 degrees, 32°F being the freezing point and 212°F the boiling point. To convert from fahrenheit to centigrade scale: 5/9(°F - 32) = °C.

Family: a group of related genera

Fermentation: incomplete oxidation of carbohydrates and carbohydrate-like compounds by microorganisms

Filamentous: composed of long threadlike structures

Filiform: having a uniform growth along the line of inoculation in streak or stab cultures

Filtrate: a liquid that has passed through a filter

Fimbriate: fringed colony growth

Fishing: the transfer of a bacterial culture from a plate to a fresh medium by touching the colony with a wire and then inoculating the medium

Flagella: whiplike processes for motility

Flocculent: containing small adherent masses of bacteria floating in or on a liquid

Fluctuating variations: temporary changes that often occur in morphology or physiology or a microorganism, usually because of environmental conditions

Focus of infection: the chief center of a morbid process

Fomite: any substance other than food that may harbor or transmit a disease; e.g., bedding, clothing, dishes

Formaldehyde: HCHO; a disinfectant gas with a pungent odor

Formalin: a 40% solution of gaseous formaldehyde

Frost count: a microscopic colony count for determining the number of bacteria in milk

Fungi: a subphylum of plants that do not contain chlorophll, including bacteria, yeasts, and molds

Fusiform: radish-shaped liquefaction of gelatin

Gamete: a sexual cell; a mature germ cell, as an unfertilized egg or mature sperm cell

Gangrene: anemic necrosis of tissue combined, usually, with invasion by saprophytic organisms

**Table 3.5.1 (continued)**

Gastroenteritis: inflammation of the stomach and intestines with symptoms similar to enteritis and dysentery, often caused by the enteric group of bacteria *Salmonella paratyphi*, *Salmonella schottmuelleri*, etc.

Genus: a group of one or more related species

Germ: a microbe or bacillus

Germicide: synonymous with disinfectant

Gingival: pertaining to the gum or gums

Glanders: a febrile acute or chronic disease of horses, transmissible to man; caused by *Malleomyces mallei*

Gonidium: a minute intracellular body that serves as a means of asexual reproduction

Gonorrhea: a contagious, catarrhal, venereal disease of the genital mucous membranes, caused by *Neisseria gonorrheae*.

Gram-negative bacteria: bacteria that lose the initial stain of the gram stain, are decolorized, and take the color of the final stain

Gram-positive bacteria: bacteria that take the initial stain of the gram stain and are not decolorized, so that they appear purple

Guarnieri bodies: cytoplasmic inclusion bodies often found in epithelial cell lesions in variola and vaccinia

H agglutinin: agglutinin specific against the (H) antigen found within the flagella of the bacterium

Haptene: a partial antigen, unable to stimulate antibody production when injected by itself but determining immunological specificity when combined with antigenic protein and reacting with the antibody so produced

Haptophore: term used by Ehrlich in his explanation of immunological reaction to denote the specific portion of the toxin or agglutinin molecule through which combinations with antitoxin or agglutinogen occurs

Hay fever: allergic acute nasal catarrh and conjunctivitis

Hemagglutinin: a specific substance in blood serum that causes agglutination of red blood corpuscles

Hematuria: the discharge of urine in which blood is present

Hemoglobin: oxygen-carrying red pigment of red blood corpuscles

Hemoslysin: an antibody that, in the presence of complement, is capable of bringing about the dissolution of foreign red blood corpuscles so that their hemoglobin is liberated

Hemolysis: the dissolution or destruction of red blood corpuscles with liberation of the hemoglobin

Hemorrhage: severe bleeding

Hemotoxin: a cytotoxin capable of destroying blood cells

Hepatitis: inflammation of the liver

Herpes simplex: a mild, acute, eruptive, vesicular virus disease of the skin and mucous membrane
Herpes zoster: shingles; an acute virus disease characterized by a vesicular dermatitis that follows a nerve trunk
Heterologous serum: serum derived from another species
Heterotrophic: living on organic matter
Homologous serum: serum derived from the same species
Hydrocarbon: any compound of hydrogen carbon
Hydrogen ion concentration: the degree of concentration of hydrogen ions in a solution used to indicate the reaction of that solution and expressed as pH, which is the logarithm of the reciprocal of the hydrogen ion concentration
Hygienic Laboratory coefficient: phenol coefficient as determined by the U.S. Hygienic Laboratory method
Hyperemia: excess of blood in any part of the body
Hyperglycemia: excess of sugar in the blood
Hypertonic: a solution that is of a concentration greater than isotonic concentration
Hyphae: branching filaments of the mold plant
Hypotonic: a solution that is of less than isotonic concentration
Idiosyncrasy: individual or peculiar susceptibility to some drug, protein, or other agent
Immunity: the ability of living individuals to resist or overcome infections
Immunization: the process of conferring immunity on an individual
Inclusion bodies: round, oval, or irregularly-shaped bodies occurring in the cytoplasm (cytoplasmic) or nucleus (intranuclear) of cells of the body in some virus diseases, such as rabies, variola, vaccinia, etc.
Incubator: an apparatus for maintaining a constant and suitable temperature for the growth and development of a bacterial culture or other materials
Indicator: a substance, usually a weak organic acid or base, that changes color when the reaction of a solution changes
Indirect stain: a stain that does not dye the organism itself but colors the background so that the organism becomes visible by contrast, e.g., nigrosin, India ink
Induration: the process of hardening; a hardened spot
Infection: invasion of body tissues by pathogenic organisms that multiply and cause disease
Infundibuliform: funnel-shaped or inverted-cone-shaped liquefaction of gelatin
Intrathecal: in the spinal column
*In utero*: within the uterus
*In vitro*: in the test tube, outside of the animal body
*In vivo*: In the living body

**Table 3.5.1 (continued)**

Involution forms: abnormal shapes taken by bacteria differing from those of the original culture
Isohemagglutinins: specific substances in blood serum that agglutinate the blood cells of other members of the same species
Isotonic: a solution equal in concentration to another solution with respect to a certain solute
Jail fever: typhus fever
Jaundice: excess of bile pigment in blood, skin, and mucous membranes with a resulting yellow appearance of the individual
Johne's disease: a chronic enteritis of cattle, caused by *Mycobacterium paratuberculosis*
Kala-azar: a fatal, infectious protozoan disease of the reticuloendothelial system, caused by *Leishmania donovani* and believed to be transmitted by the sandfly
Keratitis: inflammation of the cornea
Koplik's spots: small bluish-white spots surrounded by a reddish areola on the mucous membrane, found on the cheeks and lips during the prodromal stage of the measles
Lag phase: the early period of slow multiplication following a bacterial inoculation into a culture medium
Laryngitis: inflammation of the larynx, a condition attended by dryness and soreness of the throat, hoarseness, and cough
Leishman-Bonovan bodies: small round or oval bodies found in the spleen and liver of patients suffering from kalàazar
Leukocyte: white blood corpuscle
Locomotor-ataxia: tabes dorsalis; degeneration of the dorsal columns of the spinal cord and of the sensory nerve trunks, marked by intense pain, incoordination, loss of reflexes, etc.
Logarithmic period: the stage of maximum growth of a bacterial culture at which, if the logarithms of their numbers are plotted against the time, a straight line will be formed
Lophotrichous: having a tuft of flagella at one end
Lues: syphilis
Luminescence: the property of giving off light without manifesting an increase in heat energy
Lysin: an antibody that, in the presence of complement, dissolves or disintegrates cells
Macrogamete: the female form of the malarial parasite, which, when transferred from man to the mosquito, becomes a macrogamete
Macrophage: large wandering phagocyte, so designated by Metchnikoff
Macula: an unraised discolored spot on the skin

Maltaise: any indisposition, discomfort, or feeling of ill health

Malaria: a protozoan disease caused by members of the genus *Plasmodium* and transmitted by the bite of the anopheline mosquito; characterized by recurring fever, chills, and sweats

Malta fever: see undulant fever

Mastitis: inflammation of the mammary gland

Measles: rubeola; an acute infectious virus disease characterized by fever, catarrh, coryza, Koplik's spots on buccal membranes, and papular rash

Meningitis: inflammation of the meninges

Mesophilic bacteria: bacteria that grow best at moderate temperatures

Metabolism: the sum of all the physical and chemical processes by which the tissues are formed an maintained and energy is made available for use by the body

Metachromatic granules: deeply staining masses, irregular in size and number, that are seen in the protoplasm of various bacteria, e.g., *Corynebacterium diphtheriae*

Metastasis: the transfer of disease from one organ or part to another not directly connected with it

Microaerophilic bacteria: bacteria that grow at a low oxygen tension

Microgamete: the conjugating male element of the *Plasmodium* of malaria, which fertilizes the macrogamete in the mosquito

Microgametocyte: the male form of the malarial parasite, which is transferred from man to the mosquito

Micromicron: the millionth part of a micron (see definition for micron); designated by the symbol $\mu\mu$

Micron or micromillimeter: one-millionth part of a meter, or one-thousandth part of millimeter, 1/25,000 of an inch; designated by the Greek letter $\mu$

Microphages: small, actively motile, polynuclear leukocytes that cause phagocytosis of the bacteria of acute infections

*Microsporum*: a genus of parasitic fungi that cause skin diseases, such as ringworm and barber's itch

Millimicron: one-thousandth part of micron, or one-millionth part of a millimeter; designated by m$\mu$

Mixed culture: growth of two or more organisms in the same medium

*Monilia*: a genus of parasitic fungi that is most frequently encountered as thrush and is characterized by its fermentation of sugar with the production of gas

Monotrichous: having a single polar flagellum

Mordant: a substance used to fix a stain or dye

Moribund: in a dying state

Morphology: the science of the form and structure of organized beings

**Table 3.5.1 (continued)**

Much's granules: non-acid-fast, gram-positive granules of unknown significance observed in tuberculous lesions in which acid-fast bacilli could not be found

Mucopurulent: containing both mucus and pus

Mucous membrane: a membrane secreting mucus and lining the cavities of the body that connect with the outside air, such as the respiratory, digestive, and genitourinary tracts

Mucus: the viscid watery secretion of the mucous glands, composed of water, mucin, inorganic salts, and epithelial cells

Mutation: transmissible variation that is apt to be permanent, seemingly arising suddenly and spontaneously

Mycelium: a mass of hyphae

Mycology: the science and study of fungi

Mycosis: any disease caused by fungus

Myocarditis: inflammation of the muscular walls of the heart, or myocardium

Napiform: turnip-shaped liquefaction of gelatin

Necrosis: death of a circumscribed portion of tissue

Nergi bodies: diagnostic cytoplasmic inclusion bodies found in the affected nerve cells in rabies

Neisser-Wechsburg phenomenon: see Deviation of complement

Nephritis: inflammation of the kidney

Neufeld reaction: Method of typing the pneumococci based on the fact that the capsule swells in the presence of type-specific immune serum

Neurotropic: having an affinity for nervous tissue

Nitrification: oxidation of ammonia to nitrites and nitrates accomplished in soil by autotropic bacteria

*Nitrobacter*: a genus of bacteria that can oxidize nitrites to nitrates

Nitrogen fixation: the union of free atmospheric nitrogen with other elements to form chemical compounds, such as ammonia and nitrates or amino groups

*Nitrosomonas*: a genus of bacteria that can secure their energy by oxidizing ammonia to nitrites

Nonspecific immunity: increase of antibodies or production of immunity resulting from the injection of some nonspecific antigen

O agglutinin: agglutinin specific for the (O) antigen found within the bacterial cell

Obligate aerobe: an organism that must have free oxygen for its growth

Obligate anaerobe: an organism that can live and grow only in an environment with no or minimal amounts of free oxygen

Oocyst: an encysted oospore

Oospore: a zygote formed by the conjugation of two sexually differentiated elements

Opsonic index: the power of the blood of an individual to phagocytize any particular microorganism compared with the power of normal blood

Opsonin: an antibody in the blood that renders microorganisms or red blood cells more liable to phagocytosis

Orchitis: inflammation of a testis

Osmosis: passage through a membrane; when two solutions of unequal density are separated by a membrane that selectively prevents the passage of solute particles but is permeable to the solvent, pure solvent passes from the lesser to the greater concentration.

Otitis: inflammation of the ear

Pandemic: an epidemic involving a large population

Papilloma: an epithelial tumor in which the cells cover fingerlike processes or ridges of stroma (matrix of an organ), e.g., warts

Papule: a small, circumscribed, solid elevation of the skin

Paralysis: loss of motion or sensation in a living part of the body

Parasite: a plant or animal that lives upon or within another living organism at whose expense it grows without giving anything in return

Parenteral: not through the alimentary canal

Paraotitis: inflammation of the parotid gland; mumps

Paschen bodies: small elementary bodies demonstrable in the vesicular fluid in variola and vaccinia

Passive immunity: immunity produced by injection into one individual of serum containing antibodies formed in another

Pasteurization: the process of heating every particle of milk or milk product, maintaining the temperature for a suitable time, and then rapidly cooling, the temperatures and the time being sufficient to destroy all pathogenic organisms and to reduce the bacterial count about 90%

Pathogens: disease-producing microorganisms

Pellicle: a thin membranous film on the surface of a liquid formed as a result of bacterial growth

Penicillin: a highly effective, chemotherapeutic, nontoxic antibiotic produced by the mold *Penicillium notatum*

*Penicillium*: a genus of molds that is characterized by the development of fruiting organs resembling a broom.

Peptonization: enzymic hydrolysis of milk protein transforming the milk to a clear fluid

Pericarditis: inflammation of the pericardium (the membranous sac that contains the heart)

Peritrichous: having flagella surrounding the entire organism

**Table 3.5.1 (continued)**

Per os: by the mouth
Pertussis: whooping cough
Petechia: a small spot formed by the effusion of blood
Pfeiffer's phenomenon: when *Cholera vibrios* is injected into the peritoneal cavity of a guinea pig that has been immunized against cholera, it is found, when removing a portion of the peritoneal contents from time to time, that the vibrios lose their motility, become swollen, disintegrate, and pass into solution.
pH: see Hydrogen ion concentration
Phagocyte: any cell that destroys microorganisms or other cells by enveloping or absorbing them, e.g., endothelial cells, leukocytes
Phagocytosis: the process whereby microorganisms and other cells and substances are engulfed by phagocytes
Phenol: carbolic acid, a colorless crystalline compound, $C_6H_{11}OH$, obtained by the distillation of coal tar and having strong antiseptic and disinfectant properties
Phenol coefficient: a number indicating the relative efficiency of disinfectants. It is the quotient obtained by dividing the highest dilution of a disinfectant that kills a test organism in a fixed time by the highest dilution of phenol showing the same results.
Photodynamic sensitization: increased toxicity to bacteria of ultraviolet rays of low intensity and of visible light, resulting from the addition of certain dyes
Photogens: bacteria which produce phosphorescence or emit light
Phototaxis: the movement of organisms under the influence of light
Plague: an acute, febrile, and highly fatal epidemic disease, caused by *Yersinia pestis* and transmitted to man by fleas from rats
Plasmochin: a synthetic quinoline compound used in malaria therapy
Plasmolysis: contraction or shrinking of the protoplasm of a cell because of loss of water by osmotic action
Plating: cultivation of microorganisms in petri dishes containing a solid nutrient medium
Pleomorphism: the assumption of various distinct forms by a single organism or species
Pneumonia: inflammation of the lungs, most often of the pneumococcic origin
Polar: situated at the end or pole of a cell
Poliomyelitis: a virus disease in which there is inflammation of the gray substance of the spinal cord; commonly called infantile paralysis
Polyvalent: a term used to designate a stock vaccine made up of many strains of the same organism or different organisms
Portal of entry: the route a microorganism must take in entering the body in order to have any effect

Prausnitz-Kustner reaction: passive transfer of local hypersensitivity in many by the intradermal injection of serum of an allergic person

Precipitin: an antibody formed in response to the injection of soluble antigens and having the power to cause a precipitation or flocculation of the antigen when mixed with the latter

Precipitinogen: any substance that stimulates the formation of precipitin in the bloodstream

Predisposing cause: anything that renders a person liable to an attack of disease without actually producing it

Proagglutinoid or prozone phenomenon: this is noted when a lower dilution of a serum gives no agglutination, although agglutination occurs at higher dilutions.

Prodromal: premonitory, indicating the approach of a disease

Prognosis: a forecast of the probable result of an attack of disease; the likelihood of recovery

Prophylaxis: the prevention of disease; preventive treatment

Protozoa: unicellular, nucleated, animal organisms with diversification and specialized functions of the protoplasm

Pseudopodia: "false feet," temporary protrusions of ectoplasm serving for purposes of locomotion

Psychrophilic bacteria: cold-loving bacteria whose optimum temperature for growth is 15-20$^{o}$C or below

Puerperal sepsis: sepsis occurring after childbirth, usually due to *Staphylococcus* or *Streptococcus*.

Punctate: resembling or marked with points or dots

Punctiform colonies: pinpoint colonies, measuring less than 1 mm in diameter

Pure culture: specific bacterial growth of only one type of organism

Purpura: a disease characterized by the formation of purple patches on the skin and mucous membranes, due to subcutaneous extravasation of blood

Purulent: containing pus

Pus: a liquid inflammation product made up of dead and living leukocytes, digested bacteria, tissue debris, lymph, fibrin, serum

Pustule: a small elevation of the skin filled with pus or lymph

Putrefaction: the decomposition of animal or vegetable matter in amino acids, mercaptans, indol, skatol, and hydrogen sulfide, with an accompanying unpleasant odor

Pyemia: a general septicemia in which secondary foci of suppuration occur and multiple abscesses are formed

Pyogenic: pus producing

Pyrexia: fever

**Table 3.5.1 (continued)**

Quarantine: isolation of infected persons and of possible carriers for a period of time to prevent disease transmission

Quinine: an alkaloid of the cinchona bark that is used as a specific malaria treatment

Rabies: a fatal virus disease of man and animals characterized by extreme irritation of the central nervous system

Receptor: a term used in Ehrlich's sidechain theory of immunity to indicated that part of the cell molecule that can combine with the haptophore group of toxins of other antigens and may be attached to the cell or remain free in the bloodstream, e.g., antitoxin, agglutinin, precipitin, opsonin

Relapsing fever: any one of a group of acute infectious diseases caused by various species of *Borrelia*, and marked by alternating periods of fever and absence of fever

Remission: a diminution or abatement of the symptoms of a disease

Resistance: the ability of the individual to ward off infection

Rhinitis: inflammation of the mucous membrane of the nose

*Rhizobium*: a genus of bacteria that lives symbiotically with leguminous plants and fixes free nitrogen

Rhizoid: irregular branched or rootlike growth

Ribonucleic acid (RNA): one of two types of nucleic acids; found in both nucleus and cytoplasm and involved in protein synthesis

Rickettsiae: rickettsia bodies; gram-negative, nonmotile, intracellular parasites whose exact nature is unknown but that are generally considered intermediate between bacteria and viruses

Rubeola: measles

Saccate: sac-shaped liquefaction

Saccharose: cane sugar

Sapremia: intoxication due to presence in the blood of the products formed by the growth of saprophytes on diseased or injured tissues

Saprophytes: organisms that live on dead organic matter

*Sarcina*: a genus of Peptococcaceae characterized by occurrence of cell division in three planes, forming regular packets

Satellite phenomenon: growth of larger colonies of *Haemophilus influenzae* in the region of staphylococci and other bacteria, because of their synthesis of coenzyme

Scarlet fever: an acute, contagious, exanthematous disease caused by a strain of *Streptococcus pyogenes*

Schick test: a test for susceptibility to diphtheria by the development of an area of inflammation following intracutaneous injection of diphtheria toxin

Schizogony: asexual reproduction by segmentation

Schizomycetes: the plant class to which all bacteria belong

Schuffner's dots: coarse red granules seen in parasitized erythrocytes in certain malaria on staining with polychrome methylene blue

Schultz-Charlton phenomenon: When scarlet fever convalescent serum or antitoxin is injected intracutaneously in scarlet fever patients, there is a blanching of the rash at the site of injection.

Sensitizer: see Cytolysin

Sepis: a state of infection

Septate: divided by a septum into various compartments

Septicemia: invasion and multiplication of organisms in the bloodstream

Serum: the clear liquid that separates in the clotting of the blood from the clot and the corpuscles

Serum sickness: a form of hypersensitive reaction following the injection of foreign serum and marked by urticarial rashes, edema, fever, joint pains or swelling and pain in the lymph nodes

Shwartzman reaction: a severe, hemorrhagic, necrotic reaction produced in rabbits at the site of intradermal injection with filtrate of cultures of *Salmonella typhi* (typhosa) or certain other organisms, when this is followed within 24 hours by intravenous injection of the filtrate

Slant: solid media allowed to harden in test tubes set at a slant to increase the surface for the growth of colonies

Slime layer: carbohydrate jelly layer believed to surround cell wall of all bacteria

Smallpox: variola; a generalized febrile virus disease characterized by vesicular eruptions that become pustular and crusty, often leaving permanent pox marks

Somatic: pertaining to the body

Specific Soluble Substance: SSS; a complex, highly type-specific, polysaccharide haptene extracted from the capsular material of various bacteria, e.g., pneumococci

Spontaneous generation: the theory, now discarded, which postulated that living organisms generated from nonliving matter

Sporangia: a sac or covering for sporangiospores

Sporogony: sexual reproduction by spores

Sporozoite: a spore formed after fertilization; it is this form that is transmitted from the mosquito to man in malaria.

Stab cultures: cultures in which the organisms are inoculated far down into the solid butt of the medium to allow for possible anaerobic growth

Stain: any dye, reagent, or other material used in coloring tissues or organisms for microscopic study

Stationary phase: the stage in the growth of a bacterial culture at which multiplication of organisms gradually decreases so that as many are formed as die

**Table 3.5.1 (continued)**

Sterilization: the process of freeing completely from all living microorganisms

Stomata: openings in surfaces of leaves of plants for passage of oxygen, carbon dioxide, and water vapor

Stomatitis: inflammation of the mouth, usually attended by pain and salivation and often by fetid breath

Strangles: an infectious disease of horses caused by *Streptococcus equi* and characterized by a mucopurulent inflammation of the respiratory mucous membrane

Stratiform: liquefying to the walls of the tube at the top and then proceeding downward horizontally

Streak: inoculation of slants or plates with a streak or direct line movement across the surface of culture media

Subcutaneous: beneath the skin

Subterminal: situated near but not at the end of extremity

Sulfur bacteria: bacteria that obtain their energy from the oxidation of sulfur-containing gases to sulfuric acid

Suppuration: pus formation

Symbiosis: the living together or close association of two dissimilar organisms with mutual benefit

TAB: abbreviation for a mixed vaccine containing organisms of *Salmonella typhi, Salmonella paratyphi*, and *Salmonella schottmuelleri* (typhoid, paratyphoid A, and paratyphoid B)

Tetanus: lockjaw; an acute infectious disease caused by the toxin of *Clostridium tetani*, characterized by spasm of the voluntary muscles

Therapy: the treatment of disease

Thermal death point: the temperature that destroys all the bacteria present within a given time

Thermal death time: the length of time required to kill all the organisms in a given substance at a given temperature of 50-55°C

Thermostable: not easily affected by moderate heat

*Thiobacillus*: a genus of bacteria that obtain their energy from the oxidation of sulfides, thiosulfates, or sulfur, forming sulfur, persulfates, sulfuric acid, and sulfates

Tick: a blood-sucking arachnid parasite

Titration: volumetric determination against standard solutions of known strength

Toxemia: a general intoxication due to the absorption of bacterial products, usually toxins, formed at a local source of infections

Toxins: poisonous substances produced by bacteria. When exotoxins are injected into animals in carefully graded doses, they incite the formation of specific substances, called antitoxins, which nullify the action of the toxin.

Toxoid: anatoxin; toxin treated with formalin, etc., so that it is no longer toxic but is still capable of stimulating formation of and uniting with antitoxin

Transfer of cultures: transplanting viable bacteria from an old medium to a fresh new one

Transplant: a portion of a culture of bacteria that has been transferred from an old pure culture to a fresh new medium

Trauma: a wound or injury

Trench mouth: see Vincent's angina

*Trichophyton*: a genus of fungi consisting of flat, branched filaments and chains of spores and causing ringworm and athlete's foot

Tubercle: a nodule characteristic of tuberculosis and containing numerous bacilli; made up of small spherical cells which contain giant cells and are surrounded by a layer of spindle-shaped connective tissue cells, known as epitheloid cells

Tuberculin: a filterable substance produced in the growth of *Mycobacterium tuberculosis* in culture media. When it is injected intracutaneously in persons who have been exposed to the tuberculosis bacillus or its products, a reaction is produced in 24-48 hours consisting of infiltration and hyperemia.

Tuberculosis: an infectious disease caused by *Mycobacterium tuberculosis* and characterized by the formation of tubercles in the tissues

Tularemia: a disease of rodents, resembling plague, which may be transmitted to man; caused by *Francisella tularensis*

Twort-d'Herelle phenomenon: see Bacteriophage

Undulant fever: Malta fever; an infectious disease of man characterized by continued pyrexia, often with remissions, joint pains, skin rashes, night sweats, caused by *Brucella abortus* and *Brucella melitensis*, which cause primarily contagious abortion in cattle and goats, respectively

Univalent vaccine: a vaccine containing only one variety of organism

Urticaria: a skin disease characterized by the sudden appearance of a rash which rarely lasts longer than 2 days but may be chronic. It is characterized by smooth, elevated patches which are usually whiter than the surrounding skin and attended by severe itching

Vaccination: usually, protective inoculation against smallpox by inoculation with vaccinia virus, but also any act of protective inoculation with a virus or bacteria

Vaccine: a preparation of killed or attenuated infective agent used to produce active artificial immunity

Vaccinia: coxpox

Varicella: chickenpox

Variola: smallpox

**Table 3.5.1 (continued)**

Vector: a carrier of viruses, bacteria, or protozoa; usually refers to the animal host that carries the pathogenic organism from one human host to another; e.g., the *Anopheles* mosquito is the vector of *Plasmodium matariae*.
Venereal: due to or propagated by sexual intercourse
Vesicle: a small sac or blister containing liquid
Viable: living
Vi-antigen: an antigen found in strains of *Salmonella typhi* and some paratyphoid bacilli and believed associated with virulence
Vincent's angina: trench mouth; contagious infection of the mouth associated with the presence of *Borrelia vincenti* and the fusiform bacillus and characterized by a false membrane in the tonsillar region
Virulence: disease-producing ability
Virus: the large and heterogeneous group of microorganism that are obligate parasites and live in cells of their selected host
Viscerotropic: having a selective action on the visceral organs, including the liver and spleen
Viscid: adhering, glutinous, sticky. In bacteriology, said of a colony that strings out by clinging to a needle when it is touched to the culture and withdrawn. The sediment rises in a coherent whirl when the liquid culture is shaken.
Von Economo's disease: Japanese encephalitis type A; a chronic, sporadic, encephalitic infection of unknown etiology
Wassermann test: a complement fixation test for syphilis
Well-Felix reaction: agglutination of certain *Proteus* strains by the serum of persons with typhus fever, for which it is used as a diagnostic test
Weil's disease: infectious jaundice; an acute infectious disease characterized by jaundice, fever, muscular pain, and enlargement of the liver and spleen, caused by the spirochete *Leptospira icterohaemorrhagiae*
Wheal: a white or pinkish elevation or ridge on the skin
Whoop: the convulsive inspiration of whooping cough
Whooping cough: pertussis; an infectious disease caused by *Haemophilus pertussis* and characterized by catarrh of the respiratory tract and peculiar paraxysms of cough, ending in a prolonged crowing or whooping respiration
Widal test: an agglutination test for the diagnosis of typhoid fever
Yaws: a tropical, ulcerative, nonvenereal disease similar to syphilis and caused by *Ireponemia pertenue*
Yellow fever: a virus disease characterized by chills, fever, muscular pain, and jaundice, transmitted by the female *Aedes egypti* mosquito

Zygospore: a spore formed by the conjugation of two cells that are morphologically identical and do not show sexual differentiation
Zygote: an organism formed by the union of two germ cells
Zymogens: microorganisms that cause fermentation of carbohydrate

Source: Anonymous, 1985; Anonymous, 1991.

## 3.6. References

Anonymous. Unsafe herbs. *FDA Consumer*, 17 (October 1983): 8.

Anonymous. *FDA Analyst Operations Manual*, FDA, Washington, DC, 1985.

Anonymous. *FDA Inspections Operation Manual*, FDA, Washington, DC, 1991.

Banwart, G. J. *Basic Food Microbiology*, 2nd ed., Van Nostrand Reinhold, New York, 1989.

Bier, J. W. Protozoa and Helminths. In *Principles of Food Analysis for Filth, Decomposition, and Foreign Matter*, FDA Technical Bulletin No. 1; 2nd ed., J. R. Gorham, ed., FDA, Washington, DC, 1981.

Bryan, F. L. *Diseases Transmitted by Foods (A Classification and Summary)*, 2nd ed., USDHHS/PHS, Center for Disease Control, Atlanta, GA, 1982.

Cliver, D.O. *Foodborne Diseases*, Academic Press, San Diego, CA,1990.

Doyle, Michael P., ed., *Foodborne Bacterial Pathogens*, Marcel Dekker, Inc., New York, 1989.

Jay, J. M. *Modern Food Microbiology*, 3rd ed., Van Nostrand Reinhold, New York, 1986.

Pierson, M. D., and N. J. Stern. *Foodborne Microorganisms and Their Toxins: Developing Methodology*, Marcel Dekker, New York, 1986.

# PART B PRODUCT CATEGORIES DATA

# CHAPTER 4 FRUITS AND VEGETABLES

The data covered in this chapter include the following:

1. Selected data commonly needed in the processing of fruits and vegetables, e.g., pH, water content, commercial applications, spoilage organisms.
2. Data on the storage of fruits and vegetables. All information has been adapted from documents published by the United States Department of Agriculture (USDA) which has done more research and development in this area than any other private or public research center in the world.
3. Categories of fruits and vegetables for which the USDA has established standards for grades.
4. General descriptions for processing individual fruits, vegetables, and related produce. It must be emphasized that these descriptions include only a small number of those for which the U.S. Food and Drug Administration (FDA) has promulgated standards. Refer to Chapter 10 for such data. It should also be noted that such descriptions are very general and serve only as an introduction to food scientists and technologists who are not specialists in the processing of fruits and vegetables.

An example of the data presented in this chapter are the following three groups of information which covers the storage requirements for fruits and vegetables:

1. Vital factors: respiration rates, chilling injury, freezing injury, temperatures, relative humidity, freezing point, water content, and specific heat.
2. Mathematical calculations for the refrigeration requirements in the cold storage of fruits.
3. Requirements and observations for the controlled atmosphere storage of fruits and vegetables.

In this chapter, reference to legal government literature follows the standard format. For example, the citation "7 CFR 52" means: Volume 7 of the U.S. Code of Federal Regulations, Part 52. All U.S. legal documents are published by the U.S. Government Printing Office, Washington, D.C.

## 4.1. Processing Fruits and Vegetables

**Table 4.1.1. Major Considerations in Processing Fruits and Vegetables**

- Biology and composition
  - Composition
    - Water
    - Carbohydrates
    - Proteins and other nitrogenous substances
    - Lipids and related substances
    - Organic acids
    - Pigments
    - Minerals
    - Vitamins
    - Flavors
    - Phytates
    - Toxicants
    - Individual fruits or vegetables
  - Structure
    - Organ
    - Tissue and cell
    - Subcellular levels
  - Physiology and metabolism
    - Respiration
    - Gene expression and protein synthesis
    - Ancillary metabolic processes
    - Control mechanisms
- Postharvest handling and storage
  - Mechanical injury
  - Temperature
  - Controlled atmospheres
  - Humidity
  - Ionizing radiation
  - Other factors
- Processed fruits and vegetables
  - Methods of food preservations
    - Canning
    - Freezing
    - Concentration
    - Refrigeration
    - Dehydration
    - Freeze-drying
    - Pickling and fermenting
    - Irradiation
  - Microorganisms
    - Types
    - Regulation
    - Effects of storage in canned goods
    - Prevalence
  - Preprocessing preparation
    - Preliminary stages
    - Blanching
    - Irradiation
    - Peeling
    - Individual fruits and vegetables
  - Canning
    - Preliminary stages
    - Containers
    - Environment
    - Individual vegetables
  - Freezing
    - Basic procedures
    - Selection of fruits and vegetables
    - Frozen products: stability, quality
    - Frozen products: handling, storage, distribution
    - Individual fruits and vegetables
  - Dehydration

- Basic considerations
- Selection of method
- Types of driers
- Freeze-drying: process
- Freeze-drying: properties of products
- Dehydrated products: packaging and storage
- Quality control
- Individual fruits and vegetables

Pickling and fermenting
- Microbiology
- Types of fermented fruits and vegetables
- Pickled cucumbers
- Sauerkraut
- Olives
- Quality changes

Grades, standards, and labeling
- Food standards
- U.S. grades, standards and regulations
- Canned products
- Frozen products
- Labeling regulations
- Plant inspections

Quality control
- Concepts
- Techniques
- Statistics
- Charts
- Sampling
- Management strategy
- Individual fruits and vegetables

Sanitation and waste management
- Sanitation
- Management of residues
- Water supply
- Waste disposal
- Individual fruits and vegetables

Quality and storage
- Determining factors
- Canned products
- Dehydrated products
- Frozen products
- Long-term storage
- Stability
- Nutrition retention
- Product coding
- Individual fruits and vegetables

---

Source: Anonymous, 1985; Chan, 1983; Considine and Considine, 1982; Dennis, 1983; Goodenough and Atkin, 1981; Nelson and Tressler, 1980; Salunkhe and Desai, 1984; Ting and Rouseff, 1986; Weichman, 1987; Willis and McGlasson, 1989.

**Table 4.1.2. pH of Selected Fruits and Vegetables**

| Food Product | pH | Food Product | pH |
|---|---|---|---|
| Apple butter | 3.1-3.5 | Mushrooms | 6.2-6.6 |
| Apple sauce | 3.6-3.9 | Olives | 2.9-3.2 |
| Apples | 3.0-3.3 | Oranges | 3.2-3.8 |
| Apricots | 3.7-3.8 | Orange juice | 3.7-4.1 |
| Asparagus | 5.4-5.7 | Peaches | 3.4-3.6 |
| Cauliflower | 5.6-5.7 | Peas | 6.1-6.4 |
| Cherries | 3.4-4.0 | Pineapple juice | 3.3-3.6 |
| Corn | 6.7-7.1 | Plums, currants | 2.9-3.2 |
| Creamed corn | 6.3-6.5 | Potatoes | 5.4-5.8 |
| Cucumber | 3.0-3.5 | Prune juice | 3.7-4.1 |
| Grapefruit | 3.2-3.5 | Pumpkin | 4.1-4.4 |
| Green beans | 5.5-3.9 | Raisins | 3.6-4.2 |
| Kidney beans | 5.2-5.4 | Spinach | 5.5-5.6 |
| Legumes (alkali blanch) | 6.5-7.5 | Strawberries | 3.3-3.4 |
| Lemons | 2.3-2.6 | Tomato juice | 4.0-4.5 |
| Lima beans | 6.2-6.6 | Turnips/cabbage | 5.1-5.3 |

Source: Anonymous, 1982; Dennis, 1983; Friend, 1982; Goodenough and Atkin, 1981; Jackson and Shinn, 1979; Lopez, 1987; Nelson, 1980; Salunkhe, 1984; Ting and Rouseff, 1986; Weichman, 1987; Wills et al., 1989.

**Table 4.1.3. Fruits and Their Commercial Applications**

Apple (*Pyrus malus*)

Brandy, cakes, canned, cereals, champagne, chunks, citric acid, cocktail (alcoholic), confections, cookies, crushed, dried, essence, frozen, glacé, jelly, juice, marmalade, pectin, pie, pie filling, preserves, sauce, snack bars, and vinegar

Apricot (*Prunus armeniaca*)

Brandy, cakes, canned, cereals, chunks, confections, cookies, crushed, diet spread, dried, frozen, jam, nectar, pie, pie filling, preserves, purée, snack bars, strained, and purée

Avocado (*Persea americana*)

Crushed and purée

Banana (*Nysa saouebtyn, M. Paradisiaca)*

Dried and purée

Banana (*M. cavendeishii)*

Cereals, cookies, crushed, dried, frozen, purée, strained purée

Baobab (*Actansonia)*

Crushed, pie, and purée

Blackberries (*Rubus alleghaniensis*)

Brandy, cakes, canned, cocktail (no alcohol), cocktail (alcoholic), cookies, crushed, essence, jam, jelly, juice, nectar, pie, pie filling, preserves, purée, snack bars, strained purée, syrup, and wine

Boysenberries (*Rubus*)

Brandy, cakes, canned, cereals, cocktail (no alcohol), confections, cookies, essence, jam, jelly, juice, nectar, pie, pie filling, preserves, strained purée, syrup, and wine

Breadfruit (*Artocarpus altilis*)

Chunks and purée

Cherries, sour (*Prunus cerasus*)

Brined, cakes, canned, citric acid, cocktail (no alcohol), confections, cookies, diet spread, frozen, jam, jelly, juice, nectar, pie, pie filling, preserves, purée, and wine

Cherries, sweet (*Prunus avium*)

Brined, cakes, canned, cocktail (no alcohol), confections, cookies, frozen, nectar, pie, pie filling, preserves, and purée

Coconut (*Livistona chinensis*)

Brandy, cakes, canned, chunks, cocktail (alcoholic), confections, cookies, crushed, dried, frozen, jam, juice, nectar, pie, purée, snack bars, strained purée, and syrup

Crabapple (*Malus pumila*)

Jelly, pectin, and pickles

Cranberries (*Vaccinium macrocarpum*)

Canned, confections, crushed, glacé, jam, jelly, juice, purée, sauce, strained purée, and syrup

Currants (*Ribes vulgare*, rubrum)

Canned, cookies, crushed, essence, glacé, jam, and purée

Currants (*Ribes nigrum*)

Canned, diet spread, and purée

Elderberries (*Sambuca nigra*)

Brandy, cocktail (alcoholic), frozen, glacé, jelly, juice, nectar, and purée

Figs (*Ficus carica*)

Cakes, canned, cereals, cocktail (no alcohol), confections, cookies, crushed, diet spread, dried, frozen, glacé, leather, preserves, purée, and snack bars

Gooseberries (*Ribes hirtellum*)

Canned , cocktail (no alcohol), jelly, juice, preserves, and wine

**Table 4.1.3. (continued)**

Grapefruit (*Citrus paradisi*, pommelo)
    Canned, chunks, confections, crushed, frozen, glacé, juice, and marmalade
Grapes (*Vitis labrusca*)
    Champagne, cocktail (alcoholic), crushed, diet spread, jam, juice, sauce, vinegar, and wine
Grapes (*V. rotundifolia*)
    Champagne, jam, jelly, juice, sauce, vinegar, and wine
Grapes (*V. vinifera*)
    Champagne, cocktail (alcoholic), jam, jelly, juice, sauce, vinegar, and wine
Guava (*Psidium guajava*)
    Crushed, diet spread, jam, jelly, leather, purée, and snack bars
Honeydew melon (*Cucumic melo*)
    Crushed, diet spread, jam, jelly, leather, purée, and snack bars
Kiwifruit (*Actidia chinensis*)
    Canned, cocktail (no alcohol), crushed, and dried
Kumquat (*Fortunella margarita*)
    Jam
Lemon (*Citrus lemon*)
    Citric acid, cocktail (alcoholic), essence, frozen, glacé, juice, marmalade, and pectin
Lime (*Citrus aurantifolia*)
    Citric acid, cocktail (alcoholic), crushed, essence, frozen, juice, marmalade, and pectin
Longan (*Euphoria*)
    Citric acid and crushed
Loquat (*Eriobotrya japonica*)
    Cocktail (no alcohol) and purée
Lychee (*Litchi chinensis*)
    Canned and dried
Mango (*Mangifera indica*)
    Crushed and purée
Nectarine (*Prunun nectarina*)
    Canned, cocktail (no alcohol), confections, cookies, crushed, diet spread, frozen, jam, and purée
Olive (*Olea europala*)
    Brined and canned
Orange (*Citrus quarantium*, sinensis)
    Cereals, champagne, confections, cookies, crushed, diet spread, essence, frozen, glacé, juice, marmalade, pectin, and wine
Papaya (*Carica papaya*)

Crushed, juice, leather, purée, and snack bars

Sapote (*Pouteria sapota*)

Crushed, juice, and purée

Passion fruit (*Passiflora*)

Crushed, juice, nectar, purée, and strained purée

Peaches (*Prunus persica*)

Brandy, brined, cakes, canned, cereals, chunks, cocktail (no alcohol), confections, cookies, crushed, diet spread, dried, essence, frozen, jam, juice, leather, pickles, pie, pie filling, preserves, purée, sauce, snack bars, strained purée, syrup, and wine

Pear (*Pyrus cominunis*)

Brandy, canned, cereals, chunks, cocktail (no alcohol), confections, cookies, crushed, diet spread, dried, frozen

Persimmon (*Diospyros kaki*)

Chunks and crushed

Persimmon (*D. virginiana*)

Brandy, chunks, crushed, jam, juice, marmalade, pectin, pie, pie filling, preserves, and syrup

Pineapple (*Ananas cormosus*)

Cakes, canned, cereals, chunks, cocktail (no alcohol), confections, cookies, crushed, diet spread, frozen, glacé, jam, leather, preserves, purée, and strained purée

Plum (*Prunus domestica*)

Cereals, cocktail, confections, crushed, diet spread, juice, purée, and strained purée

Pomegranate (*Punica granatum*)

Jam, juice, leather, purée, snack bars, and strained purée

Prune (*Prunus domestica*)

Cakes, canned, cereals, cocktail (no alcohol), confections, cookies, crushed, diet spread, dried, jam, jelly, pectin, preserves, purée, vinegar, and wine

Quince (*Cydonia vulgaris*)

Glacé, jam, jelly, juice, pickles, preserves, purée, strained purée, wine

Raspberry (*Rubus idueus* or *R. stigosus*)

Cakes, canned, cereals, confections, cookies, crushed, diet spread, essence, frozen, jam, jelly, juice, nectar, pickles, pie filling, preserves, snack bars, strained purée, and syrup

Strawberry (*Fragaria chiloensis*)

Brandy, cakes, cereals, confections, cookies, crushed, frozen, and juice

Tangerine (*Citrus reticulata*)

Cocktail (no alcohol), frozen, glacé, juice, and purée

**Table 4.1.3. (continued)**

Tucuma (*Astrocaryum tucuma*)
    Crushed

Source: Anonymous, 1985; Anonymous, 1986; Chan, 1983; Considine and Considine, 1982; Duckworth, 1966; Lopez, 1987; Nelson and Tressler, 1980; Salunkhe and Desai, 1984; Ting and Rouseff, 1986; Weichmann, 1987; Wills et al., 1989.

**Table 4.1.4 Fruits, Molds, and Spoilage**

Apple
Black spot: *Alternaria, Venturia*
Blue mold rot: *Penicillium expansum*
Brown rot: *Phytophthora*
Bull's eye rot: *Cryptosporiopsis malicorticis*
Eye rot: *Nectria galligena*
Gray mold rot: *Botrytis cinerea*
Lenticel rot: *Cryptosporiopsis malicorticis*
Soft rot: *Rhizopus* spp.

Apricot
Black to brown spots: *Alternaria* spp.
Black rot: *Aspergillus* spp.
Blue mold rot: *Penicillium* sp.
Brown rot: *Monilinia fructicola*
Gray mold rot: *Botrytis cinerea*
Restricted rot with gray-black core: *Cladosporium* sp.
Watery, soft rot: *Rhizopus stolonifer*

Avocado
Anthracnose: *Collectotrichum gloeosporioides*
Stem-end rot: *Dipolida* sp.
Watery, soft rot: *Rhizopus* sp.

Banana
Anthracnose: *Collectotrichum musae*
Black rot, lesion rot: *Collectotrichum musae*
Crown rot: *Ceratocystis paradoxa, Collectotrichum musae, Fusarium roseum, Verticillium theobromae*

<u>Blueberry</u>
Gray mold rot: *Botrytis cinerea*
Mummification: *Monilinia fructicola*
Wooly growth: *Alternaria* spp.

<u>Cherry</u>
Black to brown spots: *Alternaria* spp.
Brown rot: *Monilinia fructicola*
Gray mold rot: *Botrytis cinerea*
Rhizopus rot: *Rhizopus stolonifer*

<u>Cranberry</u>
Black rot: *Centhospora lunata*
Blotch rot: *Acanthorhynchus vaccinii*
Bitter rot: *Glomerella cingulata*
Early rot: *Guignardia vaccinii*
End rot: *Godronia cassandrea*
Ripe rot: *Sporonema oxycocci*
Storage rot: *Diaporthe vaccinii*

<u>Citrus fruits</u>
Anthracnose: *Collectotrichum* spp.
Black center rot: *Alternaria* spp.
Black rot: *Aspergillus* spp.
Black spot: *Guignardia* sp.
Blue mold rot: *Penicillium italicum*
Brown rot: *Fusarium, Phytophthora*
Cocoa-brown rot: *Trichoderma* sp.
Cottony rot: *Sclerotinia* sp.
Gray mold rot: *Botrytis cinerea*
Green mold rot: *Penicillium digiatum*
Septoria spot: *Septoria sp.*
Stem-end rot: *Alternaria citri, Diplodia natalensis, Phomopsis citri*
Sooty blotch: *Stomiopeltis citri*
Sour rot: *Geotrichum candidum*

<u>Fig</u>
Black to brown spots: *Alternaria* spp.
Black rot: *Aspergillus* spp.
Blue mold rot: *Penicillium* spp.

**Table 4.1.4. (continued)**

Gray mold rot: *Botrytis cinerea*
Soft rot: *Fusarium* spp.
Spotting: *Cladosporium*
Watery soft rot: *Rhizopus* spp.

Grape
Black rot: *Cladosporium*
Blue mold rot: *Penicillium*
Gray mold rot: *Botrytis cinerea*

Grapefruit (see citrus fruits)

Guava
Anthracnose: *Collectotrichum psidii*
Fruit canker: *Pestalotia psidii*
Phoma rot: *Phoma psidii*
Soft watery rot: *Rhizopus, Botryodiplodia theobromai*

Lemon (see citrus fruits)

Lime (see citrus fruits)

Mango
Anthracnose: *Collectotrichum mangiferae*
Stem-end rot: *Diplodia* sp.

Melons (Cantaloups, Honeydews, Persians, Crenshaws, Catabas, and Summer Squash)
Alternaria rot: *Alternaria tenuis*
Blue mold rot: *Penicillium* spp.
Cladosporium (green) rot: *Cladosporium cucumerinum*
Downy mildew: *Phytophthora* sp.
Fusarium (pink to white) rot: *Fusarium* sp.
Rhizopus soft rot: *Rhizopus stolonifer*
Stem-end rot: *Diplodia* sp.

Nectarine
Black to brown spots: *Alternaria* spp.
Black rot: *Aspergillus* spp.
Blue mold rot: *Penicillium* spp.

Brown rot: *Monilinia fructicola*
Gray mold rot: *Botrytis cinerea*
Restricted rot with gray-black core: *Cladosporium* sp.
Watery, soft rot: *Rhizopus stolonifer*

Orange (see citrus fruits)

Papaya
Anthracnose: *Colletotrichum gloeosporioides*
Stem-end rot: *Diplodia* sp

Paw Paw
Anthracnose: *Colletotrichum* sp.
Ripe rot: *Botryodiplodia* sp.

Peach
Black rot: *Aspergillus* spp.
Blue rot: *Penicillium* spp.
Brown spots: *Alternaria* spp.
Gray mold rot: *Botrytis cinerea*
Restricted rot with gray-black core: *Cladosporium* sp.
Rhizopus rot: *Rhizopus stolonifer*
Sour rot: *Geotrichum candidum*
Watery, tan rot: *Diplodia* sp.

Pear
Black spots: *Venturia* sp.
Blue mold rot: *Penicillium expansum*
Bull's eye rot: *Cryptosporiopsis malicorticis*
Eye rot: *Nectria galligena*
Gray mold rot: *Botrytis cinerea*
Lenticel rot: *Cryptosporiopsis malicorticis, Phlyctaena vagabunda*
Sooty blotch: *Gloeodes* sp.
Watery, soft rot: *Rhizopus* spp.

Pineapple
Black rot: *Ceratocystis paradoxa*
Black spot, brown rot, leathery pocket: *Penicillium, Fusarium*

Plum
Black to brown spots: *Alternaria* spp.

**Table 4.1.4. (continued)**

Blue mold rot: *Penicillium* spp.
Brown rot: *Monilinia fructicola*
Gray mold rot: *Botrytis cinerea*
Restricted rot with gray-black core: *Cladosporium* sp.

Quince
Bull's eye rot: *Cryptosporiopsis*
Raspberry gray mold rot: *Botrytis cinerea*
Olive-green rot: *Cladosporium*
Rhubarb anthracnose: *Collectotrichum erumpens*
Gray mold rot: *Botrytis* spp.

Strawberry
Gray mold rot: *Botrytis cinerea*
Leathery rot: *Phytophthora* sp.
Watery, soft rot: *Rhizopus stolonifer*
Watery, white rot: *Sclerotinia* sp.

Tangerine (see citrus fruits)

Watermelon
Anthracnose: *Collectotrichum lagenarium*
Black rot: *Mycosphaerella citrullina*
Phytophthora rot: *Phytophthora capsici*
Stem-end rot: *Sclerotina sclerotiorum*
Watery, soft rot: *Sclerotinia sclerotiorium*

Source: Chan, 1983; Dennis, 1983; Goodenough et Atkin, 1981; Horst, 1990; Nel, 1985; Nyvall, 1989; Teranishi and Barrera-Benitez, 1981.

**Table 4.1.5. Water Content of Vegetables and Fruits**

| Fruit or Vegetable | % Water | Fruit or Vegetable | % Water |
|---|---|---|---|
| Apples | 75.0-85.0 | Avocados | 65.0-66.0 |
| Apricots | 85.2-85.9 | Bananas | 74.0-75.0 |
| Artichokes, globe | 83.5-84.5 | Beans, dried | 12.0-13.0 |
| Artichokes, Jerusalem | 78.5-79.5 | Beans, French | 89.0-91.0 |
| Asparagus | 92.5-93.5 | Beans, green or snap | 89.0-90.0 |

| | |
|---|---|
| Beans, lima | 65.5-66.5 |
| Beans, string | 88.0-89.0 |
| Beets, topped | 87.0-88.0 |
| Blackberries | 84.0-85.0 |
| Blueberries | 82.0-83.0 |
| Broccoli, sprouting | 89.0-90.0 |
| Brussel sprouts | 84.0-85.0 |
| Cabbage, late | 92.0-93.0 |
| Cabbage, white, fresh | 90.0-91.0 |
| Carrots, boiled | 91.0-92.0 |
| Carrots, bunch | 86.0-90.0 |
| Cauliflower | 91.0-92.0 |
| Celeriac | 88.0-89.0 |
| Celery | 93.0-84.0 |
| Cherries | 83.0-84.0 |
| Corn, dried | 10.0-11.0 |
| Corn, green | 75.0-76.0 |
| Corn, sweet | 73.0-74.0 |
| Cranberries | 87.0-88.0 |
| Cucumbers | 96.0-97.0 |
| Currants | 87.0-88.0 |
| Dates, dried | 19.0-21.0 |
| Dates, fresh | 77.0-79.0 |
| Dewberries | 87.0-88.0 |
| Eggplants | 92.0-93.0 |
| Endive | 93.0-94.0 |
| Figs, dried | 24.0-25.0 |
| Figs, fresh | 78.0-79.0 |
| Garlic, dry | 74.0-75.0 |
| Gooseberries | 88.0-89.0 |
| Grapefruit | 88.0-89.0 |
| Grapes | 81.0-92.0 |
| Kale | 86.0-87.0 |
| Kohlrabi | 90.0-91.0 |
| Leeks, green | 88.0-92.0 |
| Lemons | 89.0-90.0 |
| Lentils | 11.0-12.0 |
| Lettuce | 94.0-95.0 |
| Limes | 86.0-87.0 |
| Loganberries | 82.0-83.0 |
| Mangoes | 81.0-82.0 |
| Melons | 92.0-93.0 |
| Mushrooms, dried | 29.0-31.0 |
| Mushrooms, fresh | 90.0-92.0 |
| Nectarines | 82.0-83.0 |
| Okra | 98.0-99.0 |
| Onions | 87.0-88.0 |
| Oranges | 87.0-88.0 |
| Papaya | 90.0-91.0 |
| Parsley | 65.0-95.0 |
| Parsnips | 78.0-79.0 |
| Peaches | 86.0-90.0 |
| Pears | 82.0-83.0 |
| Peas, air dried | 13.0-14.0 |
| Peas, dried | 9.0-10.0 |
| Peas, green | 74.0-76.0 |
| Peppers, chilli, dry | 11.0-13.0 |
| Peppers, sweet | 92.0-93.0 |
| Persimmons | 78.0-79.0 |
| Pineapples | 85.0-86.0 |
| Plums and prunes, dry | 28.0-35.0 |
| Plums, fresh | 81.0-86.0 |
| Pomegranates | 85.0-86.0 |
| Popcorn, unpopped | 13.0-14.0 |
| Potatoes | 74.0-75.0 |
| Potatoes, late crop | 77.0-78.0 |
| Prunes, fresh | 81.0-86.0 |
| Pumpkins | 90.0-91.0 |
| Quinces | 85.0-86.0 |
| Radishes, spring | 93.0-94.0 |
| Radishes, spring, packaged | 93.0-94.0 |
| Raspberries, black | 80.0-81.0 |
| Raspberries, red | 84.0-85.0 |
| Rutabagas | 89.0-90.0 |
| Salsify | 79.0-80.0 |
| Sorrel | 91.0-93.0 |
| Spinach | 85.0-93.0 |
| Squash | 88.0-95.0 |
| Strawberries, fresh | 89.0-91.0 |
| Strawberries, frozen | 71.0-72.0 |
| Sweet potatoes | 68.0-69.0 |

**Table 4.1.5. (continued)**

| Fruit or Vegetable | % Water | Fruit or Vegetable | % Water |
|---|---|---|---|
| Tangerines | 87.0-88.0 | Tomatoes, mature green | 85.0-95.0 |
| Tomatoes, ripe | 94.0-95.0 | Turnips | 90.0-91.0 |

Source: Anonymous, 1982; Anonymous, 1989; Yamaguchi, 1983.

**Table 4.1.6. Microorganisms Capable of Spoiling Canned Foods**

Bacteria: Classification
1. Food Acidity; Signs or Conditions of Spoilage
2. Chemical and Physical Changes from Spoilage

*Bacillus betanigrificans:* mesophilic organism
1. pH 4.5-7.0; flat can
2. Canned beets becoming darkened

*Bacillus stearothermophilus*: thermophilic organism
1. pH 4.5-7.0; taste sour and flat
2. Byproducts include acids (acetic, lactic, and formic) but no gas.

*Bacillus subtilis:* mesophilic organism
1. pH 4.5-7.0; can swelling
2. Byproduct includes gas and spiced ham shows no putrefaction.

*Bacillus thermoacidurans*: thermophilic organism
1. pH 3.0-4.5; taste sour and flat
2. Minimal changes in pH neutral media; byproduct includes acids (lactic, acetic) but no gas.

*Byssochlamys fulva*: mesophilic organism
1. pH 3.0-4.5; taste flat
2. Canned cranberries becoming soft

*Clostridium botulinum*: mesophilic organism
1. pH 4.5-7.0; can swollen
2. Formation of gas and product putrefied

*Clostridium butyricum*: mesophilic organism
1. pH 3.0-4.5; can swollen
2. By-products include butyric acid, acetic acid, and carbon dioxide.

*Clostridium nigrificans*: thermophilic organism
1. pH 4.5-7.0; can becoming flat
2. Byproduct includes hydrogen sulfide with no acid and no gas; also product not putrefied.

*Clostridium pasteurianum*: mesophilic organism
1. pH 3.0-4.5; can swollen

2. Byproducts include butyric acid, acetic acid, and carbon dioxide.

*Clostridium putrificum*: mesophilic organism

1. pH 4.5-7.0; can swollen
2. Gas formed and product putrefied

*Clostridium sporogenes*: mesophilic organism

1. pH 4.5-7.0; can swollen
2. Gas formed and product putrefied

*Clostridium thermosaccharolyticum:* thermophilic organism

1. pH 4.5-7.0; can swollen
2. Byproducts include acids (acetic, butyric, lactic), carbon dioxide, and hydrogen. Product not putrefied.

*Lactobacillus brevis*: mesophilic organism

1. pH 3.0-4.5; can swollen
2. Byproducts include acids (acetic, lactic) and carbon dioxide.

*Lactobacillus gayoni*: mesophilic organism

1. pH 3.0-4.5; can swollen
2. Byproducts include acids (acetic and lactic) and carbon dioxide.

*Lactobacillus plantarum*: mesophilic organism

1. pH 3.0-4.5; can swollen
2. Byproducts include acid (acetic and lactic) and carbon dioxide.

*Leuconostoc mesenteroides*: mesophilic organism

1. pH 3.0-4.5; can swollen
2. Byproducts include acid (lactic and acetic) and carbon dioxide.

*Saccharomyces*: mesophilic organism

1. pH 3.0-4.5; can swollen and yeasty odor
2. Byproducts include alcohol and carbon dioxide in jams and jellies and gas in tomato products.

*Torulopsis*: mesophilic organism

1. pH 3.0-4.5; can swollen and yeasty odor
2. Byproducts include alcohol and carbon dioxide in jams and jellies and gas in tomato products.

---

Source: Anonymous, 1982; Banwart, 1989; Chan, 1983; Frazer and Westhogg, 1978; Jackson and Shinn, 1979; Jay, 1986; Lopez, 1987; Nelson and Tressler, 1980; Ting and Rouseff, 1986.

**Table 4.2.1 Respiration Rates of Fruits and Vegetables, Expressed as Rates of Carbon Dioxide Poduction (mg/kg/hour), at Various Temperatures**

| | Temperature | | | | | |
|---|---|---|---|---|---|---|
| Commodity | 0°C (32°F) | 4-5°C (40-41°F) | 10°C (50°F) | 15-16°C (59-60°F) | 20-21°C (68-70°F) | 25-27°C (77-80°F) |
| Apples, summer | 3-6 | 5-11 | 14-20 | 18-31 | 20-41 | - |
| Apples, fall | 2-4 | 5-7 | 7-10 | 9-20 | 15-25 | - |
| Apricots | 5-6 | 6-9 | 11-19 | 21-34 | 29-52 | - |
| Artichokes, globe | 15-45 | 26-60 | 55-98 | 76-145 | 135-233 | 145-300 |
| Asparagus | 27-80 | 55-136 | 90-304 | 160-327 | 275-500 | 500-600 |
| Avocados | - | 20-30 | - | 62-157 | 74-347 | 118-428 |
| Bananas, green | - | - | - | 21-23 | 33-35 | - |
| Bananas, ripening | - | - | 21-39 | 25-75 | 33-142 | 50-245 |
| Beans, lima | 10-30 | 20-36 | - | 100-125 | 133-179 | - |
| Beans, snap | 20 | 35 | 58 | 93 | 130 | 193 |
| Bean sprouts | 21-25 | 42 | 93-99 | - | - | - |
| Beets, topped | 5-7 | 9-10 | 12-14 | 17-23 | - | - |
| Beets, with leaves | 11 | 14 | 22 | 25 | 40 | - |
| Berries | | | | | | |
| Blackberries | 18-20 | 31-41 | 62 | 75 | 155 | - |
| Blueberries | 2-10 | 9-12 | 23-35 | 34-62 | 52-87 | 78-124 |

| | | | | | | |
|---|---|---|---|---|---|---|
| Cranberries | - | 4-5 | - | - | 11-18 | - |
| Gooseberries | 4-7 | 8-16 | 12-32 | 27-69 | 41-105 | - |
| Raspberries | 18-25 | 31-39 | 28-55 | 82-101 | - | - |
| Strawberries | 12-18 | 16-23 | 49-95 | 71-92 | 102-196 | 169-211 |
| Broccoli | 19-21 | 32-37 | 75-87 | 161-186 | 278-320 | - |
| Brussels sprouts | 10-30 | 22048 | 63-84 | 64-136 | 86-190 | - |
| Cabbage | 4-6 | 9-12 | 17-19 | 20-32 | 28-49 | 49-63 |
| Carrots, topped | 10-20 | 13-26 | 20-42 | 26-54 | 46-95 | - |
| Carrots, bunched | 18-35 | 25-51 | 32-62 | 55-106 | 87-121 | - |
| Cauliflower | 16-19 | 19-22 | 32-36 | 43-49 | 75-86 | 84-140 |
| Celery | 5-7 | 9-11 | 24 | 30-37 | 64 | - |
| Celeriac | 7 | 15 | 25 | 39 | 50 | - |
| Cherries, sweet | 4-5 | 10-14 | - | 25-45 | 28-32 | - |
| Cherries, sour | 6-13 | 13 | - | 27-50 | 39-50 | 53-71 |
| Citrus | | | | | | |
| Grapefruit | - | - | 7-9 | 10-18 | 13-26 | 19 |
| Lemons | - | - | 11 | 10-23 | 19-25 | 20-28 |
| Limes, Tahiti | - | - | - | 6-10 | 7-19 | 15-45 |
| Oranges | 2-5 | 4-7 | 6-9 | 13-24 | 22-34 | 25-40 |
| Cucumbers | - | - | 23-29 | 24-33 | 14-48 | 19-55 |
| Endive | 45-52 | 73 | 100 | 133 | 200 | |
| Figs, fresh | - | 11-13 | 22-23 | 49-63 | 57-95 | 85-106 |
| Garlic | 4-14 | 9-33 | 9-10 | 14-29 | 13-25 | - |

**Table 4.2.1. (continued)**

| | Temperature | | | | | |
|---|---|---|---|---|---|---|
| Commodity | 0°C (32°F) | 4-5°C (40-41°F) | 10°C (50°F) | 15-16°C (59-60°F) | 20-21°C (68-70°F) | 25-27°C (77-80°F) |
| Grapes, American | 3 | 5 | 8 | 16 | 33 | 39 |
| Grapes, vinifera | 1-2 | 3-6 | 8 | 10-12 | - | 25-30 |
| Kale | 16-27 | 34-47 | 72-84 | 120-155 | 186-265 | - |
| Kohlrabi | 10 | 16 | 31 | 49 | - | - |
| Kiwifruit | 3 | 6 | 12 | - | 16-22 | - |
| Leeks | 10-20 | 20-29 | 50-70 | 75-117 | 110 | 107-119 |
| Lettuce, head | 6-17 | 13-20 | 21-40 | 32-45 | 51-60 | 73-91 |
| Lettuce, leaf | 19-27 | 24-35 | 32-46 | 51-74 | 82-119 | 120-173 |
| Lychees | - | - | - | - | - | 75-128 |
| Mangos | - | 10-22 | - | 45 | 75-151 | 120 |
| Melons | | | | | | |
| Cantaloups | 5-6 | 9-10 | 14-16 | 34-39 | 45-65 | 62-71 |
| Honeydew | - | 3-5 | 7-9 | 12-16 | 20-27 | 26-35 |
| Watermelons | - | 3-4 | 6-90 | - | 17-25 | - |
| Mushrooms | 28-44 | 71 | 100 | - | 264-316 | - |
| Onions, dry | 3 | 3-4 | 7-8 | 10-11 | 14-19 | 27-29 |
| Onions, green | 10-32 | 17-39 | 36-62 | 66-115 | 79-178 | 98-210 |
| Okra | - | 53-59 | 86-95 | 138-143 | 248-274 | 328-362 |
| Olives | - | - | - | 27-66 | 40-105 | 56-128 |

| | | | | | | |
|---|---|---|---|---|---|---|
| Papayas | - | 4-6 | - | 15-22 | - | 39-88 |
| Parsley | 30-40 | 53-76 | 85-164 | 144-184 | 196-225 | 291-324 |
| Parsnips | 8-15 | 9-18 | 20-26 | 32-46 | - | - |
| Peaches | 4-6 | 6-9 | 16 | 33-42 | 59-102 | 81-122 |
| Pears, Bartlett | 3-7 | 5-10 | 8-21 | 15-60 | 30-70 | - |
| Pears, Lieffer | 2 | - | - | 11-24 | 15-28 | 20-29 |
| Peas, unshelled | 30-47 | 55-76 | 68-117 | 179-202 | 245-361 | 343-377 |
| Peas, shelled | 47-75 | 79-97 | - | - | 349-556 | - |
| Peppers, sweet | - | 10 | 14 | 23 | 44 | 55 |
| Persimmon, Japanese | - | 6 | - | 12-14 | 20-24 | 29-40 |
| Pineapples, mature-green | - | 2 | 4-7 | 10-16 | 19-29 | 28-43 |
| Plums, Wickson | 2-3 | 4-9 | 7-11 | 12 | 18-26 | 28-71 |
| Potatoes, immature | - | 12 | 14-21 | 14-31 | 18-45 | - |
| Potatoes, mature | - | 3-9 | 7-10 | 6-12 | 8-16 | - |
| Radishes, topped | 3-9 | 6-13 | 15-16 | 22-42 | 44-58 | 60-89 |
| Rhubarb, stalk | 9-13 | 11-18 | 25 | 31-48 | 40-57 | - |
| Romaine | - | 18-23 | 31-40 | 39-50 | 60-77 | 95-121 |
| Rutabagas | 2-6 | 5-10 | 15 | 11-28 | 41 | - |
| Spinach | 19-22 | 35-58 | 82-138 | 134-223 | 172-287 | - |
| Squash, Butternut | - | - | - | - | - | 66-121 |
| Squash, summer | 12-13 | 14-19 | 34-36 | 75-90 | 85-97 | - |
| Sweet corn, w/husks | 30-51 | 43-83 | 104-120 | 151-175 | 268-311 | 282-435 |
| Sweetpotatoes, noncured | - | - | - | 29 | - | 54-73 |

**Table 4.2.1. (continued)**

| | Temperature | | | | | |
|---|---|---|---|---|---|---|
| Commodity | 0°C (32°F) | 4-5°C (40-41°F) | 10°C (50°F) | 15-16°C (59-60°F) | 20-21°C (68-70°F) | 25-27°C (77-80°F) |
| Sweetpotatoes, cured | - | - | 14 | 20-24 | - | - |
| Tomatoes, mature-green | - | 5-8 | 12-18 | 16-28 | 28-41 | 35-51 |
| Tomatoes, ripening | - | - | 13-16 | 24-29 | 24-44 | 30-52 |
| Turnips, topped | 6-9 | 10 | 13-19 | 21-24 | 24-25 | - |
| Watercress | 15-26 | 44-49 | 91-121 | 136-205 | 302-348 | 348-438 |

Source: Hardenburg et al., 1986. Check original documents for additional details and qualifications.

**Table 4.2.2. Fruits and Vegetables Susceptible to Chilling Injury when Stored at Moderately Low but Nonfreezing Temperatures**

| Commodity | Approximate Lowest Safe Temperature °C | °F | Character of Injury when Stored between 0°C and Safe Temperature |
|---|---|---|---|
| Apples, certain cultivars | 2-3 | 36-38 | Internal browning, brown core, soggy breakdown, soft scale |
| Asparagus | 3-2 | 32-36 | Dull, gray-green, and limp tips |
| Avocados | 4.5-13 | 40-55 | Grayish-brown discoloration of flesh |
| Bananas, green | 11.5-13 | 53-56 | Dull color when ripened or ripe |
| Beans (lima) | 1-4.5 | 34-40 | Rusty brown specks, spots, or areas |
| Beans (snap) | 7 | 45 | Pitting and russeting |
| Cranberries | 2 | 36 | Rubbery texture, red flesh |
| Cucumbers | 7 | 45 | Pitting, water-soaked spots, decay |
| Eggplants | 7 | 45 | Surface scald, alternaria rot, blackening of seeds |
| Guavas | 4.5 | 40 | Pulp injury, decay |
| Grapefruit | 10 | 50 | Scald, pitting, water breakdown |
| Jicama | 13-18 | 55-65 | Surface decay, discoloration |
| Lemons | 11-13 | 52-55 | Pitting, membranous staining, red blotch |
| Limes | 7-9 | 45-48 | Pitting, turning tan with time |
| Mangoes | 10-13 | 50-55 | Grayish scaldlike discoloration of skin, uneven ripening |
| Melons | | | |
| Cantaloups | 2-5 | 36-41 | Pitting, surface decay |
| Honeydew | 7-10 | 45-50 | Reddish-tan discoloration, pitting, surface decay, failure to ripen |
| Casaba | 7-10 | 45-50 | Same as above but no discoloration |
| Crenshaw and Persian | 7-10 | 45-50 | Same as above but no discoloration |
| Watermelons | 4.5 | 40 | Pitting, objectionable flavor |

**Table 4.2.2 (continued)**

| Commodity | Approximate Lowest Safe Temperature °C | °F | Character of Injury when Stored between 0°C and Safe Temperature |
|---|---|---|---|
| Okra | 7 | 45 | Discoloration, water-soaked areas, pitting, decay |
| Olives, fresh | 7 | 45 | Internal browning |
| Oranges, California and Arizona | 7 | 45 | Pitting, failure to ripen, off-flavor, decay |
| Papayas | 7 | 45 | Pitting, failure to ripen, off-flavor, decay |
| Peppers, sweet | 7 | 45 | Sheet pitting, alternaria rot on pods and calyxes, darkening of seed |
| Pineapples | 7-10 | 45-50 | Dull green when ripened |
| Pomegranates | 4.5 | 40 | Pitting, external and internal browning |
| Potatoes | 3 | 38 | Mahogany browning (Chippewa and Sebago), sweetening |
| Pumpkins and hard-shell squashes | 10 | 50 | decay, especially alternaria rot |
| Sweetpotatoes | 13 | 55 | Decay, pitting, internal discol oration; hard core when cooked |
| Tamarillos | 3-4 | 37-40 | Surface pitting, discoloration |
| Tamatoes | | | |
| Ripe | 7-10 | 45-50 | Watersoaking and softening, decay |
| Mature-green | 13 | 55 | Poor color when ripe, alternaria rot |

Source: Hardenburg et al., 1986.

Note: Characters or symptoms of injury often apparent only after removal to warm temperatures, as in marketing. See original documents for more details.

**Table 4.2.3. Susceptibility of Fresh Fruits and Vegetables to Freezing Injury**

| Group 1: Most Susceptible | Group 2: Moderately Susceptible | Group 3: Least Susceptible |
|---|---|---|
| Apricots | Apples | Beets |
| Asparagus | Broccoli, sprouting | Brussels sprouts |
| Avocados | Cabbage, new | Cabbage, mature and savory |
| Bananas | Carrots | Dates |
| Beans, snap | Cauliflower | Kale |
| Berries (except cranberries) | Celery | Kohlrabi |
| Cucumbers | Cranberries | Parsnips |
| Eggplant | Grapefruit | Rutabagas |
| Lemons | Grapes | Salsify |
| Lettuce | Onions (dry) | Turnips |
| Limes | Oranges | |
| Okra | Parsley | |
| Peaches | Pears | |
| Peppers, sweet | Peas | |
| Plums | Radishes | |
| Potatoes | Spinach | |
| Squash, Summer | Squash, winter | |
| Sweetpotatoes | | |
| Tomatoes | | |

Source: Hardenburg et al., 1986.

**Table 4.2.4 Recommended Temperature and Relative Humidity, Approximate Storage Life, Highest Freezing Point, Water Content, and Specific Heat For Fresh Vegetables in Commercial Storage**

| Commodity | Temperature °C | Temperature °F | Relative Humidity % | Approximate Storage Life | Highest Freezing Point °C | Highest Freezing Point °F | Water Content % | Specific heat (Btu/lb·°F) |
|---|---|---|---|---|---|---|---|---|
| Artichokes, globe | 0 | 32 | 95-100 | 2-3 weeks | -1.1 | 29.9 | 83.7 | .87 |
| Artichokes, Jerusalem | -0.5-0.0 | 31-32 | 90-95 | 4-5 months | -2.2 | 28.0 | 79.8 | .84 |
| Asparagus | 0-2 | 32-35 | 95-100 | 2-3 weeks | 0.6 | 30.9 | 93.0 | .94 |
| Beans, dry | 4-10 | 40-50 | 40-50 | 6-10 months | - | - | 15.0 | .32 |
| Beans, green or snap | 4-7 | 40-45 | 95 | 7-10 days | -0.7 | 30.7 | 88.9 | .91 |
| Beans, lima | 3-5 | 37-41 | 95 | 5-7 days | -0.6 | 31.0 | 66.5 | .73 |
| Bean sprouts | 0 | 32 | 95-100 | 7-9 days | - | - | 88.8 | .91 |
| Beets, bunched | 0 | 32 | 98-100 | 10-14 days | -0.4 | 31.3 | - | - |
| Beets, topped | 0 | 32 | 98-100 | 4-6 months | -0.9 | 30.3 | 87.6 | .90 |
| Broccoli | 0 | 32 | 95-100 | 10-14 days | -0.6 | 30.9 | 89.9 | .92 |
| Brussels sprouts | 0 | 32 | 95-100 | 3-5 weeks | 0.8 | 30.5 | 84.9 | .88 |
| Cabbage, early | 0 | 32 | 98-100 | 3-6 weeks | -0.9 | 30.4 | 92.4 | .94 |
| Cabbage, late | 0 | 32 | 98-100 | 5-6 months | 0.9 | 30.4 | 92.4 | .94 |
| Cabbage, Chinese | 0 | 32 | 95-100 | 2-3 months | - | - | 95.0 | .96 |
| Carrots, bunched | 0 | 32 | 95-100 | 2 weeks | - | - | - | - |
| Carrots, mature | 0 | 32 | 98-100 | 7-9 months | -1.4 | 29.5 | 88.2 | .91 |
| Carrots, immature | 0 | 32 | 98-100 | 4-6 weeks | -1.4 | 29.5 | 88.2 | .91 |

| | | | | | | | | |
|---|---|---|---|---|---|---|---|---|
| Cassava | 0-5 | 32.41 | 85-90 | 1-2 months | - | - | - | - |
| Cauliflower | 0 | 32 | 95-98 | 3-4 weeks | -0.8 | 30.6 | 91.7 | .93 |
| Celeriac | 0 | 32 | 97-99 | 6-8 months | -0.9 | 30.3 | 88.4 | .91 |
| Celery | 0 | 32 | 98-100 | 2-3 months | -0.5 | 31.1 | 93.7 | .95 |
| Chard | 0 | 32 | 95-100 | 10-14 daysays | - | - | 91.1 | .93 |
| Chicory, witloof | 0 | 32 | 95-100 | 2-4 weeks | - | - | 95.1 | .96 |
| Collards | 0 | 32 | 95-100 | 10-14 daysays | - | | | |
| Corn, sweet | 0 | 32 | 95-98 | 5-8 days | -0.6 | 30.9 | 73.9 | .79 |
| Cucumbers | 10-13 | 50-55 | 95 | 10-14 days | -0.5 | 31.1 | 96.1 | .97 |
| Eggplants | 8-12 | 46-54 | 90-95 | 1 weeks | -0.8 | 30.6 | 92.7 | .94 |
| Endive and escarole | 0 | 32 | 95-100 | 2-3 weeks | -0.1 | 31.9 | 93.1 | .95 |
| Garlic | 0 | 32 | 65-70 | 6-7 months | -0.8 | 30.5 | 61.3 | .69 |
| Ginger | 13 | 55 | 65 | 6 months | - | - | 87.0 | .90 |
| Greens, leafy | 0 | 32 | 95-100 | 10-14 days | - | - | - | - |
| Horseradish | -1.0-0 | 30-32 | 98-100 | 10-12 months | -1.8 | 28.7 | 74.6 | .80 |
| Jicama | 13-18 | 55-65 | 65-70 | 1-2 months | - | - | - | - |
| Kale | 0 | 32 | 95-100 | 2-3 weeks | -0.5 | 31.1 | 86.6 | .89 |
| Kohlrabi | 0 | 32 | 98-100 | 2-3 months | -1.0 | 30.2 | 90.3 | .92 |
| Leeks | 0 | 32 | 95-100 | 2-3 months | -0.7 | 30.7 | 85.4 | .88 |
| Lettuce | 0 | 32 | 98-100 | 2-3 weeks | -0.2 | 31.7 | 94.8 | .96 |

**Table 4.2.4. (continued)**

| Commodity | Temperature °C | Temperature °F | Relative Humidity % | Approximate Storage Life | Highest Freezing Point °C | Highest Freezing Point °F | Water Content % | Specific heat (Btu/ lb·°F) |
|---|---|---|---|---|---|---|---|---|
| Melons | | | | | | | | |
| Cantaloup (3/4-slip) | 2-5 | 36-41 | 95 | 15 days | -1.2 | 29.9 | 92.0 | .94 |
| Cantaloup (full-slip) | 0-2 | 32-36 | 95 | 5-14 days | -1.2 | 29.9 | 92.0 | .94 |
| Casaba | 10 | 50 | 90-95 | 3 weeks | -1.0 | 30.1 | 92.7 | .94 |
| Crenshaw | 7 | 45 | 90-95 | 2 weeks | -1.0 | 30.1 | 92.7 | .94 |
| Honeydew | 7 | 45 | 90-95 | 3 weeks | -0.9 | 30.3 | 92.6 | .94 |
| Persian | 7 | 45 | 90-95 | 2 weeks | -0.8 | 30.5 | 92.7 | .94 |
| Watermelons | 10-15 | 50-60 | 90 | 2-3 weeks | -0.4 | 31.3 | 92.6 | .94 |
| Mushrooms | 0 | 32 | 95 | 3-4 days | -0.9 | 30.4 | 91.1 | .93 |
| Okra | 7-10 | 45-50 | 90-95 | 7-10 days | -1.8 | 28.7 | 89.8 | .92 |
| Onion, green | 0 | 32 | 95-100 | 3-4 weeks | -0.9 | 30.4 | 89.4 | .91 |
| Onion, dry | 0 | 32 | 65-70 | 1-8 months | -0.8 | 30.6 | 87.5 | .90 |
| Onion sets | 0 | 32 | 65-70 | 6-8 months | -0.8 | 30.6 | 87.5 | .88 |
| Parsley | 0 | 32 | 95-100 | 2-2.5 months | -1.1 | 30.0 | 85.1 | .88 |
| Parsnips | 0 | 32 | 98-100 | 4-6 months | -0.9 | 30.4 | 78.6 | .83 |
| Peas, green | 0 | 32 | 95-98 | 1-2 weeks | -0.6 | 30.9 | 74.3 | .79 |
| Peas, southern | 4-5 | 40-41 | 95 | 6-8 days | - | - | 66.8 | .73 |
| Peppers, chill (dry) | 0-10 | 32-50 | 60-70 | 6 months | - | - | 12.0 | .30 |
| Peppers, sweet | 7-13 | 45-55 | 90-95 | 2-3 weeks | -0.7 | 30.7 | 92.4 | .94 |

| | | | | | | | | |
|---|---|---|---|---|---|---|---|---|
| Potatoes, early crop | * | - | 90-95 | * | -0.6 | 30.9 | 81.2 | .85 |
| Potatoes, late crop | * | - | 90-95 | 5-10 months | -0.6 | 30.9 | 77.8 | .82 |
| Pumpkins | 10-13 | 50-55 | 50-70 | 2-3 months | -0.8 | 30.5 | 90.5 | .92 |
| Radishes, spring | - | 32 | 95-100 | 3-4 weeks | -0.7 | 30.7 | 94.5 | .96 |
| Radishes, winter | - | 32 | 95-100 | 2-4 months | - | - | - | - |
| Rhubarb | - | 32 | 95-100 | 2-4 weeks | -0.9 | 30.3 | 94.9 | .96 |
| Rutabagas | - | 32 | 98-100 | 4-6 months | -1.1 | 30.0 | 89.1 | .91 |
| Salsify | 0 | 32 | 95-98 | 2-4 months | -1.1 | 30.0 | 79.1 | .83 |
| Spinach | 0 | 32 | 95-100 | 10-14 days | -0.3 | 31.5 | 92.7 | .94 |
| Squashes, summer | 5-10 | 41-50 | 95 | 1-2 weeks | -0.5 | 31.1 | 94.0 | .95 |
| Squashes, winter | 10 | 50 | 50-70 | * | -0.8 | 30.5 | 85.1 | .88 |
| Sweetpotatoes | 13-16 | 55-60 | 85-90 | 4-7 months | -1.3 | 29.7 | 68.5 | .75 |
| Tamarillos | 3-4 | 37-40 | 85-95 | 10 weeks | - | - | - | - |
| Taro (Dasheen) | 7-10 | 45-50 | 85-90 | 4-5 months | - | - | 73.0 | .78 |
| Tomatoes, mature-green | 13-21 | 55-70 | 90-95 | 1-3 weeks | -0.6 | 31.0 | 93.0 | .94 |
| Tomatoes, firm-ripe | 8-10 | 46=50 | 90-95 | 4-7 days | -0.5 | 31.1 | 94.1 | .95 |
| Turnips | 0 | 32 | 95 | 4-5 months | -1.0 | 30.1 | 91.5 | .93 |
| Turnip greens | 0 | 32 | 95-100 | 10-14 days | -0.2 | 31.7 | 90.3 | .92 |
| Water chestnuts | 0-2 | 32-36 | 98-100 | 1-2 months | - | - | 78.3 | .83 |
| Watercress | 0 | 32 | 95-100 | 2-3 weeks | 0.3 | 31.4 | 93.3 | .95 |
| Yams | 16 | 61 | 70-80 | 6-7 months | - | - | 73.5 | .79 |

Source: Hardenburg et al., 1986.
Check the original document for additional details, especially slots marked with an *.

**Table 4.2.5 Recommended Temperature and Relative Humidity, Approximate Storage Life, Highest Freezing Point, Water Content, and Specific Heat of Fresh Fruits in Commercial Storage**

| Commodity | Temperature | | Relative Humidity | Approximate Storage Life | Highest Freezing Point | | Water Content | Specific heat |
|---|---|---|---|---|---|---|---|---|
| | °C | °F | % | | °C | °F | % | (Btu/ lb·°F) |
| Apples | -1.0-4.0 | 30-40 | 90-95 | 1-12 months | -1.5 | 29.3 | 84.1 | .87 |
| Apricots | -0.5-0.0 | 31-32 | 90-95 | 1-3 weeks | -1.0 | 30.1 | 85.4 | .88 |
| Avocados | 4.4-13 | 40-55 | 85-90 | 2-8 weeks | -0.3 | 31.5 | 76.0 | .81 |
| Bananas, green | 13-14 | 56-58 | 90-95 | * | -0.7 | 30.6 | 75.7 | .81 |
| Berries | | | | | | | | |
| Blackberries | -0.5-0 | 31-32 | 90-95 | 2-3 days | -0.7 | 30.5 | 84.8 | .88 |
| Blueberries | -0.5-0 | 31-32 | 90-95 | 2 weeks | -1.2 | 29.7 | 83.2 | .86 |
| Cranberries | 2-4 | 36-40 | 90-95 | 2-4 months | -0.8 | 30.4 | 87.4 | .90 |
| Currants | -0.5-0 | 31-32 | 90-95 | 2-3 days | -1.2 | 29.7 | 84.5 | .88 |
| Dewberries | -0.5-0 | 31-32 | 90-95 | 2-3 days | -1.2 | 29.7 | 84.5 | .88 |
| Elderberries | -0.5-0 | 31-32 | 90-95 | 1-2 weeks | - | - | 79.8 | .84 |
| Gooseberries | -0.5-0 | 31-32 | 90-95 | 3-4 weeks | -1.0 | 30.0 | 88.9 | .91 |
| Loganberries | -0.5-0 | 31-32 | 90-95 | 2-3 days | -1.2 | 29.7 | 83.0 | .86 |
| Raspberries | -0.5-0 | 31-32 | 90-95 | 2-3 days | -1.0 | 30.0 | 82.5 | .86 |
| Strawberries | 0 | 32 | 90-95 | 5-7 days | -0.7 | 30.6 | 89.9 | .92 |
| Carambola | 9-10 | 48-50 | 85-90 | 3-4 weeks | - | - | 90.4 | .92 |
| Cherries, sour | 0 | 32 | 90-95 | 3-7 days | -1.7 | 29.0 | 83.7 | .87 |
| Cherries, sweet | -1 to -0.5 | 30-31 | 90-95 | 2-3 weeks | -1.8 | 288.8 | 80.4 | .84 |
| Coconuts | 0-1.5 | 32-35 | 80-85 | 1-2 months | -0.9 | 30.4 | 46.9 | .58 |

| | | | | | | | | |
|---|---|---|---|---|---|---|---|---|
| Dates | -18 or 0 | 0 or 32 | 75 | 6-12 months | -15.7 | 3.7 | 22.5 | .38 |
| Figs, fresh | -0.5-0 | 31-32 | 85-90 | 7-10 days | -2.4 | 27.6 | 78.0 | .82 |
| Grapefruit, Calif. & Ariz. | 14-15.5 | 58-60 | 85-90 | 6-8 weeks | - | - | 87.5 | .90 |
| Grapefruit | 10-15 | 50-60 | 85-90 | 6-8 weeks | -1.0 | 30.0 | 89.1 | .91 |
| Grapes, vinifera | -1 to -0.5 | 30-31 | 90-95 | 1-6 months | -2.1 | 28.1 | 81.6 | .85 |
| Grapes, American | -0.5-0 | 31-32 | 85 | 2-8 weeks | 01.2 | 29.7 | 81.9 | .86 |
| Guavas | 5-10 | 41-50 | 90 | 2-3 weeks | - | - | 83.0 | .86 |
| Kiwifruit | -0.5-0 | 31-32 | 90-95 | 3-5 months | -1.6 | 29.0 | 82.0 | .86 |
| Lemons | * | | 85-90 | 1-6 months | -1.4 | 29.4 | 87.4 | .90 |
| Loquats | - | 32 | 90 | 3 weeks | - | - | 86.5 | .89 |
| Lychees | 1.5 | 35 | 90-95 | 3-5 weeks | - | - | 81.9 | .86 |
| Mangos | 13 | 55 | 85-90 | 2-3 weeks | -0.9 | 30.3 | 81.7 | .85 |
| Nectarines | -0.5-0 | 31-32 | 90-95 | 2-4 weeks | -0.9 | 30.4 | 81.8 | .85 |
| Olives, fresh | 5-10 | 41-50 | 85-90 | 4-6 weeks | -1.4 | 29.4 | 80.0 | .84 |
| Oranges, Calif. & Ariz. | 3-9 | 38-48 | 85-90 | 3-8 weeks | -1.2 | 29.7 | 85.5 | .88 |
| Oranges, Fla. & Texas | 0-1 | 32-34 | 85-90 | 8-12 weeks | -0.7 | 30.6 | 867.4 | .89 |
| Papayas | 7 | 45 | 85-90 | 1-3 weeks | -0.9 | 30.4 | 88.7 | .91 |
| Passion fruit | 7-10 | 45-50 | 85-90 | 3-5 weeks | - | - | 75.1 | .80 |
| Peaches | -0.5-0 | 31-32 | 90-95 | 2-4 weeks | -0.9 | 30.3 | 89.1 | .91 |
| Pears | -1.5 to -0.5 | 29-31 | 90-95 | 2-7 months | -1.5 | 29.2 | 83.2 | .87 |

**Table 4.2.5. (continued)**

| Commodity | Temperature °C | Temperature °F | Relative Humidity % | Approximate Storage Life | Highest Freezing Point °C | Highest Freezing Point °F | Water Content % | Specific heat (Btu/lb·°F) |
|---|---|---|---|---|---|---|---|---|
| Persimmons, Japanese | -1 | 30 | 90 | 3-4 months | -2.1 | 28.1 | 78.2 | .83 |
| Pineapples | 7-13 | 45-55 | 85-90 | 2-5 weeks | -1.1 | 30.0 | 85.3 | .88 |
| Plums and prunes | -0.5-0 | 31-32 | 90-95 | 2-5 weeks | -0.8 | 30.5 | 86.6 | .89 |
| Pomegranates | 5 | 41 | 90-95 | 2-3 months | -3.0 | 26.6 | 82.3 | .86 |
| Quinces | -0.5-0 | 31-32 | 90 | 2-3 months | -2.0 | 28.4 | 83.8 | .87 |
| Tangerines, mandarins, & related citrus fruits | 4 | 40 | 90-95 | 2-4 weeks | -1.0 | 30.1 | 87.3 | .90 |

Source: Hardenburg et al., 1986.
Check original document for more details, especially slots marked with an *.

**Table 4.2.6. Sample Calculations in Metric Units for Refrigeration Requirement in the Cold Storage of Pears**

---

Part A: Basic Considerations

Storage size: 15 x 15 x 4.5 m
Outside surface area, including floor: 720 $m^2$
Inside dimensions 14.7 x 14.7 x 4.2 m
Volume: 908 $m^3$
Insulation: 7.5 cm of polyurethane, with conductivity value (k) = 1.3 kJ/$m^2$/cm thickness/°C; coefficient of transmission (U) = 1.1 kJ/h/$m^2$/°C
Ambient conditions at harvest: 30°C and 50% relative humidity (RH)
Fruit temperature: at harvest, 21°C; in storage, –1.1 °C
Storage capacity: 600 bins at 500 kg fruit/bin = 300,00 kg of fruit
Bin weight: 63.5 kg; total weight of bins 38,100 kg (= 63.5 x 600 kg)
Loading rate: 200 bins, or 100,000 kg fruit/day; 3 days to fill
Cooling rate: first day, 21°C to 4.5°C; second day, 4.5°C to -1.1°C
Air changes due to door openings during cooling: six per day
Air changes due to door openings during storage: 1.8 per day
Specific heat (sp. ht.): pears, 0.86; wood bins, 0.5
Heat load in lowering air at 30°C and 50% RH to -1.1°C: 74.5 kJ/$m^3$
Heat load in lowering air from 7.2°F and 70% RH to -1.1°C: 15.3 kJ/$m^3$
Miscellaneous heat loads: lights, 2400 W/h; fans, 3 hp, two electric forklift trucks, 36,920 kJ, each for 8 h; two men working 8 h, 1000 kJ/h each

Part B: Loading During Cooling and Filling Storage

The calculations in this Part are based on the following: Temperature Difference (TD) = 30°C to -1.1°C = 31.1°C; assume 31.1°C TD on all surfaces.

| | kJ/24 h |
|---|---|
| 1. Building-transmission load: | |
| Area (720 $m^2$) x U (1.1 kJ) x TD (31.1°C) x hours (24) = | 591,149 |
| 2. Air-change load from door openings: | |
| Volume (908 $m^3$) x heat load (74.5 kJ) x air changes (6) = | 405,876 |
| 3. Product load: | |
| a. Product cooling (field heat, or sensible heat, removal): | |

**Table 4.2.6. (continued)**

| | |
|---|---|
| First day | |
| Fruit wt (100,00 kg) x sp. ht. (0.86) | |
| x TD (21° - 4.5°C) x kJ factor (4.186) = | 5,939,934 |
| Bin wt (12,700 kg) x sp. ht. (0.5) | |
| x TD (21° - 4.5°C) x kJ factor (4.186) = | 438,588 |
| Second day | |
| Fruit wt (100,000 kg) x sp. ht. (0.86) | |
| x TD (4.5°C to -1.1°C) x kJ factor (4.186) = | 2,015,977 |
| Bin wt (12,700 kg) x sp. ht. (0.5) | |
| x TD (4.5°C to -1.1°C) kJ factor (4.186) = | 148,854 |
| b. Heat of respiration during cooling (vital heat) | |
| First day | |
| (Av. temp., 13°C; resp. rate, 12,206 kJ/t/24 h) | |
| Tonne of fruit (100) x rate (12,206) = | 1,220,600 |
| Second day | |
| (Av. temp., 1.7°C; resp rate, 1741 kJ/t/24 h) | |
| Tonne of fruit (100) x rate (1741) = | 174,100 |
| Maximum accumulated in storage before cooling completed | |
| [Total fruit wt. (300,000 kg) - 2 day's loading wt (200,000 kg) = 100,000 kg, or 100 ton; resp. rate at -1.1°C, 812 kJ/t/24 h] | |
| Tonne of fruit (100) x rate (812) = | 81,200 |
| 4. Miscellaneous heat loads: | |
| Lights: Watts (2400) x kJ Watt (3.6) x hours (8) = | 69,120 |
| Fans: hp (3) x kJ hp (3112) x hours (24) = | 224,064 |
| Forklifts: 2 x 36,920 kJ per forklift for 8 h = | 73,840 |
| Labor: men (2) x kJ/h (1000) x hours (8) = | 16,000 |
| 5. Total heat load during cooling: | |
| Building transmission | 591,149 |

| | |
|---|---:|
| Air change | 405,876 |
| Product cooling | 8,543,353 |
| Production respiration | 1,475,900 |
| Miscellaneous | 383,024 |
| Subtotal | 11,399,302 |
| Add 10% for safety | 1,139,930 |
| Total required refrigeration | 12,539,232 |

Assuming that refrigeration equipment operates 18 h a day:

12,539,232 ÷ 18 = 696,624 kJ/h

Since a ton of refrigeration absorbs 12,660 kJ/24 h,

696,624 ÷ 12,660 = 55 tons of refrigeration capacity during storage.

This represents the peak capacity required.

## Part C: Load During Normal Storage Operation

The calculations in this Part are based on the following average outside ambient conditions: 7.2°C at 70% RH; storage temperature, -1.1°C; TD = 7.2°C to -1.1 °C = 8.3°C

| | kJ/24 h |
|---|---:|
| 1. Building-transmission load: Area (720 $m^2$) x U (1.1 kJ) x TD (8.3°C) x hours (24) = | 157,766 |
| 2. Air-change load from door openings: Volume (908 $m^3$) x heat load (15.3 kg) x air changes (1.8) = | 25,006 |
| 3. Product load (fruit respiration), no cooling: (Respiration rate at -1.1°C, 812 kJ/t/24 h) tonne of fruit (300) x rate (812) = | 243,600 |
| 4. Miscellaneous heat loads: | |
| Lights: Watts (2400) x kJ/Watt (3.6) x hours (4) = | 34,560 |
| Fans: hp (3) x kJ/hp (3112) x hours (24) = | 224,064 |
| Labor: men (1) x kJ/hour (1000) x hours (4) = | 4,000 |

**Table 4.2.6. (continued)**

| | |
|---|---|
| 5. Total load during storage: | |
| Building transmission | 157,766 |
| Air change | 25,006 |
| Product load (respiration) | 243,600 |
| Miscellaneous | 262-624 |
| Miscellaneous | 262,624 |
| Subtotal | 688,996 |
| Add 10% for safety | 68,899 |
| Total required refrigeration | 757,895 |

Assuming that refrigeration equipment operates 18 hours a day:

757,895 ÷ 18 = 42,105 kJ/h
42,105 ÷ 12,660 = 3.3 tons of refrigeration capacity during normal storage.

Source: Hardenburg et al., 1986.

**Table 4.2.7. Sample Calculations in English Units for Refrigeration Requirement in the Cold Storage of Pears**

Part A: Basic Considerations

Storage size: 50 x 50 x 15 ft
Outside surface area, including floor: 8000 $ft^2$
Inside dimensions: 49 x 49 x 14 ft
Volume: 33,614 $ft^3$
Insulation: 3 inches of polyurethane, with conductivity value (k) = 0.16 $Btu/ft^2$/inch thickness/°F; coefficient of transmission (U) = 0.0534 $Btu/h/ft^2$/°F1.1 $kJ/h/m^2$/°C
Ambient conditions at harvest: 85°F and 50% relative humidity (RH)
Fruit temperature: at harvest, 70°F; in storage, 30°F
Storage capacity: 600 bins at 1100 lb fruit/bin = 660,000 lb of fruit
Bin weight: 140 lb; total weight of bins 84,000 lb (= 140 x 600 kg)
Loading rate: 200 bins, or 220,000 lb fruit/day; 3 days to fill
Cooling rate: first day, 70° - 40°F; second day, 40° - 30°F
Air changes due to door openings during cooling: six per day
Air changes due to door openings during storage: 1.8 per day
Specific heat (sp. ht.): pears, 0.86; wood bins, 0.5

Heat load in lowering air at 85°F and 50% RH to 30°F: 2.0 Btu/ft$^3$
Heat load in lowering air from 45°F and 70% RH to 30°F: 0.41 Btu/ft$^3$
Miscellaneous heat loads: lights, 2400 W/h; fans, 3 hp, two electric forklift trucks, 35,000 Btu, each for 8 h; two men working 8 h, 950 Btu/h each

## Part B: Loading During Cooling and Filling Storage

The calculations in this Part are based on the following: Temperature Difference (TD) = 85°F to 30°F = 55°F; assume 55°F TD on all surfaces.

| | Btu/24 h |
|---|---|
| 1. Building-transmission load: | |
| Area (8000 ft$^2$) x U (0.534 Btu) x TD (55°F) x hours (24) = | 563,904 |
| 2. Air-change load from door openings: | |
| Volume (33,614 ft$^3$) x heat load (2.0 Btu) x air changes (6) = | 403,368 |
| 3. Product load: | |
| a. Product cooling (field heat, or sensible heat, removal): | |
| First day | |
| Fruit wt. (220,000 lb) x sp. ht. (0.86) x TD (70° - 40°F) = | 5,676,000 |
| Bin wt. (28,000 lb) x sp. ht. (0.5) x TD (70° - 40°F) = | 420,000 |
| Second day | |
| Fruit wt. (220,000 lb) x sp. ht. (0.86) x TD (40° to 30°F) = | 1,892,000 |
| Bin wt. (28,000 lb) x sp. ht. (0.5) x TD (405° to 30°F) = | 140,000 |
| b. Heat of respiration during cooling (vital heat) | |
| First day | |
| (Av. temp., 55°F; resp rate, 10,500 Btu/ton/24 h) | |
| Tons of fruit (110) x rate (10,500) = | 1,155,000 |

**Table 4.2.7.**

| | |
|---|---|
| <u>Second day</u> | |
| (Av. temp., 35$^o$F; resp rate, 1,500 Btu/ton/24 h) | |
| Tons of fruit (110) x rate (1500) = | 165,000 |
| Maximum accumulated in storage before cooling completed | |
| [Total fruit wt. (660,000 lb) - 2 day's loading wt (440,000 lb) = 220,000 lb, or 110 tons; resp. rate at 30$^o$F, 700 Btu/ton/24 h] | |
| Tons of fruit (110) x rate (1,500) = | 77,000 |
| 4. Miscellaneous heat loads: | |
| Lights: Watts (2400) x Btu/Watt (3.42) x hours (8) = | 65,664 |
| Fans: hp (3) x kJ/hp (3112) x hours (24) = | 212,400 |
| Forklifts: 2 x 36,920 kJ per forklift for 8 hours = | 70,000 |
| Labor: men (2) x kJ/h (1000) x hours (8) = | 15,200 |
| 5. Total heat load during cooling: | |
| Building transmission | 563,904 |
| Air change | 403,368 |
| Product cooling | 8,128,000 |
| Production respiration | 1,397,000 |
| Miscellaneous | 363,264 |
| Subtotal | 10,855,536 |
| Add 10% for safety | 1,085,554 |
| Total required refrigeration | 11,941,090 |

Assuming that refrigeration equipment operates 18 h a day:

$$11{,}941{,}090 \div 18 = 663{,}393 \text{ Btu/h}$$

Since a ton of refrigeration absorbs 288,000 Btu/24 h or 12,000 Btu/h,

$$663{,}393 \div 12{,}000 = 55 \text{ tons of refrigeration capacity during storage}$$

This represents the peak capacity required.

## Part C: Load During Normal Storage Operation

The calculations in this Part are based on the following average outside ambient conditions: 45°F at 70% RH; storage temperature, 30°F; TD = 45° to 30°F = 15°F]

| | Btu/24 h |
|---|---|
| 1. Building-transmission load: | |
| Area (8000 ft$^2$) x U (0.534 Btu) x TD (15°F) x hours (24) = | 153,792 |
| 2. Air-change load from door openings: | |
| Volume (33,614 ft$^3$) x heat load (0.41 Btu) x air changes (1.8) = | 24,807 |
| 3. Product load (fruit respiration), no cooling: | |
| (Respiration rate at 30°F, 700 Btu/ton/24 h) | |
| Tons of fruit (300) x rate (700) = | 231,000 |
| 4. Miscellaneous heat loads: | |
| Lights: Watts (2400) x Btu/Watt (3.42) x hours (4) = | 32,832 |
| Fans: hp (3) x Btu/hp (2950) x hours (24) = | 212,400 |
| Labor: men (1) x Btu/hour (950) x hours (4) = | 3,800 |
| 5. Total load during storage: | |
| Building transmission | 153,792 |
| Air change | 24,807 |
| Product load (respiration) | 231,000 |
| Miscellaneous | 249,032 |
| Subtotal | 658,631 |
| Add 10% for safety | 65,863 |
| Total required refrigeration | 724,494 |

Assuming that refrigeration equipment operates 18 h a day:

724,494 + 18 = 40,250 Btu/h
40,250 + 12,000 = 3.35 tons of refrigeration capacity during normal storage

Source: Hardenburg et al., 1986.

**Table 4.2.8 Summary of Controlled Atmosphere (CA) Storage Requirements for Fruits and Vegetables**

| Commodity Cultivar | Beneficial Concentration (%) | | Suitable Temperature | | Approxiamte Storage Period | Notes |
|---|---|---|---|---|---|---|
| | $O_2$ | $CO_2$ | °C | °F | | |
| | | | Fruits of the Temperate Zone | | | |
| Apple (*Malus domestica*) | | | | | | |
| Boskoop | 1.5-2.0 | <1.5 | 4 | 39 | 5-7 month | Sensitive to low temperature breakdown |
| Bramley's seedling | 2 | 6 | 3-4 | 37-39 | 7 month | CA reduces bitter pit. |
| Cortland | 2-3 | 5 | 2 | 36 | 4 to 6 months | |
| Cox Orange Pippin | 1-3 | <1 | 3 | 36 | 4 to 6 months | 2% $O_2$ 1st week, then 1-1.25% |
| Empire | 1.5-3 | 1-5 | 2 | 36 | 5 to 7 months | |
| Fuji | 2-2.5 | 1-2 | 0 | 32 | 7 to 8 months | CA reduces scald. |
| Gala | 1-2 | 1-5 | 0 | 32 | 5 months | |
| Golden Delicious | 1-3 | 1-5 | 0 | 32 | 7 to 11 months | Rapid CA is beneficial. |
| Granny Smith | 1-2 | 1-3 | 0 | 32 | 7 to 9 months | CA reduces scald. |
| Jonathan | 1-3 | 1-6 | 0-3 | 32-37 | 4 to 7 months | CA reduces Jonathan spot. |
| Newtown | 3 | 5-8 | 2-4 | 36-39 | 8 months | Susceptible to low temperature injury |
| Northern Spy | 2-3 | 2-3 | 0 | 32 | 8 months | |
| Red Delicious | 1-3 | 1-3 | 0 | 32 | 8 to 10 months | Susceptible to scald |
| Rome Beauty | 2-3 | 2-5 | 0 | 32 | 6 to 8 months | |
| Spartan | 1.5-2.5 | 1-2 | 0 | 32 | 6 to 8 months | |
| Stayman | 2-3 | 2-5 | 0 | 32 | 7 to 8 months | |

| | | | | | | |
|---|---|---|---|---|---|---|
| Worcester pearmain | 3 | 5 | 1 | 34 | 6 months | |
| Apricot (*Prunus armeniaca*) | 2-3 | 2-3 | -0.5-0 | 31-32 | 7 weeks | CA delays ripening. |
| Blackberry (*Rubus* spp.) | 5-10 | 15-20 | -0.5-0 | 31-32 | 1 week | Prompt cooling is. important. |
| Black currant (*Ribes nigrum*) | - | 25-50 | 2 | 36 | 4 weeks | 50% 1 week, then 25% |
| Blueberry (*Vaccinium* spp.) | 5-10 | 15-20 | -0.5-0 | 31.32 | 2-3 weeks | Prompt cooling is important. |
| Cherry, sweet (*Prunus avium*) | 3-10 | 10-15 | -1-0 | 30-32 | 4 weeks | High $CO_2$ reduces decay. |
| Cranberry (*Vaccinium macrocarpon*) | 1-2 | 0-5 | 3 | 37 | 2-4 months | |
| Fig (*Ficus carica*) | 5-10 | 15-20 | -1-0 | 30-32 | 2 weeks | |
| Grape (*Vitis vinifera*) | 2-5 | 1-3 | -1-0 | 30-32 | 1-6 months | |
| Kiwifruit (*Actinidia chinensis*) | 1-2 | 3-5 | 0 | 32 | 3-5 months | CA delays ripening. |
| Nectarine (*Prunus persica*) | 1-2 | 3-5 | -0.5-0 | 31-32 | 6-9 weeks | CA reduces internal breakdown. |
| Peach (*Prunus persica*) | 1-2 | 3-5 | -0.5-0 | 31-32 | 6-9 weeks | Cultivars differ in response. |
| Pear-European (*Pyrus communis*) | | | | | | |
| Anjou | 0.5-2 | 0.5-2 | -1 | 30 | 7-9 months | CA reduces scald. |
| Barlett | 1-2 | 2-3 | -1 | 30 | 3-5 months | Rapid cooling recommended. |

**Table 4.2.8. (continued)**

| Commodity Cultivar | Beneficial Concentration (%) | | Suitable Temperature | | Approxiamte Storage Period | Notes |
|---|---|---|---|---|---|---|
| | $O_2$ | $CO_2$ | °C | °F | | |
| Bosc | 1-3 | 0.5-1 | -1 | 30 | 4-6 months | Optimum maturity is critical. |
| Comice | 2 | 2-3 | -1 | 30 | 5-7 months | |
| Conference | 2 | 2 | -1 | 30 | 4-6 months | |
| Packham's Triumph | 2-3 | 1-2 | -0.5 | 31 | 8 months | Optimum maturity is critical. |
| Passe Crassane | 3-4 | 5-7 | -1 | 30 | 6-7 months | Tolerant to high $CO_2$. |
| Pear-Asian (*Pyrus serotina*) and *Pyrus bretschneideri*) | | | | | | |
| Nijiseiki (20th Century) | 3 | 1 | 0 | 32 | 9-12 months | |
| Tsu Li | 1-2 | 3 | 0 | 32 | 6-8 months | |
| Ya Li | 3-4 | 2 | 0-3 | 32-37 | 6-8 months | |
| Persimmon (*Diospyros kaki*) | 3-5 | 5-8 | -1-0 | 30-32 | 4 months | |
| Plum (*Prunus domestica*) | 1-2 | 0-5 | -0.5-0 | 31-32 | 4-5 months | CA delays ripening. |
| Raspberry (*Rubus idaeus*) | 5-10 | 15-20 | -0.5-0 | 31-32 | 1 week | Prompt cooling is important. |
| Strawberry (*Fragaria* spp.) | 5-10 | 15-20 | -0.5-0 | 31-32 | 1 week | Commercial use during transport |

| | | | | | | |
|---|---|---|---|---|---|---|
| **Subtropical and Tropical Fruits** | | | | | | |
| Avocado (*Persea americana*) | 2-5 | 3-10 | 10 | 50 | 3-6 weeks | CA reduces chilling. |
| Banana (*Musa* spp.) | 2-5 | 2-5 | 14 | 58 | 1-6 months | CA delays ripening. |
| Grapefruit (*Citrus paradisis*) | 3-10 | 5-10 | 13 | 55 | 6-8 weeks | CA reduces pitting. |
| Lemon (*Citrus limonia*) | 5-10 | 0-10 | 13 | 55 | 1-6 months | CA reduces decay. |
| Lime (*Citrus aurantifolia*) | 5-10 | 0-10 | 13 | 55 | 6-8 weeks | CA retards degreening. |
| Mango (*Mangifera indica*) | 3-5 | 5-10 | 13 | 55 | 5 weeks | CA delays ripening. |
| Olive (*Olea europaea*) | 2-3 | 0-1 | 7 | 45 | 2 months | |
| Orange (*Citrus sinensis*) | 5-10 | 0-5 | 7 | 45 | 8-12 weeks | Storage life varies with cultivars. |
| Papaya (*Carica papaya*) | 3-5 | 5-10 | 12 | 54 | 2-3 weeks | |
| Passion fruit (*Passiflora edulis*) | 5 | 5 | 7-10 | 45-50 | 6 weeks | |
| Pineapple (*Ananas comosus*) | 2-5 | 5-10 | 10 | 50 | 4 weeks | |
| **Vegetables** | | | | | | |
| Artichoke (*Cynara scolymus*) | 2-3 | 2-3 | 0 | 32 | 1 months | CA decreases discoloration. |

**Table 4.2.8. (continued)**

| Commodity Cultivar | Beneficial Concentration (%) | | Suitable Temperature | | Approxiamte Storage Period | Notes |
|---|---|---|---|---|---|---|
| | $O_2$ | $CO_2$ | °C | °F | | |
| Asparagus (*Asparagus officinalis*) | Air | 10-14 | 2 | 36 | 3 weeks | High $CO_2$ is beneficial. |
| Bean, snap (*Phaseolus vulgaris*) | 2-3 | 4-7 | 8 | 46 | 2 weeks | CA reduces color loss. |
| Bean, lima (*Phaseolus lunatus*) | >5 | 10-35 | 5-7 | 40-45 | 7-10 days | Shelled only |
| Broccoli (*Brassica oleracea*, Italian Italian group) | 1-2 | 5-10 | 0 | 32 | 1 months | CA maintains green color. |
| Brussels sprouts (*Brassica oleracea*) | 1-3 | 5-7 | 0 | 32 | 3-5 weeks | CA reduces yellowing. |
| Cabbage (*Brassica oleracea*, Capitata group) | 2-3 | 3-6 | 0 | 32 | 6-8 months use | Large scale commercial |
| Cantaloupe (*Cucumis melo*) | 3-5 | 10-20 | 8 | 46 | 2-3 weeks | CA reduces ripening. |
| Cauliflower (*Brassica oleracea*, Botrytis group) | 2-3 | 3-4 | 0 | 32 | 1 months | |
| Celery (*Apium graveolens*) | 1-4 | 3-5 | 0 | 32 | 3 months | |

| | | | | | | |
|---|---|---|---|---|---|---|
| Chinese cabbage (*Brassica campestris*) | 1-3 | 0-5 | 0 | 32 | 4-5 months | CA reduces leaf abscission. |
| Corn, sweet (*Zea mays*) | 2-4 | 5-10 | 0 | 32 | 2 weeks | CA reduces sugar loss. |
| Cucumber (*Cucumis sativus*) | 1-4 | 0 | 12 | 54 | 3 weeks | CA reduces yellowing. |
| Leek (*Allium porrum*) | 1-6 | 5-10 | 0 | 32 | 4 months | |
| Lettuce, head (*Lactuca sativa*) | 1-3 | 0 | 0 | 32 | 3-4 weeks | |
| Mushroom (*Agaricus bisporus*) | Air | 10-15 | 0 | 32 | 1-2 weeks | $CO_2$ retards cap opening. |
| Onion, dry (*Allium cepa*) | 0-1 | 0-5 | 0 | 32 | 8 months | Tolerant to low $O_2$ |
| Onion, green (*Allium cepa*) | 1 | 5 | 0 | 32 | 2 months | |
| Parsley (*Petroselinum crispum*) | 8-10 | 8-10 | 0 | 32 | 3 months | |
| Pepper, sweet (*Capsicum annuum*) | 2-5 | 0 | 12 | 54 | 3 weeks | |
| Radish (*Raphanus sativus*) | 1-2 | 2-3 | 0 | 32 | 4 months | |
| Spinach (*Spinacia oleracea*) | 7-10 | 5-10 | 0 | 32 | 3 weeks | |
| Tomato (*Lycopersicon esculentum*) | 3-5 | 2-3 | 12 | 54 | 4-6 weeks | |

Source: Courtesy of C. Y. Wang and A. E. Watada, Horticultural Crops Quality Laboratory, Product Quality and Development Institute, USDA, Beltsville, MD 20705.

## 4.3. Fruits and Vegetables with USDA Standards for Grades

**Table 4.3.1. U.S. Standards for Grades of Fresh Fruits, Vegetables, Nuts and Other Special Products**

U.S. GRADE STANDARDS FOR FRESH MARKET VEGETABLES

Anise, sweet
Artichokes
Asparagus
Beans, lima
Beans, snap
Beets
Beet greens
Broccoli, bunched Italian sprouting
Brussels sprouts
Cabbage
Carrots, bunched
Carrots, topped
Carrots with short trimmed tops
Cauliflower
Celery
Collard or broccoli greens
Corn, green
Cucumbers
Cucumbers, greenhouse
Dandelion greens
Eggplant
Endive, escarole or chicory
Garlic
Horseradish roots
Kale
Lemons
Lettuce
Lettuce, greenhouse leaf
Limes, Persian (Tahiti)
Mushrooms
Mustard greens and turnip greens
Nectarines
Okra
Onions, Bermuda-Granex-Grano types
Onions, common green
Onions, Creole
Onions (other than Bermuda-Grano and Creole types)
Onion sets
Parsley
Parsnips
Peas, fresh
Peas, southern
Peppers, sweet
Potatoes
Potatoes, seed
Radishes
Rhubarb (field grown)
Romaine
Shallots, bunched
Spinach leaves, fresh
Spinach plants
Squash, fall and winter type
Squash, summer
Sweet potatoes
Tomatoes, fresh
Tomatoes, greenhouse
Turnips or rutabagas
Watermelons

U.S. GRADE STANDARDS FOR FRESH MARKET CONSUMER VEGETABLES

Beet greens
Broccoli, Italian sprouting
Brussels sprouts

Cabbage
Carrots (bunched, topped, short trimmed tops)
Celery
Corn, green
Kale
Parsnips
Potatoes
Spinach leaves, fresh
Tomatoes
Turnips

U.S. GRADE STANDARDS FOR FRESH VEGETABLES FOR PROCESSING

Asparagus, green
Beans, lima, fresh shelled
Beans, snap
Beets
Broccoli
Cabbage
Carrots
Cauliflower
Corn, sweet
Cucumbers
Cucumbers, pickling
Mushrooms
Okra
Onions
Peas, fresh shelled, for canning or freezing
Peas, southern
Peppers, sweet
Potatoes for chipping
Potatoes
Spinach
Sweet potatoes, for canning or freezing
Tomatoes, Italian type for canning
Tomatoes, green
Tomatoes

U. S. STANDARDS FOR GRADES FOR FRESH MARKET FRUIT

Apples
Apricots
Avocados, Florida
Blueberries
Cherries, sweet
Cranberries, fresh
Dewberries and blackberries
Grapes, American (eastern type) bunch
Grapes, juice (European or vinifera type)
Grapes, table (European or vinifera type)
Grapefruit (California and Arizona)
Grapefruit (Florida)
Grapefruit (Texas, and states other than Florida, California, and Arizona)
Kiwifruit
Lemons
Limes, Persian (Tahiti)
Nectarines
Oranges (California and Arizona)
Oranges and Tangelos (Florida)
Oranges (Texas and states other than Florida, California and Arizona)
Peaches
Pears, summer and fall
Pears, winter
Pineapples
Plums and prunes, fresh
Raspberries
Strawberries
Tangerines
Tangerines (Florida)

**Table 4.3.1. (continued)**

| | |
|---|---|
| U. S. STANDARDS FOR GRADES FOR FRESH MARKET FRUIT FOR PROCESSING | Strawberries, growers' stock for manufacture |
| | Strawberries, washed and sorted for freezing |
| Apples | |
| Berries | U. S. STANDARDS FOR GRADES OF NUTS AND SPECIAL PRODUCTS |
| Blueberries | |
| Cherries, red sour, for manufacture | Almonds, shelled |
| Cherries, sweet, for cnning or freezing | Almonds in the shell |
| | Asparagus Plumosus |
| Cherries, sweet, for export for sulfur brining | Brazil nuts in the shell |
| | Filberts in the shell |
| | Mixed nuts in the shell |
| Cranberries | Peanuts, shelled Spanish type |
| Currants | Peanuts, shelled runner type |
| Grapes, bunch, American (eastern type) for processing and freezing | Peanuts, shelled Virginia type |
| | Peanuts, cleaned Virginia type in the shell |
| Grapes for processing and freezing | Pecans in the shell |
| | Pecans shelled |
| Peaches, freestone, for canning, freezing, or pulping | Tomato plants |
| | Walnuts, shelled (*Juglans regia*) |
| Pears | Walnuts (*Juglans regia*) in the shell |
| Raspberries | |

Source: 7 CFR 51, 1991.

**Table 4.3.2. U. S. Standards for Grades of Processed Fruits and Vegetables**

| | |
|---|---|
| CANNED | Beans, baked |
| | Beans, dried |
| Apples | Beans, pork and |
| Apple butter | Beans, green and wax |
| Apple juice | Beans, lima |
| Applesauce | Beets |
| Apricots, regular pack | Blackberries and other similar berries |
| Apricots, solid pack | |
| Asparagus | Carrots |

Celery
Cherries, red, tart pitted
Cherries, sweet
Chili sauce
Corn, cream style
Corn, whole kernel
Cranberry sauce
Figs, Kadota
Fruit cocktail
Fruit jelly
Fruit preserves (jams)
Fruits for salad
Grapes
Grape juice
Grapefruit
Grapefruit juice, grapefruit juice from concentrates, and concentrated grapefruit juice
Grapefruit and orange for salad
Grapefruit juice and orange juice
Hominy
Leafy greens
Lemon juice
Lemon juice for manufacture, concentrate
Mushrooms
Okra
Okra and tomatoes or vice versa
Olives, green
Olives, ripe
Onions
Orange juice, orange juice from concentrates, pasteurized orange juice, concentrated orange juice, concentrated orange juice for manufacture
Orange marmalade
Peaches, clingstone
Peaches, freestone
Pears
Peas
Peas and carrots
Peas, field and blackeye
Pickles
Pimientos
Pineapple
Pineapple juice, pineapple juice from concentrate
Plums
Potatoes, white
Prunes, canned dried
Pumpkin (squash)
Raspberries
Sauerkraut
Spinach
Squash (summer type)
Succotash
Sweetpotatoes
Tangerine juice
Tangerine juice for manufacture, concentrated
Tomatoes
Tomato catsup
Tomato juice,
Tomato juice from concentrates
Tomato juice, concentrated
Tomato paste
Tomato puree (pulp)
Tomato sauce

DRIED AND DEHYDRATED

Apples, dehydrated
Apples, dried
Apricots, dehydrated
Apricots, dried
Dates
Figs, dried
Grapefruit juice, dehydrated
Orange juice, dehydrated
Peaches, dehydrated

**Table 4.3.2. (continued)**

Peaches, dried
Pears, dried
Prunes, dehydrated
Prunes, dried
Raisins, processed
Honey, comb
Honey, extracted
Maple sirup
Molasses, sugarcane
Syrup refiners'
Syrup, sugarcane

MISCELLANEOUS

Cherries, sulfured
Olive oil
Peanut butter
Potatoes, peeled
Sauerkraut, bulk

FROZEN

Apple juice, concentrated
Apples
Apricots
Asparagus
Beans, green and wax
Beans, lima
Beans, speckled butter (lima)
Berries (black, boysen, etc.)
Blueberries
Broccoli
Brussels sprouts
Carrots
Cauliflower
Cherries, red, tart, pitted
Cherries, sweet
Corn on the cob
Corn, whole kernel
Cranberries
Grape juice, concentrated (sweetened)
Grapefruit juice, grapefruit juice from concentrate, concentrated grapefruit juice, and concentrated grapefruit juice for manufacture
Grapefruit juice and orange juice, concentrated, blended
Leafy greens, including spinach
Lemonade, concentrated
Limeade, concentrated
Melon balls
Okra
Onion rings, breaded
Orange juice: orange juice from concentrates, pasteurized orange juice, concentrated orange juice, concentrated orange juice for manufacture, and reduced-acid concentrated orange juice
Peaches
Peas
Peas and carrots
Peas, field and blackeye
Peppers, sweet
Pineapple juice, pineapple juice from concentrate
Pineapple
Plums
Potatoes, french fried
Potatoes, hash brown
Raspberries
Rhubarb
Squash (cooked)
Squash (summer)
Strawberries
Succotash
Sweetpotatoes
Tomato juice, tomato juice from concentrate

| Turnip greens with turnips | Vegetables, mixed |
|---|---|

Source: 7 CFR 52, 1991.

## 4.4. Product Descriptions

**Table 4.4.1. Descriptions for Processing Selected Fruits, Vegetables, and Related Products**

APPLES

| | |
|---|---|
| Canned apple juice | Canned apple juice is the unfermented juice obtained from sound, ripe apples, with or without parts. Concentrated apple juice may be added, provided that any added apple juice concentrate shall not contribute more than one-fourth of the total apple juice solids in the finished apple juice. No water may be added directly to the finished food. Natural volatile apple juice components (apple essence) may be restored to a level that provides a natural apple juice flavor. The apple juice may be processed with or without the addition of antioxidants and is sufficiently processed to assure preservation of the product in hermetically sealed containers. |
| Canned apples | Canned apples is the product prepared from sound, fresh apples of proper maturity and proper ripeness, which fruit is packed with or without any of the following ingredients: water, salt, spices, nutritive sweetening ingredients, and any other ingredients permissible under FDA regulations, and is sufficiently processed by heat to assure preservation of the product in hermetically sealed containers. |
| Canned applesauce | Canned applesauce is the product prepared from comminuted or chopped applies (*Malus domestica* Borkhausen), which may or may not be peeled and cored, and which may have added thereto one or more of the optional ingredients specified by the FDA. The apple ingredient is heated and, in accordance with good manufacturing practices, bruised apple particles, peel, seed, |

**Table 4.4.1 (continued)**

| | |
|---|---|
| | core material, carpel tissue, and other coarse, hard, or extraneous materials are removed. The food is sealed in containers. It is so processed by heat, either before or after sealing, as to prevent spoilage. The soluble solids content, measured by refractometer and expressed as % sucrose ($^{o}$Brix) with correction for temperature to the equivalent at 20°C (68$^{o}$F) is not less than 9% (exclusive of the solids of any added optional nutritive carbohydrate sweeteners), but without correction for invert sugar or other substances. |
| Dehydrated (low-moisture) apples | Dehydrated (low-moisture) apples are prepared from clean and sound fresh or previously dried (or evaporated) apples from which the peels and cores have been removed and which have been cut into segments. The dried (or evaporated) apple segments may be cut further into smaller segments in preparation for dehydration whereby practically all of the moisture is removed to produce a very dry texture and are prepared to assure a clean, sound, wholesome product. The sulfur dioxide content of the finished product may not exceed 1000 parts per million. No other additives may be present. |
| Dried apples | Dried apples are prepared from sound, properly ripened fruit of the common apple (*Malus pumila*) by washing, sorting, trimming, peeling, coring, and cutting into segments. The prepared apple segments are properly dried to remove the greater portion of moisture to produce a semi-dry texture. The product may be sulfured sufficiently to retard discoloration. The dried apples are sorted or cleaned, or both, to assure a clean, sound wholesome product. |
| Frozen apples | Frozen apples are prepared from sound, properly ripened fruit of *Malus sylvestris* (*Pyrus malus*); are peeled, cored, trimmed, sliced, sorted, and washed; and properly drained before filling into containers; may be packed with or without the addition of a nutritive sweetening ingredient and any other ingredient permissible under the FDA regulations; and are frozen in accordance with good commercial practice and maintained at temperatures necessary for the preservation of the product. |

Frozen concentrated apple juice

Frozen concentrated apple juice is prepared from the unfermented, unsweetened, unacidifed liquid obtained from the first pressing of properly prepared, sound, clean, mature, fresh apples, and or parts thereof by good commercial processes. The juice is clarified and concentrated to at least 22.9°Brix. The apple juice concentrate so prepared, with or without the addition of ingredients permissible under FDA regulations, is packed and frozen in accordance with good commercial practice and maintained at temperatures necessary for the preservation of the product.

## APRICOTS

Dehydrated, low-moisture apricots

Dehydrated low-moisture apricots are prepared from clean and sound fresh or previously dried apricots which are cut, chopped, or otherwise prepared into various sizes and shapes; are prepared to assure a clean, sound, wholesome product; are processed by dehydration whereby practically all of the moisture is removed to produce a very dry texture; and are packaged (including kind of container and proper closure) to assure retention of he low-moisture characteristics of the product. The product shall have been subjected to sulfur treatment sufficiently to retain a characteristic color but no other additives may be present.

Dried apricots

Dried apricots are the halved and pitted fruit of the apricot tree (*Prunus armeniaca*) from which the greater portion of moisture has been removed. Before packing, the dried fruit is processed to cleanse the fruit and may be sulfured sufficiently to retain a characteristic color.

Frozen apricots

Frozen apricots are prepared from sound, mature, fresh, peeled or unpeeled fruit of any commercial variety of apricot which are sorted, washed, and may be trimmed to assure a clean and wholesome product. The apricots are properly drained of excess water before filling into containers; may be packed with the addition of nutritive sweetening ingredients (including sirup and or sirup containing pureed apricots) and or suitable antioxidant ingredients and or any other ingredients permissible under FDA regulations. The apricots

**Table 4.4.1 (continued)**

| | |
|---|---|
| | are prepared and frozen in accordance with good commercial practice and are maintained at temperatures necessary for the preservation of the product. |
| | ASPARAGUS |
| Frozen asparagus | Frozen asparagus consists of sound and succulent fresh shoots of the asparagus plant (*Asparagus officianalis*). The product is prepared by sorting, trimming, and washing, and blanching as necessary to assure a clean and wholesome product. It is then frozen and stored at temperatures necessary for preservation. |
| | BEANS |
| Canned baked beans | Canned baked beans is the product prepared from dry mature beans used for canning; but not including soybeans. Any safe and suitable ingredients permissible under FDA regulations may be used. The product is prepared by washing, soaking, and baking by the application of dry heat in open or loosely covered containers in a closed oven at atmospheric pressure for sufficient prolonged time to produce a typical texture and flavor. It is packed in hermetically sealed containers and sufficiently processed by heat to assure preservation. |
| Canned dried beans | Canned dried beans is the product prepared from dry mature beans or peas used for canning; but not including soybeans, sweet peas, or early peas. Any safe and suitable ingredients permissible under FDA regulations may be used. The product is prepared by washing, soaking, blanching, cooking, or other processing. It is packed in hermetically sealed containers and sufficiently processed by heat to assure preservation. |
| Canned lima beans | Canned lima beans is the food prepared from seed shelled from the pods of the lima bean plant. Additional ingredients, required or optinal, may be added. Water may be added. The food is then sealed in a container and so processed by heat as to prevent spoilage. |

| | |
|---|---|
| Canned pork and beans | Canned pork and beans (canned dried white beans with pork) is the product prepared from dry mature white beans of the species *Phaseolus vulgaris* L., with pork or pork fat; and with a packing medium or sauce consisting of water, tomato products, and any other safe and suitable ingredients permissible under FDA regulations. The product is prepared by washing, soaking, blanching, or other processing. It is packed in hermetically sealed containers and sufficiently processed by heat to assure preservation. |
| Frozen green beans and frozen wax beans | Frozen green beans and frozen wax beans (frozen beans) means the frozen product prepared from the clean, sound, succulent pods of the bean plant. The pods are stemmed, washed, blanched, sorted, and properly drained. The product is then frozen in accordance with good commercial practice and maintained at temperatures necessary for the preservation of the product. |
| Frozen lima beans | Frozen lima beans means the frozen product prepared from the clean, sound, succulent seed of the lima bean plant without soaking, by shelling, washing, blanching, and properly draining, and is then frozen in accordance with good commercial practice and maintained at temperatures necessary for the preservation of the product. |
| Frozen speckled butter (lima beans) | Frozen speckled butter lima beans means the frozen product prepared from the clean, sound, freshly vined (but not seed dry) seed of the speckled butter (lima) bean plant (*Phaseolus limensis*). The skins of the seed are pigmented and the external colors range from variegated speckling of green, pink, red, and/or lavender to purple. The product is prepared by shelling the pods; by washing, blanching, and properly draining the seed which may have been sorted and blended or otherwise prepared in accordance with good commercial practice; and is frozen in accordance with good commercial practice and maintained at temperatures necessary for the preservation of the product. |

**Table 4.4.1 (continued)**

| | |
|---|---|
| | BEETS |
| Canned beets | Canned beets is the food prepared from root of the beet plant. Additional ingredients, required or optinal, may be added. Water may be added. The food is then sealed in a container and so processed by heat as to prevent spoilage. |
| | BERRIES |
| Canned cranberry sauce | Canned cranberry sauce is the jellied or semi-jellied product prepared from clean, sound, matured or fairly well-matured cranberries; a sweetening ingredient or sweetening ingredients and water. Pectin may be added but only in a quantity which reasonably compensates for deficiency, if any, of the natural pectin content of the cranberries. The mixture is concentrated and sufficiently processed by heat to assure preservation of the product in hermetically sealed containers. The soluble solids of the finished product is 35-45%. |
| Frozen berries | Frozen berries are prepared from the properly ripened fresh fruit of the plant (Genus *Rubus*); are stemmed and cleaned; may be packed with or without packing media; and are frozen and stored at temperatures necessary for the preservation of the product. |
| Frozen blueberries | Frozen blueberries are prepared from sound, properly ripened fresh fruit of the blueberry bush (Genus *Vaccinium*), including species of varieties often called "huckleberries," but not of the Genus *Gaylussacia*; are cleaned and stemmed; are properly washed; are packed with or without packing media; and are frozen and maintained at temperatures necessary for the preservation of the product. |
| Frozen cranberries | Frozen cranberries is the product prepared from the sound, mature berries of the commonly cultivated cranberry plant. The product is prepared by sorting and cleaning to assure a wholesome product. The prepared product is frozen and stored at temperatures necessary for preservation. |

## BROCCOLI

**Frozen broccoli**

Frozen broccoli is the product prepared from the fresh, clean, sound stalks or shoots of the broccoli plant by trimming, washing, blanching, sorting, and properly draining. The product is frozen in accordance with good commercial practice and maintained at temperatures necessary for its preservation.

## BRUSSELS SPROUTS

**Frozen brussels sprouts**

Frozen brussels sprouts means the frozen product prepared from the clean, sound, and succulent heads of the brussels sprouts plant by trimming, washing, blanching, and properly draining. The product is frozen in accordance with good commercial practice and maintained at temperatures necessary for its preservation.

## CARROTS

**Frozen carrots**

Frozen carrots are the clean and sound product prepared from the fresh root of the carrot plant (*Daucus carota*) by washing, sorting, peeling, trimming, and blanching, and are frozen in accordance with good commercial practice and maintained at temperatures necessary for the preservation of the product.

## CAULIFLOWER

**Frozen cauliflower**

Frozen cauliflower is prepared from the fresh flower heads of the cauliflower plant (*Brassica oleracea botrytis*) by trimming, washing, and blanching and is frozen and maintained at temperatures necessary for preservation of the product.

## CHERRIES

**Frozen red tart pitted cherries**

Frozen red tart pitted cherries is the food prepared from properly matured cherries of the domestic red sour varietal group which have been washed, pitted, sorted, and properly

**Table 4.4.1 (continued)**

| | |
|---|---|
| | drained; may be packed with or without a nutritive sweetened packing medium or any other substance permitted under FDA regulations; and are frozen and stored at temperatures necessary for the preservation of the product. |
| Frozen sweet cherries | Frozen sweet cherries are prepared from the clean, sound, properly matured fresh fruit of any commercial variety of sweet cherries, which are sorted, washed, and drained; which may be packed with or without the addition of nutritive sweetening ingredients; and which are frozen in accordance with good commercial practice and maintained at temperatures necessary for the preservation of the product. |
| Sulfured cherries | Sulfured cherries are prepared from properly matured whole cherries (*Prunus avium* or *Prunus cerasus*) of similar varietal characteristics; and are packed, with or without the addition of a hardening agent, in a solution of sulfur dioxide of sufficient strength to preserve the product. |
| | CHILI |
| Chili Sauce | Chili sauce is the product prepared from mature, clean, sound, tomatoes of the red or reddish varieties, which are peeled and chopped or crushed, or all (or a portion) of the tomatoes may be chopped, crushed, or macerated and the peelings screened out in a manner so that at least a substantial portion of the seed remains in the product, to which are added salt, spices, vinegar, nutritive sweetening ingredients, and to which may be added vegetable flavoring ingredients such as chopped onion, chopped green or red pepper, chopped green tomatoes, chopped celery, and sweet pickle relish in such quantities as will not materially alter the appearance of the product with respect to the predominance of the tomato ingredient, and any other ingredients permissible under FDA regulations. The chili sauce is processed in accordance with good commercial practice; is packed in hermetically sealed containers; and is sufficiently processed by heat, before or after sealing, to assure |

preservation of the product. The refractive index of the filtrate of the chili sauce at 20°C is not less 1.3784.

## CORN

Frozen corn-on-the-cob

Frozen corn-on-the-cob is the product prepared from sound properly matured, fresh, sweet corn ears by removing husk and silk and by sorting, trimming, and washing to assure a clean and wholesome product. The ears are blanched, then frozen, and stored at temperatures necessary for the preservation of the products.

## DATES

Dates

Dates are the properly cured fresh fruit of the date tree (*Phoenix dactylifera*) which may or may not be softened by hydration. Dates, when referred to as dry dates for processing, means that the dates are dry and have not been softened by hydration.

## FIGS

Dried figs

Dried figs are the fruit of the fig tree (*Ficus carica*) from which the greater portion of moisture has been removed. The dried figs are prepared from clean and sound fruit and are sorted and thoroughly cleaned to assure a clean, sound, wholesome product. The figs may or may not be sulfured or otherwise bleached.

## FRUITS

Canned fruits for salad

Canned fruits for salad is the product consisting of units of properly prepared apricots, yellow clingstone peaches, pears, pineapple, cherries, and/or grapes in the forms and proportions specified by regulations. The product is packed in a suitable liquid medium with or without the addition of nutritive sweetening ingredients, artificial sweetening ingredients, or other ingredients permissible under FDA regulations and is processed by heat to assure preservation of the product in hermetically sealed containers.

**Table 4.4.1 (continued)**

| | |
|---|---|
| | GRAPEFRUIT |
| Canned grapefruit and orange for salad | Canned grapefruit and orange for salad, commonly known as canned citrus salad, is prepared from sound, mature grapefruit (*Citrus paradisi* Macfadyen) and sound, mature oranges of the orange group (*Citrus sinensis*). The fruit ingredients have been properly washed, cored, with seeds and major portions of tough membrane removed. The grapefruit units are segmented and the orange units may be segmented or sliced. The product is packed with a suitable packing medium which may be water, fruit juice, nutritive carbohydrate sweeteners, artificial sweeteners, or any other safe and suitable ingredients permissible under FDA regulations. The product is sufficiently processed by heat to assure preservation in hermetically or aseptically sealed containers. |
| Frozen grapefruit | Frozen grapefruit is prepared from the matured fruit of the grapefruit tree (*Citrus paradisi*), after the fruit has been washed and peeled, and has been separated into segments by removing the core, seeds, and membrane; may be packed with or without packing media; and is frozen and stored at temperatures necessary for the preservation of the product. |
| Frozen concentrated blended grapefruit juice and orange juice | Frozen concentrated blended grapefruit juice and orange juice is the frozen product prepared from a combination of concentrated, unfermented juices obtained from sound, mature grapefruit (*Citrus paradisi*) and from sound, mature fruit of the sweet orange group (*Citrus sinensis*) and Mandarin group (*Citrus reticulata*) except tangerines. The fruit is prepared by sorting and by washing prior to extraction of the juices to a clean product. The juices may be blended upon extraction of such juices or after concentration and fresh orange juice extracted from sorted and washed fruit is admixed to the concentrate. It is recommended that the frozen concentrated blended grapefruit juice and orange juice be composed of the equivalent of not $<50\%$ orange juice in the reconstituted juice; however, in oranges yielding light-colored juices it is further recommended that as much |

as the equivalent of 75% orange juice in the reconstituted juice be used. The concentrated juice is packed in accordance with good commercial practice and is frozen and maintained at temperatures necessary for the preservation of the product.

**Grapefruit juice and orange juice**

Grapefruit juice and Orange juice is prepared from a combination of unfermented juices obtained from mature fresh grapefruit (*Citrus paradisi*) and mature sweet oranges (*Citrus sinensis*). The juice of oranges from the mandarin group (*Citrus reticulata*), however, may be added in quantities not more than 10%, by volume, of the orange juice ingredient. It is recommended that the minor juice ingredient (either orange or grapefruit) provide not <25%, by weight, of the total soluble fruit solids present in the finished product.

The fruit is prepared and the juice extracted and processed in a manner to assure a clean and wholesome product. Soluble solids, insoluble solids, Brix-acid ratios, and flavor may be adjusted by suitable manufacturing procedures.

The product is processed by appropriate physical means to assure its preservation through normal marketing channels. Such means include canning and refrigeration.

If preserved by canning, the food is processed with heat so as to assure the preservation of the juice in hermetically sealed containers.

If preserved by refrigeration, the temperature of the product is reduced so as to extend its market life. The juice may or may not have been subjected to heat prior to refrigerating. It may or may not be packed in hermetically sealed containers.

## GRAPES

**Canned grape juice**

Canned grape juice is the unfermented liquid obtained from properly matured fresh grapes. Such grape juice is prepared without concentration, without dilution, is packed with or without the addition of sweetening ingredients, and is sufficiently processed by heat to assure preservation of the

**Table 4.4.1 (continued)**

| | |
|---|---|
| | product in hermetically sealed containers (either metal or glass). |
| Frozen concentrated sweetened grape juice | Frozen concentrated sweetened grape juice is prepared from unfermented single-strength grape juice from sound, properly mature fresh grapes which, with or without aging or depectinization, is then concentrated. Single-strength grape juice or natural grape essence or a combination of single-strength grape juice and natural grape essence may be mixed to the concentrate and packed with the addition of nutritive sweetening ingredient and with or without ascorbic acid. Edible fruit acid may be added to adjust the flavor. The product is then frozen in accordance with good commercial practice and maintained at temperatures necessary for the preservation of the product. |
| | GREENS |
| Canned leafy greens | Canned leafy greens (other than spinach), means the product properly prepared from the succulent leaves of collards, kale, mustard greens, and/or turnip greens and packed with the addition of water in hermetically sealed containers and sufficiently processed by heat to prevent spoilage. The products may be acidified and or seasoned with one or more of the acidifying or seasoning ingredients and in a quantity permitted under FDA regulations. |
| Frozen leafy greens | Frozen leafy greens means the frozen product prepared from the clean, sound, succulent leaves and stems of a leafy green (including beet greens, collards, dandelion greens, endive, kale, mustard greens, spinach, swiss chard, turnip greens, and any other "market accepted" leafy green) by sorting, trimming, washing, blanching, and draining properly. The product is processed by freezing and maintained at temperatures necessary for its preservation. Any functional, optional ingredients permissible under FDA regulations may be used to acidify and/or season the product. |

## HOMINY

**Canned hominy**

Canned hominy means the canned product prepared from clean, sound field corn, either white or golden (yellow), by removal of the pericarp, and by precooking or other processing, soaking, and sorting. The product is packed in a liquid packing medium or in a jelled packing medium, in accordance with good commercial practice, and is sufficiently processed by heat to assure preservation in hermetically sealed containers.

## HONEY

**Comb honey**

Comb honey includes the following variations: comb-section honey, shallow-frame comb honey, wrapped cut-comb honey, chunk or bulk comb honey

**Extracted honey**

Extracted honey is honey that has been separated from the comb by centrifugal force, gravity, straining, or by other means.

## LEMON

**Canned lemon juice**

Canned lemon juice is the undiluted, unconcentrated, unfermented juice obtained from sound, mature lemons of one or more of the high acid varieties, which fruit is prepared by sorting, and by washing prior to extraction of the juice to assure a clean and sanitary product. The product is sufficiently processed with heat to assure preservation in hermetically sealed containers.

**Concentrated lemon juice for manu-facturing**

Concentrated lemon juice for manufacturing is the concentrated product obtained from sound, mature lemons. The fruit is prepared by sorting and by washing prior to extraction of the juice. The concentrated lemon juice is prepared and concentrated in accordance with good commercial practice. It may or may not require processing by heat, subsequent refrigeration, or freezing to assure preservation of he product. The finished product may contain added pulp, lemon oil to standardize flavor, and or

**Table 4.4.1 (continued)**

| | |
|---|---|
| | chemical preservatives permissible under provisions of FDA regulations. |
| Frozen concentrate for lemonade | Frozen concentrate for lemonade is the product prepared from lemon juice and one more nutritive sweetening ingredients. It may contain added lemon oil or concentrated lemon oil (or their extracts or emulsions) and may or may not contain water in sufficient quantities to standardize the product. The product contains not <48.0% by weight of soluble solids taken as the sucrose value determined by refractometer and corrected for acidity The lemon juice is produced from fresh, sound, ripe, and thoroughly cleansed fruit of one or more of the high acid varieties of the species *Citrus limon* (*limonia*). Such juices may be fresh or frozen or fresh concentrated or frozen concentrated. The concentrate for lemonade is processed in accordance with good commercial practice and is frozen and maintained at temperatures sufficient for the preservation of the product. If properly labeled any color material permissible under FDA regulations may be added. |
| | LIME |
| Frozen concentrate limeade | Frozen concentrate for limeade is the product prepared from lime juice and one or more nutritive sweetening ingredients to which may be added oil derived from limes for added flavor, and may or not contain water in sufficient quantities to standardize the product. The lime juice is produced from fresh, sound, mature, and thoroughly cleansed fruit of one or more of the varieties of the species *Citrus aurantifolia*. Such juice may be fresh or frozen or may be fresh juice concentrated or frozen concentrated. The concentrate for limeade is processed in accordance with good commercial practice and is frozen and maintained at temperatures sufficient for the preservation of the product. If properly labeled any artificial color permissible under FDA regulations may be added. |

## MARMALADE

**Orange marmalade** — Orange marmalade is the semisolid or gellike product prepared from orange fruit ingredients together with one or more sweetening ingredients and may contain suitable food acids, food pectins, lemon juice, or lemon peel. The ingredients are concentrated by cooking to such a point that the soluble solids of the finished marmalade is not <65%.

## MELONS

**Frozen melon balls** — Melon balls are spheres of melon flesh prepared from suitable varieties of sound, fresh cantaloupe melons and or sound, fresh honeydew melons to which melon balls prepared from sound, fresh melons of a definitely different but suitable variety may be added. The balls are prepared and washed in a manner to assure a clean and wholesome product. They are properly drained before filling into containers; may include a suitable fruit and or vegetable garnish; may be packed with the addition of nutritive or nonnutritive sweetening ingredients, including sirup and any other ingredient permissible under FDA regulations; and are prepared and frozen in accordance with good commercial practice and maintained at temperatures necessary for the preservation of the product.

## MOLASSES

**Sugarcane molasses** — Sugarcane molasses is the clean, sound, liquid product obtained by evaporating the juice of sugarcane and the removal of all or any part of the commercially crystallizable sugar.

## OKRA

**Canned okra (also see canned tomato and okra)** — Canned okra is the product prepared from clean, sound, succulent, immature pods of either the green or white varieties of the okra plant, which are properly trimmed and sorted; which may or may not have undergone partial fermentation; and is washed, packed, and processed in accordance with good commercial practice.

**Table 4.4.1 (continued)**

| | |
|---|---|
| Frozen okra | Frozen okra is the product prepared from the clean, sound, succulent, and edible fresh pods of the okra plant (*Hibiscus esculentus*) of the green variety. The product is properly prepared and properly processed and is then frozen and stored at temperatures necessary for preservation. |
| | OLIVES |
| Canned ripe olives | Canned ripe olives are prepared from properly matured olives which have first been properly treated to remove the characteristic bitterness; are packed in a solution of sodium chloride, with or without spices, and are sufficiently processed by heat in hermetically sealed containers. Canned olives which are not oxidized in processing and which possess a tan to light bronze color indicative of preparation from olives of advanced maturity and commonly referred to as tree-ripened or home-cured are not covered by the description here. |
| Green olives | Green olives are completely fermented and cured fruit of the olive tree (*Olea europea*) which have been prepared from a firm fruit of suitable maturity and variety that have been properly treated to partially remove the characteristic bitterness. The cured olives and the brine packing media have a pH of not >4 and a sodium chloride content of not <6%. Not more than a trace of reducing sugars may be present. Suitable garnishes, spices, seasoning ingredients, or any other ingredients permitted under FDA regulations may be added. |
| Olive oil | Olive oil is the edible oil obtained from the fruit of the olive tree (*Olea europaea* L.); is clarified; has a specific gravity of 0.910 to 0.915 at 25°C; has an iodine number (Hanus) of 79 to 90; has a refractive index of 1.4668 to 1.4683 at 25°C; and is packed in containers suitable for preservation of the product. |

## ONIONS

**Canned onions**

Canned onions means the canned product consisting of whole onions, properly prepared from clean, sound, succulent onion bulbs. The product is packed in accordance with good commercial practice and is sufficiently processed by heat to assure preservation of the product in hermetically sealed containers.

**Frozen breaded onion rings**

Frozen breaded onion rings is the product prepared from clean and sound, fresh onion bulbs (*Allium cepa*) from which the root bases, tops, and outer skin have been removed. The onion bulbs are sliced and separated into rings, coated with batter (or breaded) and may or may not be deep fried in a suitable fat or oil bath. The product is prepared and frozen in accordance with good commercial practice and maintained at temperatures necessary for the proper preservation of the product.

## PEACHES

**Dehydrated low-moisture peaches**

Dehydrated, low-moisture peaches are prepared from clean and sound fresh or previously dried peaches which are cut, chopped, or otherwise prepared into various sizes and shapes; are prepared to assure a clean, sound, wholesome product; are processed by dehydration whereby practically all of the moisture is removed to produce a very dry texture; and are packaged (including kind of container and proper closure) to assure retention of the low-moisture characteristics of the product. The product shall have been subjected to sulfur treatment sufficiently to retain a characteristic color but no other additives.

**Dried peaches**

Dried peaches are the halved and pitted fruit of the peach tree (*Prunes persica*) from which the greater portion of moisture has been removed. Before packing, the dried fruit is processed to cleanse the fruit and it may be sulfured sufficiently to retain a characteristic color.

**Table 4.4.1 (continued)**

| | |
|---|---|
| Frozen peaches | Frozen peaches are prepared from sound, mature, fresh peaches which are peeled, pitted, washed, cut and may be trimmed to assure a clean and wholesome product. The peaches are properly drained of excess water before filling into containers; may be packed with the addition of a nutritive sweetening ingredient, including sirup and/or sirup containing pureed peaches and any other ingredients permissible under the provisions of FDA regulations; are prepared and frozen in accordance with good commercial practice; and are maintained at temperatures necessary for the preservation of the product. |
| | PEARS |
| Dried pears | Dried pears are the halved fruit of the pear tree (*Pyrus communis*) which may or may not be cored, from which the external stems and calyx cups have been removed, and from which the greater portion of the moisture has been removed. Before packing, the dried fruit is processed to cleanse the fruit and may be sulfured sufficiently to retain a characteristic color. |
| | PEAS |
| Canned peas and carrots | Canned peas and carrots is the product which has been properly prepared from clean, sound, succulent garden peas and clean, sound carrots. The peas and carrots are packed in a suitable packing medium with or without the addition of salt, sugar, or other ingredients permissible under FDA regulations; and are sufficiently processed by heat to assure preservation of the product in hermetically sealed containers. |
| Frozen field peas and frozen black-eye peas | Frozen field peas and frozen black-eye peas (frozen peas) means the frozen product prepared from clean, sound, fresh, seed of proper maturity of the field pea plant (*Vigna sinensis*), by shelling, sorting, washing, blanching, and properly draining. The product is frozen and maintained at temperatures necessary for preservation. Frozen peas may |

contain succulent, unshelled pods (snaps) of the field pea plant as an optional ingredient used as a garnish.

Frozen peas

Frozen peas means the frozen product prepared from the clean, sound, succulent seed of the common garden pea (*Pisum sativum*) by shelling, washing, blanching, sorting, and proper draining and is frozen in accordance with good commercial practice; and maintained at temperatures necessary for the preservation of the product.

Frozen peas and carrots

Frozen peas and carrots is the product prepared from the fresh, clean, sound, immature seed of the common garden pea (*Pisum sativum*) and the fresh, clean, sound roots of the carrot plant (*Daucus carota sativa*). The peas are prepared by shelling, washing, sorting, and blanching. The prepared ingredients are properly drained and mixed; are frozen in accordance with good commercial practice: and are maintained at temperatures necessary for the preservation of the product.

## PEPPERS

Frozen sweet peppers

Frozen sweet peppers is the frozen product prepared from fresh, clean, sound, firm pods of the common commercial varieties of sweet peppers, which have been properly prepared, may or may not be blanched, and are then frozen in accordance with good commercial practice and maintained at temperatures necessary for the preservation of the product.

## PICKLES

Pickles

Pickles means the product prepared entirely or predominantly from cucumbers (*Cucumis sativus* L.). Clean, sound ingredients are used which may or may not have been previously subjected to fermentation and curing in a salt brine (solution of sodium chloride, NaCl). The prepared pickles are packed in a vinegar solution, to which may be added salt and other vegetables, nutritive sweeteners, seasonings, flavorings, spices, and other ingredients permissible under FDA regulations. The product is packed

**Table 4.4.1 (continued)**

| | |
|---|---|
| | in suitable containers and heat treated, or otherwise processed to assure preservation. |
| | PIMIENTOS |
| Canned pimientos | Canned pimientos means the canned product properly prepared from clean, sound, succulent red-ripe pods of the pimiento plant (*Capsicum annuum*). Pimientos are defined as a group of sweet peppers of the pimiento-type or Spanish pepper. This group includes Pimiento, Perfection, Truheart, and similar varieties. Pimiento fruit pods are smooth, typically conical or heart-shaped, pointed at the blossom end, thick walled, red to reddish-yellow color, and possess a mild and sweet distinct pimiento flavor.<br>The finished product is prepared from peeled pods and may contain acidifying and firming agents and any other ingredients permissible under the provisions of FDA regulations. |
| | PINEAPPLE |
| Frozen pineapple | Frozen pineapple is prepared from the properly ripened fruit of the pineapple plant (*Ananas sativus* or *Ananas comosus*); is peeled, cored, trimmed, and washed; may be packed with or without packing media; and is frozen and stored at temperatures necessary for the preservation of the product. |
| | PLUMS |
| Frozen plums | Frozen plums means the frozen product prepared from clean, sound, fresh fruit of any commercial variety of plums, exclusive of Damson variety, which are sorted, washed, drained, and pitted; which may be packed with or without the addition of a nutritive sweetening ingredient; and which are frozen in accordance with good commercial practice and maintained at temperatures necessary for the preservation of the product. |

## POTATOES

**Frozen French fried potatoes** Frozen French fried potatoes are prepared from mature, sound, white or Irish potatoes (*Solanum tuberosum*). The potatoes are washed, sorted, and trimmed as necessary to assure a clean and wholesome product. The potatoes may or may not be cut into pieces. The potatoes are processed in accordance with good commercial practice, which includes deep frying or blanching in a suitable fat or oil and which may include the addition of any ingredient permissible under FDA regulations. The prepared product is frozen and is stored at temperatures necessary for its preservation.

**Frozen hash brown potatoes** Frozen hash brown potatoes are prepared from mature, sound, white or Irish potatoes (*Solanum tuberosum*) that are washed, peeled, sorted, and trimmed to assure a clean and wholesome product. The potatoes so prepared are blanched, may or may not be fried, and are shredded or diced or chopped and frozen and stored at temperatures necessary for their preservation.

**Peeled potatoes** Peeled potatoes are clean, sound, fresh tubers of the potato plant prepared by washing, peeling, trimming, sorting, and by proper treatment, to prevent discoloration, by any means permissible under FDA regulations. The product is properly packed in suitable containers and securely closed to maintain the product in a sanitary condition.

## PRUNES

**Dehydrated, low-moisture prunes** Dehydrated, low-moisture prunes, hereinafter referred to as low-moisture prunes, are prepared from clean and sound fresh prune (plums) or clean and sound previously dried prunes; are pitted and otherwise prepared into various sizes and shapes; are prepared to assure a clean, sound, wholesome product; are processed by dehydration whereby practically all of the moisture is removed to produce a very dry texture; and are packaged (including kind of container and proper closure) to assure retention of the low-moisture characteristics of the product.

**Table 4.4.1 (continued)**

| | |
|---|---|
| Dried prunes | Dried prunes are prepared from sound, properly matured prune plums from which the greater portion of moisture is removed by drying. The dried prunes are cleaned to assure a wholesome product; they may be treated with water or steam; and a safe and suitable preservative may be added. |
| | PUMPKIN |
| Canned pumpkin and canned squash | Canned pumpkin and canned squash is the canned product prepared from clean, sound, properly matured, golden fleshed, firm shelled, sweet varieties of either pumpkins or squashed by washing, stemming, cutting, steamin and reducing to a pulp. The product is properly sieved and finished in accordance with good commercial practice and is then sufficiently processed by heat to assure preservation of the product in hermetically sealed containers. |
| | RAISINS |
| Processed raisins | Processed raisins are dried grapes of Binifera varieties, such as Thompson Seedless (*Sultanian*), Muscat of Alexandria, Muscatel Gordo Blanco, Sultana, Black Corinth, or White Corinth. The processed raisins are prepared from clean, sound dried grapes; are properly stemmed and capstemmed except for cluster or uncapstemmed raisins; are properly seeded in seeded styles; are sorted or cleaned, or both; and, except for cluster or uncapstemmed raisins, are washed in water to assure a wholesome product. Raisins with seeds that are referred to as "layer or cluster raisins" means that the raisins have not been detached from the main bunch. |
| | RASPBERRIES |
| Frozen raspberries | Frozen raspberries are prepared from the properly ripened fresh fruit of the plant (Genus *rubus*); are stemmed and cleaned; may be packed with or without packing media; and are frozen and stored at temperatures necessary for the preservation of the product. |

## SAUERKRAUT

**Bulk sauerkraut** Bulk or barreled sauerkraut is the product of characteristic acid flavor, obtained by the full fermentation, chiefly lactic, of properly prepared and shredded cabbage in the presence of 2-3% salt. It contains, upon completion of the fermentation, not <1.5% of acid, expressed as lactic acid.

**Canned sauerkraut** Canned (or packaged) sauerkraut, hereinafter referred to as canned kraut, is prepared from clean, sound, well-matured heads of he cabbage plant (*Brassica oleracea* var. *capitata* L.) which have been properly trimmed and cut; to which salt is added and which is cured by natural fermentation. The product may or may not be packed with pickled peppers, pimientos, or tomatoes or contain other flavoring ingredients to give the product specific flavor characteristics. The product (a) may be canned by processing sufficiently by heat to assure preservation in hermetically sealed containers; or (b) may be packaged in sealed containers and preserved with or without the addition of benzoate of soda or any other ingredient permissible under the provisions of FDA regulations or applicable State laws or regulations.

## SIRUP

**Refiners' sirup** Refiners' sirup means a liquid product obtained from the refining of cane or beet sugar. The total soluble nonsugar solids content of refiners' sirup exceeds 6% of the total soluble solids. All of the sirup constituents have been subjected to the processes of clarification and decolorization, or equivalent purification, and it may be partially or wholly inverted.

**Sugarcane sirup** Sugarcane sirup is the clean, sound liquid product obtained by evaporating the juice of sugarcane without the removal of any of the soluble solids or by dissolving sugarcane concrete in water and is produced and packed in accordance with good commercial practice.

**Table 4.4.1 (continued)**

| | |
|---|---|
| | SQUASH |
| Canned squash (summer type) | Canned squash (summer type) is the canned product prepared from clean, sound, succulent, immature summer type squash, called canned squash in this subpart, which has been washed, properly trimmed, and sorted. The product is packed in accordance with good commercial practice and is sufficiently processed by heat to assure preservation of the product in hermetically sealed containers. |
| Frozen cooked squash | Frozen cooked squash is the clean, sound, properly matured product made from varieties of fall- or late-type squash which have been properly prepared by washing, cutting, cleaning, steaming, reducing to a pulp, and removing seed and fiber. The product is then frozen in accordance with good commercial practice and maintained at temperatures necessary for the preservation of the product. |
| Frozen squash (summer type) | Frozen squash is the fresh, sound immature product from squash (summer type) plant which has been properly prepared and properly blanched, and is then frozen in accordance with good commercial practice and maintained at temperatures necessary for the preservation of the product. |
| | STRAWBERRIES |
| Frozen strawberries | Frozen strawberries means the frozen product prepared from sound, properly ripened fresh fruit of the strawberry plant by stemming, proper washing, sorting, and proper draining, may be packed with or without packing medium, and are then frozen in accordance with good commercial practice and maintained at temperatures necessary for the preservation of the product. |
| | SUCCOTASH |
| Frozen succotash | Frozen succotash means the frozen product prepared from fresh, clean, sound, succulent kernels of sweet corn and the immature seed of varieties of lima beans, vegetable soybeans, or fresh immature pods of green beans or wax |

beans. The ingredients are mixed and properly drained and then frozen in accordance with good commercial practice and maintained at temperatures necessary for the preservation of the product. When soybeans or green beans or wax beans are used, the product is designated as frozen soybean succotash or frozen green bean (or wax bean) succotash, as the case may be.

SWEET POTATOES

Frozen sweet potaotes

Frozen sweet potatoes is the product prepared from clean, sound root of sweet potatoes (*Ipomoea batas*) by washing, sorting, trimming, draining, and peeling, as the case may be. The product may or may not be cooked and may be prepared with the addition of suitable seasoning ingredients, sweetening ingredients, antioxidants, edible oils, spices, or other suitable additives permissible under FDA regulations. Such ingredients may or may not be admixed with the sweet potatoes or they may be contained separately from the sweet potato ingredient. When the sweet potatoes have been cooked or partially cooked prior to freezing and are packed with a high density sirup and or sugar to produce a candied effect, the frozen sweet potatoes may be considered Candied sweet potatoes for the purposes of this subpart. In lieu of peeling, the product may be prepared as unpeeled sweet potatoes in baked or stuffed form, with or without the addition of edible fat or oil to the peel. Stuffed form consists of cooked potatoes where the flesh has been removed from the peel, has been comminuted or crushed, and has been replaced in the peel or a preformed shell. The product is frozen in accordance with good commercial practice and maintained at temperatures necessary for the preservation of the product.

TANGERINE

Canned tangerine juice

Canned tangerine juice is the undiluted, unconcentrated, unfermented juice obtained from mature fresh fruit of the Mandarin orange (*Citrus reticulata*), which fruit has been properly washed; is packed with or without the addition of sweetening ingredients; is sufficiently processed by heat to

**Table 4.4.1 (continued)**

| | |
|---|---|
| | assure preservation of the product; and is packed in containers which are hermetically sealed. |
| Concentrated tangerine juice for manufacturing | Concentrated tangerine juice for manufacturing is the concentrated product obtained from sound, mature fruit of the Mandarin group (*Citrus reticulata*). The fruit is prepared by sorting and by washing prior to extraction of the juice and the extracted juice is concentrated. The concentrated tangerine juice is processed in accordance with good commercial practice and may or may not require processing by heat or subsequent refrigeration to assure preservation of the product. The finished product does not contain any additives except that cold-pressed oil to standardize flavor and chemical preservatives permissible under provisions of FDA regulations may have been added. |
| | TOMATO |
| Canned tomatoes and okra and canned okra and tomatoes | Canned tomatoes and okra (or canned okra and tomatoes) is the product prepared from clean, sound, properly matured tomatoes of red or reddish varieties, which may be canned, and the clean, sound succulent, immature pods of either the green or white varieties of the okra plant, which may be processed. The tomatoes have been prepared by trimming, washing, sorting, peeling, and coring. The okra is prepared by trimming, washing, and sorting and may be whole or cut into pieces. The tomatoes and okra may be packed with or without a small amount of tomato juice or tomato puree and with or without such other ingredients permissible under FDA regulations in accordance with good commercial practice. The product is sufficiently processed by heat to assure preservation in hermetically sealed containers. |
| Tomato sauce | Tomato sauce is the concentrated product prepared from the liquid extracted from mature, sound, whole tomatoes, the sound residue from preparing such tomatoes for canning, or the residue from partial extraction of juice, or any combination of these ingredients, to which is added salt and spices and to which may be added one or more nutritive sweetening ingredients, a vinegar or vinegars, and onion, |

| | |
|---|---|
| | garlic, or other vegetable flavoring ingredients. The refractive index of the tomato sauce at 20°C is not <1.3461. |
| | TURNIP GREENS |
| Frozen turnip greens with turnips | Frozen turnip greens with turnips is the product prepared from the clean, sound, succulent leaves and the clean, sound succulent roots of the turnip plant (*Brassica rapa*) by washing, sorting, trimming, blanching, and proper draining, and is then frozen in accordance with good commercial practice and maintained at temperatures necessary for the preservation of the product. |
| | VEGETABLES |
| Frozen mixed vegetables | Frozen mixed vegetables consist of three or more succulent vegetables, properly prepared and properly blanched; may contain vegetable (such as small pieces of sweet red peppers or sweet green peppers) added as garnish; and are frozen and maintained at temperatures necessary for the preservation of the product. |

Source: Adapted from 7 CFR 52, 1991.

## 4.5. References


Anonymous. *The A-to-Z of Container Corrosion: Another Path to Productivity*, Food Processors Institute, Washington, DC, 1982.

Anonymous. *Composition of Foods: Vegetables and Vegetable Products, Raw, Processed, and Prepared*, Agriculture Handbook No. 8-11, USDA, Washington, DC, 1984.

Anonymous. *Processed Fruits & Vegetables*, Business Trend, Commach, NY, 1985.

Anonymous. *The Time-Life Book of Vegetables and Fruits*, H. Holt & Co., New York, 1986.

Anonymous. *A Supplement to Agriculture Handbook 8*, USDA, Washington, DC, 1989.

Banwart, G. J. *Basic Food Microbiology*, 2nd ed., Van Nostrand Reinhold, New York, 1989.

Chan, H. T. Jr., ed. *Handbook of Tropical Foods*, Marcel Dekker Inc., New York, 1983.

Considine, D. M., and G. D. Considine. *Foods and Food Production Encyclopedia*, Van Nostrand Reinhold, New York, 1982.

Dennis, C., ed. *Post-harvest Pathology of Fruits & Vegetables*, Academic Press, San Diego, CA, 1983.

Duckworth, R. B. *Fruits and Vegetables*, Pergamon Press, Oxford, 1966.

Frazer, W. D., and D. C. Westhogg. *Food Microbiology*, 3rd ed., McGraw-Hill, New York, 1978.

Friend, J., ed. *Recent Advances in Biochemistry of Fruits and Vegetables*, Academic Press, San Diego, 1982.

Goodenough, P, and R. Atkin, *Quality of Stored & Processed Vegetables & Fruit*, Academic Press, San Diego, 1981.

Hardenburg, R. E., A. E. Watada, and C. Y. Wang. *The Commercial Storage of Fruits, Vegetables, and Florist and Nursery Stocks*, Agriculture Handbook 66, USDA/ARS, Washington, DC, 1986.

Horst, R. K., ed. *Westcott's Plant Disease Handbook*, 5th ed., Van Nostrand Reinhold, New York, 1990.

Jackson, J. and B. Shinn. *Fundamentals of Food Canning Technology*, Avi, New York, 1979.

Jay, J. M., *Modern Food Microbiology*, 3rd ed., Van Nostrand Reinhold, New York, 1986.

Lopez, A. *Complete Course in Canning*. Canning Trade Inc., Baltimore, MD, 1987.

Nel, P. *Deciduous Fruits and Vines: Pests and Diseases and Their Control*, B & N Imports, New York, 1985.

Nelson, P.E., and D. K. Tressler. *Fruit and Vegetable Juice Processing Technology*, 3rd ed., Van Nostrand Reinhold, New York, 1980.

Nyvall, Robert F. *Field Crop Diseases Handbook*, 2nd ed., Van Nostrand Reinhold, New York, 1989.

Salunkhe, D., and B. Desai. *Postharvest Biotechnology of Vegetables*, CRC Press, Boca Raton, FL, 1984.

Teranishi, R., and H. Barrera-Benitez, eds. *Quality of Selected Fruits & Vegetables of North America*, American Chemical Society, Washington, DC, 1981.

Ting, S. V., and R. I. Rouseff. *Citrus Fruits and Their Products*, Marcel Dekker Inc., New York, 1986.

Weichmann, Jr., ed. *Postharvest Physiology of Vegetables*, Marcel Dekker Inc., New York, 1987.

Wills, R. B. H., W. B. McGlasson, et al. *Postharvest: An Introduction to the Physiology and Handling of Fruit and Vegetables,* 3rd ed., Van Nostrand Reinhold, New York, 1989.

Yamaguchi, M. *World Vegetables: Principles, Production & Nutritive Values*. Avi, New York, 1983.

# CHAPTER 5
# MEAT AND POULTRY PRODUCTS

Unit I: Meat Products

The data presented in this chapter are mainly concerned with processed meat and poultry products. For meat products, the following categories of information are provided:

1. Some fundamental considerations in processing meats.
2. Formulas for preparing cured meats, canned meats, and sausages. The formulas have been developed from computer models, using formulations available from literature. For example, the amounts of ingredients used in each formula are given in ranges. Two types of notes are provided for each formula. Always begin with the lowest level for the ingredient salt. Since the formulas provided are in ranges, the exact legal level(s) of food additive(s) used cannot be stated. If one uses the data in this chapter, the final formula must comply with legal requirements for the usage of any food additive.
3. United States Department of Agriculture (USDA) defect criteria for meat from animal carcasses.
4. Definitions for selected terms used in labeling.
5. Legal requirements for the use of food additives in meat processing.
6. Definitions and standards promulgated by the USDA.

For poultry products, the following categories of information are provided:

1. Legal requirements for food additives used in poultry processing.
2. USDA definitions and standards for poultry products.

The citation of legal documents follows the standard format. For example, the legal citation "9 CFR 318.7" refers to Title 9 of the U.S. Code of Federal Regulations, Part 318 and Section 7.

Two government documents are cited: USDA Meat and Poultry Inspection Manual and USDA Home and Garden Bulletin 236. Both can be obtained from the USDA, Washington, D.C.

# Unit I: Meat Products

## 5.1. Processed Meats

**Table 5.1.1. General Considerations for Processed Meats**

- Curing meats
  - Sausages
  - Hams: canned and noncanned
  - Types of canned meat products
  - Other cured meats
- Raw and processed meats: components and nutrition
  - Nutritional considerations
  - Meat composition
    - Water
    - Fat
    - Minerals and ash
    - Proteins
    - Carbohydrates
    - Vitamins
- Undesirable substances in meat
  - Saturated fats and cholesterol
  - Hydrocarbons (carcinogens) formed from certain methods of cooking
  - Pathogens

Source: Compiled from Anonymous, 1953; Bechtel, 1986; Brandy et al., 1966; Gruttner, 1956; Pearson and Tauber, 1984; Price and Schweigert, 1987; Rust and Olson, 1973.

**Table 5.1.2. Cooking Meat and Cooked Meat Products**

- Considerations in cooking meat and meat products
  - Denaturation, solubility, and coagulation
  - Improving sensory attributes
  - Killing pathogens and increasing stability
  - Destruction of enzymes in meats
  - Moisture reduction at meat surface
  - Stabilizing of red pigments
- Cooking methods
  - Using dry heat
  - Using moist heat
  - Using combination methods
  - Using microwave
  - Relation to cuts of meat
  - Relation to processing techniques
  - Broiling
  - Roasting
  - Frying
  - Braising
  - Pot roasting
  - Stewing
  - Pressure cooking
  - Simmering

Source: Compiled from Bechtel, 1986; Kotschevar, 1988; Pauli, 1988; Pearson and Tauber,, 1984; Price and Schweigert, 1987; Terrell and Headlund, 1988.

**Table 5.1.3. Raw Ingredients for Preparing Processed Meats**

| | |
|---|---|
| Sausage and canned meats | Pork loins |
| Beef | Processed beef cuts |
| Pork | Raw ingredients storage |
| Veal and mutton | USDA certified pork |
| Variety meats | Meats unsuitable for processing |
| Mechanically deboned meat | Pale, soft, watery pork |
| Poultry meat | Dark, firm, and dry pork |
| Partially defatted tissue | Porcine stress syndrome |
| Smoked meats | Two-toning |
| Hams | Dark cutting beef |
| Bellies | Sex odor |
| Pork shoulders | Mutton flavor |

Source: Compiled from Anonymous, 1953; Ghinelli, 1950; Gruttner, 1956; Hoogenkamp, 1987; Karel et al., 1975; MacKenzie, 1966; Mann, 1962; Parkhurst and Mountney, 1987; Pearson and Tauber, 1984; Pearson and Dutson, 1987; Pearson et al., 1987; Rust and Olson, 1973; Woollen, 1971.

**Table 5.1.4. Types of Raw Meat Used in Comminuted Processed Meats**

Cattle

1. Boneless primal cuts: chucks, plates, flanks, and navels
2. Trimmings: from primal cuts
3. Other parts: cheeks, head meat, hearts, tripe, livers, tongues, lips, giblet meat, weasand meat

Hogs

1. Boneless primal cuts: hams, loins, shoulders, and bellies
2. Trimmings: from primal cuts
3. Other parts: cheeks, head meat, hearts, stomachs, livers, tongues, skins, snouts, piglet meat, weasand meat

Veal

1. Boneless primal cuts: yes
2. Trimmings: none
3. Other parts: hearts, cheeks, tongues

Sheep

1. Boneless primal cuts: yes
2. Trimmings: none
3. Other parts: hearts, cheeks, tongues

**Table 5.1.4. (continued)**

Source: Compiled from Gruttner, 1956; Hoogenkamp, 1987; MacKenzie, 1966; Mann, 1962; Parkhurst, 1987; Pearson and Tauber, 1984; Pearson and Dutson, 1987; Rust and Olson, 1973; Woollen, 1971.

**Table 5.1.5. Categories of Perishable Cooked and Cured Meat Products**

| Category of Meat: Examples |
|---|
| Chopped or ground: mordatella, thuringer |
| Emulsified: Bologna, franks, liver sausage, loaves, and luncheon meats |
| Jellied meats: blood, headcheese, souse, and tongue |
| Primal cuts: corned beef, Canadian bacon, ham, pastrami, smoked poultry |
| Restructed meats (sectioned and formed): dried beef, ham |

Source: Compiled from Anonymous, 1953; Ghinelli, 1950; Gruttner, 1956; MacKenzie, 1966; Mann, 1962; Parkhurst and Mountney, 1987; Pearson and Dutson, 1987; Pearson et al., 1987; Rust and Olson, 1973; Woollen, 1971.

## 5.2. Cured Meats

**Table 5.2.1. Considerations in Curing Meat Products**

- Ingredient considerations
  - Salt
  - Sugar and corn syrup solids
  - Nitrite and/or nitrate
  - Phosphates
  - Sodium ascorbate and erythorbate
  - Monosodium glutamate
  - Hydrolyzed vegetable proteins
- Color considerations
  - Muscle pigments
  - Nitrite and/or nitrate
- Methods of curing meat
  - Dry salt cure
  - Traditional dry curing
  - Pickle cure
    - Artery pumping
    - Single-needle stitch pumping
    - Multiple-needle stitch pumping
  - Curing with heat

Source: Compiled from Anonymous, 1953; Ghinelli, 1950; Gruttner, 1956; Hoogenkamp, 1987; Hugunin and Ewing, 1977; MacKenzie, 1966; Mann, 1962; Ockerman and Hansen, 1988; Pearson and Tauber, 1984; Price and Schweigert, 1987; Rust and Olson, 1973; Schmidt, 1978; Ziegler, 1952.

**Table 5.2.2. Considerations in Smoking Meat Products**

| | |
|---|---|
| Objectives of smoking meats | Smoke production |
| Smoke components | Smoke chemistry and physics |
| Phenols | Smoke deposits and meats |
| Alcohols | Smoking methods |
| Organic acids | Fuels for producing smoke |
| Carbonyls | Smoking and cooking |
| Hydrocarbons | Preparing liquid smoke |
| Gases | Adding liquid smoke |
| Smoking and nutrition | to meat products |

Source: Compiled from Anonymous, 1953; Ghinelli, 1950; Gruttner, 1956; Hall and Sleight, 1971; Maga, 1988; Pearson and Tauber, 1984; Price and Schweigert, 1987; Rust and Olson, 1973; Rutkowski, 1976; Wilson, 1961; Wistreich, 1977; Ziegler, 1952.

**Table 5.2.3. Listing of Cured and/or Smoked Meats**

| | |
|---|---|
| USDA classification of hams | Canadian bacon |
| Internal temperature reached | Wiltshire bacon |
| Permitted added substances | Beef bacon |
| Presence/absence of bone | Jowl bacon |
| Commercial processing of ham | Fat backs and heavy bellies |
| Curing | Smoked pork loin |
| Smoking and cooking | Picnic |
| Cooked ham | Shoulder butt |
| Baked ham | Corned beef |
| Country ham | Smoked fresh meat |
| Westphalian ham | Dried beef |
| Scotch ham | Smoked and cured lamb |
| Prosciutto | Smoked tongue |
| Bacon | Pickled pigs feet |

Source: Compiled from Anonymous, 1953; Ashbrook, 1973; Ghinelli, 1950; Gruttner, 1956; Hall and Sleight, 1971; Hoogenkamp, 1987; Hugunin and Ewing, 1977; Levie, 1979; MacKenzie, 1966; Maga, 1988; Mann, 1962; Ockerman and Hansen, 1988; Pearson and Tauber, 1984; Price and Schweigert, 1987; Rust and Olson, 1973; Rutkowski, 1976; Schmidt, 1978; Ziegler, 1952.

**Table 5.2.4. A Pickle Formula for Manufacturing Ham**

| Ingredients | lb | oz |
|---|---|---|
| Water | 71-73 | |
| Salt | 20-22 | |
| Sugar | 6-7 | |
| Sodium tripolyphosphate | 2-3 | |
| Sodium ascorbate | | 5-6 |
| Sodium nitrite | | 2-3 |

Source: Ranges of values are based on data from Anonymous, 1953; Barnett et al., 1965; Dunker et al., 1953; Ghinelli, 1950; MacKenzie, 1966; Pearson and Tauber, 1984; Price and Schweigert, 1987; Rust and Olson, 1973.
Always start with the lower salt indicated. The final formula must comply with legal limits of food additives used.
Note: 1 lb = ~453.6 g, 1 oz = ~28.4 g

**Table 5.2.5. A Formula for Preparing Baked-Style Ham Glaze**

| Ingredients | lb | oz |
|---|---|---|
| Corn syrup | 6-7 | |
| Brown sugar | 2-3 | |
| Gelatin | 2-3 | |
| Water | 4-6 | |
| Pineapple juice | 4-5 | |
| Finely ground paprika | | 7-8 |
| Soluble clove | | 5-6 |

Source: Ranges of values are based on data from Anonymous, 1953; Brady et al., 1949; Christian, 1960; Ghinelli, 1950; MacKenzie, 1966; Pearson and Tauber, 1984; Price and Schweigert, 1987; Rust and Olson, 1973.
Note: 1 lb = ~453.6 g, 1 oz = ~28.4 g.

Table 5.2.6. **A Pickle Formula for Pickling Boneless Ham Rolls**

| Ingredients | lb |
|---|---|
| Water | 89-91 |
| | |
| Ascorbate | 0.1-0.3 |
| Nitrite | 0.05-0.10 |
| Phosphate | 0.9-1.2 |
| Salt | 5-6 |
| Sugar | 2-3 |

Source: Ranges of values are based on data from Anonymous, 1953; Anonymous 1971; Ghinelli, 1950; Jensen and Hess, 1941; MacKenzie, 1966; Pearson and Tauber, 1984; Price and Schweigert, 1987; Reynolds et al., 1978; Rust and Olson, 1973; Skelley et al., 1964.
Always start at the lower salt level indicated. The final formula must comply with legal limits for food additives used.
Note: 1 lb = ~453.6 g, 1 oz = ~28.4 g

Table 5.2.7. **A Pickle Formula for Combination Ham and Soy Protein (66% Ham)**

| Ingredients | lb | oz |
|---|---|---|
| Water | 82-83 | |
| | | |
| Salt | 6-7 | |
| Sodium erhthorbate | | 1-2 |
| Sodium nitrite | | 0.5-0.8 |
| Sodium tripolyphosphate | | 12-14 |
| Soy protein isolate | 11-12 | |
| Sugar | 1-2 | |

Source: Ranges of values are based on data from Anonymous, 1953; Barnett, et al., 1965; Christian, 1981; Ghinelli, 1950; MacKenzie, 1966; Pearson and Tauber, 1984; Price and Schweigert, 1987; Rust and Olson, 1973; Siegel et al., 1979.
Always start at the lower salt level indicated. The final formula must comply with legal limits for food additives used.
Note: 1 lb = ~453.6 g, 1 oz = ~28.4 g

**Table 5.2.8. A Pickle Formula for Combination Roast Turkey Breast and Soy Protein Isolate Product (66% Turkey Breast)**

| Ingredients | lb | oz |
|---|---|---|
| Water | 82-83 | |
| Dextrose | 1-2 | |
| Salt | 4-5 | |
| Sodium tripolyphosphate | | 1-2 |
| Soy protein isolate | 13-14 | |

Source: Ranges of values are based on data from Anonymous, 1953; Froning, 1970; Froning and Janky, 1971; Ghinelli, 1950; MacKenzie, 1966; Pearson and Tauber, 1984; Price and Schweigert, 1987; Rust and Olson, 1973.
Always start at the lower salt level indicated. The final formula must comply with legal limits for food additives used.
Note: 1 lb = ~453.6 g, 1 oz = ~28.4 g

**Table 5.2.9. A Pickle Formula for Turkey Thigh Meat for Preparing Turkey Loaf**

| Ingredients | gal | lb |
|---|---|---|
| Water | 98-102 | |
| Nitrite | | 0.8-1.1 |
| Phosphate | | 34-36 |
| Salt | 51-53 | |

Source: Ranges of values are based on data from Anonymous, 1953; Froning, 1970; Froning, and Janky, 1971; Ghinelli, 1950; MacKenzie, 1966; Pearson and Tauber, 1984; Price and Schweigert, 1987; Rust and Olson, 1973.
Always start at the lower salt level indicated. The final formula must comply with legal limits for food additives used.
Note: 1 lb = ~453.6 g; 1 oz = ~28.4 g; 1 U.S. gal = 0.00379 $m^3$

**Table 5.2.10. A Pickle Formula for Turkey Ham**

| Ingredients | lb | gal |
|---|---|---|
| Water | | 98-102 |
| Meat | 2250-2350 | |
| | | |
| Erythorbate | 2-3 | |
| Nitrite | 1-2 | |
| Phosphate | 47-48 | |
| Salt | 11-12 | |

Source: Ranges of values are based on data from Anonymous, 1953; Froning, 1970; Froning and Janky, 1971; Ghinelli, 1950; MacKenzie, 1966; Pearson and Tauber, 1984; Reynolds et a1, 1978; Price and Schweigert, 1987; Rust and Olson, 1973; Wistreich et al., 1960.
Always start at the lower salt level indicated. The final formula must comply with legal limits for food additives used.
Note: 1 lb = ~453.6 g, 1 oz = ~28.4 g, 1 U.S. gal = 0.00379 $m^3$

## 5.3. Canned Meats

**Table 5.3.1. Basic Considerations in Canning Meats**

| | |
|---|---|
| Types of cans | Continuous agitating retorts |
| Square and Pullman base | Hydrostatic retorts |
| Pear shaped | Establishment of retort schedule |
| Round sanitary | Pasteurization |
| Drawn aluminum | Cooling, storage, shelflife |
| Canning materials | Aseptic canning and federal regulations |
| Retorts | |
| Nonagitating retorts | |

Source: Adapted from Anonymous 1953, Anonymous, 1989; Ghinelli, 1950; Karmas and Thompson, 1964; MacKenzie, 1966; Pearson and Tauber, 1984; Price and Schweigert, 1987; Rust and Olson, 1973.

**Table 5.3.2 List of Canned Meat Formulations**

| | |
|---|---|
| All canned meat | Meat stew |
| Beef stew | Pasta products with meat |
| Canned hams, pasteurized and sterile | Plastic packaged hams |
| Chili con carne | Potted meat |
| Corned beef hash | Potted products and spreads |
| Hash products | Sliced dried beef |
| Luncheon meat | Tamales |
| Meat balls with gravy | Vienna sausages |

Source: Adapted from Anonymous, 1953; Ghinelli, 1950; MacKenzie, 1966; Pearson and Tauber, 1984; Price and Schweigert, 1987; Rust and Olson, 1973.

**Table 5.3.3. A Formula for Preparing Canned Corned Beef Hash**

| Ingredients | lb | oz |
|---|---|---|
| Cooked beef | 103-105 | |
| Potatoes, dehydrated | 133-135 | |
| Beef broth | 43-46 | |
| Onions, raw | 9-11 | |
| Salt | 4-5 | |
| Ground white pepper | | 4-5 |
| Sodium nitrite | | 0.2-0.4 |

Source: Ranges of value are based on data and suggestions in the following literature: Anonymous, 1953; Anonymous, 1973; Anonymous, 1983; Anonymous, 1986; Anonymous, 1988; Ashbrook, 1973; Bloch, 1976; Ghinelli, 1950; Levie, 1979; MacKenzie, 1966; Pearson and Tauber, 1984; Price and Schweigert, 1987; Rust and Olson, 1973.
The final formula used must comply with legal limit for food additives.
Note: 1 lb = ~453.6 g, 1 oz = ~28.4 g

**Table 5.3.4. A Formula for Preparing Canned Beef Stew**

| Ingredients | lb | oz |
|---|---|---|
| Cooked beef | 103-105 | |
| Diced potatoes, rehydrated/dehydrated | 84-85 | |
| Water | 79-81 | |
| Gravy | 72-74 | |
| Diced carrots | 54-56 | |
| Chopped onions | 14-16 | |
| Flour (wheat or rice) | 4-5 | |
| Salt | 2.0-2.5 | |
| Ground black pepper | | 2-3 |

Source: Ranges of value are based on data and suggestions in the following literature: Anonymous, 1953; Anonymous, 1973; Anonymous, 1983; Anonymous, 1986; Anonymous, 1988; Ashbrook, 1973; Bloch, 1976; Ghinelli, 1950; Levie, 1979; MacKenzie, 1966; Pearson and Tauber, 1984; Price and Schweigert, 1987; Rust and Olson, 1973.
Note: 1 lb = ~453.6 g, 1 oz = ~28.4 g

**Table 5.3.5. A Formula for Preparing Canned Beef Stew Gravy**

| Ingredients | lb | oz |
|---|---|---|
| Broth from cooked meat | 79-81 | |
| Processed flour | 5-6 | |
| Salt | 2-3 | |
| Ground black pepper | | 1-2 |

Source: Ranges of value are based on data and suggestions in the following literature: Anonymous, 1953; Anonymous, 1973; Anonymous, 1983; Anonymous, 1986; Anonymous, 1988; Ashbrook, 1973; Bloch, B., 1976; Ghinelli, 1950; Levie, 1979; MacKenzie, 1966; Pearson and Tauber, 1984; Price and Schweigert, 1987; Rust and Olson, 1973.
Note: 1 lb = ~453.6 g, 1 oz = ~28.4 g

**Table 5.3.6. A Formula for Preparing Canned Chili con Carne**

| Ingredients | lb | oz |
|---|---|---|
| Beef and beef byproducts | 109-110 | |
| Beef broth or water | 98-101 | |
| Wheat flour | 8-9 | |
| Gound chili pepper | 3-4 | |
| Salt | 1.0-1.5 | |
| Ground domestic paprika | | 12-13 |
| Ground cumin | | 9-11 |
| Tomato paste | | 8-9 |
| Ground oregano | | 4-5 |
| Onion powder | | 3-4 |
| Garlic powder | | 2-3 |

Source: Ranges of value are based on data and suggestions in the following literature: Anonymous, 1953; Anonymous, 1973; Anonymous, 1983; Anonymous, 1986; Anonymous, 1988; Ashbrook, 1973; Bloch, B., 1976; Ghinelli, 1950; Levie, 1979; MacKenzie, 1966; Pearson and Tauber, 1984; Price and Schweigert, 1987; Rust and Olson, 1973.
Note: 1 lb = ~453.6 g, 1 oz = ~28.4 g

**Table 5.3.7. A Formula for Preparing Canned Vienna Sausages**

| Ingredients | lb | oz |
|---|---|---|
| Bull meat | 44-455 | |
| Beef trimmings | 19-21 | |
| Regular pork trimmings | 19-21 | |
| Beef hearts | 14-15 | |
| Crushed ice or cold water | 14-15 | |
| Salt | 2-3 | |
| Sugar | | 6-7 |
| Ground pepper | | 6-7 |
| Ground mustard | | 4-5 |
| Sodium nitrite | | 0.2-0.4 |

Source: Ranges of value are based on data and suggestions in the following literature: Anonymous, 1953; Anonymous, 1973; Anonymous, 1983; Anonymous, 1986; Anonymous, 1988; Ashbrook, 1973; Bloch, 1976; Ghinelli, 1950; Levie, 1979; MacKenzie, 1966; Pearson and Tauber, 1984; Price and Schweigert, 1987; Rust and Olson, 1973.

The final formula used must comply with legal limit for food additives.
Note: 1 lb = ~453.6 g, 1 oz = ~28.4 g

**Table 5.3.8. A Formula for Preparing Canned Meat Balls**

| Ingredients | lb | oz |
|---|---|---|
| Cow meat | 108-111 | |
| Bread crumbs | 6-7 | |
| Salt | 0.8-1.2 | |
| Onion powder | | 13-15 |
| Pepper | | 3-5 |

Source: Ranges of value are based on data and suggestions in the following literature: Anonymous, 1953; Anonymous, 1973; Anonymous, 1983; Anonymous, 1986; Anonymous, 1988; Ashbrook, 1973; Bloch, 1976; Ghinelli, 1950; Levie, 1979; MacKenzie, 1966; Pearson and Tauber, 1984; Price and Schweigert, 1987; Rust and Olson, 1973.
Note: 1 lb = ~453.6 g, 1 oz = ~28.4 g

**Table 5.3.9. A Formula for Preparing Canned Meat Balls Gravy**

| Ingredients | lb | Gal |
|---|---|---|
| Beef stock | | 15-16 |
| Flour | 7-8 | |
| Beef fat | 3-4 | |
| Salt | 1-2 | |
| Beef extract | 1-2 | |
| Onion powder | 1-2 | |
| Caramel coloring | 8-9 | |
| Celery salt | 11-12 | |
| Pepper | 8-9 | |

Source: Ranges of value are based on data and suggestions in the following literature: Anonymous, 1953; Anonymous, 1973; Anonymous, 1983; Anonymous, 1986; Anonymous, 1988; Ashbrook, 1973; Bloch, 1976; Ghinelli, 1950; Levie, 1979; MacKenzie, 1966; Pearson and Tauber, 1984; Price and Schweigert, 1987; Rust and Olson, 1973.
Note: 1 lb = ~453.6 g, 1 oz = ~28.4 g, 1 U. S. gal = ~0.00379 $m^3$

**Table 5.3.10. A Formula for Preparing Canned Luncheon Meat**

| Ingredients | lb | oz |
|---|---|---|
| Picnics | 64-66 | |
| Butts | 19-20 | |
| Shanks | 14-16 | |
| Salt | 3-4 | |
| Sugar | 2-3 | |
| Sodium nitrite | | 0.2-0.4 |

Source: Ranges of value are based on data and suggestions in the following literature: Anonymous, 1953; Anonymous, 1973; Anonymous, 1983; Anonymous, 1986; Anonymous, 1988; Ashbrook, 1973; Bloch, 1976; Ghinelli, 1950; Levie, 1979; MacKenzie, 1966; Pearson and Tauber, 1984; Price and Schweigert, 1987; Rust and Olson 1973.
The final formula used must comply with legal limit for food additives.
Note: 1 lb = ~453.6 g, 1 oz = ~28.4 g

## 5.4. Sausages and Specialty Items

### 5.4.1. Classification and Casings

**Table 5.4.1.1. Sausage Classification in the Industry (United States)**

Classification
1. Examples
2. Descriptions

Fresh sausage
1. Fresh pork sauage, bratwurst, brockwurst
2. Fresh meat (chiefly pork), uncured, comminuted, seasoned, and usually stuffed into casings; must be cooked fully before serving.

Dry and semidry sausages
1. Genoa salami, pepperoni
2. Cured meats, air dried, may be smoked before drying; served cold.

Cooked sausages
1. Frankfurters, liver sausage, braunschweiger, Bologna
2. Cured or uncured meats, comminuted, seasoned, stuffed into casings, cooked and sometimes smoked; usually served cold.

Cooked meat specialties
1. Loaves, head cheese, scrapple

2. Specially prepared meat products; cured or uncured meats, cooked but rarely smoked, often made in loaves, but generally sold in sliced, packaged form; usually served cold

Fresh, smoked sausages

1. Smoked, countrystyle pork sausage; mettwurst; kielbasa
2. Fresh meats, cured or uncured, stuffed, smoked, but not cooked; must be fully cooked before serving.

Source: Adapted from Anonymous, 1938; Gerrard, 1959; Karmas, 1977; Kutas, 1987; Ockerman, 1989; Predika, 1983; Price and Schweigert, 1987; Shannon, 1966; Tarr, 1975; Viscora, 1968.

**Table 5.4.1.2. Classifications of Sausages According to Methods of Processing**

| Methods of Preparation | Types of Sausages |
|---|---|
| Chopping | Coarsely ground |
| | Finely chopped or emulsion |
| Cooking | Cooked |
| | Uncooked |
| Curing | Cured |
| | Uncured |
| Fermentation | Fermented |
| | Unfermented |
| Moisture | Cooked: fresh and cured smoked and unsmoked |
| | Cured: smoked and unsmoked |
| | Dried: semidry and dry |
| | Fresh: unsmoked or smoked |
| | Smoked: fresh and cured |
| | Others: meat loaves and specialty items |
| Smoked | Smoked: natural smoke, flavorings, others |
| | Unsmoked |
| Water | Water added |
| | No water added |

Source: Adapted from Anonymous, 1938; Gerrard, 1959; Karmas, 1977; Kutas, 1987; Ockerman, 1989; Predika, 1983; Price and Schweigert, 1987; Shannon, 1966; Tarr, 1975; Viscora, 1968.

**Table 5.4.1.3. Classifications of Sausages According to the USDA**

| Classification | Types |
|---|---|
| Fresh sausages | Fresh pork sausage |
| | Fresh beef sausage |
| | Breakfast sausage |
| | Whole hog sausage |
| | Italian sausage products |
| Uncooked smoked sausages | Fresh smoked pork sausage |
| | Fresh smoked Kelbasa |
| Cooked sausages | Frankfurters, Bologna, knockwurst, and similar products |
| | Liver sausage and braunschweiger (liverwurst) |
| Dry and semidry sausage | Semidry sausage |
| | Dry sausages |

Source: 9 CFR 319, 1991.

**Table 5.4.1.4. Animal Casings and Their Sources**

| Casing | Part of Intestinal Tract | Animals |
|---|---|---|
| Beef bungs | Caecum (blind gut) | Cattle |
| Bladders | Urinary bladder | Cattle or hogs |
| Caps | Caecum or blind gut | Hogs |
| Hog bungs | End, 5-6 ft to anus | Hogs |
| Middles | Large intestines | Cattle, pigs |
| Rounds | Small intestine | Cattle, sheep, goats, pigs |
| Runners | Small intestines | Cattle |
| Small casings | Small intestines | Hogs, sheep, goats |
| Stomachs | Stomachs (maws) | Hog |
| Weasands | Esophagus | Cattle |

Source: Adapted from Anonymous, 1938; Gerrard, 1959; Karmas, 1977; Kutas, 1987; Mann, 1962; Ockerman, 1989; Ockerman and Hansen, 1988; Predika, 1983; Price and Schweigert, 1987; Shannon, 1966; Tarr, 1975; Viscora, 1968; Wang, 1954, 1955.

## 5.4.2. Meat and Fermentation

**Table 5.4.2.1. Basic Considerations in Meat Fermentation**

- Meat fermentation
  - Basic considerations
  - Microorganisms and starter cultures
  - Microorganisms and meat preservations
  - Microorganisms in meat
- Sausages
  - Classification and properties
  - Functional properties of meat proteins
  - Properties of raw meat
  - Formulation
  - Processing description and stages
  - Fermentation procedures
  - Heating and drying
  - Microbiology and fermentation
  - Backinoculum ("backslopping")
  - Sausage starter cultures
  - Formulations, recipes, processes
- Meat starter cultures
  - Classification and descritpion
  - Pediococci strains
  - Lactobacilli strains
  - Micrococci-type strains
  - Molds and yeasts
- Criteria determining outcomes of Meat fermentation
  - Raw meat ingredients
  - Other ingredients
    - Salt
    - Sugars
    - Complex carbohydrates
    - Spices
    - Miscellaneous
  - Processing conditions
    - Temperatures
    - Humidity
    - Smoke
    - Casing diameter
    - Buffering
    - Formulation uniformity
    - Contaminants
  - Microbial starter culture
- Fermented meat shelflife and safety factors to be considered
  - Safety
    - *Staphylococci*
    - *Salmonella*
    - *Clostridium perfringens*
    - *Trichinella spiralis*
    - *Clostridium botulinum*
    - *Staphylococcus aureus*
    - Good manufacturing practice
  - Shelf life: moisture, pH, cultures
  - Nonfermented meats and microbial additives
- Bacon
  - Nitrite, reduction, pigments, antioxidants
  - Nitrosamines
    - Formation
    - Toxicity
  - Lactobacilli cultures and nitrite
  - Starter cultures and commercial applications
  - Antibotulism
- Cured meats
  - Defintiion
  - Biochemical processes and bacterial growth
  - Bacterial cultures
  - Aging
  - Ham production

**Table 5.4.2.1. (continued)**

Source: Adapted from Anonymous, 1938; Gerrard, 1959; Karmas, 1977; Kutas, 1987; Liepe, 1983; Ockerman, 1989; Predika, 1983; Price and Schweigert, 1987; Shannon, 1966; Tarr, 1975; Viscora, 1968.

**Table 5.4.2.2. Microorganisms Used as Starter Cultures for Cured Meat Products**

Bacteria

*Achromobacter guttatus* 921; *Achromobacter* 22; *Achromobacter* X

*Aerobacter cloacae*

*Aeromonas* X; *Aeromonas* 19

*Pediococcus pentosaceus*

*Bacillus subtilis*; *Bacillus specialis*

*Flavobacterium specialis*

*Lactobacillus acidophilus*; *Lactobacillus brevis*; *Lactobacillus cucumeris*; *Lactobacillus farciminis*; *Lactobacillus fermentum*; *Lactobacillus gayoni*; *Lactobacillus hispanicus* CP-9; *Lactobacillus leichmannii*; *Lactobacillus lucanicarum* B-3; *Lactobacillus lycopersica*; *Lactobacillus maltaromicus*; *Lactobacillus mannitopoeus*; *Lactobacillus plantarum*; *Lactobacillus sake*; *Lactobacillus specialis* CP-2 P; *Lactobacillus specialis* CP-26; *Lactobacillus wehmeri*; *Lactobacillus arabinosus*; *Lactobacillus pentoaceticus*; *Lactobacillus pentosus*

*Lactobacillus* NRRL-B-5632; *Lactobacillus* L 110; *Lactobacillus* P 22; *Lactobacillus* 4669/6

*Micrococcus aquatis*; *Micrococcus aurantiacus* M 53; *Micrococcus candidus*; *Micrococcus caseolyticus*; *Micrococcus conglomeratus*; *Micrococcus epidermidis*; *Micrococcus specialis* DB 6; *Microrcoccus specialis* M 17; *Micrococcus specialis* MF-1; *Micrococcus varians; Micrococcus* M 104; *Micrococcus* M 86; *Micrococcus* M III; *Micrococcus* P 4; *Micrococcus* 199-10

*Pediococcus cerevisiae*; *Pediococcus cerevisiae* NRRL-B-652; *Pediococcus cerevisiae* CPO 18; *Pediococcus cerevisiae* B-2; *Pediococcus specialis* PS-23; *Pediococcus acidilactici*; *Pediococcus* Pc-30

*Proteus vulgaris*

*Spilirilli*

*Staphylococcus simulans* M 17; *Staphylococcus simulans* M III; *Staphylococcus xylosus*

*Streptococcus carnosus*; *Streptococcus diacetylactis*; *Streptococcus diacetylactis* 4284; *Streptococcus lactis*; *Streptococcus lactis* var. H; *Streptococcus lastis* AK-60
*Streptomyces griseus*
*Vibrio costicolus*; *Vibrio halo-denitrificans*; *Vibrio specialis*
Yeasts
*Debaryomyces hansenii*; *Debaryomyces kloeckeri* syn.; *Debaryomyces pfaffi*
*Saccharomyces specialis*
Molds
*Penicillium caseicolum (candidum)*; *Penicillium expansum*; *Penicillium nalgiovense*; *Penicillium simplicissimum*

Source: Liepe, 1983. Used with permission.

### 5.4.3. Ground Sausages

**Table 5.4.3.1. Formula for Fresh Pork Sausage**

| Ingredients | lb | oz |
|---|---|---|
| Pork trimmings (65% lean) | 79-81 | |
| Fresh belly trimmings | 19-21 | |
| Salt | 1-2 | |
| White pepper | | 4-6 |
| Sugar | | 2-3 |
| Sage | | 2-3 |
| Ginger | | 0.5-1.0 |

Source: Ranges of value are calculated from data in the following literature: Anonymous, 1938; Gerrard, 1959; 1983; Karmas, 1977; Kutas, 1987; Ockerman, 1989; Predika, 1983; Price and Schweigert, 1987; Shannon, 1966; Tarr, 1975; Viscora, 1968.
Always start the salt at the lowest level indicated.
Note: 1 lb = ~453.6 g, 1 oz = ~28.4 g

**Table 5.4.3.2. Formula for Canadian Pork or Breakfast Sausage**

| Ingredients | lb | oz |
|---|---|---|
| Jowls or regular trimmings | 54-55 | |
| Pork skirts | 14-15 | |
| Beef trimmings | 14-15 | |
| Back fat | 14-15 | |
| Ice | 14-15 | |
| Corn flour | 4-4 | |
| Dry skim milk | 3-4 | |
| Salt | 1-2 | |
| Sugar | | 8-9 |
| White pepper | | 3-5 |
| Ginger | | 0.5-1.5 |
| Mace | | 0.5-1.5 |

Source: Ranges of value are calculated from data in the following literature: Anonymous, 1938; Gerrard, 1959; Karmas, 1977; Kutas, 1987; Ockerman, 1989; Predika, 1983; Price and Schweigert, 1987; Shannon, 1966; Tarr, 1975; Viscora, 1968.

Always start the salt at the lowest level indicated.

Note: 1 lb = ~453.6 g, 1 oz = ~28.4 g

**Table 5.4.3.3. Formula for Italian Pork Sausage**

| Ingredients | lb | oz |
|---|---|---|
| Method 1 (Sweet or Nonspicy) | | |
| Pork trimmings (65% lean) | 120-125 | |
| Salt | 2-3 | |
| White pepper | 4-5 | |
| Fennel | | 4-5 |
| Paprika | | 2-3 |
| Method 2 (Hot) | | |
| Pork trimmings (65% lean) | 118-122 | |
| Salt | 2.0-2.5 | |
| White pepper | | 5-6 |
| Fennel | | 4-5 |
| Crushed red pepper (mild) | | 4-5 |
| Coriander | | 2-3 |
| Paprika | | 1-2 |

Source: Ranges of value are calculated from data in the following literature: Anonymous, 1938; Gerrard, 1959; Karmas, 1977; Kutas, 1987; Ockerman, 1989; Predika, 1983; Price and Schweigert, 1987; Shannon, 1966; Tarr, 1975; Viscora, 1968.
Always start the salt at the lowest level indicated.
Note: 1 lb = ~453.6 g, 1 oz = ~28.4 g

**Table 5.4.3.4. Formula Regular Polish Sausage**

| Ingredients | lb | oz |
|---|---|---|
| Lean beef (6% fat) | 25-27 | |
| Beef flanks (49% fat) | 10-12 | |
| Pork belly trimmings (65% fat) | 22-23 | |
| Pork butts cured (21% fat) or picnics pumped (10% fat) | 42-43 | |
| Ice | 11-13 | |
| Salt | 2-3 | |
| Sugar | 2-3 | |
| Corn syrup | 2-3 | |
| Black pepper | | 3-4 |
| Coriander | | 1-2 |
| Ginger | | 0.5-1.0 |
| Monosodium glutamate | | 1-2 |
| Garlic | | 0.2-0.3 |
| Sodium nitrite | | 0.1-0.3 |

Source: Ranges of value are calculated from data in the following literature: Anonymous, 1938; Gerrard, 1959; Karmas, 1977; Kutas, 1987; Ockerman, 1989; Predika, 1983; Price and Schweigert, 1987; Shannon, 1966; Tarr, 1975; Viscora, 1968.
Always start the salt at the lowest level indicated. The final formula must comply with the legal limit of food additives used.
Note: 1 lb = ~453.6 g, 1 oz = ~28.4 g

## 5.4.4. Summer Sausages

**Table 5.4.4.1. Formula for Thuringer**

| Ingredients | lb | oz |
|---|---|---|
| Pork trim (75% lean) | 56-58 | |
| Whole carcass beef | 42-44 | |
| Salt | 2 | 8-10 |
| Dextrose | 1-2 | |
| Ground black pepper | | 2-4 |
| Starter culture | | 1-3 |
| Whole mustard seed | | 1-3 |
| Coriander | | 1-3 |
| Sodium nitrite | | 0.1-0.3 |

Source: Ranges of value are calculated from data in the following literature: Anonymous, 1938; Gerrard, 1959; Karmas, 1977; Kutas, 1987; Ockerman, 1989; Predika, 1983; Price and Schweigert, 1987; Shannon, 1966; Tarr, 1975; Viscora, 1968.
Always start the salt at the lowest level indicated. The final formula must comply with the legal limit of food additives used.
Note: 1 lb = ~453.6 g, 1 oz = ~28.4 g

**Table 5.4.4.2. Formula for Chervelat**

| Ingredients | lb | oz |
|---|---|---|
| Beef trimmings | 74-76 | |
| Regular pork trimmings | 24-26 | |
| Salt | 2 | 13-15 |
| Sugar | 1-2 | |
| Ground black pepper | | 3-5 |
| Whole black pepper | | 1-3 |
| Sodium nitrite | | 0.1-0.3 |

Source: Ranges of value are calculated from data in the following literature: Anonymous, 1938; Gerrard, 1959; Karmas, 1977; Kutas, 1987; Ockerman, 1989; Predika, 1983; Price and Schweigert, 1987; Shannon, 1966; Tarr, 1975; Viscora, 1968.
Always start the salt at the lowest level indicated. The final formula must comply with the legal limit of food additives used.
Note: 1 lb = ~453.6 g, 1 oz = ~28.4 g

**Table 5.4.4.3. Formula for Chorizos**

| Ingredients | lb | oz |
|---|---|---|
| Pork trimmings (85% lean) | 39-41 | |
| Regular pork trimmings | 29-31 | |
| Neckbone trimmings | 29-31 | |
| Salt | 2-3 | |
| Sugar | | 8-9 |
| Sweet red peppers | | 6-7 |
| Chili powder | | 4-6 |
| Red pepper (hot) | | 2-4 |
| Sodium nitrite | | 0.1-0.3 |
| Garlic powder | | 0.1-0.3 |

Source: Ranges of value are calculated from data in the following literature: Anonymous, 1938; Gerrard, 1959; Karmas, 1977; Kutas, 1987; Ockerman, 1989; Predika, 1983; Price and Schweigert, 1987; Shannon, 1966; Tarr, 1975; Viscora, 1968.

Always start the salt at the lowest level indicated. The final formula must comply with the legal limit of food additives used.

Note: 1 lb = ~453.6 g, 1 oz = ~28.4 g

## 5.4.5. Dry Sausagees

**Table 5.4.5.1. Formula for Genoa (Beef and Pork)**

| Ingredients | lb | oz |
|---|---|---|
| Lean pork trimmings (85% lean) | 64-66 | |
| Regular pork trimmings | 34-36 | |
| Salt | 3-4 | |
| Sugar | 2-3 | |
| Burgundy wine | | 9-10 |
| White ground pepper | | 3-4 |
| Whole white pepper | | 0.5-1.5 |
| Sodium nitrate | | 0.3-0.6 |
| Garlic powder | | 0.2-0.4 |

Source: Ranges of value are calculated from data in the following literature: Anonymous, 1938; Gerrard, 1959; Karmas, 1977; Kutas, 1987; Ockerman, 1989; Predika, 1983; Price and Schweigert, 1987; Shannon, 1966; Tarr, 1975; Viscora, 1968.

**Table 5.4.5.1. (continued)**

Always start the salt at the lowest level indicated. The final formula must comply with the legal limit of food additives used.
Note: 1 lb = ~453.6 g, 1 oz = ~28.4 g

**Table 5.4.5.2. Formula for Hard Salami**

| Ingredients | lb | oz |
|---|---|---|
| Beef chucks | 39-41 | |
| Pork jowls, glands trimmed | 34-36 | |
| Regular pork trimmings | 24-26 | |
| Salt | 3-4 | |
| Sugar | 1-2 | |
| White pepper | | 3-4 |
| Sodium nitrate | | 1-3 |
| Garlic | | 0.1-0.3 |

Source: Ranges of value are calculated from data in the following literature: Anonymous, 1938; Gerrard, 1959; Karmas, 1977; Kutas, 1987; Ockerman, 1989; Predika, 1983; Price and Schweigert, 1987; Shannon, 1966; Tarr, 1975; Viscora, 1968.
Always start the salt at the lowest level indicated. The final formula must comply with the legal limit of food additives used.
Note: 1 lb = ~453.6 g, 1 oz = ~28.4 g

**Table 5.4.5.3. Formula for Pepperoni**

| Ingredients | lb | oz |
|---|---|---|
| Regular pork trimmings | 45-47 | |
| Beef chucks | 29-30 | |
| Pork hearts | 14-15 | |
| Pork cheeks | 11-12 | |
| Salt | 3-4 | |
| Sugar | 1-2 | |
| Sweet paprika | | 11-12 |
| Decorticated ground pepper | | 7-8 |
| Capsaicin | | 4-5 |
| Whole fennel seed | | 3-4 |
| Sodium nitrate | | 0.2-0.3 |

Source: Ranges of value are calculated from data in the following literature: Anonymous, 1938; Gerrard, 1959; Karmas, 1977; Kutas, 1987; Ockerman, 1989; Predika, 1983; Price and Schweigert, 1987; Shannon, 1966; Tarr, 1975; Viscora, 1968.
Always start the salt at the lowest level indicated. The final formula must comply with the legal limit of food additives used.
Note: 1 lb = ~453.6 g, 1 oz = ~28.4 g

## 5.4.6. Emulsion Sausages

**Table 5.4.6.1. Stages in Preparing Emulsion Sausages**

| | | | |
|---|---|---|---|
| 1. | Ingredients selection | 12. | Cooling or chilling |
| 2. | Size reduction: grinding | 13. | Removing skin |
| 3. | Blending and mixing | 14. | Sizing, slicing |
| 4. | Size reduction: chopping | 15. | Unit packaging |
| 5. | Blending and mixing: emulsification | 16. | Separation |
| 6. | Filling and stuffing | 17. | Reassembling |
| 7. | Closing, tying, linking | 18. | Repackaging |
| 8. | Suspension and hanging | 19. | Transporting and shipping |
| 9. | Processing, e.g., smoking | 20. | Retail display |
| 10. | Cooking, e.g., heating | 21. | Sale and marketing |
| 11. | Showering | | |

Source: Adapted from Anonymous, 1938; Gerrard, 1959; Karmas, 1977; Kutas, 1987; Ockerman, 1989; Predika, 1983; Price and Schweigert, 1987; Shannon, 1966; Tarr, 1975; Viscora, 1968.

**Table 5.4.6.2. Formula for All-Meat Bologna**

| Ingredients | lb | oz |
|---|---|---|
| Regular pork trimmings | 53-56 | |
| Beef carcass | 44-46 | |
| Ice | 24-26 | |
| Salt | 2-3 | |
| Sugar | | 9-10 |
| Ground white pepper | | 4-5 |
| Coriander | | 1-2 |
| Mace | | 1-2 |

**Table 5.4.6.2. (continued)**

| Ingredients | lb | oz |
|---|---|---|
| Sodium erythorbate | | 1-2 |
| Sodium nitrite | | 0.2-0.4 |

Source: Ranges of value are calculated from data in the following literature: Anonymous, 1938; Gerrard, 1959; Karmas, 1977; Kutas, 1987; Ockerman, 1989; Predika, 1983; Price and Schweigert, 1987; Shannon, 1966; Tarr, 1975; Viscora, 1968.
Always start the salt at the lowest level indicated. The final formula must comply with the legal limit of food additives used.
Note: 1 lb = ~453.6 g, 1 oz = ~28.4 g

**Table 5.4.6.3. Formula for Bologna from Meat and Milk Powder**

| Ingredients | lb | oz |
|---|---|---|
| Beef trimmings | 64-66 | |
| Regular pork trimmings | 44-46 | |
| Ice | 24-26 | |
| Dry skim milk | 4-5 | |
| Salt | 2-3 | |
| Sugar | 1-2 | |
| White pepper | | 2-3 |
| Coriander | | 1-2 |
| Mace | | 1-2 |
| Sodium erythorbate | | 1-2 |
| Sodium nitrite | | 0.2-0.4 |

Source: Ranges of value are calculated from data in the following literature: Anonymous, 1938; Gerrard, 1959; Karmas, 1977; Kutas, 1987; Ockerman, 1989; Predika, 1983; Price and Schweigert, 1987; Shannon, 1966; Tarr, 1975; Viscora, 1968.
Always start the salt at the lowest level indicated. The final formula must comply with the legal limit of food additives used.
Note: 1 lb = ~453.6 g, 1 oz = ~28.4 g

**Table 5.4.6.4 Formula for Meat Frankfurters**

| Ingredients | lb | oz |
|---|---|---|
| Beef carcass | 64-66 | |

| | | |
|---|---|---|
| Regular pork trimmings | 34-36 | |
| Ice | 29-31 | |
| Salt | 3-4 | |
| Corn syrup | 2-3 | |
| White pepper | | 4-5 |
| Nutmeg | | 1-2 |
| Sodium erythorbate | | 1-2 |
| Sodium nitrite | | 0.2-0.4 |

Source: Ranges of value are calculated from data in the following literature: Anonymous, 1938; Gerrard, 1959; Karmas, 1977; Kutas, 1987; Ockerman, 1989; Predika, 1983; Price and Schweigert, 1987; Shannon, 1966; Tarr, 1975; Viscora, 1968.

Always start the salt at the lowest level indicated. The final formula must comply with the legal limit of food additives used.

Note: 1 lb = ~453.6 g, 1 oz = ~28.4 g

**Table 5.4.6.5. Formula for Frankfurters from Milk Powder**

| Ingredients | lb | oz |
|---|---|---|
| Regular pork trimmings | 64-66 | |
| Pork trimmings (80% lean) | 24-26 | |
| Lean beef chucks | 9-10 | |
| Ice | 29-31 | |
| Dry skim milk | 3-4 | |
| Salt | 3-4 | |
| Pepper | | 2-3 |
| Paprika | | 2-3 |
| Capsicum | | 1-2 |
| Nutmeg | | 1-2 |
| Sodium erythorbate | | 1-2 |
| Sodium nitrite | | 0.2-0.4 |

Source: Ranges of value are calculated from data in the following literature: Anonymous, 1938; Gerrard, 1959; Karmas, 1977; Kutas, 1987; Ockerman, 1989; Predika, 1983; Price and Schweigert, 1987; Shannon, 1966; Tarr, 1975; Viscora, 1968.

Always start the salt at the lowest level indicated. The final formula must comply with the legal limit of food additives used.

Note: 1 lb = ~453.6 g, 1 oz = ~28.4 g

**Table 5.4.6.6. Formula for Bockwurst**

| Ingredients | No. | lb | oz |
|---|---|---|---|
| Pork trimmings (65% lean) | | 58-61 | |
| Veal clods | | 40-42 | |
| Fluid milk (skim or whole) | | 8-9 | |
| Ice | | 4-5 | |
| Salt | | 2-3 | |
| Sugar | | | 4-5 |
| Mace | | | 1.5 |
| Eggs | 10 | | |

Source: Ranges of value are calculated from data in the following literature: Anonymous, 1938; Gerrard, 1959; Karmas, 1977; Kutas, 1987; Ockerman, 1989; Predika, 1983; Price and Schweigert, 1987; Shannon, 1966; Tarr, 1975; Viscora, 1968.

Always start the salt at the lowest level indicated.

Note: 1 lb = ~453.6 g, 1 oz = ~28.4 g

**Table 5.4.6.7. Formula for Liver Sausage**

| Ingredients | lb | oz |
|---|---|---|
| Pork livers | 54-56 | |
| Pork jowls | 44-46 | |
| Salt | 2-3 | |
| Ground white pepper | | 4-5 |
| Coriander | | 1-2 |
| Nutmeg | | 1-2 |
| Sodium nitrite | | 0.2-0.4 |

Source: Ranges of value are calculated from data in the following literature: Anonymous, 1938; Gerrard, 1959; Karmas, 1977; Kutas, 1987; Ockerman, 1989; Predika, 1983; Price and Schweigert, 1987; Shannon, 1966; Tarr, 1975; Viscora, 1968.

Always start the salt at the lowest level indicated. The final formula must comply with the legal limit of food additives used.

Note: 1 lb = ~453.6 g, 1 oz = ~28.4 g

## 5.4.7. Other Sausages

**Table 5.4.7.1. Formula for Scrapple**

| Ingredients | lb | oz |
|---|---|---|
| Cured pork head meat | 58-62 | |
| Cured pork skins | 24-26 | |
| Pork livers | 14-16 | |
| Water | 64-66 | |
| Corn meal | 34-36 | |
| Salt | 2-3 | |
| Pepper | | 4-5 |
| Sage | | 2-3 |
| Nutmeg | | 1-2 |

Source: Ranges of value are calculated from data in the following literature: Anonymous, 1938; Gerrard, 1959; Karmas, 1977; Kutas, 1987; Ockerman, 1989; Predika, 1983; Price and Schweigert, 1987; Shannon, 1966; Tarr, 1975; Viscora, 1968.
Always start the salt at the lowest level indicated.
Note: 1 lb = ~453.6 g, 1 oz = ~28.4 g

**Table 5.4.7.2. Formula for Headcheese**

| Ingredients | lb | oz |
|---|---|---|
| Cured pork snouts (white pigs) | 24-26 | |
| Cured pork lips | 19-21 | |
| Cured pork hearts | 15-16 | |
| Cured pork cheek meat | 15-16 | |
| Cured pork skins | 15-16 | |
| Cured pork ears | 10-11 | |
| Broth from cooking water | 44-46 | |
| Onions | 5-6 | |
| Gelatin | 4-5 | |
| Vinegar | 2-3 | |
| Salt | 1-2 | |
| Mustard | | 8-9 |
| White pepper | | 4-5 |

Source: Ranges of value are calculated from data in the following literature: Anonymous, 1938; Gerrard, 1959; Karmas, 1977; Kutas, 1987; Ockerman, 1989;

**Table 5.4.7.2 (continued)**

Predika, 1983; Price and Schweigert, 1987; Shannon, 1966; Tarr, 1975; Viscora, 1968.
Always start the salt at the lowest level indicated.
Note: 1 lb = ~453.6 g, 1 oz = ~28.4 g

**Table 5.4.7.3. Formula for Beef Mortadella**

| Ingredients | lb | oz |
|---|---|---|
| Beef trimmings (85-90% lean) | 79-81 | |
| Beef navels (50% lean) | 19-21 | |
| Ice | 14-16 | |
| Salt | 2-3 | |
| Sugar | 1-2 | |
| Pepper (black) | | 4-5 |
| Ginger | | 1-2 |
| Mace | | 1-2 |
| Coriander | | 1-2 |
| Sodium nitrate | | 0.4-0.7 |
| Sodium nitrite | | 0.2-0.4 |
| Sodium ascorbate | | 0.6-0.9 |

Source: Ranges of value are calculated from data in the following literature: Anonymous, 1938; Gerrard, 1959; Karmas, 1977; Kutas, 1987; Ockerman, 1989; Predika, 1983; Price and Schweigert, 1987; Shannon, 1966; Tarr, 1975; Viscora, 1968.
Always start the salt at the lowest level indicated. The final formula must comply with the legal limit of food additives used.
Note: 1 lb = ~453.6 g, 1 oz = ~28.4 g

## 5.4.8. Specialty Items

**Table 5.4.8.1. Formula for Veal Loaf**

| Ingredients | lb | oz |
|---|---|---|
| Lean veal trimmings | 41-43 | |
| Regular pork trimmings | 34-36 | |
| Cooked tripe | 24-26 | |
| Ice | 44-46 | |

| | | |
|---|---|---|
| Dry skim milk | 11-13 | |
| Fresh chopped onions | 4-5 | |
| Salt | 2-3 | |
| Mustard flour | 1-2 | |
| White pepper | | 4-5 |
| Lemon extract | | 2-3 |
| Garlic powder | | 1-2 |
| Sodium nitrite | | 0.1-0.3 |

Source: Ranges of value are calculated from data in the following literature: Anonymous, 1938; Anonymous, 1973; Anonymous, 1983; Anonymous, 1986; Anonymous, 1988; Ashbrook, 1973; Bloch, 1976; Gerrard, 1959; Karmas, 1977, Kutas, 1987; Levie, 1979; Ockerman, 1989; Predika, 1983; Price and Schweigert, 1987; Shannon, 1966; Tarr, 1975; Viscora, 1968.
Always start the salt at the lowest level indicated. The final formula must comply with the legal limit of food additives used.
Note: 1 lb = ~453.6 g, 1 oz = ~28.4 g

**Table 5.4.8.2. Formula for Corned Beef Loaf**

| Ingredients | lb | oz |
|---|---|---|
| Lean beef | 119-121 | |
| Ice | 7-8 | |
| Salt | 4-5 | |
| Sugar | 1-2 | |
| Sodium nitrite | | 0.1-0.3 |

Source: Ranges of value are calculated from data in the following literature: Anonymous, 1973; Anonymous, 1983; Anonymous, 1986; Anonymous, 1988; Ashbrook, 1973; Bloch, 1976; Karmas, 1977; Kutas, 1987; Levie, 1979; Ockerman, 1989; Predika, 1983; Price and Schweigert, 1987; Shannon, 1966; Tarr, 1975.
Always start the salt at the lowest level indicated. The final formula must comply with the legal limit of food additives used.
Note: 1 lb = ~453.6 g, 1 oz = ~28.4 g

**Table 5.4.8.3. Formula for Kiszka**

| Ingredients | lb | oz |
|---|---|---|
| Sweet pickle ham skins | 46-48 | |
| Cured pork snouts | 19-21 | |
| Pork liver | 19-21 | |
| Cured lips | 5-6 | |
| Tripe | 5-6 | |
| Cured beef blood | 5-6 | |
| Water | 47-49 | |
| Barley | 22-24 | |
| Onions | 5-6 | |
| Salt | 1-2 | |
| Pepper | | 5-6 |
| Thyme | | 1-2 |
| Nutmeg | | 1-2 |
| Marjoram | | 1-2 |

Source: Ranges of value are calculated from data in the following literature: Anonymous, 1973; Anonymous, 1983; Anonymous, 1986; Anonymous, 1988; Ashbrook, 1973; Bloch, 1976; Karmas, 1977; Kutas, 1987; Ockerman, 1989; Predika, 1983; Price and Schweigert, 1987; Shannon, 1966; Tarr, 1975.
Note: 1 lb = ~453.6 g, 1 oz = ~28.4 g

**Table 5.4.8.4. Formula for Olive Loaf**

| Ingredients | lb | oz |
|---|---|---|
| Lean veal | 44-46 | |
| Pork trimmings (50% lean) | 44-46 | |
| Beef trimmings (85% lean) | 9-11 | |
| Ice | 44-46 | |
| Dry skim milk | 12-14 | |
| Stuffed olives | 6-7 | |
| Salt | 2-3 | |
| Sugar | 1-2 | |
| White pepper | | 4-5 |
| Marjoram | | 1-2 |
| Sodium nitrite | | 0.2-0.4 |

Source: Ranges of value are calculated from data in the following literature: Anonymous, 1973; Anonymous, 1983; Anonymous, 1986; Anonymous, 1988;

Karmas, 1977; Kutas, 1987; Ockerman, 1989; Predika, 1983; Price and Schweigert, 1987; Shannon, 1966; Tarr, 1975.
Always start the salt at the lowest level indicated. The final formula must comply with the legal limit of food additives used.
Note: 1 lb = ~453.6 g, 1 oz = ~28.4 g

**Table 5.4.8.5. Formula for Pickle and Pimiento Loaf**

| Ingredients | lb | oz |
|---|---|---|
| Pork trimmings (70% lean) | 54-56 | |
| Veal trimmings | 34-36 | |
| Beef trimmings (85% lean) | 9-11 | |
| Ice | 39-41 | |
| Dry skim milk | 14-16 | |
| Sweet pickles | 5-6 | |
| Pimientos | 5-6 | |
| Fresh onions | 5-6 | |
| Salt | 2-3 | |
| Dextrose | 1-2 | |
| White pepper | | 4-5 |
| Marjoram | | 1-2 |
| Sodium erythorbate | | 0.8-1.0 |
| Sodium nitrite | | 0.2-0.4 |

Source: Ranges of value are calculated from data in the following literature: Anonymous, 1973; Anonymous, 1983; Anonymous, 1986; Anonymous, 1988; Karmas, 1977; Kutas, 1987; Ockerman, 1989; Predika, 1983; Price and Schweigert, 1987; Shannon, 1966; Tarr, 1975.
Always start the salt at the lowest level indicated. The final formula must comply with the legal limit of food additives used.
Note: 1 lb = ~453.6 g, 1 oz = ~28.4 g

## 5.5. USDA Defect Criteria for Meat from Animal Carcasses

**Table 5.5.1. USDA Defect Criteria for Meat from Cattle, Calves, Sheep, Goats, and Equines[a]**

Blood clots
- Insignificant: less than 1-1/2" in greatest dimension
- Minor: 1-1/2" to 6" in greatest dimension
- Major: more than 6" in greatest dimension, or numerous (over 5) minor blood clots in one sample unit not seriously affecting product usability
- Critical: one or more of a number or size seriously affecting product usability

Bruises
- Insignificant: less than 1" in greatest dimension and less than 1/2" deep
- Minor: 1" to 2-1/2" in greatest dimension or 1/2" to 1" deep
- Major: more than 2-1/2" in greatest dimension or more than 1" deep, or numerous (over 5) minor bruises in one sample unit not seriously affecting product usability
- Critical: one or more of a number or size seriously affecting product usability

Bone fragments
- Insignificant defects:
  - Thin bone scrapings less than 1/32" thick x 1/8" wide x 3" long attached to muscle tissue
  - Thin flexible bone slivers, either attached to or detached from muscle tissue less than 1/4" wide and 3/4" long
  - Thin flexible fragments or chips, either attached to or detached from muscle tissue, that crumble easily and are less than 3/4" in greatest dimension
- Minor: less than 1-1/2" in greatest dimension
- Major: 1-1/2" or more in greatest dimension, or numerous (over 5) minor fragments in one sample unit not seriously affecting product usability
- Critical: one or more of a number or size seriously affecting product usability

Bone slivers (from rib)
- Minor: less than 3" long and less than 1/4" wide and flexible bone chip from a rib and more than 3/4" in greatest dimension that is thin and crumbles easily, and with or without attached muscle tissue

Detached cartilage, ligaments
- Insignificant: less than 1" long
- Minor: 1" or more long and free of muscle tissue (also see bone slivers)

Major: numerous (over 5) minor defects in one sample unit not seriously affecting product usability

Critical: defects of number seriously affecting product usability

Extraneous material

Insignificant defects:

Minute specks of dust. May affect product usability.

Pieces of plastic or paper wraps or any soft material less than 1/2"

Minor: paper or plastic wraps 1/2" to 7 square inches; a single piece covering an area equal to that of a circle 1/8" to 1/2" in diameter; a wild oat or other grass beard over 3/8" long or 3 or more pieces of wild oat or grass beards 1/8" to 3/8" long on one meat piece and without inflammation

Major defects:

Blunt piece of wood 1" or more long; paper or plastic over 7 square inches; single piece of material covering an area greater than that of a circle with a diameter exceeding 1/2"; insects without insanitation

Numerous (over 5) minor defects in a sample unit not seriously affecting product usability

Any substance causing minor bodily irritation or discomfort (Chemicals, hard objects, etc.)

Critical defects:

Any substance causing injury or illness (poisonous or toxic chemicals, sharp pieces of metal, glass, hard plastic, etc.)

Large insects, insects associated with insanitation, or any material of number or size seriously affecting product usability

Hair, hide, wool

Minor defects

Hide (with or without hair) or wool less than 1/2" in greatest dimension. A total of five to 10 single strands of hair or wool.

Total number of hairs, divide by 10, and round off to nearest whole number to determine total hair defects. For example: 34 hairs equal 3 defects and 35 hairs equal 4 defects. When a second step is necessary, total number of hairs in step one and two, divide by 10, and round off to nearest whole number as described above.

A cluster of hair (strands too numerous to count) in one area.

**Table 5.5.1. (continued)**

Hide (with or without hair) or wool 1/2" or more in greatest dimension; numerous (over 25) single strands of hair in one sample unit; numerous (over 5) clusters of hair in one sample unit, provided none of the above seriously affects product usability

Critical: hair, hide or wool of amount seriously affecting product usability

Ingesta

Minor: amount equal to area of a circle 1/2" or less in diameter

Critical: amount equal to area of a circle more than 1/2" in diameter

Off condition

Critical: any off condition

Parasitic lesions

Minor: parasites not transmissible to man; one, two, or three closely associated lesions on one piece of meat (score as one lesion, ovine only); first lesion found in a sample.

Major: each succeeding parasitic lesion in the sample

Pathologic lesions

Major: any lesion (not evident on postmortem inspection) not seriously affecting product acceptability

Major: any lesion unless excepted by USDA regulations

Stains, discolored areas

Insignificant: very light stains of any size or stains covering an area less than that of a circle 1/2" in diameter

Minor: equal to area of a circle 1/2" to 1-1/2"

Major: equal to area of a circle greater than 1-1/2" in diameter; numerous (over 5) minor stains in one sample unit (12 pounds) not seriously affecting product usability

Critical: minor or major areas of a number seriously affecting product usability

Other

Minor: defect that individually or in aggregate affects product appearance, but not its usability

Major: defect that individually or in aggregate materially affects product usability

Critical: defect that individually or in aggregate seriously affects appearance or usability of product

Source: USDA Meat and Poultry Inspection Manual, 1991.

[a]Each defect can be classified in one of four ways: insignificant, minor, major, or critical. For each classification, there can be more than one type of defects.

**Table 5.5.2. USDA Defect Criteria for Meat from Swine Carcasses[a]**

Blood clots
- Insignificant: less than 1-1/2" in greatest dimension
- Minor: 1-1/2" to 6" in greatest dimension
- Major: more than 6" in greatest dimension, or numerous (over 5) minor blood clots in one sample unit not seriously affecting product usability
- Critical: one or more of a number or size seriously affecting product usability

Bruises
- Insignificant: less than 1" in greatest dimension and less than 1/2" deep
- Minor: 1" to 2-1/2" in greatest dimension or 1/2" to 1" deep
- Major: more than 2-1/2" in greatest dimension or more than 1" deep, or numerous (over 5) minor bruises in one sample unit not seriously affecting product usability
- Critical: one or more of a number or size seriously affecting product usability

Bone fragments
- Insignificant:
  - Thin bone scrapings less than 1/32" thick x 1/8" wide x 3" long attached to or detached from muscle tissue
  - Thin flexible bone slivers, either attached to or detached from muscle tissue and less than 1/4" wide and 3/4" long
  - Thin flexible fragments or chips either attached to or detached from muscle tissue that crumble easily and are less than 3/4" in greatest dimension
- Minor: less than 1-1/2" in greatest dimension
- Major: 1-1/2" or more in greatest dimension, or numerous (over 5) minor fragments in one sample unit not seriously affecting product usability
- Critical: one or more of a number or size seriously affecting product usability

Bone slivers (from rib)
- Minor: less than 3" long and less than 1/4" wide and flexible bone chip from a rib and more than 3/4" in greatest dimension that is thin and crumbles easily, and with or without attached muscle tissue

Detached cartilage, ligaments
- Insignificant: less than 1" long
- Minor: 1" or more long and free of muscle tissue (also see bone slivers)
- Major: numerous (over 5) minor defects in one sample unit not seriously affecting product usability
- Critical: defects of number seriously affecting product usability

**Table 5.5.2. (continued)**

Extraneous material

- Insignificant:
  - Minute specks of dust. May affect product usability.
  - Pieces of plastic or paper wraps or any soft material less than 1/2"
- Minor: paper or plastic wraps 1/2" to 7 sq. in.; a single piece covering an area equal to that of a circle 1/8" to 1/2" in diameter; a wild oat or other grass beard over 3/8" long or 3 or more pieces of wild oat or grass beards 1/8" to 3/8" long on one meat piece and without inflammation
- Major:
  - Blunt piece of wood 1" or more long; paper or plastic over 7 square inches; single piece of material covering an area greater than that of a circle with a diameter exceeding 1/2"; insects without insanitation
  - Numerous (over 5) minor defects in a sample unit not seriously affecting product usability
  - Any substance causing minor bodily irritation or discomfort (Chemicals, hard objects, etc.)
- Critical:
  - Any substance causing injury or illness (poisonous or toxic chemicals, sharp pieces of metal, glass, hard plastic, etc.)
  - Large insects, insects associated with insanitation, or any material of number or size seriously affecting product usability

Skin, hair, hair roots

- Insignificant: skin (with or without hair or visible hair roots) individually or in aggregate less than 1 square inch
- Minor: skin (with or without hair or visible hair roots) individually or in the aggregate 1 sq. in. to 3 sq. in. A total of 2 or 3 single strands of hair or 5-10 visible hair roots. Total number of hairs or visible hair roots in sample divided by 3 for hairs or 10 for visible hair roots and round off to nearest whole number. For example, 10 hairs equal 3 defects, 38 visible hair roots equal 4 defects. When a second step is necessary, total the hair or visible hair roots from both steps. Also, cluster of hair or visible hair roots (strands too numerous to count) in one area.
- Major: skin with or without hair or visible hair roots individually or in aggregate over 3 square inches; numerous (over 13), provided none of above seriously affect product usability
- Critical: hair, skin, or visible hair roots seriously affecting product usability

Ingesta

- Minor: amount equal to area of a circle 1/2" or less in diameter

Critical: amount equal to area of a circle more than 1/2" in diameter

Off condition

Critical: any off condition

Lips, ear canals, teeth, kidney, liver

Major: any sample unit containing tooth or teeth, ear canal(s), lip with or without teeth marks, piece (s) of kidney or liver

Pathologic lesions

Major: any lesion (not evident on postmortem inspection) not seriously affecting product acceptability

Critical: any lesion unless excepted by USDA regulations

Stains, discolored areas

Insignificant: very light stains of any size or stains covering an area less than that of a circle 1/2" in diameter

Minor: equal to area of a circle 1/2" to 1-1/2"

Major: equal to area of a circle greater than 1-1/2" in diameter; numerous (over 5) minor stains in one sample unit (12 pounds) not seriously affecting product usability

Critical: minor or major areas of a number seriously affecting product usability

Lung tissue

Critical: any amount

Other

Minor: defect that individually or in aggregate affects product appearance, but not its usability

Major: defect that individually or in aggregate materially affects product usability

Critical: defect that individually or in aggregate seriously affects appearance or usability of product

---

Source: USDA Meat and Poultry Inspection Manual, 1991.

[a]Each defect can be classified in one of four ways: insignificant, minor, major, or critical. For each classification, there can be more than one type of defects.

## 5.6. Definitions for Meat And Poultry Labeling

**Table 5.6.1. Meat and Poultry Labeling: Geographical and Seasonal Terms**

Geographical Terms

1. Terms having geographical significance with reference to a locality other than that in which the product is prepared may appear on the label only when qualified by the word "style," "type," or"brand," as the case may be.
2. They must be accompanied by a prominent qualifying statement identifying the country, state, territory, or locality in which the product is prepared, using terms appropriate to effect the qualification.
3. When the word "style" or "type" is used, there must be a recognized style or type of product identified with and peculiar to the area represented by the geographical term and the product must possess the characteristics of such style or type.
4. A geographical term that has come into general usage as a trade name may be used without the qualifications.
5. Such terms include "Frankfurter," "Vienna," "Bologna," "Lebanon Bologna," "Braunschweiger," "Thuringer," "Genoa," "Leona," "Berliner," "Holstein," "Goteborg," "Milan," "Polish," "Italian," and "Irish" as applied to stews and the term "Boston" as applied to pork shoulder butts.
4. They need not be accompanied with the word "style," "type," or "brand," or a statement identifying the locality in which the product is prepared.
5. Such terms as "farm" or "country" cannot be used on labels in connection with products unless such products are actually prepared on the farm or in the country. However, if the product is prepared in the same way as on the farm or in the country these terms, if qualified by the word "style," may be used. Also, the term "farm" may be used as part of a brand designation when qualified by the word "brand," and followed with a statement identifying the locality in which the product is prepared. However, sausage containing cereal should not be labeled "farm style" or "country style," and lard not rendered in an open kettle should not be designated as "farm style" or "country style."

Spring Lamb or Genuine Spring Lamb

The term "spring lamb" or "genuine spring lamb" is applicable only to carcasses of new-crop lambs slaughtered during the period beginning in March and terminating not beyond the close of the week containing the first Monday in October.

Source: 9 CFR 317.8, 1991.

**Table 5.6.2. Meat and Poultry Labeling: General Definitions**

Agar-agar: When agar-agar is used in canned jellied meat food products, there should appear on the label in a prominent manner, contiguous to the product name, a statement to indicate the use of agar-agar.

Baked: It applies only to product that has been cooked by the direct action of dry heat and for a sufficient time to permit the product to assume the characteristics of a baked article, such as the formation of a brown crust on the surface, rendering out of surface fat, and the caramelization of the sugar if applied. Baked loaves shall be heated to a temperature of at least 160°F and baked pork cuts shall be heated to an internal temperature of at least 170°F.

Blood: Products that contain blood from livestock as permitted by regulations should be labeled with a name that includes the term "blood," and the specific kind of blood should be declared in the ingredient statement, e.g., "swine blood," in the manner required by regulations.

Bones: Cooked, cured, or pickled pigs feet, pigs knuckles, and similar products should be labeled to show that the bones remain in the product, if such is the case. The designation "semi-boneless," should not be used if less than 50% of the total weight of bones has been removed.

Browned: When products such as loaves are browned by dipping in hot edible oil or by a flame, the label should state such a fact, e.g., by the words "browned in hot cottonseed oil" or "browned by a flame," as the case may be, appearing as part of the product name.

Chemical additives and dry sausages: Sausage of the dry varieties treated with potassium sorbate or propylparaben (propyl *p*-hydroxybenzoate) should be marked or labeled with a statement disclosing such treatment and the purpose thereof, such as "dipped in a potassium sorbate solution to retard mold growth."

Chemical additives and fresh porks: When ascorbic acid, erythorbic acid, citric acid, sodium ascorbate, and sodium citrate, singly or in combination, are added to fresh pork cuts, there should appear on the label of that product, in letters of the same style and type and not less than one fourth the size of letters in the product name, contiguous to the name of the product, a statement identifying the specific approved substance(s) by its common name and the purpose for which it is added, such as, "sprayed with a solution of water, ascorbic acid and citric acid to maintain color."

Chemical additives and restructured meat: When sodium alginate, calcium carbonate, lactic acid and calcium carbonate (or glucono delta-lactone) are used together in a dry binding matrix in restructured, formed meat food products, there should appear on the label, contiguous to the product

**Table 5.6.2. (continued)**

name, a statement to indicate the use of sodium alginate, calcium carbonate and lactic acid and calcium carbonate (or glucono delta-lactone).

Chitterlings: This term should apply to the large intestines of swine, or young bovine animals when preceded with the word "calf" or "veal." Meat food products that contain chitterlings or calf or veal chitterlings should be identified with product names that refer to such ingredients, as for instance, "chitterling loaf," "chitterling pie," or "calf chitterlings and gravy," and should be packed in containers having a capacity of 3 pounds or less and of a kind usually sold at retail intact and bearing such other required information.

Date: A calendar date may be shown on labeling when declared in accordance with the following:

1. The calendar date should express the month of the year and the day of the month for all products and also the year in the case of products hermetically sealed in metal or glass containers, dried or frozen products, or other required products.
2. Immediately adjacent to the calendar date should be a phrase explaining the meaning of such date, in terms of "packing" date, "sell by" date, or "use before" date, with or without a further qualifying phrase, e.g., "for maximum freshness" or "for best quality," and such phrases must be approved before use.

Dimethylpolysiloxan: When dimethylpolysiloxan is added as an antifoaming agent to rendered fats, its presence should be declared on the label contiguous to the name of the product. Such declaration should read "Dimethylpolysiloxan added."

Fat or oil additives:

1. When <20% beef fat, mutton fat, oleo stearin, vegetable stearin, or hardened vegetable fat is mixed with lard or with rendered pork fat, there should appear on the label the words "beef fat added," "mutton fat added," "oleo stearin, added," "vegetable stearin added," or "hardened vegetable fat added," as the case may be.
2. If >20% is added, the product name should refer to the particular animal fat or fats used, such as, "lard and beef fat." The designation "vegetable fat" is applicable to vegetable oil, vegetable stearin, or a combination of such oil and stearin, whereas the designations "vegetable oil" and "vegetable stearin" should be applicable only to the oil and the stearin respectively, when used in meat food products.

Fatty acid esters: When monoglycerides, diglycerides, and/or polyglycerol esters of fatty acids are added to rendered animal fat or a combinaiton of such fat and vegetable fat, there should appear on the label in a prominent manner and contiguous to the name of the product a statement such as "with monoglycerides and diglycerides added," or with diglycerides and monoglycerides," or "with polyglycerol esters of fatty acids" as the case may be.

Fresh: It should not be used on labels to designate product that contains any sodium nitrate, sodium nitrite, potassium nitrate, or potassium nitrite, or that has been salted for preservation.

Gelatin: It means (i) the jelly prepared in official establishments by cooking pork skins, tendons, or connective tissue from inspected and passed product, and (ii) dry commercial gelatin or the jelly resulting from its use.

Goat: Meat of goats should be identified as goat meat or chevon.

Ham: The term "ham" without any prefix indicating the species of animal from which derived, should be used in labeling only in connection with the hind legs of swine. Ham shanks as such or ham shank meat as such or the trimmings accruing in the trimming and shaping of hams should not be labeled "ham" or "ham meat" without qualification. When used in connection with a chopped product the term "ham" or "ham meat" should not include the skin.

Inorganic chlorides: When approved inorganic chlorides are used on steaks or other raw meat cuts there should appear on the label in a prominent manner, contiguous to the product name, the statement, "tenderized with [names of approved inorganic chloride(s)] to indicate the use of such inorganic chlorides. Any other approved substance that may be in the solution should also be included in the statement.

Leaf lard: It is lard prepared from fresh leaf fat.

Loaf: If a noncanned product contains the term "loaf" in its label, its shape must be one of the following:

1. If distributed from the official establishment in consumer size containers, it may be in any shape.
2. If distributed in a container of a size larger than that sold intact at retail the product should be prepared in rectangular form, or other approved form.
3. If labeled as an "old-fashioned loaf," it should be prepared in a traditional form, such as rectangular with rounded top or circular with flat bottom and rounded top.

Meat extract or extract of beef: Such terms without qualification should not be used on labels in connection with products prepared from organs or other parts of the carcass, other than fresh meat. Extracts prepared from any parts of the carcass other than fresh meat may be properly labeled as extracts with the true name of the parts from which prepared. In the case

**Table 5.6.2. (continued)**

of extract in fluid form, the word "fluid" should also appear on the label, as, for example, "fluid extract of beef."

Packaging materials: When any product is enclosed in a container along with a packing substance such as brine, vinegar, or agar jelly, a declaration of the packing substance should be printed prominently on the label as part of the name of the product, as for example, "frankfurts packed in brine," "lamb tongue packed in vinegar," or "beef tongue packed in agar jelly," as the case may be. The packing substance should not be used in such a manner as will result in the container being so filled as to be misleading.

Propionate (calcium or sodium): When pizzas are formulated with crust containing calcium propionate or sodium propionate, there should appear on the label contiguous to the name of the product the statement "_____added to retard spoilage of crust," preceded by the name of the preservative.

Proteolytic enzymes: When approved proteolytic enzymes are used on steaks or other raw meat cuts, there should appear on the label, in a prominent manner, contiguous to the product name, the statement, "tenderized with (approved enzyme)," to indicate the use of such enzymes. Any other approved substance that may be used in the solution should also be included in the statement.

Rendered: When lard or hardened lard is mixed with rendered pork fat or hardened rendered pork fat, the mixture should be designated "rendered pork fat" or "hardened rendered pork fat," as the case may be. Oil, stearin, or stock obtained from beef or mutton fats rendered at a temperature above 170°F should not be designated as "oleo oil," "oleo stearin," or "oleo stock," respectively.

Shankless and hockless: The terms "shankless" and "hockless" should apply only to hams and pork shoulders from which the shank or hock has been completely removed, thus eliminating the entire tibia and fibula, or radius and ulna, respectively, together with the overlying muscle, skin, and other tissue.

Single ingredient: The terms "all," "pure," "100%," and terms of similar connotation should not be used on labels for products to identify ingredient content, unless the product is prepared solely from a single ingredient.

Vegetable additives for bockwurst: When bread, cereal, vegetable starch, starchy vegetable flour, soy flour, soy protein concentrate, dry or dried whey, reduced-lactose whey, reduced-minerals whey, whey protein concentrate, or isolated soy protein is added in bockwurst, there should appear on the label in a prominent manner and contiguous to the product name, the name of such added ingredient, e.g., "bread added," "cereal added," or "soy protein concentrate added," as the case may be.

Vegetable additives for sausages: When cereal, vegetable starch, starchy vegetable flour, soy flour, soy protein concentrate, isolated soy protein, dried milk, nonfat dry milk, dry or dried whey, reduced-lactose whey, reduced-minerals whey, whey protein concentrate, or calcium-reduced dried skim milk is added to sausage within the prescribed limits, there should appear on the label in a prominent manner, contiguous to the name of the product, the name of each such added ingredient, as, for example, "cereal added," "with cereal," "potato flour added," "cereal and potato flour added," "soy flour added," "soy protein concentrate added," "isolated soy protein added," "nonfat dry milk added," "calcium reduced dried skim milk added," or "cereal and nonfat dry milk added," as the case may be.

Source: 9 CFR 317.8, 1991.

## 5.7. Legal Requirements for Food Additives Used in Meat Processing

**Table 5.7.1. Legal Requirements for Acidifiers, Anticoagulants, and Antifoaming Agents Permitted in Meat Processing**

Specific Chemical or Group of Chemicals: Purpose
Products: Amounts

Acidifiers

Acetic acid: to adjust acidity
Various products: sufficient to purpose

Citric acid: to adjust acidity
Various products: sufficient to purpose

Lactic acid: to adjust acidity
Various products: sufficient to purpose

Phosphoric acid: to adjust acidity
Various products: sufficient to purpose

Tartaric acid: to adjust acidity
Various products: sufficient to purpose

Anticoagulants

Citric acid, sodium citrate: to prevent clotting
Fresh blood of livestock: 0.2% with or without water. When water is used to make a solution of citric acid or sodium citrate added to blood of

**Table 5.7.1. (continued)**

| Specific Chemical or Group of Chemicals: Purpose<br>Products: Amounts |
|---|
| livestock, not more than 2 parts of water to 1 part of citic acid or sodium citrate shall be used. |
| Antifoaming Agents |
| Methyl polysilicone: to retard foaming |
| Soups, rendered fats: 10 ppm |
| Curing pickle: 50 ppm |

Source: 9 CFR 318.7, 1991.

**Table 5.7.2. Legal Requirements for Antioxidants and Oxygen Interceptors Permitted in Meat Processing**

| Specific Chemical or Group of Chemicals: Purpose<br>Products: Amounts |
|---|
| BHA (butylated hydroxyanisole): to retard rancidity |
| Dry sausage: (1) 0.003% based on total weight; (2) 0.006% in combination |
| BHT (butylated hydroxytoluene): to retard rancidity |
| Dry sausage: 0.003% based on total weight |
| Propyl gallate: to retard rancidity |
| Dry sausage: 0.003% based on total weight |
| TBHQ (tertiary butylhydroquinone): to retard rancidity |
| Dry sausage: 0.003% based on total weight |
| BHA (butylated hydroxyanisole): to retard rancidity |
| Rendered animal fat or a combination of such fat and vegetable fat: (1) 0.01%; (2) 0.02% in combination |
| BHT (butylated hydroxyanisole): to retard rancidity |
| Rendered animal fat or a combination of such fat and vegetable fat: (1) 0.01%; (2) 0.02% in combination |
| Glycine: to retard rancidity |
| Rendered animal fat or a combination of such fat and vegetable fat: (1) 0.01%; (2) 0.02% in combination |
| Propyl gallate: to retard rancidity |
| Rendered animal fat or a combination of such fat and vegetable fat: (1) 0.01%; (2) 0.02% in combination |
| Resin guaiac: to retard rancidity |

Rendered animal fat or a combination of such fat and vegetable fat: (1) 0.01%; (2) 0.02% in combination

TBHQ (tertiary butylhydroquinone): to retard rancidity

Rendered animal fat or a combination of such fat and vegetable fat: (1) 0.01%; (2) 0.02% in combination only with BHA and/or BHT

Tocopherols: to retard rancidity

Rendered animal fat or a combination of such fat and vegetable fat: 0.03%. A 30% concentration of tocopherols in vegetable oils shall be used when added as an antioxidant to products designatged as "lard" or "rendered pork fat."

BHA (butylated hydroxyanisole): to retard rancidity

Fresh pork, sausage, brown and serve sausages, fresh Italian sausage products, pregrilled beef patties, fresh sausage made from beef or beef and pork, cooked or raw pizza topping, and cooked or raw meatballs: (1) 0.01% based on fat content; (2) 0.02% in combination based on fat content

BHT (butylated hydroxytoluene): to retard rancidity

Fresh pork, sausage, brown-and-serve sausages, fresh Italian sausage products, pregrilled beef patties, fresh sausage made from beef or beef and pork, cooked or raw pizza topping, and cooked or raw meatballs: 0.01% based on fat content

Propyl gallate: to retard rancidity

Fresh pork, sausage, brown-and-serve sausages, fresh Italian sausage products, pregrilled beef patties, fresh sausage made from beef or beef and pork, cooked or raw pizza topping, and cooked or raw meatballs: 0.01% based on fat content

TBHQ (tertiary butylhydroquinone): To retard rancidity

Fresh pork, sausage, brown and serve sausages, fresh Italian sausage products, pregrilled beef patties, fresh sausage made from beef or beef and pork, cooked or raw pizza topping, and cooked or raw meatballs: (1) 0.01% based on fat content; (2) 0.02% in combination only with BHA and/or BHT based on fat content

BHA (butylated hydroxyanisole): to retard rancidity

Dried meats: (1) 0.01% based on total weight; (2) 0.01% in combination

BHT (butylated hydroxytoluene): to retard rancidity

Dried meats: 0.01% based on total weight

BHT (butylated hydroxytoluene): to retard rancidity

Dried meats: 0.01% based on total weight

Propyl gallate: to retard rancidity

Dried meats: 0.01% based on total weight

TBHQ (tertiary butylhydroquinone): to retard rancidity

**Table 5.7.2. (continued)**

| Specific Chemical or Group of Chemicals: Purpose<br>Products: Amounts |
|---|
| Dried meats: (1) 0.01% based on total weight; (2) 0.01% in combination only with BHA and/or BHT |
| BHA (butylated hydroxyanisole): to retard rancidity |
| Margarine or oleomargarine: 0.02% (by weight of the finished product) individually or in combination with other antioxidants approved for use in margarine |
| BHT (butylated hydroxytoluene):to retard rancidity |
| Margarine or oleomargarine: 0.02% (by weight of the finished product) individually or in combination with other antioxidants approved for use in margarine |
| Octyl gallate: to retard rancidity |
| Margarine or oleomargarine: 0.02% (by weight of the finished product) individually or in combination with other antioxidants approved for use in margarine |
| Propyl gallate: to retard rancidity |
| Margarine or oleomargarine: 0.02% (by weight of the finished product) individually or in combination with other antioxidants approved for use in margarine |
| Dodecyl gallate: to retard rancidity |
| Margarine or oleomargarine: 0.02% (by weight of the finished product) individually or in combination with other antioxidants approved for use in margarine |
| Ascorbyl palmitate: to retard rancidity |
| Margarine or oleomargarine: 0.02% (by weight of the finished product) individually or in combination with other antioxidants approved for use in margarine |
| Ascorbyl stearate: to retard rancidity |
| Margarine or oleomargarine: 0.02% (by weight of the finished product) individually or in combination with other antioxidants approved for use in margarine |
| TBHQ (tertiary butylhydroquinone): to retard rancidity |
| Margarine or oleomargarine: 0.02% alone or in combination only with BHA and/or BHT based on fat or oil content |

Source: 9 CFR 318.7, 1991.

**Table 5.7.3. Legal Requirements for Binders Permitted in Meat Processing**

Specific Chemical or Group of Chemicals: Purpose
Products: Amounts

Agar-agar: to stabilize and thicken
Thermally processed canned jellied meat food products: 0.25% of finished product

Algin: to extend and stabilize product
Breading mix, sauces: sufficient for purpose

A mixture of sodium alginate, calcium carbonate, lactic acid, and calcium lactate: to bind meat pieces
Restructured meat food products: Sodium alginate at 1.0% calcium carbonate at 0.2% and lactic acid and calcium lactate at 0.3% of product formulation. Added mixture may not exceed 1.5% of product at formulation. Ingredients of mixture must be added dry.

Carrageenan: to bind meat pieces
Restructured meat food products: Sodium alginate at 1.0% calcium carbonate at 0.2% and lactic acid and calcium lactate at 0.3% of product formulation. Added mixture may not exceed 1.5% of product at formulation. Ingredients of mixture must be added dry.

Carboxymethylcellulose (cellulose gum): to bind meat pieces
Baked pies: Sodium alginate at 1.0% calcium carbonate at 0.2% and lactic acid and calcium lactate at 0.3% of product formulation. Added mixture may not exceed 1.5% of product at formulation. Ingredients of mixture must be added dry.

Enzyme (rennet)-treated calcium-reduced dried skim milk and calcium lactate: to bind and extend product
Sausages, as provided for in 9 CFR 319: 3.5% total finished product. Calcium lactate required at a rate of 10% of binder.
Imitation sausages, nonspecific loaves, soups, stews: sufficient for purpose. Calcium lactate required at a rate of 25% of binder.

Enzyme (rennet)-treated sodium caseiante and calcium lactate: to bind and extend product
Imitation sausages, nonspecific loaves, soups, stews: sufficient for purpose. Calcium lactate required at rate of 25% of binder.

Gums, vegetables: to bind and extend product
Egg roll: sufficient for purpose. Calcium lactate required at rate of 25% of binder.

Methylcellulose: to extend and to stabilize product (also carrier)
Meat and vegetable patties: 0.15%

**Table 5.7.3. (continued)**

| Specific Chemical or Group of Chemicals: Purpose<br>Products: Amounts |
|---|
| Isolated soy protein: to bind and extend product<br>Sausage, as provided for in 9 CFR 319: 2% |
| Isolated soy protein: to bind and extend product<br>Imitation sausage, nonspecific loaves, soups, stews: sufficient for purpose |
| Sodium caseiante: to bind and extend product<br>Imitation sauage, nonspecific loaves, soups, stews: sufficient for purpose |
| Dry or dried whey: to bind or thicken<br>Sausage as provided for in 9 CFR 319, bockwurst: 3.5% individually or collectively |
| Reduced-lactose whey: to bind or thicken<br>Sausage as provided for in 9 CFR 319, bockwurst: 3.5% individually or collectively |
| Reduced-minerals whey: to bind or thicken<br>Sausage as provided for in 9 CFR 319, bockwurst: 3.5% individually or collectively |
| Whey protein concentrate: to bind or thicken<br>Sausage as provided for in 9 CFR 319, bockwurst: 3.5% individually or collectively |
| Dry or dried whey: to bind or thicken<br>Imitation sausage soups, stews, nonspecific loaves: sufficient for purpose |
| Reduced-lactose whey: To bind or thicken<br>Sausage as provided for in 9 CFR 319, bockwurst: 3.5% individually or collectively |
| Reduced-minerals whey: to bind or thicken<br>Sausage as provided for in 9 CFR 319, bockwurst: 3.5% individually or collectively |
| Whey protein concentrate: to bind or thicken<br>Sausage as provided for in 9 CFR 319, bockwurst: 3.5% individually or collectively |
| Dry or dried whey: to bind or thicken<br>Chili con carne, pork or beef with barbecue sauce: 8% individually or collectively with other binders |
| Reduced-lactose whey: to bind or thicken<br>Chili con carne, pork or beef with barbecue sauce: 8% individually or collectively with other binders |
| Reduced-mineral whey: to bind or thicken<br>Chili con carne, pork or beef with barbecue sauce: 8% individually or collectively with other binders |

Whey protein concentrate: to bind or thicken
  Chili con carne, pork or beef with barbecue sauce: 8% individually or collectively with other binders
Xanthan gum: to maintain uniform viscosity, suspension of particulate matter, emulsion stability, and freeze-thaw stability
  Meat sauces, gravies or sauces and meats, canned or frozen, and/or refrigerated meat salads, canned or frozen meat stews, canned chili or chili with beans, pizza topping mixes, and batter or breading mixes: 8% individually or collectively with other binders

Source: 9 CFR 318.7, 1991.

**Table 5.7.4. Legal Requirements for Bleaching, Catalytic, and Coloring Agents Permitted in Meat Processing**

Specific Chemical or Group of Chemicals: Purpose
  Products: Amounts

Bleaching Agents

Hydrogen peroxide: to remove color
  Tripe (substance must be removed from product by rinsing with clear water): sufficient for purpose

Catalysts
[substances must be eliminated during process]

Nickel: to accelerate chemical reaction
  Rendered animal fats or a combination of such fats and vegetable fats: sufficient for purpose
Sodium amide: rearrangement of fatty acid and radicals
  Rendered animal fats or a combination of such fats and vegetable fats: sufficient for purpose

Coloring Agents [Natural]

Alkanet, annatto, carotene, cochineal, green chlorophyll, saffron, and tumeric:
  To color casings or rendered fats; marking and branding product
  Sausage casings, oleomargrine, shortening; marking or branding ink on product: sufficient for purpose (may be mixed with approved artificial dyes or harmless inert material such as common salt and sugar).

**Table 5.7.4 (continued)**

Specific Chemical or Group of Chemicals: Purpose
Products: Amounts

Coloring Agents [Artificial]

Color additives listed in 21 CFR 74 and 82 (operator must furnish evidence to inspector in charge that color additive has been certified for use in connection with foods by the FDA): to color casings or rendered fats, marking and branding product.
Sausage casings, oleomargrine, shortening; marking or branding ink on product: sufficient for purpose (may be mixed with approved natural coloring matters or harmless inert material such as common salt or sugar).
Titanium dioxide: to color casings or rendered fats; marking and branding product
Canned ham salad spread and creamed-type canned products: 0.5%.

Source: 9 CFR 318.7, 1991.

**Table 5.7.5. Legal Requirements for Curing Accelerator Agents Permitted in Meat Processing**

Specific Chemical or Group of Chemicals: Purpose
Products: Amounts

Curing Accelerators
[must be used only in combination with curing agents]

Ascorbic acid: to accelerate color fixing or preserve color during storage
Cured pork and beef cuts, cured comminuted meat food product: 75 oz to 100 gallon pickle at 10% pump level, 3/4 oz to 100 lb meat or meat byproduct; 10% solution to surfaces of cured cuts prior to packaging. The use of such solution shall not result in the addition of a significant amount of moisture to the product.
Erythorbic acid: to accelerate color fixing or preserve color during storage
Cured pork and beef cuts, cured comminuted meat food product: 75 oz to 100 gallon pickle at 10% pump level, 3/4 oz to 100 lb meat or meat byproduct; 10% solution to surfaces of cured cuts prior to packaging. The use of such solution shall not result in the addition of a significant amount of moisture to the product.
Fumaric acid: To accelerate color fixing

Cured, comminuted meat or meat food products: 0.065% (or 1 oz to 100 lb) of the weight of the meat or meat byproducts, before processing

Glucone delta lactone: to accelerate color fixing

Cured, comminuted meat or meat food product: 8 oz to each 100 lb of meat or meat byproduct

Genoa salami: 16 oz to 100 lb of meat (1.0%)

Sodium acid pyrophosphate: to accelerate color fixing

Frankfurters, wieners, Vienna, Bologna, garlic Bologna: not to exceed, alone or in combination with other curing accelerators, the following: 8 oz in 100 lb of the meat, or meat and meat byproducts, content of the formula, nor 0.5% in the finished product

Sodium ascorbate: to accelerate color fixing or preserve color during storage

Cured pork and beef cuts, cured comminuted meat food product: 87.5% oz to 100 gallon pickle at 10% pump level; 7/8 oz to 100 lb meat or meat byproduct; 10% solution to surfaces of cured cuts prior to packaging. (The use of such solutions shall not result in the addition of a significant amount of moisture to the product.)

Sodium erythorbate: to accelerate color fixing or preserve color during storage

Cured pork and beef cuts, cured comminuted meat food product: 87.5% oz to 100 gallon pickle at 10% pump level; 7/8 oz to 100 lb meat or meat byproduct; 10% solution to surfaces of cured cuts prior to packaging. (The use of such solution shall not result in the addition of a significant amount of moisture to the product.)

Citric acid or sodium citrate: to accelerate color fixing or preserve color during storage

Cured pork and beef cuts, cured comminuted meat food product: May be used in cured products or in 10% solution used to spray surfaces of cured cuts prior to packaging to replace up to 50% of the ascorbic acid, erythorbic acid, sodium ascorbate, or sodium erythorbate that is used.

## Curing Agents

Sodium or potassium nitrate: Source of nitrite

Cured products other than bacon; nitrates may not be used in baby, junior, and toddler foods: 7 lb to 100 gallon pickle; 3-1/2 oz to 100 lb meat (dry cure); 2-3/4 oz to 100 lb chopped meat

Sodium or potassium nitrite: Supplies of sodium nitrite and potassium nitrite and mixtures containing them must be kept securely under the care of a responsible employee of the establishment; the specific nitrate content of such supplies must be known and clearly marked accordingly: to fix color

**Table 5.7.5. (continued)**

| Specific Chemical or Group of Chemicals: Purpose<br>Products: Amounts |
|---|
| Cured products; nitrites may not be used in baby, junior, or toddler foods: 2 lb to 100 gallon pickle at 10% pump level; 1 oz to 100 lb meat (dry cure); 1/4 oz to 100 lb chopped meat and/or meat byproduct. The use of nitrites, nitrates, or combination shall not result in more than 200 ppm of nitrite, calculated as sodium nitrate, in finished product, except that nitrites may be used in bacon only in accordance with 9 CFR 318.7(b). |

Source: 9 CFR 318.7, 1991.

**Table 5.7.6. Legal Requirements for Denuding Agents[a] Permitted in Meat Processing**

| Specific Chemical or Group of Chemicals: Purpose<br>Products: Amounts |
|---|
| Lime (calcium oxide, calcium hydroxide): to denude mucous membranes<br>Tripe: sufficient for purpose |
| Sodium carbonate: to denude mucous membranes<br>Tripe: sufficient for purpose |
| Sodium hydroxide: to denude mucous membranes<br>Tripe: sufficient for purpose |
| Sodium persulfate: to denude mucous membranes<br>Tripe: sufficient for purpose |
| Sodium silicates (ortho, meta, and sesqui): to denude mucous membranes<br>Tripe: sufficient for purpose |
| Trisodium phosphate: to denude mucous membranes<br>Tripe: sufficient for purpose |

Source: 9 CFR 318.7, 1991.

[a]May be used in combination. Must be removed from tripe by rinsing with potable water.

**Table 5.7.7. Legal Requirements for Emulsifying Agents and Film-Forming Agents Permitted in Meat Processing**

| Specific Chemical or Group of Chemicals: Purpose<br>Products: Amounts |
|---|
| Emulsifying Agents |
| Acetylated monoglycerides: to emulsify product<br>Shortening: sufficient for purpose |
| Diacetyl tartaric acid esters of mono- and diglycerides: to emulsify product<br>Rendered animal fat or a combination of such fat with vegetable: sufficient for purpose |
| Lecithin: to emulsify product (also as an antioxidant)<br>Oleomargarine, shortening, various meat food products: 0.5% in oleomargarine; use in other products: sufficient amount for emulsification |
| Mono- and diglycerides (glycerol palmitate, etc.): to emulsify product<br>Rendered animal fat or a combination of such fat with vegetable fat; oleomargarine: sufficient for purpose in lard and shortening, 0.5% in oleomargerine |
| Mono- and diglycerides of fatty acids esterified with any of the following acids: acetic, acetyltartaric, citric, lactic, tartaric, and their sodium and calcium salts; the sodium sulfoacetate derivatives of these mono- and diglycerides: to emulsify product<br>Margerine or oleomargarine: 0.5% |
| Polyglycerol esters of fatty acids [polyglycerol esters of fatty acids are restricted to those up to and including the decaglycerol esters and otherwise meeting the requirements of 172.854(a) of the the general food additive regulations]: to emulsify product<br>Rendered animal fat or a combination of such fat with vegetable fat when use is not precluded by standards of identity or composition; oleomargarine: sufficient for purpose for rendered animal fat or combination with vegetable fat; 0.5% for oleomargarine |
| 1,2-propylene glycol esters of fatty acids: to emulsify product<br>Margarine or oleomargarine: 2.0% |
| Polysorbate 80 [polyoxyethylene (20) sorbitan monooleate]: to emulsify product<br>Shortening for use in nonstandardized baked goods, baking mixes, icings, fillings, and toppings and in the frying of foods: 1% when used alone. If used with polysorbate 60 the combined total shall not exceed 1%. |

**Table 5.7.7. (continued)**

Specific Chemical or Group of Chemicals: Purpose
  Products: Amounts

Propylene glycol mono- and diesters of fats and fatty acids: to emulsify product
  Rendered animal fat or a combination of such fat with vegetable fat: sufficient for purpose.
Polysorbate 60 [polyoxyethylene (20) sorbitan monostearate]: to emulsify product
  Shortening for use in nonstandardized baked goods, baking goods, baking mixes, icings, fillings, and toppings and in the frying of foods: 1% when used alone. If used with polysorbate 80 the combined total shall not exceed 1%.
Stearyl-2-lactylic acid: to emulsify product
  Shortening to be used for cake icings and fillings: 3.0%
Stearyl monoglyceridyl citrate: to emulsify product
  Shortening: sufficient for purpose

Film-Forming Agents

A mixture consisting of water, sodium alginate, calcium chloride, sodium carboxymethylcellulose, and corn syrup solids: to reduce cooler shrinkage and help protect surface
  Freshly dressed meat carcasses. Such carcasses must bear a statement "Protected with a film of water, corn syrup solids, sodium alginate, calcium chloride and sodium carboxymethylcellulose": formulation may not exceed 1.5% of hot carcass weight when applied. Chilled weight may not exceed hot weight.

Source: 9 CFR 318.7, 1991.

**Table 5.7.8. Legal Requirements for Flavoring Agents, Protectors, and Developers**

Specific Chemical or Group of Chemicals: Purpose
  Products: Amounts

Program approved artificial smoke flavoring: to flavor product
  Various: formulation may not exceed 1.5% of hot carcass weight when applied. Chilled weight may not exceed hot weight.
Program approved smoke flavoring: to flavor product

Various: formulation may not exceed 1.5% of hot carcass weight when applied. Chilled weight may not exceed hot weight.

Autolyzed yeast extract: to flavor product.

Various: formulation may not exceed 1.5% of hot carcass weight when applied. Chilled weight may not exceed hot weight.

Harmless bacteria starters of the *Acidophilus* type, lactic acid starter, or culture of *Pediococcus cerevisiae*: to develop flavor

Dry sausage, pork roll, thuringer, Lebanon Bologna, cervelat, and salami: 0.5%

Harmless lactic acid-producing bacteria: to prevent growth of *Clostridium botulinum*

Bacon: sufficient for purpose

Benzoic acid (sodium, potassium, and calcium salts): to retard flavor reversion

Margarine or oleomargarine: 0.1% individually, or if used in combination or with sorbic acid and its salts, 0.2% (expressed as the acids in the weight of the finished foods).

Calcium lactate: to protect flavor

Cooked semidry and dry products, including sausage, imitation sausage, and nonspecific meat food sticks: 0.6% in product formulation

Citirc acid: flavoring

Chili con carne: sufficient for purpose

Corn syrup solids, corn syrup, glucose syrup: to flavor

Sausage, hamburger, meat loaf, luncheon meat, chopped or pressed ham: 2.0% individually or collectively, calculated on a dry basis

Dextrose: to flavor product

Sausage, ham, and cured products: sufficient for purpose

Diacetyl: to flavor product

Oleomargine: sufficient for purpose

Disodium guanylate: to flavor product

Various: sufficient for purpose

Disodium inosinate: to flavor product

Various: sufficient for purpose

Hydrolyzed plant protein: to flavor product

Various: sufficient for purpose

Isopropyl citrate: to protect flavor

Oleomargarine: 0.02%

Malt syrup: to flavor product

Cured products: 2.5%

Milk protein hydrolyzate: to flavor product

Various: sufficient for purpose

**Table 5.7.8. (continued)**

| Specific Chemical or Group of Chemicals: Purpose<br>Products: Amounts |
|---|
| Monosodium glutamate: to flavor product<br>Various: sufficient for purpose |
| Monoammonium glutamate: to flavor product<br>Various: sufficient for purpose |
| Sodium sulfoacetate derivative of mono- and diglycerides: to flavor product<br>Various: 0.5% |
| Sodium tripolyphosphate: to help protect flavor<br>"Fresh beef," "beef for further cooking," "cooked beef," beef patties, meat loaves, meat toppings, and similar products derived from pork, lamb, veal, mutton, and goat meat that are cooked or frozen after processing: 0.5% of total product |
| Mixtures of sodium tripolyphosphate and sodium metaphosphate, insoluble; and sodium polyphosphates, glassy: to help protect flavor<br>"Fresh beef," "beef for further cooking," "cooked beef," beef patties, meat loaves, meat toppings, and similar products derived from pork, lamb, veal, mutton, and goat meat that are cooked or frozen after processing: 0.5% of total product |
| Sorbitol: to flavor, to facilitate the removal of casings from product, and to reduce caramelization and charring<br>Cooked sausage labeled frankfurter, frank, furter, wiener, knockwurst: $\leq$2% of the weight of the formula, excluding the formula weight of water or ice; not permitted in combination with corn syrup, and/or corn syrup solids |
| Starter distillate: to help protect flavor<br>Oleomargarine: sufficient for purpose |
| Stearyl citrate: to protect flavor<br>Oleomargarine: 0.15% |
| Sugars (sucrose and dextrose): to flavor product<br>Various: sufficient for purpose |

Source: 9 CFR 318.7, 1991.

**Table 5.7.9. Legal Requirements for Gases and Hog Scald Agents Permitted in Meat Processing**

| Specific Chemical or Group of Chemicals: Purpose<br>Products: Amounts |
|---|
| Gases |
| Carbon dioxide, solid (dry ice): to cool product<br>Chopping of meat, packaging of product: sufficient for purpose |
| Nitrogen: to exclude oxygen<br>Sealed container: sufficient for purpose |
| Hog Scald Agents<br>[must be removed by subsequent cleaning operations] |
| Caustic soda: to remove hair<br>Hog carcasses: sufficient for purpose |
| Dimethylpolysiloxane: to remove hair<br>Hog carcasses: sufficient for purpose |
| Diocytl sodium sulfosuccinate: to remove hair<br>Hog carcasses: sufficient for purpose |
| Disodium calcium ethylenediamine tetraacetate: to remove hair<br>Hog carcasses: sufficient for purpose |
| Disodium phosphate: to remove hair<br>Hog carcasses: sufficient for purpose |
| Ethylenediamine tetraacetic acid (sodium salts): to remove hair<br>Hog carcasses: sufficient for purpose |
| Lime (calcium oxide, calcium hydroxide): to remove hair<br>Hog carcasses: sufficient for purpose |
| Propylene glycol: to remove hair<br>Hog carcasses: sufficient for purpose |
| Soap (prepared by the reaction of calcium, potassium, or sodium with rosin or fatty acids of natural fats and oils): to remove hair<br>Hog carcasses: sufficient for purpose |
| Sodium acid pyrophosphate: to remove hair<br>Hog carcasses: sufficient for purpose |
| Sodium carbonate: to remove hair<br>Hog carcasses: sufficient for purpose |
| Sodium dodecylbenzene sulfonate: to remove hair<br>Hog carcasses: sufficient for purpose |

**Table 5.7.9. (continued)**

| Specific Chemical or Group of Chemicals: Purpose<br>Products: Amounts |
|---|
| Sodium gluconate: to remove hair<br>Hog carcasses: sufficient for purpose |
| Sodium hexametaphosphate: to remove hair<br>Hog carcasses: sufficient for purpose |
| Sodium lauryl sulfate: to remove hair<br>Hog carcasses: sufficient for purpose |
| Sodium mono- and dimethylnaphthalene sulfonate (molecular weight 245-260): to remove hair<br>Hog carcasses: sufficient for purpose |
| Sodium *n*-alkylbenzene sulfonate (alkyl group predominantly $C_{12}$ and $C_{13}$ and not less 95% $C_{10}$ and $C_{16}$): to remove hair<br>Hog carcasses: sufficient for purpose |
| Sodium pyrophosphate: to remove hair<br>Hog carcasses: sufficient for purpose |
| Sodium silicates (ortho, meta, and sesqui): to remove hair<br>Hog carcasses: sufficient for purpose |
| Sodium sulfate: to remove hair<br>Hog carcasses: sufficient for purpose |
| Sodium tripolyphosphate: to remove hair<br>Hog carcasses: sufficient for purpose |
| Sucrose: to remove hair<br>Hog carcasses: sufficient for purpose |
| Triethanolamine dodecylbenzene sulfonate: to remove hair<br>Hog carcasses: sufficient for purpose |
| Trisodium phosphate: to remove hair<br>Hog carcasses: sufficient for purpose |

Source: 9 CFR 318.7, 1991.

**Table 5.7.10. Legal Requirements for Miscellaneous Agents Permitted in Meat Processing**

| Specific Chemical or Group of Chemicals: Purpose<br>Products: Amounts |
|---|
| Ascorbic acid, erythorbic acid, citric acid, sodium ascorbate, and sodium citrate, singly or in combination, under an approved, partial quality control (PQC) program [9 CFR 318.4(d) & (e)]: to maintain color |

Fresh pork cuts: not to exceed either 500 ppm or 1.8 mg/sq. in. of surface of ascorbic acid, erythorbic acid, or sodium ascorbate, singly or in combination; and/or not to exceed either 250 ppm or 0.9 mg/sq. in. of surface of citric acid or sodium citrite, singly or in combination

*d*- and di-$\alpha$-tocopherol: to inhibit nitrosamine formation

Pump-cured bacon: 500 ppm; by injection or surface application

Potassium sorbate: to retard mold growth

Dry sausage: 10% in water solution may be applied to casings after stuffing or casings may be dipped in a 10% water solution prior to stuffing.

Silicon dioxide: processing aid/dispersant

Tocopherol-containing bacon curing premixes: at level not to exceed 4.0% in the dry mix

Sorbic aicd (sodium, potassium, and calcium salts): to preserve product and to retard mold growth

Margarine or oleomargarine: 0.1% individually, or if used in combination or with benzoic acid or its salts, 0.2% (expressed as the acids) in the weight of the finished foods

Calcium disodium, EDTA (calcium disodium ethylenediamine tetraacetate): to preserve product and to protect flavor

Margarine or oleomargarine: 75 parts per million by weight of the finished oleomargarine or margarine

Propyl paraben (propyl *p*-hydroxybenzoate): to retard mold growth

Dry sausage: 3.5% in water solution may be applied to casings after stuffing, or casings may be dipped in solution prior to stuffing.

Sodium bicarbonate: to neutralize excess acidity, clearing vegetables

Rendered fats, soups, curing pickle: sufficient for purpose

Calcium propionate: to retard mold growth

Pizza crust: 0.32% alone or in combination based on weight of the flour brace used

Sodium propionate: to retard mold growth

Pizza crust: 0.32% alone or in combination based on weight of the flour brace used

Sodium hydroxide: to decrease the amount of cooked-out juices

Meat food products containing phosphates: May be used only in combination with phosphates in a ratio not to exceed one part sodium hydroxide to four parts phosphate; the combination shall not exceed 5% in pickle at 10% pump level; 0.5% in product.

Disodium phosphate: to decrease the amount of cooked-out juices

Meat food products except where otherwise prohibited by the Federal meat inspection regulations: 5% of phosphate in pickle at 10% pump

**Table 5.7.10. (continued)**

| Specific Chemical or Group of Chemicals: Purpose<br>Products: Amounts |
|---|
| level; 0.5% of phosphate in product (only clear solution may be injected into product). |
| Monosodium phosphate: to decrease the amount of cooked-out juices |
| Meat food products except where otherwise prohibited by the Federal meat inspection regulations: 5% of phosphate in pickle at 10% pump level; 0.5% of phosphate in product (only clear solution may be injected into product). |
| Sodium metaphosphate, insoluble: to decrease the amount of cooked-out juices |
| Meat food products except where otherwise prohibited by the Federal meat inspection regulations: 5% of phosphate in pickle at 10% pump level; 0.5% of phosphate in product (only clear solution may be injected into product). |
| Sodium polyphosphate, insoluble: to decrease the amount of cooked-out juices |
| Meat food products except where otherwise prohibited by the Federal meat inspection regulations: 5% of phosphate in pickle at 10% pump level; 0.5% of phosphate in product (only clear solution may be injected into product). |
| Sodium tripolyphosphate: to decrease the amount of cooked-out juices |
| Meat food products except where otherwise prohibited by the Federal meat inspection regulations: 5% of phosphate in pickle at 10% pump level; 0.5% of phosphate in product (only clear solution may be injected into product). |
| Sodium pyrophosphate: to decrease the amount of cooked-out juices |
| Meat food products except where otherwise prohibited by the Federal meat inspection regulations: 5% of phosphate in pickle at 10% pump level; 0.5% of phosphate in product (only clear solution may be injected into product). |
| Sodium acid pyrophosphate: to decrease the amount of cooked-out juices |
| Meat food products except where otherwise prohibited by the Federal meat inspection regulations: 5% of phosphate in pickle at 10% pump level; 0.5% of phosphate in product (only clear solution may be injected into product). |
| Dipotassium phosphate: to decrease the amount of cooked-out juices |
| Meat food products except where otherwise prohibited by the Federal meat inspection regulations: 5% of phosphate in pickle at 10% pump level; 0.5% of phosphate in product (only clear solution may be injected into product). |
| Monopotassium phosphate: to decrease the amount of cooked-out juices |

Meat food products except where otherwise prohibited by the Federal meat inspection regulations: 5% of phosphate in pickle at 10% pump level; 0.5% of phosphate in product (only clear solution may be injected into product).

Potassium tripolyphosphate: to decrease the amount of cooked-out juices

Meat food products except where otherwise prohibited by the Federal meat inspection regulations: 5% of phosphate in pickle at 10% pump level; 0.5% of phosphate in product (only clear solution may be injected into product).

Potassium pyrophosphate: to decrease the amount of cooked-out juices

Meat food products except where otherwise prohibited by the Federal meat inspection regulations: 5% of phosphate in pickle at 10% pump level; 0.5% of phosphate in product (only clear solution may be injected into product).

Citric acid (sodium and potassium salts): to acidity

Margarine or oleomargarine: sufficient for purpose

Lactic acid (sodium and potassium salts): to acidity

Margarine or oleomargarine: sufficient for purpose

L-Tartaric aicd (sodium and potassium salts): to acidity

Margarine or oleomargarine: sufficient for purpose

Adipic acid: to acidity

Margarine or oleomargarine: sufficient for purpose

Phosphoric acid: to acidity

Margarine or oleomargarine: sufficient for purpose

Hydrochloric acid: to acidity

Margarine or oleomargarine: sufficient for purpose

Sodium bicarbonate: to alkalize

Margarine or oleomargarine: sufficient for purpose

Sodium carbonate: to alkalize

Margarine or oleomargarine: sufficient for purpose

Sodium hydroxide: to alkalize

Margarine or oleomargarine: sufficient for purpose

Potassium carbonate: to alkalize

Margarine or oleomargarine: sufficient for purpose

Potassium bicarbonate: to alkalize

Margarine or oleomargarine: sufficient for purpose

---

Source: 9 CFR 318.7, 1991.

**Table 5.7.11. Legal Requirements for Proteolytic Enzymes Permitted in Meat Processing**

Specific Chemical or Group of Chemicals: Purpose
Products: Amounts

*Aspergillus oryzae*: to soften tissues
Raw meat cuts: solutions consisting of water and approved proteolytic enzymes applied or injected into raw meat cuts shall not result in a gain of more than 3% above the weight of the untreated product.
*Aspergillus flavus oryzae* group: to soften tissues
Raw meat cuts: solutions consisting of water and approved proteolytic enzymes applied or injected into raw meat cuts shall not result in a gain of more than 3% above the weight of the untreated product.
Bromelin: to soften tissues
Raw meat cuts: solutions consisting of water and approved proteolytic enzymes applied or injected into raw meat cuts shall not result in a gain of more than 3% above the weight of the untreated product.
Ficin: to soften tissues
Raw meat cuts: solutions consisting of water and approved proteolytic enzymes applied or injected into raw meat cuts shall not result in a gain of more than 3% percent above the weight of the untreated product.
Papain: to soften tissues
Raw meat cuts: solutions consisting of water and approved proteolytic enzymes applied or injected into raw meat cuts shall not result in a gain of more than 3% above the weight of the untreated product.

Source: 9 CFR 318.7, 1991.

**Table 5.7.12. Legal Requirements for Refining and Rendering Agents Permitted in Meat Processing**

Specific Chemical or Group of Chemicals: Purpose
Products: Amounts

Refining Agents
[must be eliminated during process of manufacture]

Acetic acid: to separate fatty acids and glycerol
Rendered fats: sufficient for purpose
Bicarbonate of soda: to separate fatty acids and glycerol

Rendered fats: sufficient for purpose
Carbon (purified charcoal): to aid in refining of animal fats
Rendered fats: sufficient for purpose
Caustic soda (sodium hydroxide): to refine fats
Rendered fats: sufficient for purpose
Diatomaceous earth (fuller's earth): to refine fats
Rendered fats: sufficient for purpose
Sodium carbonate: to refine fats
Rendered fats: sufficient for purpose
Tannic acid: to refine fats
Rendered fats: sufficient for purpose

Rendering Agents

Tricalcium phosphate: to aid rendering
Animal fats: sufficient for purpose
Trisodium phosphate: to aid rendering
Animal fats: sufficient for purpose

Source: 9 CFR 318.7, 1991.

### 5.7.13. Legal Requirements for Radiation Permitted in Meat Processing

Specific Chemical or Group of Chemicals: Purpose
Products: Amounts

Ionizing radiation limited to gamma rays from cobalt-60 or cesium-137: to control *Trichinella spiralis*
Pork carcasses, or fresh or previously frozen cuts or pork carcasses, that have not been cured or heat-processed: minimum absorbed dose of 0.3 kiloGray (30 kilorads) to a maximum absorbed dose of 1 kiloGray (100 kilorads)

Source: 9 CFR 318.7, 1991.

**Table 5.7.14 Legal Requirements for Artificial Sweeteners and Synergists Permitted in Meat Processing**

| Specific Chemical or Group of Chemicals: Purpose<br>Products: Amounts |
|---|
| Artificial Sweeteners |
| Sacccharin: to sweeten product<br>Bacon: 0.01%. |
| Synergists<br>[used in combination with antioxidants] |
| Citric acid: to increase effectiveness of antioxidants<br>Lard and shortening: 0.01% alone or in combination with antioxidants in lard or shortening |
| Citric acid: to increase effectiveness of antioxidants<br>Dry sausage: 0.003% in dry sausage in combination with antioxidants |
| Citric acid: to increase effectiveness of antioxidants<br>Fresh pork sausage: 0.01% on basis of fat content, in combination with antioxidants |
| Citric acid: to increase effectiveness of antioxidants<br>Dried meats: 0.01% on basis of total weight in combination with antioxidants |
| Malic acid: to increase effectiveness of antioxidants<br>Lard and shortening: 0.01% on basis of total weight in combination with antioxidants |
| Monoglyceride citrate: to increase effectiveness of antioxidants<br>Lard, shortening, fresh pork sausage, dried meats: 0.02% |
| Monoisopropyl citrate: to increase effectiveness of antioxidants<br>Lard, shortening, oleomargarine, fresh pork sausage, dried meats: 0.02% |
| Phosphoric acid: to increase effectiveness of antioxidants<br>Lard and shortening: 0.01% |

Source: 9 CFR 318.7, 1991.

**Table 5.7.15. Legal Requirements for Tenderizing Agents Permitted in Meat Processing**

| Specific Chemical or Group of Chemicals: Purpose<br>Products: Amounts |
|---|
| *Aspergillus oryzae*: to soften tissue |
| Raw meat cuts: solutions consisting of water and approved proteolytic enzymes applied or injected into raw meat cuts shall not result in a gain of more than 3% above the weight of the untreated product. |
| *Aspergillus flavus oryzae* group: to soften tissue |
| Raw meat cuts: solutions consisting of water and approved proteolytic enzymes applied or injected into raw meat cuts shall not result in a gain of more than 3% above the weight of the untreated product. |
| Bromelin: to soften tissue |
| Raw meat cuts: solutions consisting of water and approved proteolytic enzymes applied or injected into raw meat cuts shall not result in a gain of more than 3% above the weight of the untreated product. |
| Calcium chloride: to soften tissue |
| Raw meat cuts: not more than 3% of a 0.8 molar solution |
| Ficin: to soften tissue |
| Raw meat cuts: solutions consisting of water and approved proteolytic enzymes applied or injected into raw meat cuts shall not result in a gain of more than 3% above the weight of the untreated product. |
| Papain: to soften tissue |
| Raw meat cuts: solutions consisting of water and approved proteolytic enzymes applied or injected into raw meat cuts shall not result in a gain of more than 3% above the weight of the untreated product. |
| Potassium chloride: to soften tissue |
| Raw meat cuts: not more than 3% of a 2.0 molar solution |
| Magnesium chloride: to soften tissue |
| Raw meat cuts: not more than 3% of a 0.8 molar solution |
| Potassium, magnesium or calcium chloride: to soften tissue |
| Raw meat cuts: A solution of approved inorganic chlorides injected into or applied to raw meat cuts shall not result in a gain of more than 3% above the weight of the untreated product. |

Source: 9 CFR 318.7, 1991.

## 5.8. Definitions and Standards Promulgated by the USDA for Meat Products

### Table 5.8.1. Definition for Mechanically Separated (Species)

Mechanically separated (species) product is any finely comminuted product resulting from the mechanical separation and removal of most of the bone from attached skeletal muscle of livestock carcasses and meeting other legal requirements. Examples of such product are "mechanically separated beef," "mechanically separated veal," "mechanically separated pork," and "mechanically separated lamb."

1. At least 98% of the bone particles present in such product have a maximum size no greater than 0.5 mm in their greatest dimension and there are no bone particles larger than 0.85 mm in their greatest dimension.
2. The product has a calcium content exceeding 0.75%, as a measure of a bone solids content of <3%, and has a minimum PER of 2.5 (with some exceptions).
3. Such product has a protein content of ≥14% and a fat content of <30%.
4. An essential amino acid content of at least 33% of the total amino acids present in "mechanically separated (species)" is evidence of compliance with the protein quality requirement. Essential amino acid content includes isoleucine, leucine, lysine, methionine, phenylalanine, threonine, and valine content, and the total amino acids present include isoleucine, leucine, lysine, methionine, phenylalanine, threonine, valine, tyrosine, arginine, histidine, alanine, aspartic acid, glutamic acid, glycine, proline, serine, and hydroxyproline.
5. A prerequisite for label approval for products consisting of "mechanically separated (species)" is that such product has been produced by an establishment under an approved plant quality control system.

Source: 9 CFR 319.

### Table 5.8.2. Limitations on Applications of Mechanically Separated (Species)

1. Food products required to be prepared from one species cannot contain mechanically separated (species) products of any other species.
2. Mechanically separated (species) products with a protein content of not less than 14% and a fat content of not more than 30% may constitute up to

20% of the livestock and poultry product portion of most meat food products with some exceptions.

3. Mechanically separated (species) products for processing may constitute up to 20% of the livestock and poultry product portion of any meat food product with a definition and standard of identity or composition which establishes a maximum limit on the fat content.
4. Mechanically separated (species) product for processing may be used in baby, junior, or toddler foods, ground beef, hamburger, fabricated steaks, barbecued meats, roast beef–parboiled and steam roasted, corned (cured) beef cuts, certain cured pork products, tripe with milk, lima beans with ham and similar products, beef with gravy and gravy with beef, and meat pies.

Source: 9 CFR 319, 1991.

**Table 5.8.3. Listing for Meat Products with USDA Definitions and Standards**

Raw meat products
- Miscellaneous beef products
  - Chopped beef, ground beef
  - Hamburger
  - Beef patties
  - Fabricated steak
  - Partially defatted beef fatty tissue
- Miscellaneous pork products
  - Partially defatted pork fatty tissue

Cooked meats
- Barbecued meats
- Roast beef–parboiled and steam roasted

Cured meats, unsmoked and smoked
- Bacon
- Corned beef
- Corned beef brisket
- Corned beef round and other corned beef cuts
- Cured beef tongue
- Cured pork products
- "Country ham," "country-style ham," "dry cured ham," "country pork shoulder," "country-style pork shoulder," and "dry cured pork shoulder"
- "Ham patties," "chopped ham," "pressed ham," "spiced ham," and similar products

Sausage generally: fresh sausage
- Breakfast sausage
- Fresh beef sausage
- Fresh pork sausage
- Italian sausage products
- Sausage
- Whole hog sausage

Uncooked, smoked sausage
- Smoked pork sausage

Cooked sausage
- Cheesefurters and similar products
- Frankfurters, frank, furter, hotdog, weiner, Vienna, Bologna, garlic Bologna, knockwurst, and similar products
- Liver sausage and braunschweiger

**Table 5.8.3. (continued)**

Luncheon meat, loaves, and jellied products
- Luncheon meat
- Meat loaf

Meat specialties, puddings and nonspecific loaves
- Bockwurst
- Scrapple

Canned, frozen, or dehydrated meat food products
- Beans with frankfurters in sauce, sauerkraut with wieners and juice, and similar products
- Beef with gravy and gravy with beef
- Chili con carne
- Chili con carne with beans
- Chow mein vegetables with meat, and chop suey vegetables with meat
- Corned beef hash
- Hash
- Lima beans with ham in sauce, beans with ham in sauce, beans with bacon in sauce, and similar products
- Meat stews
- Pork with barbecue sauce and beef with barbecue sauce
  - Spaghetti sauce with meat
  - Spaghetti with meat balls and sauce, spaghetti with meat and sauce, and similar products
- Tamales
- Tripe with milk

Meat food entree products, pies, and turnovers
- Meat pies

Meat snacks, hors d'oeuvres, pizza, and specialty items
- Pizza

Fats, oils, shortenings
- Lard, leaf lard
- Margarine or oleomargarine
- Mixed fat shortening
- Rendered animal fat or mixture thereof

Meat soups, soup mixes, broths, stocks, and extracts
- Fluid extract of meat
- Meat extract

Meat salads and meat spreads
- Deviled ham, deviled tongue, and similar products
- Ham spread, tongue spread, and similar products
- Potted meat food product and deviled meat food products

Miscellaneous
- Breaded products
- Liver meat food products

Source: 9 CFR 319, 1991.

**Table 5.8.4. Definition and Standard for Raw Meat Products**

Miscellaneous Beef Products

Chopped Beef, Ground Beef

1. It contains chopped fresh and/or frozen beef with or without seasoning and without the addition of beef fat.
2. It contains not more than 30% fat.
3. It does not contain added water, phosphates, binders, or extenders.
4. When beef cheek meat (trimmed beef cheeks) is used in the preparation of chopped or ground beef, the amount of such cheek meat is limited to 25%; and if in excess of natural proportions, its presence is declared on the label in the required ingredient statement and contiguous to the name of the product.

Hamburger

1. It consists of chopped fresh and/or frozen beef with or without the addition of beef fat as such and/or seasoning.
2. It does not contain more than 30% fat.
3. It does not contain added water, phosphates, binders, or extenders.
4. Beef cheek meat (trimmed beef cheeks) may be used in the preparation of hamburger only in accordance with prescribed conditions.

Beef Patties

1. It consists of chopped fresh and/or frozen beef with or without the addition of beef fat as such and/or seasonings.
2. It contains binders or extenders; mechanically separated (species) products, and/or partially defatted beef fatty tissue may be used without added water or with added water only in amounts such that the product characteristics are essentially that of a meat pattie.

Fabricated Steak

Fabricated steak includes fabricated beef steaks, veal steaks, beef and veal steaks, or veal and beef steaks, and similar products, such as those labeled "beef steak, chopped, shaped, frozen," "minute steak, formed, wafer sliced, frozen," "veal steaks, beef added, chopped–molded–cubed–frozen, hydrolyzed plant protein, and flavoring." Characteristics of such products are:

1. They should be prepared by comminuting and forming the product from fresh and/or frozen meat, with or without added fat, of the species indicated on the label.

**Table 5.8.4. (continued)**

2. They should not contain more than 30% fat and should not contain added water, binders, or extenders.
3. Beef cheek meat (trimmed beef cheeks) may be used in the preparation of fabricated beef steaks only in accordance with the conditions prescribed under "chopped beef."

Partially Defatted Beef Fatty Tissue

It is a beef byproduct derived from the low-temperature rendering (not exceeding 120°F) of fresh beef fatty tissue. It should have a pinkish color and a fresh odor and appearance.

Miscellaneous Pork Products

Partially Defatted Pork Fatty Tissue

It is a pork byproduct derived from the low-temperature rendering (not exceeding 120°F) of fresh pork fatty tissue, exclusive of skin. It should have a pinkish color and a fresh odor and appearance.

Source: 9 CFR 319, 1991.

**Table 5.8.5. Definition and Standard for Cooked Meats**

Barbecued Meats

Barbecued meats, such as product labeled "beef barbecue" or "barbecued pork," are cooked by the direct action of dry heat resulting from the burning of hard wood or the hot coals therefrom for a sufficient period to assume the usual characteristics of a barbecued article, which include the formation of a brown crust on the surface and the rendering of surface fat. The product may be basted with a sauce during the cooking process. The weight of barbecued meat does not exceed 70% of the weight of the fresh uncooked meat.

Roast Beef Parboiled and Steam Roasted

This is prepared so that the weight of the finished product, excluding salt and flavoring material, does not exceed 70% of the fresh beef weight. Beef cheek meat and beef head meat, from which the overlying glandular and connective tissues have been removed, and beef heart meat, exclusive of the heart cap, may be used individually or collectively to the extent of 5% of the meat ingredients in the preparation of such canned products. If this is the case, their presence should be reflected in the required statement of ingredients.

Source: 9 CFR 319, 1991.

**Table 5.8.6. Definition and Standard for Cooked Meats, Unsmoked and Smoked: Corned Beef, Corned Beef Brisket, Corned Beef Round and Other Corned Beef Cuts, Cured Beef Tongue, Bacon**

Corned Beef

This is prepared from beef briskets, navels, clods, middle ribs, rounds, rumps, or similar cuts using one or a combination of specified curing ingredients. Canned product labeled "corned beef" should be prepared so that the weight of the finished product, excluding cure, salt, and flavoring material, does not exceed 70% of the fresh beef weight. Corned beef other than canned should be cured in pieces weighing not less than 1 lb, and if cooked, its weight does not exceed the weight of the fresh uncured beef. Beef cheek meat, beef head meat, and beef heart meat may be used to the extent of 5% of the meat ingredient in preparation of this product when trimmed as specified. If this is the case, their presence should be reflected in the required statement of ingredients. The application of curing solution to beef cuts, other than briskets, which are intended for bulk corned beef should not result in an increase in the weight of the finished cured product of more than 10% over the weight of the fresh uncured meat.

Corned Beef Brisket

In preparing "corned beef brisket," the application of curing solution to the beef brisket should not result in an increase in the weight of the finished cured product of more than 20% over the weight of the fresh uncured brisket. If the product is cooked, the weight of the finished product should not exceed the weight of the fresh uncured brisket.

Corned Beef Round and Other Corned Beef Cuts

In preparing "corned beef round" and other corned beef cuts, except "corned beef briskets," the curing solution should be applied to pieces of beef weighing not less than 1 lb, and such application should not result in an increased

**Table 5.8.6. (continued)**

weight of the cured beef product of more than 10% over the weight of the fresh uncured beef cut. If the product is cooked, the weight of the finished product should not exceed the weight of the fresh uncured beef cut.

Cured Beef Tongue

In preparing "cured beef tongue," the application of curing solution to the fresh beef tongue should not result in an increase in the weight of the cured beef tongue of more than 10% over the weight of the fresh uncured boof tongue.

Bacon

The weight of cured pork bellies ready for slicing and labeling as "bacon" should not exceed the weight of the fresh uncured pork belles.

Source: 9 CFR 319, 1991.

**Table 5.8.7. Definition and Standard for Cooked Meats, Unsmoked and Smoked: Cured Pork Products**

A. Cured pork products, including hams, shoulders, picnics, butts and loins, should comply with the minimum meat protein fat free (PFF) percentage requirements. The percentage should be the minimum meat protein which is indigenous to the raw unprocessed pork expressed as a percent of the nonfat portion of the finished product.

The minimum meat PFF percentage and product name (qualifying statements) for each type of cured pork product is as follows.

Cooked Ham, Loin[a]

1. 20.5%; common and usual name
2. 18.5%; common and usual name (with natural juices)
3. 17.0%; common and usual name (water added)
4. <17.0%; common and usual name (and water product–x% of weight is added ingredients).[b]

Cooked Shoulder, Butt, Picnic[a]

1. 20.0%; common and usual name
2. 18.0%; common and usual name (with natural juices)

3. 16.5%; common and usual name (water added)
4. <16.5%; common and usual name (and water product–x% of weight is added ingredients.)[b]

Uncooked Cured Ham, Loin

1. 18.0%; uncooked common and usual name
2. <18.0%; uncooked common and usual name (and water product–x% of weight is added ingredients.)[b]

Uncooked Cured Shoulder, Butt, Picnic

1. 17.5%; uncooked common and usual name
2. <17.5%; uncooked common and usual name (and water product–x% of weight is added ingredients.)[b]

B. Cured pork products for shall bear a required statement as part of the product name in lettering not less than 3/8 inch in height, or in lettering not less than one-third the size of the largest letter in the product name if it is in the same color and style of print and on the same color background as the product name.

---

Source: 9 CFR 319, 1991.

[a]The term "cooked" is not appropraite for use on labels of cured pork products heated only for the purpose of destruction of possible live trichinae.

[b]Processors may immediately follow this qualifying statement with a list of the ingredients in descending order of predominance rather than having the traditional ingredients statement. In any case, the maximum percentage of added substances in the finished product on a total weight percentage basis would be inserted as the x value; for example, " ham and water product–20% of weight is added ingredients."

**Table 5.8.8. Definition and Standard for Cooked Meats, Unsmoked and Smoked: "Ham Patties," "Chopped Ham," "Pressed Ham," "Spiced Ham," and Similar Products**

---

A. Finely divided (chopped, ground, flaked, chipped) cured ham products such as "ham patties," "chopped ham," "pressed ham," and "spiced ham" should comply with minimum meat protein fat free (PFF) percentage requirements. The minimum meat PFF percentage is the minimum meat protein indigenous to the raw, unprocessed pork expressed as a percent of

**Table 5.8.8. (continued)**

the nonfat portion of the finished product. The percentage and product name (qualifying statments) for each type of cured pork product are:

1. "Ham patties," "chopped ham," "pressed ham," "spiced ham": 19.5%; common and usual name
2. "Ham patties," "chopped ham," "pressed ham," "spiced ham": 17.5%; common and usual name (with natural juices)
3. "Ham patties," "chopped ham," "pressed ham," "spiced ham": 16.0%; common and usual name (water added)
4. "Ham patties," "chopped ham," "pressed ham," "spiced ham": <16,0%; common and usual name (and water product–x% of weight is added ingredients). Processors may immediately follow this qualifying statement with a list of the ingredients in descending order of predominance rather than having the traditional ingredients statement. In any case, the maximum percentage of added substances in the finished product on a total weight percentage basis would be inserted as the x value; for example, "ham and water product–20% of weight is added ingredients."

B. Cured pork products except "ham patties" may contain finely chopped ham shank meat to the extent of 25% over that normally present in boneless ham. Mechanically separated (species) product may be used.

C. "Ham patties" may not contain more than 35% fat by analysis.

Source: 9 CFR 319, 1991.

**Table 5.8.9. Definition and Standard for Cooked Meats, Unsmoked and Smoked: "Country Ham," "Country-Style Ham," "Dry Cured Ham," "Country Pork Shoulder," "Country-Style Pork Shoulder," and "Dry Cured Pork Shoulder"**

1. They are the uncooked, cured, dried, smoked or unsmoked meat food products made respectively from a single piece of meat conforming to the definition of "ham," or from a single piece of meat from a pork shoulder. They are prepared in accordance with prescribed criteria by the dry application of salt (NaCl) or by the dry application of salt (NaCl) and one or more of the optional ingredients. They may not be injected with curing solution nor placed in curing solutions.
2. The product must be treated for the destruction of possible live trichinae.

3. The entire exterior of the ham or pork should shoule be coated by the dry application of salt or salt combined with other permitted ingredients. Additional salt, or salt mixed with other permitted ingredients, may be reapplied to the product as necessary to insure complete penetration.
4. When sodium or potassium nitrate, or sodium or potassium nitrite, or a combination thereof is used, the application of salt should be in sufficient quantity to insure that the finished product has an internal salt content of at least 4%.
5. When no sodium nitrate, potassium nitrate, sodium nitrite, potassium nitrite or a combination thereof is used, the application of salt should be in sufficient quantity to insure that the finished product has a brine concentration of not less than 10% or a water activity of not more than 0.92.
6. For hams or pork shoulders labeled "country" or "country style," the combined period for curing and salt equalization should not be less than 25 days for pork shoulders; the total time for curing, salt equalization, and drying should not be less than 70 days for hams and should not be less than 50 days for pork shoulders. During the drying and smoking period, the internal temperature of the product must not exceed 95°F. Such temperature requirement does not apply to product dried or smoked under natural climatic conditions.
7. For hams or pork shoulders labeled "dry cured," the combined period for curing and salt equalization should not be less than 45 days for hams and should not be less than 25 days for pork shoulders; and the total time for curing, salt equalization, and drying should not be less than 55 days for hams and should not be less than 40 days for pork shoulders.
8. The weight of the finished hams and pork shoulders should be at least 18% less than the fresh uncured weight of the article.
9. The optional ingredients for products are:
   a. Nutritive sweeteners, spices, seasonings, and flavorings
   b. Sodium or potassium nitrate and sodium or potassium nitrite if used as prescribed

Source: 9 CFR 319, 1991.

**Table 5.8.10. Definition and Standard for a Sausage and Fresh Sausages**

Sausage

1. Sausage is the coarse or finely comminuted meat food product prepared from one or more kinds of meat or meat and meat byproducts, containing

**Table 5.8.10. (continued)**

various amounts of water as permitted, usually seasoned with condimented proportions of condimental substances, and frequently cured.

2. Certain sausage may contain binders and extenders; e.g., cereal, vegetable starch, starchy vegetable flour, soy flour, soy protein concentrate, isolated soy protein, nonfat dry milk, dry or dried whey, reduced-lactose whey, reduced-minerals whey, whey protein concentrate, calcium-reduced dried skim milk, enzyme (rennet)-treated calcium-reduced dried skim milk, and calcium lactate or dried milk.
3. The finished product should contain no more than 3.5% of these additives individually or collectively. Two percent of isolated soy protein should be deemed equivalent to 3 1/2% of any one or more of these binders.
4. Sausage may not contain phosphates except that certain phosphates may be used in cooked sausage.
5. To facilitate chopping or mixing or to dissolve the usual curing ingredients, water or ice may be used in the preparation of uncooked sausage in an amount not to exceed 3% of the total ingredients in the formula.
6. Cooked sausages such as Polish sausage, cotto salami, braunschweiger, liver sausage, and similar cooked sausage products may contain no more than 10% of added water in the finished product.
7. Sausage may contain mechanical separated (species) products.

Fresh Pork Sausage

1. Fresh pork sausage is sausage prepared with fresh pork or frozen pork or both, but not including pork byproducts, and may contain mechanically separated (species) product and may be seasoned with permitted condimental substances.
2. The finished product shall not contain more than 50% fat.
3. To facilitate chopping or mixing, water or ice may be used in an amount not to exceed 3% of the total ingredients used.

Fresh Beef Sausage

1. Fresh beef sausage is sausage prepared with fresh beef or frozen beef, or both, but not including beef byproducts, and may contain mechanically separated (species) product and may be seasoned with permitted condimental substances.
2. The finished product shall not contain more than 30% fat.
3. To facilitate chopping or mixing, water or ice may be used in an amount not to exceed 3% of the total ingredients used.

Breakfast Sausage

1. Breakfast sausage is sausage prepared with fresh and/or frozen meat, or fresh and/or frozen meat and meat byproducts, and may contain mechanically separated (species) product and may be seasoned with permitted condimental substances.
2. The finished product shall not contain more than 50% fat.
3. To facilitate chopping or mixing, water or ice may be used in an amount not to exceed 3% of the total ingredients used.
4. Extenders or binders may be used to the extent of 3 1/2% of the finished sausage.

Whole Hog Sausage

1. "Whole hog sausage" is sausage prepared with fresh and/or frozen meat from swine in such proportions as are normal to a single animal, and may include any mechanically separated (species) product from the animal and may be seasoned with permitted condimental substances.
2. The finished product should not contain more than 50% fat.
3. To facilitate chopping or mixing, water or ice may be used in an amount not to exceed 3% of the total ingredients used.

Italian Sausage Products

1. Italian sausage products are cured or uncured sausages containing at least 85% meat, or combination of meat and fat, with the total fat content constituting not more than 35% of the finished product. They are prepared according to legal requirements and should contain salt, pepper, and either fennel or anise, or a combination of fennel and anise. Such products may contain any or all of the permitted optional ingredients.
2. "Italian sausage" are be prepared with fresh or frozen pork, or pork and pork fat, and may contain mechanically separated (Species) product.
3. "Italian sausage with beef," "Italian sausage with veal," or "Italian sausage with beef and veal" are prepared so that fresh or frozen pork constitutes the major portion of the meat content requirement. Mechanically separated (species) product may be used. When pork muscle tissue is combined with beef or veal, or both, in the preparation of bulk-packed products, or patties, it is treated to destroy possible live trichinae.
4. "Italian beef sausage" or "kosher Italian beef sausage" should be prepared with fresh or frozen beef or beef and beef fat. "Italian veal sausage" or

**Table 5.8.10. (continued)**

"kosher Italian veal sausage" should be prepared with fresh or frozen veal or veal and veal fat. Mechanically separated (species) product may be used.

5. Italian sausage products may contain sodium nitrite or potassium nitrite in amounts not to exceed those allowed. Such products are labeled with the word "cured" in the product name, such as "cured Italian sausage." The word "cured" should be displayed on the product lable in the same size and style of lettering as other words in the product name.
6. Optional ingredients permitted in Italian sausage products include:

   a. Spices (including paprika) and flavorings
   b. Water or ice to facilitate chopping or mixing, but not to exceed 3% of the total weight of all ingredients including the water
   c. Red or green peppers, or both
   d. Dehydrated or fresh onions, garlic, and parsley
   e. Sugar, dextrose, corn syrup, corn syrup solids, and glucose syrup
   f. Monosodium glutamate and antioxidants

7. If Italian sausage products are cooked or smoked, requirements for compliance are based on the uncooked or unsmoked product. The product before cooking or smoking should contain no more than 3% water as specified. Cooked product should be labeled with the word "cooked" in the product name, such as "cooked Italian sausage" or "cooked cured Italian sausage." or "smoked cured Italian sausage." The words "cooked" and "smoked" should be displayed on the product label in the same size and style of lettering as other words in the product name.

Source: 9 CFR 319, 1991.

**Table 5.8.11. Definition and Standard for Uncooked, Smoked Pork Sausage**

1. "Smoked pork sausage" is pork sausage that is smoked with hardwood or other approved nonresinous materials.
2. It may be seasoned with permitted condimental substances. The finished product should not contain more than 50% fat.
3. To facilitate chopping or mixing, water or ice may be used in an amount not to exceed 3% of the total ingredients used.

Source: 9 CFR 319, 1991.

**Table 5.8.12. Definition and Standard for Cooked Sausage: Frankfurter, Frank, Furter, Hotdog, Weiner, Vienna, Bologna, Garlic Bologna, Knockwurst, and Similar Products**

Frankfurter, Frank, Furter, Hotdog, Weiner, Vienna, Bologna, Garlic Bologna, Knockwurst, and Similar Cooked Sausages

1. They are comminuted, semisolid sausages prepared from one or more kinds of raw skeletal muscle meat or raw skeletal muscle meat and raw or cooked poultry meat, and seasoned and cured, using one or more of the permitted curing agents. They may or may not be smoked.
2. The finished products should not contain more than 30% fat.
3. Water or ice, or both, may be used to facilitate chopping or mixing or to dissolve the curing ingredients but the sausage should contain no more than 40% of a combination of fat and added water.
4. These sausage products may contain only approved phosphates.
5. They may contain raw or cooked poultry meat not in excess of 15% of the total ingredients, excluding water, in the sausage, and mechanically separated (species) products. Such poultry meat ingredients should be designated in the ingredient statement on the label.
6. They are labeled by its generic name, e.g., frankfurter, frank, furter, hotdog, wiener, Vienna, Bologna, garlic Bologna, or knockwurst. When such sausage products are prepared with meat from a single species of cattle, sheep, swine, or goats they should be labeled with the term designating the particular species in conjunction with the generic name, e.g., "beef frankfurter," and when such sausage products are prepared in part with mechanically separated (species) product, they should be labeled accordingly.

Frankfurter, Frank, Furter, Hotdog, Weiner, Vienna, Bologna, Garlic Bologna, Knockwurst, and Similar Cooked Sausages that Contain Byproducts

1. When labeled with the phrase "with byproducts" or "with variety meats" in the product name, the products are comminuted, semisolid sausages consisting of not less than 15% of one or more kinds of raw skeletal muscle meat with
    1. Raw meat byproducts, or
    2. Raw meat byproducts and raw or cooked poultry products
2. They should be seasoned and cured, using one or more of the permitted curing ingredients. They may or may not be smoked.
3. Partially defatted pork fatty tissue or partially defatted beef fatty tissue, or a combination of both, may be used in an amount not exceeding 15% of

**Table 5.8.12. (continued)**

the meat and meat byproducts or meat, meat byproducts, and poultry products ingredients.

4. The finished products should not contain more than 30% fat.
5. Water or ice, or both, may be used to facilitate chopping or mixing to dissolve the curing and seasoning ingredients and the sausage should contain no more than 40% of a combination of fat and added water.
6. These sausage products may contain only phosphates approved by regulations.
7. These sausage products may contain poultry products, individually or in combination, not in excess of 15% of the total ingredients, excluding water, in the sausage, and may contain mechanically separated (species) product in accordance with requirements. Such poultry products shall not contain kidneys or sex glands. The amount of poultry skin present in the sausage must not exceed the natural proportion of skin present on the whole carcass of the kind of poultry used in the sausage. The poultry products used in the sausage shall be designated in the ingredient statement on the label of such sausage. Meat byproducts used in the sausage should be designated individually in the ingredient statement on the label for such sausage.
8. They should be labeled by its generic name, e.g., frankfurter, frank, furter, hotdog, wiener, Vienna, Bologna, garlic Bologna, or knockwurst, in conjunction with the phrase "with byproducts" or "with variety meats," with such supplemental phrase shown in a prominent manner directly contiguous to the generic name and in the same color on an identical background.

"Frankfurter, Calcium-reduced Dried Skim Milk Added,"
or "Bologna, with Byproducts
(or Variety Meats), Soy Flour Added"

1. When labeled as such, one or more of the following binders may be used in cooked sausage otherwise complying with other requirements:

    a. Dried milk
    b. Calcium-reduced dried skim milk
    c. Enzyme (rennet)-treated calcium-reduced dried skim milk and calcium lactate
    d. Nonfat dry milk
    e. Dry or dried whey
    f. Reduced-lactose whey
    g. Reduced-minerals whey
    h. Whey protein concentrate
    i. Cereal
    j. Vegetable starch
    k. Starchy vegetable flour

l. Soy flour
m. Soy protein concentrate
n. Isolated soy protein

However, such ingredients, individually or collectively, should not exceed 3 1/2% of the finished product, except that 2% of isolated soy protein should be deemed to be the equivalent of 3 1/2% of any one or more of the other binders.

General Requirements

1. Cooked sausages should not be labeled with terms such as "all-meat" or "all-(species)," or otherwise to indicate they do not contain nonmeat infredients or are prepared only from meat.
2. Poultry meat means deboned chicken meat or turkey meat, or both, without skin or added fat; poultry products mean chicken or turkey, or chicken meat or turkey meat, or poultry byproducts.
3. Meat byproducts (or variety meats), mean pork stomachs or snouts; beef, veal, lamb, or goat tripe; beef, veal, lamb, goat, or pork hearts, tongues, fat, lips, weasands, and spleens; and partially defatted pork fatty tissue, or partially defatted beef fatty tissue.

Source: 9 CFR 319, 1991.

**Table 5.8.13. Definition and Standard for Cooked Sausage: Cheesefurters and Similar Products**

1. They are products in casings which resemble frankfurters except that they contain sufficient cheese to give definite characteristics to the finished article.
2. They may contain cereal, vegetable starch, starchy vegetable flour, soy flour, soy protein concentrate, isolated soy protein, nonfat dry milk, dry or dried whey, reduced-lactose whey, reduced-minerals whey, whey protein concentrate, calcium-reduced dried skim milk, enzyme (rennet)-treated calcium-reduced dried skim milk and calcium lactate, or dried milk.
3. The finished product shall contain no more than 3.5% of these additives, individually and collectively, exclusive of the cheese constituent. In determining the maximum amount of the ingredients specified in this subparagraph which may be used, individually and collectively, in a product, 2% of isolated soy protein shall be considered the equivalent of 3.5% of any other ingredient specified in this subparagraph. When any such additive is added to these products, there shall appear on the label in a prominent manner, contiguous to the name of the product, the name of each such added ingredient, as for example, "cereal added," "with cereal,"

**Table 5.8.13. (continued)**

"potato flour added," "cereal and potato flour added," "soy flour added," "nonfat dry milk added," "cereal and nonfat dry milk added," as the case may be. These products shall contain no more than 40% of a combination of fat and added water, and no more than 30% fat and shall comply with the other provisions for cooked sausages[a].

Source: 9 CFR 319, 1991.
[a]See Tables 5.8.10 to 5.8.15 in this chapter.

**Table 5.8.14. Definition and Standard for Cooked Sausage: Braunschweiger and Liver Sausage or Liverwurst**

Braunschweiger

1. It is a cooked sausage made from fresh, cured, and/or frozen pork, beef, and/or veal and at least 30% pork, beef, and/or veal livers computed on the weight of the fresh livers.
2. It may also contain pork and/or beef fat. Mechanically separated (species) product may be used.
3. Binders and extenders may be used as permitted.
4. The product may have a smoked taste characteristic, which may be imparted by use of smoked meats, smoke flavoring, or smoking.
5. If prepared from components of a single species, the product name may reflect the species, e.g., "beef braunschweiger." Braunschweiger may also be labeled as any of the following: "braunschweiger–a liver sausage," "braunschweiger–a liverwurst," or "braunschweiger (liver sausage)" or "braunschweiger (liverwurst)."

Liver Sausage orr Liverwurst

1. It is a cooked sausage made from fresh, cured, and/or frozen pork, beef, and/or veal and at least 30% pork, beef, veal, sheep, and/or goat livers computed on the weight of the fresh livers.
2. It may also contain pork and/or beef byproducts.
3. Mechanically separated (species) product may be used in accordance with requirements.
4. Binders and extenders maybe used as permitted.
5. If prepared from components of a single species, the product name may reflect that species, e.g., "pork liver sausage."

Source: 9 CFR 319, 1991.

**Table 5.8.15. Definition and Standard for Cooked Sausage: Luncheon Meat, Loaves, and Jellied Products**

Luncheon Meat

1. It is a cured, cooked meat food product made from comminuted meat.
2. Mechanically separated (species) product may be used.
3. To facilitate chopping or mixing or to dissolve the usual curing ingredients, water or ice may be used in the preparation of luncheon meat in an amount not to exceed 3% of the total ingredients.

Meat Loaf

1. It is a cooked meat food product in loaf form made from comminuted meat.
2. Mechanically separated (species) product may be used in accordance with 9 CFR 319.6. To facilitate chopping or mixing, water or ice may be used in an amount not to exceed 3% of the total ingredients used.

Source: 9 CFR 319, 1991.

**Table 5.8.16. Definition and Standard for Meat Specialties, Puddings, and Nonspecific Loaves**

Scrapple

1. It contains not less than 40% meat and/or meat byproducts computed on the basis of the fresh weight, exclusive of bone.
2. Mechanically separated (species) product may be used in accordance with requirements. The meal or flour used may be derived from grain and/or soybeans.

Bockwurst

1. It is an uncured, comminuted meat food product that may or may not be cooked.
2. It contains meat; milk, or water, or a combination thereof; eggs; vegetables; and any of the optional ingredients.

**Table 5.8.16. (continued)**

3. Meat should constitute not less than 70% of the total weight of the product and should consist of pork or a mixture of pork and veal; pork and beef; or pork, veal, and beef. Such meat should be fresh or fresh frozen meat.
4. Pork may be omitted when the specie or species of meat used in the product is identified in the product name (e.g., veal bockwurst, beef bockwurst, or beef and veal bockwurst).
5. Mechanically separated (species) product may be used in accordance with requirements.
6. Specifications for other nonoptional ingredients are:

   a. The milk may be fresh whole milk, dried milk, nonfat dry milk, calcium-reduced dried skim milk, enzyme (rennet)-treated calcium-reduced dried skim milk and calcium lactate, or any combination.
   b. Eggs refer to whole eggs that are fresh, frozen, or dried.
   c. Vegetables refer to onions, chives, parsley, and leeks, alone or in any combination.

7. Bockwurst may contain one or more of the following optional ingredients:

   a. Pork fat
   b. Celery, fresh or dehydrated
   c. Spices, flavorings
   d. Salt
   e. Egg whites, fresh, frozen, or dried
   f. Corn syrup solids, corn syrup, or glucose syrup with a maximum limit of 2% individually or collectively, calculated on a dry basis. The maximum quantities of such ingredients should be computed on the basis of the total weight of the ingredients.
   g. Autolyzed yeast extract, hydrolyzed plant protein, milk protein hydrolyzate, and monosodium glutamate
   h. Sugars (sucrose and dextrose)
   i. Cereal, bread, vegetable starch, starchy vegetable flour, soy flour, soy protein concentrate, dry or dried whey, reduced-lactose whey, reduced-minerals whey, whey protein concentrate, and isolated soy protein, provided such ingredients, individually or collectively, do not exceed 3 1/2% of total weight of all the ingredients, except that 2% of isolated soy protein should be deemed to be the equivalent of 3 1/2% of any one or more of the other ingredients permitted. Bockwurst containing any of the permitted ingredients should be labeled.

8. If bockwurst is cooked or partially cooked, the composition of the raw mix from which it is prepared should be used in determining whether it meets the requirements of this section.

Source: 9 CFR 319, 1991.

**Table 5.8.17. Definition and Standard for Canned, Frozen, or Dehydrated Meat Food Products**

Chili Con Carne

1. It should contain not less than 40% of meat computed on the weight of the fresh meat.
2. Mechanically separated (species) product may be used in accordance with requirements.
3. Head meat, cheek meat, and heart meat exclusive of the heart cap may be used to the extent of 25% of the meat ingredients under specific declaration on the label. The mixture may contain not more than 8%, individually or collectively, of cereal, vegetable starch, starchy vegetable flour, soy flour, soy protein concentrate, isolated soy protein, dried milk, nonfat dry milk, dry or dried whey, reduced-lactose whey, reduced-minerals whey, whey protein concentrate, or calcium-reduced dried skim milk.

Chili Con Carne With Beans

1. It should contain not less than 25% of meat computed on the weight of the fresh meat.
2. Mechanically separated (species) product may be used in accordance with requirements.
3. Head meat, cheek meat, or heart meat exclusive of the heart cap may be used to the extent of 25% of the meat ingredient, and its presence shall be reflected in the required statement of ingredients.

Hash

1. It should contain not less than 35% of meat computed on the weight of the cooked and trimmed meat. The weight of the cooked meat used in this calculation shall not exceed 70% of the weight of the uncooked fresh meat.
2. Mechanically separated (species) product may be used in accordance with requirements.

**Table 5.8.17. (continued)**

Corned Beef Hash

1. It is the semisolid food product in the form of a compact mass, which is prepared with beef, potatoes, curing agents, seasonings, and any of the optional ingredients.
2. Either fresh beef, cured beef, or canned corned beef or a mixture of two or more of these ingredients may be used, and the finished product should contain not less than 35% of beef computed on the weight of the cooked and trimmed beef. The weight of the cooked meat used in this calculation should not exceed 70% of the weight of the uncooked fresh meat.
3. Potatoes refers to fresh potatoes, dehydrated potatoes, cooked dehydrated potatoes, or a mixture of two or more of these ingredients.
4. The curing agents that may be used are salt, sodium nitrate, sodium nitrite, potassium nitrate, or potassium nitrite, or a combination of two or more of these ingredients. When sodium nitrate, sodium nitrite, potassium nitrate, or potassium nitrite is used it should be used in amounts not exceeding those specified.
5. The seasonings that may be used, singly or in combination, are salt, sugar (sucrose or dextrose), spice, and flavoring, including essential oils, oleoresins, and other spice extractives.
6. Corned beef hash may contain one or more of the following optional ingredients:

   a. Beef cheek meat and beef head meat from which the overlying glandular and connective tissues have been removed, and beef heart meat, exclusive of the heart cap, may be used individually or collectively to the extent of 5% of the meat ingredients,
   b. Onions, including fresh onions, dehydrated onions, or onion powder
   c. Garlic, including fresh garlic, dehydrated garlic, or garlic powder
   d. Water
   e. Beef broth or beef stock
   f. Monosodium glutamate
   g. Hydrolyzed plant protein
   h. Beef fat
   i. Mechanically separated (species) product when derived from carcasses of cattle may be used in accordance with regulations.
7. The finished product should not contain more than 15% fat nor more than 72% moisture.
8. When any of beef organ meat is used, the label should bear the following applicable statement: "Beef cheek meat constitutes 5% of the meat

ingredient," or "beef head meat constitutes 5% of the meat ingredient," or "beef heart meat consitutes 5% of the meat ingredient." When two or more of the ingredients are used, the words "constitutes 5% of meat ingredient" need only appear once. Whenever the words "corned beef hash" are featured on the label so conspicuosly as to identify the contents, the prescribed statements should immediately and conspicuously precede or follow such name without intervening written, printed, or other graphic matter.

## Meat Stews

1. Meat stews such as "beef stew" or "lamb stew" should contain not less than 25% of meat of the species named on the label, computed on the weight of the fresh meat.
2. Mechanically separated (species) product may be used in accordance with regulations.

## Tamales

1. Tamales should be prepared with at least 25% meat computed on the weight of the uncooked fresh meat in relation to all ingredients of the tamales. When tamales are packed in sauce or gravy, the name of the product should include a prominent reference to the sauce or gravy; for example, "tamales with sauce" or "tamales with gravy." Product labeled "tamales with sauce" or "tamales with gravy" should contain not less than 20% meat, computed on the weight of the uncooked fresh meat in relation to the total ingredients making up the tamales and sauce or the tamales and gravy. Mechanically separated (species) product may be used in accordance with 9 CFR 319.6.

## Spaghetti with Meatballs and Sauce, Spaghetti with Meat and Sauce, and Similar Products

1. They should contain not less than 12% of meat computed on the weight of the fresh meat.
2. Mechanically separated (species) product may be used in accordance with regulations.
3. The presence of the sauce or gravy constituent should be declared prominently on the label as part of the name of the product.
4. Meatballs may be prepared with not more than 12%, singly and collectively, of farinaceous material, soy flour, soy protein concentrate,

**Table 5.8.17 (continued)**

isolated soy protein, nonfat dry milk, calcium-reduced dried skim milk, and similar substances.

Spaghetti Sauce with Meat

1. It should contain not less than 6% of meat computed on the weight of the fresh meat.
2. Mechanically separated (species) product may be used in accordance with regulations.

Tripe with Milk

It should be prepared so that the finished canned article, exclusive of the cooked-out juices and milk, contains at least 65% tripe. The product should be prepared with not less than 10% milk.

Beans with Frankfurters in Sauce, Sauerkraut with Wieners and Juice, and Similar Products

They should contain not less than 20% frankfurters of wieners computed on the weight of the smoked and cooked sausage prior to its inclusion with the beans or sauerkraut.

Lima Beans with Ham in Sauce, Beans with Ham in Sauce, Beans with Bacon in Sauce, and Similar Products

They should contain not less than 12% ham or bacon computed on the weight of the smoked ham or bacon prior to its inclusion with the beans and sauce.

Chow Mein Vegetables with Meat, and Chop Suey Vegetables with Meat

They should contain not less than 12% meat computed on the weight of the uncooked fresh meat prior to its inclusion with the other ingredients. Mechanically separated (species) product may be used in accordance with regulations.

Pork with Barbecue Sauce and Beef with Barbecue Sauce

1. They should contain not less than 50% meat of the species specified on the label, computed on the weight of the cooked and trimmed meat.
2. Mechanically separated (species) product may be used.
3. The weight of the cooked meat used in this calculation should not exceed 70% of the uncooked weight of the meat.
4. If uncooked meat is used in formulating the products, they should contain at least 72% meat computed on the weight of the fresh uncooked meat.

Beef with Gravy and Gravy with Beef

They should not be made with beef that, in the aggregate for each lot, contains more than 30% trimmable fat, that is, fat which can be removed by thorough, practicable trimming and sorting.

Source: 9 CFR 319, 1991.

**Table 5.8.18. Definition and Standard for Meat Pies and Pizza**

Meat Pies

Meat pies such as "beef pie," "veal pie," and "pork pie" should contain meat of the species specified on the label, in an amount not less than 25% of all ingredients including crust, and should be computed on the basis of the fresh uncooked meat.

Pizza

1. "Pizza with meat" is a bread base meat food product with tomato sauce, cheese, and meat topping. It should contain cooked meat made from not less than 15% raw meat. Mechanically separated (species) product may be used.
2. "Pizza with sausage" is a bread base meat food product with tomato sauce, cheese, and not less than 12% cooked sausage or 10% dry sausage; e.g., pepperoni. Mechanically separated (species) product may be used.

Source: 9 CFR 319, 1991.

**Table 5.8.19. Definition and Standard for Meat Extract and Fluid Extract of Meat**

Meat Extract

Meat extract (e.g., "beef extract") should contain not more than 25% of moisture.

Fluid Extract of Meat

Fluid extract of meat (e.g., "fluid extract of beef") should contain not more than 50% of moisture.

Source: 9 CFR 319, 1991.

**Table 5.8.20. Definition and Standard for Meat Salads and Meat Spreads**

Deviled Ham, Deviled Tongue, and Similar Products

1. "Deviled ham" is a semiplastic cured meat food product made from finely comminuted ham and containing condiments. Mechanically separated (species) product may be used. It may contain added ham fat. However, the total fat content should not exceed 35% of the finished product. The moisture content of deviled ham should not exceed that of the fresh unprocessed meat.
2. The moisture content of "deviled tongue" and similar products should not exceed that of the fresh, unprocessed meat.

Potted Meat Food Product and Deviled Meat Food Product

They should not contain cereal, vegetable flour, nonfat dry milk, or similar substances. The amount of water added should be limited to that necessary to replace moisture lost during processing.

Ham Spread, Tongue Spread, and Similar Products

1. They should contain not less than 50% of the meat ingredient named, computed on the weight of the fresh meat.
2. Other meat and fat may be used to give the desired spreading consistency provided it does not detract from the character of the spreads named.
3. Mechanically separated (species) product may be used.

Source: 9 CFR 319, 1991.

**Table 5.8.21. Definition and Standard for Breaded Products and Liver Meat Food Products**

Breaded Products

The amount of batter and breading used as a coating for breaded product should not exceed 30% of the weight of the finished breaded product.

Liver Meat Food Products

Meat food products characterized and labeled as liver products, such as liver loaf, liver cheese, liver spread, liver mush, liver paste, and liver pudding, should contain not less than 30% of pork, beef, sheep, or goat livers computed on the fresh weight of the livers.

Source: 9 CFR 319, 1991.

**Table 5.8.22. USDA Required Percentage of Meat in Meat Products**

| Meat Products | Percentage of Meat, Fat, Water |
|---|---|
| Baby food | |
| High meat dinner | At least 26% meat |
| Meat and broth | At least 61% meat |
| Vegetable with meat | At least 8% meat |
| Bacon (cooked) | Weight of cooked bacon is 40% of uncooked, cured, smoked bacon |
| Bacon and tomato spread | At least 20% cooked bacon |
| Bacon dressing | At least 8% cured, smoked bacon |
| Barbecue sauce with meat | At least 35% meat (cooked basis) |
| Barbecued meat | Weight of meat when barbecued cannot exceed 70% of the fresh uncooked meat; must have barbecued (crusted) appearance and be prepared over burning or smouldering hardwood or its sawdust; if cooked by other dry heat means, product name must mention the method of cooking. |
| Beans with bacon or ham in sauce | At least 12% bacon or ham (cooked basis) |

**Table 5.8.22. (continued)**

| Meat Products | Percentage of Meat, Fat, Water |
|---|---|
| Beans with frankfurters in sauce | At least 20% franks |
| Beans with meat in sauce | At least 12% meat |
| Beans with meatballs in sauce | At least 20% meatballs |
| Beef à la king | At least 20% beef (cooked basis) |
| Beef à la mode | At least 50% beef |
| Beef almondine with vegetables | At least 18% beef (cooked basis); product must contain almonds |
| Beef and dumplings with gravy or beef and gravy with dumplings | At least 25% beef |
| Beef burgundy | At least 50% beef; enough wine to characterize the sauce |
| Beef carbonade | At least 50% beef |
| Beef roulade | At least 50% beef (cooked basis) |
| Beef sausage (raw) | No more than 30% fat; no byproducts, no extenders, and no more than 3% water |
| Beef Stroganoff | At least 45% fresh, uncooked beef or 30% cooked beef and one of the following:<br>At least 10% sour cream or<br>A combination of at least 78% sour cream and 5% wine or 98% whole milk, 2% sour cream, and 28% wine |
| Beef with barbecue sauce | At least 50% beef (cooked basis) |
| Beef with gravy | At least 50% beef (cooked basis) |
| Breaded steaks, chops, etc. | Breading cannot exceed 30% of finished product weight. |
| Breakfast (frozen product containing meat) | At least 15% cooked meat based on total net weight of breakfast |
| Breakfast sausage | No more than 50% fat; may contain 38% binders and extenders and 3% water. |
| Brown and serve sausage | No more than 35% fat and no more than 10% added water |

| | |
|---|---|
| Brunswick stew | At least 25% meat, made up of at least two kinds of meat, including poultry meat; must contain corn as one of the vegetables. |
| Burgundy sauce with beef and noodles | At least 25% beef (cooked basis) and up to 20% noodles; enough wine to characterize the sauce |
| Cabbage rolls with meat in sauce | At least 12% meat |
| Cannelloni with meat and sauce | At least 10% meat |
| Cappelletti with meat in sauce | At least 12% meat |
| Cheesefurter | Shall contain sufficient cheese to characterize the product |
| Chili con carne | At least 40% meat |
| Chili con carne with beans | At least 25% meat |
| Chili hot dog with meat | At least 40% meat in chili |
| Chili Max | At least 16% meat; must be qualified with true product name: "beans, macaroni, and beef in sauce." |
| Chili sauce with meat | At least 6% meat |
| Chop suey (American style) with macaroni and meat | At least 25% meat |
| Chop suey vegetables with meat | At least 12% meat |
| Chopped ham | Must be prepared from fresh, cured, or smoked ham, plus certain kinds of curing agents and seasonings; may contain dehydrated onions, dehydrated garlic, corn syrup, and not more than 3% water to dissolve the curing agents. |
| Chow mein vegetables with meat | At least 12% meat |
| Chow mein vegetables with meat and noodles | At least 8% meat and the noodles must equal no more than one-third of the product |
| Corn dog | Must be accompanied by true product name, "batter wrapped franks on a stick;" limited to 65% batter and a minimum of 35% frankfurter. |
| Corned beef and cabbage | At least 25% corned beef (cooked basis) |

**Table 5.8.22. (continued)**

| Meat Products | Percentage of Meat, Fat, Water |
|---|---|
| Corned beef hash | At least 35% beef (cooked basis); must contain potatoes, curing agents, and seasonings; may contain onions, garlic, beef broth, beef fat, or others; no more than 15% fat; no more than 72% moisture. |
| Country ham | A dry-cured product frequently coated with spices; minimum 4% salt content |
| Creamed meat products or creamed sauce with meat products (chipped beef, cooked beef, cured beef, ham, franks, meatballs, etc.) | At least 18% meat product (cooked basis) |
| Crepe with meat | Based on total net weight of product; at least 20% meat (cooked basis) if filling has no other major characterizing ingredient, or 10% meat (cooked basis) if one other major characterizing ingredient (crepe with meat and cheese, for example) |
| Croquettes | At least 35% meat (cooked basis); 50% fresh basis |
| Curried sauce with meat and rice | At least 35% meat (cooked basis) in the sauce and meat part; no more than 50% cooked rice |
| Deviled ham | No more than 35% fat; no added moisture; no cereal |
| Dinner (frozen product meat) | At least 25% meat or meat food product containing (cooked basis) figured on total meal minus appetizer, bread, and dessert; consumer package must weigh at least 10 ounces (284 grams). |
| Dumplings with meat in sauce | At least 18% meat |
| Egg foo young with meat | At least 12% meat |
| Egg roll with meat | At least 10% meat |
| Egg roll with meat and seafood | At least 5% meat |
| Egg Benedict | At least 18% cured smoked ham |
| Enchilada with meat | At least 15% meat |
| Entree: meat or meat food product and one vegetable | At least 50% meat or meat food product (cooked basis) |
| Frankfurter, Bologna, | May contain only skeletal meat; no more than 30% |

| | |
|---|---|
| and similar cooked sausage | fat, 10% added water, and 2% corn syrup; no more than 15% poultry meat (exclusive of water in formula). |
| Frankfurter, Bologna, and similar cooked sausage with by-products or variety meats | Same limitations as above on fat, added water, and corn syrup; must contain at least 15% skeletal meat. These products must be specifically labeled, such as "frankfurters with byproducts," and each byproduct or variety meat must be specifically named in the list of ingredients. These include heart, tongue, spleen, tripe, and stomach. |
| Frankfurter, Bologna, and similar cooked sausage with byproducts or variety meats, and which also contain nonmeat binders | Product made with the above formulas and also containing up to 3 1/2% nonmeat binders (or 2% isolated soy protein). These products must be distinctively labeled, such as, "frankfurters with byproducts, nonfat dry milk added." The binders must be named in the list of ingredients. |
| Fried rice with meat | At least 10% meat |
| Fritter | At least 35% meat; no more than 65% breading |
| German style potato salad with bacon | At least 14% bacon (cooked basis) |
| Goulash | At least 25% meat |
| Gravy | At least 25% meat stock or broth, or at least 6% meat |
| Gravy and sauerbraten | 35% meat (cooked basis) |
| Gravy and Swiss steak | At least 35% meat (cooked basis) |
| Gravy and Yankee pot roast | At least 35% meat (cooked basis) |
| Gravy with beef | At least 35% beef (cooked basis) |
| Ham (canned) | Limited to 8% total weight gain after processing |
| Ham, cooked or cooked and smoked (not canned) | Must not weigh more after processing than the fresh ham weighs before curing and smoking; if contains up to 10% added weight, must be labeled "ham, water added." |
| Ham à la king | At least 20% ham (cooked basis) |
| Ham and cheese spread | At least 25% ham (cooked basis) |
| Ham chowder | |
| Ready-to-eat | At least 5% ham (cooked basis) |
| Condensed | At least 10% ham (cooked basis) |
| Ham salad | At least 35% ham (cooked basis) |

**Table 5.8.22. (continued)**

| Meat Products | Percentage of Meat, Fat, Water |
|---|---|
| Ham spread | At least 50% ham |
| Hamburger, hamburg, burger, ground beef, or chopped beef | No more than 30% fat; no extenders |
| Hash | At least 35% meat (cooked basis) |
| Hors d'oeuvre | At least 15% meat (cooked basis) or 10% bacon (cooked basis) |
| Jambalaya with meat | At least 25% meat (cooked basis) |
| Knish | At least 15% meat (cooked basis) |
| Kreplach | At least 20% meat |
| Lasagna with meat and sauce, or cheese lasagna with meat | At least 12% meat |
| Lasagna with meat sauce | At least 6% meat |
| Lasagna with sauce, cheese, and dry sausage | At least 8% dry sausage |
| Lima beans with ham or bacon in sauce | At least 12% ham or bacon (cooked basis) |
| Liver products, such as liver loaf, liver paste, liver pâté, liver cheese, liver spread, liverwurst, Braunschweiger, and liver sausage | At least 30% liver |
| Macaroni and beef in in sauce | At least 12% beef |
| Macaroni and cheese with ham | At least 12% ham (cooked basis) |
| Macaroni and meat | At least 25% meat |
| Macaroni salad with ham or beef | At least 12% meat (cooked basis) |
| Manicotti with meat in sauce (contains a meat filling) | At least 10% meat |
| Margarine (or oleo-) | If product is entirely of animal fat or contains some animal fat, it is processed under federal inspection; |

| | |
|---|---|
| | must contain–individually or in combination–pasteurized cream, cow's milk, skim milk, combination of nonfat dry milk and water or finely ground soybeans and water; may contain butter, salt, artificial coloring, vitamins A and D, and permitted functional substances; finished product must contain at least 80% fat from animal or vegetable Source: 9 CFR 319.s; labels must clearly state which types of fat are used. |
| Meat and dumplings in sauce | At least 25% meat |
| Meat and vegetables | At least 50% meat |
| Meat casserole | At least 25% fresh, uncooked meat or 18% cooked meat |
| Meat curry | At least 50% meat |
| Meat loaf (baked or oven-ready) | At least 65% meat and no more than 12% cereal products |
| Meat pasty | At least 25% meat |
| Meat pie or vegetable meat pie | At least 25% meat |
| Meat ravioli | At least 10% meat in ravioli |
| Meat ravioli in sauce | At least 10% meat in ravioli; at least 50% ravioli in total product |
| Meat salad | At least 35% meat (cooked basis) |
| Meat sauce | At least 6% meat |
| Meat soup | |
| Ready-to-eat | At least 5% meat |
| Condensed | At least 10% meat |
| Meat spread | At least 50% meat |
| Meat stew | At least 25% meat |
| Meat taco | At least 15% meat |
| Meat taco filling | At least 40% meat |
| Meat turnover | At least 25% meat |
| Meat Wellington | At least 50% cooked tenderloin spread with liver pâté or similar coating and covered with not more than 30% pastry |
| Meatballs | No more than 12% extenders, including textured vegetable protein; at least 65% meat |
| Meatballs in sauce | At least 50% meatballs (cooked basis) |
| Meatball Stroganoff | At least 45% meatballs (cooked basis) |
| Mince meat | At least 12% meat |

**Table 5.8.22. (continued)**

| Meat Products | Percentage of Meat, Fat, Water |
|---|---|
| Moussaka | At least 25% meat; must be qualified on label as "eggplant and meat casserole." |
| New England boiled dinner | At least 25% cooked corned beef |
| Omelet with bacon | At least 9% bacon (cooked basis) |
| Omelet with dry sausage | At least 12% dry sausage |
| Omelet with ham | At least 18% ham (cooked basis) |
| Omelet with meat food product, such as creamed chipped beef or corned beef hash | At least 25% meat food product |
| Omelet, western | At least 18% cooked ham; contains onions and green and/or red bell peppers. |
| Pâté de foie gras | At least 30% liver |
| Pepper steak (Chinese) | At least 30% thin, braised strips of beef (cooked basis) |
| Peppers and Italian sausage in sauce | At least 20% sausage (cooked basis) |
| Pizza with meat | At least 15% meat |
| Pizza with sausage | At least 12% sausage (cooked basis) or 10% dry sausage, such as pepperoni |
| Pork sausage | Not more than 50% fat or 3% water; may contain no byproducts or extenders. |
| Pork with barbecue sauce | At least 50% pork (cooked basis) |
| Pork with dressing | At least 50% pork (cooked basis) |
| Pork with dressing and gravy | At least 30% pork (cooked basis) |
| Proscuitto | A flat, dry-cured ham coated with spice |
| Quiche Lorraine | At least 8% bacon or ham (cooked basis) and 10% swiss or gruyere cheese |
| Rice with meat | At least 12% meat |
| Salisbury steak | At least 65% meat and no more than 12% extenders, including textured vegetable protein |
| Sandwich, meat | At least 35% meat in total sandwich; bread component may not exceed 50% of the sandwich. |
| Sauerbraten | At least 50% beef (cooked basis) |

| | |
|---|---|
| Sauerkraut balls with meat | At least 30% meat |
| Sauerkraut with wieners and juice | At least 20% wieners |
| Sausage with sauer-kraut in sauce | At least 40% sausage (cooked basis) |
| Scalloped potatoes and ham or sausage | At least 20% ham or sausage (cooked basis) |
| Scallopini ("veal scallopini," for example) | At least 35% meat (cooked basis) |
| Scrambled eggs with ham in a pancake | At least 9% ham (cooked basis) |
| Scrapple | At least 40% meat and/or meat byproducts |
| Shepherd's pie | At least 25% meat; no more than 50% mashed potatoes |
| Sloppy joe | At least 35% meat (cooked basis); must be qualified with true product name, "barbecue sauce with beef". |
| Snack | At least 15% meat (cooked basis) or 10% bacon (cooked basis) |
| Spaghetti sauce with meat | At least 6% meat |
| Spaghetti with meat or meat-balls in sauce | At least 12% meat |
| Spanish rice with meat | At least 20% meat (cooked basis) |
| Stuffed cabbage with meat in sauce | At least 12% meat |
| Stuffed pepper with meat in sauce | At least 12% meat |
| Sukiyaki | At least 30% meat |
| Sweet and sour meat | At least 25% meat and at least 16% fruit |
| Swiss steak with gravy | At least 50% meat (cooked basis) |
| Tamale | At least 25% meat |
| Tamale with sauce or gravy | At least 20% meat |
| Tamale pie | At least 20% meat; filling must be at least 40% of total product |
| Taquito | At least 15% meat |
| Tongue spread | At least 50% tongue |

**Table 5.8.22. (continued)**

| | |
|---|---|
| Tortellini with meat | At least 10% meat |
| Tortellini with meat in sauce | At least 50% cooked meat tortellini |
| Veal and peppers in sauce | At least 30% meat (cooked basis) |
| Veal bird | At least 60% meat and no more than 40% stuffing |
| Veal cordon bleu | At least 60% veal, 5% ham, and containing Swiss, gruyere, mozzarella, or pasteurized process Swiss cheese |
| Veal fricassee | At least 40% meat |
| Veal parmigiana | At least 40% breaded veal in sauce |
| Veal scallopini | At least 35% veal (cooked basis) |
| Veal steak | Chopped, shaped, cubed, frozen. Beef can be added up to 20% with product name show as, "veal steaks, beef added, chopped, shaped, and cubed"; if more than 20% beef, must be labeled, "veal and beef steak, chopped, shaped, and cubed"; no more than 30% fat in total product. |
| Vegetable and meat casserole | At least 25% meat |
| Vegetable and meat pie | At least 25% meat |
| Won ton soup | At least 5% meat |

Source: Adapted from USDA Home and Garden Bulletin Number 236, 1986.

## Unit II Poultry Products

### 5.9. Legal Requirements for Food Additives Permitted in Poultry Processing

**Table 5.9.1. Legal Requirements for Acidifiers, Antifoaming Agents, Antioxidants, and Oxygen Interceptors Permitted in Poultry Processing**

Specific Chemical or Group of Chemicals: Purpose
Products: Amounts

Acidifiers

Acetic acid: to adjust acidity
Various: sufficient for purpose

Citric acid: to adjust acidity
  Various: sufficient for purpose
Glucono delta-lactone: to adjust acidity
  Various: sufficient for purpose
Lactic acid: to adjust acidity
  Various: sufficient for purpose
Phosphoric acid: to adjust acidity
  Various: sufficient for purpose
Tartaric acid: to adjust acidity
  Various: sufficient for purpose

Antifoaming Agent

Methyl polysilicone: to retard foaming
  Soups: 10 ppm
  Rendered fats: 10 ppm
  Curing pickle: 50 ppm

Antioxidants and Oxygen Interceptors

BHA (butylated hydroxyanisole): to retard rancidity
  Various: 0.01% based on fat content (0.02% in combination with any other antioxidant listed in this table based on fat content)
BHT (butylated hydroxytoluene): to retard rancidity
  Various: 0.01% based on fat content (0.02% in combination with any other antioxidant listed in this table based on fat content)
Propyl gallate: to retard rancidity
  Various: 0.01% based on fat content (0.02% in combination with any other antioxidant listed in this table based on fat content)
TBHQ (tertiary butylhydroquinone): to retard rancidity
  Various: 0.01% based on fat content (0.02% in combination only with BHA and/or BHT based on fat content)
Tocopherols: to retard rancidity
  Various: 0.03% based on fat content (0.02% in combination with any other antioxidant listed in this table, except TBHQ, based on fat content)

Source: 9 CFR 381.147, 1991.

**Table 5.9.2. Legal Requirements for Binders and Extenders Permitted in Poultry Processing**

| Specific Chemical or Group of Chemicals: Purpose<br>Products: Amounts |
|---|
| Algin: to extend and stabilize product<br>Various: sufficient for purpose |
| Carrageenan: to extend and stabilize product<br>Various: sufficient for purpose |
| Carboxymethylcellulose (cellulose gum): to extend and stabilize product<br>Various: sufficient for purpose |
| Enzyme (rennet)-treated calcium-reduced dried skim milk and calcium lactate: to bind and extend product<br>Various: sufficient for purpose (calcium lactate required at rate of 10% of binder) |
| Enzyme (rennet)-treated sodium caseinate and calcium lactate: to bind and extend product<br>Various: sufficient for purpose (calcium lactate required at rate of 25% percent of binder) |
| Gums, vegetable: to bind and extend product<br>Various: sufficient for purpose (calcium lactate required at rate of 25% of binder) |
| Methylcellulose: to extend and to stabilize product (also carrier)<br>Various: 0.15% |
| Isolated soy protein: to bind and extend product<br>Various: sufficient for purpose |
| Sodium caseinate: to bind and extend product<br>Various: sufficient for purpose |
| Whey: to bind and extend product<br>Various: sufficient for purpose |
| Xanthan gum: to maintain uniform viscosity; to maintain suspension of particulate matter; to maintain emulsion stability; to maintain freeze-thaw stability<br>Various, except uncooked products or sausages of other products with a moisture limitation established by regulations |

Source: 9 CFR 381.147, 1991.

**Table 5.9.3. Legal Requirements for Chilling Media and Coloring Agents Permitted in Poultry Processing**

| Specific Chemical or Group of Chemicals: Purpose<br>Products: Amounts |
|---|
| Chilling Media |
| Salt (NaCl): to aid in chilling<br>Raw poultry products: 700 lb to 10,000 gal of water |
| Coloring Agents (Natural) |
| Annatto, carotene: to color products<br>Various: sufficient for purpose |
| Coloring Agents (Artificial) |
| Coal tar dyes (FD&C certified), titanium dioxide: to color products; to whiten products<br>Salads and spreads: 0.05% |

Source: 9 CFR 381.147, 1991.

**Table 5.9.4. Legal Requirements for Curing Accelerators Permitted in Poultry Processing[a]**

| Specific Chemical or Group of Chemicals: Purpose<br>Products: Amounts |
|---|
| Ascorbic acid: to accelerate color fixing<br>Cured poultry; cured, comminuted poultry products: 75 oz to 100 gal pickle at 1% pump level; 3/4 oz to 100 lb of poultry product; 10% solution to surfaces of the product prior to packaging. (The use of such solution shall not result in the addition of a significant amount of moisture to the product.) |
| Erythorbic acid: to accelerate color fixing<br>Cured poultry; cured, comminuted poultry products: 75 oz to 100 gal pickle at 10% pump level; 3/4 oz to 100 lb of poultry product; 10% solution to surfaces of the product prior to packaging. (The use of such solution shall not result in the addition of a significant amount of moisture to the product.) |

**Table 5.9.4. (continued)**

Fumaric acid: to accelerate color fixing.
  Cured, comminuted poultry or poultry products: 0.065% (or 1 oz to 100 lb) of the weight of the poultry of poultry byproducts, before processing
Sodium ascorbate: to accelerate color fixing
  Cured, comminuted poultry or poultry products: 87.5 oz to 100 gal pickle at 10% pump level; 7/8 oz to 100 lb of poultry product; 10% solution to surfaces of product prior to packaging. (The use of such solution shall not result in the addition of a significant amount of moisture to the product.)
Sodium erythorbate: to accelerate color fixing
  Cured, comminuted poultry or poultry products: 87.5 oz to 100 gal pickle at 10% pump level; 7/8 oz to 100 lb of poultry product; 10% solution to surfaces of product prior to packaging. (The use of such solution shall not result in the addition of a significant amount of moisture to the product.)
Citric acid or sodium citrate: to accelerate color fixing
  Cured, comminuted poultry or poultry products: May be used in cured products to replace up to 50% of the ascorbic acid or sodium ascorbate that is used.

Source: 9 CFR 381.147, 1991.
[a]All curing accelerators must be used only in combination with curing agents.

**Table 5.9.5 Legal Requirements for Curing Agents Permitted in Poultry Processing**

Specific Chemical or Group of Chemicals: Purpose
  Products: Amounts

Sodium or potassium nitrate: source of nitrite
  Cured, comminuted poultry or poultry products: 7 lb to 100 gal pickle; 3 1/2 oz to 100 lb or poultry product (dry cure); 2 3/4 oz to 100 lb of chopped poultry meat
Sodium or potassium nitrite. (Supplies of sodium nitrite and potassium nitrite and mixtures containing them must be kept securely under the care of a responsible employee of the establishment. The specific nitrite content of such supplies must be known and clearly marked accordingly.): to fix color
  Cured products: 2 lb to 100 gal pickle at 10% pump level; 1 oz to 100 lb of poultry product (dry cure); 1.4 oz to 100 lb chopped poultry meat. The use of nitrites, nitrates, or combination shall not result in more

than 200 ppm of nitrite, calculated as sodium nitrite, in finished product.

Source: 9 CFR 381.147, 1991.

**Table 5.9.6. Legal Requirements for Emulsifying Agents Permitted in Poultry Processing**

| Specific Chemical or Group of Chemicals: Purpose<br>Products: Amounts |
|---|
| Acetylated monoglycerides: to emulsify product<br>Various: sufficient for purpose |
| Diacetyl tartaric acid esters of mono- and diglycerides: to emulsify product<br>Rendered poultry fat or a combination of such fat with vegetable fat: sufficient for purpose. |
| Glycerol-lacto stearate, oleate or palmitate: to emulsify product<br>Rendered poultry fat or a combination of such fat with vegetable fat: sufficient for purpose |
| Lecithin: to emulsify product (also as antioxidant)<br>Various: sufficient for purpose |
| Mono- and diglycerides (glycerol palmitate, etc).: to emulsify product<br>Various: sufficient for purpose |
| Polysorbate 80 [polyoxyethylene (20) sorbitan monooleate]: to emulsify product<br>Various: 1% when used alone. If used with polysorbate 60, the combined total shall not exceed 1%. |
| Propylene glycol mono- and diesters of fats and fatty acids: to emulsify product<br>Rendered poultry fat or a combination of such fat with vegetable fat: sufficient for purpose |
| Polysorbate 60 [polyoxyethylene (20) sorbitan monostearate]: to emulsify product<br>Rendered poultry fat or a combination of such fat with vegetable fat: 1% when used alone. If used with polysorbate 80, the combined total shall not exceed 1%. |

Source: 9 CFR 381.147, 1991.

**Table 5.9.7. Legal Requirements for Flavoring Agents Permitted in Poultry Processing**

| Specific Chemical or Group of Chemicals: Purpose<br>Products: Amounts |
|---|
| Approved artificial smoke flavorings: to flavor product<br>Various: sufficient for purpose |
| Approved smoke flavoring to flavor product<br>Various: sufficient for purpose |
| Autolyzed yeast extract: to flavor product<br>Various: sufficient for purpose |
| Citric acid: to protect flavor: to flavor product<br>Various: sufficient for purpose |
| Corn syrup solids; corn syrup; glucose syrup: to flavor product<br>Various: sufficient for purpose |
| Disodium inosinate: to flavor product<br>Various: sufficient for purpose |
| Disodium guanylate: to flavor product<br>Various: sufficient for purpose |
| Hydrolyzed plant protein: to flavor product<br>Various: sufficient for purpose |
| Malt syrup: to flavor product<br>Various: sufficient for purpose |
| Milk protein hydrolyzate: to flavor product<br>Various: sufficient for purpose |
| Monosodium glutamate: to flavor product<br>Various: sufficient for purpose |
| Monoammonium glutamate: to flavor product<br>Various: sufficient for purpose |
| Sodium sulfoacetate derivative of mono- and diglycerides: to flavor product<br>Various: sufficient for purpose |
| Sugars approved (sucrose and dextrose): to flavor product<br>Various: sufficient for purpose |

Source: 9 CFR 381.147, 1991.

**Table 5.9.8. Legal Requirements for Gases Permitted in Poultry Processing**

| Specific Chemical or Group of Chemicals: Purpose<br>Products: Amounts |
|---|

Carbon dioxide solid (dry ice): to cool product or facilitate chopping or packaging
 Various: sufficient for purpose
Carbon dioxide liquid: contact freezing
 Various: sufficient for purpose
Nitrogen: to exclude oxygen from sealed containers
 Various: sufficient for purpose
Nitrogen liquid: contact freezing
 Various: sufficient for purpose

Source: 9 CFR 381.147, 1991.

**Table 5.9.9. Legal Requirements for Miscellaneous Agents Permitted in Poultry Processing**

| Specific Chemical or Group of Chemicals: Purpose<br>Products: Amounts |
|---|
| Sodium bicarbonate: to neutralize excess acidity; cleaning vegetables<br>Rendered fat, soups, curing pickle: sufficient for purpose |
| Calcium propionate: to retard mold growth<br>Fresh pie dough: 0.3% of calcium propionate or sodium propionate alone, or in combination, based on weight of the flour used |
| Sodium hydroxide: to decrease the amount of cooked-out juices<br>Poultry food products containing phosphates: may be used only in combination with phosphate in a ratio not to exceed one part sodium hydroxide to four parts phosphate |
| Sodium propionate: to retard mold growth<br>Fresh pie dough: 0.3% of calcium propionate or sodium propionate alone, or in combination, based on weight of the flour used |
| Disodium phosphate: to decrease the amount of cooked-out juices<br>Poultry food products except where otherwise prohibited by the poultry products inspection regulations: 0.5% of total product |
| Monosodium phosphate: to decrease the amount of cooked-out juices<br>Poultry food products except where otherwise prohibited by the poultry products inspection regulations: 0.5% of total product |
| Sodium metaphosphate, insoluble: to decrease the amount of cooked-out juices<br>Poultry food products except where otherwise prohibited by the poultry products inspection regulations: 0.5% of total product |
| Sodium polyphosphate, glassy: to decrease the amount of cooked-out juices<br>Poultry food products except where otherwise prohibited by the poultry products inspection regulations: 0.5% of total product |

**Table 5.9.9. (continued)**

Sodium tripolyphosphate: to decrease the amount of cooked-out juices
    Poultry food products except where otherwise prohibited by the poultry products inspection regulations: 0.5% of total product
Sodium pyrophosphate: to decrease the amount of cooked-out juices
    Poultry food products except where otherwise prohibited by the poultry products inspection regulations: 0.5% of total product
Sodium acid pyrophosphate: to decrease the amount of cooked-out juices
    Poultry food products except where otherwise prohibited by the poultry products inspection regulations: 0.5% of total product
Dipotassium phosphate: to decrease the amount of cooked-out juices
    Poultry food products except where otherwise prohibited by the poultry products inspection regulations: 0.5% of total product
Potassium tripolyphosphate: to decrease the amount of cooked-out juices
    Poultry food products except where otherwise prohibited by the poultry products inspection regulations: 0.5% of total product
Potassium pyrophosphate: to decrease the amount of cooked-out juices
    Poultry food products except where otherwise prohibited by the poultry products inspection regulations: 0.5% of total product

Source: 9 CFR 381.147, 1991.

**Table 5.9.10. Legal Requirements for Scalding Agents Permitted in Poultry Processing[a]**

Specific Chemical or Group of Chemicals: Purpose
    Products: Amounts

Alpha-hydro-omega-hydroxy poly(oxyethylene) poly(oxypropylene)(minimum 15 moles) poly(oxyethylene) block copolymer (polyoxamer): to remove feathers
    Poultry carcasses: not to exceed 0.05% by weight in scald water
Dimethylpolysiloxane: to remove feathers: to remove feathers
    Poultry carcasses: sufficient for purpose
Dioctyl sodium sulfosuccinate: to remove feathers
    Poultry carcasses: sufficient for purpose
Dipotassium phosphate: to remove feathers
    Poultry carcasses: sufficient for purpose
Ethylenediamine tetraacetic acid (sodium salts): to remove feathers
    Poultry carcasses: sufficient for purpose
Lime (calcium oxide, calcium hydroxide): to remove feathers

Poultry carcasses: sufficient for purpose
Polyoxyethylene (20) sorbitan monooleate: to remove feathers
Poultry carcasses: not to exceed 0.0175% in scald water
Potassium hydroxide: to remove feathers
Poultry carcasses: sufficient for purposes
Propylene glycol: to remove feathers
Poultry carcasses: sufficient for purposes
Sodium acid phosphate: to remove feathers
Poultry carcasses: sufficient for purposes
Sodium bicarbonate: to remove feathers
Poultry carcasses: sufficient for purposes
Sodium carbonate: to remove feathers
Poultry carcasses: sufficient for purposes
Sodium dodecylbenzene sulfonate: to remove feathers
Poultry carcasses: sufficient for purposes
Sodium 2-ethylhexyl sulfate: to remove feathers
Poultry carcasses: sufficient for purposes
Sodium hexametaphosphate: to remove feathers
Poultry carcasses: sufficient for purposes
Sodium hydroxide: to remove feathers
Poultry carcasses: sufficient for purposes
Sodium lauryl sulfate: to remove feathers
Poultry carcasses: sufficient for purposes
Sodium phosphate (mono-, di-, tribasic): to remove feathers
Poultry carcasses: sufficient for purposes
Sodium pyrophosphate: to remove feathers
Poultry carcasses: sufficient for purposes
Sodium sesquicarbonate: to remove feathers
Poultry carcasses: sufficient for purposes
Sodium sulfate: to remove feathers
Poultry carcasses: sufficient for purposes
Sodium tripolyphosphate: to remove feathers
Poultry carcasses: sufficient for purposes
Tetrasodium pyrophosphate: to remove feathers
Poultry carcasses: sufficient for purposes
Sodium tripolyphosphate: to remove feathers
Poultry carcasses: sufficient for purposes
Sodium pyrophosphate: to remove feathers
Poultry carcasses: sufficient for purposes
Sodium acid pyrophosphate: to remove feathers
Poultry carcasses: sufficient for purposes

**Table 5.9.10. (continued)**

Source: 9 CFR 381.147, 1991.
[a]Poultry scalding agents must be removed by subsequent cleaning operations.

**Table 5.9.11. Legal Requirements for Proteolytic Enzymes Permitted in Poultry Processing**

| Specific Chemical or Group of Chemicals: Purpose<br>Products: Amounts |
|---|
| *Aspergillus oryzae*: to soften tissue<br>Raw poultry muscle tissue of hen, cock, mature turkey, mature duck, mature goose, and mature guinea: Solutions consisting of water and approved proteolytic enzyme applied or injected into raw poultry tissue shall not result in a gain of more than 3% above the weight of the untreated product. |
| *Aspergillus flavus oryzae* group: to soften tissue<br>Raw poultry muscle tissue of hen, cock, mature turkey, mature duck, mature goose, and mature guinea: Solutions consisting of water and approved proteolytic enzyme applied or injected into raw poultry tissue shall not result in a gain of more than 3% above the weight of the untreated product. |
| Bromelin: to soften tissue<br>Raw poultry muscle tissue of hen, cock, mature turkey, mature duck, mature goose, and mature guinea: Solutions consisting of water and approved proteolytic enzyme applied or injected into raw poultry tissue shall not result in a gain of more than 3% above the weight of the untreated product. |
| Ficin: to soften tissue<br>Raw poultry muscle tissue of hen, cock, mature turkey, mature duck, mature goose, and mature guinea: Solutions consisting of water and approved proteolytic enzyme applied or injected into raw poultry tissue shall not result in a gain of more than 3% above the weight of the untreated product. |
| Papain: to soften tissue<br>Raw poultry muscle tissue of hen, cock, mature turkey, mature duck, mature goose, and mature guinea: Solutions consisting of water and approved proteolytic enzyme applied or injected into raw poultry tissue shall not result in a gain of more than 3% above the weight of the untreated product. |

Source: 9 CFR 381.147, 1991.

**Table 5.9.12. Legal Requirements for Synergists and Tenderizing Agents Permitted in Poultry Processing**

Specific Chemical or Group of Chemicals: Purpose
  Products: Amounts

Synergists [Used in Combination with Antioxidants]

Citric acid: to increase effectiveness of antioxidants
  Poultry fats: 0.01% alone or in combination with antioxidants in poultry fats

Malic acid: to increase effectiveness of antioxidants
  Poultry fats: 0.01% alone or in combination with antioxidants in poultry fats

Monoisopropyl citrate: to increase effectiveness of antioxidants
  Poultry fats: 0.01% poultry fats

Tenderizing Agents

Phosphoric acid: to increase effectiveness of antioxidants
  Poultry fats: 0.01%

Monoglyceride citrate: to increase effectiveness of antioxidants
  Poultry fats: 0.02%

*Aspergillus oryzae*: to soften tissue
  Raw poultry muscle tissue of hen, cock, mature turkey, mature duck, mature goose, and mature guinea: Solutions consisting of water and approved proteolytic enzymes applied or injected into raw poultry tissue shall not result in a gain of more than 3% above the weight of the untreated product.

*Aspergullus flavus oryzae* group: to soften tissue
  Raw poultry muscle tissue of hen, cock, mature turkey, mature duck, mature goose, and mature guinea: Solutions consisting of water and approved proteolytic enzymes applied or injected into raw poultry tissue shall not result in a gain of more than 3% above the weight of the untreated product.

Bromelin: to soften tissue
  Raw poultry muscle tissue of hen, cock, mature turkey, mature duck, mature goose, and mature guinea: Solutions consisting of water and approved proteolytic enzymes applied or injected into raw poultry tissue shall not result in a gain of more than 3% above the weight of the untreated product.

**Table 5.9.12. (continued)**

Ficin: to soften tissue

Raw poultry muscle tissue of hen, cock, mature turkey, mature duck, mature goose, and mature guinea: Solutions consisting of water and approved proteolytic enzymes applied or injected into raw poultry tissue shall not result in a gain of more than 3% above the weight of the untreated product.

Papain: to soften tissue

Raw poultry muscle tissue of hen, cock, mature turkey, mature duck, mature goose, and mature guinea: Solutions consisting of water and approved proteolytic enzymes applied or injected into raw poultry tissue shall not result in a gain of more than 3% above the weight of the untreated product.

Potassium chloride: to soften tissue

Raw poultry muscle tissue of hen, cock, mature turkey, mature duck, mature goose, and mature guinea: not more than 3% of a 2.0 molar solution

Magnesium chloride: to soften tissue

Raw poultry muscle tissue of hen, cock, mature turkey, mature duck, mature goose, and mature guinea: not more than 3% of a 0.8 molar solution

Calcium chloride: to soften tissue

Raw poultry muscle tissue of hen, cock, mature turkey, mature duck, mature goose, and mature guinea: not more than 3% of a 0.08 molar solution

Potassium, magnesium, or calcium chloride: to soften tissue

Raw poultry muscle tissue of hen, cock, mature turkey, mature duck, mature goose, and mature guinea: A solution of approved inorganic chlorides alone or in combination, applied or injected into raw poultry muscle tissue shall not result in a gain of more than 3% above the weight of the untreated product.

Source: 9 CFR 381.147, 1991.

## 5.10. USDA Definitions and Standards for Poultry Products

**Table 5.10.1. Listing of Poultry Products with USDA Definitions and Standards of Identity or Composition**

| |
|---|
| Percentage of poultry meat as an ingredient in a poultry product |
| Content of light and dark meat for specific poultry product |
| Canned boned poultry food |
| Content for poultry dinners (frozen) and pies |
| Poultry rolls |
| Poultry burgers, patties, à la Kiev, steak, fillet, baked, roasted, barbecued, and breaded |
| Contents of poultry dishes and specialty items |
| Maximum percentage of skin in certain poultry products |
| Requirements for adding solutions to ready-to-cook poultry products |
| Requirements for kinds, classes, and cuts of raw poultry |
| Requirements for specific cuts of poultry |
| Turkey ham |
| Minimal or maximal percentage of poultry meat in processed poultry products |

Source: 9 CFR 381.155-381.171, 1991.

**Table 5.10.2. Definition and Percentage of Poultry Meat as An Ingredient in a Poultry Product**

| |
|---|
| Poultry meat: derived from poultry processed, cooked, and cooled in a prescribed manner, e.g., with or without use of liquid or moisture in direct contact with the poultry meat following the cooking and cooling of the poultry |
| Percentage: If, following cooking and cooling of poultry meat to be used in poultry products, liquid or moisture is used in direct contact with the meat and the percentage of solids, excluding salt, in the poultry meat is found to be below 34%, the processor must raise the percentage in one of two ways. The percentage of poultry meat should be increased in proportion to the deficiency, or the meat should be so processed as to raise the solids content, excluding salt, to 34%. |

Source: 9 CFR 381.156, 1991.

**Table 5.10.3. Required Content of Light and Dark Meat for Specific Poultry Product**

| Product Label | Light Meat % | Dark Meat % |
|---|---|---|
| Natural proportions | 50-65 | 50-35 |
| Light or white meat | 100 | 0 |
| Dark meat | 0 | 100 |
| Light and dark meat | 51-65 | 49-35 |
| Dark and light meat | 35-49 | 65-51 |
| Mostly white meat | 66 or more | 34 or less |
| Mostly dark meat | 34 or less | 66 or more |

Source: 9 CFR 381.156, 1991.

**Table 5.10.4. Definition for Canned Boned Poultry Food**

1. Canned boned poultry should be prepared from cooked deboned poultry meat and may contain skin and fat not in excess of natural whole carcass proportions.
2. Gelatin stabilizers, or similar solidifying or emulsifying agents should not be added to product labeled "boned (kind of poultry)–solid pack" but may be added in quantities not in excess of a total of 0.5% of the total ingredients in the preparation of other canned boned poultry products. If so, the common name of the substance should be included in the name of the product, e.g., "boned chicken with broth–gelatin added."
3. The minimum percent of cooked, deboned poultry meat of kind indicated, with skin, fat, and seasoning, is as follows and the maximum percentage liquid that may be added is in parenthesis:

    a. Boned (kind of poultry)–solid pack: 95% (5%)
    b. Boned (kind of poultry): 90% (10%)
    c. Boned (kind of poultry) with broth: 80% (20%)
    d. Boned (kind of poultry) (–) percent broth: 50% (50%). Total amount of liquid added should be included in the name of the product, e.g., "boned chicken with 25 percent broth."

    When liquid is added, it may be in the form of, but is not limited to, broth or extractives. The product under "c" and "d" may be prepared from raw boned poultry in combination with cooked boned poultry as long as the product complies with the specified standard. The percentages are calculated on the basis of the total ingredients used in the preparation of the product.

4. Canned boned poultry with natural juices ["boned (kind of poultry) with natural juices"] should be prepared from either raw boned poultry or a mixture of raw boned poultry and cooked boned poultry and should have no liquid added during the preparation of the product.
5. Canned shredded poultry [shredded (kind of poultry)] consists of poultry meat reduced to a shredded appearance, from the kind of poultry indicated, with meat, skin, and fat not in excess of the natural whole carcass proportions. Canned shredded poultry from specific parts may include skin or fat in excess of the proportions normally found on a whole carcass, but not in excess of the proportions of skin and fat normal to the particular part or parts; and such product should be labeled accordingly.
6. Poultry products intended for infant or geriatric use and represented as having a "high meat" content must comply with specifications. The minimum percentage of cooked, deboned poultry meat of kind indicated, with seasoning is as follows and the maximum percentage liquid that may be added is in parenthesis:

    a. Strained or chopped (kind) with broth: 43% (57)
    b. High-meat dinner: 18.75%

    When liquid is added, it may be in the form of, but is not limited to, broth or extractives. Alternatively, the product may be prepared from raw boned poultry in combination with cooked boned poultry as long as the product complies with the specified standard. Label must indicate in some manner that product is for infant or geriatric servings.

Source: 9 CFR 381.157, 1991.

**Table 5.10.5. Required Content for Poultry Dinners (Frozen) and Pies**

| | (Kind) Pies | | (Kind) Dinners | |
|---|---|---|---|---|
| | Percent | Weight | Percent | Weight |
| Minimum cooked deboned poultry meat of kind indicated | 14 or | 1 1/8 oz per 8 oz pie[a] | 18 or | 2 oz[a, b] |
| Minimum raw deboned poultry meat of kind indicated | 25 or | 2 oz per 8 oz pie[a] | 25 or | 2 oz per 8 oz pie[a] |

Source: 9 CFR 381.158, 1991.

[a]Use the percentage or weight, whichever is greater. For example: 14% or 1 1/8 oz, 25% or 2 oz; 18% or 2 oz, 25% or 2 oz.

[b]Excluding weight of appetizers, desserts, etc. A minimum of 45% or 5 oz per dinner, whichever is greater, of cooked poultry including bone and breading may be used in lieu of minimum 18% or 2 oz of cooked deboned poultry meat and the cooked poultry including bone and breading should not contain more than 30% breading.

**Table 5.10.6. Definition for Poultry Rolls**

1. Binding agents, including but not limited to gelatin and wheat gluten, may be added in quantities not in excess of a total of 3% for cooked rolls and 2% for raw rolls, based on the total ingredients used in the preparation of the product, without affecting the name of the product. However, when such agents are added in excess of 3% or 2%, whichever is applicable, the common name of the agent or the term "binders added" should be included in the name of the product; e.g., "turkey roll–gelatin added."
2. With respect to heat-processed rolls, 2% or less liquid based on the weight of the finished product without liquid may remain with or be returned to product labeled as "(kind of poultry) roll."
3. Heat-processed rolls that have more than 2% liquid remaining with or returned to the product should be labeled as "(kind of poultry) roll with natural juices." If more than 2% of any liquid other than natural cookout juices is added, the product must be labeled to indicate that fact; e.g., "turkey roll with broth." Liquid should not be returned or added to product in excess of the amount normally cooked out during preparation.

Source: 9 CFR 159, 1991.

**Table 5.10.7 Definitions for Poultry Burgers, Patties, à la Kiev, Steak, Fillet, Baked, Roasted, Barbecued, and Breaded**

| Product: Definition |
|---|
| (Kind of poultry) burgers and (kind of poultry) patties: 100% poultry of the kind indicated, with skin and fat not in excess of natural proportions. Product containing fillers or binders should be named "(kind of poultry) patties." |
| (Kind of poultry) à la Kiev: poultry meat of the kind indicated, stuffed with butter which may be seasoned and the product may be wrapped in sufficient skin to cover the meat. It may be dipped in batter, fried, and frozen. |
| (Kind of poultry) steak or fillet: a boneless slice or strip of poultry meat of the kind indicated |
| (Kind of poultry) baked or (kind of poultry) roasted: ready-to-cook poultry of the kind indicated that has been cooked in dry source heat, e.g., oven roasted or oven baked |
| (Kind of poultry) barbecued: ready-to-cook poultry of the kind indicated that has been cooked in dry heat and basted with a seasoned sauce |
| (Kind of poultry) barbecued prepared with moist heat: ready-to-cook poultry of the kind indicated that has been cooked by the action of moist heat in a barbecue sauce |
| Breaded products: any poultry product coated with breading or a batter and breading in an amount not to exceed 30% of the weight of the finished breaded product |

Source: 9 CFR 381.160-381.166, 1991.

**Table 5.10.8. Contents of Poultry Dishes and Specialty Items**

The minimum percentage cooked deboned poultry meat of kind indicated for poultry specialty dishes and items should meet the percentage requirements:

1. (Kind) ravioli: 2%
2. (Kind) soup: 2%
3. Chop suey with (kind): 2%
4. (Kind) chop suey: 4%
5. (Kind) chow mein without noodles: 4%
6. (Kind) tamales: 6%

**Table 5.10.8. (continued)**

7. Noodles or dumplings with (ind): 6%. This standard also applies to products named (kind) with rice or similar starches.
8. (Kind) stew: 12%
9. (Kind) noodles or dumplings: 15%. This standard also applies to products named (ind) with rice or similar starches.
10. (Kind) with vegetables: 15%
11. Gravy with sliced (kind): 15%
12. (Kind) tetrazzini: 15%
13. (Kind) chili with beans: 17%
14. Creamed (kind): 20%
15. (Kind) cacciatore: 20%
16. (Kind) fricassee: 20%
17. (Kind) à-la-king: 20%
18. (Kind) croquettes: 25%
19. Sliced (kind) with gravy and dressing: 25%
20. (Kind) salad: 25%. The 25% includes poultry meat plus proportions of skin and fat natural to the poultry used.
21. (Kind) chili: 28%
22. (Kind) hash: 30%
23. Sliced (kind) with gravy: 35%
24. Minced (kind) barbecue: 40%

Also, for poultry specialty dishes and items, the minimum percentage cooked poultry of the kind indicated, indicating bone, should meet the percentage requirements:

1. (Kind) fricassee of wings: 40%
2. (Kind) noodles or dumplings: 30%
3. (Kind) cacciatore: 40%
4. (Kind) fricassee: 40%

For all items listed in this table, the following should be noted:

1. The percentages are calculated on a ready-to-serve basis, except that soup bases in institutional packs which are prepared for sale to institutional users should have a minimum of 15% cooked deboned poultry meat based on the weight of the soup base product.
2. The product name may contain other appropriate descriptive terms such as "noodle," e.g., "chicken noodle soup."

Source: 9 CFR 381.167, 1991.

**Table 5.10.9. Maximum Percentage of Skin in Certain Poultry Products**

| Product Name | Percentage Skin Raw | Percentage Skin Cooked |
|---|---|---|
| Boneless turkey breast or boneless turkey breast roll | 14 | |
| Boneless turkey thigh or boneless turkey thigh roll | 8 | |
| Boneless turkey or turkey roll | 15 | |
| Boneless chicken breast or boneless chicken breast roll | 18 | 20 |
| Boneless chicken or chicken roll | 20 | 25 |

Source: 9 CFR 381.168, 1991.

**Table 5.10.10. Requirements for Adding Solutions to Ready-to-Cook Poultry Products**

1. Butter alone, or solutions of poultry broth, poultry stock, water, or edible fats, or mixtures therof, in which are included functional substances such as spices, flavor enhancers, emulsifier, phosphates, coloring materials, or other substances, may be introduced by injection into the thick muscles (breast and legs) of ready-to-cook poultry carcasses and may be introduced by injection or marinating into any separate bone-in part therefrom, for the purpose of providing a basting medium or similar function.
2. The ingredients of the added materials and the manner of addition to the products must have government approval. The introduction of the added materials should increase the weight of the processed product by approximately 3% over the weight of the raw product after washing and chilling in compliance with established regulations.
3. Approval for this practice depends upon the ability of the processor to control the finished product, within a range of three-tenths of 1% accuracy, so that the average percentage of basting material in each outgoing lot is not greater than 3.3% or less than 2.7% of basting material when tested by an approved plant control procedure would be in compliance.
4. The control procedures to be eligible for approval are summarized below:

**Table 5.10.10 (continued)**

a. Assure compliance with all labeling requirements.
b. Control the variability of the amount of added approved solution with the limits defined above.
c. Provide for the disposition of all products not in compliance with requirements.
d. Incorporate a system of raw weight identification of a sufficient number of poultry and/or poultry parts to allow effective monitoring of the system by concerned parties.

Source: 9 CFR 381.169, 1991.

**Table 5.10.11. Standards and Requirements for Kinds, Classes, and Cuts of Raw Poultry**

| Kinds of Raw Poultry<br>Classes of Raw Poultry |
|---|
| Chickens |
| Rock Cornish game hen or Cornish game hen: a young immature chicken (usually 5-6 weeks of age) weighing not more than 2 pounds ready-to-cook weight, which was prepared from a Cornish chicken or the progeny of a Cornish chicken crossed with another breed of chicken. |
| Rock Cornish fryer, roaster, or hen: the progeny of a cross between a purebred Cornish and a purebred Rock chicken, without regard to the weight of the carcass involved; however, the term "fryer," "roaster," or "hen" should apply only if the carcasses are from birds with ages and characteristics that qualify them for such designation. |
| Broiler or fryer: a young chicken (usually under 13 weeks of age), of either sex, that is tender-meated with soft, pliable, smooth-textured skin and flexible breastbone cartilage |
| Roaster or roasting chicken: a young chicken (usually 3-5 months of age), of either sex, that is tender-meated with soft, pliable, smooth-textured skin and breastbone cartilage that may be somewhat less flexible than that of a broiler or fryer |
| Capon: a surgically unsexed male chicken (usually under 8 months of age) that is tender-meated with soft, pliable, smooth-textured skin |
| Hen, fowl, for baking or stewing: a mature female chicken (usually more than 10 months of age) with meat less tender than that of a roaster or roasting chicken and a nonflexible breastbone tip |

Cock or rooster: a mature male chicken with coarse skin, toughened and darkened meat, and a hardened breastbone tip

Turkeys

Fryer–roaster turkey; a young immature turkey (usually under 16 weeks of age), of either sex, that is tender-meated with soft, pliable, smooth-texture skin, and flexible breastbone cartilage

Young turkey: a turkey (usually under 8 months of age) that is tender-meated with soft, pliable, smooth-textured skin, and breastbone cartilage that is somewhat less flexible than in a fryer–roaster turkey. Sex designation is optional.

Yearling turkey: a fully matured turkey (usually under 15 months of age) that is reasonably tender-meated and with reasonably smooth-textured skin. Sex designation is optional.

Mature turkey or old turkey (hen or tom): an old turkey of either sex (usually in excess of 15 months of age) with coarse skin and toughened flesh

Ducks

Broilers duckling or fryer duckling: a young duck (usually under 8 weeks of age), of either sex, that is tender-meated and has a soft bill and soft windpipe

Roaster duckling: a young duck (usually under 16 weeks of age), of either sex, that is tender-meated and has a bill that is not completely hardened and a windpipe that is easily dented

Mature duck or old duck: a duck (usually over 6 months of age), or either sex, with toughened flesh, hardened bill, and hardened windpipe

Geese

Young goose: may be of either sex, tender-meated, with a windpipe that is easily dented

Mature goose or old goose: may be of either sex and has toughened flesh and hardened windpipe

Guineas

Young guinea: may be of either sex, is tender-meated, and has a flexible breastbone cartilage

Mature guinea or old guinea: may be of either sex, has toughened flesh, and a hardened breastbone

---

Source: 9 CFR 170, 1991.

**Table 5.10.12. Requirements for Specific Cuts of Poultry**

| Cuts of Poultry: Requirements |
|---|
| Breasts: They should be separated from the back at the shoulder joint and by a cut running backward and downward from that point along the junction of the vertebral and sternal ribs. The ribs may be removed from the breasts, and the breasts may be cut along the breastbone to make two approximately equal halves; or the wishbone portion may be removed before cutting the remainder along the breastbone to make three parts. Pieces cut in this manner may be substituted for lighter or heavier pieces for exact weight-making purposes and the package may contain two or more of such parts without affecting the appropriateness of the labeling as, e.g., "chicken breasts." Neck skin should not be included with the breasts, except that "turkey breasts" may include neck skin up the whisker. |
| Breasts with ribs: They should be separated from the back at the junction of the vertebral ribs and back. Breasts with ribs may be cut along the breastbone to make two approximately equal halves; or the wishbone portion may be removed before cutting the remainder along the breastbone to make three parts. Pieces cut in this manner may be substituted for lighter or heavier pieces for exact weight-making purposes and the package may contain two or more of such parts without affecting the appropriateness of the labeling as "breasts with ribs." Neck skin should not be included, except that "turkey breasts with ribs" may include neck skin up to the whisker. |
| Wishbones (pulley bones): They, with covering muscle and skin tissue, should be severed from the breast approximately halfway between the end of the wishbone (hypocledium) and front point of the breastbone (cranial process of the sternal crest) to a point where the wishbone joins the shoulder. Neck skin should not be included with the wishbone. |
| Drumsticks: They should be separated from the thigh by a cut through the knee joint (femorotibial and patellar joint) and from the hock joint (tarsal joint). |
| Thighs: They should be disjointed at the hip joint and may include the pelvic meat but should not include the pelvic bones. Back skin should not be included. |
| (Kind of poultry) legs: They should be the poultry product which includes the thigh and the drumstick, i.e., the whole leg, and may include the pelvic meat but should not include the pelvic bones. Back skin should not be included. |
| Wings: They should include the entire wing with all muscle and skin tissue intact, except that the wingtip may be removed. |
| Backs: They should include the pelvic bones and all the vertebrae posterior to the shoulder joint. The meat should not be peeled from the pelvic bones. |

The vertebral ribs and/or scapula may be removed or included without affecting the appropriateness of the name. Skin should be substantially intact.

Stripped backs: They should include the vertebrae from the shoulder joint to the tail and include the pelvic bones. The meat may be stripped off of the pelvic bones.

Necks: They, with or without neck skin, should be separated from the carcass at the shoulder joint.

Halves: They are prepared by making a full-length back and breast split of an eviscerated poultry carcass so as to produce approximately equal right and left sides.

Quarters: They consist of the entire eviscerated poultry carcass, which has been cut into four equal parts, but excluding the neck.

Breast quarter: It consists of half a breast with the wing and a portion of the back attached.

Breast quarter without wing: It consists of a front quarter of a poultry carcass from which the wing has been removed.

Leg quarter: It consists of a poultry thigh and drumstick, with a portion of the back attached.

Thigh with back portion: It consists of a poultry thigh with back portion attached.

Legs with pelvic bone: It consists of a poultry leg with adhering meat and skin and pelvic bone.

Wing drummette: It consists of the humerus of a poultry wing with adhering skin and meat attached.

Wing portion: It consists of a poultry wing except that the drummette has been removed.

Cutup poultry: It is any cut-up or disjointed portion of poultry or any edible part thereof.

Giblets: It consists of approximately equal numbers of hearts, gizzards, and livers, as determined on a count basis.

---

Source: 9 CFR 381.170, 1991.

**Table 5.10.13. Definition and Standard for Turkey Ham**

---

1. It should be fabricated from boneless, turkey thigh meat, with skin and the surface fat attached to the skin removed.
2. The product may or may not be smoked and should be cured using one or more of the approved curing agents. The product may also contain cure accelerators, phosphates, and flavoring agents; common salt, sugars, spices,

**Table 5.10.13. (continued)**

spice extractives, dehydrated garlic, and dehydrated onions; and water for purpose of dissolving and dispersing the substances specified above.

3. The cooked finished product weight should be no more than the original weight of the turkey thigh meat used prior to curing.
4. The product name should be qualified with the statement "cured turkey thigh meat."
5. If the product is fabricated from pieces of turkey thigh meat that result from the cutting through the muscle (as opposed the whole thighs intact or whole thighs with some incidental separation of muscle tissue during removal of the bone), the product name should be further qualified by a descriptive statement. The product name of product fabricated from such pieces of turkey thigh meat equivalent in size to a one-half inch cube or greater should be further qualified to specify that the product is "chunked and formed." The product name of product fabricated from such pieces of turkey thigh meat smaller than the equivalent of a one-half inch cube should be further qualified to specify that the product is "ground and formed" or "chopped and formed" as appropriate.

Source: 9 CFR 381.171, 1991.

**Table 5.10.14 USDA Required Minimal or Maximal Percentage of Poultry Meat in Processed Poultry Products**

| Name | Minimum or Maximum Percentage |
|---|---|
| Poultry à la king | At least 20% poultry meat |
| Poultry barbecue | At least 20% poultry meat |
| Poultry burgers | 100% meat with skin and fat |
| Poultry, breaded | No more than 30% breading |
| Poultry cacciatore | At least 20% poultry meat or 40% with bone |
| Poultry canneloni | At least 7% poultry meat |
| Poultry chili | At least 28% poultry meat |
| Poultry chili with beans | At least 17% poultry meat |
| Poultry chop suey | At least 4% poultry meat |
| Chop suey with poultry | At least 2% poultry meat |
| Poultry creole with rice | At least 35% cooked meat in poultry and sauce portion; not more than 50% cooked rice in total product |

| | |
|---|---|
| Poultry chow mein with or without noodles | At least 4% poultry meat |
| Poultry croquettes | At least 25% poultry meat |
| Poultry, creamed | At least 20% poultry meat |
| Poultry dinners | At least 18% poultry meat |
| Poultry fricassee | At least 20% poultry meat |
| Poultry fricassee with wings | At least 40% poultry wings (cooked basis with bone) |
| Poultry gizzards and gravy | At least 35% cooked gizzards |
| Poultry hash | At least 30% poultry meat |
| Poultry meat loaf | A minimum of 65% raw poultry or 50% poultry meat and a maximum of 12% extenders |
| Poultry with noodles or dumplings | At least 15% meat or 30% with bone |
| Noodles or dumplings with poultry | At least 6% poultry meat |
| Poultry paella | At least 35% poultry meat or 35% poultry meat and other meat (cooked basis), no more than 35% cooked rice, must contain seafood |
| Poultry parmagiana | At least 40% breaded poultry |
| Poultry pies | At least 14% poultry meat |
| Poultry ravioli | At least 2% poultry meat |
| Poultry salad mix | At least 45% poultry meat |
| Poultry salad | At least 25% poultry meat |
| Poultry soup | At least 2% poultry meat |
| Poultry-flavored soup | No minimum requirement (less than 2% poultry meat) |
| Poultry spread | At least 30% poultry meat |
| Poultry stew | At least 12% poultry meat |
| Poultry stew with dumplings | At least 8.4% cooked poultry meat (based on 70% of stew requirement) |
| Poultry tamales | At least 6% poultry meat |
| Poultry tetrazzini | At least 15% poultry meat |
| Poultry turnover | At least 14% poultry meat |
| Poultry with gravy | At least 35% poultry meat |
| Poultry with poultry | At least 15% poultry meat |
| Poultry gravy plus dressing | At least 25% poultry meat |
| Poultry with rice | At least 15% poultry meat |

**Table 5.10.14. (continued)**

| Name | Minimum or Maximum Percentage |
|---|---|
| Poultry scrapple | At least 30% poultry; same standard as scrapple made with meat |
| Poultry with vegetables | At least 15% poultry meat |

Source: Adapted from USDA Home and Garden Bulletin Number 236, 1986.

## 5.11. References


Anonymous. *Sausage and Ready-To-Serve Meats*, American Meat Institute, Chicago, IL, 1938.

Anonymous. *Meat Products*, Bull. No 804, American Dry Milk Institute, Chicago, IL, 1953.

Anonymous. *The All Beef Cookbook*, Benjamin Co., Elmsford, NY, 1973.

Anonymous. *Variety Meats*, Time Life Books, New York, 1983.

Anonymous. *Famous Brands Meat Cook Book*, Sammis Publishers, New York, 1986.

Anonymous. *Meat and Game Cooking*, Ortho, San Francisco, CA, 1988.

Anonymous. *Canned Foods: Principles of Thermal Process Control, Acidification, and Container Closure Evaluation*, 5th ed., The Food Processors Institute, Washington, DC, 1989.

Ashbrook, G. E. *Butchering, Processing, and the Preservation of Meat*, Van Nostrand Reinhold, New York, 1973.

Barnett, H. W., H. R. Nordin, et al. A study of factors affecting the flavour of cured ham, In *Proc. 11th European Meeting of Meat Research Workers*, Belgrade, Yugoslavia, 1965.

Bechtel, P. J. *Muscle as Food*. Academic Press, San Diego, CA, 1986.

Bloch, B. *Meat Board Meat Book*. Benjamin Co., Elmsford, NY, 1976.

Brady, D. E., F. H. Smith, et al. Characteristics of country-style hams as related to sugar content of curing mixture. *Food Res. 14*, 303: (1949).

Brady, D. E., F. H. Smith, et al. The characteristics of country style hams as related to sugar content of curing mixture. *Food Res.*, *14*, 303: (1953).

Brandy, P. J., G. Migake, et al. *Meat Hygiene*, 3rd Ed., Lea & Febiger, Philadelphia, PA, 1966.

Christian, J. A. *Curing Hams Country Style*. N. Carolina State Coll., Agric. Ext. Serv. Cir. 405, North Carolina State Univ., Raleigh, NC, 1960.

Christian, J. A. *Curing Georgia Hams--Country Style*. Georgia Cooperative Extension Service Bull. No. 627, 1981.

Cross, H. R., G. C. Smith, et al. Effect of quality attributes upon processing and palatability characteristics of commercially cured hams. *J. Food Sci.*, *36*, 982: (1971).

Dunker, C.F., M. Berman, et al. Quality and nutritive properties of different types of commercially cured hams. III. Vitamin content, biological value of protein and bacteriology. *Food Technol*, 7, 288: (1953).

Fields, M. D. and C. F.Dunker. Quality and nutritive properties of different types of commercially cured hams. I. Curing methods and chemical composition. *Food Technol.*, 6, 329: (1952).

Froning, G. W. Poultry meat sources and their emulsifying characteristics as related to processing variables. *Poultry Sci*, *49*, 1625: (1970).

Froning, G. W., and D. Janky. Effect of pH and salt preblending on emulsifying characteristics of mechanically deboned turkey frame meat. *Poultry Sci.*, *50*, 1206: (1971).

Gerrard, F. *Sausage and Small Goods Production*. Leonard Hill, London, 1959.

Ghinelli, I. *The Preservation of Meat*, Casanova, Parma, Italy, 1950.

Gruttner, F. *Handbook of Meat Processing*, 5th Ed., Serger & Hempel, Braunschweig, Germany, 1956.

Hall, R. and J. Sleight. *Home Book of Smoke Cooking Meats, Fish, and Game*, Stackpole, Harrisburg, PA, 1971.

Hoogenkamp, H. W. *Milk Proteins in Meat and Poultry Products*, DMV Campina, Inc., Stone Mountain, GA, 1987.

Hugunin, A. G., and N. L. Ewing. *Dairy Based Ingredients for Food Products*, Dairy Research Inc., Rosemont, IL, 1977.

Jensen, L. B. and W. T. Hess. A study of ham curing. *Food Res. 6*, 273: (1941).

Karel, M., O. Fennema and D.B. Lund. *Principles of Food Science, Part II: Physical Principles of Food Preservation*. Marcel Dekker, New York, 1975.

Karmas, E. *Sausage Products Technology*, Noyes Data Corporation, Park Ridge, NJ, 1977.

Karmas, E., and J.E. Thompson. Certain properties of canned hams as influenced by conditions of thermal processing. *Food Technol.*, *18*, 126: (1964).

Kotschevar, L. H. *Standards, Principles, and Techniques in Quantity Food Production*. 4th ed., Van Nostrand Reinhold, New York, 1988.

Kutas, R. *Great Sausage Recipes and Meat Curing*. Macmillan, New York, 1987.

Lawrie, R. A. Meat Science, 4th ed. Pergamon Press, London, 1985.

Levie, A. *Meat Handbook*, 4th ed., Van Nostrand Reinhold, New York, 1979.

Liepe, H. U. Starter Cultures in Meat Production. In *Biotechnology, Volume 5: Food and Feed Production with Microorganisms*, G. Reed, ed., VCH, New York, 1983.

MacFarlane, J. J., G. R. Schmidt, et al. Binding of meat pieces: A comparison of myosin, actomyosin, and sarcoplasmic proteins as binding agents. *J. Food Sci.*, *43*,. 123 (1977).

MacKenzie, D. S. *Prepared Meat Product Manufacturing*, 2nd ed., American Meat Institute, Chicago, IL, 1966.

Maga, J. A. *Smoke in Food Processing*, CRC Press, Boca Raton, FL, 1988.

Mandigo, R. W., and R. L. Hendrickson. Influence of hot processing pork carcasses on cured ham. *Food Technol.*, *20*, 538: (1966)

Mann, I. *Processing and Utilization of Animal By-products*, Food and Agriculture Development Paper 75, Food and Agriculture Organization, United Nations, Rome, Italy, 1962.

Methven, B. *Microwaving Meats*. Cy De Cosse, Minneapolis, MN, 1979.

Ockerman, H. W. *Sausage and Processed Meat Formulations*, Van Nostrand Reinhold, New York, 1989.

Ockerman, H. W., and C. L. Hansen. *Animal By-product Processing*, VCH, New York, 1988.

Parkhurst, C. R., and G. J. Mountney. *Poultry, Meat, and Egg Production*, Van Nostrand Reinhold, New York, 1987.

Pauli, E. *Classical Cooking: the Modern Way*, 2nd ed., Van Nostrand Reinhold, New York, 1988.

Pearson, A. M., and T. R. Dutson, eds. *Advances in Meat Research, Volume 2: Meat and Poultry Microbiology*, Van Nostrand Reinhold, New York, 1986.

Pearson, A. M., and T. R. Dutson, eds. *Advances in Meat Research, Volume 3: Restructured Meat and Poultry Products*, Van Nostrand Reinhold, New York, 1987.

Pearson, A. M., and F. W. Tauber. *Processed Meats*, Van Nostrand Reinhold, New York, 1984.

Pearson, A. M., T. R. Dutson, and A. Bailey, eds. *Advances in Meat Research, Volume 4: Collagen*, Van Nostrand Reinhold, New York, 1987.

Pomeranz, Y. *Functional Properties of Food Components*, Academic Press, San Diego, CA,1985.

Predika, J. *The Sausage-Making Cookbook*, Stackpole, Harrisburg, PA, 1983.

Price, J. F., and B. S. Schweigert, eds. *The Science of Meat and Meat Products*, 3rd ed., Food and Nutrition Press, Bridgeport, CT, 1987.

Reynolds, J. B., D. B. Anderson, et al. The effects of ultrasonic treatment on binding strength in cured ham rolls. *J. Food Sci.* 43, 866: (1978).

Rust, R. E., and D. G. Olson. *Meat Curing Principles and Modern Practices*. Kock Supplies, Inc. , Kansas City, MO, 1973.

Rutkowski, A. *Advances in Smoking of Foods*, Pergamon Press, New York, 1976.

Saffle, R. L. Meat emulsions. *Adv. Food Res.*, *16*, 105: (1968).

Schmidt, G. R. Sectioned and formed meat. *Proc. Recip. Meat Conf.*, *31*, 18: (1978).

Shannon, W. J. Modern sausage production. *Meat Process, 6*, 33: (1966).

Siegel, D. G., D. M. Theno, et al. Meat massaging: The effects of salt, phosphate and massaging on the presence of specific skeletal muscle proteins in the exudate of sectioned and formed ham. *J. Food Sci.*, *43*, 327: (1978A).

Siegel, D. G., D. M. Theno, et al. Meat massaging: The effects of salt, phosphate and massaging on cooking loss, binding strength and exudate composition in sectioned and formed ham. *J. Food Sci.*, 43, 431: (1978B).

Siegel, D. G., W. B. Tuley, et al. Sensory, textural and yield properties of a combination ham extended with isolated soy protein. *J. Food Sci.*, 44, 1049: (1979A).

Skelley, G. C , J. D. Kenp, et al. Quick aging of hams. *J. Animal Sci., 23*: (1964).

Tarr, Y. Y. *The Super-Easy-Step-by-Step Sausagemaking*. Random House, New York,, 1975.

Terrell, M. E., and D. B. Headlund. *Large Quantity Recipes*, 4th ed., Van Nostrand Reinhold, New York, 1988.

Viscora. *Automatic and Continuous Processes of Making Small Sausages*, Bull. No. 6, Viscora, Paris, 1968.

Wang, H. *Histology of Beef Casings*. Bull. No. 17, American Meat Institute Foundation, Chicago, IL, 1954.

Wang, H. *Histology of Pork Casings*, Bull. No. 25, American Meat Institute Foundation, Chicago, IL, 1955.

Wilson, G. D. A Summary of the Technical Aspects of Smoking Meats, Special Rept. No. 26, American Meat Institute Foundation, Chicago, IL, 1961.

Wistreich, H. Smoking of meats. In *Proc. Meat Indus. Res. Conf.*, American Meat Institute Foundation, Washington, DC, 1977.

Wistreich, H. E., R. E. Morse, et al. Curing of hams: A study of sodium chloride accumulation. I. Method, effect of temperature, cations, muscles and solution concentration. *Food Technol.*, *13*, 65:(1959).

Wistreich, H. E., R. E. Morse, et al. Curing of ham: A study of sodium chloride accumulation. II. Combined effects of time, solution concentration, and solution name. *Food Technol*, *14*, 549: (1960).

Woollen, A. Growing interest in meat tumbling. *Food Manufacture, 46*, 35: (1971).

Ziegler, P. T. *The Meat We Eat*, 3rd ed., Interstate Printers & Publishers, Danville, IL, 1952.

# CHAPTER 6
# BAKERY PRODUCTS

This chapter provides four groups of information on the processing of bakery products:

1. General data such as pH, bread faults, baking times and temperatures.
2. Formulas for selected rolls, cakes, cookies, and others.
3. Manufacturing processes for sour doughs.
4. Enrichment standards for flour, bread and related products.

The formulas for making bakery products have been developed from a computer model using available formulations from literature. For example, the amounts of ingredients used in each formula are given in ranges. Two types of notes are provided for each formula. Always begin with the lowest level for the ingredient salt. Since the formulas provided are in ranges, the exact legal level(s) of food additive(s) used cannot be stated. If one uses the data in this chapter, the final formula must comply with legal requirements for the usage of any food additive.

The citation for federal legal documents follows the standard format. For example, "21 CFR 136" refers to U.S. Code of Federal Regulations, Volume 21 and Part 136. All such documents are available from the Government Printing Office, Washington, DC.

## 6.1 General Information

**Table 6.1.1. Fundamentals in the Manufacture of Baked Goods**

- History and definitions
- Nutritional values
- Raw materials
  - Flour
  - Water
  - Salt
  - Other ingredients
  - Leavening agents
    - Biological
      - Yeasts
      - Sour dough starter cultures
    - Chemical leavening agents
    - Physical leavening agents
  - Additives affecting the processing characteristics of flour and doughs (dough conditioners)
    - Regulation of water absorption in doughs
    - Improvement of the properties of doughs and baked goods
    - Effect of dough fermentation
    - Dough acidification
  - Formulated additives
- Formation of the dough
  - Preparation of raw materials
  - Kneading of doughs
    - Mixers
    - The mixing process
    - Dough formation
- Leavening of doughs
  - Processing of wheat doughs
    - Straight dough process
    - Sponge dough process
    - Liquid preferment process
    - Continuous mix process
    - No time doughs
- Rye dough (sour dough) processes
  - Seed
  - Sour dough process with several stages
  - Two-stage sour dough process
  - Single-stage sour doughs
  - Salt sour processes
  - Freeze-dried sour dough
  - Effect of processing conditions on the sour dough fermention
  - Effect of processing conditions on the degradation of carboxylic acids
  - Effect of processing conditions on the hydrolysis of proteins
- Processing of fermented doughs
  - Fermentation
  - Dough makeup
  - Proofing
- Baking
  - Baking ovens
  - Processing conditions during baking
  - Physicochemical changes
  - Formation of aroma substances
- Bread quality and its preservation
  - Bread defects
  - Bread staling
  - Microbial spoilage
  - Retention of bread quality
    - Storage
    - Packaging
    - Preservation
      - Physical processes
      - Chemical processes

Source: Spicher, 1983. Used with permission.

## 6.1.2. pH Values of Bakery Products

| Product | pH | Product | pH |
|---|---|---|---|
| Angel food cake | 5.1-6.3 | Clear | 5.9-6.4 |
| Apples | 2.8-3.2 | Rye | 6.3-6.5 |
| Apricots (sweetened) | 3.7-4.1 | Cake | 5.0-5.8 |
| Baking powder | 7.0 | Cookie and pie | 4.9-5.9 |
| Baking soda | 8.4-8.7 | Fruit cake | 4.3-5.1 |
| Biscuits (baking powder) | 6.7-7.4 | Gelatin | 3.9-4.3 |
| Blackberries | 3.1-3.3 | Gooseberries | 2.8-3.3 |
| Blueberries | 3.1-3.3 | Huckleberries | 2.6-3.1 |
| Bran | 6.1-6.9 | Jam, fruit | 3.4-4.1 |
| Bread | | Jellies, fruit | 3.1-3.5 |
| Rye | 4.2-5.4 | Lemon juice | 2.1-2.5 |
| Rye (sour) | 4.3-4.6 | Lime juice | 2.2-2.5 |
| White | 4.9-5.8 | Loganberries | 2.6-3.1 |
| Whole wheat | 5.7-5.9 | Milk | |
| Cocoa | | Buttermilk powder | 4.0-5.3 |
| Natural and chocolate | 5.3-5.9 | Evaporated | 6.0-6.3 |
| Light dutched | 5.9-6.5 | Powdered (sweet) | 6.4-6.9 |
| Medium dutched | 5.9-6.9 | Sweet buttermilk | 6.3-6.5 |
| Heavy dutched | 6.7-7.9 | Sweetened condensed | 6.3-6.7 |
| Cookies | 6.6-6.1 | Malt syrup | 4.6-5.0 |
| Cheese (baker's) | 3.9-4.3 | Malt, dry | 5.1-6.1 |
| Cherries, red sour | 3.1-3.9 | Molasses | 4.9-5.2 |
| Crackers | 7.0-8.5 | Orange juice | 3.3-4.1 |
| Dates | 6.1-6.5 | Peaches | 3.4-3.9 |
| Devil's food cake | 7.5-8.1 | Pears | 3.7-4.7 |
| Doughnuts | 6.3-7.9 | Plums | 3.1-4.1 |
| Doughnut flour | 5.9-7.0 | Pound cake | 6.5-7.2 |
| Eggs | | Prunes | 3.7-4.1 |
| Whites | 8.8-9.1 | Pumpkin | 4.7-5.3 |
| Whites (after aging) | 5.2-5.5 | Raspberries, black | 3.1-4.3 |
| Whole | 6.3-8.3 | Raspberries, red | 3.0-3.5 |
| Yolks | 6.2-6.7 | Rhubarb | 3.0-3.4 |
| Fermenting dough | 4.6-5.1 | Sponge cake | 7.2-7.7 |
| Figs | 4.3-4.9 | Strawberries | 2.9-4.0 |
| Flour | | Sugar (cane or beet) | 6.3-6.9 |
| Bread, unbleached | 5.8-6.1 | Sugar (dextrose) | 4.7-6.1 |
| White, bleached | 5.7-6.0 | Syrup (invert) | 2.3-4.5 |

**Table 6.1.2. (continued)**

| Product | pH | Product | pH |
|---|---|---|---|
| Vinegar | 2.5-3.5 | Yeast | 4.9-6.1 |
| Water, distilled | 6.8-7.0 | Yellow layer cake | 6.8-7.2 |
| White layer cake | 7.0-7.5 | | |

Source: Amendols, 1983; Barrows, 1984; Cunningham, 1984; De Villers and Van der Berg, 1985; Fance, 1983; Field, 1985; Gisslen, 19083; Pomeranmz, 1985; Pyler, 1988; Spicher, 1983

### 6.1.3. Baking Times and Temperatures for Various Baked Goods

| | Time | °C |
|---|---|---|
| Cookies | 8 min | 350 |
| Crisp (flat) bread, knäckebrot | 8 min | 340 |
| Hamburg black bread | 4-5 h | 200-210 |
| Pumpernickel | 16-35 h | 100-180 |
| Rhenish black bread | 2-4 h | 210-230 |
| Rye bread (1 kg) | 40-60 min | 220-230 |
| Rye bread (2 kg) | 75 min | 250-270 |
| Rye bread (3 kg) | 90 min | 250-270 |
| Waffels | 2-9 min | 200-400 |
| Wheat bread and wheat mixed grain bread (1 kg) | 35-40 min | 220-230 |
| Wheat bread (2 kg) | 50 min | 220-230 |
| Westphalian black bread | 8-10 h | 180-200 |

Source: Amendols, 1983; Barrows, 1984; Cunningham, 1984; De Villers and Van der Berg, 1985; Fance, 1983; Field, 1985; Gisslen, 1983; Pomeranz, 1985; Pyler, 1988; Spicher, 1983

### 6.1.4. Factors Causing External Bread Faults

Blisters: Under Crust

1. Prebaking procedures
   - Mixing inadequate
   - Dough not mature
   - Moulding not careful
   - Final proof too much steam

Overproofing
2. Baking procedures
    Rough handling at oven
    Oven steam (condensation) excessive
    Depanner (separation) too much vacuum

Crust: Thick
1. Ingredients
    Sugar and shortening in formula not adequate
    Milk in formula not adequate
2. Prebaking procedures
    Diastatic activity inadequate
    Dough too old
    Humidity in final proof not adequate
    Final proof too much steam
3. Baking procedures
    Oven temperature low
    Overbaking

Crust Color: Dark
1. Ingredients
    High sugar content in formula
    High milk product content in formula
2. Prebaking procedures
    Immature dough
3. Baking procedures
    Over too hot
    Oven top heat high
    Baking excessive

Crust Color: Pale
1. Ingredients
    Residual sugar not enough
    Mineral yeast food too much
2. Prebaking procedures
    Diastatic activity insufficient
    Fermentation temperature too high
    Dough old
    Humidity inadequate for final proof
3. Baking procedures
    Oven not hot enough

**Table 6.1.4. (continued)**

Oven top heat low
Baking insufficient

Loaf: Volume Inadequate

1. Ingredients
   - Flour immature or green
   - Flour wrong type
   - Yeast insufficient
   - Yeast poor quality or mishandled
   - Salt too much
   - Shortening not enough
2. Prebaking procedures
   - Diastatic activity too much
   - Absorption not enough
   - Mixing ingredients inadequate or excessive
   - Dough fermentation insufficient or excessive
   - Dough temperature not high enough
   - Pan proof not adequate
3. Baking procedures
   - Pan size and dough weight not matching
   - Oven too hot

Loaf: Volume Too Large

1. Ingredients
   - Salt not enough
2. Prebaking procedures
   - Overaged dough
   - Overproofing
3. Baking procedures
   - Overscaling for pan size
   - Oven not hot enough

Shred and Wild Break: Present

1. Prebaking procedures
   - Dough slightly overaged
   - Mixing ingredients excessive
   - Oxidation not sufficient
   - Humidity in final proof not adequate
   - Moulding (low pressure board) not proper

<u>Shred and Break: Absent</u>

1. Prebaking procedures
   - Flour weak
   - Mixing not adequate
   - Dough age improper (too young or too old)
   - Diastatic activity too much
   - Oxidation too much
   - Final proof excessive
2. Baking procedures
   - Oven too hot

<u>Tops: Shell</u>

1. Ingredients
   - Flour immature
2. Prebaking procedures
   - Diastatic activity low
   - Oxidation not sufficient
   - Dough young
   - Humidity in final proof not sufficient
   - Underproofing
   - Moulding (low pressure board) too tight
3. Baking procedures
   - Top heat in oven too high
   - Steam in oven not adequate

---

Source: Amendols, 1983; Anonymous, 1981; Barrows, 1984; Cunningham, 1984; De Villers and Van der Berg, 1985; Fance, 1983; Field, 1985; Gisslen, 1983; Orcutt and Taylor, 1981; Pomeranz, 1985; Pyler, 1988; Spicher, 1983; Sultan, 1976.

### 6.1.5. Factors Causing Internal Bread Faults

---

<u>Color: Dull Gray</u>

1. Prebaking procedures
   - Diastatic activity too much
   - Mixing improper (excessive or inadequate)
   - Dough young (round, thick cells)
   - Dough old (open cell structure)
   - Sheeting and moulding not done properly
   - Dough underweight for pan size
   - Proof too much

**Table 6.1.5. (continued)**

2. Baking procedures
    - Oven not hot enough

<u>Crumb: Streaked</u>

1. Prebaking procedures
    - Dough mixing not proper
    - Dusting flour excessive
    - Trough grease or divider oil excessive
    - Shortening not adequate
    - Dough crusting during fermentation and final proof
    - Moulder setting not proper
    - Rework dough used improperly
    - Pan grease too much

<u>Flavor: Poor</u>

1. Ingredients
    - Quality poor
    - Storage unsatisfactory
    - Formula imbalance
    - Salt low
2. Prebaking procedures
    - Dough too old or too young
    - Baking plant unsanitary
    - Trough old and pan greasy
3. Baking procedures
    - Baking under- or overdone

<u>Grain: Poor</u>

1. Ingredients
    - Weak flour
2. Prebaking procedures
    - Dough too stiff
    - Dough too slack
    - Dough too young
    - Moulding not properly done
    - Dough weight inadequate for pan size
3. Baking procedures
    - Oven not hot enough

<u>Holes: Present in Bread</u>

1. Ingredients
   - Floor weak or immature
   - Formula not balanced
2. Prebaking procedures
   - Mixing not done properly (under or excessive mixing)
   - Doughs too old or too young
   - Sponge or dough crusting during fermentation
   - Dusting flour or divider oil excessive
   - Moulding not proper
   - Final proof temperature too high
   - Oven steam not adequate

<u>Keeping Quality: Poor</u>

1. Ingredients
   - Poor quality
   - Formula imbalance
   - Formula lean
2. Prebaking procedures
   - Mixing not properly done
   - Fermentation excessive
   - Fermentation inadequate
   - Proofing excessive
3. Baking procedures
   - Oven temperature low
   - Excessive cooling before wrapping

<u>Texture: Poor</u>

1. Prebaking procedures
   - Dough too stiff
   - Mixing not properly done
   - Diastatic activity too much
   - Shortening not adequate
   - Dough too old
   - Sponge or dough crusting during fermentation
   - Dough crusting during final proof
   - Proofing excessive
   - Dough weight inadequate for pan size
2. Baking procedures
   - Oven not hot enough

**Table 6.1.5. (continued)**

Source: Amendols, 1983; Anonymous, 1981; Barrows, 1984; Cunningham, 1984; De Villers and Van der Berg, 1985; Fance, 1983; Field, 1985; Gisslen, 1983; Orcutt and Taylor, 1981; Pomeranz, 1985; Pyler, 1988; Spicher, 1983; Sultan, 1976.

## 6.2. Formulas for Some Rolls, Cakes, Cookies, and Variety Baked Goods

**Table 6.2.1. Soft Roll Formula**

| Ingredients | Percent |
|---|---|
| Flour | 100 |
| Nonfat dry milk | 1-2 |
| Salt | 1.75-2.25 |
| Shortening | 4.5-6.5 |
| Sugar | 10-12 |
| Water (variable) | 57-61 |
| Yeast | 3-3.5 |

Source: Compiled from Amendols, 1983; Barrows, 1984; Cunningham, 1984; Fance, 1983; Field, 1985; Matz, 1987; Miller, 1981; Orcutt and Taylor, 1981.

**Table 6.2.2. Formula for Devil's Food and Chocolate Cakes**

| Ingredients | Devil's Food Cake, % | Chocolate Cake, % |
|---|---|---|
| Flour | 14.5-15.5 | 17.0-18.0 |
| Sugar | 25.5-26.5 | 28.0-29.0 |
| Shortening | 11.0-12.0 | 8.0-9.0 |
| Eggs (whole) | 13.0-14.0 | 10.0-11.0 |
| Cocoa (natural) | 7.0-8.0 | - |
| Chocolate liquor | - | 8.0-9.0 |
| Nonfat dry milk solids | 3.0-4.0 | 3.0-4.0 |
| Water | 22.0-23.0 | 22.0-23.0 |
| Salt | 0.4-0.8 | 0.5-0.8 |
| Soda | 0.5-0.6 | 0.1-0.2 |
| Baking powder | - | 0.5-1.0 |

Source: Compiled from Amendols, 1983; Anonymous, 1981; Anonymous, 1982; Barrows, 1984; Cunningham, 1984; Field, 1985; Gisslen, 1983; Goodbody, 1986; Matz, 1987; Miller, 1981.
The final formula must comply with the legal limit of any food additive used.

**Table 6.2.3. Formula for Angel Food Cake**

| Ingredient | Percent |
|---|---|
| Egg white | 40.0-42.0 |
| Flour | 14.0-16.0 |
| Sugar | 40.0-42.0 |
| Salt | 0.4-0.6 |
| Cream of tartar | 0.6-0.7 |
| Vanilla | 0.7-0.9 |

Source: Compiled from Amendols, 1983; Anonymous, 1981; Anonymous, 1982; Barrows, 1984; Cunningham, 1984; Field, 1985; Gisslen, 1983; Goodbody, 1986; Matz, 1987; Miller, 1981.
The final formula must comply with the legal limit of any food additive used.

**Table 6.2.4. Formulas for Solid- and French-Type Cheese Cake**

| | Solid Type, % | French Type, % |
|---|---|---|
| Baker's cheese | 40.0-42.0 | 34.0-35.0 |
| Bread flour | - | 2.0-3.0 |
| Butter | - | 2.5-3.5 |
| Cake flour | - | 2.0-3.0 |
| Egg whites | - | 7.0-8.0 |
| Egg yolks | 3.5-4.5 | - |
| Emulsifier shortening | 16.0-17.0 | - |
| Nonfat dry milk | 0.4-0.6 | 2.0-3.0 |
| Powdered lemon juice | 0.0.1-0.3 | 0.2-0.5 |
| Salt | 0.2-0.6 | 0.2-0.5 |
| Starch | 1.0-2.0 | - |
| Sugar | 17.0-18.0 | 34.0-35.0 |
| Vanilla extract | 5.0-6.0 | 26.0-27.0 |
| Whole eggs | 12.0-13.0 | 7.5-8.5 |

Source: Compiled from Amendols, 1983; Anonymous, 1981; Anonymous, 1982; Barrows, 1984; Cunningham, 1984; Field, 1985; Gisslen, 1983; Goodbody, 1986; Matz, 1987; Miller, 1981.
The final formula must comply with the legal limit of any food additive used.

**Table 6.2.5. Formulas for Fudge and Milk Chocolate Cakes**

| Ingredients | Fudge Cake, % | Milk Chocolate Cake, % |
|---|---|---|
| Flour | 14.0-15.0 | 16.5-17.5 |
| Sugar | 28.5-29.5 | 27.0-28.0 |
| Shortening | 9.0-10.0 | 9.0-10.0 |
| Eggs (whole) | 11.5-12.5 | 10.0-11.0 |
| Cocoa (natural) | 7.0-8.0 | - |
| Chocolate liquor | - | 4.0-5.0 |
| Nonfat dry milk solids | 2.5-3.5 | 8.0-9.0 |
| Water | 22.0-23.0 | 22.0-23.0 |
| Salt | 0.5-0.6 | 0.6-0.7 |
| Soda | 0.15-0.25 | 0.1-0.2 |
| Baking powder | 0.4-0.5 | 0.5-1.0 |

Source: Compiled from Amendols, 1983; Anonymous, 1981; Anonymous, 1982; Barrows, 1984; Cunningham, 1984; Field, 1985; Gisslen, 1983; Goodbody, 1986; Matz, 1987; Miller, 1981.
The final formula must comply with the legal limit of any food additive used.

**Table 6.2.6. Formula for White Cake**

| Ingredients | Percentage |
|---|---|
| Sugar | 25.0-26.0 |
| Flour | 25.0-26.0 |
| Shortening, regular hydrogenation | 12.5-13.5 |
| Egg whites | 17.0-18.0 |
| Milk | 15.0-16.0 |
| Baking powder | 0.6-1.0 |
| Salt | 0.3-0.7 |
| Flavor | 0.3-0.8 |

Source:Compiled from Amendols, 1983; Anonymous, 1981; Anonymous, 1982; Barrows, 1984; Cunningham, 1984; Field, 1985; Gisslen, 1983; Goodbody, 1986; Matz, 1987; Miller, 1981.
The final formula must comply with the legal limit of any food additive used.

**Table 6.2.7. Formula for a Yellow Cake**

| Ingredients | Percentage |
|---|---|
| Sugar | 25.0-26.0 |
| Flour | 29.0-30.0 |
| Shortening, regular hydrogenation | 13.0-14.0 |
| Whole eggs | 14.0-15.0 |
| Milk | 14.5-15.5 |
| Baking powder | 0.5-1.0 |
| Salt | 0.3-0.8 |
| Flavor | 0.2-0.9 |

Source:compiled from Amendols, 1983; Anonymous, 1981; Anonymous, 1982; Barrows, 1984; Cunningham, 1984; Field, 1985; Gisslen, 1983; Goodbody, 1986; Matz, 1987; Miller, 1981.
The final formula must comply with the legal limit of any food additive used.

**Table 6.2.8. Formula for a Chiffon Cake**

| Ingredients | Percentage |
|---|---|
| Cake flour | 14.0-15.0 |
| Cream of tartar | 0.1-0.2 |
| Egg whites | 31.5-32.5 |
| Egg yolk | 10.0-11.0 |
| Granulated sugar | 15.5-16.5 |
| Oil | 10.0-11.0 |
| Powdered sugar | 15.5-16.5 |
| Salt | 0.3-0.4 |

Source: Compiled from Amendols, 1983; Anonymous, 1981; Anonymous, 1982; Barrows, 1984; Cunningham, 1984; Field, 1985; Gisslen, 1983; Goodbody, 1986; Matz, 1987; Miller, 1981.
The final formula must comply with the legal limit of any food additive used.

**Table 6.2.9. Formula for Rotary Molded Cookies**

| Ingredients | Parts by Weight |
|---|---|
| Flour | 100 |
| Egg solids | 1.5-2.52 |
| Flavor | 1.0-2.0 |
| Invert sugar | 1.5-2.5 |
| Nonfat dry milk solids | 1.5-2.5 |
| Salt | 1.0-2.0 |
| Shortening | 30-45 |
| Sodium bicarbonate | 0.4-0.9 |
| Sugar | 35-55 |
| Water | 7-9 |

Source: Compiled from Amendols, 1983; Anonymous, 1981; Anonymous, 1982; Barrows, 1984; Cunningham, 1984; Field, 1985; Gisslen, 1983; Goodbody, 1986; Matz, 1987; Miller, 1981.
The final formula must comply with the legal limit of any food additive used.

**Table 6.2.10. Formula for Oatmeal Cookies**

| Ingredients | Parts by Weight |
|---|---|
| Oatmeal and flour (50/50) | 100 |
| Chocolate chips, ground raisins, or ground dates | 28-32 |
| Cinnamon | 0.1-0.4 |
| Egg solids | 1.5-2.5 |
| Flavor | 0.5-1.2 |
| Invert sugar | 1.5-2.5 |
| Nonfat dry milk solids | 1.5-2.5 |
| Salt | 1.5-2.5 |
| Shortening | 45-55 |
| Sodium bicarbonate | 0.5-1.5 |
| Sugar | 55-65 |
| Water | 14-15 |

Source: Compiled from Amendols, 1983; Anonymous, 1981; Anonymous, 1982; Barrows, 1984; Cunningham, 1984; Field, 1985; Gisslen, 1983; Goodbody, 1986; Matz, 1987; Miller, 1981.
The final formula must comply with the legal limit of any food additive used.

**Table 6.2.11. Formula for Chocolate Chip Cookies**

| Ingredients | Parts by Weight |
|---|---|
| Flour | 100 |
| Chocolate chips | 50-60 |
| Egg solids | 6-8 |
| Flavor | 0.5-1.5 |
| Salt | 1.0-2.0 |
| Shortening | 55-65 |
| Sodium bicarbonate | 0.3-0.9 |
| Sugar | 70-80 |
| Water | 16-18 |

Source: Compiled from Amendols, 1983; Anonymous, 1981; Anonymous, 1982; Barrows, 1984; Cunningham, 1984; Field, 1985; Gisslen, 1983; Goodbody, 1986; Matz, 1987; Miller, 1981.
The final formula must comply with the legal limit of any food additive used.

**Table 6.2.12. Formula for Wire-Cut Cookies**

| | Parts by Weight |
|---|---|
| Flour | 100 |
| Invert sugar | 7.0-24.0 |
| Lecithin | 0.2-0.6 |
| Milk | 2.0-6.0 |
| Salt | 0.8-1.5 |
| Shortening | 28-43 |
| Sodium bicarbonate | 0.4-1.2 |
| Sugar | 28-42 |
| Water | 8.0-23.0 |
| Whole eggs | 6.5-11.0 |

Source: Compiled from Amendols, 1983; Anonymous, 1981; Anonymous, 1982; Barrows, 1984; Cunningham, 1984; Field, 1985; Gisslen, 1983; Goodbody, 1986; Matz, 1987; Miller, 1981.
The final formula must comply with the legal limit of any food additive used.

**Table 6.2.13. Formula for a Croissant**

| Ingredients | Parts by Weight |
|---|---|
| Bread flour | 100 |
| Margarine | 1.5-2.5 |
| Nonfat dry milk | 2.5-3.5 |
| Salt | 1.0-2.0 |
| Sugar | 9.5-10.5 |
| Water | 45-60 |
| Yeast | 6-8 |
| Yeast food | 0.4-0.7 |

Source: Compiled from Amendols, 1983; Anonymous, 1981; Barrows, 1984; Cunningham, 1984; De Villers and Van der Berg, 1985; Fance, 1983; Field, 1985; Gisslen, 1983; Green, 1984; Matz, 1987; Miller, 1981; Pyler, 1988; Sultan, 1976.
The final formula must comply with the legal limit of any food additive used.

**Table 6.2.14. Formula for Yeast-Raised Doughnuts**

| | Parts by Weight |
|---|---|
| Flour | 100 |
| Egg yolk | 0-6 |
| Nonfat dry milk | 1.5-5.5 |
| Salt | 0.5-2.5 |
| Shortening | 7.5-15.5 |
| Sweetener, sucrose and/or dextrose | 5.5.-11.5 |
| Water | 50.0-60.0 |
| Yeast | 3-7 |

Source: Compiled from Amendols, 1983; Barrows, 1984; Clancy, 1975; Cunningham, 1984; De Villers and Van der Berg, 1985; Fance, 1983; Field, 1985; Gisslen, 1983; Matz, 1987; Miller, 1981; Pyler, 1988; Sultan, 1976.
The final formula must comply with the legal limit of any food additive used.

**Table 6.2.15. Formula for Pizza Crust**

| Ingredients | Parts Where Applicable |
|---|---|
| Basic | |
| Baking powder | 0.4-4.5 |
| Calcium propionate | 0.08-0.32 |
| Flour | 100 |
| Salt | 0.5-2.5 |
| Shortening | 3.5-14.5 |
| Sugar | 0.5--5.5 |
| Water | 50.0-75.0 |
| Yeast | 0.4-5.5 |
| Optional | |
| Corn meal | 5.0-21.0 |
| L-Cysteine or $NaHSO_3$ | 4.0-95.0 ppm |
| Flavoring | Variable |
| Proteolytic enzyme | Variable |
| Sodium stearoyl 2-lactylate | 0.2-0.6 |
| Sours | 0.5-3.5 |
| Vinegar | 0.2-1.3 |
| Vital wheat gluten | 0.5-2.5 |

Source: Compiled from Amendols, 1983; Barrows, 1984; Clancy, 1975; Cunningham, 1984; De Villers and Van der Berg, 1985; Fance, 1983; Field, 1985; Gisslen, 1983; Matz, 1987; Miller, 1981; Pyler, 1988; Sultan, 1976.
The final formula must comply with the legal limit of any food additive used.

**Table 6.2.16. Formula for Doughnut Mix**

| Ingredients | Parts by Weight |
|---|---|
| Bread flour | 34-36 |
| Cake flour | 60-70 |
| Defatted soy flour | 6-8 |
| Dextrose | 3-4 |
| Egg yolk solids | 6-8 |
| Finely granulated sugar | 35-45 |
| Mace | 0.5-1.0 |
| Nonfat dry milk | 9.5-10.5 |
| Salt | 1.0-2.0 |
| Shortening | 5.0-6.0 |

Source: Compiled from Amendols, 1983; Barrows, 1984; Clancy, 1975; Cunningham, 1984; De Villers, 1985; Fance, 1983; Field, 1985; Gisslen, 1983; Matz, 1987; Miller, 1981; Pyler, 1988; Sultan, 1976.
The final formula must comply with the legal limit of any food additive used.

**Table 6.2.17. Formula for Wheat Flour Tortilla**

| Ingredeints | Parts by Weight |
|---|---|
| Wheat flour (high protein) | 100 |
| Water (tempered to 100°F) | 45-50 |
| Shortening | 7-9 |
| Salt | 1-2 |
| Baking powder | 0.3-0.5 |
| Dough conditioner | 0.3-0.7 |
| Mold inhibitor | 0.5-1.0 |

Source: Compiled from Amendols, 1983; Barrows, 1984; Clancy, 1975; Cunningham, 1984; De Villers and Van der Berg, 1985; Fance, 1983; Field, 1985; Gisslen, 1983; Matz, 1987; Miller, 1981; Pyler, 1988; Sultan, 1976.
The final formula must comply with the legal limit of any food additive used.

**Table 6.2.18. Formula for Pies Crust Mix**

| Ingredients | Parts |
|---|---|
| Soft wheat flour | 100 |
| Hydrogenated shortening | 29-31 |
| Nonfat dry milk solids | 2-3 |
| Salt | 2-3 |
| Sodium propionate | 0.3-0.6 |
| Soy flour | 5.5-6.5 |
| Sugar | 5.5-6.5 |

Source: Compiled from Amendols, 1983; Anonymous, 1981; Barrows, 1984; Choate, 1984; Cunningham, 1984; De Villers and Van der Berg, 1985; Fance, 1983; Field, 1985; Fobel, 1988; Gisslen, 1983; Matz, 1987; Miller, 1981.
The final formula must comply with the legal limit of any food additive used.

**Table 6.2.19. Formula for English Muffin (Straight Dough)**

| Ingredients | Parts |
|---|---|
| Flour (13% protein) | 100 |
| Water | 79-81 |
| Yeast | 0.7-1.2 |
| Salt | 1-2 |
| Sugar | 2-2.5 |
| Shortening | 1.0-1.5 |
| Calcium propionate | 0.6-1.0 |
| Fungal protease | 0.8-1.1 |

Source: Compiled from Amendols, 1983; Barrows, 1984; Cunningham, 1984; De Villers and Van der Berg, 1985; Fance, 1983; Field, 1985; Gisslen, 1983; Matz, 1987; Miller, 1981; Prange and Pauli, 1984; Sultan, 1976.
The final formula must comply with the legal limit of any food additive used.

**Table 6.2.20. Dough Formulas for Bread and Danish**

| | Parts by Weight | |
|---|---|---|
| Ingredients | Bread dough | Danish dough |
| Patent flour | 99-101 | 74-76 |
| Pastry flour | - | 24-26 |
| Water | 60-64 | 42-44 |
| Sugar | 3.5-4.5 | 17-19 |
| Shortening | 2.5-3.5 | 14-16 |
| Salt | 1.8-2.2 | 1.5-2.5 |
| Nonfat dry milk solids | - | 5.5-6.5 |
| Whey solids | 3.5-4.5 | - |
| Whole eggs | - | 11-13 |
| Yeast | 2-3 | 9-11 |
| Yeast food | 0.4-0.6 | - |

Source: Compiled from Amendols, 1983; Anonymous, 1981; Barrows, 1984; Casella, 1968; Cunningham, 1984; De Villers and Van der Berg, 1985; Field, 1985; Gisslen, 1983; Matz, 1987; Miller, 1981; Orcutt and Taylor, 1981.
The final formula must comply with the legal limit of any food additive used.

## 6.3 Enrichment Standards

**Table 6.3.1. United States Enrichment Standards for Cereals and Flours**

| Product | Thiamine | Riboflavin | Niacin | Iron | Calcium | Vitamin A |
|---|---|---|---|---|---|---|
| Enriched flour | 2.9 | 1.8 | 24 | 20 | (960)[a] | |
| Enriched self-rising flour | 2.9 | 1.8 | 24 | 20 | 960 | |
| Enriched farina | 2.0-2.5 | 1.2-1.5 | 16-20 | 13 | (500)[a] | |
| ASCS domestic wheat flour | 2.9 | 1.8 | 24 | 20 | 500-625 | 4000-6000 |
| ASCS export wheat flour | 2.9 | 1.8 | 24 | 20 | 500-625 | 4000-6000 |
| Enriched corn meal and grits | 2.0-3.0 | 1.2-1.8 | 16-24 | 13-26 | (500-750)[a] | |
| Enriched self-rising corn meal | 2.0-3.0 | 1.2-1.8 | 16-24 | 13-26 | (500-750)[a] | |
| ASCS domestic corn meal | 2.0-3.0 | 1.2-1.8 | 16-24 | 13-26 | | |
| ASCS domestic corn grits | 2.0-3.0 | 1.2-1.8 | 16-24 | 21-26 | | |
| ASCS corn masa flour | 2.0 | 1.2 | 16 | 13-26 | | |
| ASCS export corn meal and soy-fortified corn meal | 2.0-3.0 | 1.2-1.8 | 16-24 | 13-26 | 500-750 | 4000-6000 |
| Enriched rice | 2.0-4.0 | 1.2-2.4[b] | 16-32 | 13-26 | (500-1000)[a] | |
| Enriched macaroni and noodle products[c] | 4.0-5.0 | 1.7-2.2 | 27-34 | 13.0-16.5 | (500-625)[a] | |
| Enriched bread[d] | 1.8 | 1.1 | 15 | 12.5 | (600)[a] | |
| U.S. RDA (mg or IU/day) | 1.5 | 1.7 | 20 | 18 | 1000 | 5000 |

Source: Adapted from 21 CFR 136, 137, and 139 (1991); Pomeranz, 1987 (used with permission). This table shows the final nutrient levels required for enriched cereal products as specified by the U.S. Food and Drug Administration and the USDA Agriculture Stabilization and Conservation Service (ASCS). All figures are in milligrams per pound of product except for vitamin A, which is in International Units (IU) per pound. When two figures are shown it indicates a minimum-maximum range. Where one number is shown it indicates the minimum level with overages left to good manufacturing practice. Standards in parentheses are optional.

[a]No claim of calcium enrichment can be made when calcium is present for technological reasons at levels less than the minimum value shown except as required by nutritional labeling (21 CFR 101.9).

[b]The riboflavin standard for rice has been constant for many years.

[c]Enriched pasta products are normally made from semolina, durum flour, or wheat flour enriched to these same levels. There are, however, no official standards for enriched cereals used in pasta production.

[d]Bread containing 62% or more of flour enriched to the standards for enriched flour will meet the standards for enriched bread. Bread can be additionally fortified to meet the standards proposed by the Food and Nutrition Board of the National Academy of Sciences. In addition to the above, the flour would contain, in mg/lb: 0.3 folic acid, 2.0 pyridoxine, 900 calcium, 200 magnesium, 100 zinc, and 5000 IU/lb vitamin A. Such bread is to be labeled "special formula enriched bread."

Note: 1 lb = ~453.6 g.

**6.4.1. Relationship Between Sour Dough Starter Lactobacillus spp. and Bread Quality[a]**

| | Sour Dough | | | Bread Quality | | |
|---|---|---|---|---|---|---|
| *Lactobacillus* spp. | pH | Acidity | pH | Acidity[b] | Crumb | Flavor |
| *L. acidophilus* | 4.50 | 7.25 | 4.60 | 6.40 | 2-3 | 2 |
| | 4.20-5.00 | 5.00-8.20 | 4.35-4.75 | 5.80-6.80 | | |
| *L. casei* | 4.25 | 7.00 | 4.70 | 6.20 | 3 | 2 |
| | 4.10-4.30 | 6.60-7.50 | 4.65-4.80 | 6.10-6.50 | | |
| *L. plantarum* | 4.05 | 7.90 | 4.50 | 6.50 | 4 | 2 |
| | 3.95-4.15 | 7.60-8.20 | 4.45-4.60 | 6.20-6.60 | | |
| *L. farciminis* | 3.95 | 8.10 | 4.40 | 6.60 | 4 | 2 |
| | 3.95-4.00 | 8.00-8.30 | 4.35-4.50 | 6.40-6.80 | | |
| *L. brevis* | 4.20 | 9.95 | 4.40 | 8.00 | 3-4 | 3 |
| | 4.10-4.40 | 9.00-11.20 | - | 7.90-8.10 | | |
| *L. brevis* var. | 3.90 | 11.50 | 4.30 | 8.90 | 5 | 4 |
| *lindneri* I | 3.80-4.00 | 10.60-12.30 | 4.20-4.30 | 8.30-9.20 | | |
| *L. brevis* var. | 3.90 | 11.65 | 4.25 | 9.00 | 5 | 4 |
| *lindneri* II | 3.80-4.00 | 11.20-12.00 | 4.10-4.30 | 8.90-9.20 | | |
| *L. buchneri* | 4.35 | 9.,10 | 4.60 | 7.70 | 3 | 3 |
| | 4.30-4.40 | 8.50-9.60 | 4.55-4.70 | 7.60-7.90 | | |
| *L. fermentum* | 4.30 | 8.46 | 5.70 | 6.80 | 2-3 | 2 |
| | 4.20-4.50 | 7.50-9.70 | 4.65-4.75 | 6.40-7.10 | | |
| *L. fructivorans* | 4.25 | 9.30 | 4.30 | 8.50 | 3-4 | 2 |
| | 4.10-4.50 | 8.20-11.10 | 4.10-4.50 | 8.20-8.70 | | |

| | | | | | | |
|---|---|---|---|---|---|---|
| Starter (B)[c] | 3.80 | 12.55 | 4.30 | 8.50 | 5 | 4-5 |
| | 3.70-3.90 | 11.60-13.40 | - | 8.10-9.00 | | |
| Spontaneous sour | 4.60 | 6.10 | 4.80 | 6.10 | 2 | 0 |

Source: Adapted from Spicher, 1983. Used with permission.
[a]Bread quality is defined by crumb and flavor. Crumb: elasticity of crumb on a scale of 1-5. Flavor: a scale of 1-5.
[b]Acidity: mL 0.1 *N* NaOH per 10 g.
[c]Commercial culture.

### 6.4.2. Scheme of a Three-Stage Sour Dough Process for Production of Rye Mixed-Grain Bread (Basic Sour Overnight) Consisting of 70% Rye Flour and 30% Wheat Flour

| | Seed Sour | Basic Sour | Full Sour | Dough[a] |
|---|---|---|---|---|
| Duration clock time | 14-20 | 20-04 | 4-7 | 7 |
| Sour dough (kg) | 0.3 | 2.0 | 19.0 | 68.5 |
| Rye flour (kg) | 1.0 | 11.0 | 24.0 | 34.0 |
| Wheat flour (kg) | -- | -- | -- | 34.0 |
| Water (L) | 1.0 | 6.0 | 26.5 | 35.5 |
| Dough yield[b] | 200 | 158 | 190 | 168 |
| Sour dough temperature ($^{o}$C) | 25-26 | 23-27 | 28-31 | 38 |

Source: Adapted from Spicher, 1983. Used with permission.
[a]Yeast, 1.0 kg; salt, 1.8 kg; dough rest 5 min.
[b]In relation to 100 parts of flour.

## 6.4.3. Detmold One-Stage Sour Dough Process for the Production of Mixed-Grain Bread

| | Ingredients | | | Process Conditions | |
|---|---|---|---|---|---|
| | Rye Mixed Grain[a] Bread | Wheat Mixed Grain[b] Bread | Processing Conditions or Processing Criteria | Rye Mixed Grain[a] Bread | Wheat Mixed Grain[b] Bread |
| Full sour in dough (% of total flour) | 25% | 15% | | | |
| Detmold One-stage sour | | | | | |
| Rye flour | 25.0 kg | 15.0 kg | Dough yield: 180% | | |
| Seed sour | 0.5 kg | 0.3 kg | Temperature at start: 31°C | | |
| Water | 20.0 L | 12.0 L | Temperature at end: 23°C | | |
| Total without starter | 45.0 kg | 27.0 kg | Fermentation time: 15-24 h | | |
| Dough | | | | | |
| Detmold One-stage sour | 45.0 kg | 27.0 kg | | | |
| Rye flour | 45.0 kg | 15.0 kg | | | |
| Wheat flour | 30.0 kg | 70.0 kg | Dough yield (theory) | 168% | 163% |
| Yeast | 1.8 kg | 2.0 kg | Dough temperature | 28°C | 26°C |
| Salt | 1.8 kg | 2.0 kg | Dough rest time | 5 min | 5 min |
| Water, about | 48.0 L | 51.0 L | | | |
| Total | 171.6 kg | 167.0 kg | | | |

Source: Adapted from Spicher, 1983. Used with permission.
[a]Rye mixed-grain bread: 70% rye flour, 30% wheat flour.
[b]Wheat mixed-grain bread: 70% wheat flour, 30% rye flour.

### 6.4.4 Berlin Short (One-Stage) Sour Dough Process for the Production of Mixed-Grain Bread

| | Ingredients | | | Process Conditions | |
|---|---|---|---|---|---|
| | Rye Mixed Grain[a] Bread | Wheat Mixed Grain[b] Bread | Processing Conditions or Processing Criteria | Rye Mixed Grain[a] Bread | Wheat Mixed Grain[b] Bread |
| Full sour in dough (% of total flour) | 40% | 20% | | | |
| Berlin short sour | | | | | |
| Rye flour | 40.0 kg | 20.0 kg | Dough yield: 190% | | |
| Seed sour | 8.0 kg | 4.0 kg | Temperature: 35°C | | |
| Water | 36.0 L | 18.0 L | Fermentation time: 3-4 h | | |
| Total without starter | 76.0 kg | 38.0 kg | | | |
| Dough | | | | | |
| Berlin Short sour | 76.0 kg | 38.0 kg | | | |
| Rye flour | 30.0 kg | 10.0 kg | | | |
| Wheat flour | 30.0 kg | 70.0 kg | Dough yield (theory) | 168% | 163% |
| Yeast | 2.0 kg | 2.0 kg | Dough temperature | 28°C | 26°C |
| Salt | 2.0 kg | 2.0 kg | Dough rest time | 5 min | 5 min |
| Water, about | 32.0 L | 45.0 L | | | |
| Total | 172.6 kg | 167.0 kg | | | |

Source: Adapted from Spicher, 1983. Used with permission.
[a]Rye mixed-grain bread: 70% rye flour, 30% wheat flour.
[b]Wheat mixed-grain bread: 70% wheat flour, 30% rye flour.

**6.4.5 Monheim Salt-Sour One Stage Sour Dough Process for the Production of Mixed-Grain Bread**

| | Ingredients | | | Process Conditions | |
|---|---|---|---|---|---|
| | Rye Mixed Grain[a] Bread | Wheat Mixed Grain[b] Bread | Processing Conditions or Processing Criteria | Rye Mixed Grain[a] Bread | Wheat Mixed Grain[b] Bread |
| Full sour in dough (% of total flour) | 30% | 20% | | | |
| Monheim salt-sour | | | | | |
| Rye flour | 30.0 kg | 20.0 kg | Dough yield: 200% | | |
| Seed sour | 6.0 kg | 4.0 kg | Temperature, start: 30-35°C | | |
| Salt | 0.6 kg | 0.4 kg | Temperature, end: 20-25°C | | |
| Water | 30.0 L | 20.0 L | Fermentation time: 18-24 h | | |
| Total without starter | 60.6 kg | 40.0 kg | | | |
| Dough | | | | | |
| Salt-sour | 60.6 kg | 40.4 kg | | | |
| Rye flour | 40.0 kg | 10.0 kg | | | |
| Wheat flour | 30.0 kg | 70.0 kg | Dough yield (theory) | 168% | 163% |
| Yeast | 2.0 kg | 2.0 kg | Dough temperature | 28°C | 26°C |
| Salt | 1.2 kg | 1.6 kg | Dough rest time | 5 min | 5 min |
| Water, about | 38.0 L | 43.0 L | | | |
| Total | 171.6 kg | 167.0 kg | | | |

Source: Adapted from Spicher, 1983. Used with permission.
[a]Rye mixed-grain bread: 70% rye flour, 30% wheat flour.
[b]Wheat mixed-grain bread: 70% wheat flour, 30% rye flour.

**6.4.6. Two-Stage Sour Dough Process for the Production of Mixed-Grain Bread with a Fermentation Time of 2.5-3.5 Hours for the Full Sour**

| | Ingredients | | | | |
|---|---|---|---|---|---|
| | Rye Mixed Grain[a] Bread | Wheat Mixed Grain[b] Bread | Processing Conditions or Processing Criteria | Process Conditions | |
| Full sour in dough (% of total flour) | 40% | 20% | | | |
| Rye flour in sour dough | 40.0 kg | 20.0 kg | | | |
| Seed flour 2.5% of rye flour in dough | 1.0 kg | 0.5 kg | | | |
| First stage: basic sour | | | | | |
| Seed sour | 1.0 kg | 0.5 kg | Dough yield: 150% | | |
| Rye flour (16 x) | 16.0 kg | 8.0 kg | Temperature: 22-26°C | | |
| Water (8 x) | 8.0 L | 4.0 L | Fermentation time: 15-24 h | | |
| Total without starter | 24.0 kg | 12.0 kg | | | |
| Second stage: full sour | | | | | |
| Basic sour | 24.0 kg | 12.0 kg | Dough yield: 180% | Fermentation Time | Taste of Bread |
| Flour (1 x) | 24.0 kg | 12.0 kg | Temperature | | |
| Water (1 x) | 24.0 kg | 12.0 kg | 33°C | 2.5 h | Milder |
| | | | 31°C | 2.25 h | More acid |
| Total | 72.0 kg | 36.0 kg | 29°C | 3 h | |
| | | | 27°C | 3.5 h | |

**Table 6.4.6. (continued)**

| | Ingredients | | | | |
|---|---|---|---|---|---|
| | Rye Mixed Grain[a] Bread | Wheat Mixed Grain[b] Bread | Processing Conditions or Processing Criteria | Process Conditions | |
| Dough | | | | | |
| Full sour | 72.0 kg | 36.0 kg | | Rye Mixed Grain Bread | Wheat Mixed Grain Bread |
| Rye flour | 40.0 kg | 20.0 kg | | | |
| Wheat flour | 20.0 kg | 60.0 kg | Dough yield (theory) | 168% | 163% |
| Yeast | 1.8 kg | 2.0 kg | Dough temperature | 28ºC | 26ºC |
| Salt | 1.8 kg | 2.0 kg | Dough rest time: | 5 min | 5 min |
| Water | 36.0 L | 47.0 L | | | |
| Total | 171.6 kg | 167.0 kg | | | |

Source: Adapted from Spicher 1983. Used with permission.
[a]Rye mixed-grain bread: 70% rye flour, 30% wheat flour.
[b]Wheat mixed-grain bread: 70% wheat flour, 30% rye flour.

## 6.5. References


Amendols, J. *Understanding Baking*, Van Nostrand Reinhold, New York, 1983.

Anonymous. *Time-Life Books: Pies and Pastries*, Time-Life, New York, 1981.

Anonymous. *Time-Life Books: Cakes*, Time-Life, New York, 1982.

Barrows, A. B. *Bakery Specialities*, Elsevier, New York, 1984.

Casella, D. *World of Baking*, D. White, Port Washington, New York, 1968.

Choate, J. *The Great Americn Pie Book*, Yankee Books, Dublin, NH, 1984.

Clancy, J. *The John Clancy Baking Book*, Clancy's Kitchen, New York, 1975.

Cunningham, M. *The Fannie Farmer Baking Book*, Knopf, New York, 1984

De Villers, S. J., and E. Van der Berg. *Breads and Biscuits*, Stevens, Milwaukee, WI, 1985.

Fance W. J. The Students' *Technology of Breadmaking and Flour Confectionery*, 2nd ed., Routledge, Chapman and Hall, New York, 1983.

Field, C. *The Italian Baker*, Harper and Row, New York, 1985.

Fobel, A. *Old-Fashioned Baking Book*, Ballantine, New York, 1988.

Gisslen, W. *Professional Baking*, Wiley, New York, 1983.

Goodbody, M. *Pretty Cakes*, Harper and Row, New York, 1986.

Green, M. *The Little Croissant Cookbook*, Gastar, Reading, MA, 1984.

Heatter, M. *Maida Heatter's Book of Great Cookies*, Knopf, New York, 1977.

Matz, S. *Formulas and Processes for Bakers*, Pan Tech. International, McAllen, Texas, 1987.

Miller, B. S., ed. *Variety Breads in the United States*, American Association of Cereal Chemists, St. Paul, MN, 1981.

Orcutt, G, and S. Taylor, *Breads, Rolls, & Pastries*, Yankee Books, Dublin, NH, 1981.

Pomeranz, Y. *Functional Properties of Food Components*, Academic Press, San Diego, CA,1985.

Pomeranz, Y. *Modern Cereal Science and Technology*, VCH, New York, 1987.

Prange, K, and J. Pauli. *Muffin Mania*, Liberty, Deerfield, Fl, 1984.

Pyler, E. J. *Baking Science & Technology*, 3rd ed., Sosland, Merriam, KS, 1988.

Spicher, G. Baked goods, In *Biotechnology, Volume 5: Food and Feed Production with Microorganisms*, G. Reed, ed. VCH, New York, 1983.

Sultan, W. *Practical Baking Manual*, Avi, New York, 1976.

# CHAPTER 7
# FISH LIST

*The Fish List* is a compilation of existing acceptable market names for imported and domestically available food fish. The list was developed by the Food and Drug Administration (FDA), in cooperation with the National Marine Fisheries Service (NMFS). Its purpose is to promote uniformity in the use of FDA acceptable market names by the fish industry and to provide consistent advice on these names. The list represents an extensive, although not necessarily complete, listing of food fish commonly sold in the United States.

In determining which fish species would be included, and what market names would be acceptable, the following criteria were used:

1. The fish species is currently sold in interstate commerce in the U.S. or has strong potential for sale.
2. The fish species is not listed as endangered.
3. The fish species is not prohibited by law from sale in interstate commerce.
4. Common and/or market names already prescribed by Federal regulations have been incorporated into the list.

The references used in determining scientific and common names included the American Fisheries Society's *List of Common and Scientific Names of Fishes from the United States and Canada* and *World List of Fishes Important to North America, Exclusive of Species from Continental Waters of the United States and Canada*; Food and Agriculture Organization (FAO) species catalogs and/or species identification sheets and source country references for species limited to that country.

Market names have been determined by common usage in the U.S. market place. When more than one name is in use for a species, the acceptable market name has been decided based on the above references and in consultation with FDA and NMFS.

The List is divided into two parts. Part I lists species by "acceptable market name" alphabetically with "scientific name", "common name" and "regional name". This list is provided in Table 7.1.

Part II provides the "scientific names" alphabetically, and the "acceptable market name" for each. This is presented in Table 8.2

The data presented in this book should serve as a useful guide for the following groups or individuals:

1. Fish processors. Although most fish processors have a copy of the original document, for various reasons many do not. These data will save the latter much inconvenience associated with getting a government document through normal channels including interlibrary loan.

2. Food scientists and technologists. Many food processors that deal with fish products are unaware of the existing of this document. By being included in a reference book for food scientists and technologists, the data have a better chance of being used by these professionals. This facilitates the government's effort to disseminate information and assists a processor to comply with federal recommendations.
3. Foreign exporters and domestic importers. Since more than 100 countries export seafood to the United States, the data should serve as a useful reference for foreign exporters and domestic importers in labeling their fish products. Foreign exporters frequently are unaware of relevant U.S. government recommendations and have trouble with obtaining such documents even if they are familiar with their existence. This communication gap is common in all areas of foreign trades. Labeling of imported fish products has been an issue of industrial dispute for many years.

According to the FDA and NMFS, it is advisable to use either the acceptable market name or the common name in labeling fish products. This will help assure that identity labeling of the fish will comply with the regulations and recommendations of these two agencies. Use of the regional name is not encouraged, and may cause the fish to be misbranded. The listing of regional names has been included for information only, and to help reference the acceptable market name.

**Table 7.1. Alphabetical Listing of Acceptable Market Name for Food Fish and Their Corresponding Scientific, Common, and Regional Names**

| Market Name | Scientific Name | Common Name | Regional Name |
|---|---|---|---|
| Aholehole | *Kuhlia marginata* | Flagtail, spotted | Damagan |
| Aholehole | *Kuhlia rupestris* | Flagtail, rock | Buffalo bream/perch |
| Aholehole | *Kuhlia sandvicensis* | Aholehole, Hawaiian | |
| Alewife/river herring | *Alosa pseudoharengus* | Alewife | Gaspegeau/alewife/branch herring/ gray herring |
| Alfonsin | *Beryx decacactylus* | Bream, red | Alfonsin a la casta larga |
| Alfonsin | *Beryx splendens* | Alfonsin a casta | Roadbars/goldeneye perch |
| Alfonsin | *Trachichthodes affinis* | Alfonsin | Red snapper/golden snapper/king snapper |
| Amberjack | *Seriola dumerili* | Amberjack, greater | Yellowtail/Atlantic amberjack/madregal/bonito |
| Amberjack | *Seriola fasciata* | Amberjack, lesser | Madregal/bonito |
| Amberjack | *Seriola zonata* | Rudderfish, banded | Amberjack/rudderfish/jackfish/ banded mackerel |
| Amberjack/yellowtail | *Seriola lalandei* | Yellowtail | Great amberjack/horseeye bonito/amberjack/coronado |
| Amberjack/yellowtail | *Seriola quinqueradiata* | Amberjack, king | Hamachi/buri/racing tuna |
| Anchovy | *Anchoa cayorum* | Anchovy, key | Manjua |
| Anchovy | *Anchoa compressa* | Anchovy, deepbody | |
| Anchovy | *Anchoa delicatissima* | Anchovy, slough | |
| Anchovy | *Anchoa duodecim* | Anchovy, New Jersey | |
| Anchovy | *Anchoa hepsetus* | Anchovy, striped | Manjua/anchovy fry |
| Anchovy | *Anchoa lyolepis* | Anchovy, dusky | |

| | | | |
|---|---|---|---|
| Anchovy | *Anchoa mitchilli* | Anchovy, bay | Common anchovy/whitebait |
| Anchovy | *Anchoa nasuta* | Anchovy, longnose | |
| Anchovy | *Anchoviella estauqual* | Anchovy Camiguana | |
| Anchovy | *Anchoviella letidentostole* | Anchovy | |
| Anchovy | *Anchoviella perfasciata* | Anchovy, flat | Grubber/broadhead |
| Anchovy | *Cetengraulis mysticetus* | Anchoveta | |
| Anchovy | *Engraulis australis* | Anchovy, southern | Australian anchovy/whitebait/ smig/kokowhawa |
| Anchovy | *Engraulis encrasicolus* | Anchovy, European | |
| Anchovy | *Engraulis eurystole* | Anchovy, silver | |
| Anchovy | *Engraulis japonica* | Anchovy, Japanese | Halfmouthed sardine |
| Anchovy | *Engraulis mordax* | Anchovy, northern | North Pacific anchovy/California anchovy |
| Anchovy | *Engraulis ringens* | Anchoveta | |
| Anchovy | *Stolephorus buccaneeri* | Anchovy, round headed | |
| Anchovy | *Stolephorus miarcha* | Anchovy, slim | |
| Anchovy | *Stolephorus purpureus* | Anchovy | Nehu |
| Angelfish | *Holacanthus bermudensis* | Angelfish, blue | Common angelfish/queen angelfish/ Isabelita |
| Angelfish | *Holacanthus ciliaris* | Angelfish, queen | Yellow angelfish/golden angel/ blue angel/Isabelita |
| Angelfish | *Holacanthus tricolor* | Rock beauty | Catalineta/black angel |
| Angelfish | *Pomacanthus arcuatus* | Angelfish, gray | Black angelfish/black angel/chirivita/Portugaise |
| Angelfish | *Pomacanthus paru* | Angelfish, French | Paru/Indian fish/flatfish |
| Argentine | *Argentina elongata* | Argentine, silverside | Nigisu |
| Armourhead | *Pentaceros richardsoni* | Armourhead, pelagic | Southern boarfish/boarfish |
| Barracuda | *Sphyraena ensis* | Vicuda | |

**Table 7.1 (continued)**

| Market Name | Scientific Name | Common Name | Regional Name |
|---|---|---|---|
| Barracuda | *Sphyraena argentea* | Barracuda, Pacific | California barracuda/barracouta |
| Barracuda | *Sphyraena barracuda* | Barracuda, great | Short barracuda/picuda/becuna/ seapike/pike |
| Barracuda | *Sphyraena guachancho* | Guaguache | Guaguanche pelon |
| Barracuda | *Sphyraene picudilla* | Sennet, southern | Picuda/picudilla/barracuda |
| Bass | *Ambloplites cavifrons* | Bass, roanoke | Southern rock bass |
| Bass | *Ambloplites constellatus* | Bass, Ozark | |
| Bass | *Ambloplites rupestris* | Bass, rock | Redeye/goggle-eye/sunfish/sun perch/white bass |
| Bass | *Lepomis gulosus* | Warmouth | Goggle-eye/bigmouth/black sunfish/bream/buffalo bass |
| Bass | *Micropterus dolomieui* | Bass, smallmouth | Smallmouth/black bass/bass hogfish/ black perch |
| Bass | *Micropterus notius* | Bass, suwannee | |
| Bass | *Micropterus punctulatus* | Bass, spotted | |
| Bass | *Micropterus salmoides* | Bass, largemouth | Black bass/largemouth osego bass/chub/trout/jumper |
| Bass | *Micropterus treculi* | Bass, guadalupe | |
| Bass | *Morone chrysops* | Bass, white | White lake bass/silver bass |
| Bass | *Morone chrysops x saxatilis* | Bass, sunshine | White and striped bass hybrid (aquaculture) |
| Bass | *Morone mississippiensis* | Bass, yellow | |
| Bass | *Morone saxatilis* | Bass, striped | Rockfish/rock/striper/linesides |
| Bass | *Synagrops bellus* | Bass, blackmouth | |
| Bass, sea | *Acanthistius brasilianus* | Bass, Argentine sea | |

| | | | |
|---|---|---|---|
| Bass, sea | *Centropristis ocycurus* | Bass, bank sea | Gulf sea bass/tallywag |
| Bass, sea | *Centropristis philadelphica* | Bass, rock sea | Redfish |
| Bass, sea | *Centropristis striata* | Bass, Black Sea | Blackfish/rock bass/black bass bluefish/tallywag |
| Bass, sea | *Dicentrarchus labrax* | Bass, European sea | Bass |
| Bass, sea | *Epinephelus merra* | Grouper, speckled dwarf | Spotted grouper/honey comb rockcod/honey comb cod |
| Bass, sea | *Epinephelus tauvina* | Bass, Black sea | Estuary rockcod/spotted river cod/ groper/cod |
| Bass, sea | *Lateolabrax japonicus* | Bass, Japan sea | Japanese seaperch/seaperch/ perch/common sea bass |
| Bass, sea | *Paralabrax callaensis* | Bass, Peruvian sea | Southern rock bass |
| Bass, sea | *Paralabrax clathratus* | Bass, kelp | Rock bass/California kelp bass/cabrilla |
| Bass, sea | *Paralabrax maculatofasciatus* | Bass, spotted sand | Redspotted rock bass/spotted cabrilla |
| Bass, sea | *Paralabrax nebulifer* | Bass, barred sand | California sand bass/sand bass/Johnny verde/kelp bass |
| Bass sea | *Paranthias furcifer* | Creole-fish | Southern creole-fish/barber |
| Bass, sea | *Polyprion americanus* | Wreckfish | Bass/groper/moeone/toti |
| Bass, sea | *Polyprion moene* | Wreckfish | Common bigeye/catalufa/blear-eye/ glass-eye |
| Bigeye | *Priacanthus arenatus* | Bigeye | Little bigeye/deep bigeye |
| Bigeye | *Pristigenys alta* | Bigeye, short | Tailor/snapper |
| Bluefish | *Pomatomus saltatrix* | Bluefish | Bluegill sunfish/blue sunfish/sunfish/roach/bream |
| Bluegill | *Lepomis macrochirus* | Bluegill | Bonita/stoneye/blue bream |
| Bluenose | *Hyperoglyphe antarctica* | Cutlerfish, Antarctic | |
| Boarfish | *Antigonia capros* | Boarfish, deepbody | |
| Boarfish | *Antigonia combatia* | Boarfish, shortspine | Sowfish/common boarfish |

**Table 7.1 (continued)**

| Market Name | Scientific Name | Common Name | Regional Name |
|---|---|---|---|
| Boarfish | *Paristiopterus labiosus* | Boarfish, giant | Ladyfish/tarpon/bananafish/grubber/macabi |
| Bonefish | *Albula vulpes* | Bonefish | Sanducha |
| Bonito | *Sarda chiliensis* | Bonito, Pacific | California bonito/skip jack belted bonito/oriental bonito/skip jack |
| Bonito | *Sarda orientalis (S. velox)* | Bonito, striped | Belted bonito/pelamid/shortfinned tunny |
| Bonito | *Sarda sarda* | Bonito, Atlantic | Red baitfish/redbait/red sea-haarder |
| Bonnetmouth | *Emmelichthys nitidus* | Bonnetmouth | Rubyfish/red perch/rocksalmon/salmon |
| Bonnetmouth | *Plagiogenion rubiginosus* | Bonnetmouth | Bonnetmouth |
| Bowfin | *Amia calva* | Bowfin | Bowfish/dogfish/mudfish/grindle/cyprus trout |
| Bream | *Argyrops bleekeri* | Tai, Taiwan | |
| Bream | *Argyrops spinifer* | Bream, long-spined red | Redfin/king soldierbream/longspined snapper |
| Bream | *Pentapodus macrurus* | Bream, threadfin | |
| Bream | *Sparus auratus* | Bream, gilt-headed | |
| Bream, threadfin | *Nemipterus japonicus* | Bream, Japanese threadfin | Sesugo/melon-coat/longtailed nemipterid |
| Bream/bogue | *Boops boops* | Bream/bogue | |
| Brotula | *Brotula barbarta* | Brotula, bearded | |
| Buffalo | *Ictiobus bubalus* | Buffalo, smallmouth | Razorbacked buffalo/suckermouth/buffalofish |
| Buffalo | *Ictiobus cyprinellus* | Buffalo, bigmouth | Redmouth buffalofish/largemouth buffalofish |

| | | | |
|---|---|---|---|
| Buffalo | *Ictiobus niger* | Buffalo, black | |
| Bullhead | *Ictalurus natalis* | Bullhead, yellow | Yellow cat |
| Bullhead | *Ictalurus platycephalus* | Bullhead, flat | Mud cat/brown cat |
| Bullhead/catfish | *Ictalurus melas* | Bullhead, black | Small catfish |
| Bullhead/catfish | *Ictalurus nebulosus* | Bullhead, brown | Common bullhead/American catfish/ mud cat/mudpout |
| Bumper | *Chloroscombrus chrysurus* | Bumper, Atlantic | Little bumper/Casebe/yellowtail |
| Bumper | *Chloroscombrus orqueta* | Bumper, Pacific | Orqueta |
| Burbot | *Lota lota* | Burbot | Ellpout/ling/freshwater cod |
| Butterfish | *Peprilus alepidotus* | Harvestfish | Starfish/star/butterfish/diamond |
| Butterfish | *Peprilus burti* | Butterfish, Gulf | |
| Butterfish | *Peprilus simillimus* | Pompany, Pacific | California pompano/Pacific butterfish |
| Butterfish | *Peprilus triacanthus* | Butterfish | Dollarfish/skip jack/ harvesterfish/sheepshead |
| Buterflyfish | *Chaetodon ocellatus* | Butterflyfish, spotfin | |
| Butterflyfish | *Chaetodon striatus* | Butterflyfish, banded | |
| Cabrilla | *Epinephelus analogus* | Cabrilla, spotted | Cabrilla Pinta/Gag |
| Capelin | *Mallotus villosus* | Capelin | Caplin/Lodde/Moiva |
| Cardinalfish | *Epigonus telescopus* | Cardinalfish, bigeye | |
| Carp | *Cyprinus carpio* | Carp, common | European carp/German carp/mirror carp |
| Catfish | *Brachyplatysome vaillanti* | Catfish, Brazilian | |
| Catfish | *Ictalurus catus* | Catfish, white | White cat/channel cat |
| Catfish | *Ictalurus furcatus* | Catfish, blue | Blue cat/bullhead/chuckleheaded cat/Mississippi cat |
| Catfish | *Ictalurus pricei* | Catfish, Yaqui | |
| Catfish | *Ictalurus punctatus* | Catfish, channel | Channel cat/spotted cat/white cat/ lake catfish |

**Table 7.1 (continued)**

| Market Name | Scientific Name | Common Name | Regional Name |
|---|---|---|---|
| Catfish | *Pinirampus* spp. | Catfish, flatwhiskered | Barba-chata |
| Catfish | *Platynematichthys notatus* | Coroata | |
| Catfish | *Pseudoplatystoma tigrinum* | Caparari | Tiger catfish/cabezona |
| Catfish | *Pylodictis olivaris* | Catfish, flathead | |
| Catfish, ocean | *Anarhichas denticulatus* | Wolffish, northern | Bullhead catfish/broadheaded catfish/ Arctic wolffish |
| Catfish, ocean | *Anarhichas lupus* | Wolffish, Atlantic | Wolffish/catfish/seacat/ocean whitefish/rock salmon |
| Catfish, ocean | *Anarhichas minor* | Wolffish, spotted | Spotted catfish/lesser catfish/smaller catfish |
| Catfish, ocean | *Anarhichas orientalis* | Wolffish, Bering | Wolffish |
| Catfish, sea | *Ariopsis felis* | Catfish, hardhead | Sea catfish/sea cat |
| Catfish, sea | *Bagre marinus* | Catfish, Gafftopsail | Sea catfish |
| Char | *Salvelinus alpinus* | Char, Arctic | Common char/alpine trout/alpine char/sea trout |
| Chimaera | *Harriotta raleighana* | Chimera, longnosed | Ratfish/azumaginzame/ratfish |
| Chimaera | *Hydrolagus novaezelandiae* | Ratfish | Dark ghost shark/pearl fillets |
| Chimaera | *Hydrolagus* spp. | Ratfish | Atlantic, Pacific, African chimarera/elephantfish |
| Chub | *Coregonus kiyi* | Kiyi | Waterbelly/mooneye/Kiyi |
| Chub | *Kyphosus incisor* | Chub, yellow | |
| Chub | *Kyphosus sectatrix* | Chub, Bermuda | Spotted Bermuda chub/stripped rudderfish/chopa |
| Cisco/chub | *Coregonus reighardi* | Cisco, shortnose | Bloater/grayback |
| Cisco/chub | *Coregonus alpenae* | Cisco, longjaw | |

| | | | |
|---|---|---|---|
| Cisco/chub | *Coregonus zenithicus* | Cisco, shortjaw | |
| Cisco,/Tullibee | *Coregonus artedii* | Cisco/lake herring | Shallow-water cisco/grayback/ Erie herring |
| Cobia | *Rachycentron canadum* | Cobia | Crabeater/sergeantfish/bonito/ cabio/cubby yew |
| Cod | *Arctogadus borisovi* | Cod, toothed | East Siberian cod |
| Cod | *Arctogadus glacialis* | Cod, polar | Arctic Greenland cod |
| Cod | *Boreogadus saida* | Cod, Arctic | Polar cod/Oikardirsg saika |
| Cod | *Eleginus gracilis* | Cod, saffron | Wachna cod/northern cod/navaga |
| Cod | *Gadus ogac* | Cod, Greenland | Greenland codfish/fjord cod/ pilot/ogac/uvak |
| Cod, Morid | *Lotella rhacina* | Cod, rock | Beardie |
| Cod, Morid | *Mora pacifica* | Cod, Morid | Goodly-eyed cod/ribaldo/deepsea cod/white cod |
| Cod, Morid | *Pseudophycis breviusculus* | Cod, Morid | Bastard red cod/red cod/cloudy bay cod/hoka |
| Cod/Alaska cod | *Gadus macrocephalus* | Cod, Pacific | Alaska codfish/grey cod/true cod/treska |
| Cod/codfish | *Gadus morhua* | Cod, Atlantic | Rock cod/codling/scrod cod |
| Corvina | *Cynoscion parvipinnis* | Corvina, shortfin | California bluefish/shortfin seabass |
| Cowfish | *Lactophyrs polygonia* | Cowfish, honeycomb | |
| Cowfish | *Lactophyrs quadricornis* | Cowfish, scrawled | |
| Crappie | *Pomoxis annularis* | Crappie, white | Silver bass/white bass/speckled perch/bachelor |
| Crappie | *Pomoxis nigromaculatus* | Crappie, black | Calico bass/Oswego bass/strawberry bass/grass bass |
| Croaker | *Argyrosomus argentatus* | Croaker, white | Silver jewfish/white Chinese croaker/Roncador |

**Table 7.1 (continued)**

| Market Name | Scientific Name | Common Name | Regional Name |
|---|---|---|---|
| Croaker | *Argyrosomus hololepidotus* | Cob | Kob |
| Croaker | *Argyrosomus japonicus* | Croaker, Japanese | Green perch |
| Croaker | *Argyrosomus macrophthalmus* | Croaker, white | |
| Croaker | *Argyrosomus nibe* | Croaker, black | Black Chinese croaker/blackmouth croaker |
| Croaker | *Bairdiella batabana* | Croaker, blue | |
| Croaker | *Bairdiella sanctaeluciae* | Croaker, striped | |
| Croaker | *Cheilotrema saturnum* | Croaker, black | |
| Croaker | *Genyonemus lineatus* | Croaker, white | Little roncador/kingfish |
| Croaker | *Johnius* spp. | Jewfish, spotted | Black jew/jew/twostriped Jewfish/croaker/codfish |
| Croaker | *Micropogonias furnieri* | Drummer, whitemouth | Verrugato/corvina brasileira |
| Croaker | *Micropogonias undulatus* | Croaker, Atlantic | Hardhead |
| Croaker | *Nebris microps* | Croaker, smalleye | |
| Croaker | *Nibea* spp. | Croaker | Sharpnose croaker/spotted croaker |
| Croaker | *Odontoscion dentex* | Croaker, reef | Corvina |
| Croaker | *Pachypops* spp. | Croaker | Brazilian croaker/Guyanan croaker |
| Croaker | *Pachyurus* spp. | Croaker | Parana croaker/Colombian croaker/ San Francisco croaker |
| Croaker | *Paralonchurus* spp. | Croaker | Coco/suco |
| Croaker | *Plagioscion* spp. | Croaker | Pacora/pescada depiaui/pescada branca |
| Croaker | *Pseudotolithes* spp. | Captainfish | Longneck croaker/weakfish |
| Croaker | *Pterotolithus* spp. | Croaker | Bigmouth croaker/false seatrout |
| Croaker | *Roncador stearnsi* | Croaker, spotfin | Roncador |

| | | | |
|---|---|---|---|
| Croaker | *Umbrina roncador* | Croaker, yellowfin | |
| Croaker/corvina | *Cynoscion orthonopterus* | Corvina, Gulf | Seatrout |
| Croaker/corvina | *Isopisthus parvipinnis* | Corfina, bigtooth | Shortfin corvina |
| Croaker/corvina | *Micropogonias opercularis* | Corvina | Corvina blanca |
| Croaker/shadefish | *Argyrosomus regius* | Shadefish, Atlantic | Maigre/shadefish/croaker/jewfish/ salmon bass |
| Croaker/yellowfish | *Pseudosciaena* spp. | Croaker, yellow | Yellow flower fish/large yellow croaker |
| Cunner | *Tautogolabrus adspersus* | Cunner | Biergall/perch/sea perch/blue perch/bergall/nipper |
| Cusk | *Brosme brosme* | Cusk | Tusk/torsk/European cusk/brosmius |
| Cusk-eel | *Lepophidium cervinum* | Cusk-eel, fawn | |
| Cusk-eel | *Lepophidium graellsi* | Cusk-eel blackedge | |
| Cutlassfish | *Aphanopuys carbo* | Scabbardfish, black | Espada preta |
| Cutlassfish | *Lepidopus caudatus* | Scabbardfish, | Scalefoot/ribbonfish/buttersnoek/ frostfish/hikau |
| Cutlassfish | *Lepidopus caudatus* | Scabbardfish, silver | Ribbonfish/buttersnoek/espada branca/southern frostfish |
| Cutlassfish | *Trichiurus lepturus* | Cutlassfish, Atlantic | Hairtail/ribbonfish/scabbardfish/ snakefish |
| Cutlassfish | *Trichiurus nitens* | Cutlassfish, Pacific | |
| Cutlassfish | *Trichiurus savaka* | Hairtail, smallhead | |
| Dogfish | *Centrophorus squamosus* | Catshark | |
| Dogfish | *Centroscymnus fabricii* | Dogfish, black | |
| Dogfish | *Mustelus antarcticus* | Dogfish | Lemonfish |
| Dogfish | *Mustelus canis* | Dogfish, smooth | Smooth hound/nurse shark/smooth grayfish/grayfish |
| Dogfish | *Mustelus lenticulatus* | Shark, dogfish | Rig/spotted dogfish/lemonfish |

**Table 7.1 (continued)**

| Market Name | Scientific Name | Common Name | Regional Name |
|---|---|---|---|
| Dogfish | *Scyliorhinus canicula* | Dogfish, lessor spotted | Small spotted catshark/common dogfish/rock salmon |
| Dogfish | *Scyliorhinus stellaris* | Dogfish, large spotted | Greater spotted dogfish/ nursehound/catfish/huss |
| Dogfish | *Squalus acanthias* | Dogfish, spiny/spring | Spiked dogfish/grayfish/spur dog/ picked dogfish |
| Dogfish | *Squalus blainvillei* | Dogfish, northern | |
| Dory | *Cyttus novae-zelandiae* | Dory, silver | Pink-finned dory/silbaadoori |
| Dory | *Zenopsis conchifer* | Dory, buckler | |
| Dory | *Zenopsis nebbulosa* | Dory, mirror | Kagamidai/ginbatou/mirror perch |
| Dory | *Zeus faber* | John dory, European | Girtie/dory/St. Peter's fish/peterfish/doorkeeper |
| Driftfish | *Hyperoglyphe bythites* | Driftfish, black | |
| Driftfish | *Hyperoglyphe perciformis* | Barrelfish | |
| Drum | *Eqquetus punctatus* | Drum, spotted | Serrana Hispana |
| Drum | *Larimus* spp. | Drum | Bastard perch/banded croaker/ banded larimus |
| Drum | *Pogonias cromis* | Drum, black | Oyster drum/sea drum/gray drum/ drumfish |
| Drum | *Stellifer* spp. | Drum | Smalleye drum/yellow drum/rake drum/rake stardrum |
| Drum | *Umbrina corides* | Drum, sand | Roncador/corb/pargo/striped drum/ Catalina croaker |
| Drum, freshwater | *Aplodinotus grunniens* | Drum, freshwater | Sheepshead/gaspergou/croaker/ jewelhead/crocus |

| | | | |
|---|---|---|---|
| Drum, cubbyu | *Equetus umbrosus* | Cubbyu | |
| Drum/lion fish | *Collichthys* spp. | Lion fish | |
| Drum/meagre | *Sciaena* spp. | Drum/meagre | Deepwater drum/brown meagre/corb |
| Drum/queenfish | *Seriphus politus* | Queenfish | White croaker |
| Drum/redfish | *Sciaenops ocellatus* | Drum, red | Channel bass/spottail/spotted bass/bull redfish |
| Drummer | *Kyphosus vaigiensis* | Drummer, lowfinned | Brass bream/drummer/quili/pilotfish |
| Eel | *Anguilla anguilla* | Eel, European | Common eel/glass eel/elver/yellow eel |
| Eel | *Anguilla australis* | Eel, short-fin | Shortfinned freshwater eel/yellow eel/ silver eel |
| Eel | *Anguilla dieffenbachii* | Eel, long-finned | Yellow/silver eel/kuwharuwharu |
| Eel | *Anguilla japonicus* | Eel, Japanese | Unagi |
| Eel, conger | *Ariosoma balearicum* | conger, bandtooth | |
| Eel, conger | *conger cinereus* | Eel, gray conger | Indian conger eel/Pacific white eel/ black conger eel |
| Eel, conger | *conger conger* | Eel, conger | Sea eel/silver eel/southern conger/koiro/ngoio |
| Eel, conger | *conger marginatus* | Eel, white | Puhi uha |
| Eel, conger | *conger oceanicus* | Eel, conger | Sea eel/American eel/white eel |
| Eel, conger | *conger oligoporus* | Eel, conger | Puhi |
| Eel, conger | *conger triporiceps* | conger, manytooth | |
| Eel, conger | *Congrina aequoreus* | Eel, conger | Puhi |
| Eel, conger | *Gnathophis catalinensis* | conger, Catalina | |
| Eel, conger | *Hildebrandia flava* | Conger, yellow | |
| Eel, conger | *Hildebrandia gracilior* | Conger, whiptail | |
| Eel, conger | *Paraconger caudilimbatus* | Conger, margintail | |
| Eel, freshwater | *Anguilla rostrata* | Eel, American | Common eel/freshwater eel/Atlantic eel/silver eel |

**Table 7.1 (continued)**

| Market Name | Scientific Name | Common Name | Regional Name |
|---|---|---|---|
| Eel, snake/keoghfish | *Ophichthus rex* | Snake eel, giant | |
| Eelpout | *Zoarces viviparus* | Eelpout | Guffer eel/viviparous blenny |
| Eelpout/ocean pout | *Macrozoarces americanus* | Pout ocean | Mutton fish/eelpout/conger eel/ mother-of-eels/ling |
| Elephant fish | *Callorhynchus millii* | Elephant fish | Silver trumpeter/goast shark/ elephantfish |
| Emperor | *Lethrinus* spp. | Emperor | Longnosed pigfaced bream/mu/ rosefaced pigface bream |
| Fanfish | *Pteraclis aesticola* | Fanfish, Pacific | |
| Filefish | *Aluterus monocers* | Filefish, unicorn | Unicorn or yellowfinned leatherjacket/andora |
| Filefish | *Aluterus schoepfi* | Filefish, orange | Unicornfish/sunfish/foolfish/lija/ turbot/hog fish |
| Flounder | *Ancylopsetta dilecta* | Flounder, three-eye | |
| Flounder | *Arnoglossus scapha* | Flounder | Witch/megrim/mehue/hirame |
| Flounder | *Atheresthes evermanni* | Flounder, kamchatka | Arrowtooth halibut/Arctic arrowtooth/turbot/paltus |
| Flounder | *Bothyus lunatus* | Flounder, peacock | Plate-fish |
| Flounder | *Bothus mancus* | Flounder, tropical | Maimed flounder/paki'i |
| Flounder | *Bothus ocellatus* | Flounder, eyed | Eyed platefish |
| Flounder | *Bothus pantherinus* | Flounder, panther | Mottled flounder/leopard flounder/ paku |
| Flounder | *Colistium guntheri* | Brill, New Zealand | Brill |
| Flounder | *Colistium nudipinnis* | Brill | Turbot |
| Flounder | *Hippoglossoides robustus* | Flounder, Bering | |

| | | | |
|---|---|---|---|
| Flounder | *Limanda ferruginea* | Flounder, yellowtail | Yellowtail/rusty dab/sand dab |
| Flounder | *Liopsetta glacialis* | Flounder, Arctic | Smoothback flounder/polar flounder |
| Flounder | *Paralichthys albigutta* | Flounder, gulf | |
| Flounder | *Paralichthys oblongus* | Flounder, fourspot | Fourspotted flounder or fluke |
| Flounder | *Paralichthys olivaceus* | Flounder, olive | Bastard halibut/false halibut/ left mouth |
| Flounder | *Paralichthys patagonicus* | Flounder, patagonian | |
| Flounder | *Paralichthys squamilentus* | Flounder, Broad | |
| Flounder | *Pelecanichthys crumenalis* | Flounder, longjawed | |
| Flounder | *Pelotretis flavilatus* | Sole, New Zealand lemon | Karea |
| Flounder | *Peltorhampus novaezeelandiae* | Sole, New Zealnd | New Zealand common sole/ patikirors |
| Flounder | *Platichthys stellatus* | Flounder, starry | Great flounder/California flounder/grindstone/emerywheel |
| Flounder | *Psettodes erumei* | Flounder, Indian Ocean | Queensland halibut/arrowtooth flounder/turbot |
| Flounder | *Pseudorhombus arsius* | Flounder, largetoothed | Smoothscaled brill/false brill/bastard halibut |
| Flounder | *Pseudorhombus jenynsii* | Flounder, small-toothed | Sole |
| Flounder | *Pseudorhombus pentophthalmus* | Flounder, fivespot | Tama-ganzodirane/Tama ganzobirame |
| Flounder | *Rhombosolea leporina* | Flounder, yellowbelly | Yellow flounder/Patiki-totora |
| Flounder | *Rhombosolea plebeia* | Flounder, sand | Dab/Patiki |
| Flounder | *Rhombosolea retiara* | Flounder, black | Freshwater flounder/river flounder |
| Flounder | *Rhombosolea tapirina* | Flounder, greenback | Karei |
| Flounder | *Samariscus triocellatus* | Flounder | Ali |
| Flounder | *Scophthalmus aquosus* | Windowpane | Brill/turbot/spotted flounder/sand dab/sand flounder |

**Table 7.1 (continued)**

| Market Name | Scientific Name | Common Name | Regional Name |
|---|---|---|---|
| Flounder | *Scophthalmus rhombus* | Brill | |
| Flounder, arrowtooth | *Atheresthes stomias* | Flounder, arrowtooth | Arrowtooth halibut/turbo/paltus |
| Flounder/dab | *Limanda limanda* | Dab, common | |
| Flounder/dab | *Limanda proboscidea* | Dab, longhead | Snout sole/longsnouted flounder |
| Flounder/fluke | *Paralichthys dendtatus* | Flounder, summer | Plaicefish/plaice/northern fluke |
| Flounder/fluke | *Paralichthys lethostigma* | Flounder, southern | |
| Flounder/fluke | *Platichthys flesus* | Flounder, European | White fluke/North Atlantic flounder;/Sandberg/pallus |
| Flounder/sole | *Pseudopleuronectes americanus* | Flounder, winter/ lemon sole | Blackback flounder/blackback/ Georges bank flounder |
| Flounder/whiff | *Lepidorhombus boscii* | Scaldfish, fourspot | Fourspot megrim |
| Flounder/whiff | *Lepidorhombus whiffiagonis* | Megrim | Sailfluke |
| Flyingfish | *Cypselurus atrisignis* | Flyingfish, Jenkins | |
| Flyingfish | *Cypselurus californicus* | Flyingfish, California | Great flyingfish |
| Flyingfish | *Cypselurus cyanopterus* | Flyingfish, margined | |
| Flyingfish | *Cypselurus exsiliens* | Flyingfish, bandwing | |
| Flyingfish | *Cypselurus furcatus* | Flyingfish, spotfin | |
| Flyingfish | *Cypselurus hubbsi* | Flyingfish, blotchwing | |
| Flyingfish | *Cypselurus melanurus* | Flyingfish, Atlantic | |
| Flyingfish | *Cypselurus simus* | Flyingfish, shortnosed | Malolo |
| Flyingfish | *Cypselurus speculiger* | Flyingfish, Cuvier's | |
| Flyingfish | *Cypselurus spilonopterus* | Flyingfish, spotted wing | |
| Flyingfish | *Cyspelurus comatus* | Flyingfish, clearwing | Fourwinged flyingfish |
| Flyingfish | *Exocoetus obtusirostris* | Flyingfish, ocean twowing | |

| | | | |
|---|---|---|---|
| Flyingfish | *Exocoetus volitans* | Flyingfish, tropical twowing | |
| Flyingfish | *Fodiator acutus* | Flyingfish, sharpchin | |
| Flyingfish | *Hirundichthys affinis* | Flyingfish, fourwing | |
| Flyingfish | *Hirundichthys rondeleti* | Flyingfish, blackwing | |
| Flyingfish | *Oxyporhamphus micropterus* | Flyingfish, smallwing | |
| Flyingfish | *Parexocoetus brachypterus* | Flyingfish, sailfin | |
| Flyingfish | *Prognichthys gibbifrons* | Flyingfish, bluntnose | |
| Gar | *Lepisosteus osseus* | Gar, longnose | Common gar pike/gar pike/Ohio gar/garfish/billfish |
| Gar | *Lepisosteus spatula* | Gar, alligator | Mississippi alligator gar |
| Gemfish | *Epinnula magistralis* | Domine | |
| Gemfish | *Lepidocybium flavobrunneum* | Escolar | |
| Gemfish | *Nesiarchus nasutus* | Gemfish, black | |
| Gemfish/Barracouta | *Rexea solandri* | Genfish, silver | Southen kingfish/takati |
| Gemfish/Barracouta | *Thyrsites atun* | Mackerel, snake | Atun/snoek/manga |
| Gemfish/caballa | *Thyrsites lepidopoides* | Mackerel, white snake | Caballa blanca/cavalinha |
| Goatfish | *Mulloidichthys* spp. | Goatfish | King mullet/weke a a/vanikolo goatfish/weke ula |
| Goatfish | *Mullus auratus* | Goatfish, red | Red mullet/northern goatfish/ surmullet |
| Goatfish | *Parupeneus* spp. | Goatfish | Moano kea/moano/kumu/munu/ sanmoiti |
| Goatfish | *Pseudupeneus* spp. | Goatfish, spotted | Red goatfish/salmonete |
| Goatfish | *Upeneichthys lineatus* | Goatfish, bluespotted | Red mullet/goatfish/ahuruhuru |
| Goatfish | *Upeneus* spp. | Goatfish | Wele puep/yellowstriped goatfish/ kakunir |

**Table 7.1 (continued)**

| Market Name | Scientific Name | Common Name | Regional Name |
|---|---|---|---|
| Goldeye | *Hiodon alosoides* | Goldeye | Winnipeg goldeye/naccaysh/ la quesche |
| Goldfish | *Carassius auratus* | Goldfish | |
| Grayling | *Thymallus arcticus* | Grayling, Arctic | American or North American grayling |
| Greenbone | *Coridodax pullus* | Greenbone | Butterfish/kelp salmon/kelpie/koae/ marari |
| Greenling | *Hexagrammos decagrammus* | Greenling, kelp | Rock trout/sea trout/boregat/bodieron |
| Greenling | *Hexagrammos lagocephalus* | Greenling, rock | Red rock trout |
| Greenling | *Hexagrammos octogrammus* | Greenling, masked | Alaska Greenling/greenfish |
| Greenling | *Hexagrammos stelleri* | Greenling, whitespotted | Common greenling |
| Grenadier | *Coryphaenoides pectoralis* | Rattail, giant | |
| Grenadier | *Coryphaenoides rupestris* | Grenadier, rock | Round nose grenadier |
| Grenadier | *Lepidorhynchus denticulatus* | Whiptail, deepsea | Rattail |
| Grenadier | *Macrourus* spp. | Grenadier | Rattail/popeye/roughhead |
| Grenadier | *Nezumia bairdi* | Marlin-spike | Rattail/grenadier |
| Grenadier | *Trachyrhynchus murrayi* | Grenadier, roughnose | Rattail |
| Grouper | *Caprodon schlegelii* | Grouper | Pinkfish/longfin/pink maomao |
| Grouper | *Cephalopholis acanthistius* | Coney, gulf | |
| Grouper | *Cephalopholis argus* | Grouper, purplespotted | Bluespotted argus or grouper/black rockcod |
| Grouper | *Cephalopholis taeniops* | Grouper, spotted | Blue spotted sea bass |
| Grouper | *Cephalopholis urodelus* | Grouper, chevron tailed | Mata'cle/flagtail(cd) rockcod/ bandtailed coral cod |
| Grouper | *Diplectrum formosum* | Perch, sand | Sandfish/squirrelfish/serrano |
| Grouper | *Epinephelus aeneus* | Grouper, white | Striped grouper |

| | | | |
|---|---|---|---|
| Grouper | *Epinephelus areolatus* | Rockcod, yellowspotted | Greenspotted rockcod/areolated rockcod or reefcod |
| Grouper | *Epinephelus chlorostigma* | Rockcod, brownspotted | Browspotted reefcod |
| Grouper | *Epinephelus cruentatus* | Graysby | |
| Grouper | *Epinephelus fasciatus* | Grouper, blacktip | Blacktip rockcod/banded reefcod/ sea bass |
| Grouper | *Epinephelus flavolimbatus* | Grouper, yellowedge | Yellowfinned grouper |
| Grouper | *Epinephelus fulva* | Coney | |
| Grouper | *Epinephelus fuscoguttatus* | Grouper, mottled | Flowery cod/carpet cod/black rockcod/rockcod |
| Grouper | *Epinephelus guaza* | Grouper, dusky | |
| Grouper | *Epinephelus inermis* | Grouper, marbled | |
| Grouper | *Epinephelus morio* | Grouper, red | Cherna Americana/negr |
| Grouper | *Epinephelus mystacinus* | Grouper, misty | Black grouper |
| Grouper | *Epinephelus nigritus* | Grouper, Warsaw | Black jewfish/black grouper |
| Grouper | *Epinephelus niveatus* | Grouper, snowy | Spotted grouper |
| Grouper | *Epinephelus striatus* | Grouper, Nassau | Hamlet/cherna criolla/rockfish |
| Grouper | *Mycteroperca bonaci* | Grouper, black | Rockfish/marbled rockfish/runner/ springer |
| Grouper | *Mycteroperca interstitialis* | Grouper, yellowmouth | Yellowfin grouper/princess rockfish |
| Grouper | *Mycteroperca jordani* | Grouper, Gulf | Baya |
| Grouper | *Mycteroperca rubra* | Grouper, comb | Scirenga |
| Grouper | *Mycteroperca tigris* | Grouper, tiger | Tiger rockfish/gag/ragtailed rockfish/bonacigato |
| Grouper | *Mycteroperca venenosa* | Grouper, yellowfin | Red rockfish/monkey rockfish |
| Grouper | *Mycteroperca xenarcha* | Grouper, broomtail | |
| Grouper/gag | *Mycteroperca microlepis* | Gag | Velvet rockfish/small-scaled rockfish |

**Table 7.1 (continued)**

| Market Name | Scientific Name | Common Name | Regional Name |
|---|---|---|---|
| Grouper/hind | *Epinephelus guttatus* | Hind, red | Spotted jewfish/strawberry grouper/black grouper |
| Grouper/jewfish | *Epinephelus itajara* | Jewfish | Giant Atlantic jewfish/guasa/mero/cabrilla/hind |
| Grunion | *Leuresthes tenius* | Grunion, California | Smelt |
| Grunt | *Anisotremus interruptus* | Grunt, burrito | Mojarron/sargo |
| Grunt | *Conodon nobilis* | Grunt, barred | |
| Grunt | *Haemulon bonariense* | Grunt, black | Ronco prieto |
| Grunt | *Haemulon carbonarium* | Grunt, Caesar | Black Grunt/Blacktail grunt |
| Grunt | *Haemulon chrysargyreum* | Grunt, smallmouth | |
| Grunt | *Haemulon flavolineatum* | Grunt, French | Yellowstriped grunt/yellow grunt/white grunt |
| Grunt | *Haemulon macrostomum* | Grunt, Spanish | Striped grunt/gray grunt/streaked grunt/sow grunt |
| Grunt | *Haemulon parrai* | Sailors choice | Ronco/ronco blanco/ronco prieto/bastard margate |
| Grunt | *Haemulon plumieri* | Grunt, white | Black grunt/common grunt/ronco ronco/boar grunt |
| Grunt | *Haemulon sciurus* | Grunt, bluestriped | Yellow grunt/boar grunt/humpback grunt |
| Grunt | *Haemulon scudderii* | Prieta mojarra | |
| Grunt | *Haemulon steindachneri* | Grunt, Latin | Roncador raiado |
| Grunt | *Haemulon striatum* | Grunt, striped | Common tomtate/white grunt |
| Grunt | *Orthopristis chrysoptera* | Pigfish | Hogfish/sailors choice/redmouthed grunt |

| | | | |
|---|---|---|---|
| Grunt | *Pomadasys crocro* | Grunt, burro | |
| Grunt/Catalina | *Anisotremus taeniatus* | Catalina | |
| Grunt/cottonwick | *Haemulon melanurum* | Cottonwick | Ceaser grunt/jeniguana/comical grunt |
| Grunt/margate | *Anisotremus surinamensis* | Margate, black | Pompon |
| Grunt/margate | *Haemulon album* | Margate | Margate fish/sailors choice/Margaret grunt/bream |
| Grunt/porkfish | *Anisotremus virginicus* | Porkfish | Catalineta |
| Grunt/sargo | *Anisotremus davidsoni* | Sargo | Sargo raiado |
| Grunt/sweetlips | *Plectorhynchus chaetodonides* | Sweetlips, harlequin | Manyspotted sweetlips/multispotted sweetlips |
| Grouper/sweetlips | *Plectorhynchus nigrus* | Sweetlips, brown | Brown sweetlips/blubberlip bream black sweetlip |
| Grunt/sweetlips | *Plectorhynchus lineatus* | Sweetlips, yellowbanded | |
| Grunt/sweetlips | *Plectorhynchus orientalis* | Sweetlips, oriental | |
| Grunt/sweetlips | *Plectorhynchus pictus* | Sweetlips, painted | Morwong/blackall/bluey/thicklipped grunt |
| Grunt/tomtate | *Haemulon aurolineatum* | Tomtate | Yellow tomtate/white grunt/jeniguano |
| Grunt/tomtate | *Haemulon rimator* | Tomtate | Redmouthed grunt/white grunt/ceaser/redmouth grunt |
| Guitarfish | *Platyrhinoidis triseriata* | Thornback | |
| Guitarfish | *Rhinobatos lentiginosus* | Guitarfish, Atlantic | |
| Guitarfish | *Rhinobatos productus* | Guitarfish, shovelnose | |
| Guitarfish | *Zapteryx exasperata* | Guitarfish, banded | |
| Haddock | *Melanogrammus aeglefinus* | Haddock | Scrod haddock |
| Hake | *Phycis chesteri* | Hake, longfin | |
| Hake | *Urophycis brasiliensis* | Hake, Brazilian | West Atlantic forkbeard/squirrel codling |

**Table 7.1 (continued)**

| Market Name | Scientific Name | Common Name | Regional Name |
|---|---|---|---|
| Hake | *Urophycis chuss* | Hake, red | Squirrel hake/mud hake/squirrel ling/ white hake |
| Hake | *Urophycis cirrata* | Hake, Gulf | |
| Hake | *Urophycis earlli* | Hake, Carolina | |
| Hake | *Urophycis floridana* | Hake, southern | |
| Hake | *Urophycis regia* | Hake, spotted | Spotted codling/cod |
| Hake | *Urophycis tenius* | Hake, white | White hake/codling/squirrel hake/ling/steakfish |
| Halfmoon | *Medialuna californiensis* | Halfmoon | California halfmoon/medialuna/ blue perch |
| Halibut | *Hippoglossus hippoglossus* | Halibut, Atlantic | Common halibut |
| Halibut | *Hippoglossus stenolepis* | Halibut, Pacific | Common halibut/whitesided paltus |
| Halibut/California halibut | *Paralichthys californicus* | Halibut, California | Bastard halibut/Monterey halibut |
| Hamlet | *Epinephelus afer* | Hamlet, mutton | |
| Herring | *Etrumeus teres* | Herring, round | |
| Herring | *Ilisha africans* | Ilisha, African | West African ilisha |
| Herring | *Ilisha elongata* | Ilisha, pugnose | Pacific ilisha/Chinese herring |
| Herring | *Ilisha furthi* | Ilisha, Pacific | Sardine/machete |
| Herring | *Ilisha megaloptera* | Ilisha, Bigeye | |
| Herring | *Ilisha melastoma* | Ilisha, Indian | |
| Herring | *Ilisha pristigastroides* | Ilish, Javan | |
| Herring | *Opisthopterus tardoore* | Tardoore | Longfinned Herring |
| Herring | *Pellona ditchela* | Pellona, Indian | Ditchellee/toothed shad/bony herring |
| Herring, thread | *Opisthonema oglinum* | Thread herring, Atlantic | Hairy-back/shad-herring/sprat/bristle |

| | | | |
|---|---|---|---|
| | | | herring/sargo |
| Herring, river herring | *Alosa aestivalis* | Herring, blueback | Blue herring/summer herring/glut herring |
| Herring/river herring | *Alosa chrysochloris* | Herring, skipjack | Skip jack/blue herring/green herring/golden shad |
| Herring/sea herring/slid | *Clupea harengus harengus* | Herring, Atlantic | Common herring/Labrador herring/sardine/sperling |
| Herring/sea herring/slid | *Clupea harengus pallasi* | Herring, Pacific | California herring/easter herring/seld |
| Hind | *Epinephelus adscensionis* | Hind, rock | Ascension rockcod/rockcod/hind/polka dot |
| Hind | *Epinephelus drummondhayi* | Hind, speckled | John paw |
| Hogfish | *Bodianus pulchellus* | Hogfish, spotfin | |
| Hogfish | *Bodianus rufus* | Hogfish, Spanish | Spanish ladyfish/ladyfish |
| Hogfish | *Lachnolaimus maximus* | Hogfish | Common hogfish/hogsnapper/capitan/perro perro |
| Houndfish | *Tylosurus corcodilus* | Houndfish | Hound/guardfish/garfish/crocodile needlefish |
| Icefish | *Champsocephalus gunnari* | Icefish, mackerel | |
| Icefish | *Notothenia moariensis* | Icefish | Black cod/maori chief |
| Icefish | *Salangichthys* spp. | Icefish, Japanese | Whitebait/noodlefish/whitefish/salanx |
| Jack | *Caranx bartholomaei* | Jack, yellow | Cibi amarillo |
| Jack | *Caranx caballus* | Jack, green | Common green jack/cocinero/cocinero dorado/jurel |
| Jack | *Caranx helvolus* | Jack, whitemouth | Omaka/dusky jack/whitetongued crevally |
| Jack | *Caranx hippos* | Jack, crevalle | Toro/cavally/common jack/jack runner/rudderfish |
| Jack | *Caranx latus* | Jack, horse-eye | Jurel/xurel/yellow jack |

**Table 7.1 (continued)**

| Market Name | Scientific Name | Common Name | Regional Name |
|---|---|---|---|
| Jack | *Caranx lugubris* | Jack, black | Black ulua/tarakito/tinosa |
| Jack | *Caranx ruber* | Jack, bar | Little skip jack/green jack/cibi/ carbonero/skip jack |
| Jack | *Caranx vinctus* | Jack, mazatlan striped | Cocinera |
| Jack | *Oligoplites saurus* | Leatherjacket | Leathercoat/zapatero/runner/ quiebra/sauteur |
| Jack | *Selene peruviana* | Moonfish, Pacific | |
| Jack | *Selene setapinnis* | Moonfish, Atlantic | |
| Jack | *Selene vomer* | Lookdown | Atlantic lookdown/moonfish/ horsehead/dollarfish |
| Jack | *Seriola rivoliana* | Jack, almaco | |
| Jack | *Uraspis secunda* | Jack, cottonmouth | Segundo/volantin |
| Jack/blue runner | *Caranx crysos* | Runner, blue | Hardtail/hardtail jack/yellow jack/ jack crevalle |
| Jack/rainbow runner | *Elegatis bipinnulata* | Runner, rainbow | Kamanu/runner/yellowtail/skip jack/shoemaker |
| Jack/roosterfish | *Nematistius pectoralis* | Roosterfish | |
| Jack/trevally | *Caranx georgianus* | Trevally, white | Jackfish/araara/silver trevally/skip jack |
| Jack/trevally | *Caranx ignobilis* | Trevally, giant | Lowly trevally/yellowfin trevally/giant kingfish |
| Jack/trevally | *Caranx melampygus* | Jack, spotted | Starjack/blue ulua/blue jack/spotted trevally |
| Jackfish/crevalle | *Alectis indicus* | Jack, threadfin | Diamond trevally/mirrorfish/old maid/ kagami ulua |
| Jobfish | *Aphareus furcatus* | Jobfish, smalltooth | |

| | | | |
|---|---|---|---|
| Jobfish | *Aphareus rutilans* | Jobfish, rusty | |
| Jobfish | *Aprion virescens* | Jobfish, green | |
| Jobfish | *Pristipomoides flavipinnis* | Snapper/jobfish goldeneye | Yelloweye opakapaka/silk snapper |
| Jobfish | *Pristipomoides sieboldii* | Jobfish, lavender | Kali kali |
| Kahawai | *Arripis trutta* | Kahawai | Salmon trout/Australian salmon/sake |
| Kelpfish | *Chironemus marmoratus* | Kelpfish | Marbled morwong/hiwihiwi/ngakoikoi |
| Killifish | *Fundulus confluentus* | Killifish, marsh | Atlantic killifish |
| Killifish | *Fundulus heteroclitus* | Mummichog | Mudminnow/pike minnow/mudfish/killifish/killy |
| Killifish | *Fundulus majalis* | Killifish, striped | Striped mummichog/gudgeon/bull minnow/mayfish |
| Kingfish | *Menticirrhus americanus* | Kingfish, southern | Sea mullet/Carolina whiting/sand whiting/roundhead |
| Kingfish | *Menticirrhuys littoralis* | Kingfish, gulf | King whiting/surf whiting/silver whiting/Virginia mullet |
| Kingfish | *Menticirrhus saxatilis* | Kingfish, northern | Northern whiting/king whiting/sea mullet/hake/barb |
| Kingfish/corbina | *Menticirrhus undulatus* | Corbina, California | California whiting/sand sucker |
| Kingklip | *Genypterus blacodes* | Kingklip, golden | Ling/cloudy bay cod/Australian rockling |
| Kingklip | *Genypterus capensis* | Kingklip, South African | African kingklip |
| Kingklip | *Genypterus chilensis* | Kingklip, red | Ling/hoka/hokarari/congrio Colorado |
| Kingklip | *Genypterus maculatus* | Kingklip, black | Chacha/congrio negro |
| Ladyfish | *Elops affinis* | Machete | |
| Ladyfish | *Elops hawaiensis* | Ladyfish | Hawaiian tarpon/giant herring/tenpounder/awa awa |
| Ladyfish | *Elops saurus* | Ladyfish | Tenpounder/bony fish/bigeye herring/bonefish/chiro |

**Table 7.1 (continued)**

| Market Name | Scientific Name | Common Name | Regional Name |
|---|---|---|---|
| Ling | *Molva dypterygia* | Ling, blue | Lesser ling/trade ling |
| Ling | *Molva molva* | Ling | European ling/ling cod/molva/sea pike/ drizzie |
| Ling, Mediterranean | *Molva macrophthalus* | Ling, Spanish | |
| Lingcod | *Ophiodon elongatus* | Lingcod | Cultus cod/blue cod/buffalo cod/ling |
| Lizardfish | *Synodus foetens* | Lizardfish, inshore | Galliwasp/lagarto/soapfish/providence whiting |
| Lizardfish | *Synodus intermedius* | Sand diver | Galliwasp |
| Louvar | *Luvarus imperialis* | Louvar | |
| Lumpfish | *Cyclopterus lumpus* | Lumpfish | Lumpsucker/lump/henfish/sea owl/paddle/ sea hen |
| Mackerel | *Auxis thazard* | Mackerel, frigate | Plain bonito/judeu/Melva/ke oke o/auxide |
| Mackerel | *Grammatorcynus bicaneatus* | Mackerel, double-lined | Thazard-kisarakus/carite cazon pintado |
| Mackerel | *Rastrelliger kanagurta* | Mackerel, Indian | Rakegilled mackerel/longjawed mackerel/gurukuma |
| Mackerel | *Scomber japonicus* | Mackerel, chub | Common mackerel/Japanese mackerel/ Pacific mackerel |
| Mackerel | *Scomber scombrus* | Mackerel, Atlantic | Common mackerel/Boston mackerel/ caballa |
| Mackerel | *Scombermorus commerson* | Mackerel, narrow-barred | Seer |
| Mackerel | *Scomberomorus concolor* | Sierra, Gulf | Monterey Spanish mackerel/ Thazard de Monterey |
| Mackerel | *Scomberomorus maculatus* | Mackerel, Spanish | Spotted cybium/bay mackerel/spotted mackerel |
| Mackerel | *Scomberomorus regalis* | Cero | Painted mackerel/kingfish/pintado/ |

| | | | |
|---|---|---|---|
| | | | Sierra/king mackerel |
| Mackerel | *Scomberomorus sierra* | Sierra, Pacific | Mexican Sierra/Sierra mackerel/Thazard Sierra |
| Mackerel | *Scomberomorus tritor* | Mackerel, Spanish | |
| Mackerel, Atka | *Pleurogrammus monopterygius* | Mackerel, Atka | Atkafish/yellowfish |
| Mackerel, chub | *Scomber australasicus* | Mackerel, spotted chub | Common mackerel/slimy mackerel/blue mackerel |
| Mackerel, Jack | *Trachurus declivis* | Mackerel, Jack | Hature/horse mackerel/scad/bung/yellowtail |
| Mackerel, Jack | *Trachurus japonicus* | Scad, Japanese | Aji/maaji |
| Mackerel, Jack | *Trachurus lathami* | Scad, rough | |
| Mackerel, Jack | *Trachurus murphyi* | Scad, Inca | |
| Mackerel, Jack | *Trachurus symmetricus* | Mackerel, Jack | Saurel/rough scad/California horse mackerel/Pacific Jack mackerel |
| Mackerel, king | *Scomberomorus cavalla* | Mackerel, king | Cavalla/cero/Sierra |
| Mackerel/snake | *Ruvettus pretiosus* | Oilfish | |
| Mackerel/scad | *Trachurus novaeze-landiae* | Mackerel, Jack | Hature/horse mackerel/cowanyoung |
| Mahi-mahi | *Coryphaena equisetis* | Dolphin, pompano | Dorado |
| Mahi-mahi | *Coryphaena hippurus* | Dolphin | Corado/common dolphinfish/dolphin |
| Manta | *Manta birostris* | Manta, Atlantic | |
| Manta | *Manta hamiltoni* | Manta, Pacific | |
| Manta | *Mobula hypostoma* | Ray, devil | |
| Manta | *Mobula japanica* | Mobula, spinetail | |
| Manta | *Mobula lucasana* | Mobula, smoothtail | |
| Marlin | *Makaira indica* | Marlin, black | Indian spearfish/shirokajiki/spearfish/swordfish |

**Table 7.1 (continued)**

| Market Name | Scientific Name | Common Name | Regional Name |
|---|---|---|---|
| Marlin | *Makaira nigricans* | Marlin, blue | Sailfish/spikefish/boohoo/agika prieta/ kurokajiki |
| Marlin | *Tetrapturus albidus* | Marlin, white | Atlantic white marlin/jiki |
| Marlin | *Tetrapturus audax* | Marlin, striped | Marlinspike/barred marlin/spearfish/takaketonga |
| Menhaden | *Brevoortia gunteri* | Menhaden, finescale | |
| Menhaden | *Brevoortia patronus* | Menhaden, Gulf | Largescale menhaden/alewife/shad |
| Menhaden | *Brevoortia smithi* | Menhaden, yellowfin | Yellowfin shad |
| Menhaden | *Brevoortia tyrannus* | Menhaden, Atlantic | American menhaden/mossbunker/ skip jack/shad/pogy/bunker |
| Milkfish | *Chanos chanos* | Milkfish | Awa/bangos/bandang |
| Mojarra | *Diapterus auratus* | Pompano, Irish | Muttonfish |
| Mojarra | *Diapterus plumieri* | Mojarra, striped | |
| Monkfish | *Lophius americanus* | Goosefish | Monktails/angler/fishing frog/allmouth/molligut |
| Monkfish | *Lophius piscatorius* | Monkfish | Monk/anglerfish/goosefish/allmouth/ frogfish |
| Mooneye | *Hiodon tergisus* | Mooneye | Toothed herring/river whitefish |
| Moray | *Muraena retifera* | Moray, recticulate | |
| Morwong | *Aplodactylus meandratus* | Marblefish | Granite trout/keke |
| Morwong | *Cheilodactylus gibbosus* | Morwong, magpie | Banded morwong/black and white morwong |
| Morwong | *Cheilodactylus macropterus* | Tarakihi | Silver perch/jackass morwong/ocean bream/silver bream |
| Morwong | *Cheilodactylus spectabilis* | Morwong, brownband | Red moki |

| | | | |
|---|---|---|---|
| Morwong | *Nemadactylus carponotatus* | Morwong, blue | Queen snapper (Australia) |
| Morwong | *Nemadactylus douglasi* | Morwong | Porae (New Zealand) |
| Mullet | *Agonostomus monticola* | Mullet, mountain | Trucha |
| Mullet | *Aldrichetta forsteri* | Mullet, yelloweye | Herring/aua/awa/matakawhiti |
| Mullet | *Crenimugil crenilabis* | Mullet, wartynosed | Fringelip mullet |
| Mullet | *Mugil cephalus* | Mullet, striped | Sea mullet/gray mullet/black mullet/ river mullet |
| Mullet | *Mugil curema* | Mullet, white | Blueback mullet/Liza |
| Mullet | *Mugil gaimardianus* | Mullet, redeye | |
| Mullet | *Mugil liza* | Liza | Blueback mullet/fantail mullet/ stripped mullet |
| Mullet | *Mugil trichodon* | Mullet, fantail | |
| Mullet | *Mullus brabatus* | Mullet, red | Surmullet/unstriped red mullet |
| Mullet | *Mullus surmuletus* | Mullet, red | Striped mullet |
| Mullet | *Neomyxus chaptalii* | Mullet | Uouoa |
| Muskellunge | *Esox masquinongy* | Muskellunge | Muscalonga/tiger musky/musky/luge/ great pike |
| Opah | *Lampris guttatus* | Opah | Moonfish |
| Opaleye | *Firella nigricans* | Opaleye | Greenfish/California bluefish/ mangrove fish/parore |
| Oreo dory | *Allocyttus* spp. | Oreo dory, black | |
| Oreo dory | *Pseudocyttus maculatus* | Oreo dory, smooth | Smooth oreo/small-spined oreo |
| Oscar | *Astronotus ocellatus* | Oscar | Peacock bass/peacock cichlid/velvet cichlid |
| Pacu | *Pacu* spp. | Pacu | |
| Paddlefish | *Polyodon spathuyla* | Paddlefish | Spoonbilled sturgeon/spoonbill/ buckbill cat |
| Pargo | *Hoplopagrus guntheri* | Pargo, striped | Pargo raisero |

**Table 7.1 (continued)**

| Market Name | Scientific Name | Common Name | Regional Name |
|---|---|---|---|
| Parrotfish | *Scarus coelestinus* | Parrotfish, midnight | Purple parrotfish/indigo parrotfish |
| Parrotfish | *Sparisoma viride* | Parrotfish, stoplight | |
| Perch | *Bairdiella chrysoura* | Perch, silver | Mademoiselle/yellowtail/ground croaker/sand perch |
| Perch | *Hermosilla azurea* | Perch, zebra | |
| Perch, Nile | *Lates nilotica* | Perch, Nile | |
| Perch, ocean | *Sebastes alutus* | Ocean perch, Pacific | |
| Perch, ocean | *Sebastes fasciatus* | Redfish, Labrador | |
| Perch, ocean | *Sebastes marinus* | Redfish/ocean perch | Rosefish/redfish/snapper/sea perch/ redbream |
| Perch, ocean | *Sebastes mentella* | Redvish, deepwater | Atlantic redfish/deepwater rosefish |
| Perch, ocean | *Sebastes viviparus* | Redfish, Norway | Rosefish/Norway haddock |
| Perch, pile | *Rhacochilus vacca* | Perch, pile | |
| Perch, white | *Morone americana* | Perch, white | |
| Perch, yellow/ lake perch | *Perca flavescens* | Perch, yellow | Ring perch/striped perch/Jack perch/redfin |
| Picarel | *Spicara maena* | Picarel, blotched | |
| Pickerel | *Esox americanus americanus* | Pickerel, redfin | Grass pickerel/barred pickerel/smaller pickerel |
| Pickerel | *Esox americanus vermiculatus* | Pickerel, grass | Little pickerel/grass pike |
| Pickerel | *Esox niger* | Pickerel, chain | Eastern pickerel/mud pickerel/grass pickerel/pike |
| Pike | *Esox lucius* | Pike, northern | Jack/snake/jackpike/jackfish/grass pike |
| Pilchard/sardine | *Sardina pilchardus* | Pilchard, European | True sardine |
| Pilchard/sardine | *Sardinops neopilchardus* | Sardine, Australian | Australian pilchard/smid/maray |

| | | | |
|---|---|---|---|
| Pilchard/sardine | *Sardinops sagax melanosticta* | Pilchard, Japanese | Japanese sardine/true sardine/ spotlined sardine |
| Pilchard/sardine | *Sardinops sagax ocellata* | Pilchard, South African | South African sardine |
| Pipefish | *Syngnathus fuscus* | Pipefish, northern | Banded pipefish/green pipefish/billfish |
| Plaice | *Pleuronectes quadrituberculatus* | Plaice, Alaska | Yellowbellied flounder |
| Plaice | *Plueronected platessa* | Plaice, European | Sanddab |
| Plaice/dab | *Hippoglossoides platessoides* | Plaice, American | Sea dab/Canadian plaice/sand dab/blackdab/sole |
| Pollock | *Pollachius pollachius* | Pollack | Lythe/European Pollock/saithe |
| Pollock | *Pollachius virens* | Pollock | Saithe/coalfish/coley/green cod/ Boston bluefish |
| Pollock/Alaska pollock | *Theragra chalcogramma* | Pollock, walleye | Whiting/Pacific tomcod/Pacific pollock |
| Pomfret | *Brama brama* | Pomfret, Atlantic | Bream/Jim bream/Ray's bream |
| Pomfret | *Brama japonica* | Pomfret, Pacific | Bream/Ray's bream |
| Pomfret | *Taractes rubescens* | Pomfret | |
| Pompano | *Alectis ciliaris* | Pompano, African | Threadfin/threadfish/hairfish/ cobblerfish/sunfish |
| Pompano | *Trachinotus blochi* | Pompano, tropical | Bloch's dart/ladyfish/pampano/ pamphlet |
| Pompano | *Trachinotus carolinus* | Pompano, Florida | Common pompano/Atlantic pompano/sunfish/butterfish |
| Pompano | *Trachinotus paitensis* | Pompano, paloma | |
| Pompano/palometa | *Trachinotus goodei* | Palometa | Permit/great pompano/longfinned pompano/alewife |
| Pompano/permit | *Trachinotus falcatus* | Permit | Round pompano/pompano palmeta/ ovate pompano |

**Table 7.1 (continued)**

| Market Name | Scientific Name | Common Name | Regional Name |
|---|---|---|---|
| Pompano/permit | *Trachinotus kennedyi* | Permit, palometta | Pompano |
| Pompano/pompanito | *Trachinotus rhodopus* | Pompano, gafftopsail | Pampanito/domingo |
| Porgy | *Calamus* spp. | Porgy | Shad porgy/grass porgy/bluebone porgy/sheephead porgy |
| Porgy | *Chrysophrys auratus* | Porgy, red Hawaiian | Snapper/schnapper/red bream/ brim/squire |
| Porgy | *Dentex gibbosus* | Dentex | |
| Porgy | *Diplodus* spp. | Porgy | Silver bream/spottail pinfish/bream/sargo/puntazzo |
| Porgy | *Lagodon rhomboides* | Pinfish | Bream/sailors choice/sargo/chop spina/Spanish porgy |
| Porgy | *Pagrus pagrus* | Porgy | Seabream/pink porgy/silver snapper/besugo/red porgy |
| Porgy | *Stenotomus caprinus* | Porgy, longspine | |
| Porgy/scup | *Stenotomus chrysops* | Scup | Porgy/maiden/fair maid/ironsides/ northern porgy |
| Puffer | *Lagocephalus laevigatus* | Puffer, smooth | Jugfish/rabbitfish/tambor/swellfish/ silvery puffer |
| Puffer | *Lagocephalus lagocephalus* | Puffer, oceanic | |
| Puffer | *Sphoeroides annulatus* | Puffer, bullseye | |
| Puffer | *Sphoeroides dorsalis* | Puffer, marbled | |
| Puffer | *Sphoeroides lobatus* | Puffer, longnose | |
| Puffer | *Sphoeroides maculatus* | Puffer, northern | Sea squab/blowfish/swellfish/ ballonfish/globefish |
| Puffer | *Sphoeroides mephelus* | Puffer, southern | |

| | | | |
|---|---|---|---|
| Puffer | *Sphoeroides pachygaster* | Puffer, blunthead | |
| Puffer | *Sphoeroides parvus* | Puffer, least | |
| Puffer | *Sphoeroides spengleri* | Puffer, bandtail | |
| Puffer | *Sphoeroides testudineus* | Puffer, checkered | Tambor/globefish/blowfish/northern puffer |
| Racehorse | *Congiopodus leucopaecilus* | Pigfish, southern | Marbled pigfish/puramorua |
| Ray,bat | *Myliobatis californica* | Ray, bat | |
| Ray, bullnose | *Myliobatis freminvillei* | Ray, bullnose | |
| Ray, cownose | *Rhinoptera bonasus* | Ray, cownose | American cownose/cowfish |
| Ray, eagle | *Aetobatus narinari* | Ray, spotted eagle | |
| Ray, eagle | *Myliobatis goodei* | Ray, southern eagle | |
| Ray, electric | *Narcine brasiliensis* | Ray, lesser electric | |
| Ray, electric | *Torpedo califonica* | Ray, Pacific electric | |
| Rockfish | *Helicolenus papillosus* | Rockfish | Scarpee/jock stewart/seaperch |
| Rockfish | *Scorpaena cardinalis* | Rockfish, red | Red rock cod/matua-whapukui |
| Rockfish | *Sebastes aleutianus* | Rockfish, rougheye | |
| Rockfish | *Sebastes atrovirens* | Rockfish, kelp | Garrupa |
| Rockfish | *Sebastes auriculatus* | Rockfish, brown | |
| Rockfish | *Sebases aurora* | Rockfish, aurora | |
| Rockfish | *Sebastes babcocki* | Rockfish, redbanded | |
| Rockfish | *Sebastes boreaclis* | Rockfish, shortraker | |
| Rockfish | *Sebastes brevispinis* | Rockfish, silvergray | |
| Rockfish | *Sebastes carnatus* | Rockfish, gopher | Flesh-colored rockcod |
| Rockfish | *Sebastes caurinus* | Rockfish, copper | |
| Rockfish | *Sebastes chlorostictus* | Rockfish, greenspotted | |
| Rockfish | *Sebastes chrysomelas* | Rockfish, black & yellow | |
| Rockfish | *Sebastes ciliatus* | Rockfish, dusky | |
| Rockfish | *Sebastes constellatus* | Rockfish, starry | Spotted rockfish |

**Table 7.1 (continued)**

| Market Name | Scientific Name | Common Name | Regional Name |
|---|---|---|---|
| Rockfish | *Sebastes crameri* | Rockfish, darkblotched | Blackmouth rockfish |
| Rockfish | *Sebastes dalli* | Rockfish, calico | |
| Rockfish | *Sebastes diploproa* | Rockfish, splitnose | Lobejawed rockfish |
| Rockfish | *Sebastes elongatus* | Rockfish, greenstriped | |
| Rockfish | *Sebastes emphaeus* | Rockfish, puget sound | |
| Rockfish | *Sebastes ensifer* | Rockfish, swordspine | |
| Rockfish | *Sebastes entomelas* | Rockfish, widow | Pacific red snapper (Cal-Wash) |
| Rockfish | *Sebastes eos* | Rockfish, pink | Pink rockfish |
| Rockfish | *Sebastes flavidus* | Rockfish, yellowtail | Pacific red snapper (Cal-Wash)/ green rockfish |
| Rockfish | *Sebastes gilli* | Rockfish, bronzespotted | |
| Rockfish | *Sebastes goodei* | Chilipepper | Pacific red snapper (Cal-Wash) |
| Rockfish | *Sebastes helvomaculatus* | Rockfish, rosethorn | Swordspine/flyfish |
| Rockfish | *Sebastes hopkinsi* | Rockfish, squarespot | |
| Rockfish | *Sebastes jordani* | Rockfish, shortbelly | Pacific red snapper (Cal-Wash) |
| Rockfish | *Sebastes lentiginosus* | Rockfish, freckled | |
| Rockfish | *Sebastes levis* | Cowcod | Pacific red snapper(Cal-Wash) |
| Rockfish | *Sebastes macdonaldi* | Rockfish, Mexican | Coralred rockfish |
| Rockfish | *Sebastes maliger* | Rockfish, quillback | Yellowback rockcod/brown rockcod/orangespotted rockcod |
| Rockfish | *Sebastes melanops* | Rockfish, black | Pacific red snapper (Cal-Wash) |
| Rockfish | *Sebastes melanosema* | Rockfish, semaphore | |
| Rockfish | *Sebastes melanostomus* | Rockfish, blackgill | |
| Rockfish | *Sebastes miniatus* | Rockfish, vermillion | Pacific red snapper (Cal-Wash)/ rasciera/rasher |

| | | | |
|---|---|---|---|
| Rockfish | *Sebastes mystinus* | Rockfish, blue | Black rockfish/rockcod/priestfish |
| Rockfish | *Sebastes nebulosus* | Rockfish, China | Yellowspotted rockcod |
| Rockfish | *Sebastes nigrocinctus* | Rockfish, tiger | Blackbanded rockcod |
| Rockfish | *Sebastes ovalis* | Rockkfish, speckled | Pacific red snapper (Cal-Wash)/ widowfish//viura |
| Rockfish | *Sebastes paucispinis* | Bocaccio | Pacific red snapper (Cal-Wash)/ jack/merou |
| Rockfish | *Sebastes phillipsi* | Rockfish, chameleon | |
| Rockfish | *Sebastes pinniger* | Rockfish, canary | Pacific red snapper(Cal-Wash)/ orange rockfish |
| Rockfish | *Sebastes polyspinis* | Rockfish, northern | Multispined bass |
| Rockfish | *Sebastes proriger* | Rockfish, redstripe | Redstriped rockfish |
| Rockfish | *Sebastes rastrelliger* | Rockfish, grass | |
| Rockfish | *Sebastes reedi* | Rockfish, yellowmouth | |
| Rockfish | *Sebastes rosaceus* | Rockfish, rosy | Corsair |
| Rockfish | *Sebastes rosenblatti* | Rockfish, greenblotched | |
| Rockfish | *Sebastes ruberrimus* | Rockfish, yelloweye | Pacific red snapper (Cal-Wash)/ rasphead rockfish |
| Rockfish | *Sebastes rubrivinctus* | Rockfish, flag | Spanish flag |
| Rockfish | *Sebastes rufinanus* | Rockfish, dwarf-red | |
| Rockfish | *Sebastes rufus* | Rockfish, bank | Pacific red snapper (Cal-Wash) |
| Rockfish | *Sebastes saxicola* | Rockfish, stripetail | Olivebacked rockfish |
| Rockfish | *Sebastes semicinctus* | Rockfish, halfbanded | |
| Rockfish | *Sebastes serranoides* | Rockfish, olive | Pacific red snapper (Cal-Wash) |
| Rockfish | *Sebastes serriceps* | Treefish | |
| Rockfish | *Sebastes simulator* | Rockfish, pinkrose | |
| Rockfish | *Sebastes umbrosus* | Rockfish, honeycomb | |
| Rockfish | *Sebastes wilsoni* | Rockfish, pygmy | Wilson's rockfish |

**Table 7.1 (continued)**

| Market Name | Scientific Name | Common Name | Regional Name |
|---|---|---|---|
| Rockfish | *Sebastes zacentrus* | Rockfish, sharpchin | Bigeyed rockfish |
| Rockling | *Enchelyopus cimbrius* | Rockling, fourbeard | |
| Rosefish | *Helicolenus dactylopterus* | Rosefish, blackbelly | Redfish/serran imperial/fanegal/cardonniera |
| Roughy | *Hoplostethus mediterraneus* | Roughy, silver | Mediterranean redfish/rosy soldierfish/flint-perch |
| Roughy | *Paratrachichthys trailli* | Roughy, pinkfinned | Sandpaper fish/common roughy |
| Roughy, orange | *Hoplostethus atlanticus* | Roughy, orange | |
| Rudderfish | *Kyphosus cinerascens* | Rudderfish | Bluefish/blue chub/nenue/tropical drummer |
| Ruff, black | *Centrolophus niger* | Ruff, black | Rudderfish/black-fish/boarlace |
| Sablefish | *Anoplopoma fimbria* | Sablefish | Black cod/butterfish/skilfish/skil/beshow/coalfish |
| Sailfish | *Istiophorus gladius* | Sailfish, Indo-Pacific | Bashokajiki |
| Sailfish | *Istiophorus platypterus* | Sailfish | |
| Salmon, Atlantic | *Salmo salar* | Salmon, Atlantic | Sea salmon/kennebec/sebago salmon/black salmon |
| Salmon, cherry | *Onhcorhynchus masou* | Salmon, cherry | Masu salmon/masu/salmon trout |
| Salmon, chinook/king/spring | *Oncorhynchus tshawytscha* | Salmon, Chinook | Tee salmon/winter salmon/quinnat salmon |
| Salmon, chum/keta | *Oncorhynchus keta* | Salmon, chum | Dog salmon/calico salmon/chub/dog/salmon |
| Salmon, coho/silver/medium red | *Oncorhynchus kisutch* | Salmon, coho | Hoopid salmon/white salmon |

| | | | |
|---|---|---|---|
| Salmon, pink/ humpback | *Oncorhynchus gorbuscha* | Salmon, pink | Dog salmon/gorbusch/haddo/holia |
| Salmon, sockeye/ blueback/red | *Oncorrhynchus nerka* | Salmon, sockeye | Redfish/red salmon/spring-run salmon/summer sockeye |
| Sanddab | *Citharichthys sordidus* | Sanddab, Pacific | Mottled sanddab/soft flounder |
| Sandlance | *Ammodytes hexapterus* | Sanddab, Pacific | |
| Sandperch | *Mugiloides chilensis* | Sandperch, Chilean | Sea salmon |
| Sandperch | *Parapercis* spp. | Weever | Grubfish/sand smelt/sand perch/ whiting/blue cod |
| Sardine | *Harengula clupeola* | Pilchard, false | Cailleu/sardina escamuda/sprat |
| Sardine | *Harengula humeralis* | Sardine, redear | Sprat/sardine/redear pilchard/ pilchard/sardina |
| Sardine | *Harengula jaguana* | Sardine, scaled | |
| Sardine | *Sardinella anchovia* | Sardine, Spanish | False sardine/sardina/bang |
| Sardine | *Sardinella aurita* | Sardine, Spanish | Smooth sardine/gilt sardine or pilchard |
| Sardine | *Sardinella brasiliensis* | Sardine, orangespot | Spanish sardine/sardinia/sardinha verdadeira |
| Sardine | *Sardinella fimbriata* | Sardine, fringescale | Tunsoy/roundbodied sardine/ fimbriated sardine |
| Sardine | *Sardinella longiceps* | Sardine, oil | Tamban |
| Sardine | *Sardinella perforata* | Sardine, perforated-scale | Deepbodied sardine/deep body herring/lapad |
| Sardine/pilchard | *Sardinops sagax* | Sardine, Pacific | Cal. pilchard/So. Am. sardine/ Chilean pilchard |
| Sauger | *Stizostedion canadense* | Sauger | Eastern sauger/sand pike/sand pickerel/gray pike/Canadian pike |
| Saury | *Cololabis saira* | Saury, Pacific | |

**Table 7.1 (continued)**

| Market Name | Scientific Name | Common Name | Regional Name |
|---|---|---|---|
| Saury | *Scomberesox saurus* | Saury, Atlantic | Needlefish/ocean piper |
| Sawfish | *Pristis pectinata* | Sawfish, smalltooth | Largetooth sawfish/common sawfish |
| Sawfish | *Pristis perotteti* | Sawfish, largetooth | |
| Scad | *Caranx mate* | Scad, yellowtail | Black ulua/'omaka/deep trevally/ finlet scad |
| Scad | *Decapterus koheru* | Koheru | Hature |
| Scad | *Decapterus macarellus* | Scad, mackerel | Antonino/ocean robin |
| Scad | *Decapterus muroadsi* | Scad, amberstripe | Brownstriped mackerel scad/horse-scad mackerel |
| Scad | *Decapterus punctatus* | Scad, round | Cigarfish/round robin/dotted scad/ quia-quia |
| Scad | *Decapterus russelli* | Scad, northern mackerel | Russell's mackerel scad/mackerel scad/round scad |
| Scad | *Decapterus scombrinus* | Scad, Mexican | |
| Scad | *Decapterus tabl* | Scad, redtail | |
| Scad | *Selar crumenophthalmus* | Scad, bigeye | Gogle eye/cigarfish/goggle eye jack/ chicharro |
| Scad | *Trachurus picturatus* | Scad, blue | |
| Scad | *Trachurus trachurus* | Mackerel, Atlantic horse | Rough scad/saurel/jack mackerel/ Japanese horse mackerel |
| Scamp | *Mycteroperca phenax* | Scamp | Bacalao |
| Schoolmaster | *Litjanus apodus* | Schoolmaster | Caji/pargo amarillo |
| Scorpionfish | *Scorpaena scrofa* | Scorpionfish, orange | |
| Sculpin | *Hemitripterus americanus* | Sea raven, Atlantic | Raven/yellow sculpin/red sculpin/ sea sculpin |

| | | | |
|---|---|---|---|
| Sculpin | *Myoxocephalus polyacanthocephalus* | Sculpin, great | Kalog |
| Sculpin/cabezon | *Scorpaenichthys marmoratus* | Cabezon | Giant marbled sculpin/marbled sculpin |
| Seabream | *Archosargus rhomboidalis* | Bream, sea | Sheepshead/sheepshead porgy |
| Seabream | *Pagellus* spp. | Seabream | |
| Searobin | *Chelidonichthys kumu* | Searobin, bluefin | Grondin rouget/kumu/kumu gurnard/kumukumu/gurnard |
| Searobin | *Chelidonichthys lucerna* | Gurnard, tub | |
| Searobin | *Peristedion miniatum* | Searobin, armored | |
| Searobin | *Prionotus carolinus* | Searobin, northern | Gurnard/robin/flying fish/green-eye/carolina robin |
| Searobin | *Pterygotrigla picta* | Gaurnard, spotted | |
| Seatrout | *Cynoscion arenarius* | Seatrout, sand | Sand trout/white trout |
| Seatrout | *Cynoscion nebulosus* | Seatrout, spotted | Speckled trout/squeteague/gator trout/spotted weakfish |
| Seatrout | *Cynoscion nobilis* | Seabass, white | Corbina/California white seabass |
| Seatrout | *Cynoscion nothus* | Seatrout, silver | White seatrout/squeteague/bastard weakfish |
| Shad | *Alosa alabamae* | Shad, Alabama | Ohio shad/Gulf shad |
| Shad | *Alosa alosa* | Shad, Allis | |
| Shad | *Alosa fallax* | Shad, twaite | |
| Shad | *Alosa mediocris* | Shad, hickory | Hickory jack/hick/bonejack/shad herring/fall herring |
| Shad | *Alosa sapidissima* | Shad, American | Atlantic shad/Potomac shad/north silver shad |
| Shad | *Dorosoma cepedianum* | Shad, gizzard | Atlantic shad/mud shad/hickory shad/lake shad/sawbelly |

**Table 7.1 (continued)**

| Market Name | Scientific Name | Common Name | Regional Name |
|---|---|---|---|
| Shad | *Dorosoma petenense* | Shad, threadfin | |
| Shark | *Carcharhinus albimarginatus* | Shrark, silver tip | Whiptain reef shark |
| Shark | *Carcharhinus altimus* | Shark, bignose | Knopp's shark |
| Shark | *Carcharhinus amblyrhynchus* | Shark, gray reef | Graceful shark/longnose blacktail shark/black shark |
| Shark | *Carcharhinus brachyurus* | Shark, narrowtoothed | Bronze whaler shark/toiki/Cooper shark/river shark |
| Shark | *Carcharhinus brevipinna* | Shark, spinner | Longnosed blackfin shark |
| Shark | *Carcharhinus falciformis* | Shark, silky | |
| Shark | *Carcharhinus isodon* | Shark, finetooth | Smoothtooth shark |
| Shark | *Carcharhinus leucas* | Shark, bull | Cub shark/roundnose shark/pigeye shark/fish shark |
| Shark | *Carcharhinus limbatus* | Shark, blacktip | Blacktip/small blacktip shark/grey shark/soupfin |
| Shark | *Carcharhinus longimanus* | Shark, whitetip | Oceanic whitetip |
| Shark | *Carcharhinus melanopterus* | Shark, blacktip reef | Blacktip/black shark/blackfin reef shark |
| Shark | *Carcharhinus obscurus* | Shark, dusky | Dusky shark/bay shark/brown shark/shovelnose |
| Shark | *Carcharhinus perezi* | Shark, reef | |
| Shark | *Carcharhinus plumbeus* | Shark, sandbar | Atlantic gray shark |
| Shark | *Carcharhinus porosus* | Shark, smalltail | Sharpnosed shark/cuero duro |
| Shark | *Carcharhinus signatus* | Shark, night | |
| Shark | *Carcharodon carcharias* | Shark, white | Great white/white pointer/ |

| | | | |
|---|---|---|---|
| | | | maneater/blue pointer |
| Shark | *Cetorhinus maximus* | Shark, basking | |
| Shark | *Galeocerdo cuviere* | Shark, tiger | Spotted shark/leopard shark |
| Shark | *Galeorhinus galeus* | Shark, tope | Grey shark/soupfin/greyboy/flake |
| Shark | *Hexanchus griseus* | Shark, sixgill | Cow shark/bullshark/bulldog/grey shark/mud shark |
| Shark | *Lamna ditropis* | Shark, salmon | Mackerel shark/porbeagle |
| Shark | *Negaprion brevirostris* | Shark, lemon | Yellow shark |
| Shark | *Notorynchus cepedianus* | Shark, sevengill | Cow shark/broadsnouted or roundsnout sevengill shark |
| Shark | *Prionace glauca* | Shark, blue | Blue dog/blue whaler/blue pointer/great blue shark |
| Shark | *Triaenodon obesus* | Shark, whitetip reef | Blunthead shark/whitetip/smooth dogfish/houndshark |
| Shark | *Triakis semifasciata* | Shark, leopard | Catshark |
| Shark, angel | *Squatina californica* | Shark, Pacific angel | California angel shark/monkfish/northern angel shark |
| Shark, angel | *Squatina dumerili* | Shark, Atlantic angel | Angelfish/nurse fish/monkfish/sand devil |
| Shark, mako | *Isurus oxyrinchus* | Shark, shortfin mako | Blue pointer/bonito shark/Atlantic & Pacific mako shark |
| Shark, mako | *Isurus paucus* | Shark, longfin mako | Longfin mako shark |
| Shark, thresher | *Alopias pelagicus* | Shark, pelagic thresher | Whiptail shark/fox shark |
| Shark, thresher | *Alopias superciliosus* | Shark, bigeye thresher | Bigeye |
| Shark, thresher | *Alopias vulpinus* | Shark, common thresher | Fox shark/sea fox/swingletail/whiptail or thintail shark |
| Shark/bonnethead | *Sphyrna tiburo* | Shark/bonnethead | Bonnet shark/bonnet/bonnetnose shark/shoverhead |

**Table 7.1 (continued)**

| Market Name | Scientific Name | Common Name | Regional Name |
|---|---|---|---|
| Shark/hammerhead | *Sphyrna lewini* | Shark, scalloped hammerhead | Hammerheaded shark/bronze hammerhead |
| Shark/hammerhead | *Sphyrna mokarran* | Shark, great hammerhead | Great hammerhead shark |
| Shark/hammerhead | *Sphyrna tudes* | Shark, smalleye hammerhead | Great hammerhead/squat-headed hammerhead |
| Shark/hammerhead | *Sphyrna zygaena* | Shark, smooth hammerhead | Common hammerhead/ roundheaded hammerhead |
| Shark/porbeagle | *Lamna nasus* | Shark, porbeagle | Swordfin/porpoise shark/blue shark/mackerel shark |
| Shark/smoothhound | *Mustelus californicus* | Smoothhound, grey | |
| Shark/smoothhound | *Mustelus henlei* | Smoothhound, brown | Drab brown smoothhound |
| Sheephead | *Semicossyphus pulcher* | Sheephead, California | |
| Sheepshead | *Archosargus probatocephalus* | Sheepshead | Sheepshead porgy/convict fish |
| Silverside | *Atherinopsis affinis* | Topsmelt | Bay smelt/jack/least smelt |
| Silverside | *Atherinopsis californiensis* | Jacksmelt | California blue smelt/Cal. smelt/ pecado del rey |
| Silverside | *Basilichthys australis* | Silverside, So. American | |
| Silverside | *Menidia menidia* | Silverside, Atlantic | Northern silverside/dotted silverside/spearling |
| Skate | *Raja aleutica* | Skate, Aleutian | |
| Skate | *Raja binoculata* | Skate, big | |
| Skate | *Raja eglanteria* | Skate, clearnose | |
| Skate | *Raja erinacea* | Skate, little | |
| Skate | *Raja garmani* | Skate, rosette | |

| | | | |
|---|---|---|---|
| Skate | *Raja inornata* | Skate, California | |
| Skate | *Raja interrupta* | Skate, Bering | |
| Skate | *Raja kincaidi* | Skate, sandpaper | |
| Skate | *Raja laevis* | Skate, barndoo | |
| Skate | *Raja ocellata* | Skate, winter | |
| Skate | *Raja olseni* | Skate, spreadfin | |
| Skate | *Raja parmifera* | Skate, Alaska | |
| Skate | *Raja radiata* | Skate, thorny | |
| Skate | *Raja rhina* | Skate, longnose | |
| Skate | *Raja rosispinis* | Skate, flathead | |
| Skate | *Raja senta* | Skate, smooth | |
| Skate | *Raja spinicauda* | Skate, spinytail | |
| Skate | *Raja* spp. | Skate | Starry, blackbellied, roughskinned skate/swhai/suei |
| Skate | *Raja stellulata* | Skate, starry | |
| Skate | *Raja texana* | Skate, roundel | |
| Skate | *Raja trachura* | Skate, roughtail | |
| Skilfish | *Erilepis zonifer* | Skilfish | Giant skilfish/fat priest |
| Smelt | *Allosmerus elongatus* | Smelt, whitebait | |
| Smelt | *Argentina semifasciata* | Smelt, deep sea | Japanese argentine/halfbanded argentine |
| Smelt | *Argentina silus* | Smelt, great silver | Herring smelt/Atlantic argentine |
| Smelt | *Argentina sphyraena* | Smelt, lesser silver | Argentine/sildsmelt/glassauge/ stromsild |
| Smelt | *Hypomesus pretiosus* | Smelt, surf | Whitebait |
| Smelt | *Hypomesus transpacificus* | Smelt, delta | |
| Smelt | *Osmerus dentex* | Smelt, Arctic | White sea smelt |
| Smelt | *Osmerus eperlanus* | Smelt, European | Sparling |

**Table 7.1 (continued)**

| Market Name | Scientific Name | Common Name | Regional Name |
|---|---|---|---|
| Smelt | *Retropinna retropinna* | Smelt, common | Whitebait/cucumber fish/waikato/paraki/Australian smelt |
| Smelt | *Spirinchus starksi* | Smelt, night | |
| Smelt | *Spirinchus thaleichthys* | Smelt, longfin | Sacamento smelt |
| Smelt | *Thaleichthys pacificus* | Smelt, eulachon | Eurachon/candlefish/oolachon/Pacific smelt |
| Smelt/American smelt | *Osmerus mordax* | Smelt, rainbow | Spirling/saltwater smelt/freshwater smelt/icefish |
| Snapper,black | *Apsilus dentatus* | Snapper, black | Arnillo |
| Snapper, black and white | *Macolor niger* | Snapper, black and white | |
| Snapper, blackfin | *Lutjanus buccanella* | Snapper, blackfin | Boucanello/red snapper |
| Snapper, blacktail | *Lutjanus fulvus* | Snapper, blacktail | Flame-colored snapper/toau/red-margined seaperch |
| Snapper, blubberlip | *Lutjanus rivulatus* | Snapper, blubberlip | |
| Snapper, bluestriped | *Lutjanus kasmira* | Snapper, bluestriped | Bluebanded snapper/ta'ape/savani/funai/snapper |
| Snapper, cardinal | *Pristipomoides macrophthalmus* | Snapper, cardinal | |
| Snapper, Colorado | *Lutjanus colorado* | Snapper, Colorado | Red snapper/pargo Colorado |
| Snapper, crimson | *Pristipomoides filamentosus* | Snapper, crimson | Opakapaka |
| Snapper, cubera | *Lutjanus cyanopterus* | Snapper, cubera | Cuban snapper/cubera |
| Snapper, dog | *Lutjanus jocu* | Snapper, dog | |
| Snapper, emperor | *Lutjanus sebae* | Snapper, emperor | Red emperor |
| Snapper, golden | *Lutjanus inermis* | Snapper, golden | |

| | | | |
|---|---|---|---|
| Snapper, gray | *Lutjanus griseus* | Snapper, gray | Mangrove snapper/lawyer/cabellerote/pargo prieto |
| Snapper, humpback | *Lutjanus gibbus* | Snapper, humpback | Paddletail/hunched snapper/mala'l/red snapper |
| Snapper, lane | *Lutjanus synagris* | Snapper, lane | Spot snapper/redtail snapper/silk snapper |
| Snapper, mahogany | *Lutjanus mahogoni* | Snapper, mahogany | Ojanco |
| Snapper, malabar | *Lutjanus malabaricus* | Snapper, malabar | Scarlet sea perch/red bream/malabar red snapper |
| Snapper, midnight | *Macolor macularius* | Snapper, midnight | |
| Snapper, mullet | *Lutjanus aratus* | Snapper, mullet | |
| Snapper, mutton | *Lutjanus analis* | Snapper mutton | Muttonfish/pargo/pargo criollo |
| Snapper, onespot | *Lutjanus monostigma* | Snapper, onespot | Kakaka/red snapper/blackspot mayamaya |
| Snapper, Pacific | *Lutjanus peru* | Snapper, Pacific | Pacific red snapper |
| Snapper, Pacific dog | *Lutjanus novemfasciatus* | Snapper, Pacific dog | Argo prieto/pargo moreno |
| Snapper, queen | *Etelis oculatus* | Sanpper, queen | Bleareyed snapper/night snapper/cachucho |
| Snapper, red | *Lutjanus campechanus* | Sanpper, red | Caribbean red snapper/Mexican snapper |
| Snapper, red Caribbean | *Lutjanus purpureus* | Snapper, Caribbean red | |
| Snapper, ruby | *Etelis carbunculus* | Snapper, ruby | Queen snapper/ehu/onaga/'ula'ula kuae/palu |
| Snapper, rufous | *Lutijanus jordani* | Snapper, rufous | Hhuachinango/red snapper/Jordan's snapper |
| Snapper, sailfin | *Symphorichthys spilurus* | Snapper, sailfin | |

**Table 7.1 (continued)**

| Market Name | Scientific Name | Common Name | Regional Name |
|---|---|---|---|
| Snapper, scarlet | *Lutjanus sanguineus* | Snapper, blood | Saddletailed sea perch/red bream/red jaw |
| Snapper, silk | *Lutjanus vivanus* | Snapper, silk | West Indian snapper/day snapper/ longfin red snapper |
| Snapper, spotted rose | *Lutjanus guttatus* | Snapper, spotted rose | Mutton snapper/flamenco/pargo chibato |
| Snapper, twinspot | *Lutjanus bohar* | Snapper, twinspot | Twospot snapper/whitespotted snapper/mu/tagafi |
| Snapper, vermilion | *Rhomboplites aurorubens* | Sanpper, vermilion | Beeliner/clubhead snapper/night snapper |
| Snapper, wenchman | *Pristipomoides aquilonaris* | Wenchman | |
| Snapper, yellow-striped | *Etelis coruscans* | Snapper, yellowstriped | Ruby snapper |
| Snapper, yellowtail | *Ocyurus chrysurus* | Snapper, yellowtail | Palu-I'usama |
| Snook | *Centropomus ensiferus* | Snook, swordspine | |
| Snook | *Centropomus nigrescens* | Snook | Robalo prieto |
| Snook | *Centropomus ornatus* | Snook | |
| Snook | *Centropomus parallelus* | Snook, fat | Little snook |
| Snook | *Centropomus pectinatus* | Snook, tarpon | |
| Snook | *Centropomus pedemacula* | Snook | |
| Snook | *Centropomus robalito* | Constintino | |
| Snook | *Centropomus undecimalis* | Snook | Robalo/saltwater pike/common snook/common marine snook |
| Snook | *Centropomus vindis* | Snook | |
| Sole | *Aseraggodes kobensis* | Sole, Kobe | |

| | | | |
|---|---|---|---|
| Sole | *Aseraggodes macleayanus* | Sole, narrowbanded | |
| Sole | *Austroglossus microlepis* | Sole | West coast sole/tong |
| Sole | *Austroglossus pectoralis* | Sole | East coast sole/tong |
| Sole | *Lyopsetta exilis* | Sole, slender | |
| Sole | *Microchirus variegatus* | Sole, thickback | |
| Sole | *Microstomus kitt* | Sole, lemon | Lemon dab/smear dab |
| Sole | *Parophrys vetulus* | Sole, English | Lemon sole |
| Sole | *Solea vulgaris* | Sole, European | Common sole |
| Sole | *Synaptura orientalis* | Sole, oriental black | Black sole/oriental sole |
| Sole, Dover | *Microstomus pacificus* | Sole, Dover | Slime sole/slippery sole |
| Sole/flounder | *Clidoderma asperrimum* | Sole, roughscale | Roughskinned sole |
| Sole/flounder | *Embassichthys bathybius* | Sole, deepsea | |
| Sole/flounder | *Eopsetta jordani* | Sole, petrale | California sole/brill/petral |
| Sole/flounder | *Glyptocephalus cynoglossus* | Flounder, witch/ gray sole | Craig fluke/pole flounder/flet/ lemon sole/pole dab |
| Sole/flounder | *Glyptocephalus zachirus* | Sole, rex | Longfin sole/witch |
| Sole/flounder | *Hippoglossina stomata* | Sole, bigmouth | |
| Sole/flounder | *Hippoglossoides elassodon* | Sole, flathead | Flathead flounder |
| Sole/flounder | *Isopsetta isolepis* | Sole, butter | |
| Sole/flounder | *Lepidopsetta bilineata* | Sole, rock | Rock flounder/twolined flounder/ whitebellied flounder |
| Sole/flounder | *Limanda aspera* | Sole, yellowfin | Alaska dab/yellowfinned flounder |
| Sole/flounder | *Pleuronichthys coenosus* | Sole, c-o | Popeye/c-o turbot |
| Sole/flounder | *Pleuronichthys decurrens* | Sole, curlfin | Curlfin turbot |
| Sole/flounder | *Psettichthys melanostictus* | Sole, sand | |
| Sole/flounder | *Xystreurys liolepis* | Sole, fantail | |
| Spadefish | *Chaetodipterus faber* | Spadefish, Atlantic | Porgy/moonfish/angelfish/tripletail paguala |

**Table 7.1 (continued)**

| Market Name | Scientific Name | Common Name | Regional Name |
|---|---|---|---|
| Spadefish | *Chaetodipterus zonatus* | Spadefish, Pacific | |
| Spearfish | *Tetrapturus angustirostris* | Spearfish, shortbill | Japanese spearfish/shortnose spearfish |
| Spearfish | *Tetrapturus pfluegeri* | Spearfish, longbill | |
| Spinefoot | *Siganus* spp. | Spinefoot | Rabbitfish |
| Spot | *Leiostomus xanthurus* | Spot | Norfolk spot/goody/post croaker/oldwife/porgy |
| Sprat/brisling | *Sprasttus* spp. | Sprat | Sardine/New Zealand herring/kupae |
| Squirrelfish | *Holocentrus ascensionis* | Squirrelfish | Soldierfish/longjaw squirrelfish/ squirrel/welshman |
| Squirrelfish | *Holocentrus rufus* | Squirrelfish, longspine | |
| Squirrelfish | *Myripristis argyromus* | Squirrelfish | 'U' a manpachi |
| Squirrelfish | *Myripristis berndti* | Squirrelfish | 'U' u menpachi |
| Squirrelfish | *Sargocentrol lacteoguttatum* | Squirrelfish | Ala ihi/malau-faiumu |
| Squirrelfish | *Sargocentron spiniferum* | Squirrelfish, scarlet | Ala ihi mama |
| Stargazer | *Genyagnus monopterygius* | Stargazer, spotted | Kourepoua/catfish |
| Stargazer | *Kathetostoma averruncus* | Stargazaer, smooth | |
| Stargazer | *Kathetostoma giganteum* | Stargazer, giant | Monkfish/bulldog/flathead |
| Sturgeon | *Acipenser gueldenstaedti* | Sturgeon, Russian | Nisetru/waxdick |
| Sturgeon | *Acipenser medirostris* | Sturgeon, green | Saghalien/sturgeonsakhalin sturgeon/Japanese sturgeon |
| Sturgeon | *Acipenser oxyrhynchus* | Sturgeon, Atlantic | American Atlantic sturgeon/sea sturgeon/sharpnosed sturgeon |
| Sturgeon | *Acipenser stellatus* | Sturgeon, star | Pastruga/sveriuga |
| Sturgeon | *Acipenser transmontanus* | Sturgeon, white | Oregon or Sacramento sturgeon |

| | | | |
|---|---|---|---|
| Sturgeon | *Huso huso* | Sturgeon, European | Great sturgeon/hausen/giant beluga |
| Sucker | *Carpiodes carpio* | Carpsucker, river | Common river carpsucker |
| Sucker | *Carpiodes cyprinus* | Quillback | Quillback carpsucker/eastern carpsucker/carpsucker |
| Sucker | *Catostomus commersoni* | Sucker, white | Black sucker/brood sucker/ freshwater mullet |
| Sucker | *Cycleptus elongatus* | Sucker, blue | Gourd-seed sucker/Missouri sucker/blackhorse |
| Sucker/redhorse | *Moxostoma macrolepidotum* | Redhorse, shorthead | Carolina redhorse/neuse redhorse/ northern redhorse |
| Sunfish | *Archoplites interruptus* | Perch, Sacramento | Sacramento sunfish |
| Sunfish | *Lepomis auritus* | Sunfish, redbreast | Redbreast/yellowbelly sunfish/ redbellied sunfish |
| Sunfish | *Lepomis cyanellus* | Sunfish, green | Bluespotted sunfish/little redeye/ blackeyes |
| Sunfish | *Lepomis gibbosus* | Pumpkinseed | Common sunfish/sunny/tabaccobox/ sand perch/bream |
| Sunfish | *Lepomis microlophus* | Sunfish, redear | Shellcracker/yellow bream |
| Sunfish | *Lepomis punctatus* | Sunfish, spotted | Stumpknocker |
| Surfperch | *Amphistichus argenteus* | Surfperch, barred | Surffish |
| Surfperch | *Amphistichus koelzi* | Surfperch, calico | |
| Surfperch | *Amphistichus rhodoterus* | Surfperch, redtail | |
| Surfperch | *Cymatogaster aggregata* | Surfperch, shiner | Shiner perch/shiner/yellow shiner/perch/sparada |
| Surfperch | *Embiotoca jacksoni* | Surfperch, black | Black seaperch/common surffish/ black surffish |
| Surfperch | *Embiotoca lateralis* | Surfperch, striped | Blue surfperch/blue perch |
| Surfperch | *Hypersopon argenteum* | Surfperch, walleye | Walleye seaperch/white perch |

**Table 7.1 (continued)**

| Market Name | Scientific Name | Common Name | Regional Name |
| --- | --- | --- | --- |
| Surfperch | *Rhacochilus toxotes* | Seaperch, rubberlip | Bigmouth surffish/alfione |
| Swordfish | *Xiphias gladius* | Swordfish | Broadbilled swordfish/ broadbill/espada/emperado |
| Tang | *Acanthurus chirugus* | Doctorfish | |
| Tang | *Acanthurus coeruleus* | Tang, blue | Blue surgeon/blue doctorfish/ surgeonfish/barbero |
| Tarpon | *Megalops atlanticus* | Tarpon, Atlantic | Silverfish/tarpum/silver king/sabalo/ savalle |
| Tautog | *Tautoga onitis* | Tautog | Blackfish/oysterfish/black porgy/sea dog/cub |
| Thornyhead | *Sebastolobus alascanus* | Thornyhead, shortspine | Shortspine channel rockfish/ spinycheek rockfish |
| Thornyhead | *Sebastolobus altivelis* | Thornyhead, longspine | Longspine channel rockfish/ spinycheek rockfish |
| Tilapia | *Tilapia aurea* | Tilapia, blue | |
| Tilapia | *Tilapia galilaea* | Tilapia, mango | Galilee mouthbrooder/St. Peters fish |
| Tilapia | *Tilapia macrochir* | Tilapia, longfin | Greenhead bream/greenhead tilapia |
| Tilapia | *Tilapia melanotheron* | Tilapia, blackchin | Blackchin mouthbrooder/African sunfish/black mouthbrooder |
| Tilapia | *Tilapia mossambica* | Tilapia, Mozambique | Mozambique mouthbrooder/African mouthbrooder |
| Tilapia | *Tilapia nilotica* | Tilapia, Nile | Nile mouthbrooder/Nile perch |
| Tilapia | *Tilapia, rendalli* | Tilapia, redbreast | Readbreasted bream/covgo tilapia |
| Tilefish | *Caulolatilus chrysops* | Tilefish, goldface | Blackline tilefish/blanquill |
| Tilefish | *Caulolatilus microps* | Tilefish, blueline | Gray tilefish |

| | | | |
|---|---|---|---|
| Tilefish | *Caulolatilus princeps* | Whitefish, ocean | Whitefish/blanquillo |
| Tilefish | *Lopholatilus chamaeleonticeps* | Tilefish, golden | Blue tilefish |
| Tilefish | *Malacanthus plumieri* | Tilefish, sand | Blanquillo/sandfish |
| Toadfish | *Opsanus beta* | Toadfish, gulf | |
| Toadfish | *Opsanus tau* | Toadfish, oyster | Sapo/slimer/oysterfish/scorpion/opossum toadfish |
| Tomcod | *Microgadus proximus* | Tomcod, Pacific | California tomcod |
| Tomcod | *Microgadus tomcod* | Tomcod, Atlantic | Frostfish/snig/tommy cod |
| Tonguesole | *Cynoglossus* spp. | Tonguesole | |
| Torpedo | *Torpedo nobiliana* | Torpedo, Atlantic | Black or dark electric ray/skate/crampfish |
| Trevally | *Caranx dentex* | Trevally, white | Buta ulua/butaguchi/underjaw kingfish |
| Trevally | *Caranxx sexfasciatus* | Trevally, bigeye | Six-banded trevally/pake ulua/great trevally |
| Triggerfish | *Balistes capriscus* | Triggerfish, gray | Spotted triggerfish/leatherjacket turbot/cucuyo |
| Triggerfish | *Balistes vetula* | Triggerfish, queen | Oldwife/old wench/bluestriped triggerfish/concino |
| Triggerfish | *Cantherdermis sufflamen* | Triggerfish, ocean | Sobaco |
| Triggerfish | *Melichthys niger* | Filefish, fringed | Black durgon/humuhumu eleele |
| Triggerfish | *Navodon convexirostris* | Triggerfish | |
| Triggerfish | *Navodon scabra* | Triggerfish | Parika scab |
| Tripletail | *Datinoides quadrifasciatus* | Tripletail, fourband | Blackbanded tripletail |
| Tripletail | *Lobotes pacificus* | Tripletail, west coast | Berrugate |
| Tripletail | *Lobotes surinamensis* | Tripletail | Falsher/chobie/lumpfish/strawberry bass/black perch |

**Table 7.1 (continued)**

| Market Name | Scientific Name | Common Name | Regional Name |
|---|---|---|---|
| Trout | *Salmo aguabonita* | Trout, golden | |
| Trout | *Salmo gilae* | Trout, gila | |
| Trout | *Salmo trutta* | Trout, brown | Sea trout/salmon trout |
| Trout, brook | *Salvelinus fontinalis* | Trout, brook | Splake/Alsation charr |
| Trout, cutthroat | *Salmo clarki* | Trout, cutthroat | Columbia river trout/blackspotted trout |
| Trout, lake | *Salvelinus namaycush* | Trout, lake | Siscowet/splake/gray trout/salmon trout |
| Trout, rainbow/ steelhead | *Salmo gairdneri* | Trout, rainbow | Coastal rainbow trout/silver trout salmon/brook trout |
| Trout/dolly varden | *Salvelinus malma* | Trout, dolly varden | Dolly varden trout/malma trout/ Oregon char/bull trout |
| Trout/iconnu | *Stenodus leucichthys* | Inconnu | Fighting sheefish/Caspian inconnu sheefish |
| Trumpeter | *Latridopsis ciliaris* | Trumpeter, bastard | Blue moki/moki trumpeter/moki |
| Trumpeter | *Latris lineata* | Trumpeter, striped | Real trumpeter/kohikohi/Tasmanian trumpeter |
| Trunkfish | *Lactophyrs triqueter* | Trunkfish, smooth | |
| Tuna | *Auxis rochei* | Tuna, bullet | Frigate tuna/plain bonito/bonitou/ melvera |
| Tuna | *Euthynnus affinis* | Kawakawa | Little tunny/black skip jack/mackerel tuna/bonita |
| Tuna | *Euthynnus alletteratus* | Tunny, little | Little tuna/false albacore/bonito |
| Tuna | *Euthynnus lineatus* | Skipjack, black | |
| Tuna | *Katsuwonus pelamis* | Tuna, skip jack | Ocean bonito/lesser tuna/aku |

| | | | |
|---|---|---|---|
| Tuna | *Thunnus alalunga* (*T. germo*) | Albacore | Germon/longfinned tuna/albecor |
| Tuna | *Thunnus albacares* | Tuna, yellowfin | Ahi |
| Tuna | *Thunnus atlanticus* | Tuna, blackfin | |
| Tuna | *Thunnus maccoyii* | Tuna, southern bluefin | Southern bluefin tuna/southern tunny/tunny |
| Tuna | *Thunnus obesus* | Tuna, bigeye | Big eye/ahi-B |
| Tuna | *Thunnus thynnus* | Tuna, bluefin | Tunny/albacore/giant bluefin/northern bluefin |
| Tuna | *Thunnus tonggol* | Tuna, longtain | Bluefin tuna/northern bluefin tuna/bonito |
| Turbot | *Thpsopsetta guttulata* | Turbot, diamond | Diamond flounder |
| Turbot | *Pleuronichthys ritteri* | Turbot, spotted | |
| Turbot | *Pleuronichthys verticalis* | Turbot, hornyhead | |
| Turbot | *Psettodes belcheni* | Turbot, spottedtail | Spring turbot |
| Turbot | *Psettodes bennetti* | Turbot, spring | |
| Turbot | *Scophthalmus maximus* | Turbot | Butt/breet/britt |
| Turbot, Greenland | *Reinhardtius hippoglossoides* | Halibut, Greenland | Newfoundland turbot/turbot/blue halibut |
| Wahoo | *Acanthocybium solanderi* | Wahoo | Kinkfish/peto/guarapucu/ono/thazard batard |
| Walleye | *Stizostedion vitreum* | Pike, walleye | Yellow pike/yellow walleye/pickerel/pikeperch/dory |
| Warehou | *Seriolella brama* | Warehou, blue | Common warehou/whario |
| Warehou | *Seriolella caerulee* | Warehou, white | Deepsea warehou |
| Warehou | *Seriolella punctata* | Warehou, silver | Spotted warehou/ibodai/silverfish |
| Weakfish | *Cynoscion* spp. | Weakfish | Seatrout/squeteague/gray trout/saltwater trout |

**Table 7.1 (continued)**

| Market Name | Scientific Name | Common Name | Regional Name |
|---|---|---|---|
| Weakfish | *Macrodon ancylodon* | Weakfish, king | Pescadilla-real/pescadinha-real |
| Whitefish | *Coregonus clupeaformis* | Whitefish, lake | Inland whitefish/Labrador white fish |
| Whitefish | *Coregonus pidschian* | Whitefish, humpback | Siberian whitefish/Arctic whitefish |
| Whitefish | *Prosopium cylindraceum* | Whitefish, round | Pilotfish/memominee whitefish/ frostfish |
| Whiting | *Merlangius merlangus* | Whiting, European | English whiting/scotch haddock |
| Whiting | *Merluccius australis* | Hake, New Zealand | Hake/Antarctic queen/southern hake/ English hake |
| Whiting | *Merluccius bilinearis* | Hake, silver | American hake/silverfish/stockfish/ winter trout |
| Whiting | *Merluccius capensis* | Hake, cape | Stockfish/shallowwater hake/ South African hake |
| Whiting | *Merluccius gayi* | Hake, Chilean | English hake |
| Whiting | *Merluccius hubbsi* | Hake, Argentine | Southwest Atlantic hake/ patagonian whiting |
| Whiting | *Merluccius merluccius* | Hake, European | |
| Whiting | *Merluccius polylepis* | Hake, Patagonian | Merluza |
| Whiting, blue | *Micromesistius australis* | Whiting, southern blue | Southern poutassou |
| Whiting, blue | *Micromesistius poutassou* | Poutassou | Couch's whiting |
| Whiting, New Zealand | *Macruronus novaezealandiae* | Hoki | |
| Whiting/Pacific whiting | *Merluccius productus* | Hake, Pacific | California hake |

Source: *The Fish List: FDA Guide to Acceptable Market Names for Food Fish Sold in Interstate Commerce.* Food and Drug Administration, Washington, D.C., 1988.

**Table 7.2 Alphabetical Listing of Food Fish in the United States by Scientific Names and Market Names**

| Scientific Names | Market Names |
|---|---|
| *Acanthistius brasilianus* | Bass, sea |
| *Acanthocybium solanderi* | Wahoo |
| *Acanthurus chirurgus* | Tang |
| *Acanthurus coeruleus* | Tang |
| *Acipenser gueldenstaedti* | Sturgeon |
| *Acipenser medirostris* | Sturgeon |
| *Acipenser oxyrhynchus* | Sturgeon |
| *Acipenser stellatus* | Sturgeon |
| *Acipenser transmontanus* | Sturgeon |
| *Aetobatus narinari* | Ray, eagle |
| *Agnostomus monticola* | Mullet |
| *Albula vulpes* | Bonefish |
| *Aldrichetta forsteri* | Mullet |
| *Alectis ciliaris* | Pompano |
| *Alectis indicus* | Jackfish/crevalle |
| *Allocyttus* spp. | Oreo dory |
| *Allosmerus elongatus* | Smelt |
| *Alopias pelagicus* | Shark, thresher |
| *Alopias superciliosus* | Shark, thresher |
| *Alopias vulpinus* | Shark, thresher |
| *Alosa aestivalis* | Herring/river herring |
| *Alosa alabamae* | Shad |
| *Alosa alosa* | Shad |
| *Alosa chrysochloris* | Herring/river herring |
| *Alosa fallax* | Shad |
| *Alosa mediocris* | Shad |
| *Alosa pseudoharengus* | Alewife/river herring |
| *Alosa sapidissima* | Shad |
| *Aluterus monocers* | Filefish |
| *Aluterus schoepfi* | Filefish |
| *Ambloplites cavifrons* | Bass |
| *Ambloplites constellatus* | Bass |
| *Ambloplites rupestris* | Bass |
| *Amia calva* | Bowfin |
| *Ammodytes hexapterus* | Sandlance |
| *Amphistichus argenteus* | Surfperch |
| *Amphistichus koelzi* | Surfperch |

**Table 7.2 (continued)**

| Scientific Names | Market Names |
|---|---|
| *Amphistichus rhodoterus* | Surfperch |
| *Anarhichas denticulatus* | Catfish, ocean |
| *Anarhichas lupus* | Catfish, ocean |
| *Anarhichas minor* | Catfish, ocean |
| *Anarhichas orientalis* | Catfish, ocean |
| *Anchoa cayorum* | Anchovy |
| *Anchoa compressa* | Anchovy |
| *Anchoa delicatissima* | Anchoyy |
| *Anchoa duodecim* | Anchovy |
| *Anchoa hepsetus* | Anchovy |
| *Anchoa lyolepis* | Anchovy |
| *Anchoa mitchilli* | Anchovy |
| *Anchoa nasuta* | Anchovy |
| *Anchoviella estauqual* | Anchovy |
| *Anchoviella letidentostole* | Anchovy |
| *Anchoviella perfasciata* | Anchovy |
| *Ancylopsetta dilecta* | Flounder |
| *Anguilla anguilla* | Eel |
| *Anguilla australis* | Eel |
| *Anguilla dieffenbachii* | Eel |
| *Anguilla japonicus* | Eel |
| *Anguilla rostrata* | Eel, freshwater |
| *Anisotremus davidsoni* | Grunt/sargo |
| *Anisotremus interruptus* | Grunt |
| *Anisotremus surinamensis* | Grunt/margate |
| *Anisotremus taeniatus* | Grunt/Catalina |
| *Anisotremus virginicus* | Grunt/porkfish |
| *Anoplopoma fimbria* | Sablefish |
| *Antigonia capros* | Boarfish |
| *Antigonia combatia* | Boarfish |
| *Aphanopus carbo* | Cutlassfish |
| *Aphareus furcatus* | Jobfish |
| *Aphareus rutilans* | Jobfish |
| *Aplodactylus meandratus* | Morwong |
| *Aplodinotus grunniens* | Drum, freshwater |
| *Aprion virescens* | Jobfish |
| *Apsilus dentatus* | Snapper, black |
| *Archoplites interruptus* | Sunfish |
| *Archosargus probatocephalus* | Sheepshead |

| | |
|---|---|
| *Archosargus rhomboidalis* | Seabream |
| *Arctogadus borisovi* | Cod |
| *Arctogadus glacialis* | Cod |
| *Argentina elongata* | Argentine |
| *Argentina semifasciata* | Smelt |
| *Argentina silus* | Smelt |
| *Argentina sphyraena* | Smelt |
| *Argyrops bleekeri* | Bream |
| *Argyrops spinifer* | Bream |
| *Argyrosomus argentatus* | Croaker |
| *Argyrosomus hololepidotus* | Croaker |
| *Argyrosomus japonicus* | Croaker |
| *Argyrosomus macrophthalmus* | Croaker |
| *Argyrosomus nibe* | Croaker |
| *Argyrosomus regius* | Croaker/shadefish |
| *Ariopsis felis* | Catfish, sea |
| *Ariosoma balearicum* | Eel, conger |
| *Arnoglossus scapha* | Flounder |
| *Arripis trutta* | Kahawai |
| *Aseraggodes kobensis* | Sole |
| *Aserggodes macleayanus* | Sole |
| *Astronotus ocellatus* | Oscar |
| *Atheresthes evermanni* | Flounder |
| *Atheresthes stomias* | Flounder, arrowtooth |
| *Atherinopsis affinis* | Silverside |
| *Atherinopsis californiensis* | Silverside |
| *Austroglossus microlepsis* | Sole |
| *Austroglossus pectoralis* | Sole |
| *Auxis rochei* | Tuna |
| *Auxis thazard* | Mackerel |
| *Bagre marinus* | Catfish, sea |
| *Bairdiella batabana* | Croaker |
| *Bairdiella chrysoura* | Perch |
| *Bairdiella sanctaeluciae* | Croaker |
| *Balistes capriscus* | Triggerfish |
| *Balistes vetula* | Triggerfish |
| *Basilichthys australis* | Silverside |
| *Beryx decadactylus* | Alfonsin |
| *Beryx splendens* | Alfonsin |
| *Bodianus pulchellus* | Hogfish |
| *Bodianus rufus* | Hogfish |
| *Boops boops* | Bream/bogue |

**Table 7.2 (continued)**

| Scientific Names | Market Names |
|---|---|
| *Boreogadus saida* | Cod |
| *Bothus lunatus* | Flounder |
| *Bothus mancus* | Flounder |
| *Bothus ocellatus* | Flounder |
| *Bothus pantherinus* | Flounder |
| *Brachyplatysoma vaillanti* | Catfish |
| *Brama brama* | Pomfret |
| *Brama japonica* | Pomfret |
| *Brevoortia gunteri* | Menhaden |
| *Brevoortia patronus* | Menhaden |
| *Brevoortia smithi* | Menhaden |
| *Brevoortia tyrannus* | Menhaden |
| *Brosme brosme* | Cusk |
| *Brotula barbarta* | Brotula |
| *Calamus* spp. | Porgy |
| *Callorhynchus millii* | Elephant fish |
| *Cantherdermis sufflamen* | Triggerfish |
| *Caprodon schlegelii* | Grouper |
| *Caranx bartholomaei* | Jack |
| *Caranx caballus* | Jack |
| *Caranx crysos* | Jack/blue runner |
| *Caranx dentex* | Trevally |
| *Caranx georgianus* | Jack/trevally |
| *Caranx helvolus* | Jack |
| *Caranx hippos* | Jack |
| *Caranx ignobilis* | Jack/trevally |
| *Caranx latus* | Jack |
| *Caranx lugubris* | Jack |
| *Caranx mate* | Scad |
| *Caranx melampygus* | Jack/trevally |
| *Caranx ruber* | Jack |
| *Caranx sexfasciatus* | Trevally |
| *Caranx vinctus* | Jack |
| *Carassius auratus* | Goldfish |
| *Carcharhinus albimarginatus* | Shark |
| *Carcharhinus altimus* | Shark |
| *Carcharhinus amblyrhynchus* | Shark |
| *Carcharhinus brachyurus* | Shark |
| *Carcharhinus brevipinna* | Shark |

| | |
|---|---|
| *Carcharhinus falciformis* | Shark |
| *Carcharhinus isodon* | Shark |
| *Carcharhinus leucas* | Shark |
| *Carcharhinus limbatus* | Shark |
| *Carcharhinus longimanus* | Shark |
| *Carcharhinus melanopterus* | Shark |
| *Carcharhinus obscurus* | Shark |
| *Carcharhinus perezi* | Shark |
| *Carcharhinus plumbeus* | Shark |
| *Carcharhinus porosus* | Shark |
| *Carcharhinus signatus* | Shark |
| *Carcharodon carcharias* | Shark |
| *Carpiodes carpio* | Sucker |
| *Carpiodes cyprinus* | Sucker |
| *Catostomus commersoni* | Sucker |
| *Caulolatilus chrysops* | Tilefish |
| *Caulolatilus microps* | Tilefish |
| *Caulolatilus princeps* | Tilefish |
| *Centrolophus niger* | Ruff, black |
| *Centrophorus squamosus* | Dogfish |
| *Centropomus ensiferus* | Snook |
| *Centropomus nigrescens* | Snook |
| *Centropomus ornatus* | Snook |
| *Centropomus parallelus* | Snook |
| *Centropomus pectinatus* | Snook |
| *Centropomus pedemacula* | Snook |
| *Centropomus robalito* | Snook |
| *Centropomus undecimalis* | Snook |
| *Centropomus vindis* | Snook |
| *Centropristis ocyurus* | Bass, sea |
| *Centropristis philadelphica* | Bass, sea |
| *Centropristis striata* | Bass, sea |
| *Centroscymnus fabricii* | Dogfish |
| *Cephalopholis acanthistius* | Grouper |
| *Cephalopholis argus* | Grouper |
| *Cephalopholis taeniops* | Grouper |
| *Cephalopholis urodelus* | Grouper |
| *Cetengraulis mysticetus* | Anchovy |
| *Cetorhinus maximus* | Shark |
| *Chaetodipterus faber* | Spadefish |
| *Chaetodipterus zonatus* | Spadefish |
| *Chaetodon ocellatus* | Butterflyfish |

**Table 7.2 (continued)**

| Scientific Names | Market Names |
|---|---|
| *Chaetodon striatus* | Butterflyfish |
| *Champsocephalus gunnari* | Icefish |
| *Chanos chanos* | Milkfish |
| *Cheilodactylus gibbosus* | Morwong |
| *Cheilodactylus macropterus* | Morwong |
| *Cheilodactylus spectabilis* | Morwong |
| *Cheilotrema saturnum* | Croaker |
| *Chelidonichthys kumu* | Searobin |
| *Chelidonichthys lucerna* | Searobin |
| *Chironemus marmoratus* | Kelpfish |
| *Chloroscombrus chrysurus* | Bumper |
| *Chloroscombrus orqueta* | Bumper |
| *Chrysophrys auratus* | Porgy |
| *Citharichthys sordidus* | Sanddab |
| *Clidoderma asperrimum* | Sole/flounder |
| *Clupea harengus harengus* | Herring/sea herring/sild |
| *Clupea harengus pallasi* | Herring/sea herring/sild |
| *Colistium guntheri* | Flounder |
| *Colistium nudipinnis* | Flounder |
| *Collichthys* spp. | Drum/lion fish |
| *Coloabis saira* | Saury |
| *Conger cinereus* | Eel, conger |
| *Conger conger* | Eel, conger |
| *Conger marginatus* | Eel, conger |
| *Conger oceanicus* | Eel, conger |
| *Conger oligoporus* | Eel, conger |
| *Conger triporiceps* | Eel, conger |
| *Congipodus leucopaecilus* | Racehorse |
| *Congrina aequoreus* | Eel, conger |
| *Conodon nobilis* | Grunt |
| *Coregonus alpenae* | Cisco/chub |
| *Coregonus artedii* | Cisco/tullibee |
| *Coregonus clupeaformis* | Whitefish |
| *Coregonus kiyi* | Chub |
| *Coregonus pidschian* | Whitefish |
| *Coregonus reighardi* | Cisco |
| *Coregonus zenithicus* | Cisco/chub |
| *Coridodax pullus* | Greenbone |
| *Coryphaena equisetis* | Mahi-mahi |

| | |
|---|---|
| *Coryphaena hippurus* | Mahi-mahi |
| *Coryphaenoides pectoralis* | Grenadier |
| *Coryphaenoides rupertris* | Grenadier |
| *Crenimugil crenilabis* | Mullet |
| *Cycleptus elongatus* | Sucker |
| *Cycleptus lumpus* | Lumpfish |
| *Cymatogaster aggregata* | Surfperch |
| *Cynoglossus* spp. | Tonguesole |
| *Cynoscion arenarius* | Seatrout |
| *Cynoscion nebulosus* | Seatrout |
| *Cynoscion nobilis* | Seatrout |
| *Cynoscion nothus* | Seatrout |
| *Cynoscion orthonopterus* | Croaker/corvina |
| *Cynoscion parvinnis* | Corvina |
| *Cynoscion* spp. | Weakfish |
| *Cyprinus carpio* | Carp |
| *Cypselurus atrisignis* | Flyingfish |
| *Cypselurus californicus* | Flyingfish |
| *Cypselurus cyanopterus* | Flyingfish |
| *Cypselurus exsiliens* | Flyingfish |
| *Cypselurus furcatus* | Flyingfish |
| *Cypselurus hubbsi* | Flyingfish |
| *Cypselurus melanurus* | Flyingfish |
| *Cypselurus simus* | Flyingfish |
| *Cypselurus speculiger* | Flyingfish |
| *Cypselurus spilonopterus* | Flyingfish |
| *Cyspelurus comatus* | Flyingfish |
| *Cyttus novae-zelandiae* | Dory |
| *Datinoides quadrifasciatus* | Tripletail |
| *Decapterus koheru* | Scad |
| *Decapterus macarellus* | Scad |
| *Decapterus muroadsi* | Scad |
| *Decapterus punctatus* | Scad |
| *Decapterus russelli* | Scad |
| *Decapterus scombrinus* | Scad |
| *Decapterus tabl* | Scad |
| *Dentex gibbosus* | Porgy |
| *Diapterus auratus* | Mojarra |
| *Diapterus plumieri* | Mojarra |
| *Dicentrarchus labrax* | Bass, sea |
| *Diplectrum formosum* | Grouper |
| *Diplodus* spp. | Porgy |

**Table 7.2 (continued)**

| Scientific Names | Market Names |
|---|---|
| *Dorosoma cepedianum* | Shad |
| *Dorosoma petenese* | Shad |
| *Elegatis bipinnulata* | Jack/rainbow runner |
| *Eleginus gracilis* | Cod |
| *Elops affinis* | Ladyfish |
| *Elops hawaiensis* | Ladyfish |
| *Elops saurus* | Ladyfish |
| *Embassichthys bathybius* | Sole/flounder |
| *Embiotoca jacksoni* | Surfperch |
| *Embiotoca lateralis* | Surfperch |
| *Emmelichthys nitidus* | Bonnetmouth |
| *Enchelyopus cimbrius* | Rockling |
| *Engraulis australis* | Anchovy |
| *Engraulis encrasicolus* | Anchovy |
| *Engraulis eurystole* | Anchovy |
| *Engraulis japonica* | Anchovy |
| *Engraulis mordax* | Anchovy |
| *Engraulis ringens* | Anchovy |
| *Eopsetta jordani* | Sole/flounder |
| *Epigonus telescopus* | Cardinalfish |
| *Epinephelus adscensionis* | Hind |
| *Epinephelus aeneus* | Grouper |
| *Epinephelus afer* | Hamlet |
| *Epinephelus analogus* | Cabrilla |
| *Epinephelus areolatus* | Grouper |
| *Epinephelus chlorostigma* | Grouper |
| *Epinephelus cruentatus* | Grouper |
| *Epinephelus drummondhayi* | Hind |
| *Epinephelus fasciatus* | Grouper |
| *Epinephelus flavolimbatus* | Grouper |
| *Epinephelus fulva* | Grouper |
| *Epinephelus fuscoguttatus* | Grouper |
| *Epinephelus guaza* | Grouper |
| *Epinephelus guttatus* | Grouper/hind |
| *Epinephelus inermis* | Grouper |
| *Epinephelus itajara* | Grouper/jewfish |
| *Epinephelus merra* | Bass, sea |
| *Epinephelus morio* | Grouper |
| *Epinephelus mystacinus* | Grouper |

| | |
|---|---|
| *Epinephelus nigritus* | Grouper |
| *Epinephelus niveatus* | Grouper |
| *Epinephelus striatus* | Grouper |
| *Epinephelus tauvina* | Bass, sea |
| *Epinnula magistralis* | Gemfish |
| *Equetus punctatus* | Drum |
| *Equetus umbrosus* | Drum/cubbyu |
| *Erilepis zonifer* | Skilfish |
| *Esox americanus americanus* | Pickerel |
| *Esox americanus vermiculatus* | Pickerel |
| *Esox lucius* | Pike |
| *Esox masquinongy* | Muskellunge |
| *Esox niger* | Pickerel |
| *Etelis carbunculus* | Snapper, ruby |
| *Etelis coruscans* | Snapper, yellowstriped |
| *Etelis oculantus* | Snapper, queen |
| *Etrumeus teres* | Herring |
| *Euthynnus affinis* | Tuna |
| *Euthynnus alletteratus* | Tuna |
| *Euthynnus lineatus* | Tuna |
| *Exocoetus obtusirostris* | Flyingfish |
| *Exocoetus volitans* | Flyingfish |
| *Fodiator acutus* | Flyingfish |
| *Fundulus confluentus* | Killifish |
| *Fundulus heteroclitus* | Killifish |
| *Fundulus majalis* | Killifish |
| *Gadus macrocephalus* | Cod/Alaska cod |
| *Gadus morhua* | Cod/codfish |
| *Gadus ogac* | Cod |
| *Galeocerdo cuviere* | Shark |
| *Galeorhinus galeus* | Shark |
| *Genyagnus monopterygius* | Stargazer |
| *Genyonemus lineatus* | Croaker |
| *Genypterus blafodes* | Kingklip |
| *Genypterus caspensis* | Kingklip |
| *Genypterus chilensis* | Kingklip |
| *Genypterus maculatus* | Kingklip |
| *Girella nigricans* | Opaleye |
| *Glyptocephgalus cynoglossus* | Sole/flounder |
| *Gnathophis catalinensis* | Eel, conger |
| *Grammatorcynis bicaneatus* | Mackerel |
| *Gylptocephalus zachirus* | Sole/flounder |

**Table 7.2 (continued)**

| Scientific Names | Market Names |
|---|---|
| *Haemulon album* | Grunt/margate |
| *Haemulon aurolineatum* | Grunt/tomtate |
| *Haemulon bonariense* | Grunt |
| *Haemulon carbonarium* | Grunt |
| *Haemulon chrysargyreum* | Grunt |
| *Haemulon flavolineatum* | Grunt |
| *Haemulon macrostomum* | Grunt |
| *Haemulon melanurum* | Grunt/cottonwick |
| *Haemulon parrai* | Grunt |
| *Haemulon plumieri* | Grunt |
| *Haemulon rimator* | Grunt/tomatate |
| *Haemulon sciurus* | Grunt |
| *Haemulon scudderii* | Grunt |
| *Haemulon steindachneri* | Grunt |
| *Haemulon striatum* | Grunt |
| *Harengula clupeola* | Sardine |
| *Harengula humeralis* | Sardine |
| *Harengula jaguana* | Sardine |
| *Hariotta raleighana* | Chimaera |
| *Helicolenus dactylopterus* | Rosefish |
| *Helicolenus papillosus* | Rockfish |
| *Hemitripterus americanus* | Sculpin |
| *Hermosilla azurea* | Perch |
| *Hexagrammos decagrammus* | Greenling |
| *Hexagrammos lagocephalus* | Greenling |
| *Hexagrammos octogrammus* | Greenling |
| *Hexagrammos stelleri* | Greenling |
| *Hexanchus griseus* | Shark |
| *Hildebrandia flava* | Eel, conger |
| *Hildebrandia gracilior* | Eel, conger |
| *Hiodon alosoides* | Goldeye |
| *Hiodon tergisus* | Mooneye |
| *Hippoglossina stomata* | Sole/flounder |
| *Hippoglossoides elassodon* | Sole/flounder |
| *Hippoglossoides platessoides* | Plaice/dab |
| *Hippoglossoides robusts* | Flounder |
| *Hippoglossus hippoglossus* | Halibut |
| *Hippoglossus stenolepis* | Halibut |
| *Hirundichthys affinis* | Flyingfish |

| | |
|---|---|
| *Hirundichthys rondeleti* | Flyingfish |
| *Holacanthus bermudensis* | Angelfish |
| *Holacanthus ciliaris* | Angelfish |
| *Holacanthus tricolor* | Angelfish |
| *Holocentrus ascensionis* | Squirrelfish |
| *Holocentrus rufus* | Squirrelfish |
| *Hoplopagrus guntheri* | Pargo |
| *Hoplostethus atlanticus* | Roughy, orange |
| *Hoplostethus mediterraneus* | Roughy |
| *Huso huso* | Sturgeon |
| *Hydrolagus novaezelandiae* | Chimaera |
| *Hydrolagus* spp. | Chimaera |
| *Hyperoglyphe antarctica* | Bluenose |
| *Hyperoglyphe bythites* | Driftfish |
| *Hyperoglyphe perciformis* | Driftfish |
| *Hypersopon argenteum* | Surfperch |
| *Hypomesus pretiosus* | Smelt |
| *Hyspomesus transpacificus* | Smelt |
| *Hyposetta guttulata* | Turbot |
| *Ictalurus catus* | Catfish |
| *Ictalurus furcatus* | Catfish |
| *Ictalurus melas* | Bullhead/catfish |
| *Ictalurus natalis* | Bullhead |
| *Ictalurus nebulosus* | Bullhead/catfish |
| *Ictalurus platycephalus* | Bullhead |
| *Ictalurus pricei* | Catfish |
| *Ictalurus punctatus* | Catfish |
| *Ictiobus bubalus* | Buffalo |
| *Ictiobus cyprinellus* | Buffalo |
| *Ictiobus niger* | Buffalo |
| *Ilisha africans* | Herring |
| *Ilisha elongata* | Herring |
| *Ilisha furthi* | Herring |
| *Ilisha megaloptera* | Herring |
| *Ilisha melastoma* | Herring |
| *Ilisha pristigastroides* | Herring |
| *Isopisthus parvipinnis* | Croaker/corvina |
| *Isopsetta isolepis* | Sole/flounder |
| *Istiophorus gladius* | Sailfish |
| *Istiophorus platypterus* | Sailfish |
| *Isurus oxyrinchus* | Shark, mako |
| *Isurus paucus* | Shark, mako |

**Table 7.2 (continued)**

| Scientific Names | Market Names |
|---|---|
| *Johnius* spp. | Croaker |
| *Kathetostoma averruncus* | Stargazer |
| *Kathetostoma giganteum* | Stargazer |
| *Katsuwonus pelamis* | Tuna |
| *Kuhlia marginata* | Aholehole |
| *Kuhlia rupestris* | Aholehole |
| *Kuhlia sandvicensis* | Aholehole |
| *Kyphosus cinerascens* | Rudderfish |
| *Kyphosus incisor* | Chub |
| *Kyphosus sectatrix* | Chub |
| *Kyphosus vaigiensis* | Drummer |
| *Lachnolaimus maximus* | Hogfish |
| *Lactophyrs polygonia* | Cowfish |
| *Lactophyrs quadricornis* | Cowfish |
| *Lactophyrs triqueter* | Trunkfish |
| *Lagocephalus laevigatus* | Puffer |
| *Lagocephalus lagocephalus* | Puffer |
| *Lagodon rhomboides* | Porgy |
| *Lamna ditropis* | Shark |
| *Lamna nasus* | Shark/porbeagle |
| *Lampris guttatus* | Opah |
| *Larimus* spp. | Drum |
| *Lateolabrax japonicus* | Bass, sea |
| *Lates nilotica* | Perch, Nile |
| *Latridopsis cilliaris* | Trumpeter |
| *Latris lineata* | Trumpeter |
| *Leiostomus xanthurus* | Spot |
| *Leipidocybium flavobrunneum* | Gemfish |
| *Lepidopsetta bilineata* | Sole/flounder |
| *Lepidopus caudatus* | Cutlassfish |
| *Lepidopus caudatus* | Cutlassfish |
| *Lepidorhombus boscii* | Flounder/whiff |
| *Lepidorhombus whiffiiagonis* | Flounder/whiff |
| *Lepidorhynchus denticulatus* | Grenadier |
| *Lepisosteus osseus* | Gar |
| *Lepisosteus spatula* | Gar |
| *Lepomis auritus* | Sunfish |
| *Lepomis cyanellus* | Sunfish |
| *Lepomis gibbosus* | Sunfish |

| | |
|---|---|
| *Lepomis gulosus* | Bass |
| *Lepomis macrochirus* | Bluegill |
| *Lepomis microlophus* | Sunfish |
| *Lepomis punctatus* | Sunfish |
| *Lepophidium cervinum* | Cusk-eel |
| *Lepophidium graellsi* | Cusk-eel |
| *Lethrinus* spp. | Emperor |
| *Leuresthes tenuis* | Grunion |
| *Limanda aspera* | Sole/flounder |
| *Limanda ferruginea* | Flounder |
| *Limanda limanda* | Flounder/dab |
| *Limanda proboscidea* | Flounder/dab |
| *Liopsetta glacialis* | Flounder |
| *Lobotes pacificus* | Tripletail |
| *Labotes surinamensis* | Tripletail |
| *Lophius americanus* | Monkfish |
| *Lophius piscatorius* | Monkfish |
| *Lopholatilus chamaeleonticeps* | Tilefish |
| *Lota lota* | Burbot |
| *Lotella rhacina* | Cod, morid |
| *Lutjanus analis* | Snapper, mutton |
| *Lutjanus apodus* | Schoolmaster |
| *Lutjanus aratus* | Snapper, mullet |
| *Lutjanus bohar* | Snapper, twinspot |
| *Lutjanus buccanella* | Snapper, blackfin |
| *Lutjanus campechanus* | Snapper, red |
| *Lutjanus colorado* | Snapper, Colorado |
| *Lutjamus cyanopterus* | Snapper, cubera |
| *Lutjanus fulvus* | Snapper, blacktail |
| *Lutjanus gibbus* | Snapper, humpback |
| *Lutjanus griseus* | Snapper, gray |
| *Lutjanus guttatus* | Snapper, spotted rose |
| *Lutjanus inermis* | Snapper, golden |
| *Lutjanus jocu* | Snapper, dog |
| *Lutjanus jordani* | Snapper, rufous |
| *Lutjanus kasmira* | Snapper, bluestriped |
| *Lutjanus mahogoni* | Snapper, mahogany |
| *Lutjanus malabaricus* | Snapper, malabar |
| *Lutjanus monostigma* | Snapper, onespot |
| *Lutjanus novemfasciatus* | Snapper, Pacific dog |
| *Lutjanus peru* | Snapper, Pacific |
| *Lutjanus purpureus* | Snapper, red Caribbean |

**Table 7.2 (continued)**

| Scientific Names | Market Names |
|---|---|
| *Lutjanus rivulatus* | Snapper, blubberlip |
| *Lutjanus sanguineus* | Snapper, scarlet |
| *Lutjanus sebae* | Snapper, emperor |
| *Lutjanus synagris* | Snapper, lane |
| *Lutjanus vivanus* | Snapper, silk |
| *Luvarus imperialis* | Louvar |
| *Lyopsetta exilis* | Sole |
| *Macolor macularius* | Snapper, midnight |
| *Macolor niger* | Snapper, black and white |
| *Macrodon ancylodon* | Weakfish |
| *Macrourus* spp. | Grenadier |
| *Macrozoarces americanus* | Eelpout/ocean pout |
| *Macruronus novaezealandiae* | Whiting, New Zealand |
| *Makaira indica* | Marlin |
| *Makaira nigricans* | Marlin |
| *Malacanthus plumieri* | Tilefish |
| *Mallotus villosus* | Capelin |
| *Manta birostris* | Manta |
| *Manta hamiltoni* | Manta |
| *Medialuna californiensis* | Halfmoon |
| *Megalops atlanticus* | Tarpon |
| *Melanogrammus aeglefinus* | Haddock |
| *Melichthys niger* | Triggerfish |
| *Menidia menidia* | Silverside |
| *Menticirrhus americanus* | Kingfish |
| *Menticirrhus littoralis* | Kingfish |
| *Menticirrhus saxatilis* | Kingfish |
| *Menticirrhus undulatus* | Kingfish/corbina |
| *Merlangius merlangus* | Whiting |
| *Merluccius australis* | Whiting |
| *Merluccius bilinearis* | Whiting |
| *Merluccius capensis* | Whiting |
| *Merluccius gayi* | Whiting |
| *Merluccius hubbsi* | Whiting |
| *Merluccius merluccius* | Whiting |
| *Merluccius polylepis* | Whiting |
| *Merluccius productus* | Whiting/Pacific whiting |
| *Microchirus variegatus* | Sole |
| *Microgadus proximus* | Tomcod |

| | |
|---|---|
| *Microgadus tomcod* | Tomcod |
| *Micromesistius australis* | Whiting, blue |
| *Micromesistius poutassou* | Whiting, blue |
| *Micropogonias furnieri* | Croaker |
| *Micropogonias opercularis* | Croaker/corvina |
| *Micropogonias undulatus* | Croaker |
| *Micropterus dolomieui* | Bass |
| *Micropterus notius* | Bass |
| *Micropterus punctulatus* | Bass |
| *Micropterus salmoides* | Bass |
| *Micropterus treculi* | Bass |
| *Microstomus kitt* | Sole |
| *Microstomus pacificus* | Sole, dover |
| *Mobula hypostoma* | Manta |
| *Mobula japanica* | Manta |
| *Mobula lucasana* | Manta |
| *Molva dypterygia* | Ling |
| *Molva macrophthalus* | Ling, Mediterranean |
| *Molva molva* | Ling |
| *Mora pacifica* | Cod, morid |
| *Morone americana* | Perch, white |
| *Morone chrysops* | Bass |
| *Morone chrysops x saxatilis* | Bass |
| *Morone mississippiensis* | Bass |
| *Morone saxatilis* | Bass |
| *Moxostoma macrolepidotum* | Sucker/redhorse |
| *Mugil cephalus* | Mullet |
| *Mugil curema* | Mullet |
| *Mugil gaimardianus* | Mullet |
| *Mugil liza* | Mullet |
| *Mugil trichodon* | Mullet |
| *Muigiloides chilensis* | Sandperch |
| *Mulloidichthys* spp. | Goatfish |
| *Mullus auratus* | Goatfish |
| *Mullus brabatus* | Mullet |
| *Mullus surmeletus* | Mullet |
| *Muraena retifera* | Moray |
| *Mustelus antarcticus* | Dogfish |
| *Mustelus californicus* | Shark/smoothhound |
| *Mustelus canis* | Dogfish |
| *Mustelus henlei* | Shark/smoothhound |
| *Mustelus lenticulatus* | Dogfish |

**Table 7.2 (continued)**

| Scientific Names | Market Names |
|---|---|
| *Mycteroperca bonaci* | Grouper |
| *Mycteroperca interstitialis* | Grouper |
| *Mycteroperca jordani* | Grouper |
| *Mycteroperca microlepis* | Grouper/gag |
| *Mycteroperca phenax* | Scamp |
| *Mycteroperca rubra* | Grouper |
| *Mycteroperca tigris* | Grouper |
| *Mycteroperca venenosa* | Grouper |
| *Mycteroperca xenarcha* | Grouper |
| *Myliobatis californica* | Ray, bat |
| *Myliobatis freminvillei* | Ray, bullnose |
| *Myliobatis goodei* | Ray, eagle |
| *Myoxocephalus polyacanthocephalus* | Sculpin |
| *Myripristis argyromus* | Squirrelfish |
| *Myripristis berndti* | Squirrelfish |
| *Narcine brasiliensis* | Ray, electric |
| *Navodon convexirostris* | Triggerfish |
| *Navodon scabra* | Triggerfish |
| *Nebris microps* | Croaker |
| *Negaprion brevirostris* | Shark |
| *Nemadactylus carponotatus* | Morwong |
| *Nemadactylus douglasi* | Morwong |
| *Nematistius pectoralis* | Jack/roosterfish |
| *Nemipterus japonicus* | Bream, threadfin |
| *Neomyxus chaptalii* | Mullet |
| *Nesiarchus nasutus* | Gemfish |
| *Nezumia bairdi* | Grenadier |
| *Nibea* spp. | Croaker |
| *Notorynchus cepedianus* | Shark |
| *Notothenia moariensis* | Icefish |
| *Ocyurus chrysurus* | Sanpper, yellowtail |
| *Odontoscion dentex* | Croaker |
| *Oligoplites saurus* | Jack |
| *Oncorhynchus gorbuscha* | Salmon, pink/humpback salmon |
| *Oncorhynchus keta* | Salmon, chum/keta |
| *Oncorhynchus kisutch* | Salmon, coho/silver/med. red |
| *Oncorhynchus nerka* | Salmon, sockeye/blueback/red |
| *Oncorhynchus tshawytscha* | Salmon, chinook/king/spring |
| *Onhcorhynchus masou* | Salmon, cherry |

| | |
|---|---|
| *Ophichthus rex* | Eel, snake/keoghfish |
| *Ophiodon elongatus* | Lingcod |
| *Opisthonema oglinum* | Herring, thread |
| *Opisthopterus tardoore* | Herring |
| *Opsanus beta* | Toadfish |
| *Opsanus tau* | Toadfish |
| *Orthopristis chrysoptera* | Grunt |
| *Osmerus dentex* | Smelt |
| *Osmerus eperlanus* | Smelt |
| *Osmerus mordax* | Smelt/American smelt |
| *Oxyporhamphus micropterus* | Flyingfish |
| *Pachypops* spp. | Croaker |
| *Pachyurus* spp. | Croaker |
| *Pacu sp.* | Pacu |
| *Pagellus* spp. | Seabream |
| *Pagrus pagrus* | Porgy |
| *Paraconger caudilimbatus* | Eel, conger |
| *Paralabrax callaensis* | Bass, sea |
| *Paralabrax clathratus* | Bass, sea |
| *Paralabrax maculatofasciatus* | Bass, sea |
| *Paralabrax nebulifer* | Bass, sea |
| *Paralichthys albigutta* | Flounder |
| *Paralichthys californicus* | Halibut/California halibut |
| *Paralichthys dentatus* | Flounder/fluke |
| *Paralichthys lethostigma* | Flounder/fluke |
| *Paralichthys oblongus* | Flounder |
| *Paralichthys olivaceus* | Flounder |
| *Paralichthys patagonicus* | Flounder |
| *Paralichthys squamilentus* | Flounder |
| *Paralonchurus* spp. | Croaker |
| *Paranthias furcifer* | Bass, sea |
| *Parapercis* spp. | Sandperch |
| *Paratrachichthys trailli* | Roughy |
| *Parexocoetus brachypterus* | Flyingfish |
| *Paristiopterus labiosus* | Boarfish |
| *Parophrys vetulus* | Sole |
| *Parupeneus* spp. | Goatfish |
| *Pelecanichthys crumenalis* | Flounder |
| *Pellona ditchela* | Herring |
| *Pelotretis flavilatus* | Flounder |
| *Peltorhampus novaezeelandiae* | Flounder |
| *Pentaceros richardsoni* | Armourhead |

**Table 7.2 (continued)**

| Scientific Names | Market Names |
|---|---|
| *Pentapodus macrurus* | Bream |
| *Peprilus alepidotus* | Butterfish |
| *Peprilus burti* | Butterfish |
| *Peprilus simillimus* | Butterfish |
| *Peprilus triacanthus* | Butterfish |
| *Perca flavescens* | Perch, yellow/lake perch |
| *Peristedion miniatum* | Searobin |
| *Phycis chesteri* | Hake |
| *Pinirampus* spp. | Catfish |
| *Plagiogenion rubiginosus* | Bonnetmouth |
| *Plagioscion* spp. | Croaker |
| *Platichthys flesus* | Flounder/fluke |
| *Platichthys stellatus* | Flounder |
| *Platynematichthys notatus* | Catfish |
| *Platyrhinoidis triseriata* | Guitarfish |
| *Plectorhynchus chaetodonoides* | Grunt/sweetlips |
| *Plectorhynchus lineatus* | Grunt/sweetlips |
| *Plectorhynchus nigrus* | Grunt/sweetlips |
| *Plectorhynchus orientalis* | Grunt/sweetlips |
| *Plectorhynchus pictus* | Grunt/sweetlips |
| *Pleurogrammus monopterygius* | Mackerel, atka |
| *Pleuronectes quadrituberculatus* | Plaice |
| *Pleuronichthys coenosus* | Sole/flounder |
| *Pleuronichthys decurrens* | Sole/flounder |
| *Pleuronichthys ritteri* | Turbot |
| *Pleuronichthys verticalis* | Turbot |
| *Pleuronectes platessa* | Plaice |
| *Pogonias cromis* | Drum |
| *Pollachius pollachius* | Pollock |
| *Polhachius virens* | Pollock |
| *Polyodon spathula* | Paddlefish |
| *Polyprion americanus* | Bass, sea |
| *Polyprion moene* | Bass, sea |
| *Pomacanthus arcuatus* | Angelfish |
| *Pomacanthus paru* | Angelfish |
| *Pomadasys crocro* | Grunt |
| *Pomatomus saltatrix* | Bluefish |
| *Pomoxis annularis* | Crappie |
| *Pomoxis nigromaculatus* | Crappie |

| | |
|---|---|
| *Priancanthus arenatus* | Bigeye |
| *Prionace glauca* | Shark |
| *Prionaotus carolinus* | Searobin |
| *Pristigenys alta* | Bigeye |
| *Pristipomoides aquilonaris* | Snapper, wenchman |
| *Pristipomoides filamentosus* | Snapper, crimson |
| *Pristopomoides flavipinnis* | Jobfish |
| *Pristipomoides macrophthalmus* | Snapper, cardinal |
| *Pristipomoides sieboldii* | Jobfish |
| *Pristis pectinata* | Sawfish |
| *Pristis perotteti* | Sawfish |
| *Prognichthys gibbifrons* | Flyingfish |
| *Pfosopium cylindraceum* | Whitefish |
| *Psettichthys mealnostictus* | Sole/flounder |
| *Psettodes belcheni* | Turbot |
| *Psettodes bcenetti* | Turbot |
| *Psettodes erumei* | Flounder |
| *Pseudocyttus maculatus* | Oreo dory |
| *Pseudophycis breviusculus* | Cod, morid |
| *Pseudoplatystoma tigrinum* | Catfish |
| *Pseudopleuronectes americanus* | Flounder/sole |
| *Pseudorhombus arsius* | Flounder |
| *Pseudorhombus jenynsii* | Flounder |
| *Pseudorhombus pentophthalmus* | Flounder |
| *Pseudosciaena* spp. | Croaker/yellowfish |
| *Pseudotolithes* | Croaker |
| *Pseudupeneus* spp. | Goatfish |
| *Pteraclis aesticola* | Fanfish |
| *Pterotolithus* spp. | Croaker |
| *Pterygotrigla picta* | Searobin |
| *Pylodictis olivaris* | Catfish |
| *Rachycentron canadium* | Cobia |
| *Raja aleutica* | Skate |
| *Raja binoculate* | Skate |
| *Raja eglanteria* | Skate |
| *Raja erinacea* | Skate |
| *Raja garmani* | Skate |
| *Raja inornata* | Skate |
| *Raja interrupta* | Skate |
| *Raja kincaidi* | Skate |
| *Raja laevis* | Skate |
| *Raja ocellata* | Skate |

**Table 7.2 (continued)**

| Scientific Names | Market Names |
|---|---|
| *Raja olseni* | Skate |
| *Raja parmifera* | Skate |
| *Raja radiata* | Skate |
| *Raja rhina* | Skate |
| *Raja rosispinis* | Skate |
| *Raja senta* | Skate |
| *Raja spinicauda* | Skate |
| *Raja* spp. | Skate |
| *Raja stellulata* | Skate |
| *Raja texana* | Skate |
| *Raja trachura* | Skate |
| *Rastrelliger kanagurta* | Mackerel |
| *Reinhardtius hippoglossoides* | Turbot, Greenland |
| *Retropinna retropinna* | Smelt |
| *Rexea solandri* | Gemfish/barracouta |
| *Rhacochilus toxotes* | Surfperch |
| *Rhacochilus vacca* | Perch, pile |
| *Rhinobatos lentiginosus* | Guitarfish |
| *Rhinobatos productus* | Guitarfish |
| *Rhinoptera bonasus* | Ray, cownose |
| *Rhomboplites aurorubens* | Snapper, vermilion |
| *Rhombosolea leporina* | Flounder |
| *Rhombosolea plebeia* | Flounder |
| *Rhombosolea retiaria* | Flounder |
| *Rhombosolea tapirina* | Flounder |
| *Roncador stearnsi* | Croaker |
| *Ruvettus pretiosus* | Mackerel, snake |
| *Salangichthys* spp. | Icefish |
| *Salmo aguabonita* | Trout |
| *Salmo clarki* | Trout, cutthroat |
| *Salmo gairdneri* | Trout, rainbow/steelhead |
| *Salmo gilae* | Trout |
| *Salmo salar* | Salmon, Atlantic |
| *Salmo trutta* | Trout |
| *Salvelinus alpinus* | Char |
| *Salvelinus fontinalis* | Trout, brook |
| *Salvelinus malma* | Trout/Dolly Varden |
| *Salveninus namaycush* | Trout, lake |
| *Samariscus triocellatus* | Flounder |

| | |
|---|---|
| *Sarda chiliensis* | Bonito |
| *Sarda orientalis (S. velox)* | Bonito |
| *Sarda sarda* | Bonito |
| *Sardina pilchardus* | Pilchard/sardine |
| *Sardinella anchovia* | Sardine |
| *Sardinella aurita* | Sardine |
| *Sardinella brasiliensis* | Sardine |
| *Sardinella fimbriata* | Sardine |
| *Sardinella longiceps* | Sardine |
| *Sardinella perforata* | Sardine |
| *Sardinops neopilchardus* | Pilchard/sardine |
| *Sardinops sagax* | Sardine/pilchard |
| *Sardinops sagax melanosticta* | Pilchard/sardine |
| *Sardinops sagax ocellata* | Pilchard/sardine |
| *Sargocentron lacteoguttatum* | Squirrelfish |
| *Sargocentron spiniferum* | Squirrelfish |
| *Scarus coelestinus* | Parrotfish |
| *Sciaena* spp. | Drum/meagre |
| *Sciaenops ocellatus* | Drum/redfish |
| *Scomber austrlasicus* | Mackerel, chub |
| *Scomber japonicus* | Mackerel |
| *Scomber scombrus* | Mackerel |
| *Scomberesox saurus* | Saury |
| *Scombermorus commerson* | Mackerel |
| *Scomberomorus cavella* | Mackerel, king |
| *Scomberomorus concolor* | Mackerel |
| *Scomberomorus maculatus* | Mackerel |
| *Scomberomorus regalis* | Mackerel |
| *Scomberomorus sierra* | Mackerel |
| *Scomberomorus tritor* | Mackerel |
| *Scophthalmus aquosus* | Flounder |
| *Scophthalmus maximus* | Turbot |
| *Scophthalmus rhombus* | Flounder |
| *Scorpaena cardinalis* | Rockfish |
| *Scorpaena scrofa* | Scorpionfish |
| *Scorpaenichthys marmoratus* | Sculpin/cabezon |
| *Scyliorhinus canicula* | Dogfish |
| *Schliorhinus stellaris* | Dogfish |
| *Sebastes aleutianus* | Rockfish |
| *Sebastes alutus* | Perch, ocean |
| *Sebastes atrovirens* | Rockfish |
| *Sebastes auriculatus* | Rockfish |

**Table 7.2 (continued)**

| Scientific Names | Market Names |
|---|---|
| *Sebastes aurora* | Rockfish |
| *Sebastes babcocki* | Rockfish |
| *Sebastes borealis* | Rockfish |
| *Sebastes brevispinis* | Rockfish |
| *Sebastes carnatus* | Rockfish |
| *Sebastes caurinus* | Rockfish |
| *Sebastes chlorostictus* | Rockfish |
| *Sebastes chrysomelas* | Rockfish |
| *Sebastes ciliatus* | Rockfish |
| *Sebastes constellatus* | Rockfish |
| *Sebastes crameri* | Rockfish |
| *Sebastes dalli* | Rockfish |
| *Sebastes diploproa* | Rockfish |
| *Sebastes elongatus* | Rockfish |
| *Sebastes emphaeus* | Rockfish |
| *Sebastes ensifer* | Rockfish |
| *Sebastes entomelas* | Rockfish |
| *Sebastes eos* | Rockfish |
| *Sebastes fasciatus* | Perch, ocean |
| *Sebastes flavidus* | Rockfish |
| *Sebastes gilli* | Rockfish |
| *Sebastes goodei* | Rockfish |
| *Sebastes helvomaculatus* | Rockfish |
| *Sebastes hopkinsi* | Rockfish |
| *Sebastes jordani* | Rockfish |
| *Sebastes lentiginosus* | Rockfish |
| *Sebastes levis* | Rockfish |
| *Sebastes macdonaldi* | Rockfish |
| *Sebastes maliger* | Rockfish |
| *Sebastes marinus* | Perch, ocean |
| *Sebastes melanops* | Rockfish |
| *Sebastes melanosema* | Rockfish |
| *Sebastes melanostomus* | Rockfish |
| *Sebastes mentella* | Perch, ocean |
| *Sebastes miniatus* | Rockfish |
| *Sebastes mystinus* | Rockfish |
| *Sebastes nebulosus* | Rockfish |
| *Sebastes nigrocinctus* | Rockfish |
| *Sebastes ovalis* | Rockfish |

| | |
|---|---|
| *Sebastes paucispinis* | Rockfish |
| *Sebastes phillipsi* | Rockfish |
| *Sebastes pinniger* | Rockfish |
| *Sebastes polyspinis* | Rockfish |
| *Sebastes proriger* | Rockfish |
| *Sebastes rastrelliger* | Rockfish |
| *Sebastes reedi* | Rockfish |
| *Sebastes rosaceus* | Rockfish |
| *Sebastes rosenblatti* | Rockfish |
| *Sebastes ruberrimus* | Rockfish |
| *Sebastes rubrivinctus* | Rockfish |
| *Sebastes rufinanus* | Rockfish |
| *Sebastes rufus* | Rockfish |
| *Sebastes saxicola* | Rockfish |
| *Sebastes semicinctus* | Rockfish |
| *Sebastes serranoides* | Rockfish |
| *Sebastes serriceps* | Rockfish |
| *Sebastes simulator* | Rockfish |
| *Sebastes umbcrosus* | Rockfish |
| *Sebastes viviparus* | Perch, ocean |
| *Sebastes wilsoni* | Rockfish |
| *Sebastes zacentrus* | Rockfish |
| *Sebastolobus alascanus* | Thornyhead |
| *Sebastolobus altivelis* | Thornyhead |
| *Selar crumenophthalmus* | Scad |
| *Selene peruviana* | Jack |
| *Selene setapinnis* | Jack |
| *Selene vomer* | Jack |
| *Semicossyphus pulcher* | Sheephead |
| *Seriola dumerili* | Amberjack |
| *Seriola fasciata* | Amberjack |
| *Seriola lalandei* | Amberjack/yellowtail |
| *Seriola quinqueradiata* | Amberjack/yellowtail |
| *Seriola rivoliana* | Jack |
| *Seriola zonata* | Amberjack |
| *Seriolella brama* | Warehou |
| *Seriolella caerulee* | Warehou |
| *Seriolella punctatda* | Warehou |
| *Seriphus politus* | Drum/queenfish |
| *Siganus* spp. | Spinefoot |
| *Solea vulgaris* | Sole |
| *Sparisoma viride* | Parrotfish |

**Table 7.2 (continued)**

| Scientific Names | Market Names |
|---|---|
| *Sparus auratus* | Bream |
| *Sphlyraena ensis* | Barracuda |
| *Sphoeroides annulatus* | Puffer |
| *Sphoeroides dorsalis* | Puffer |
| *Sphoeroides lobatus* | Puffer |
| *Sphoeroides maculatus* | Puffer |
| *Sphoeroides mephelus* | Puffer |
| *Sphoeroides pachygaster* | Puffer |
| *Sphoeroides parvus* | Puffer |
| *Sphoeroides spengleri* | Puffer |
| *Sphoeroides testudineus* | Puffer |
| *Sphyraena argentea* | Barracuda |
| *Sphyraena barracuda* | Barracuda |
| *Sphyraena guachancho* | Barracuda |
| *Sphyraena picudilla* | Barracuda |
| *Sphyrna lewini* | Shark/hammerhead |
| *Sphyrna mokarran* | Shark/hammerhead |
| *Sphyrna tiburo* | Shark/bonnethead |
| *Sphyrna tudes* | Shark/hammerhead |
| *Sphyrna zygaena* | Shark/hammerhead |
| *Spirinchus maena* | Picarel |
| *Spirinchus starksi* | Smelt |
| *Spirinchus thaleichthys* | Smelt |
| *Sprattus* spp. | Sprat/brisling |
| *Squalus acanthias* | Dogfish |
| *Squalus blainvillei* | Dogfish |
| *Squatina californica* | Shark, angel |
| *Squatina dumerili* | Shark, angel |
| *Stellifer* spp. | Drum |
| *Stenodus leucichthys* | Trout/iconnu |
| *Stenotomus caprinus caprinus* | Porgy |
| *Stenotomus chrysops* | Porgy/scup |
| *Stizostedion canadense* | Sauger |
| *Stizostedion vitreum* | Walleye |
| *Stolephorus buccaneeri* | Anchovy |
| *Stolephorus miarcha* | Anchovy |
| *Stolephorus purpureus* | Anchovy |
| *Symphorichthys spilurus* | Snapper, sailfin |
| *Synagrops bellus* | Bass |

| | |
|---|---|
| *Synaptura orientalis* | Sole |
| *Syngnathus fuscus* | Pipefish |
| *Synodus foetens* | Lizardfish |
| *Synodus intermedius* | Lizardfish |
| *Taractes rubescens* | Pomfret |
| *Tautoga onitis* | Tautog |
| *Tautogolabrus adspersus* | Cunner |
| *Tetrapturus albidus* | Marlin |
| *Tetrapturus angustirostris* | Spearfish |
| *Tetrapturus audax* | Marlin |
| *Tetrapturus pfluegeri* | Spearfish |
| *Thaleichthys pacificus* | Smelt |
| *Theragra chalcogramma* | Pollock/Alaska pollock |
| *Thunnus alalunga* | Tuna |
| *Thunnus albacares* | Tuna |
| *Thunnus atlanticus* | Tuna |
| *Thunnus maccoyii* | Tuna |
| *Thunnus obesus* | Tuna |
| *Thunnus thynnus* | Tuna |
| *Thunnus tonggol* | Tuna |
| *Thymallus arcticus* | Grayling |
| *Thyrsites atun* | Gemfish/barracouta |
| *Thyrsites lepidopoides* | Gemfish/caballa |
| *Tilapia aurea* | Tilapia |
| *Tilapia galilaea* | Tilapia |
| *Tilapia macrochir* | Tilapia |
| *Tilapia melanotheron* | Tilapia |
| *Tilapia mossambica* | Tilapia |
| *Tilapia nilotica* | Tilapia |
| *Tilapia rendalli* | Tilapia |
| *Torpedo califonica* | Ray, electric |
| *Torpedo nobiliana* | Torpedo |
| *Trachichthodes affinis* | Alfonsin |
| *Trachinotus blochi* | Pompano |
| *Trachinotus carolinus* | Pompano |
| *Trachinotus falcatus* | Pompano/permit |
| *Trachinotus goodei* | Pompano/palometa |
| *Trachinotus Kennedyi* | Pompano/permit |
| *Trachinotus paitensis* | Pompano |
| *Trachinotus rhodopus* | Pompano/pompanito |
| *Trachurus declivis* | Mackerel, jack |
| *Trachurus japonicus* | Mackerel, jack |

**Table 7.2 (continued)**

| Scientific Names | Market Names |
|---|---|
| *Trachurus lathami* | Mackerel, jack |
| *Trachurus murphyi* | Mackerel, jack |
| *Trachurus novaeze-landiae* | Mackerel/scad |
| *Trachurus picturatus* | Scad |
| *Trachurus symmetricus* | Mackerel, jack |
| *Trachurus trachurus* | Scad |
| *Trachyrhynchus murrayl* | Grenadier |
| *Triaenodon obesus* | Shark |
| *Triakis semifasciata* | Shark |
| *Trichiurus lepturus* | Cutlassfish |
| *Trichiurus nitens* | Cutlassfish |
| *Trichiurus savaka* | Cutlassfish |
| *Tylosurus crocodilus* | Houndfish |
| *Umbrina corides* | Drum |
| *Umbrina roncador* | Croaker |
| *Upeneichthys lineatus* | Goatfish |
| *Upeneus* spp. | Goatfish |
| *Uraspis secunda* | Jack |
| *Urophycis brasiliensis* | Hake |
| *Urophycis chusss* | Hake |
| *Urophycis cirrata* | Hake |
| *Urophycis earlli* | Hake |
| *Urophycis floridana* | Hake |
| *Urophycis regia* | Hake |
| *Urophycis tenuis* | Hake |
| *Xiphias gladius* | Swordfish |
| *Xystreurys liolepis* | Sole/flounder |
| *Zapteryl exasperata* | Guitarfish |
| *Zenopsis conchifer* | Dory |
| *Zenopsis neubulosa* | Dory |
| *Zeus faber* | Dory |
| *Zoarces viviparus* | Eelpout |

Source: The fish list: FDA guide to acceptable market names for food fish sold in interstate commerce. Food and Drug Administration, Washington, D.C., 1988.

# CHAPTER 8
# ALCOHOLIC BEVERAGES

This chapter provides data on the following:

1. Beer. Coverage includes: fundamentals of manufacture, types of beer, legal definitions and standards in the United States, quality control.
2. Wine. Coverage includes: legal definitions and standards in the United States, chemicals, materials, and processes authorized for the treatment of wine, juice, and distilling materials in the United States.
3. Liquor. Coverage includes: legal definitions and standards for distilled beverage spirits.
4. Flavorings. Coverage includes: usage and limitations for natural flavoring substances permitted in alcoholic beverages.

Although many food scientists and technologists are aware that there are legal standard and definitions for alcoholic beverages, few are familiar with the fact that there are stringent restrictions on the use of chemical substances as flavoring agents. Further, the U.S. Bureau of Alcohol, Tobacco, and Firearms (BATF) has described in details the chemicals, materials, and processes authorized for the treatment of wine and related substances. Therefore, this chapter provides an opportunity for interested individuals to locate both groups of information in one commercial source without referring to the original legal document. Most legal documents are difficult and inconvenient to use.

The use of federal legal documents follows the standard format. For example, "27 CFR 240.1051a" refers to the U.S. Code of Federal Regulations, Volume 27, Part 240, Section 1051a. All federal documents are available from the Superintendent of Document, Government Printing Office, Washington, D.C.

Two other abbreviations used in this chapter include: FDA, Food and Drug Administration; GRAS, Generally Recognized As Safe.

## 8.1. Beer

**Table 8.1.1. Basic Data for Beer Manufacture**

- History
  - Man's first alcoholic drink
  - Man's earliest brewing
    - Mesopotamia and Egypt
    - Greece and Rome
    - European tribes
    - Africa
    - China
    - India
    - South America
  - Ancient brewing and nutrition
  - Sanitary considerations
  - Large-scale brewing
  - Early brewing science
- Beer types in the world
  - Classical beer types
  - Beerlike beverages
  - Beer production in the world
- Beer constituents
  - Water
  - Alcohol
  - Carbohydrates
  - Nitrogen compounds
  - Inorganic constituents
  - Organic acids
  - Carbon dioxide
  - Other compounds
- Materials used in brewing
  - Water
    - Sources and purity
    - Minerals
    - Heavy metals
  - Barley and malt
    - Barley
      - Harvest and storage
      - Weathering
    - Malt
      - Earliest malt
      - Modern malting process
  - Brewing adjuncts
  - Hops
    - Earliest use of hops
    - The hop family
    - Hop utilization
    - Hop chemistry
  - Brewer's yeast
    - Characteristics and classification
    - Cell morphology and physiology
    - Yeast metabolism
      - Carbohydate metabolism
      - Metabolism of nitrogenous compounds
      - Lipid synthesis
      - Sterol synthesis
      - Sulfur compounds
      - Other metabolic reactions
- The brewing process
  - Brewhouse operation
    - Milling
      - Malt milling
      - Wet milling
      - Adjunct milling
    - Mashing
      - Infusion mashing
      - Decoction mashing
      - Malt conversion
      - Adjunct conversion
      - Enzyme activity during mashing
    - Lautering
      - The lauter tub
      - Run off
      - Sparging

- Draft beer
  - Cooperage
  - Racking
- Quality assurance
  - Physical and chemical measurements
    - Cleaning and sanitation
    - Raw materials acceptability
    - Biological survey of beer in progress
    - Analysis of the finished beer
    - Flavor analysis of the finished beer
  - Flavor measurements
    - Tasting beer
      - What we taste with
      - How we taste
      - The value of tasting
    - The quality assurance flavor panel
      - Screening for flavor panel trainees
      - Training a flavor panel
      - The work of a flavor panel
  - Tastable beer defects
    - Diacetyl
    - Metallic tastes
    - High-air beer
    - Light-struck beer
    - Old, oxidized beer
    - Medicinal odors
    - Grainy, harsh, astringent, bitter
    - Flavor depression

Source: Adapted from Hardwick, 1983. Used with permission.

**Table 8.1.2. Classical Beer Types Brewed in the World**

| Type | Character | Origin | Alcohol (vol %) | Flavor Features |
|---|---|---|---|---|
| Bottom Fermented | | | | |
| Bock | Lager | Bavaria, Canada, U.S. | 6 | Full bodied |
| Doppelbock | Lager/ale | Bavaria | 7-13 | Full bod., estery, winey |
| Dortmunder | Lager | Dortmund | 5+ | Light hops, dry, estery |
| Light beers | Lager | U.S.A. | 4.2-5 | Light bodied, light hops |
| Munchener | Lager/ale | Munich | 4-4.8 | Malty, dry, mod. bitter |
| Pilsner | Lager | Pilsen | 4.5-5 | Full bodied, hoppy |
| Vienna (Marzen) | Lager | Vienna | 5.5 | Full bodied, hoppy |
| Top Fermented | | | | |
| Ales | Ale | U.K., U.S., Canada, Australia | 2.5-5 | Hoppy, estery, bitter |
| Alt | Ale | Dusseldorf | 4 | Estery, bitter |
| Barley wine | Ale/wine | U.K. | 8-12 | Rich, full, estery |

| | | | | |
|---|---|---|---|---|
| Kolsch | Ale | Cologne | 4.4 | Light, estery, hoppy |
| Porter | Stout | London, Canada, U.S. | 5.7-5 | Very malty, rich |
| Provisie | Ale | Belgium | 6 | Sweet, alelike |
| Saisons | Ale | Belgium, France | 5 | Light, hoppy, estery |
| Stout (Bitter) | Stout | Ireland | 4-7 | Dry, bitter |
| Stout (Mackeson) | Stout | U.K. | 3.7-4 | Sweet, mild, lact. sour |
| Strong/old ale | Ale | U.K. | 6-8.4 | estery, heavy, hoppy |
| Trappiste | Ale | Belgium, Dutch Abbeys | 6-8 | Full bodied, estery |
| Wheat-Malt Beers | | | | |
| Berliner, Sweisse | Lager | Berlin | 2.5-3 | Light flavored |
| Gueuze-Lambic | Acid ale | Brussels | 5+ | Acidic |
| Hoegards wit | Ale | E. of Brussels | 5 | Full bodied, bitter |
| South Germany, Weizenbeer | Lager/ale | Bavaria | 5-6 | Full bodied, low hops |

Source: Adapted from Hardwick, 1983. Used with permission.

**Table 8.1.3. Legal Class and Type of Beer in the United States**

Malt Beverage: Products containing less than one-half of 1 percent (.5%) of alcohol by volume should bear the class designation "malt beverage," or "cereal beverage," or "near beer." If the designation "near beer" is used, both words must appear in the same size and style of type, in the same color of ink, and on the same background. No product containing less than one-half of 1 percent of alcohol by volume should bear the class designations "beer," "lager beer," "lager," "ale," "porter," or "stout," or any other class or type designation commonly applied to malt beverages containing one-half of 1 percent or more of alcohol by volume.

Concentrated and unconcentrated malt beverages: Malt beverages which have been concentrated by the removal of water and reconstituted by the addition of water and carbon dioxide should be labeled in the same manner as malt beverages which have not been concentrated and reconstituted, except that there should appear in direct conjunction with, and as a part of, the class designation the statement "PRODUCED FROM____________

CONCENTRATE" (the blank to be filled in with the appropriate class designation). All parts of the class designation should appear in lettering of substantially the same size and kind.

Half and half: No product shall be designated as "half and half" unless it is in fact composed of equal parts of two classes of malt beverages the names of which are conspicuously stated in conjunction with the designation "half and half".

Class designations: No product other than a malt beverage fermented at comparatively high temperature, possessing the characteristics generally attributed to "ale," "porter," or "stout" and produced without the use of coloring or flavoring materials (other than those recognized in standard brewing practices) should bear any of the above class designations.

Geographical names: Geographical names for distinctive types of malt beverages should not be applied to malt beverages produced in any place other than the particular region indicated by the name unless:

1. In direct conjunction with the name there appears the word "type" or the word "American", or some other statement indicating the true place of production in lettering substantially as conspicuous as such name, and
2. The malt beverages to which the name is applied conform to the type so designated.

The following are examples of distinctive types of beer with geographical names that have not become generic; Dortmund, Dortmunder, Vienna, Wein, Weiner, Bavarian, Munich, Munichner, Salvator, Kulmbacker, Wurtzburger, Pilsen (Pilsener and Pilsner). However, beer which is produced in the United States may be designated as "Pilsen," Pilsener," or "Pilsner" without further modification, if it conforms to such type.

Generic names: Only such geographical names for distinctive types of malt beverages as the Bureau of Alcohol, Tobacco, and Firearms finds have by usage and common knowledge lost their geographical significance to such an extent that they have become generic should be deemed to have become generic, e.g., India pale ale.

---

Source: 27 CFR 7, 1991.

**Table 8.1.4. Legal Definitions of Major Terms Used in the Manufacture of Beer in the United States**

Balling: the percent by weight of dissolved solids at 60°F present in wort and beer usually determined by a balling saccharometer

Barrel: when used as a unit of measure, the quantity equal to 31 U.S. gallons; when used as a container, a consumer package or keg containing 1/8, 1/6, 1/4, 1/3, or 1/2 barrel, a whole barrel, or other size authorized by the BATF

Beer: beer, ale, porter, stout, and other similar fermented beverages (including sake or similar products) of any name or description containing one-half of one percent or more of alcohol by volume, brewed or produced from malt, wholly or in part, or from any substitute for male

Cereal beverage: a beverage, produced either wholly or in part from malt (or a substitute for malt), and either fermented or unfermented, which contains, when ready for consumption, less than 1/2% by volume of alcohol by volume

Gallon: the liquid measure containing 231 $in^3$

Racking: the filling of kegs or barrels

Wort: the product of brewing before fermentation which results in beer

Source: 27 CFR 7, 25, 1991.

**Table 8.1.5. Flow Chart for Malting of Barley in a Malt House**

---

BARLEY

↓

Scalper ——▷ Dust, chaff

↓

STORAGE

↓

Cleaner ——▷ Dust, chaff, corn, and weed seeds

↓

Separator ——▷ Oats, wheat, cracked barley, weed seeds, etc.

↓

Grader ——▷ Undersized kernels
(Small, Medium, Large)

↓

Steep Tank

|

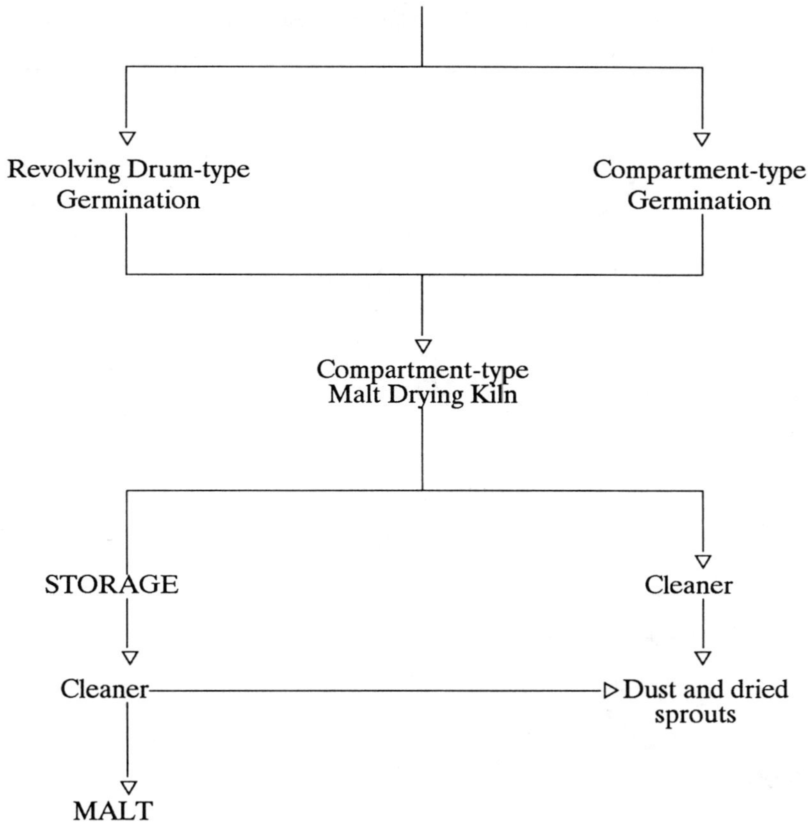


Source: Adapted from Hardwick, 1983. Used with permission.

**Table 8.1.6. Flow Chart for Brewing**

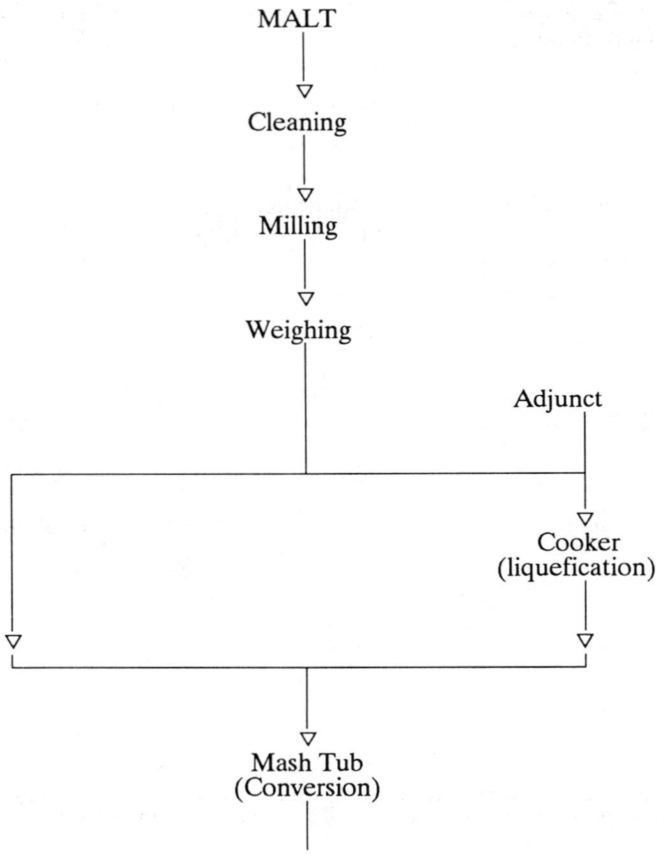

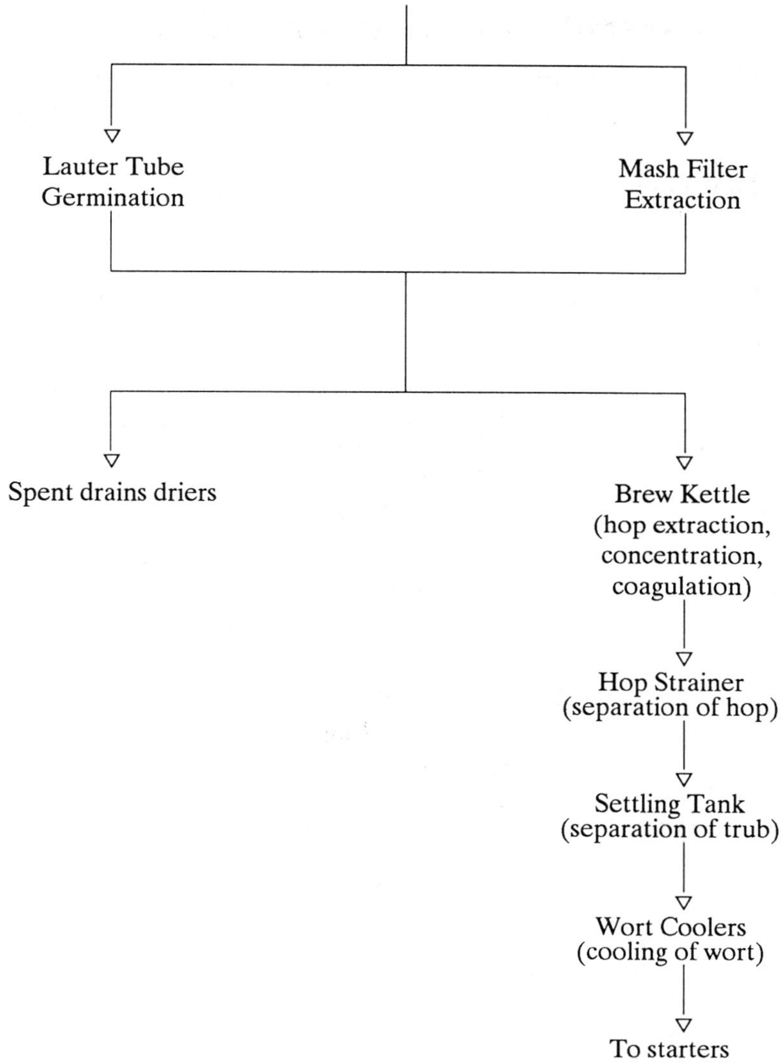


Source: Adapted from Hardwick, 1983. Used with permission.

**Table 8.1.7 Flow Chart for Fermentation and Finishing**

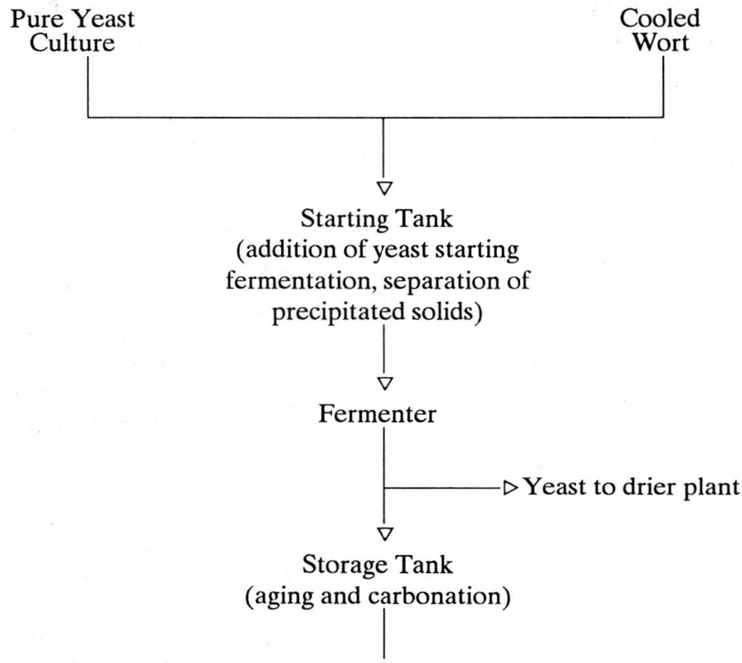

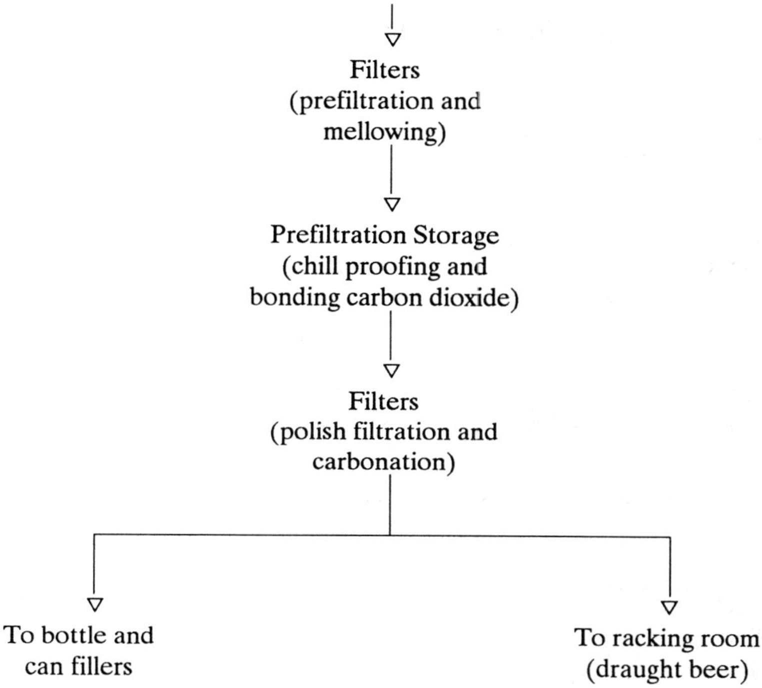


Source: Adapted from Hardwick, 1983. Used with permission.

**Table 8.1.8. Components of a Typical Beer**

| Substance | Percentage | No. of Chemicals | Source |
|---|---|---|---|
| Water | 90 | 1 | |
| Alcohol | 4 | 1 | Yeast, malt |
| Carbohydrate | 4 | 16 | Malt, adjuncts |
| Inorganic salts | 0.8 | 10 | Wort, malt |
| Nitrogen compounds | 0.3 | 35 | Malt |
| Organic acids | 0.2 | 13 | Yeast, malt |
| $CO_2$ | 0.5 | 1 | Yeast |
| Other compounds | 0.1 | $\pm 750$ | Malt, yeast, hops |

Source: Adapted from Hardwick, 1983. Used with permission.

**Table 8.1.9. Concentration of Carbohydrates in Regular Beers**

| Carbohydrates | g/100 mL |
|---|---|
| Monosaccharides | 0.1-0.2 |
| Disaccharides | 0.4-1.0 |
| Trisaccharides | 0.2-0.3 |
| Polysaccharides, 4-24 glucose units | 2.6-3.2 |

Source: Adapted from Hardwick, 1983. Used with permission.
Note: 1 L = ~0.001 $m^3$

**Table 8.1.10. Adjuncts Used in Brewing**

| Cereal Source | Form of Product |
|---|---|
| Barley | Dehusked, milled whole grain |
| Corn | Dry milled grits: germ- and skin- free |
| | Pregelatinized grits or flakes |
| | Refined grits (starch) |
| Corn sugar | Glucose from starch conversion |
| Corn syrup | Can add at end of kettle boil. |
| Rice | Rice grits, broken grains; flakes |
| | Pregelatinized rice grits |
| Sorghum grits | From kaffir corn and milo maize |
| Wheat | Low-protein wheat flakes |

Source: Adapted from Hardwick, 1983. Used with permission.

**Table 8.1.11. Relationship Between Certain Chemicals In Commercial Hops and Speific Beer Flavor**

| Substance | Percentage in Hops | Effect on Beer Flavor |
|---|---|---|
| Bitter resins | 6-17 | |
| Alpha acids | | |
| Humulone | | |
| Cohumulone | 5-8 | Bitter |
| Adhynulone | | |
| Beta acids | | |
| Lupulone | | |
| Colupulone | 7-10 | Bitter |
| Adlupulone | | |
| Hydrocarbons | 0.2-0.6 | |
| Myrcene | 0.1-0.3 | |
| Pinenes | 0.003-0.005 | |
| Caryophyllene | 0.05-0.08 | Aromatic |
| Farnescene | 0.07-0.1 | Terpene-like |
| Humulene | 0.08-0.15 | |
| Selinenes | 0.005-0.015 | |
| Oxygenated esters | 0.05-0.1 | Floral |
| Tannins | 3-5 | Harsh, astringent |
| Lipids and wax | 2-4 | Dulling |

Source: Adapted from Hardwick, 1983. Used with permission.

**Table 8.1.12. Differentiating Features in the Characteristics of *Saccharomyces cerevisiae* and *Saccharomyces uvarum***

| | *S. cerevisiae* | *S. uvarum* |
|---|---|---|
| Glyceraldehyde fermentation | (−) | (+) |
| $H_2S$ production | (+) | (+++) |
| Melibiose fermentation | (−) | (+) |
| Respiratory activity | (+++) | (+) |
| Sporulation | (+) | (−) |
| Yeast flocculance | (−) | (+) |
| Yeast head | (+) | (−) |

Source: Adapted from Hardwick, 1983. Used with permission.

**Table 8.1.13. Concentration of Some Amino Acids in a Wort and Its Beer**

| Amino Acid | ppm In Wort | ppm In Beer |
|---|---|---|
| Aspartic acid | 50 | 10 |
| Asparagine; glutamine | 140 | |
| Glutamic acid | 45 | 8 |
| Alanine | 88 | 24 |
| Isoleucine | 50 | 7 |
| Phenylalanine | 90 | 40 |
| Cystein + cystine | Trace | Trace |
| Arginine | 100 | 9 |
| Threonine | 39 | 7 |
| Proline | 210 | 220 |
| Valine | 88 | 18 |
| Leucine | 111 | 13 |
| Lysine | 40 | 3 |
| Serine | 50 | 10 |
| Glycine | 28 | 13 |
| Methionine | 30 | 3 |
| Tyrosine | 80 | 29 |
| Histidine | 23 | 3 |
| Tryptophan | 30 | 10 |

Source: Adapted from Hardwick, 1983. Used with permission.

**Table 8.1.14. Accepted Names for Gram-Negative Bacteria Associated with Brewing**

| Approved name | Synonyms |
|---|---|
| *Citrobacter freundii* | *Escherichia freundii* |
| *Enterobacter aerogenes* | *Aerobacter aerogenes* |
| *Enterobacter agglomerans* | *Erwinia herbicola* |
| *Enterobacter cloacae* | |
| *Hafnia alvei* | *Enterobacter alvei, Enterobacter hafniae* |
| *Klebsiella pneumoniae* | *Aerobacter aerogenes* |
| *Obesumbacterium proteus* | *Flavobacterium proteus, Hafnia protea* |

Source: Adapted from Hardwick, 1983. Used with permission.

**Table 8.1.15. Brewery Production Stage, Infection, and Microorganisms**

| Production Stage: Organisms |
|---|
| 1. Mashing/sweet wort: thermophylic lactics |
| 2. Cool wort to pitching: *Obesumbacterium, Lactobacillus, Acetobacter, Gluconobacter*, Enterobacteriaceae |
| 3. Fermentation: *Obesumbacterium, Acetobacter, Gluconobacter*, Enterobacteriaceae, *Lactobacillus, Pediococcus* |
| 4. Package in trace: *Acetobacter, Gluconobacter, Zymomonas, Lactobacillus, Pediococcus* |

Source: Adapted from Hardwick, 1983. Used with permission.

**Table 8.1.16. Some Wild Yeasts Reported to Spoil Beer**

*Bretanomyces bruxellensis, B. lambicus, B. anomolus, B. clausenii, B. intermedius*
*Candida utilis, C. lambica*
*Debaryomyces* spp.
*Dekkera bruxellensis, D. intermedia*
*Hanseniaspora* spp.
*Hansenula anomala*
*Kloeckera apiculata*
*Pichia membranaefaciens*
*Rhodotorula* spp.
*Saccharomyces cerevisiae* var. *ellipsoideus*
*Saccharomyces cerevisiae* var. *turbidaus*
*Saccharomyces bayanus (pastorianus), S. diastaticus, S. inusitatus*
*Torulopsis colliculosa, T. inconspicua*

Source: Adapted from Hardwick, 1983. Used with permission.

**Table 8.1.17. Raw Materials, Analytical Procedures, and Quality Assurance in a Brewery**

- Malt
  - Percentage moisture
  - Percentage extract
  - Laboratory wort
    - Soluble nitrogen
    - Free amino nitrogen
  - Diastatic power
  - Alpha-amylase
  - Protein
- Adjuncts
  - Adjunct cereals
    - Moisture
    - Oil
    - Enzyme conversion yield
    - Protein
    - Ash
  - Adjunct syrups
    - Clarity
    - Color
    - Percentage solubles
    - Percentage fermentable
    - pH
    - Protein
    - Ash
- Hop materials
  - Hops
    - Insect count
    - Percentage moisture
    - Alpha acids
    - Beta acids
  - Hop extracts
    - Alpha acids
    - Beta acids
  - Isomerized hop extracts
    - Iso-alpha acids
    - Rho-iso-alpha acids
- Water
  - Turbidity
  - Conductivity
  - Hardness
  - Alkalinity
  - Dissolved solids
  - pH
  - Individual minerals
    - Organic matter
  - Trihalomethanes
  - Chloroform
  - Nitrates

Source: Adapted from Hardwick, 1983. Used with permission.

**Table 8.1.18. "In Process" Wort and Beer, Analytical Procedure, and Quality Assurance**

| Chemical Tests for Wort | | |
|---|---|---|
| Specific gravity | Total acidity | Free amino acids |
| Extract | Color | Viscosity |
| Iodine test | Protein | |
| pH | Reducing sugars | |

Chemical Tests for Beer

| | |
|---|---|
| Specific gravity | Protein |
| Degree of fermentation | Fermentable sugars |
| Total acidity | Free amino nitrogen |
| Color | Dissolved oxygen |
| pH | |

Microbiological Tests

| | |
|---|---|
| Beer | Yeast |
| Cultured samples | Microscopic |
| Bacterial cultures | Cell morphology |
| Wild yeast cultures | Viability stains |
| Direct yeast counts | Bacteria search |
| Cleaning checks | Percentage budding |
| Air sample cultures | Culturing |
| Stained slides | Differential (bacteria) |
| Fermenter beer | Differential (yeast) |
| Filtered beer | Measuring |
| Suspect cultures | Total cell concentration |
| | Total yeast solids |

Source: Adapted from Hardwick, 1983. Used with permission.

**Table 8.1.19. Finished Beer, Analytical Procedures, and Quality Assurance**

| | |
|---|---|
| Air ($O_2$) content | Original extract |
| Alcohol content | Final extract |
| Calcium/oxalate content | Final degree of fermentation |
| Clarity, fresh | Foam retention |
| Carbon dioxide | Iodine reaction |
| Clarity, finished | Iron |
| Color | Iso-alpha acids |
| Copper | pH |
| Diacetyl | Protein |
| Final specific gravity | Sodium |
| Flavor evaluation | Sulfur dioxide |

Source: Adapted from Hardwick, 1983. Used with permission.

## 8.2. Wine

**Table 8.2.1. Fundamentals of Wine and Brandy Manufacture**

- Definitions
- Yeasts and alcoholic fermentation
  - Yeasts
    - Taxonomy, ecology
    - Industrially important yeasts
    - Killer yeasts
    - Effect of yeasts on the organoleptic character of wines
  - Yeast nutrition in grape musts
    - Composition of grape musts
    - Nutritional requirements of yeasts and their provision in musts
  - Growth of yeasts and alcoholic fermentation
    - Growth cycle of yeasts and kinetics of the fermentation
    - Fermentation problems and their causes
    - Stimulation of the fermentation
    - Concept of the survival factor
  - Biochemistry of the fermentation of grape must
    - Primary and secondary products
    - Volatile substances contributing to the aroma of wine
- Lactic acid bacteria and the Malolactic fermentation
  - Lactic acid bacteria of wines
    - Taxonomy
    - Ecology
    - The role of lactic acid bacteria in vinification
  - Bacterial growth and malolactic fermentation
    - Development of lactic acid bacteria during vinification; kinetics and biochemistry of the malolactic fermentation
    - Parameters affecting the development of lactic acid bacteria in wines
    - Stimulation of bacterial growth and of the malolactic fermentation
- Bacterial spoilage of wines
  - Spoilage by lactic acid bacteria
  - Spoilage by acetic acid bacteria
    - Taxonomy, ecology
    - Physiological properties
    - Effect of the metabolism of acetic acid bacteria on the quality of musts and wines
  - Wine technology
    - Grapes and corrective measures for the vintage
    - Classical fermentation
      - White wine production
      - Red wine production
    - Biological stabilization of wines; the role of sulfur dioxide and sorbic acid
  - Special processes
    - Sparkling wines
    - Sherry and port
    - Brandy
  - Production volumes of wines and brandy

Source: Adapted from Lafon-Lafourcade, 1983. Used with permission.

**Table 8.2.2. Legal Definitions for Major Terms Used in the Manufacture of Wine in the United States**

Agricultural wine: wine made from suitable agricultural products other than the juice of grapes, berries, or other fruits

Allied products: commercial fruit products and byproducts not taxable as wine.

Amelioration: the addition to juice or wine, before, during, and after fermentation, of either water or pure sugar, or a combination of water and pure sugar, or liquid sugar or invert sugar syrup, to adjust the acid content or to develop alcohol by fermentation

Artificially carbonated wine: effervescent wine artificially charged with carbon dioxide

Effervescent wine: Sparkling wine and artificially carbonated wine and does not include still wine as legally defined.

Fold: The ratio of the volume of the fruit mash or juice to the volume of the volatile fruit-flavor concentrate produced from such fruit mash or juice; e.g., 1 gal of concentrate of 100-fold would be the product from 100 gal of fruit mash or juice.

Fruit wine: wine made from the juice of sound, ripe fruit (including berries) other than grapes

Gallon or wine gallon: A United States gallon of liquid measure equivalent to the volume of 231 $in^3$.

Heavy bodied blending wine: wine made from fruit without added sugar, with or without added wine spirits, and conforming to the definition of natural wine in all respects except as to maximum total solids content

Invert sugar syrup: a solution of invert sugar that has been prepared by recognized methods of inversion from pure sugar. It should be a substantially colorless solution and contain not less than 60% sugar by weight (60°Brix).

Juice (or must): the unfermented juice of fruit, berries and authorized agricultural products, exclusive of pulp, skins, or seeds

Lees: the settlings of wine

Liter or litre: a metric unit of capacity equal to 1000 cubic centimeters of aloholic beverage, and equivalent to 33.814 U.S. fluid ounces. A liter is subdivided into 1000 milliliters (mL).

Liquid sugar: a substantially colorless pure sugar and water solution containing not less than 60% pure sugar by weight (60°Brix).

Proof: the ethyl alcohol content of a liquid at 60°F, stated as twice the percentage of ethyl alcohol by volume

Proof gallon: the alcoholic equivalent of a United States gallon at 60°F, containing 50% of ethyl alcohol by volume

**Table 8.2.2. (continued)**

Pure sugar: pure refined sugar, suitable for human consumption, having a dextrose equivalent of not less than 95% on a dry basis, and produced from cane, beets, or fruit, or from grain or other sources of starch. However, the invert sugar syrup produced from such pure sugar by reocgnized methods of inversion may may be used to prepare any sugar syrup, or solution of water and pure sugar.

Same kind of fruit: in the case of grapes, all of the several species and varieties of grapes. In the case of fruits other than grapes, this term includes all of the several species and varieties of any given kind. However, this does not preclude a more precise identification of the composition of the product for the purpose of its designation.

Sparkling wine or champagne: effervescent wine charged with carbon dioxide, resulting from fermentation of the wine within a closed container or bottle

Special natural wine: a product, such as vermouth, made according to an approved formula in accordance with regulations issued by the BATF

Specially sweetened natural wine: a product having a total solids content in excess of 17% by weight and an alcohol content of not more than 14% by volume, made in accordance with prescribed regulations

Standard agricultural wine: agricultural wine made with prescribed regulations

Standard wine: This includes natural wine, specially sweetened natural wine, special natural wine, and standard agricultural wine.

Still wine: noneffervescent wine, including those wines containing carbon dioxide

Sugar: pure sugar, liquid sugar, and invert sugar syrup

Total solids: the degrees Brix of the dealcoholized wine

Volatile fruit-flavor concentrate: any volatile fruit-flavor concentrate produced by any process that includes evaporations from any fruit mash or juice at a concentrate plant

Wine: "Wine," when used without qualification, includes all still wines, champagne, and other sparkling wines, artificially carbonated wine, and special natural wine or other wine products produced on bonded wine cellar premises.

Wine spirits: They include brandy or wine spirits produced in a distilled spirits plant (with or without the use of water to facilitate extraction and ditillation) exclusively from:

1. Fresh or dried fruit, or their residues or
2. The wine or wine residues therefrom or
3. Special natural wine. However, in the production of natural wine or special natural wine, if sugar has been used, the wine or the residuum

may not be used if the unfermented sugars have been refermented. Such wine spirits spirits should not be reduced with water from the distillation proof or be distilled at less than 140 degrees of proof. However, commercial brandy aged in wood for a period of not less than 2 years, and barreled at not less than 100 degrees of proof, should be deemed wine spirits.

Source: 27 CFR 4, 240 (1991).
Note: 1 U.S. gal = ~0.00379 $m^3$; $^oF = {^oC} \times 1.8 + 32$; 1 L = 0.001 $m^3$

**Table 8.2.3. Legal Classes and Standards of Identity for Wine in the United States**

Class 1: Grape Wine

"Grape wine" is wine produced by the normal alcoholic fermentation of the juice of sound, ripe grapes (including restored or unrestored pure condensed grape must) with or without the addition, after fermentation, of pure condensed grape must, and with or without added grape brandy or alcohol, but without other addition or abstraction except as may occur in cellar treatment. However, the product may be ameliorated before, during, or after fermentation by one of three methods.

1. Methods of amelioration:
   a. By adding, separately or in combination, dry sugar or such an amount of sugar and water solution as will not increase the volume of the resulting product more than 35%; but in no event should any product so ameliorated have an alcoholic content, derived by fermentation, of more than 13% by volume, or a natural acid content, if water has been added, of less than 5 parts per thousand, or a total solids content of more than 22 grams per 100 cubic centimeters.
   b. By adding, separately or in combination, not more than 20% by weight of dry sugar, or not more than 10% by weight of water.
   c. In the case of domestic wine, the BATF has prescribed a special method.
2. Maximum volatile acidity. The maximum volatile acidity, calculated as acetic acid and exclusive of sufur dioxide, should not be, for natural red wine, more than 0.14 g, and for other grape wine, more than 0.12 g/100 cc (20$^o$C). Grape wine deriving its characteristic color or lack of color from the presence or absence of the red coloring matter of the skins, juice, or pulp of grapes may be designated as "red wine," "pink (or rose) wine,"

**Table 8.2.3. (continued)**

"amber wine," or "white wine" as the case may be, and may be further designated as "natural."

3. "Table wine" is grape wine having an alcoholic content not in excess of 14% by volume. Such wine may also be designated as "light wine," "red table wine," "light white wine," "sweet table wine," etc., as the case may be.
4. "Dessert wine" is grape wine having an alcoholic content in excess of 14% but not in excess of 24% by volume. Dessert wine having the taste, aroma, and characteristics generally attributed to sherry and an alcoholic content, derived in part from added grape brandy or alcohol, of not less than 17% by volume, may be designated as "sherry." Dessert wine having the taste, aroma, and characteristics generally attributed to angelica, madeira, muscatel, and port and an alcoholic content, derived in part from added grape brandy or alcohol, of not less than 18% by volume, may be designated as "angelica," "madeira," "muscatel," or "port," respectively. Dessert wines having the taste, aroma, and characteristics generally attributed to any of the above products and an alcoholic content, derived in part from added grape brandy or alcohol, in excess of 14% by volume but, in the case of sherry, less than 17%, or, in other cases, less than 18% by volume may be designated as "light sherry," "light angelica," "light madeira," "light muscatel" or light port," respectively.

Class 2: Sparkling Grape Wine

1. "Sparkling grape wine" (including "sparkling wine," "sparkling red wine," and "sparkling white wine") is grape wine made effervescent with carbon dioxide resulting solely from the fermentation of the wine within a closed container, tank, or bottle.
2. "Champagne" is a type of sparkling light wine that derives its effervescence solely from the secondary fermentation of the wine within glass containers of not greater than 1 gallon capacity, and other characteristics attributed to champagne as made in the Champagne district of France.
3. A sparkling light wine having the taste, aroma, and characteristics generally attributed to champagne but not otherwise conforming to the standard for "champagne" may, in addition to but not in lieu of the class designation "sparkling wines" be further designated as "champagne style" or "champagne type" or "American (or New York State, California, etc.) champagne–bulk process"; all the words in such further designation should appear in lettering of substantially the same size and such lettering should not be substantially larger than the words "sparkling wine."

4. "Crackling wine," "petillant wine," "frizzante wine" (including dremant, perlant, reciotto, and other similar wine) is sparkling light wine normally less effervescent than champagne or other similar sparkling wine, but containing sufficient carbon dioxide in solution to produce, upon pouring under normal conditions, after the disappearance of air bubbles, a slow and steady effervescence evidenced by the formation of gas bubbles flowing through the wine. Crackling wine that derives its effervescence from secondary fermentation in containers greater than 1-gal capacity should be designated "crackling wine–bulk process," and the words "bulk process" should appear in lettering of substantially the same size as the words "crackling wine."

### Class 3: Carbonated Grape Wine

"Carbonated grape wine" (including "carbonated wine," "carbonated red wine," and "carbonated white wine") is grape wine made effervescent with carbon dioxide other than that resulting solely from the secondary fermentation of the wine within a closed container, tank, or bottle.

### Class 4: Citrus Wine

"Citrus wine" or "citrus fruit wine" is wine produced by the normal alcoholic fermentation of the juice of sound, ripe citrus fruit (including restored or unrestored pure condensed citrus must), with or without the addition, after fermentation, of pure condensed citrus must, and with or without added citrus brandy or alcohol, but without any other addition or abstraction except as may occur in cellar treatment. However, a domestic product may be ameliorated or sweetened in accordance with prescribed regulations and any product other than domestic may be ameliorated before, during, or after fermentation by adding, separately or in combination, dry sugar or such an amount of sugar and water solution as will not increase the volume of the resulting product more than 35%, but in no event should any product so ameliorated have an alcoholic content, derived by fermentation, of more than 13% by volume, or a natural acid content, if water has been added, of less than 5 parts per thousand, or a total solids content of more than 22 g/100 cc.

1. The maximum volatile acidity, calculated as acetic acid and exclusive of sulfur dioxide, should not be, for natural citrus wine, more than 0.14 gram, and for other citrus wine, more than 0.12 g/100 cc (10°C).
2. Any citrus wine containing no added brandy or alcohol may be further designated as "natural."

**Table 8.2.3. (continued)**

3. "Citrus table wine" or "citrus fruit table wine" is citrus wine having an alcoholic content not in excess of 14% by volume. Such wine may also be designated "light citrus wine," "light citrus fruit wine," "light sweet citrus fruit wine," etc., as the case may be.
4. "Citrus dessert wine" or "citrus fruit dessert wine" is citrus wine having an alcoholic content in excess of 14% but not in excess of 24% by volume.
5. Citrus wine derived wholly (except for sugar, water, or added alcohol) from one kind of citrus fruit, should be designated by the word "wine" qualified by the name of such citrus fruit, e.g., "orange wine," "grapefruit wine." Citrus wine not derived wholly from one kind of citrus fruit should be designated as "citrus wine" or "citrus fruit wine" qualified by a truthful and adequate statement of composition appearing in direct conjunction therewith. Citrus wine rendered effervescent by carbon dioxide resulting solely from the secondary fermentation of the wine within a closed container, tank, or bottle should be further designated as "sparkling"; and citrus wine rendered effervescent by carbon dioxide otherwise derived should be further designated as "carbonated."

Class 5: Fruit Wine

"Fruit wine" is wine (other than grape wine or citrus wine) produced by the normal alcoholic fermentation of the juice of sound, ripe fruit (including restored or unrestored pure condensed fruit must), with or without the addition, after fermentation, of pure condensed fruit must, and with or without added fruit brandy or alcohol, but without other addition or abstraction except as may occur in cellar treatment. However, a domestic product may be ameliorated or sweetened in accordance with the prescribed regulations and any product other than domestic may be ameliorated before, during, or after fermentation by adding, separately or in combination, dry sugar, or such an amount of sugar and water solution as will increase the volume of the resulting product, in the case of wines produced from loganberries, currants, or gooseberries, having a normal acidity of 20 parts or more per thousand, not more than 60%, and in the case of other fruit wines, not more than 35%, but in no event should any product so ameliorated have an alcoholic content, derived by fermentation, of more than 13% by volume, or a natural acid content, if water has been added, of less than 5 parts per thousand, or a total solids content of more than 22 g/100 cc.

1. The maximum volatile acidity, calculated as acetic acid and exclusive of sulfur dioxide, should not be, for natural fruit wine, more than 0.14 gram, and for other fruit wine, more than 0.12 g/100 cc (20$^{o}$C).

2. Any fruit wine containing no added brandy or alcohol may be further designated as "natural."
3. "Berry wine" is fruit wine produced from berries.
4. "Fruit table wine" or "berry table wine" is fruit or berry wine having an alcoholic content not in excess of 14% by volume. Such wine may also be designated "light fruit wine," or "light berry wine."
5. "Fruit dessert wine" or "berry dessert wine" is fruit or berry wine having an alcoholic content in excess of 14% but not in excess of 24% by volume.
6. Fruit wine derived wholly (except for sugar, water, or added alcohol) from one kind of fruit should be designated by the word "wine" qualified by the name of such fruit, e.g., "peach wine," "blackberry wine." Fruit wine not derived wholly from one kind of fruit should be designated as "fruit wine" or "berry wine," as the case may be, qualified by a truthful and adequate statement of composition appearing in direct conjunction therewith. Fruit wines that are derived wholly (except for sugar, water, or added alcohol) from apples or pears may be designated "cider" and "perry," respectively, and should be so designated if lacking in vinous taste, aroma, and characteristics. Fruit wine rendered effervescent by carbon dioxide resulting solely from the secondary fermentation of the wine within a closed container, tank, or bottle should be further designated as "sparkling"; and fruit wine rendered effervescent by carbon dioxide otherwise derived should be further designated as "carbonated."

### Class 6: Wine From Other Agrricultural Products

Wine of this class is wine (other than grape wine, citrus wine, or fruit wine) made by the normal alcoholic fermentation of sound fermentable agricultural products, either fresh or dried, or of the restored or unrestored pure condensed must thereof, with the addition before or during fermentation of a volume of water not greater than the minimum necessary to correct natural moisture deficiencies in such products, with or without the addition, after fermentation, of pure condensed must, and with or without added alcohol or such other spirits as will not alter the character of the product, but without other addition or abstraction except as may occur in cellar treatment. However, a domestic product may be ameliorated or sweetened in accordance with prescribed regulations, and any product other than domestic may be ameliorated before, during, or after fermentation by adding, separately or in combination, dry sugar or such an amount of sugar and water solution as will not increase the volume of the resulting product more than 35%, but in no event should any product so ameliorated have an alcoholic content, derived by fermentation of more than

**Table 8.2.3. (continued)**

13% by volume, or a natural acid content, if water has been added, of less than 5 parts per thousand, or a total solids content of more than 22 g/100 cc.

1. The maximum volatile acidity, calculated as acetic acid and exclusive of sulfur dioxide, should not be, for natural wine of this class, more than 0.14 gram, and for other wine of this class, more than 0.12 g/100 cc (20$^{o}$C).
2. Wine of this class containing no added alcohol or other spirits may be further designated as "natural."
3. "Table wine" of this class is wine having an alcoholic content not in excess of 14% by volume. Such wine may also be designated as "light."
4. "Dessert wine" of this class is wine having an alcoholic content in excess of 14% but not in excess of 24% by volume.
5. "Raisin wine" is wine of this class made from dried grapes.
6. "Sake" is wine of this class produced from rice in accordance with the commonly accepted method of manufacture of such product.
7. Wine of this class derived wholly (except for sugar, water, or added alcohol) from one kind of agricultural product should, except in the case of "sake," be designated by the word "wine" qualified by the name of such agricultural product, e.g., "honey wine," "raisin wine," "dried blackberry wine." Wine of this class not derived wholly from one kind of agricultural product should be designated as "wine" qualified by a truthful and adequate statement of composition appearing in direct conjunciton therewith. Wine of this class rendered effervescent by carbon dioxide resulting solely from the secondary fermentation of wine within a closed container, tank, or bottle should be further designated as "sparkling"; and wine of this class rendered effervescent by carbon dioxide otherwise derived should be further designated as "carbonated."

Class 7: Apertif Wine

"Aperitif wine" is wine having an alcoholic content of not less than 15% by volume, compounded from grape wine containing added brandy or alcohol, flavored with herbs and other natural aromatic flavoring materials, with or without the addition of caramel for coloring purposes, and possessing the taste, aroma, and characteristics generally attributed to aperitif wine and should be so designated unless designated as "vermouth" under prescribed regulations.

1. "Vermouth" is a type of aperitif wine compounded from grape wine, having the taste, aroma, and characteristics generally attributed to vermouth, and should be so designated.

## Class 8: Imitation and Substandard Wine

"Imitation wine" should bear as a part of its designation the word "imitation," and should include:

1. Any wine containing synthetic materials.
2. Any wine made from a mixture of water with residue remaining after thorough pressing of grapes, fruit, or other agricultural products.
3. Any class or type of wine the taste, aroma, color, or other characteristics of which have been acquired in whole or in part, by treatment with methods or materials of any kind, if the taste, aroma, color, or other characteristics of normal wines of such class or type are acquired without such treatment.
4. Any wine made from must concentrated at any time to more than 80$^{o}$ (Valling).

"Substandard wine" should bear as a part of its designation the word "substandard," and should include:

1. Any wine having a volatile acidity in excess of the maximum prescribed by regulations.
2. Any wine for which no maximum volatile acidity is prescribed, having a volatile acidity, calculated as acetic acid and exclusive of sulfur dioxide, in excess of 0.14 g/100 cc (20$^{o}$C).
3. Any wine for which a standard of identity is prescribed, which, through disease, decomposition, or otherwise, fails to have the composition, color, and clean vinous taste and aroma of normal wines conforming to such standard.
4. Any "grape wine," "citrus wine," "fruit wine," or "wine from other agricultural products" to which has been added sugar and water solution in an amount which is in excess of the limitations prescribed in the standards of identity for these products, unless, in the case of "citrus wine," "fruit wine" and "wine from other agricultural products" the normal acidity of the material from which such wine is produced is 20 parts or more per thousand and the volume of the resulting product has not been increased more than 60% by such addition.

## Class 9: Retsina Wine

"Retsina wine" is grape table wine fermented or flavored with resin.

---

Source: 27 CFR 4, 240 (1991).
Note: 1 U.S. gal = ~0.00379 $m^3$; $^{o}F = {}^{o}C \times 1.8 + 32$; 1 L = 0.001 $m^3$

**Table 8.2.4. Chemical Substances Authorized for the Treatment of Wine and Juice in the United States**

| Nutrients | Use[a] | Reference or Limitation[b] |
|---|---|---|
| Acacia (gum arabic) | To clarify and to stabilize wine | The amount used shall not exceed 2 lb/1000 gal (0.24 g/L) of wine. 21 CFR 184.12330 (GRAS) |
| Activated carbon | To assist precipitation during fermentation | 27 CFR 240.361, 240.365, 240.401, and 240.405 (GRAS) |
| | To clarify and to purify wine | The amount used to clarify and to purify wine shall be included in the total amount of activated carbon used to remove excessive color in wine. 27 CFR 240.524, 240.527, and 240.527a (GRAS) |
| | To remove excessive color in white wine | The amount used shall not exceed 9 lb/1000 gal (1.08 g/L) of wine. If the amount used exceeds this limit, a notice is required. |
| | To remove excessive color in pale dry sherry or cocktail sherry | The amount used shall not exceed 25 lb/1000 gal. (3.0 g/L) of wine. 27 CFR 240.527 (GRAS) |
| | To reduce color in the juice from red and black grapes | The amount used to reduce color must not exceed 4 lb/1000 gal. (0.48 g/L) of juice. Wine made from juice so treated shall have a color of not more than 95% transmittance. |
| Albumen (egg white) | Fining agent for wine | May be prepared in a light brine: 1 oz (28.35 grams) potassium chloride, 2 lbs (907.2 grams) egg white, 1 gal (3.785 L) of water. Usage not to exceed 1.5 gals of solution per 1000 gals of wine. (GRAS) |

| | | |
|---|---|---|
| Aluminosilicates (hydrated), e.g., bentonite (Wyoming clay) and kaolin | To clarify wine | 21 CFR 182.2727, 182.2729, 184.1155 (GRAS), and 186.1256. The sodium content of the wine shall not be increased by such treatment. Request for GRAS advisory opinion pending. |
| Ammonium carbonate | Yeast nutrient to facilitate fermentation in wine | The natural fixed acids shall not be reduced below 5 g/L. The amount used shall not exceed 2 lb/1000 gal (0.24 g/L). 21 CFR 184.1137 (GRAS) |
| Ammonium phosphate (mono- and dibasic) | Yeast nutrient in wine production and to start secondary fermentation in the production of sparkling wines | As a yeast nutrient in wine production, the amount shall not exceed 1.7 lb/1000 gal (0.20 g/L). In the production of sparkling wines, the amount used shall not exceed 8 lb/1000 gal (1.0 g/L) of wine. 21 CFR 184.1141 (GRAS) |
| Ascorbic acid, isoascorbic acid (erythorbic acid) | To prevent oxidation of color and flavor components of juice and wine | May be added to fruit, grapes, berries, and other materials used in wine production, to the juice of such materials, or to the wine, within limitations which do not alter the class or type of the wine. 21 CFR 182.3013 and 182.3041 (GRAS) |
| Autolyzed yeast | Yeast nutrient to facilitate fermentation in the production of grape or fruit wine | 21 CFR 172.896, 184.1983 (GRAS) |
| Calcium carbonate (with or without calcium salts of tartaric and malic acids) | To reduce the excess natural acid in high-acid wine | The natural or fixed acids shall not be reduced below 5 g/L. 21 CFR 184.1069, 184.1099, and 184.1191 (GRAS) |

**Table 8.2.4. (continued)**

| Nutrients | Use[a] | Reference or Limitation[b] |
|---|---|---|
| | A fining agent for cold stabilization | The amount used shall not exceed 30 lb/1000 gal (3.59 g/L) of wine. |
| Calcium sulfate (gypsum) | To lower pH in sherry wine | The calcium sulfate content of the finished wine shall not exceed 16.69 lb/1000 gal (2.0 g/L). 27 CFR 240.385; 21 CFR 184.1230 (GRAS) |
| Carbon dioxide | To stabilize and to preserve wine | 27 CFR 240.531 through 240.535; 21 CFR 184.1240 (GRAS) |
| | To maintain counterpressure during the transfer of sparkling wines | The carbon dioxide content of the finished wine shall not be increased during the transfer operation. 27 CFR 240.510 |
| Casein, potassium salt of casein | To clarify wine | GRAS, 27 CFR 240.528. |
| Citric acid | To correct natural acid deficiencies in wine | 27 CFR 240.364, 240.404, and 240.512; 21 CFR 182.1033 (GRAS) |
| | To stabilize wine other than citrus wine | The amount of citric acid shall not exceed 5.8 lb/1000 gal (0.7 g/L). 27 CFR 240.526, 240.539; 21 CFR 182.1033 (GRAS) |
| Copper sulfate | To remove hydrogen sulfide and/or mercaptans from wine | The quantity of copper sulfate added (calculated as copper) shall not exceed 0.5 part copper per million parts of wine with the residual level of copper not to be in excess of 0.2 part per million. GRAS |

| | | |
|---|---|---|
| Defoaming agents: [polyoxy ethylene (40) monostearate, silicon dioxide, dimethylpoly-siloxane, and sorbitan monostearate] | To control foaming | Defoaming agents that are 100% active may be used in amounts not exceeding 0.15 lb/1000 gal (0.018 g/L) of wine. Defoaming agents that are 30% active may be used in amounts not exceeding 0.5 lb/1000 gal (0.06 g/L) of wine. Silicon dioxide shall be completely removed by filtration. The amount of silicon remaining in the wine shall not exceed 10 parts per million and no soluble residue in excess of 25 parts per million shall remain in the wine. 21 CFR 173.340 |
| Enzymatic activity | Specific for individual group of enzymes | The enzyme preparation used shall be prepared from nontoxic and non-pathogenic microorganisms in accordance with good manufacturing practice and be approved for use in food by either FDA regulation or by FDA advisory opinion. |
| Carbo hydrase ($\alpha$-Amylase) | To convert starches to fermentable carbohydrates | The amylase enzyme activity shall be derived from *Aspergillus niger*, *Aspergillus oryzae*, *Bacillus subtilis*, or barley malt per FDA advisory opinion of 8/18/83 or from *Rhizopus oryzae* per 21 CFR 173.130 or from *Bacillus licheniformis* per 21 CFR 184.1027. |
| Carbo hydrase ($\beta$-amylase) | To convert starches to fermentable carbohydrates | The amylase enzyme activity shall be derived from barley malt per FDA advisory opinion dated 8/18/83. |
| Carbohydrase (gluco-amylase, amylogluco-sidase) | To convert starches to fermentable carbohydrates | The amylase enzyme activity shall be derived from *Aspergillus niger* or *Aspergillus oryzae* per FDA advisory opinion dated 8/18/83 or from *Rhizopus oryzae* per 21 CFR 173.110. |

**Table 8.2.4. (continued)**

| Nutrients | Use[a] | Reference or Limitation[b] |
|---|---|---|
| Catalase | To clarify and to stabilize wine | The enzyme activity used shall be derived from *Aspergillus niger* or bovine liver. |
| Cellulase | To clarify and to stabilize wine and to facilitate separation of the juice from the fruit | The enzyme activity used shall be derived from *Aspergillus niger* (GRAS). |
| Glucose oxidase | To clarify and to stabilize wine. | The enzyme activity used shall be derived from *Aspergillus niger* (GRAS). |
| Pectinase | To clarify and to stabilize wine and to facilitate separation of juice from the fruit | The enzyme activity used shall be derived from *Aspergillus niger* (GRAS) |
| Protease (general) | To reduce or to remove heat-labile proteins | The enzyme activity used shall be derived from *Aspergillus niger*, *Bacillus subtilis*, *Bacillus licheniformis* per 21 CFR 184.1027 (GRAS). |
| Protease (bromelin) | To reduce or to remove heat-labile proteins. | The enzyme activity used shall be derived from *Ananus comosus* or *Ananus bracteatus* (L.) (GRAS). |
| Protease (ficin) | To reduce or to remove heat-labile proteins | The enzyme activity used shall be derived from *Ficus* spp. (GRAS). |
| Protease (papain) | To reduce or to remove heat-labile proteins | The enzyme activity used shall be derived from *Carica papaya* (L.) per 21 CFR 184.1585 (GRAS). |
| Protease (pepsin) | To reduce or to remove heat-labile proteins | The enzyme activity used shall be derived from porcine or bovine stomachs (GRAS). |

| | | |
|---|---|---|
| Protease (trypsin) | To reduce or to remove heat-labile proteins | The enzyme activity used shall be derived from porcine or bovine pancreas (GRAS). |
| Ethyl maltol | To smooth wine | Use limited to not more than 100 mg/L 21 CFR 172.515. |
| Ferrocyanide compounds (sequestered complexes) | To remove trace metal from wine and to remove objectionable levels of sulfide and mercaptans from wine | No insoluble or soluble residue in excess of 1 part per million shall remain in the finished wine and the basic character of the wine shall not be changed by such treatment. (GRAS). |
| Ferrous sulfate | To clarify and to stabilize wine | The amount used shall not exceed 3 oz/1000 gal (0.22 g/L) of wine. 21 CFR 182.8315 (GRAS) |
| Fumaric acid. | To correct natural acid deficiencies in grape wine | Use must not exceed 25 lb/1000 gal (3.0 g/L). 27 CFR 240.364, 240.404, and 240.512; 21 CFR 172.350. The fumaric acid content of the finished wine shall not exceed 3.0 g/L. |
| | To stabilize wine | Use must not exceed 25 lb/1000 gal (3.0 g/L). 27 CFR 240.526, 240.539; 21 CFR 172.350. The fumaric acid content of the finished wine shall not exceed 3.0 g/L. |
| Gelatin (food grade) | To clarify juice or wine | GRAS |
| Granular cork | To treat wine | The amount used shall not exceed 10 lb/1000 gal of wine (1.2 g/L). Request for GRAS advisory opinion pending. |

**Table 8.2.4. (continued)**

| Nutrients | Use[a] | Reference or Limitation[b] |
|---|---|---|
| Hydrogen peroxide | To facilitate secondary fermentation in the production of sparkling wine | Amount used shall not exceed 3 parts per million. The finished wine shall contain no residual hydrogen peroxide (GRAS). |
| | To remove color from the juice of red and black grapes | The amount used shall not exceed 500 parts per million. The use of hydrogen peroxide is limited to oxidizing color pigment in the juice of red and black grapes. 21 CFR 182.1366 (GRAS) |
| Isinglass | To clarify wine | Request for GRAS advisory opinion pending. |
| Lactic acid | To correct natural acid deficiencies in grape wine | 27 CFR 240.364 and 2240.512; 21 CFR 182.1061 (GRAS) |
| Malic acid | To correct natural acid deficiencies in juice or wine | 27 CFR 240.364, 240.404, and 240.512. 21 CFR 184.1069 (GRAS) |
| Malolactic bacteria | To stabilize grape wine | Malolactic bacteria of the type *Leuconostoc oenos* may be used in treating wine. Request for GRAS advisory opinion pending. |
| Maltol | To smooth wine | Use limited to not more than 250 mg/L. 21 CFR 172.515 |

| | | |
|---|---|---|
| Nitrogen gas | To maintain pressure during filtering and bottling or canning of wine and to prevent oxidation of wine | 21 CFR 184.1540 (GRAS) |
| Oak chips or particles, uncharred and untreated | To smooth wine | 21 CFR 172.510 |
| Oxygen and compressed air | In baking or maturing wine and aeration of sherry | May be used provided it does not cause changes in the wine other than those occurring during the usual storage in wooden cooperage over a period of time. |
| Parabens (*n*-alkyl esters of *p*-hydroxybenzoic acid: heptyl-, propyl-, and methyl- | To preserve wine | Use separately or in combination must not exceed 0.1 percent by weight (w/w). Use of *n*-heptyl paraben must not exceed 12 parts per million. 21 CFR 172.145, 184.1490, 184.1670 (GRAS) |
| Polyvinyl-polypyrrolidone (PVPP) | To clarify and to stabilize wine | The amount used shall not exceed 6 lb/1000 gal (0.72 g/L) of wine. Material shall be removed during filtration. 21 CFR 173.50. |
| Polyvinyl-pyrrolidone (PVP) | To clarify wine | The residual level of PVP in the finished wine shall not exceed 60 parts per million. 21 CFR 173.55 |
| Potassium benzoate | To preserve wine | Use separately or in combination with other approved antimicrobial agents must not exceed 0.1% of the volume of juice and/or wine so treated. GRAS advisory opinion pending. |

**Table 8.2.4. (continued)**

| Nutrients | Use[a] | Reference or Limitation[b] |
|---|---|---|
| Potassium bitartrate | To stabilize grape wine | The amount used shall not exceed 35 lb/1000 gal (4.129 g/L) of grape wine. 21 CFR 184.1077 (GRAS) |
| Potassium carbonate and/or potassium bicarbonate | To reduce excess natural acidity in wine | The natural or fixed acids shall not be reduced below 5 parts per thousand (5 g/L). 21 CFR 184.1613 and 184.1619 (GRAS) |
| Potassium citrate | pH control agent and sequestrant in treatment of citrus wines | The amount of potassium citrate shall not exceed 25 lb/1000 gal (3.0 g/L) of finished wine. 27 CFR 240.404; 21 CFR 182.1625 and 182.16625 (GRAS) |
| Potassium metabisulfite | To sterilize and to preserve wine | The sulfur dioxide content of the finished wine shall not exceed the limitations prescribed in 27 CFR 4.22. 21 CFR 182.3637 (GRAS) |
| Propylene glycol | Solvent for parabens | The amount used shall not exceed 40 parts per million in wine. 21 CFR 184.1666 (GRAS) |
| Silica gel (colloidal silicon dioxide) | To clarify wine | Use shall not exceed the equivalent of 20 lb colloidal silicon dioxide at a 30% concentration per 1000 gal of wine (2.4 g/L). Silicon dioxide shall be completely removed by filtration (GRAS). |
| Sorbic acid and potassium salt of sorbic acid | To sterilize and to preserve wine; to inhibit mold growth and secondary fermentation | The finished wine shall contain more than 300 mg of sorbic acid per liter of wine. 21 CFR 183.3089 and 182.3640 (GRAS) |

| | | |
|---|---|---|
| Soy flour (defatted) | Yeast nutrient to facilitate fermentation of wine | The amount used shall not exceed 2 lb/1000 gal (0.24 g/L) of wine. GRAS |
| Sulfur dioxide | To sterilize and to preserve wine | The sulfur dioxide content of the finished wine shall not exceed the limitations prescribed in 27 CFR 4.22(b)(1). 21 CFR 182.3862 (GRAS) |
| Tannin | To adjust tannin content in apple juice or wine | The residual amount of tannin shall not exceed 3.0 g/L, calculated as gallic acid equivalents (GAE). GRAS |
| | To clarify or to correct tannin deficiencies in grape juice or wine | The residual amount of tannin, calculated in gallic acid equivalents, shall not exceed 0.8 g/L in white wine and 3.0 g/L in red and rose wines. Only tannin that does not impart color may be used in the cellar treatment of juice or wine. GRAS |
| Tartaric acid | To correct natural acid deficiencies in grape juice or grape wine | Use as prescribed in 27 CFR 240.364 and 240.512, and 21 CFR 184.1099 (GRAS) |
| Thiamine hydrochloride | Yeast nutrient to facilitate fermentation of wine | The amount used shall not exceed 0.005 lb/1000 gal (0.6 mg/L) of wine or juice. 21 CFR 184.1875 (GRAS) |
| Urea. | Yeast nutrient to facilitate fermentation of wine. | The amount used shall not exceed 2 lb/1000 gal (0.24 g/L) of wine. 21 CFR 184.1923 (GRAS) |

Source: 27 CFR 240.1051a, 1991.

[a]To stabilize: to prevent or to retard unwanted alteration of chemical and/or physical properties.

[b]GRAS: an acronym for "generally recognized as safe." The term means that the treating material has an FDA listing in 21 CFR 182, 184, or is considered to be generally recognized as safe by the U.S. Food and Drug Administration.

Note: 1 lb = ~453.6 g, 1 L = ~0.001 $m^3$, 1 U.S. gal = ~0.379 $m^3$

**Table 8.2.5. Materials Authorized for Treatment of Distilling Material in the United States**

| Materials | Use | Reference or Limitation[a] |
|---|---|---|
| Ammonium phosphate (mono- and dibasic) | Yeast nutrient in distilling material | The amount used shall not exceed 10 lb/1000 gal (1.2 g/L). 21 CFR 184.1141 (GRAS) |
| Amylase enzyme activity | | The enzyme preparation used shall be prepared from nontoxic and non-pathogenic microorganisms in accordance with good manufacturing practice and be approved for use in food by either FDA regulation or by FDA advisory opinion. |
| Carbohydrase ($\alpha$-amylase) | To convert starches to fermentable carbohydrates | The amylase enzyme activity shall be derived from *Aspergillus niger*, *Aspergillus oryzae*, *Bacillus subtilis*, or barley malt, or from *Rhizopus oryzae* (21 CFR 173.130), *Bacillus licheniformis* (21 CFR 184.1027). |
| Carbohydrase ($\beta$-amylase) | To convert starches to fermentable carbohydrates | The amylase enzyme activity shall be derived from barley malt. |
| Carbohydrase (glucomylase, amylo-glucosidase) | To convert starches to fermentable carbohydrates | The amylase enzyme activity shall be derived from *Aspergillus niger, Aspergillus oryzae, Rhizopus oryzae,* or from *Rhizopus niveus* (21 CFR 173.110). |
| Copper sulfate | To eliminate hydrogen sulfide and mercaptans | The finished brandy or wine spirits produced from distilling material to which copper sulfate has been added shall not contain more than 2 parts per million (2 mg/L) residual copper. GRAS |

| | | |
|---|---|---|
| Hydrogen peroxide | To reduce the bisulfite aldehyde complex in distilling material | The amount used shall not exceed 200 parts per million. 21 CFR 182.1366 (GRAS) |
| Potassium permanganate | Oxidizing agent | The finished brandy or wine spirits produced from distilling material to which potassium permanganate has been added must be free of chemical residue resulting from such treatment (GRAS). |
| Sodium benzoate | To prevent fermentation of the sugar in special natural wine being accumulated as distilling material | The amount used shall not exceed 0.1% (w/w). 21 CFR 184.1733 (GRAS) |
| Sodium hydroxide | Acid neutralizing agent | The finished brandy or wine spirits produced from distilling material to which sodium hydroxide has been added must be free of chemical residue resulting from such treatment. 21 CFR 184.1763 (GRAS) |
| Sulfuric acid | To effect favorable yeast development in distilling material; to prevent fermentation of the sugar in special natural wine being accumulated as distilling material; | 27 CFR 240.486; 21 CFR 184.1095 (GRAS) |

**Table 8.2.5. (continued)**

| Materials | Use | Reference or Limitation |
|---|---|---|
| | to lower pH to 2.5 to prevent putrefaction and/or ethyl acetate development | |

Source: 27 CFR 1051a, 1991.
[a]GRAS: an acronym for "generally recognized as safe." The term means that the treating material has an FDA listing in 21 CFR 182, 184, or is considered to be generally recognized as safe by the U.S. Food and Drug Administration.
Note: 1 lb = ~453.6 g, 1 L = ~0.001 $m^3$, 1 U.S. gal = ~0.379 $m^3$

**Table 8.2.6. Processes Authorized for the Treatment of Wine, Juice, and Distilling Material in the United States**

| Processes | Use | Reference or Limitation |
|---|---|---|
| Ion exchange | Various applications in the treatment of wine | Anion, cation, and nonionic resins, except those anionic resins in the mineral acid state, may be used in batch or continuous column processes as total or partial treatment of wine, provided that with regard to the finished wine:<br><br>1. Such treatment does not alter the vinous character of the wine.<br>2. The treatment does not reduce the color of the wine to less than that normally contained in such wine.<br>3. The treatment does not increase inorganic anions in the wine by more than 10 mg/L.<br>4. The treatment does not reduce the metallic cation concentration in the wine to less than 300 mg/L. |

| | | |
|---|---|---|
| | | 5. The treatment does not reduce natural or fixed acid in grape wine below 4 g/L for red table wines, 3 g/L for white table wines, 2.5 g/L for all other grape wine.<br>6. Treatment does not reduce the pH of the wine to less than pH 3 or increase the pH to more than pH 4.5.<br>7. The resins used have not imparted to the wine any material or characteristic (incidental to the resin treatment) that may be prohibited under any other section of the regulations in this part. The winemaker may employ conditioning and/or regenerating agents consisting of water, fruit acids common to the wine being treated, and inorganic acids, salts, and/or bases provided the conditioned or regenerated resin is rinsed with water until the resin and container are essentially free from unreacted (excess) conditioning or regenerating agents prior to the introduction of the wine. 21 CFR 173.25 |
| Thermal gradient processing | To separate standard wine into low-alcohol and high-alcohol wine fractions | The fractions derived from such processing must retain the color, taste, and other vinous characteristics of the wine prior to treatment. Such treatment must not reduce the ethyl alcohol content of the low-alcohol fraction to less than 3% by volume nor increase the alcohol content of the high-alcohol fraction to more than 24% by volume. Blending is restricted to wine of the proprietor's own production. Reconstitution, i.e., the addition of water other than that originally present in the wine prior to processing, is prohibited. |

**Table 8.2.6. (continued)**

| Processes | Use | Reference or Limitation |
|---|---|---|
| Thin-film evaporation under reduced pressure | To separate standard wine into a low-alcohol wine fraction and into a higher alcohol distillate | Wine so treated must retain the color, flavor, and other vinous characteristics of the wine prior to treatment. Such treatment must not reduce the ethyl alcohol content of the low-alcohol fraction to less than 3% by volume. Blending is restricted to wine of the proprietor's own production. Reconstitution, i.e., the addition of water other than that originally present in the wine prior to processing, is prohibited. Water lost in processing may be recovered by refluxing in a closed continuous system and returned to the wine from which it was extracted. |
| Elimination of sulfur dioxide by physical process | To reduce the sulfur dioxide content of juice | Use of a physical process to remove sulfur dioxide from juice must not alter the basic character of the juice so treated. |

Source: 27 CFR 240.1051b, 1991.
Note: 1 lb = ~453.6 g, 1 L = ~0.001 $m^3$, 1 U.S. gal = ~0.379 $m^3$

## 8.3. Distilled Beverage Spirits

**Table 8.3.1. Fundamentals of Manufacturing Distilled Beverage Spirits**

- History
  - The Coffey still
  - Definition of products
- Product manufacture
  - Products from cereal grains
    - Bourbon whiskey
      - Raw materials and their handling
        - Blending
      - Grain preparation for fermentation
      - Yeasting
      - Fermentation
      - Distillation
      - Recovering byproduct
      - Processing distillate
        - Maturation
    - Tequila

- Filtration
- Neutral spirits
- Other whiskeys
  - Scotch whiskey
  - Irish whiskey
  - Rye whiskey
  - Corn whiskey
  - Light whiskey
  - Blended whiskey
  - Canadian whiskey
- Products from nongrains
  - Rum
  - Vodka

- Microorganisms
  - Yeast
    - Alcohol production
    - Cogener production
  - Bacteria
  - Lactic souring of yeast mash
  - Indigenous microflora in the fermentation
    - Grain
    - Cooking and starch conversion
    - Equipment
    - Fermentation

Source: Adapted from Bluhm, 1983. Used with permission.

**Table 8.3.2. Legal Definitions for Some Major Terms Used in the Manufacture of Distilled Beverage Spirits in the United States**

Age: The period during which, after distillation and before bottling, distilled spirits have been stored in oak containers. "Age" for bourbon whiskey, rye whiskey, wheat whiskey, malt whiskey, or rye malt whiskey, and straight whiskies other than straight corn whiskey, means the period the whiskey has been stored in charred new oak containers.

Distilled spirits: Ethyl alcohol, hydrated oxide of ethyl, spirits of wine, whiskey, rum, brandy, gin, and other distilled spirits, including all dilutions and mixtures thereof, for nonindustrial use. They exclude mixtures containing wine, bottled at 48 degrees of proof or less, if the mixture contains more than 50% wine on a proof gallon basis.

Gallon: U.S. gallon of 231 in$^3$ of alcoholic beverage at 60°F.

Liter or litre: A metric unit of capacity equal to 1000 cc of distilled spirits at 15.56°C (60°F), and equivalent to 33.814 U.S. fluid oz. A liter is subdivided into 1000 mL.

Proof of distillation: The composite proof of the spirits at the time the production gauge is made, or, if the spirits have been reduced in proof prior to the production gauge, the proof of the spirits prior to such reduction, unless the spirits are subsequently redistilled at a higher proof than the proof prior to reduction.

Proof gallon: A gallon of liquid at 60°F which contains 50% by volume of ethyl alcohol having a specific gravity fo 0.7939 at 60°F referred to water at 60°F as unity, or the alcoholic equivalent thereof.

**Table 8.3.2. (continued)**

Source: 27 CFR 5, 19 (1991).
Note: $^{o}$F = $^{o}$C x 1.8 + 32; 1 L = 0.001 $m^3$; 1 U.S. gal = ~0.00379 $m^3$; 1 lb = ~453.6 g; 1 oz = ~28.4 g

**Table 8.3.3. Legal Classifications of Liquor in the United States**

Class 1: Neutral Spirits or Alcohol

"Neutral spirits" or "alcohol" are distilled spirits produced from any material at or above 190$^{o}$ proof, and, if bottled, bottled at not less than 80$^{o}$ proof.

1. "Vodka" is neutral spirits so distilled, or so treated after distillation with charcoal or other materials, as to be without distinctive character, aroma, taste, or color.
2. "Grain spirits" are neutral spirits distilled from a fermented mash of grain and stored in oak containers.

Class 2: Whiskey

"Whiskey" is an alcoholic distillate from a fermented mash of grain produced at less than 190$^{o}$ proof in such manner that the distillate possesses the taste, aroma, and characteristics generally attributed to whiskey, stored in oak containers (except that corn whiskey need not be so stored), and bottled at not les than 80$^{o}$ proof, and also includes mixtures of such distillates for which no specific standards of identity are prescribed.

1. "Bourbon whiskey", "rye whiskey", "wheat whiskey", "malt whiskey", or "rye malt whiskey" is whiskey produced at not exceeding 160$^{o}$ proof from a fermented mash of not less than 51% corn, rye, wheat, malted barley, or malted rye grain, respectively, and stored at not more than 125$^{o}$ proof in charred new oak containers; and also includes mixtures of such whiskies of the same type.
2. "Corn whiskey" is whiskey produced at not exceeding 160$^{o}$ proof from a fermented mash of not less than 80% corn grain, and if stored in oak containers stored at not more than 125$^{o}$ proof in used or uncharred new oak containers and not subjected in any manner to treatment with charred wood and also includes mixtures of such whiskey.
3. Whiskies conforming to the standards prescribed for bourbon and corn whiskies, which have been stored in the type of oak containers prescribed,

for a period of 2 years or more should be further designated as "straight"; for example, "straight bourbon whiskey", "straight corn whiskey." Whiskey conforming to the standards prescribed for bourbon whiskey, except that it was produced from a fermented mash of less than 51% of any one type of grain, and stored for a period of 2 years or more in charred new oak containers, should be designated merely as "straight whiskey". No other whiskies may be designated "straight." "Straight whiskey" includes mixtures of straight whiskies of the same type produced in the same State.

3. "Whiskey distilled from bourbon (rye, wheat, malt, or rye malt) mash" is whiskey produced in the United States at not exceeding $160^{o}$ proof from a fermented mash of not less than 51% corn, rye, wheat, malted barley, or malted rye grain, respectively, and stored in used oak containers; and also includes mixtures of such whiskies of the same type. Whiskey conforming to the standard of identity for corn whiskey must be designated corn whiskey.
4. "Light whiskey" is whiskey produced in the United States at more than $160^{o}$ proof, on or after January 26, 1968, and stored in used or uncharred new oak containers; and also includes mixtures of such whiskies. If "light whiskey" is mixed with less than 20% of straight whiskey on a proof gallon basis, the mixture shall be designated "blended light whiskey" (light whiskey—blend).
5. "Blended whiskey" (whiskey—a blend) is a mixture that contains straight whiskey or a blend of straight whiskies at not less than 20% on a proof gallon basis, excluding alcohol derived from added harmless coloring, flavoring or blending materials, and, separately or in combination, whiskey or neutral spirits. A blended whisky containing not less than 51% on a proof gallon basis of one of the types of straight whiskey should be further designated by that specific type of straight whiskey; for example, "blended rye whiskey" (rye whiskey—a blend).
6. "A blend of straight whiskies" (blended straight whiskies) is a mixture of straight whiskies that does not conform to the standard of identity for "straight whiskey." Products so designated may contain harmless coloring, flavoring, or blnding materials as set forth by regulations.
   a. "A blend of straight whiskies" (blended straight whiskies) consisting entirely of one of the types of straight whiskey, and not conforming to the standard for straight whiskey, should be further designated by that specific type of straight whiskey. For example, "a blend of straight whiskies" consisting entirely of one of the types of straight whiskey should include straight whiskey of the same type that was produced in the same State or by the same proprietor within the same State.

**Table 8.3.3. (continued)**

However, such whiskey must contain harmless coloring, flavoring, or blending materials as required by regulations.

b. The harmless coloring, flavoring, or blending materials allowed should not include neutral spirits or alcohol in their original state. Neutral spirits or alcohol may only appear in a "blend of straight whiskies" or in a "blend of straight whiskies consisting entirely of one of the types of straight whiskey" as a vehicle for recognized flavoring of blending material.

7. "Spirit whiskey" is a mixture of neutral spirits and not less than 5% on a proof gallon basis of whiskey, or straight whiskey, or straight whiskey and whiskey, if the straight whiskey component is less than 20% on a proof gallon basis.
8. "Scotch whiskey" is whiskey that is a distinctive product of Scotland, manufactured in Scotland in compliance with the laws of the United Kingdom regulating the manufacture of Scotch whiskey for consumption in the United Kingdom. However, if such product is a mixture of whiskies, such mixture is "blended Scotch whiskey" (Scotch whiskey—a blend).
9. "Irish whiskey" is whiskey that is a distinctive product of Ireland, manufactured either in the Republic of Ireland or in Northern Ireland, in compliance with their laws regulating the manufacture of Irish whiskey for home consumption. However, if such product is a mixture of whiskies, such mixture is "blended Irish whiskey" (Irish whiskey—a blend).
10. "Canadian whiskey" is whiskey that is a distinctive product of Canada, manufactured in Canada in compliance with the laws of Canada regulating the manufacture of Canadian whiskey for consumption in Canada. However, if such product is a mixture of whiskies, such mixture is "blended Canadian whiskey" (Canadian whiskey—a blend).

Class 3: Gin

"Gin" is a product obtained by original distillation from mash, or by redistillation of distilled spirits, or by mixing neutral spirits, with or over juniper berries and other aromatics, or with or over extracts derived from infusions, percolations, or maceration of such materials, and includes mixtures of gin and neutral spirits. It should derive its main characteristic flavor from juniper berries and be bottled at not less than $80^o$ proof. Gin produced exclusively by original distillation or by redistillation may be further designated as "distilled." "Dry gin" (London dry gin), "Geneva gin" (Hollands gin), and "Old Tom gin" (Tom gin) are types of gin known under such designations.

## Class 4: Brandy

"Brandy" is an alcoholic distillate from the fermented juice, mash, or wine of fruit, or from the residue thereof, produced at less than 190° proof in such manner that the distillate possesses the taste, aroma, and characteristics generally attributed to the product, and bottled at not less than 80° proof. Brandy, or mixtures thereof, not conforming to any of the standards prescribed should be designated as "brandy," and such designation should be immediately followed by a truthful and adequate statement of composition.

1. "Fruit brandy" is brandy distilled solely from the fermented juice or mash of whole, sound, ripe fruit or from standard grape, citrus, or other fruit wine with or without the addition of not more than 20% by weight of the pomace of such juice or wine or 30% by volume of the lees of such wine or both (calculated prior to the addition of water to facilitate fermentation or distillation). Fruit brandy should include mixtures of such brandy with not more than 30% (calculated on a proof gallon basis) of lees brandy. Fruit brandy derived from grapes should be designated as "grape brandy" or "brandy." However, in the case of brandy (other than neutral brandy, pomace brandy, marc brandy, or grappa brandy) distilled from the fermented juice, mash, or wine of grapes, or the residue thereof, which has been stored in oak containers for less than 2 years, the statement of class and type should be immediately preceded, in the same size and kind of type, by the word "immature." Fruit brandy, other than grape brandy, derived from one variety of fruit, should be designated by the word "brandy" qualified by the name of such fruit (for example, "peach brandy"). However, "apple brandy" may be designated "applejack." Fruit brandy derived from more than one variety of fruit should be designated as "fruit brandy" qualified by a truthful and adequate statement of composition.
2. "Cognac," or "Cognac (grape) brandy," is grape brandy distilled in the Cognac region of France, which is entitled to be so designated by the laws and regulations of the French Government.
3. "Dried fruit brandy" is brandy that conforms to the standard for fruit brandy except that it has been derived from sound dried fruit or from the standard wine of such fruit. Brandy derived from raisins, or from raisin wine, should be designated as "raisin brandy." Other brandies should be designated in the same manner as fruit brandy from the corresponding variety or varieties of fruit except that the name of the fruit should be qualified by the word "dried."

**Table 8.3.3. (continued)**

4. "Lees brandy" is brandy distilled from the lees of standard grape, citrus, or other fruit wine and should be designated as "lees brandy," qualified by the name of the fruit from which such lees are derived.
5. "Pomace brandy," or "marc brandy," is brandy distilled from the skin and pulp of sound, ripe grapes, citrus, or other fruit, after the withdrawal of the juice or wine therefrom, and should be designated as "pomace brandy" or "marc brandy" qualified by the name of the fruit from which it has been derived. Grape pomace brandy may be designated as "grappa" or "grappa brandy."
6. "Residue brandy" is brandy distilled wholly or in part from the fermented residue of fruit or wine, and should be designated as "residue brandy" qualified by the name of the fruit from which it has been derived. Brandy distilled wholly or in part from residue materials, which conform to any of the standards prescribed, may regardless of such fact be designated "residue brandy," but the use of such designation should be conclusive, precluding any later change of designation.
7. "Neutral brandy" is brandy produced at more than 170$^{o}$ proof and should be designated in accordance with the standards, except that the designation should be qualified by the word "neutral citrus residue brandy."
8. "Substandard brandy" should bear as a part of its designation the word "substandard" and should include:
   a. Any brandy distilled from fermented juice, mash, or wine having a volatile acidity, calculated as acetic acid and exclusive of sulfur dioxide, in excess of 0.2 g/100 cc (20$^{o}$C); measurements of volatile acidity should be calculated exclusive of water added to facilitate distillation.
   b. Any brandy that has been distilled from unsound, moldy, diseased, or decomposed juice, mash, wine, lees, pomace, or residue or that shows in the finished product any taste, aroma, or characteristic associated with products distilled from such material

Class 5: Blended Applejack

"Blended applejack" (applejack—a blend) is a mixture that contains at least 20% of apple brandy (applejack) on a proof gallon basis, stored in oak containers for not less than 2 years, and not more than 80% of neutral spirits on a proof gallon basis is such mixture at the time of bottling is not less than 80$^{o}$ proof.

## Class 6: Rum

"Rum," is an alcoholic distillate from the fermented juice of sugar cane, sugar cane syrup, sugar cane molasses, or other sugar cane by-products, produced at less than 190° proof in such manner that the distillate possesses the taste, aroma and characteristics generally attributed to rum, and bottled at not less than 80° proof; and also includes mixtures solely of such distillates.

## Class 7: Tequila

"Tequila" is an alcoholic distillate from a fermented mash derived principally from the *Agave tequilana* Weber ("blue" variety), with or without additional fermentable substances, distilled in such a manner that the distillate possesses the taste, aroma, and characteristics generally attributed to tequila and bottled at not less than 80° proof, and also includes mixtures solely of such distillates. Tequila is a distinctive product of Mexico, manufactured in Mexico in compliance with the laws of Mexico regulating the manufacture of tequila for consumption in that country.

## Class 8: Cordials and Liqueurs

Cordials and liqueurs are products obtained by mixing or redistilling distilled spirits with or over fruits, flowers, plants, or pure juices therefrom, or other natural flavoring materials, or with extracts derived from infusions, percolation, or maceration of such materials, and containing sugar, dextrose, or levulose, or a combination thereof, in an amount not less than 2 1/2% by weight of the finished product.

1. "Sloe gin" is a cordial or liqueur with the main characteristic flavor derived from sloe berries.
2. "Rye liqueur," "bourbon liqueur" (rye, bourbon cordial) are liqueurs, bottled at not less than 60° proof, in which not less than 51%, on a proof gallon basis, of the distilled spirits used are, respectively, rye or bourbon whiskey, straight rye or straight bourbon whiskey, or whiskey distilled from a rye or bourbon mash, and which possess a predominant characteristic rye or bourbon flavor derived from such whiskey. Wine, if used, must be within the 2 1/2% limitation prescribed for coloring, flavoring, and blending materials.
3. "Rock and rye," "rock and bourbon," "rock and brandy," "rock and rum" are liqueurs, bottled at not less than 48° proof, in which, in the case of rock and rye and rock and bourbon, not less than 51%, on a proof gallon basis,

**Table 8.3.3. (continued)**

of the distilled spirits used are, respectively, rye or bourbon whiskey, straight rye or straight bourbon whiskey, or whiskey distilled from a rye or bourbon mash, and, in the case of rock and brandy and rock and rum, the distilled spirits used are all grape brandy or rum, respectively; containing rock candy or sugar syrup, with or without the addition of fruit, fruit juices, or other natural flavoring materials, and possessing, respectively, a predominant characteristic rye, bourbon, brandy, or rum flavor derived from the distilled spirits used. Wine, if used, must be within the 2 1/2% limitation prescribed for harmless coloring, flavoring, and blending materials.

4. "Rum liqueur," "gin liqueur," "brandy liqueur," are liqueurs, bottled at not less than 60 proof, in which the distilled spirits used are entirely rum, gin, or brandy, respectively, and that possess, respectively, a predominant characteristic rum, gin, or brandy flavor derived from the distilled spirits used. In the case of brandy liqueur, the type of brandy must be stated in accordance with prescribed regulations. However, liqueurs made entirely with grape brandy may be designated simply as "brandy liqueur." Wine, if used, must be within the 2 1/2% limitation prescribed for harmless coloring, flavoring, and blending materials.
5. The designation of a cordial or liqueur may include the word "dry" if the sugar, dextrose, or levulose, or a combination thereof, are less than 10% by weight of the finished product.
6. Cordials and liqueurs should not be designated as "distilled" or "compound."

Class 9: Flavored Brandy, Flavored Gin, Flavored Rum, Flavored Vodka, and Flavored Whiskey

"Flavored brandy," "flavored gin," "flavored rum," "flavored vodka," and "flavored whiskey" are brandy, gin, rum, vodka, and whiskey, respectively, to which have been added natural flavoring materials, with or without the addition of sugar, and bottled at not less than $70^{0}$ proof. The name of the predominant flavor should appear as a part of the designation. If the finished product contains more than 2 1/2% by volume of wine, the kinds and percentages by volume of wine must be stated as a part of the designation, except that a flavored brandy may contain an additional 12 1/2% by volume of wine, without label disclosure, if the additional wine is derived from the particular fruit corresponding to the labeled flavor of the product.

## Class 10: Imitations

Imitations should bear, as a part of the designation, the word "imitation" and should include the following:

1. Any class or type of distilled spirits to which has been added coloring or flavoring material of such nature as to cause the resultant product to simulate any other class or type of distilled spirits
2. Any class or type of distilled spirits (other than distilled spirits required to bear a distinctive or fanciful name and a truthful and adequate statement of composition) to which has been added flavors considered to be artificial or imitation. In determining whether a flavor is artificial or imitation, recognition will be given to what is considered to be "good commercial practice" in the flavor manufacturing industry.
3. Any class of type of distilled spirits (except cordials, liqueurs and specialties marketed under labels which do not indicate, or imply, that a particular class or type of distilled spirits was used in the manufacture thereof) to which has been added any whiskey essence, brandy essence, rum essence, or similar essence or extract that simulates or enhances, or is used by the trade or in the particular product to simulate or enhance, the characteristics of any class or type of distilled spirits
4. Any type of whiskey to which beading oil has been added
5. Any rum to which neutral spirits or distilled spirits other than rum have been added
6. Any brandy made from distilling material to which has been added any amount of sugar other than the kind and amount of sugar expressly authorized in the production of standard wine
7. Any brandy to which neutral spirits or distilled spirits other than brandy have been added, except that this provision should not apply to any product conforming to the standard of identity for blended applejack.

## Class 11: Geographical Designations

Geographical names for distictive types of distilled spirits should not be applied to distilled spirits produced in any other place than the particular region indicated by the name, unless: (1) In direct conjunction with the name there appears the word "type" or the word "American" or some other adjective indicating the true place of production, in lettering substantially as conspicuous as such name; and (2) the distilled spirits to which the name is applied conform to the distilled spritis or that particular region.

**Table 8.3.3. (continued)**

1. Examples. The following are examples of distinctive types of distilled spirits with geographical names that have not become generic: Eau de Vie do Dantzig (Danziger Goldwasser), Ojen, Swedish punch.
2. Standards of identity. Goegraphical names for distinctive types of distilled spirits should be used to designate only distilled spirits conforming to the standard of identity, if any, for such type specified in this section, or if no such standard is so specified, then in accordance with the trade understanding of that destinctive type.
3. Generic names. Only such geographical names for distilled spirits as the BATF finds have by usage and common knowledge lost their geographical significance to such extent that they have become generic should be deemed to have become generic. Examples are London dry gin, Geneva (Hollands) gin.
4. Nondistinctive types. Geographical names that are not names for distinctive types of distilled spirits, and that have not become generic, should not be applied to distilled spirits produced in any other place than the particular place or region indicated in the name. Examples are Cognac, Armagnac, Greek brandy, Pisco brandy, Jamaica rum, Puerto Rico rum, Demerara rum.
5. The words "Scotch," "Scots," "highland," or "highlands" and similar words connoting, indicating, or commonly associated with Scotland should not be used to designate any product not wholly produced in Scotland.

Class 12: Products Without Geographical Designations
But Distinctive of a Particular Place

The whiskies of the types prescribed by the BATF are distinctive products of the United States and if produced in a foreign country should be designated by the applicable designation prescribed, together with the words "American type" or the words "produced (distilled, blended) in ______," the blank to be filled in with the name of the foreign country. However, the word "bourbon" should not be used to describe any whiskey or whiskey-based distilled spirits not produced in the United States. If whiskey of any of these types is composed in part of whiskey or whiskies produced in a foreign country there should be stated, on the brand label, the percentage of such whiskey and the country of origin thereof.

1. The name for other distilled spirits that are distinctive products of a particular place or country, an example is Habanero, should not be given to the product of any other place or country unless the designation for

such product includes the word "type" or an adjective such as "American," or the like, clearly indicating the true place of production. The provision for place of production should not apply to designations that by usage and common knowledge have lost their geographical significance to such an extent that the BATF finds they have become generic. Examples are Slivovitz, Zubrovka, Aquavit, Arrack, and Kirschwasser.

Source: 27 CFR 5, 19 (1991).
Note: $^{o}F$ = $^{o}C$ x 1.8 + 32; 1 L = 0.001 $m^3$; 1 U.S. gal = ~0.00379 $m^3$; 1 lb = ~453.6 g; 1 oz = ~28.4 g

## 8.4 Natural Flavoring Substances Permitted In Alcoholic Beverages

According to the U. S. Food and Drug Administration, natural flavoring substances and natural adjuvants may be safely used in food in accordance with the following conditions

1. They are used in the minimum quantity required to produce their intended physical or technical effect and in accordance with all the principles of good manufacturing practice.
2. In the appropriate forms (plant parts, fluid and solid extracts, concretes, absolutes, oils, gums, balsams, resins, oleoresins, waxes, and distilaltes) they consist of one or more the items listed in Table 8.4.1, used alone or in combination with flavoring substances and adjuvants generally recognized as safe in food, previously sanctioned for such use or regulated by prescribed requirements.

**Table 8.4.1 Natural Flavoring Substances And Natural Substances Used In Conjunction With Flavors**

| Common Name | Scientific Name | Limittaions |
|---|---|---|
| Aloe | *Aloe perryi* Baker, *A. barbadensis* Mill., *A. ferox* Mill., and hybrids of this sp. with *A. africana* Mill. and *A. spicata* Baker | None stated |
| Althea root and flowers | *Althea officinalis* L. | None stated |
| Amyris (West Indian sandalwood) | *Amyris balsamifera* L. | None stated |

**Table 8.4.1. (continued)**

| | | |
|---|---|---|
| Angola weed | *Rocella fuciformis* Ach | In alcoholic beverages only |
| Arnica flowers | *Arnica montana* L., *A. fulgens* Pursh, *A. sororia* Greene, or *A. cordifolia* Hooker | In alcoholic beverages only |
| Artemisia (wormwood) | *Artemisia* spp. | Finished food thujone free |
| Artichoke leaves | *Cynara scolymus* L | In alcoholic beverages only |
| Benzoin resin | *Styrax benzoin* Dryander, *S. paralleloneurus* Perkins, *S. tonkinensis* (Pierre) Craib ex Hartwich, or other spp. of the Section *Anthostyrax* of the genus *Styrax* | None stated |
| Blackberry bark | *Rubus*, section *Eubatus* | None stated |
| Boldus (boldo) leaves | *Peumus boldus* Mol | In alcoholic beverages only |
| Boronia flowers | *Boronia megastigma* Nees | None stated |
| Byronia root | *Bryonia alba* L., or *B. diocia* Jacq | In alcoholic beverages only |
| Buchu leaves | *Barosma betulina* Bartl. et Wendl., *B. crenulata* (L.) Hook, or *B. serratifolia* Wild | None stated |
| Buckbean leaves | *Menyanthes trifoliata* L. | In alcoholic beverages only |
| Cajeput | *Melaleuca leucadendron* L. and other *Melaleuca* spp. | None stated |
| Calumba root | *Jateorhiza palmata* (Lam.) Miers | In alcoholic beverages only |
| Camphor tree | *Cinnamomum camphora* (L.) Nees et Eberm | Safrole free |
| Cascara sagrada | *Rhamnus purshiana* DC | None stated |
| Cassie flowers | *Acacia farnesiana* (L.) Wild | None stated |

| | | |
|---|---|---|
| Castor oil | *Ricinus communis* L | None stated |
| Catechu, black | *Acacia catechu* Wild | None stated |
| Cedar, white (aborvitae), leaves and twigs | *Thuja occidentalis* L. | Finished food thujone free |
| Centuary | *Centaurium umbellatum* Gilib | In alcoholic beverages only |
| Cherry pits | *Prunus avium* L. or P. *cerasus* L. | Not to exceed 25 ppm prussic acid |
| Cherry-laurel leaves | *Prunus laurocerasus* L. | Not to exceed 25 ppm prussic acid |
| Chestnut leaves | *Castanea dentata* (Marsh.) Borkh | None stated |
| Chirata | *Swertia chirata* Buch.-Ham | In alcoholic beverages only |
| Cinchona, red, bark | *Cinchona succirubra* Pav. or its hybrids | In beverages only; not more than 83 ppm total cinchona alkaloids in finished beverage |
| Cinchona, yellow, bark | *Cinchona ledgeriana* Moens, *C. calisaya* Wedd., or hybrids of these with other spp. of *Cinchona* | In beverages only; not more than 83 ppm total cinchona alkaloids in finished beverage |
| Copaiba | South American spp. of *Copaifera* L. | None stated |
| Cork, oak | *Quercus suber* L., or *Q. occidentalis* F. Gay | In alcoholic beverages only |

**Table 8.4.1. (continued)**

| | | |
|---|---|---|
| Costmary | *Chrysanthemum balsamita* L | In alcoholic beverages only |
| Costus root | *Saussurea lappa* Clarke | None stated |
| Cubeb | *Piper cubeba* L. | None stated |
| Currant, black, buds and leaves | *Ribes nigrum* L | None stated |
| Damiana leaves | *Turnera diffusa* Willd | None stated |
| Davana | *Artemisia pallens* Wall. | None stated |
| Dill, Indian | *Anethum sowa* Roxb. (*Peucedanum graveolens* Benth et Hook., *Anethum graveolens* L.) | None stated |
| Dittany (fraxinella) roots | *Dictamnus aslbus* L | In alcoholic beverages only |
| Dittany of Crete | *Origanum dictamnus* L. | None stated |
| Dragon's blood (dracorubin) | *Daemonorops* spp | None stated |
| Elder tree leaves | *Sambucus nigra* L. | In alcoholic beverages only; not to exceed 25 ppm prussic acid in the flavor |
| Elecampane rhizome and roots | *Inula helenium* L. | In alcoholic beverages only |
| Elemi | *Canarium commune* L. or *C. luzonicum* Miq. | None stated |
| Erigeron | *Erigeron canadensis* L. | None stated |
| Eucalyptus globulus leaves | *Eucalyptus globulus* Labill | None stated |
| Fir ("pine") needles and twigs | *Abies sibirica* Ledeb., *A. alba* Mill., *A. sachalinesis* Masters or *A. mayriana* Miyabe et Kudo | None stated |
| Fir, balsam, needles and twigs | *Abies balsamea* (L.) Mill | None stated |

| | | |
|---|---|---|
| Galanga, greater | *Alpinia galanga* Willd | In alcoholic beverages only |
| Galbanum | *Ferula galbaniflua* Boiss. et Buhse and other *Ferula* spp. | None stated |
| Gambir (catechu, pale) | *Uncaria gambir* Roxb | None stated |
| Genet flowers | *Spartium junceum* L. | None stated |
| Gentian rhizome and roots | *Gentiana lutea* L. | None stated |
| Gentian, stemless | *Gentiana acaulis* L. | In alcoholic beverages only |
| Germander, chamaedrys | *Teucrium chamaedrys* L. | In alcoholic beverages only |
| Germander, golden | *Teucrium polium* L. | In alcoholic beverages only |
| Guaiac | *Guaiacum officinale* L., *G. santum* L., *Bulnesia sarmienti* Lor. | None stated |
| Guarana | *Paullinia cupana* HBK | None stated |
| Haw, black, bark | *Viburnum prunifolium* L. | None Stated |
| Hemlock needles and twigs | *Tsuga canadensis* (L.) Carr. or *T. heterophylla* (Raf.) Sarg. | None stated |
| Hyacinth flowers | *Hyacinthus orientalis* L. | None stated |
| Iceland moss | *Cetraria islandica* Ach | In alcoholic beverages only |
| Imperatoria | *Peucedanum ostruthium* (L.). Koch (*Imperatoria ostruthium* L.) | None stated |
| Iva | *Achillea moschata* Jacq | In alcoholic beverages only |
| Labdanum | *Cistus* spp. | None stated |
| Lemon-verbena | *Lippia citriodora* HBK | In alcoholic beverages only |

**Table 8.4.1. (continued)**

| | | |
|---|---|---|
| Linaloe wood | *Bursera delpechiana* Poiss. and other *Bursera* spp., | None stated |
| Linden leaves | *Tillia* spp | In alcoholic beverages only |
| Lovage | *Levisticum officinale* Koch | None stated |
| Lungmoss (lungwort) | *Sticta pulmonacea* Ach. | None stated |
| Maidenhair fern | *Adiantum capillus-veneris* L. | In alcoholic beverages only |
| Maple, mountain | *Acer spicatum* Lam | None stated |
| Mimosa (black wattle) flowers | *Acacia decurrens* Wild. var. *dealbata* | None stated |
| Mullein flowers | *Verbascum phlomoides* L.or *V. thapsiforme* Schrad. | In alcoholic beverages only |
| Myrrh | *Commiphora molmol* Engl., *C. abyssinica* (Berg) Engl., or other *Commiphora* spp. | None stated |
| Myrtle leaves | *Myrtus communis* L. | In alcoholic beverages only |
| Oak, English, wood | *Quercus robur* L. | In alcoholic beverages only |
| Oak, white, chips | *Quercus alba* L. | None stated |
| Oak moss | *Evernia prunastri* (L.) Ach., E. *furfuracea* (L.) Mann, and other lichens | Finished food thujone |
| Olibanum | *Boswellia carten* Birdw. and other *Boswellia* spp. | None stated |
| Opopanax (bisabolmyrrh) | *Opopanax chironium* Koch (true opopanax) of *Commiphora erythraea* Engl. var. *Llabrescens* | None stated |
| Orris root | *Iris germanica* L. (including its variety *florentina* Dykes) and *I. pallida* Lam. | None stated |

| | | |
|---|---|---|
| Pansy | *Viola tricolor* L | In alcoholic beverages only |
| Passion flower | *Passiflora incarnata* L. | None stated |
| Patchouly | *Pogostemon cablin* Benth. and *P. heyneanus* Benth. | None stated |
| Peach leaves | *Prunus persica* (L.) Batsch | In alcoholic beverates only; not to exceed 25 ppm prussic acid in the flavor |
| Pennyroyal, American | *Hedeoma pulegioides* (L.) Pers | None stated |
| Pennyroyal, European | *Mentha pulegium* L. | None stated |
| Pine, dwarf, needles and twigs | *Pinus mugo* Turra var. *pumilio* (Haenke) Zenari | None stated |
| Pine, Scotch, needles and twigs | *Pinus sylvestris* L. | None stated |
| Pine, white, bark | *Pinus strobus* L. | In alcoholic beverages only |
| Pine, white oil | *Pinus palustris* Mill., and other *Pinus* spp. | None stated |
| Poplar buds | *Populus balsamifera* L. (*P. tacamahacca* Mill.), *P. candicans* Alt., or *P. nigra* L. | In alcoholic beverages only |
| Quassia | *Picrasma excelsa* (Sw.) Planch, or *Quassia amara* L. | None stated |
| Quebracho bark | *Aspidosperma quebracho-blanco* Schlecht, or (*Quebrachia lorentzii* (Griseb)) *Schinopsis lorentzii* (Griseb.) Engl. | |
| Quillaia (soapbark) | *Quillaja saponaria* Mol | None stated |
| Red saunders (red sandalwood) | *Pterocarpus san alinus* L. | In alcoholic beverages only |

**Table 8.4.1. (continued)**

| | | |
|---|---|---|
| Rhatany root | *Krameria triandra* Ruiz et Pav. or K. *argentea* Mart. | None stated |
| Rhubarb, garden root | *Rheum rhaponticum* L. | In alcoholic beverages only |
| Rhubarb root | *Rheum officinale* Baill., *R. palmatum* L., or other spp. (excepting *R. rhaponticum* L.) or hybrids of *Rheum* grown in China | None stated |
| Roselle | *Hibiscus sabdariffa* L | In alcoholic beverages only |
| Rosin (colophony) | *Pinus palustris* Mill., and other *Pinus* spp. | In alcoholic beverages only |
| St. Johnsword leaves, flowers, and caulis | *Hypericum perforatum* L. | Hypericin-free alcohol distillate form only; in alcoholic beverages only |
| Sandalwood, white (yellow, or East Indian). | *Santalum album* L. | None stated |
| Sandarac | *Tetraclinis articulata* (Vahl.), Mast | In alcoholic beverages only |
| Sarsaparilla | *Smilax aristolochiaefolia* Mill., (Mexican sarsaparilla), *S. regelii* Killip et Morton (Honduras sarsaparilla), *S. febrifuga* Kunth (Ecuadorean sarsaparilla), or undetermined *Smilax* spp. (Ecuadorean or Central American sarsaparilla) | None stated |
| Sassafras leaves | *Sassafras albidum* (Nutt.) Nees | Safrole free |
| Senna, Alexandria | *Cassia acutifolia* Delile | None stated |

| | | |
|---|---|---|
| Serpentaria (Virginia snakeroot) | *Aristolochia serpentaria* L | In alcoholic beverages only |
| Simaruba bark | *Simaruba amara* Aubl | In alcoholic beverages only |
| Snakeroot, Canadian (wild ginger) | *Asarum canadense* L. | None stated |
| Spruce needles and twigs | *Picea glauca* (Moench) Voss or *P. mariana* (Mill.) BSP | None stated |
| Storax (styrax) | *Liquidambar orientalis* Mill. or *L. styraciflua L* | None stated |
| Tagetes (marigold) | *Tagetes patula* L., *T erecta* L., or *T. minuta* L. *(T. glandulifera* Schrank*)* | As oil only |
| Tansy | *Tanacetum vulgare* L. | In alcoholic beverages only; finished alcoholic beverage thujone free |
| Thistle, blessed (holy thistle) | *Onicus benedictus* L. | In alcoholic beverages only |
| Thymus capitatus (Spanish "origanum") | *Thymus capitatus* Hoffmg. et Link | None stated |
| Tolu | *Myroxylon balsamum* (L.) Harms | None stated |
| Turpentine | *Pinus palustris* Mill. and other *Pinus* spp. which yield terpene oils exclusively | None stated |
| Valerian rhizome and roots | *Valeriana officinalis* L. | None stated |
| Veronica | *Veronica officinalis* L. | In alcoholic beverages only |
| Vervain, European | *Verbena officinalis* L. | In alcoholic beverages only |
| Violet, Swiss | *Viola calcarata* L. | None stated |

**Table 8.4.1. (continued)**

| | | |
|---|---|---|
| Walnut husks (hulls), leaves, and green nuts | *Juglans nigra* L. or *J. regia* L. | None stated |
| Woodruff, sweet | *Asperula odorata* L. | In alcoholic beverages only |
| Yarrow | *Achillea millefolium* L. | In beverages only; finished beverage thujone free |
| Yerba santa | *Eriodictyon californicum* (Hook, et Arn.) Torr | None stated |
| Yucca, Joshua-tree | *Yucca brevifolia* Engelm | None Stated |
| Yucca, Mohave | *Yucca schidigera* Roezl ex Ortgies *(Y. mohavensis* Sarg.) | None stated |

Souce: 21 CFR 172.515, 1991.

## 8.5. References

Lafon-Lafourcade, S. Wine and brandy. In Biotechnology, Vol. 5, *Food and Feed Production with Microorganisms*, G. Reed, ed., VCH, New York, 1983.

Hardwick, W. A. Beer, In Biotechnology, Vol. 5, *Food and Feed Production with Microorganisms*, G. Reed, ed., VCH, New York, 1983.

Bluhm, L. Distilled beverages, In Biotechnology, Vol. 5, *Food and Feed Production with Microorganisms*, G. Reed, ed., VCH, New York, 1983.

# PART C
# SAFETY, LAWS, AND REGULATIONS

# CHAPTER 9
# SAFETY AND SANITATION

Recently, two issues in public health have assumed increasing importance. One is the safety of a worker in a food processing plant and the other the safety of the food consumed by the public. This chapter provides six categories of data. The first one is directly concerned with the safety of workers in food processing plants. The remaining five provide reference data on the sanitation of food processing, which has a direct bearing on the safety of the finished products.

Data on safety of workers in a food processing plant cover the following:

1. Types of hazards and controls.
2. Occupational hazards associated with the processing of individual groups of products.
3. Agents harmful to a worker.
4. Exposure standards for chemicals in a food processing plant.

Data on sanitation cover the following:

1. Analyzing hazards associated with certain critical steps in food processing.
2. Cleaning and sanitizing operations.
3. Foods and their defects.
4. Foods and poisonous contaminants.
5. Legal applications of certain food additives.

The citation of government legal documents follows the standard format. For example, 21 CFR 172.515 refers to the U.S. Code of Federal Regulations, Volume 21, Part 172 and Section 515. All documents issued by federal agencies can be obtained from one or both of the following:

1. Government Printing Office, Washington, D.C.
2. The agency.

The abbreviations used include:

FDA: U.S. Food and Drug Administration.
GMP: Good Manufacturing Practice.
GRAS: General Recognized As Safe.
USDA: U.S. Department of Agriculture.

## 9.1. Safety of Workers in a Food Processing Plant

The U.S. Occupational Safety and Health Administration has issued numerous documents (e.g., interal and public reports, books, contract studies) on different aspects of occupational safety in food processing plants. Several tables in this chapter have been drawn from one such contract study: *Information Profiles on Potential Occupational Hazards: The Food Processing Industry*, U.S. Contract Number 210-78-0019, prepared for National Institute for Occupational Safety and Health (Food and Drug Administration and Centers for Disease Control) in February 1980 by the Center for Chemical Hazard Assessment, Syracuse Research Corporation, Merrill Lane, Syracuse, NY 13210. The corporation's file code is SRC TR-80-525. For reference purpose, this document will be cited as "Information Profiles."

**Table 9.1.1. Food Processing: Safety Hazards and Controls**

Meat Processing

Handling live, immobile, and slaughtered animals
- Potential hazard: strains, contusions
- Control: mechanization, training

Cutting and use of sharp tools
- Potential hazard: lacerations, loss of body members
- Control: protective clothing, gloves, guards, training

Wet flooring, platforms, decks
- Potential hazard: falls, sprains
- Control: Drains, shields

Steam
- Potential hazard: burns, scalds
- Control: Shields, reliefs

Animal-borne microorganisms
- Potential hazard: brucellosis, dermatitis
- Control: inspection

Dairy Processing

Handling churns, homogenizers, plasticizers, evaporators, freezers
- Potential hazard: lacerations, contusions, etc., from moving machine parts
- Control: guards, shields, layout, clothing, insulation

Handling in-process materials, products

**Table 9.1.1. (continued)**

Potential hazard: strains and contusions
  Control: mechanization, training

Food Preservation Processes

Cleaning, cutting, screening, peeling raw fruit and vegetables
  Potential hazard: lacerations, bruises, pinches in operating and maintaining the tools and machines
  Control: guards, shields, clothing, layout, training
Blanching, cooking, pasteurizing, curing, freezing products
  Potential hazard: burns, scalds, extreme temperatures
  Control: insulation, shields
  Storing, packaging, shipping
Potential hazard: cuts, bruises from packaging machines, sprains
  Control: guards, gloves, shields, mechanization

Grain Mill Processes

Operating and servicing breaking rolls, sieves, conveying and elevating equipment, man-lifts
  Potential hazard: bruises, contusions, pinches, lacerations, falls
  Control: guards, barriers, training
Handling feed, in-process material products
  Potential hazard: body strains
  Control: Mechanization, training
Dust, noise, vibration
  Potential hazard: Respiratory effects, hearing
  Control: Ventilation, insulation

Bakery Processes

Mixing, kneading, and forming machinery/conveyors
  Potential hazard: injuries from moving parts
  Control: guards, shields, layout
Baking ovens
  Potential hazard: burns, hot working environments
  Control: insulation, clothing, air conditioning
Handling in-process materials, products
  Potential hazard: strains
  Control: mechanization, training

Sugar and Confectionary Processes

Cleaning, grinding, shredding, and extraction machinery
Potential hazard: lacerations, contusions from moving machine parts
Control: guards, protective clothing, layout, ventilation, drains, overflows
Purifiers and chemicals
Potential hazard: lime, sulfur dioxide, chlorine dioxide, formaldehyde
Control: controls, ventilation
Concentrators, crystallizers
Potential hazard: burns, spills, leaks
Control: overflows, drains
Centrifuges, filters, dryers under operating and maintenance conditions
Potential hazard: lacerations, contusions, burns
Control: controls, maintenance, insulation/sizing

Fat and Oil Recovery Processes

Extracting oil and fat from animal and vegetable processes by steam distillation, mechanical expression, solvent extraction
Potential hazard: burns and scalds from steam and liquor leaks, spills; breaks and leaks from presses; vapors and gases from extractors
Control: insulation, barriers, layout, controls against overloads and spills, ventilation, and monitoring
Cleaning, grinding, shredding feeds
Potential hazard: machine injuries
Control: guards, training
Purification, hydrogenation processing
Potential hazard: chemical effects
Control: ventilation

Beverages

Handling in-processing materials
Potential hazard: body straining from lifting
Control: increasing mechanization, training
Broken glass
Potential hazard: lacerations
Control: protective clothing, gloves

Source: Adapted from "Information Profiles."

**Table 9.1.2. Occupational Hazards Associated with Agricultural Services: Migrant Laborers**

| Hazards | Item | Location |
|---|---|---|
| Disease | Site drainage, proximity to livestock feeding, sanitation | Labor camp |
| Safety and health | Adequate shelter | Labor camp |
| Disease | Adequate water supply | Labor camp |
| Disease | Toilet facilities | Labor camp |
| Disease | Sewage disposal | Labor camp |
| Disease | Washing/bathing facilities | Labor camp |
| Safety and health | Lighting | Labor camp |
| Health | Refuse disposal | Labor camp |
| Health | Approved kitchen facilities | Labor camp |
| Safety and health | First aid supplies and procedures | Labor camp |

Source: Adapted from *Industrial Data Report*, North Carolina Department of Labor, Occupational Safety and Health Administration, Raleigh, NC, 1976.

**Table 9.1.3 Occupational Hazards Associated with Agricultural Services: Contract Sorting, Grading, and Packing Fruit and Vegetables**

| Hazards | Item | Location |
|---|---|---|
| Major Hazards | | |
| Heavy boxes falling from lift trucks and transportation trucks crushing employees | Mechanical handling equipment, chocks, dock boards | Docks |
| Amputations and mangled limbs | Chains, sprockets, gears, V-belts, flat belts | Conveyors |
| Slips and falls. leg, back, and other body injuries | Water from floating systems | Floors |
| Body injuries, mashed fingers, mashed toes, bruises | Forklift trucks, hand trucks | Packing and shipping |

| | | |
|---|---|---|
| Other Hazards | | |
| Possible eye and puncture injuries | Box staple machines | Packing |
| Poor stacking, leaning boxes, mechanical equipment not working in aisles | Aisles marked, boxes stacked, blocked, interlocked | Warehouse |
| Water on floors, slips and falls; fruits and vegetables on floors, trips, and falls | Poor housekeeping | Throughout |
| Carbon monoxide | Monoxide fumes released from lift trucks | Throughout |

Source: Adapted from *Industrial Data Report*, North Carolina Department of Labor, Occupational Safety and Health Administration, Raleigh, NC, 1976.

**Table 9.1.4. Occupational Hazards Associated with Grain Mill Products**

| Hazards | Item | Location |
|---|---|---|
| Major Hazards | | |
| Oxygen-deficient atmospheres, combustible dusts, long high ladders. Dust may cause respiratory problems. | Silos, bins, hoppers | Storage |
| Other Hazards | | |
| Hearing loss | Noise | Throughout |

Source: Adapted from *Industrial Data Report*, North Carolina Department of Labor, Occupational Safety and Health Administration, Raleigh, NC, 1976.

**Table 9.1.5. Occupational Hazards Associated with Dairy Products**

| Hazards | Item | Location |
|---|---|---|
| Major Hazards | | |
| Falls, strains and foot injuries | Raw milk | Receiving |
| Falls, burns | Milk | Pasteurization |
| Falls, burns | Milk | Homogenization |
| Machine injuries (point of operation and shafts | Dairy products | Mixing or curing |
| Machine injuries, conveyors (point of operation-shafts and gears), cuts, dermatitis | Milk or dairy products | Bottling or packaging |
| Loss of hearing | Noise | Throughout |
| Falls, strains | Wet floors | Throughout |
| Eye injuries, hand injuries, electrocution | Shop equipment and hand tools | Maintenance |
| Other Hazards | | |
| Falling objects, strains, falls, cuts | Cartons, containers, bottles | Warehouse |
| Falls, strains | Dairy products | Distribution |

Source: Adapted from *Industrial Data Report*, North Carolina Department of Labor, Occupational Safety and Health Administration, Raleigh, NC, 1976.

**Table 9.1.6. Occupational Hazards Associated with Meat Products**

| Hazards | Item | Location |
|---|---|---|
| Major Hazards | | |
| Strains, falls and hand and arm amputations | Trolley conveyors | Hanging |
| Cuts, gun wounds | Hand knives and guns | Killing |
| Hand and arm amputations, cuts, stab wounds, tenosynovitis | Hand knives, saws; mechanical saws, knives, and cutters | Butchering |
| Falls, strains | Slick, wet floors | Throughout |

| | | |
|---|---|---|
| Electrocution | Electrical devices | Throughout |
| Amputations-hand, arm, and leg | Mechanical grinders, choppers, mixers, and conveyors | Mixing, grinding, chopping meat or ice |
| Burns, falls, noise | Steam vats, cookers, oven cookers, canners | Cooking, smoking, canning |
| Bacterial infection | Anthrax, Q fever, brucellosis, erysipelas | Throughout |
| Other Hazards | | |
| Falls, eye and skin burns, amputations | Chemical, steam cleaning of plant and equipment | Throughout |
| Infections and dermatitis | Wet, greasy conditions, bone and knife cuts | Throughout |
| Falls, falling objects, dermatitis | Improper clothing and protective equipment | Throughout |
| Falls, fire, explosion, asphyxiation | Ice, floors, ammonia, refrigerant | Freezers, refrigerators |
| Falls, collisions, strains | Forks and hand trucks | Shipping |
| Electrical shock, amputations | Electrical-mechanical devices | Maintenance |

Source: Adapted from *Industrial Data Report*, North Carolina Department of Labor, Occupational Safety and Health Administration, Raleigh, NC, 1976.

**Table 9.1.7. Occupational Hazards Associated with Beverages, Beer or Ale, Wines, Soft Drinks**

| Hazards | Item | Location |
|---|---|---|
| Major Hazards | | |
| Fire and explosion, respiratory illness | Grain dust | Grainery |
| Pedestrian and driver injuries | Lift trucks | Warehouse |
| Exposed points of operation, | Machinery | Manufacturing |

**Table 9.1.7. (continued)**

| Hazards | Item | Location |
|---|---|---|
| power transmission apparatus, rotating parts and in run ning nip points | | |
| Unconsciousness and hysteria | Carbon dioxide | Manufacturing |
| Unconsciousness and hysteria | Ammonia | Manufacturing |
| Burns and respiratory problems | Caustics | Manufacturing |
| Suffocation and death during cleaning | Tanks | Manufacturing |
| Fire and explosion, electrocution from faulty wiring and connections | Electricity | Plant |
| Falling on feet and legs, crushed hands | Large cartons and crates of beer, wine, and soft drinks | Loading and unloading docks |
| Flash fire and explosion from ignition of dusts by electrical sources | Combustible dust | Grain storage area |
| Items stacked too high, uneven, not interlocked; slick floors | Stacking heavy cartons and crates | Storage warehouse, lines, and conveyors |
| Danger of hands in V-belts, conveyors, pulleys, gears; burns from hot water; slips and falls | Mixers, hot water, electrical hazards, ungrounded machinery, wet floors | Work area |
| Openings, housekeeping, slips and falls | Floor holes | Receiving |
| Nip points | Screw conveyors | Receiving |
| Crushing by rolling or moving vehicles | Railroad cars | Receiving |
| Other Hazards | | |
| Body injury from falls | Slippery floors | Plant |
| Lifting and crushing | Heavy containers | Warehouse |
| Hearing loss | Noise | Manufacturing |

| | | |
|---|---|---|
| Exploding bottles, cuts and eye injuries | Bottles | Plant |
| Eye and skin irritation | Air cleaning | Plant |
| Unsafe operation of fork lifts (horseplay), unqualified operator, improper dock plates | Material handling equipment and tractor trailer | Shipping and receiving |
| Oxygen-deficient atmospheres, combustible dusts | Bins, hoppers, | Storage |
| Slipping, tripping | Housekeeping | Preparation area |
| Amputation and injured limbs from contact with gears, shafts, pulleys, belts, chains, and sprockets | Mechanical power transmission apparatus | Throughout plant |
| Falls | Man lifts | Throughout plant |
| Affects body system | Organic vapors | Throughout plant |
| Sprains, abrasions, puncture wounds, eye injuries | Improper tools, punches, chisels, etc. | Shop |

Source: Adapted from *Industrial Data Report*, North Carolina Department of Labor, Occupational Safety and Health Administration, Raleigh, NC, 1976.

**Table 9.1.8. Occupational Hazards Associated with Bread, Cake, Cookies, Crackers, and Related Products**

| Hazards | Item | Location |
|---|---|---|
| Major Hazards | | |
| Amputation and mangled limbs from contact with gears, shafts, pulleys, belts, chains, and sprockets | Mechanical power transmission apparatus | Throughout plant |
| Slipping, tripping, and falling hazards | Housekeeping | Throughout plant |
| Amputation and mangled limbs from nip points and sharp blades | Point of operation | Throughout plant |
| Electrocution from inadequate | Electrical | Throughout |

**Table 9.1.8. (continued)**

| Hazards | Item | Location |
|---|---|---|
| grounding | connections | plant |
| Burns from hot pipes and hot fat splashes; inhalation of carbon monoxide | Ovens and open fat kettles | Throughout plant |
| Other Hazards | | |
| Broken chain links and pulleys causing mixing bowls to fall on employees | Cranes and hoists | Mixing |
| Back strains and pulled muscles | Lifting | Mixing and baking areas |
| Explosion or fire | Combustible dusts | Storage |

Source: Adapted from *Industrial Data Report*, North Carolina Department of Labor, Occupational Safety and Health Administration, Raleigh, NC, 1976.

**Table 9.1.9. Occupational Hazards Associated with Canned and Preserved Vegetables**

| Hazards | Item | Location |
|---|---|---|
| Major Hazards | | |
| Falling objects from stacked materials and forklift trucks | Boxes, cases | Receiving, storing and shipping |
| Overturning, striking person, striking other vehicles | Forklift trucks | Cleaning area |
| Carbon monoxide poisoning, asphyxiation | Forklift trucks | Cleaning area |
| Burns from steam, hot kettles or caustics, foreign objects in eyes | Cleaning agent, dust and dirt | Cleaning area |
| Slips, trips, falls | Wet floors | Throughout plant |
| Unguarded nip points in conveyor | Conveyors | Throughout |

**Table 9.1.9. (continued)**

| Hazards | Item | Location |
|---|---|---|
| chains and belts and on chain and belt drives | | plant |
| Hearing loss | Noise | Throughout plant |
| Unguarded moving parts causing hand injuries | All machines | Canning room |
| Other Hazards | | |
| Back injuries, hernia, hand and foot injuries from manual loading pallets | Encasing and palletizing | Storage preparation |
| Fire, dermatitis, inhalation of toxic vapors | Glue | Labelling area |
| Inhalation of toxic vapors | Acetic acid | Seasoning area |

Source: Adapted from *Industrial Data Report*, North Carolina Department of Labor, Occupational Safety and Health Administration, Raleigh, NC, 1976.

**Table 9.1.10. Occupational Hazards Associated with Canned Seafoods**

| Hazards | Item | Location |
|---|---|---|
| Major Hazards | | |
| Back injuries and hernia from lifting and handling heavy objects | Boxes and baskets | Receiving and shipping |
| Cuts to fingers and hands from use of opening, scaling, and dressing tools | Hand tools | Processing room |
| Unguarded moving parts injure hands and arms | Canning machine | Canning room |
| Hearing loss | Noise | Canning room |
| Slips and falls, especially when handling heavy objects | Wet floors | Throughout plant |
| Burns to exposed portions of | Hot pipes | Cooking |

**Table 9.1.10. (continued)**

| Hazards | Item | Location |
|---|---|---|
| body on contact with hot metal insulation | and vessels | room (seldom used) |

Source: Adapted from *Industrial Data Report*, North Carolina Department of Labor, Occupational Safety and Health Administration, Raleigh, NC, 1976.

**Table 9.1.11. Occupational Hazards Associated with Chemical Products**

| Hazards | Item | Location |
|---|---|---|
| Major Hazards | | |
| Carbon monoxide poisoning | Toxic | Garages |
| Asphyxiation, damage to body systems, burns | Tanks or vats of toxic materials | Dye houses |
| Fire, explosion, damage to body systems | Toxic organic paint vapors | Paint |
| Skin burns, inhalation of caustics, carcinogens | Caustics, dyes | Drug room |
| Amputations and mangled limbs from contact with gears, shafts, pulleys, belts, chains, and sprockets | Mechanical power transmission apparatus (no lockouts) | Dye houses |
| Other Hazards | | |
| Skin irritation, burns | Toxic materials leaking from lines and fittings | Receiving and shipping |
| Slips and falls | Wet surfaces, ladders | Receiving and shipping |
| Falls, falling objects | Loading platform (dockboards) insecure storage | Warehouse |
| Burns | Ruptured containers | Warehouse |

| | | |
|---|---|---|
| Burns, skin irritation, dermatitis | Solvents used for cleaning | Garages |
| Dermatitis | Solvents used for cleaning | Paint booths |
| Eye injuries | Lack of eye protection | Dye houses |
| Slips and falls | Wet floors | Dye houses |
| Dermatitis | Chemical contact | Dye houses |
| Bronchial and lung irritation | Lack of respirators or inadequate ventilation | Dye houses |
| Burns | Exposed steam lines | Dye houses |

Source: Adapted from *Industrial Data Report*, North Carolina Department of Labor, Occupational Safety and Health Administration, Raleigh, NC, 1976.

**Table 9.1.12. Occupational Hazards Associated with Farm Machinery and Equipment**

| Hazards | Item | Location |
|---|---|---|
| Major Hazards | | |
| Mashed toes, hand laceration, falling on material, back injuries and hernia | Casting, sheetmetal, steel shapes, paints, lumber | Receiving |
| Amputation and lacerations | Points of operation such as grinding, sawing, milling, drilling, tapping, rolling, braking, shearing | Manufacturing, throughout |
| Amputation and laceration from contact with gears, pulleys and chairs | Mechanical power transmission apparatus | Throughout |
| Hearing loss | Noise | Throughout |
| Eye and face burns | Welding | Throughout |
| Explosion and fire | Spraying painting | Throughout |
| Other Hazards | | |
| Foreign bodies in eyes | Metal chips and welding slag | Drilling, cutting |

**Table 9.1.12 (continued)**

| Hazards | Item | Location |
|---|---|---|
| | | punching, finishing, assembly |
| Toxic fumes and gases from welding, fire hazard and toxic vapors from solvents and paint operations | Ventilation | Finishing, welding |
| Fire and explosion | Vapors and residue | Finishing |

Source: Adapted from *Industrial Data Report*, North Carolina Department of Labor, Occupational Safety and Health Administration, Raleigh, NC, 1976.

**Table 9.1.13. Occupational Hazards Associated with Public Warehousing**

| Hazards | Item | Location |
|---|---|---|
| Major Hazards | | |
| Shifting or falling material crushing | Mechanical handling equipment, wheel chocks, dockboards, stored material | Docks, bays, and racks |
| Vehicle-vehicle accidents, vehicle-employee accidents | Mechanical handling equipment | Throughout |
| Burns and explosions | Acid and fumes | Battery charging area |
| Respiratory ailments, headaches, dizziness, mental confusion, and death; explosions | Carbon monoxide | Forklift area |
| Slips and falls | Floors-wet and ice | Refrigerated storage area |
| Other Hazards | | |
| Being trapped by fire or harmful gases or fumes | Suitable numbers and marking | Exits |
| Fire and tripping | Housekeeping | Throughout |
| Mildly irritating to eyes and upper air passage | Freon | Refrigerated storage areas |

Source: Adapted from *Industrial Data Report*, North Carolina Department of Labor, Occupational Safety and Health Administration, Raleigh, NC, 1976.

**Table 9.1.14 Occupational Hazards Associated with Groceries and Related Products**

| Hazards | Item | Location |
|---|---|---|
| Major Hazards | | |
| Falling objects, striking employees | Stored goods | Racks in warehouse and aisle |
| Dizziness, headaches, and death; explosions | Carbon monoxide | Fuel-powered fork trucks |
| Fire and fumes and/or Vapors | Exits | Throughout |
| Other Hazards | | |
| Falling objects, crushing overturned equipment | Mechanical handling equipment, chocks and dockboards | Docks |
| Fire, tripping and falling | Aisles, clearance and markings, housekeeping | Docks |
| Burns, toxic fumes | Acids | Battery charging area |
| Damage to body systems, respiratory ailments | Refrigerant gas | Refrigerated areas |

Source: Adapted from *Industrial Data Report*, North Carolina Department of Labor, Occupational Safety and Health Administration, Raleigh, NC, 1976.

**Table 9.1.15 Occupational Hazards Associated with Retailed Bakery Products**

| Hazards | Item | Location |
|---|---|---|
| Major Hazards | | |
| Amputation and mangled limbs from pulleys and nip point | V-belt drive | Air compression area |
| Slipping and falling | Dockboard | Receiving area |
| Cuts from broken glass; strains from bending and lifting | Showcases | Sales floor |
| Amputated fingers and cuts | Bread slicer | Sales counter |
| Other Hazards | | |
| Freezing and suffocation from entrapment | Chill boxes | Refrigeration area |
| Slipping and tripping | Floors | Throughout store |

Source: Adapted from *Industrial Data Report*, North Carolina Department of Labor, Occupational Safety and Health Administration, Raleigh, NC, 1976.

**Table 9.1.16 Occupational Hazards Associated with Eating Places**

| Hazards | Item | Location |
|---|---|---|
| Major Hazards | | |
| Falling on feet and legs, mashed fingers | Large cartons of food products | Unloading dock |
| Crushing by falling objects, falls and slips | Stacking heavy cartons, conveyors | Storage warehouse, refrigerator lines and conveyors |
| Amputation of fingers, hands from saws and knives contact with V-belts, pulleys, gears, dermatitis | Saws, knives, mixers, slicers, ovens | Preparation area |

| | | |
|---|---|---|
| Burns from water and pipes, electrical shock, severe cuts from knives, falls, burns, dermatitis | Hot water, electrical hazards, ungrounded machinery, broken glasses, dishes, wet and slick floors, hot grease | Kitchen |
| Other Hazards | | |
| Tripping, slipping, falls | Housekeeping | Warehouse storage area, preparation area, kitchen |

Source: Adapted from *Industrial Data Report*, North Carolina Department of Labor, Occupational Safety and Health Administration, Raleigh, NC, 1976.

**Table 9.1.17. Potential Harmful Effects to Workers from Agents Present or Used in a Food Processing Plant**

| Agents | A Worker May Suffer Dermatitis (D), or Systemic (S) or Local (L) Toxic Effect in a Plant That Processes | | | | | | | |
|---|---|---|---|---|---|---|---|---|
| | Meat | Dairy Products | Preserved Foods | Sugar & Grain Milling | Bakery Products | Confect-ioneries | Fats & Oils | Beverages |
| Biological | | | | | | | | |
| Bacteria | D, S | D | D | | | | | |
| Fungi | D, S | D | D | | | | | |
| | L | | | | | | | |
| Mites | D | | | | | | | |
| "Monilia" | | | D | | | | | |
| Parasites | D | | D | D | | | | |
| Viruses | D | D | | | | | | |
| Physical | | | | | | | | |
| Cold | D, S | | | | | | | |
| Heat | L | | L | | D, L | | L | |
| Infrared radiation | | S, L | S, L | | | | | |
| Ionizing radiation | | D, S | | | | | | |

| | | | | | | | | |
|---|---|---|---|---|---|---|---|---|
| Chemical | | | | | | | | |
| Antibiotics | D | D | | | | | | |
| Bleaches | | | D | D | | | | |
| Brine | D | | D | L | | | | |
| Detergents | D | D | | | | | | |
| Dusts | | | D, S, L | D, S, L | | | | |
| Dyes | | D | | | D | | | |
| Enzymes | D | | D | | | | | |
| Moisture | D | | D | | D | | | |
| Organic | | | D | | | D | | |
| Pesticides | | | | D | | | | |
| Waxes | | | D | | | | | |
| Specific chemicals | | | | | | | | |
| Acetonitrile | | | S, L | | | | S, L | |
| Amyl alcohol | | | | | | | S, L | |
| Benzoly peroxide | | | | | D | | | |
| Carbon dioxide | | | | S, L | | | | S, L |
| Carbon disulfide | | | | | | | S, L | |
| Chloride, organic | | | | | | | S, L | |
| Dichloroethyl ether | | | | | | | S, L | |
| Dioxane | | | | | | | S, L | |
| Ethyl alcohol | | | | | | | | S, L |
| Ethylene glycol | | | | | | | S, L | |
| Ethylene oxide | | | | S, L | | | S, L | |
| Formates | | | | S, L | | | | |

**Table 9.1.17. (continued)**

| Agents | Meat | Dairy Products | Preserved Foods | Sugar & Grain Milling | Bakery Products | Confect-ioneries | Fats & Oils | Beverages |
|---|---|---|---|---|---|---|---|---|
| | A Worker May Suffer Dermatitis (D), or Systemic (S) or Local (L) Toxic Effect in a Plant That Processes | | | | | | | |
| Hydrocarbon | | | | | | | S | |
| Hydrogen chloride | | | S, L | | | | | |
| Hydorgen sulfide | S, L | | | | | S, L | | S, L |
| Lime | | | | | | S, L | | |
| Propyl alcohol | | | L | | | | | |
| Sulfur chloride | | | L | | | L | | |
| Sulfur dioxide | | | S, L | | | | | S, L |
| Sulfurous acid | | | S, L | | | | | |

Source: Adapted from "Information Profiles."

**Table 11.1.18 Exposure Standards for Chemicals Associated with the Food Processing Industry**

| Chemical | TLV | | OSHA | | NIOSH |
|---|---|---|---|---|---|
| | ppm | $mg/m^3$ | ppm | $mg/m^3$ | |
| Acetonitrile | 40 | 70 | 40 | 70 | |
| Benzoyl peroxide | -- | 5 | - | 5 | 0.5 ppm |

| | | | | | |
|---|---|---|---|---|---|
| Carbon dioxide | 5000 | 9000 | 5000 | 9000 | 10000 ppm, 30000 ppm 10 min ceiling |
| Carbon disufide | 20 | 60 | 20 30C 100/30 min | 360 | 1 ppm 10 ppm, 15 min |
| Dichloroethyl | 5 | 30 | 15C | 90C | |
| Dioxane | 50 | 180 | 100 | 360 | 1 ppm, 30 min ceiling |
| Ethyl alcohol | 1000 | 1900 | 1000 | 1900 | |
| Ethylene glycol | | | | | |
| Particulate | -- | 10 | | | |
| Vapor | 100 | 260 | | | |
| Ethylene oxide | 50 | 90 | 50 | 90 | |
| Hydrogen chloride | 5C | 7C | 5C | 7C | |
| Hydrogen sulfide | 10 | 15 | 20C 50/ 10 min | -- | 15 mg/m$^3$, 10 min ceiling |
| Propyl alcohol | 200 | 500 | 200 | 500 | |
| Sulfur dioxide | 5 | 13 | 5 | 13 | 0.5 ppm |
| Sulfur monochloride | 1 | 6 | 1 | 6 | |

Source: Adapted from "Information Profiles." Consult the original document for more details.
Note: All values listed are 8-hr time-weighted average concentrations except as indicated.
Abbreviations: TLV = Threshold Limit Values; OSHA = Occupational Safety and Health Administration; NIOSH = National Institute of Occupational Safety and Health; C = Acceptable ceiling concentration.

## 9.2. Hazard Analysis and Critical Control Point System

On November 1989, the U.S. National Advisory Committee on Microbiological Criteria for Foods (U.S. Department of Agriculture) adopted the document *HACCP Principles for Food Production*. This document defines hazard analysis and critical control point (HACCP) as a systematic approach to be used in food production as a means to assure food safety. Seven basic principles underlie the concept. These principles include an assessment of the inherent risks that may be present from harvest through ultimate consumption. Six hazard characteristics and a ranking schematic are used to identify those points throughout the food product and distribution system, whereby control must be exercised in order to reduce or eliminate potential risks. A guide for HACCP plan development and critical control point (CCP) identification is noted. Further, the document points out the additional areas that are to be included in the HACCP plan–the need to establish critical limits that must be met at each CCP, appropriate monitoring procedures, corrective action procedures to be taken if a deviation is encountered, recordkeeping, and verification activities

This section presents a slightly modified version of this document which will be referred to as "HACCP Document." This document can be obtained from the U.S. Department of Agriculture, Washington, D.C.

**Table 9.2.1. Contents of the Document on HACCP**

| |
|---|
| Preamble |
| Definitions |
| Purpose and principles |
| Explanation of principles |
| Principle No. 1 |
| Principle No. 2 |
| Principle No. 3 |
| Principle No. 4 |
| Principle No. 5 |
| Principle No. 6 |
| Principle No. 7 |
| Implementation guide |

Source: The "HACCP Document."

**Table 9.2.2. Preamble**

The National Advisory Committee on Microbiological Criteria for Foods (Committee) endorses the hazard analysis and critical control point (HACCP) System as an effective and rational approach to the assurance of food safety. In the application of HACCP, the use of microbiological testing is seldom an effective means of monitoring critical control points (ccp) because of the time required to obtain results. In most instances, monitoring of CCP can best be accomplished through the use of physical and chemical tests, and through visual observations. Microbiological criteria do, however, play a role in verifying that the overall HACCP system is working.

The Committee believes that the HACCP principles should be standardized to create uniformity in its work, and in training and applying the HACCP system by industry and regulatory authorities. In accordance with the National Academy of Sciences recommendation, the HACCP system must be developed by each food establishment and tailored to its individual product, processing, and distribution conditions.

Source: The "HACCP Document."

**Table 9.2.3. Definitions**

Continuous monitoring: uninterrupted recording of data such as a recording of temperature on a strip chart

Control point: any point in a specific food system where loss of control does not lead to an unacceptable health risk

Critical control point: any point or procedure in a specific food system where loss of control may result in an unacceptable health risk

Critical defect: a defect that may result in hazardous or unsafe conditions for individuals using and depending upon the product

Critical limit: one or more prescribed tolerances that must be met to insure that a critical control point effectively controls a microbiological health hazard

Deviation: failure to meet a required critical limit for a critical control point

HACCP plan: the written document which delineates the formal procedures to be followed in accordance with these general principles

HACCP system: the result of the implementation of the HACCP principles

Hazard: any biological, chemical, or physical property that may cause an unacceptable consumer health risk

Monitoring: a planned sequence of observations or measurements of critical limits designed to produce an accurate record and intended to insure that the critical limit maintains product safety

**Table 9.2.3. (continued)**

Risk: an estimate of the likely occurrence of a hazard or danger
Risk category: one category: one of six categories prioritizing risk based on food hazards
Sensitive ingredient: any ingredient historically associated with a known microbiological hazard
Significant risk: posing moderate likelihood of causing an unacceptable health risk
Spot check: supplemental tests performed on a random basis.
Verification: methods, procedures and tests used to determine whether the HACCP system is in compliance with the HACCP plan.

Source: The "HACCP Document."

**Table 9.2.4. Purpose and Principles**

HACCP is a systemic approach to food safety, consisting of the seven following principles:

1 Assess hazards and risks associated with growing, harvesting, raw materials and ingredients, processing, manufacturing, distribution, marketing, preparation, and consumption of the food
2 Determine CCP required to control the identified hazards.
3 Establish the critical limits that must be met at each identified CCP.
4 Establish procedures to monitor CCP.
5 Establish corrective action to be taken when there is a deviation identified by monitoring a CCP.
6 Establish effective recordkeeping systems that document the HACCP plan.
7 Establish procedures for verification that the HACCP system is working correctly.

Source: The "HACCP Document."

**Table 9.2.5. Principle No. 1: Definition and Description**

Definition

Assess hazards associated with growing, harvesting, raw materials and ingredients, processing, manufacturing, distribution, marketing, preparation, and consumption of the food.

Description

It provides for a systematic evaluation of a specific food and its ingredients or components to determine the risk from hazardous microorganisms or their toxins. Hazard analysis is most useful for guiding the safe design of a food product and defining the CCPs that eliminate or control hazardous microorganisms or their toxins at any point during the entire production sequence. The hazard assessment is a two-part process consisting of ranking a food according to six hazard characteristics, followed by the assignment of risk category which is based upon the ranking.

Ranking according to hazard characteristics is based on assessing a food in terms of (a) whether the product contains microbiologically sensitive ingredients, (b) whether the process does not contain a controlled processing step that effectively destroys harmful microorganisms, (c) whether there is significant risk of postprocessing contamination with harmful microorganisms or their toxins, and (d) whether there is substantial potential for abusive handling in distribution or in consumer handling or preparation that could render the product harmful when consumed, or (e) whether there is no terminal heat process after packaging or when cooked in the home. Ranking according to these six characteristics results in the assignment of risk categories based on how many of the characteristics are present.

The risk categories are utilized for recognizing the hazard risk for ingredients and how they must be treated or processed to reduce the risk for the entire food production and distribution sequence.

The hazard assessment procedure is ideally conducted after developing a working description of the product, establishing the types of raw materials and ingredients required for preparation of the product, and preparing a diagram for the food production sequence. The two-part assessment of hazard analysis and assignment of risk categories is conducted according to the next two tables.

Source: The "HACCP Document."

**Table 9.2.6. Principle No. 1: Hazard Analysis and Assignment of Risk Categories**

Ranking Food According to Hazard Characteristics

Explanation

Rank the food according to hazard characteristics A through F, using a plus (+) to indicate a potential hazard. The number of pluses will determine the risk category. A model diagram outlining this concept is provided in Table 9.2.7. As indicated, if the product falls under hazard class A, it should automatically be considered Risk Category VI.

Hazards Ranking

Hazard A: A special class that applies to nonsterile products designated and intended for consumption by at-risk populations, e.g., infants, the aged, the infirm, or immunocompromised individuals

Hazard B: The product contains "sensitive ingredients" in terms of microbiological hazards.

Hazard C: The process does not contain a controlled processing step that effectively destroys harmful microorganisms.

Hazard D: The product is subject to recontamination after processing before packaging.

Hazard E: There is substantial potential for abusive handling in distribution or in consumer handling that could render the product harmful when consumed.

Hazard F: There is no terminal heat process after packaging or when cooked in the home.

Note: Hazards can also be stated for chemical or physical hazards, particularly if a food is subject to them.

Assignment Of Risk Category (Based On Ranking By Hazard Characteristics)

Category VI: A special category that applies to nonsterile products designated and intended for consumption by at-risk populations, e.g., infants, the aged, the infirm, or immunocompromised individuals. All six hazard characteristics must be considered.

Category V: Food products subject to all five general hazard characteristics; hazard class B, C, D, E, F

Category IV: Food products subject to four of the general hazard characteristics
Category III: Food products subject to three of the general hazard characteristics
Category II: Food products subject to two of the general hazard characteristics
Category I: Food products subject to one of the general hazard characteristics
Category O: Hazard class-No hazard

Note: Ingredients are treated in the same manner in respect to how they are received at the plant, before processing. This permits determination of how to reduce risk in the food system.

Source: The "HACCP Document."

**Table 9.2.7. Principle No. 1: A Model Chart To Assess A Food by Hazard Characteristic and Risk Category**

| Food Ingredient or Product | Hazard Characteristics (A, B, C, D, E, F) | Risk Category (VI, V, IV, III, II, I, 0) |
|---|---|---|
| T | A + (special category)[a] | VI |
| U | Five +'s (B through F) | V |
| V | Four +'s (B through F) | IV |
| W | Three +'s (B through F) | III |
| X | Two +'s (B through F) | II |
| Y | One + (B through F) | I |
| A | No +'s | 0 |

Source: The "HACCP Document."

[a]Hazard characteristic A automatically is risk category VI, but any combination of B through F may also be present. It is recommended that a chart be utilized that provides assessment of a food by hazard characteristic and risk category. A format for this chart is as shown in the table.

**Table 9.2.8 Principle No. 2: Definition and Description**

Definition

Principle No. 2 determines CCP required to control the identified hazards.

Description

A CCP is defined as any point or procedure in a specific food system where loss of control may result in an unacceptable health risk. A CCP must be established where control can be exercised. All hazards identified by the hazard analysis must be controlled at some point(s) in the food production sequence, from harvesting and growing raw materials to the ultimate consumption of the food.

CCPs are located at any point in a food sequence where hazardous microorganisms need to be destroyed or controlled. For example, a specified heat process, at a given time and temperature to destroy a specified microbiological pathogen, is a CCP. Likewise, refrigeration required to prevent hazardous organisms from growing or the adjustment of a food to a pH necessary to prevent toxin formation is a CCP.

Types of CCP may include, but are not limited to: cooking, chilling, sanitizing, formulation control, prevention of cross contamination, employee hygiene, and environmental hygiene.

CCPs must be carefully developed and documented. In addition, they must be used only for purposes of product safety. They should not be confused with control points that do not control safety. For comparison, a control point is defined as any point in a specific food system where loss of control does not lead to an unacceptable health risk.

Source: The "HACCP Document."

**Table 9.2.9. Principle No. 3: Definition and Description**

Definition

Principle No. 3 establishes the critical limits that must be met at each identified CCP.

Description

A critical limit is defined as one or more prescribed tolerances that must be met to insure that a CCP effectively controls a microbiological health hazard. There may be more than one critical limit for a CCP. If any one of those critical limits is out of control, the CCP will be out of control and a potential hazard can exist. The criteria most frequently utilized for critical limits are temperature, time, humidity, moisture level ($A_w$), pH, titratable acidity, preservatives, salt concentration, available chlorine, viscosity, and in some cases sensorial information such as texture, aroma, and visual appearance. Many different types of limit information may be needed for safe control of a CCP.

For example, the cooking of meat patties should be designed to eliminate the most heat-resistant vegetative pathogen that could reasonably be expected to be in the product. The critical limits must be specified for temperature, time, and meat patty thickness. Technical development of these critical limits requires accurate information on the probable maximum numbers of these microorganisms in the meat, use of additional ingredients and the potential for recontamination.

The relationship between the CCP and its critical limits for the meat patty example is shown as follows:

| Critical Control Point | Critical Limits |
|---|---|
| Meat patty cooked to destroy the most heat-resistant pathogen, based on lethality tests. The minimum lethal cook will usually be designated "to reach an internal patty temperature of x for time y." | Minimum operating temperature of cooker to achieve microbiological lethality at center of coldest patty. Time to achieve lethality (belt speed expressed at rpm). Patty thickness. Other possible critical limits:<br>* oven humidity    * cooker sanitation<br>* patty composition    * etc. |

Note: This example illustrates that the type and number of critical limits will vary depending on the type of cooking system and equipment used and for meat patties.

Source: The "HACCP Document."

**Table 9.2.10. Principle No. 4: Definition and Description**

Definition

Principle No. 4 establishes procedures to monitor CCPs.

Description

Monitoring is the scheduled testing or observation of a CCP and its limits. Monitoring results must be documented. From the monitoring standpoint, failure to control a CCP is a critical defect.

A critical defect is defined as a defect that may result in hazardous or unsafe conditions for individuals using and depending upon the product. Because of the potentially serious consequences of a critical defect, monitoring procedures must be extremely effective.

Ideally, monitoring should be at the 100% level. Continuous monitoring is possible with many types of physical and chemical methods. For example, the temperature and time for the scheduled thermal process of low-acid canned foods is recorded continuously on temperature recording charts. If the temperature falls below the scheduled temperature or the time is insufficient, as recorded on the chart, the retort load is restrained as a process deviation. Likewise, pH measurement may be done continually in fluids or by testing of a batch before processing. There are many ways to monitor CCP limits on a continuous or batch basis and record the data on charts. The high reliability of continuous monitoring is always preferred when feasible. It requires careful calibration of equipment.

When it is not possible to monitor a critical limit on a full-time basis, it is necessary to establish that the monitoring interval will be reliable enough to indicate that the hazard is under control. Statistically designed data collection systems or sampling systems lend themselves to this purpose. However, statistical procedures are most useful for measuring and reducing the variation in food formulations, manufacturing equipment, and measuring devices. Thus, they increase the reliability of the system.

When using statistical process control, it is important to recognize that there is no tolerance for exceeding a critical limit. For example, when a pH of 4.6 or less is required for product safety, no single product unit may have a pH above 4.6. To compensate for variation, the maximum of the product may be targeted at a pH below 4.6. Statistical process control can be applied to understand variation in the system and to assure that no unit exceeds a pH of 4.6. Statistical audits can be based on this concept.

Most monitoring procedures for CCP will need to be done rapidly because they relate to on-line processes and there will not be time for lengthy analytical testing. Microbiological testing is seldom effective for monitoring CCP due to its time-consuming nature. Therefore, physical and chemical measurements are preferred because they may be done rapidly and can indicate microbiological control of the process.

Measurements

Physical and chemical measurements that may be utilized for monitoring include:

- Temperature
- Time
- pH
- Sanitations at CCP
- Specific preventive measures for cross contamination
- Specific food handling procedures
- Moisture level

Spot checks are useful for supplementing the monitoring of certain CCP and their respective limits. They may be used to check incoming precertified ingredients and to assess equipment and environmental sanitation, airborne contamination, cleaning and sanitizing of gloves, and any place where followup is needed. Spot checks may consist of physical and chemical tests and, where needed, microbiological tests.

With certain foods, microbiologically sensitive ingredients, or imports, there may be no alternative to microbiological testing. However, a sampling frequency that is adequate for reliable detection of low levels of pathogens is seldom possible because of the large number of samples needed. For this reason, microbiological testing has limitations in a HACCP system but is valuable as a means of establishing and randomly verifying the effectiveness of control at CCP (challenge tests, spot checking, or for troubleshooting.)

All records and documents associated with CCP monitoring must be signed by the person doing the monitoring and signed by a responsible official of the company.

---

Source: The "HACCP Document."

**Table 9.2.11. Principle No. 5: Definition and Description**

Definition

Principle No. 5 establishes corrective action to be taken when there is a deviation identified by monitoring of a CCP.

Description

Actions taken must eliminate the actual or potential hazard that has been created by deviation from the HACCP plan and assure safe disposition of the product involved. Because of the variations in CCP for different foods and the diversity of possible deviations, specific corrective actions must be developed for each CCP in the HACCP plan. The actions must demonstrate that the CCP has been brought under control. Deviation procedures must be documented in the HACCP plan and agreed to by the appropriate regulatory agency prior to approval of the plan.

Should a deviation occur, the plant will place the product on hold pending completion of appropriate corrective actions and analyses. In instances where it may be difficult to determine the safety of the product, then the testing and final disposition must be agreed to by the government. In instances not associated with safety, government consultation is not required.

Identification of deviant lots and corrective actions taken to assure safety of these lots must be noted in the HACCP record and remain on file for a reasonable period after the expiration date or expected shelf life of the product.

Source: The "HACCP Document."

**Table 9.2.12. Principle No. 6: Definition and Description**

Definition

Principle No. 6 establishes effective record-keeping systems that document the HACCP plan.

Description

The HACCP plan must be on file at the food establishment. Additionally, it is to include documentation relating to CCP and any action on critical deviations and disposition of product. These materials are to be made available to government inspectors upon request. The HACCP plan clearly designates records that will be available for government inspection. Certain records that deal with the functioning of the HACCP system and proprietary information are not necessarily available to regulatory agencies.

Generally, the types of records utilized in the total HACCP system will include those listed in Table 9.2.13.

Note: Only those records pertaining to CCP must be made available to regulatory agencies.

Source: The "HACCP Document."

**Table 9.2.13 Principle No. 6: Record Keeping System**

Records for Ingredients

1. Supplier certification documenting compliance with processor's specifications
2. Processor audit records verifying supplier compliance
3. Storage temperature record for temperature-sensitive ingredients
4. Storage time records of limited shelf life ingredients

Records Relating To Product Safety

1. Sufficient data and records to establish the efficacy of barriers in maintaining product safety
2. Sufficient data and records establishing the safe shelf life of the product
3. Documentation of the adequacy of the processing procedures from a knowledgeable process authority

Records Relating to Processing

1. Records from all monitored CCP
2. System records verifying the continued adequacy of the processes

Records Relating To Packaging

1. Records indicating compliance with specifications of packaging materials
2. Records indicating compliance with scaling specifications

**Table 9.2.13 (continued)**

Records Relating to Storage and Distribution

1. Temperature records
2. Records showing no product shipped after shelf life date on temperature sensitive products

Other Records

1. Deviation file
2. Modification to the HACCP plan file indicating approved revisions and changes in ingredients, formulations, processing, packaging, and distribution control, as needed

Source: The "HACCP Document."

**Table 9.2.14. Principle No. 7: Definition and Description**

Definition

Principle No. 7 establishes procedures for verification that the HACCP system is working correctly.

Description

Verification consists of methods, procedures, and tests used to determine that the HACCP system is to compliance with the HACCP plan. Both the producer and the regulatory agency have a role in verifying HACCP plan compliance. Verification confirms that all hazards were identified in the HACCP plan when it was developed. Verification measures may include physical, chemical, and sensory methods and testing for conformance with microbiological criteria when established.

Activities

Examples of verification activities include but are not limited to:

1. Establishment of appropriate verification inspection schedules
2. Review of the HACCP plan
3. Review of the CCP records
4. Review deviations and dispositions

5. Visual inspections of operations to observe whether CCP are under control
6. Random sample collection and analysis
7. Written record of verification inspections, which certifies compliance with the HACCP plan or deviations from the plan and the corrective actions taken

Verification Inspections

Verification inspections should be conducted when:

1. Routinely, or on an unannounced basis to assure selected CCP are under control
2. It is determined that intensive coverage of a specific commodity is needed because of new information on food safety issues requiring assurance that the HACCP plan remains effective
3. Foods produced have been implicated as a vehicle of a foodborne disease
4. Requested on a consultative basis or established criteria have not been met

Verification Inspection Reports

Elements that must be included in verification inspection reports:

1. Existence of an approved HACCP plan and designation of a person(s) responsible for administering and unpdating the HACCP plan
2. All records and documents associated with CCP monitoring must be signed by the person monitoring and the approved by a responsible official of the firm.
   a. Direct monitoring data of the CCP while in operation
   b. Certification that monitoring equipment is properly calibrated and in working order
   c. Deviation procedures
3. Any sample analysis for attributes confirming that CCP are under control is to include physical, chemical, microbiological or organoleptic methods.

Source: The "HACCP Document."

**Table 9.2.15. HACCP System: Implementation Guide**

A guide for HACCP plan implementation for a specific food includes:

1. Describe the food and its intended use.
2. Develop a flow diagram for the production of the food.
3. Perform a hazard assessment (Principle 1).
   a. Ingredients prior to any processing step.
   b. End product
4. Select the CCP (Principle 2).
   a. Enter on flow diagram in numerical order.
   b. List CCP number and description.
5. Establish critical limits (Principle 3).
6. Establish monitoring requirements (Principle 4).
7. Establish corrective action to be taken when there is a deviation identified by monitoring of a CCP (Principle 5).
8. Establish effective recordkeeping systems that document the HACCP plan (Principle 6).
9. Establish procedures for industrial and governmental verification that the HACCP system is working properly. Verification measure may include physical, chemical, and sensory methods and, when needed, establishment of microbiological criteria (Principle 7).

Source: The "HACCP Document."

## 9.3. Cleaning and Sanitizing Operations in a Food Service or Food Processing Establishment

Effective cleaning of equipment in the food industry reduces chances for contamination of food during preparation/processing, storage, and serving. Clean equipment is essential for subsequent sanitizing treatment which minimizes the chances of transmitting disease organisms to the consumer. Cleaning minimizes attraction of vermin, increases life of equipment, improves employee morale and efficiency, and is important from other esthetic considerations.

In 1983, H. Haverland, Director of State Training Branch of the U.S. Food and Drug Administration, presented a series of lectures on the topic as part of a training course. The information in this section has been adapted from the lectures.

**Table 9.3.1. Definitions of Terms**

Cleaning: A process that will remove soil and prevent accumulation of food residues which may decompose or support the growth of disease causing organisms or the production of toxins.

Sanitizing: A process that destroys disease-causing organisms which may be present on equipment and utensils after cleaning. Chemical sanitizer used shall meet the requirements of 21 CFR 178.1010.

Soil: Soil has been appropriately defined as "matter out of place"; grease on a gear, for example, is a lubricant, but that same grease on a table top becomes soil.

**Table 9.3.2. Water Used**

The primary constituent of all food processing plant cleaners is water. Basic water requirements common to all food processing operations are that it must be free from disease-producing organisms, toxic metal ions, and objectionable odors and taste. Pure water presents no problems, but no food processing establishment has an ideal water supply. Therefore, the cleaning compounds must be tailored to the individual water supply and type of operation.

Water impurities affecting cleaning:

A. Suspended matter must be kept to a minimum to avoid deposits on clean equipment surfaces.

B. Soluble iron and manganese salts: Concentrations above 0.3 ppm will cause colored deposits on equipment surfaces. Suspended matter and soluble iron and manganese can be removed only by treatment.

C. Water hardness: Water hardness due to salts of calcium and magnesium presents a major problem in the use of cleaners by reducing effectiveness and by forming surface deposits.

  1. Types of hardness
     a. Carbonate hardness (formerly called temporary hardness) is due to calcium and magnesium carbonates and bicarbonates and can be removed by heating.
     b. Noncarbonate hardness (formerly called permanent hardness) is due to calcium sulfates, calcium chloride, magnesium sulfate, and magnesium chloride, which cannot be removed by heating.
  2. The hardness of waters varies considerably from place to place. In general, surface waters are softer than ground waters.

**Table 9.3.2. (continued)**

3. Hardness classification

| Class | Grains per Gallon |
|---|---|
| Soft | 0-3.5 |
| Moderately hard | 3.5-7.0 |
| Hard | 7.0-10.5 |
| Very Hard | Over 10.5 |

4. In terms of savings and effectiveness, the ideal water for general food-plant cleaning purposes is completely softened water. However, a degree of hardness is often preferable in some food processing, such as canning of certain vegetables.

Note: 1 grain per gallon = 17.1 ppm

**Table 9.3.3. Fundamentals of Cleaning**

A. To bring a cleaning agent into intimate contact with the soil
B. To displace the soil from the surface to be cleaned
C. To disperse the soil in the solvent
D. To prevent redepositing of the dispersed soil back onto the clean surface

**Table 9.3.4. Definitions of Functions Performed by Cleaning Agents**

Deflocculation or dispersion: the action in which groups or clumps of particles are broken up into individual particles and spread out, suspended, in the solution

Dissolving: the reaction that produces water-soluble materials from the water-insoluble soil

Emulsification: a process wherein fats are broken up into tiny globules and are suspended in the cleaning solution

Penetration: the action of liquids entering porous materials through cracks, pinholes, or small channels

Peptization: physical formation of colloidal solutions from partially soluble materials

Saponification: action of alkali on fats resulting in the formation of soap

Suspension: the action in which insoluble particles are held in solution and not allowed to settle out onto the utensils

Rinsability: the ability to break the surface tension of the water in the solution and permit the utensil to drain dry

Water softening

1. Precipitation: soften water by precipitating out the hardness
2. Sequestration: the action of an inorganic compound attaching itself to the water hardness particles and inactivating them so they do not combine with other material in the water and precipitate out
3. Chelation: the same as sequestration except that an organic compound is used

Wetting: Action of water in contacting all soil; helps to reduce surface tension (wetting agents usually do a good job of emulsification)

Synergism: A chemical used as a builder with a soap or detergent, results in a detergency that is greater than the total detergency of the chemical or the soap used independently

**Table 9.3.5. Factors Affecting Cleaning Efficiency**

The proper selection of the *cleaner* to do the job, *concentration* of cleaner, *time* in contact with the surface, *force* or *velocity*, and *temperature* are all important to have good cleaning. Each of the above can be varied independently to adjust a cleaning operation to a particular problem or plant operating practice. These factors will vary from hand cleaning to circulation cleaning and will depend upon the type and condition of soil to be removed. The functions of various factors are:

A. The cleaner
B. The temperature: Increasing temperature: (a) decreases the strength of bond between soil and surface, (b) decreases viscosity, (c) increases solubility of soluble materials, and (d) increases chemical reaction rate.
C. Velocity or force: In hand cleaning force is applied by "elbow grease," whereas fluid flow is used to apply cleaning force in clean in place (CIP) systems. Increased turbulence provides more effective removal of film from surfaces. However, efficiency is less effected by turbulence as the physical-chemical effectiveness of the detergent increases. CIP cleaning velocities of 5 ft/sec are recommended to ensure adequate turbulence.
D. Time: All other factors remaining constant, cleaning efficiency can be increased by utilizing longer times.
E. Concentration: Increase in concentration increases the reaction rate. It is the least effective variable to change in cleaning (time, temperature, and velocity are other variables).

**Table 9.3.6. Desirable Properties of Good Cleaners**

1. Quick and complete solubility
2. Good wetting or penetrating action
3. Dissolving action on food solids
4. Emulsifying action on fat
5. Deflocculating, dispersing, or suspending action
6. Good rinsing properties
7. Complete water softening power
8. Noncorrosive on metal surfaces
9. Germicidal action
10. Economy in use

**Table 9.3.7. Cleaning Methods**

A. Removal of gross food particles
   Loose material should be removed before the application of cleaning solutions. This may be accomplished by flushing the equipment surface with cold or warm water under moderate pressure. Very hot water or steam should not be used because it may make cleaning more difficult.
B. Application of cleaning compounds
   There are many methods of subjecting the surfaces of equipment to cleaning compounds and solutions. Effectiveness and the economy of the method generally dictates its use.
   1. Soaking: Small equipment or fittings or valves may be immersed in cleaning solutions in a sink while larger vessels such as vats and tanks may be partially filled with a predissolved cleaning solution. The cleaning solution should be hot (125°F) and the equipment permitted to soak for 15-30 minutes before being manually or mechanically scrubbed. One relatively recent approach is the ultrasonic cleaning tanks in which equipment is immersed in a cleaning solution and cleaned by the scrubbing action of microscopic bubbles caused by high frequency vibrations (20,000-40,000 cycles per second).
   2. Spray methods: Cleaning solutions may be sprayed on equipment surfaces by use of either fixed or portable spraying units using either hot water or steam. These methods are extensively used in the food industry.
   3. Clean-in-place (CIP) systems: This method is an automated cleaning system generally used in conjunction with permanent-welded pipeline

systems. In CIP cleaning, fluid turbulence in pipelines is considered to be the major source of energy required for soil removal.

4 Abrasive cleaning: Abrasive type powders and pastes are still available and used for removing difficult soil. Complete rinsing is necessary and care should be taken to avoid scratching stainless steel surfaces. Scouring pads should not be used on food-contact surfaces because small metal pieces from the pads may serve as focal points for corrosion or may be picked up in the food.

C Rinsing

All equipment surfaces should be thoroughly rinsed with clean potable water immediately after being cleaned in order to remove all traces of the cleaning solution. Very hot water may be desirable to decrease drying times.

Note: $^{\circ}F = {^{\circ}C} \times 1.8 + 32$

**Table 9.3.8. Sanitizing and Chemical Sanitizing Agents**

The primary reason for the application of effective sanitizing procedures is to destroy those disease organisms which may be present on equipment or utensils after cleaning, and thus prevent the transfer of such organisms to the ultimate consumer. In addition, sanitizing procedures may prevent spoilage of foods or prevent the interference of microorganisms in various industrial processes that depend on pure cultures.

There are a wide variety of known chemicals whose properties destroy or inhibit the growth of microorganisms. Many of these chemicals, however, are not suitable for use on food-contact surfaces because they may corrode, stain, or leave a film on the surface. Others may be highly toxic or too expensive for practical use. Therefore, the discussion on chemical sanitizing agents will be restricted to those agents in common use in the food industry.

**Table 9.3.9. Chlorine as a Chemical Sanitizing Agent**

Chlorine: Chlorine and its compounds combine indiscriminately with any and all proteins and protoplasms. The mode of bactericidal action is thought to be the reaction of chlorine with certain oxidizable groups in vital enzyme systems.

A. Inorganic chlorine products
   1. Types in common use

a. Calcium hypochlorite: generally in powder form of 70% available chlorine
b. Sodium hypochlorites: generally in aqueous solution of 2-15% available chlorine

2. Characteristics and limitations
   a. Effective sanitizer if high enough residual used.
   b. Organic matter may cause a substantial reduction in bactericidal effectiveness.
   c. Temperature and pH may exert marked influence upon the bactericidal effectiveness.
   d. Relatively unaffected by water hardness
   e. No film left on surface, but they may leave odor or taste.
   f. Bactericidal activity is good against a wide variety of microorganisms.

B. Organic chlorine products
   1. Types
      a. "Chloramine-T": in powdered form of 25% available chlorine
      b. Dichlorocyanuric and trichlorocyanuric acids: in powdered form of 70-90% available chlorine
      c. Characteristics and limitations
         i. Slower bactericidal action than hypochlorites
         ii. Factors affecting hypochlorites similarly affect organic chlorine compounds.
         iii. Relatively nonirritating to skin

**Table 9.3.10. Iodophors as a Sanitizing Agent**

Iodophors are soluble complexes of iodine combined usually with nonionic surface-active agents, loosely bound. Their characteristics are:

1. Rapid bactericidal action in acid pH range in cold or hard water
2. Less affected by organic matter than hypochlorites
3. Nontoxic in ordinary concentration
4. Noncorrosive, nonirritating to skin
5. Yellow or amber color of solution is proportional to concentration
6. Less effective against bacterial spores than hypochlorites
7. Do not stain and only minimal taste and odor

**Table 9.3.11. Quaternary Ammonium Compounds as Sanitizing Agents**

Quaternary ammonium compounds are synthetic surface-active agents. The most common ones are the cationic detergents, which are poor detergents but excellent germicides. In these compounds, the organic radical is the cation and the anion is usually chlorine. The mechanisms of germicidal action is not completely understood but is associated with enzyme inhibition and leakage of cell constituents.

A. Types of compounds
   1. Alkyl dimethyl benzylammonium chloride; alkyl dimethyl ethylbenzylammonium chloride: These compounds are effective in water ranging from 500 to 1100 ppm hardness without added sequestering agents.
   2. Diisobutyl phenoxy ethoxy ethyl dimethylbenzylammonium chloride; methyl dodecylbenzyl trimethylammonium chloride: These compounds require sodium tripolyphosphate to raise hard water levels to a minimum of 500 ppm.

B. Characteristics and limitations
   1. Require high dilution for germicidal or bacteriostatic action.
   2. Very selective in destruction or inhibition of various types of organisms
   3. Form bacteriostatic film on surface after treatment.
   4. Noncorrosive and nonirritating to skin
   5. No taste or odor in use dilutions
   6. More stable in presence of organic matter than some other chemical sanitizers
   7. Incompatible with soap, anionic detergents and inorganic polyphosphate
   8. Difficult to measure residual accurately

**Table 9.3.12. Factors Affecting Action of Chemical Sanitizers**

A. Contact of sanitizer: In order for a chemical to react with microorganisms, it must achieve intimate contact.

B. Selectivity of sanitizer: Certain sanitizers are nonselective in their ability to destroy a wide variety of microorganisms, while others demonstrate a degree of selectivity. Chlorine is relatively nonselective; however, both iodophors and quaternary compounds have a selectivity that may limit their application.

**Table 9.3.12. (continued)**

C. Concentration of sanitizer: In general, the more concentrated a sanitizer, the more rapid and certain its actions. Increases in concentration are usually related to exponential increases in effectiveness until a certain point, when they accomplish less noticeable effectiveness.

D. Temperature of solution: All of the common sanitizers increase in activity as the solution temperature increases. This is partly based on the principle that chemical reactions in general are speeded up by raising the temperature. However, a higher temperature also generally lowers surface tension, increases pH, decreases viscosity, and effects other changes that may enhance its germicidal action. It should be noted that chlorine compounds are more corrosive at high temperatures and iodine tends to sublime at temperatures above 120°F.

E. pH of solution: The pH of the solution exerts a very pronounced influence on most sanitizers. Quaternary compounds present a varied reaction to pH depending on the type of organism being destroyed. Chlorine and iodophors generally decrease in effectiveness with an increase in pH.

F. Time of exposure: Sufficient time must be allowed for whatever chemical reactions that occur to destroy the microorganism. The required time will depend not only on the preceding factors, but on microorganism populations and the populations of cells having varied susceptibility to the sanitizer due to cell age, spore formation, and other physiological factors of the microorganisms.

Note: °F = °C x 1.8 + 32.

**Table 9.3.13. Minimum Concentration of Chemical Sanitizing Agents**

| Chemical agent | Minimum concentration (ppm) |
|---|---|
| Hypochlorites | 50 |
| Chloramine-T | 200 |
| Quaternary Ammonium Compounds | 200 |
| Iodophors | 12.5 |

Note: For sanitization of assembled equipment, the solution strength should be checked at the outlet end. When it is in excess of the above minimum concentrations, the solution should be pumped through the entire equipment for at least 1 minute.

**Table 9.3.14. Field Tests for Sanitizer Strength**

A. Chlorine sanitizers

1. Starch iodide method (iodometric): a titration test in which chlorine displaces iodine from potassium iodide in an acid solution and forms a blue color with starch. This is decolorized by the addition of standard sodium thiosulfate. This test is generally used to measure high residuals.
2. Orthotolidine colorimetric comparison: a test in which a colorless solution of orthotolidine added to a chlorine solution produces an orange-brown colored compound proportional to its concentration. It is compared to a standardized color.
3. Indicator paper test: a rapid test of limited accuracy in which test papers usually impregnated with starch–iodide are immersed and the developed color compared with a standard

B. Quaternary compounds

There are several satisfactory tests for determining quaternary concentration. Some reagents are in tablet form and others use test papers by which a color comparison is made.

| Solution pH | Temperature (°F) | Time min |
|---|---|---|
| 7.5-9.5 | 75 or above | 1 |
| 7.0 or below | 75 or above | 2 |
| 5.0 or above | 75 or above | 1/2 |
| 5.0 or below | 75 - 120 | 1 |

C. Iodophors

Although iodophors have a builtin color indicator that is relatively accurate, color comparative kits and other kits are available for testing.

**Table 9.3.15. Physical Sanitizing Agents: Water Used**

A. Heat

1. Moist heat
   a. Hot water: an effective, nonselective sanitization method for food-contact surfaces; however, spores may remain alive even after an hour of boiling temperatures. The microbicidal action is thought to be the coagulation of some protein molecules in the cell. The use of hot water has several advantages in that is readily available, inexpensive, and nontoxic. Sanitizing can be

**Table 9.3.15. (continued)**

accomplished by either pumping the water through assembled equipment or immersing equipment into the water. When pumping it through equipment, the temperature should be maintained to at least 170°F. for at least 5 min as checked at the outlet end of the equipment. When immersing equipment the water should be maintained at a temperature of at least 17°F for 1-5 min depending on the size of the equipment.

b. Steam: an excellent agent for treating food equipment. Treatment on heavily contaminated surfaces may cake on the organic residues and prevent lethal heat from penetrating to the microorganism. Steam flow in cabinets should be maintained long enough to keep the thermometer reading above 170°F for at least 15 min or above 200°F. for at least 5 minutes. When steam is used on assembled equipment, the temperature should be maintained at 200°F. for at least 5 min as checked at the outlet end of the assembled equipment.

2. Dry Heat
Hot air ovens and chambers are not generally used because the method requires longer times and higher temperatures. When such cabinets are used, the temperature must be at least 180°F for a holding period of at least 20 min.

B. Ultraviolet Radiation
Low-pressure mercury vapor lamps, which produce effective bactericidal action by the emission of radiation at a wavelength of around 2500Å, have had limited use in the food industry. Major application has been with disinfection of air; however, installations of lamps have been reported on bread-slicing machines, over open vats in breweries, and in chill rooms for meats. Bacterial resistance will highly influence the lethal exposure time. Moreover, the light rays must actually strike the microorganisms because the rays are absorbed by dust, thin films of grease, and opaque or turbid solutions.

## 9.4. FDA Food Defect Action Levels

Many inquiries are received by the FDA as to permitted variations from absolute cleanliness or soundness in foods. The Food, Drug, and Cosmetic Act does not explicitly provide for "tolerances" for filth or decomposition in foods. It states that a food is adulterated if it consists *in whole or in part* of a filthy, putrid, or decomposed substance. Some foods, even when produced under current good manufacturing and/or processing practices, contain natural or unavoidable defects at lower levels that are not hazardous to health.

This does not mean that a food must be condemned because of the presence of foreign matter in amounts below the irreducible minimum after all possible precautions have been taken. FDA recognizes that it is not possible to grow, harvest, and process crops that are totally free of natural defects. The alternative, to increase the use of chemicals to control insects, rodents, and other sources of contamination, is not acceptable because of the potential health hazards from chemical residues. To resolve the problem, the FDA has established maximum levels for such defects in foods produced under good manufacturing and/or processing practices and uses these levels for recommending regulatory actions. Currently, the FDA has published a list of "defect action levels," stating the amounts of contamination that will subject the food to enforcement action. Such food defect action levels are set on the basis of no hazard to health and are changed from time to time. Any products that may be harmful to consumers are subject to action based on their health hazard, whether or not they exceed the "defect" levels. In addition, poor manufacturing processes by a manufacturer will result in regulatory action by FDA whether or not the product exceeds the defect action level. The levels are not averages. Actually the average levels are much lower and FDA continues to lower the action levels as industry performance improves.

Defect action levels are established for products whenever it is necessary and feasible. Such levels are subject to change upon the development of new technology or the availability of new information.

Compliance with defect action levels does not excuse failure to observe either the requirement in Section 402(a)(4) of the Federal Food, Drug, and Cosmetic Act that food may not be prepared, packed, or held under insanitary conditions or the other requirements in this section that food manufacturers must observe current good manufacturing practices. Evidence obtained through factory inspection indicating such a violation renders the food unlawful, even though the amounts of natural or unavoidable defects are lower than the currently established action levels. The manufacturer of food must at all times utilize quality control procedures that will reduce natural or unavoidable defects to the lowest level currently feasible.

The mixing of a food containing defects above the current defect action level with another lot of food is not permitted and renders the final food unlawful regardless of the defect level of the final food.

It will not be advisable to reproduce such defect action levels in view of frequent changes. Instead, the next table provides a list of foods and the types of defects for which the FDA has established action levels. An interested party may write to the FDA and obtain the brochure that provides the latest defect action levels for such foods.

**Table 9.4.1. Foods and Definitions for Defects**

| Food Products | Defects |
|---|---|
| Allspice | |
| Whole | Percentage moldy berries by weight |
| Ground | Number of insect fragments |
| | Number of rodent hairs |
| Apple butter | Percentage mold count |
| | Number of rodent hairs |
| | Number of whole or equivalent insects |
| Apricot, peach, and pear nectars and purees | Percentage mold counts |
| Apricots, canned | Percentage insect infested or insect damaged |
| Asparagus, canned or frozen | Percentage by count of spears or pieces infected with beetle eggs and/or sacs |
| | Number of thrips |
| | Number of insects (whole or equivalent) per 100 g |
| | Number of insects |
| Bay (laurel) leaves | Percentage moldy pieces by weight |
| | Percentage of insect infested pieces by weight |
| | Milligrams excreta after processing |
| Beets, canned | Percentage by weight of pieces with dry rot |
| Berries | |
| Drupelet, canned and frozen (blackberries, raspberries, etc.) | Percentage mold count |
| | Number of insect larvae |
| | Number of whole insects or equivalent (excluding thrips, aphids, and mites) |
| Lingon, canned | Number of insect larvae |
| Multer, canned | Number of thrips per No. 2 can |
| Broccoli, frozen | Number of aphids, thrips, mites |

| | |
|---|---|
| Brussels sprouts, frozen | Number of aphids, thrips |
| *Capsicum*: | |
| Pods | Percentage by weight of insect infested and moldy pods |
| | Milligrams mammalian excreta |
| Ground Capsicum (excluding paprika) | Percentage moldy count |
| | Number of insect fragments |
| | Number of rodent hairs |
| Ground paprika | Percentage moldy count |
| | Number of insect fragments |
| | Number of rodent hairs |
| Cassia or cinnamon bark (whole) | Percentage moldy pieces by weight |
| | Percentage insect infested pieces by weight |
| | Milligrams mammalian excreta |
| Cinnamon, ground | Number of insect fragments per 50 g |
| | Number of rodent hairs per 50 g |
| Cherries | |
| Brined and maraschino | Percentage rejects due to maggot |
| Fresh, canned, or frozen | Percentage defects due to insects other than maggots |
| Cherry jam | Percentage mold count |
| Chocolate and chocolate liquor | Number of microscopic fragments |
| | Number of rodent hairs |
| | For chocolate liquor, % shells on the basis of alkali-free nibs |
| Citrus fruit juices, canned | Percentage microscopic mold count |
| | Number of *Drosophilia* and other fly eggs |
| | Percentage of maggots |
| Cloves | Percentage stems by weight |
| Cocoa beans | Percentage mold by count |
| | Percentage by count insect infested, including insect damaged |
| | Percentage moldy and insect infested cocoa beans |
| | Milligrams mammalian excreta |
| Cocoa powder, press cake | Number of microscopic fragments |
| | Number of rodent hairs |
| | Percentage shell calculated on the basis of alkali-free nibs |

**Table 9.4.1. (continued)**

| Food Products | Defects |
|---|---|
| Coffee beans, green | Percentage by count insect infested or insect damaged<br>Number of live or dead insects (whole or parts) in immediate containers, immediate proximity of the lot, in cloth or burlap bag containers<br>Percentage mold by count |
| Coffee beans graded green | Beans are poorer than Grade 8 of the New York Green Coffee Association |
| Condimental seeds other than fennel seeds and sesame seeds | Milligrams mammalian excreta |
| Corn: sweet corn, canned | Number of larvae, cast, skins, larval or casts skin fragments of corn ear worm or corn borer |
| Corn husks for tamales | Percentage by weight insect infested and insect damaged<br>Percentage by weight moldy |
| Cornmeal | Number of whole insects or equivalent<br>Number of insect fragments<br>Number of rodent hairs<br>Number of rodent excreta fragments |
| Cranberry sauce | Percentage mold count |
| Cumin seed | Percentage ash or acid-insoluble ash |
| Currant jam, black | Percentage mold count |
| Currants | Percentage by count wormy currants |
| Curry powder | Number of insect fragments<br>Number of rodent hairs |
| Date material (chopped, sliced, or macerated) | Number of dead insects (whole or equivalent)<br>Number of pits and/or pit fragments |
| Dates, pitted | Percentage by count reject dates (moldy, dead insects, insect excreta, sour, dirty, and/or worthless)<br>Number of pits and/or pit fragments |
| Dates, whole | Percentage by count reject dates (moldy, dead insects, insect excreta, sour, dirty, and/or worthless) |

| | |
|---|---|
| Eggs and other egg products | Number of cans decomposed and the extent of microscopic counts of bacteria |
| Fennel seed | Percentage of insects infested<br>Percentage containing mammalian excreta<br>Milligrams mammalian excreta |
| Fig paste | Number insect heads per 100 grams of fig paste |
| Figs | Percentage by count rejected<br>Percentage by count insect infested, moldy, dirty fruit, and/or pieces of fruit |
| Fish, fresh or frozen: (applies only to fish or fillets weighing 3 pounds or less) | Percentage with Class 3 decomposition<br>Percentage with Class 2 decomposition<br>Extent of Class 2 versus Class 3 decomposition<br><br>Classes of decomposition:<br>1. No odor of decomposition<br>2. Slight odor of decomposition<br>3. Definite odor of decomposition |
| Tullibees, ciscoes, inconnus, chubs, and whitefish | Number of parasitic cysts in whole or fillets<br>Number of fish infested |
| Blue fin and other freshwater herring | Number of cysts per whole or fillets<br>Number of fish infested |
| Red fish and ocean perch | Percentage fillets containing copepods accompanied by pus pockets |
| Food storage and warehousing | Facility rodent infested:<br>Number of bags in lot rodent gnawed<br>Number of bags in lot bearing either rodent urine stains or rodent pellets<br>Food in container(s) in lot containing rodent-gnawed material, or rodent excreta or urine<br><br>Facility may or may not be rodent infested:<br>Number of bags bearing rodent urine stains<br>Number of bags rodent gnawed<br>Number of bags bearing either rodent urine or rodent pellets<br>Food in bag in lot containing rodent-gnawed material or rodent excreta or urine |

**Table 9.4.1. (continued)**

| Food Products | Defects |
|---|---|
| | Number of bags bearing rodent urine stains or rodent pellets |
| | Insect filth: |
| | Number of insects (whole/part, live/dead) in immediate containers |
| | Number of insects (whole/part, live/dead) in immediate proximity of the lot |
| Ginger, whole | Percentage or pieces by weight of moldy or insect infested ginger |
| | Milligrams mammalian excreta |
| Greens, canned | Percentage leaves, by count or weight, showing mildew |
| Hops | Number of aphids |
| Macaroni and noodle products | Grams insect fragments |
| | Grams rodent hairs |
| Mace | Percentage or pieces by weight of moldy and/or insect infested pieces |
| | Milligrams mammalian excreta |
| | Percentage foreign matter |
| Marjoram | |
| Whole (unprocessed) | Percentage insect infested moldy pieces by weight |
| | Milligrams mammalian excreta |
| Ground | Number of insect fragments |
| | Number of rodent hairs |
| Unground (processed) | Number of insect fragments |
| | Number of rodent hairs |
| Mushrooms, canned and dried | Number of maggots |
| | Number of mites |
| | Percentage decomposed |
| Nutmeg | |
| Whole | Percentage insect infested and/or moldy pieces |
| Ground | Number of insect fragments |
| | Number of rodent hairs |
| Nuts, tree | Reject nuts (insect infested, rancid, moldy, gummy, and shriveled or empty shells) |
| Olives: | |
| Pitted | Percentage by count of olives with |

| | |
|---|---|
| | whole pits and/or pit fragments |
| Imported green | Percentage by count showing damage by olive fruit fly |
| Salad | Percentage by weight showing damage due to the olive fruit fly |
| Salt-cured | Percentage by count of olives with scale insects<br>Percentage by count of moldy olives |
| Imported black | Percentage by count showing damage by olive fruit fly |
| Oregano | |
| Whole (unprocessed) | Percentage insect infested and/or moldy pieces<br>Milligrams mammalian excreta |
| Ground | Number of insect fragments<br>Number of rodent hairs |
| Crushed | Number of insect fragments<br>Number of rodent hairs |
| Peaches, canned and frozen | Percentage wormy or moldy by count<br>Number of larval fragments |
| Peanut butter | Number of insect fragments<br>Number of rodent hairs<br>Gritty taste<br>Milligrams insoluble inorganic residue |
| Peanuts, shelled | Percentage reject kernels (insect infested, moldy, rancid, otherwise decomposed, blanks, and shriveled) Number whole insects or equivalent |
| Peanuts, unshelled | Percentage reject peanuts (insect infested, moldy, rancid, otherwise decomposed, blanks, and shriveled) |
| Peas: black-eyed, cowpeas, field peas, dried | Percentage by count insect damaged by lygus bug |
| Peas: cowpeas, black-eyed peas (succulent), canned | Number cowpea curculio larvae or the equivalent |
| Peas and beans, dried | Percentage insect infested pieces by count and/or insect damaged by storage insects |
| Pepper, | |
| whole | Percentage insect infested and/or moldy pieces<br>Milligram mammalian excreta<br>Percentage pickings and siftings by weight |

**Table 9.4.1. (continued)**

| Food Products | Defects |
|---|---|
| Ground | Number of insect fragments |
| | Number of rodent hairs |
| Pineapple, canned | Percentage microscopic mold count |
| Pineapple juice | Percentage microscopic mold count |
| Plums, canned | Percentage with rot spots |
| Popcorn | Number rodent excreta pellets |
| | Number of rodent hairs |
| | Percentage gnawed grains |
| | Percentage by weight of field corn |
| Potato chips | Percentage by weight of rot |
| Prunes: dried and dehydrated, low-moisture | Percentage by count with whole pits and/or pit fragments |
| Prunes, pitted | Percentage by count with whole pits and/or pit fragments |
| Raisins | Percentage by count of natural raisins showing molds |
| | Number sand and grit per natural or golden bleached raisins |
| | Number whole or equivalent insects |
| | Number *Drosophila* eggs |
| Sage | |
| Whole (unprocessed | Percentage insect infested and/or moldy pieces by weight |
| | Milligrams mammalian excreta |
| Ground | Number of insect fragments |
| Salmon, canned | A defective can is defined as one that contains Class 2 or Class 3 decomposition.<br><br>Classes of decomposition:<br>1. No odor of decomposition<br>2. Slight odor of decomposition<br>3. Definite odor of decomposition |
| Sesame seeds | Percentage insect infested and/or moldy pieces by weight |
| | Percentage decomposed by weight |
| | Milligrams mammalian excreta |
| | Percentage foreign matter by weight |
| Shrimp: Fresh or frozen, | Percentage Class 3 and/or Class 2 |

| | |
|---|---|
| (peeled or breaded) | decomposition as raw, headless as determined by organoleptic examination |
| | Classes of decomposition:<br>1. No odor of decomposition<br>2. Slight odor of decomposition<br>3. Definite odor of decomposition |
| Shrimp: imported canned or cooked/frozen | Indole levels |
| Spices, leafy, other than bay leaves | Percentage insect-infested and/or moldy pieces<br>Milligrams mammalian excreta |
| Spinach, canned or frozen | Number aphids, thrips<br>Number larvae and/or larval fragments of spinach worms (caterpillars)<br>Number leaf miners<br>Percentage leaves by count or weight showing areas affected by mildew or other type of decomposition |
| Strawberries: frozen, whole or sliced | Percentage mold count<br>Gritty taste |
| Tomato catsup | Percentage mold count |
| Tomato juice | Number fly eggs<br>Number maggots<br>Percentage mold count |
| Tomato paste or puree | Number mold count<br>Percentage mold count |
| Tomato powder, (spray-dried) | Percentage mold count |
| Tomatoes, canned, with or without added tomato juice | Percentage mold count |
| Tomatoes, canned, packed in tomato puree juice | Percentage mold count |
| Paste, pizza, and other sauces | Number fly eggs<br>Number maggots |

**Table 9.4.1. (continued)**

| Food Products | Defects |
|---|---|
| Pizza sauce (based on 6% total tomato solids after pulping) | Percentage mold count |
| Tomato sauce (undiluted) | Percentage mold count |
| Tomato soup and tomato products | Percentage mold count |
| Tuna, canned: albacore, skip-jack, and yellowfin | Histamine content |
| Wheat | Percentage insect-damaged kernels<br>Milligrams rodent excreta pellets and/or pellet fragments |
| Wheat flour | Number of insect fragments<br>Number of rodent hairs |

## 9.5. FDA Food Action Levels for Poisonous or Deleterious Substances

An unavoidable poisonous or deleterious substance in food may be added or naturally occurring. A "naturally occurring poisonous or deleterious substance" is one that is an inherent natural constituent of a food and is not the result of environmental, agricultural, industrial, or other contamination. An "added poisonous or deleterious substance" is one that is not a naturally occurring poisonous or deleterious substance. When a naturally occurring poisonous or deleterious substance is increased to abnormal levels through mishandling or other intervening acts, it is an added poisonous or deleterious substance to the extent of such an increase.

Action levels for poisonous or deleterious substances in human food and animal feed are established by the U.S. Food and Drug Administration to control levels of contaminants. Action levels and tolerances represent limits at or above which FDA will take legal action to remove adulterated products from the market. Where no established action level or tolerance exists, FDA may take legal action against the product at the minimal detectable level of the contaminant. Action levels and tolerances are established based on the unavoidability of the poisonous or deleterious substance and do not represent permissible levels of contamination where it is avoidable. The blending of a food or feed containing a substance in excess of an action level or tolerance with another food or feed is not permitted, and the final product resulting from blending is unlawful, regardless of the level of the contaminant.

It will not be advisable to reproduce such food action levels in view of frequent changes. Instead, the next table provides a list of contaminants and those foods for which the FDA has established action levels for such contaminants. An interested party may write to the FDA and obtain a current listing that provides the latest information on action levels, contaminants, and food commodities.

**Table 9.5.1. A Listing of Contaminants and Those Foods for Which There Are Action Levels**

- Aflatoxin
  - Brazil nuts
  - Cottonseed meal (for nondairy animals)
  - Finished feeds and feed ingredients
  - Foods
  - Milk (aflatoxin M-1)
  - Peanuts and peanut products
  - Pistachio nuts
- Aldrin and Dieldrin
  - Alfalfa
  - Animal feed, processed
  - Artichokes
  - Asparagus
  - Bananas
  - Beets (garden and sugar)
  - Beet tops (garden and sugar)
  - Broccoli
  - Brussels sprouts
  - Bulb vegetables
  - Cabbage
  - Carrots
  - Cauliflower
  - Cereal grains (except buckwheat, millet, teosinte, and wild rice)
  - Celery
  - Clover
  - Collards
  - Cowpea hay
  - Cucumbers
  - Eggplant
  - Eggs
  - Endive (escarole)
  - Fats and oils (animal feed)
  - Figs
  - Fish (edible portion)
  - Forage, fodder, and straw of cereal grains (except those of buckwheat, millet, teosinte, and wild rice)
  - Grapefruit
  - Hay
  - Horseradish
  - Kale
  - Kohlrabi
  - Legume vegetables (except guar, jackbeans, lablab beans, and lentils)
  - Lemons
  - Lespedeza
  - Lettuce
  - Limes

**Table 9.5.1. (continued)**

Mangoes
Melons
Milk (fat basis)
Mustard greens
Oranges
Parsnips
Pea hay
Peaches
Peanuts
Peanut hay
Peppers
Pimentos
Pineapple
Pome fruits (except crabapples and loquats)
Potatoes
Radishes
Radish tops
Rutabagas
Salisfy roots
Salsify tops
Small fruits and berries
Soybean hay
Spinach
Squash
Stone fruits (except Chickasaw, Damson, and Japanese plums, and peaches)
Sugarbeet pulp (animal feed)
Sweet potatoes
Swiss chard
Tangerines
Tomatoes
Turnips
Turnip tops

Benzene hexachloride (BHC)
Animal feed, processed
Apples
Asparagus
Avocados
Beans
Brassica (cole) leafy vegetables (except broccoli raab, rape greens)
Celery
Carrots
Cereal grains (except buckwheat, millet, popcorn, teosinte, wild rice)
Citrus fruits
Cocoa beans
Cucurbit vegetables (except Balsam pears, Chinese waxgourds, gherkins, gourds)
Eggplant
Eggs
Endive
Figs
Frog legs (edible portion)
Guavas
Hays (animal feed)
Lettuce
Mangoes
Milk and dairy products (fat basis)
Okra
Onions
Paprika
Pears
Peas
Pecans
Peppers
Pineapples
Quinces
Rabbits (fat basis)
Root and tuber vegetables (except carrots)
Small fruits and berries
Spinach
Swiss chard
Stone fruits (except Chickasaw, Damson, and Japanese plums)
Tomatoes

Turnip greens
Cadmium
Pottery (ceramics): flatware (avg of units) μg/mL of leaching solution
Pottery (ceramics): small hollowware (any one of units) μg/mL of leaching solution
Pottery (ceramics) large hollowware (any one of units) μg/mL of leaching solution
Chlordane
Animal fat, rendered
Animal feed, processed
Asparagus
Bananas
Beans
Beets (with or without tops)
Beet greens
Brassica (cole) leafy vegetables (except broccoli raab, Chinese mustard cabbage, and rape greens)
Carrots
Celery
Citrus fruits
Corn
Cucumbers
Eggplant
Fish (edible portion)
Lettuce
Melons
Okra
Onions
Papayas
Parsnips
Peanuts
Peas
Peppers
Pineapple
Pome fruits (except crabapples and loquats)
Potatoes
Radishes (with or without tops)
Radish tops
Rutabagas (with or without tops)
Rutabaga tops
Small fruits and berries (except cranberries, currants, elderberries, gooseberries, and olallie berries)
Spinach
Squash
Stone fruits (except Chickasaw, Damson, and Japanese plums)
Sweet potatoes
Swiss chard
Tomatoes
Turnips (with or without tops)
Turnip greens
Chlordecone [Decachloroctahydro-1,3,4-methano-2*H*-cyclobuta(*cd*)pentalen-2-one]
Crabmeat
Fish and shellfish (edible portion)
*Crotalaria* Seeds (*C. spectabilis, C. saquittalis, C. striata*)
Grains and feeds average of whole seed/pound
Dicofol [1,1 bis(*p*-chlorophenyl)-2,2,2,-trichloroethanol]
Animal feed processed
DDT, DDE, and TDE,
Animal feed, processed
Artichokes
Asparagus
Avocados
Beets (roots and tops)
Brassica (cole) leafy vegetables (except broccoli raab, Chinese mustard cabbage, and rape greens)
Carrots

**Table 9.5.1. (continued)**

- Cereal grains (except buchwheat, fresh sweetcorn, millet, popcorn, teosinte, and wild rice)
- Celery
- Citrus fruits
- Cocoa beans
- Corn, fresh sweet
- Cottonseed
- Cucumbers
- Eggplant
- Eggs
- Endive (escarole)
- Fish (edible portion)
- Grapes
- Guavas
- Hay
- Hops
- Legume vegetables (except guar, jackbeans, lablab beans, and lentils)
- Lettuce
- Mangoes
- Melons
- Milk (fat basis)
- Mushrooms
- Okra
- Onions (dry bulb)
- Papayas
- Parsnips (roots and tops)
- Peanuts
- Peppermint hay
- Peppermint oil
- Peppers
- Pineapples
- Pome fruits (except crabapples and loquats)
- Potatoes
- Radishes (roots and tops)
- Rutabagas (roots and tops)
- Small fruits and berries (except elderberries, grapes, and olallie berries)
- Soybean oil (crude)
- Spearmint hay
- Spearmint oil
- Spinach
- Squash
- Stone fruits (except Chickasaw, Damson, and Japanese plums)
- Sweet potatoes
- Swiss chard
- Tomatoes
- Tomato pomace
- Turnips (roots and tops)

Dimethylnitrosamine (nitrosodimethylamine)

- Barley malt
- Malt beverages

Endrin

- Animal feed, processed
- Asparagus
- Beans
- Citrus fruits
- Corn, fresh sweet
- Eggs
- Figs
- Fish and shellfish (edible portion)
- Fish byproducts, including fish oil (animal feed)
- Guavas
- Leafy vegetables
- Mangoes
- Melons
- Milk and dairy products (fat basis)
- Oilseed meal (animal feed)
- Okra
- Peas
- Pimentoes
- Pineapples
- Pome fruits

Pumpkins
Root vegetables
Small fruits
Stone fruits
Vegetable oils and fats, including soapstock (animal feed)

Ethylene dibromide (EDB)
Grain products, intermediate (milled), must be cooked prior to consumption
Grain products, ready-to-eat (cooked), require no cooking prior to consumption
Honey, ready-to-eat (will not undergo further processing prior to consumption)

Heptachlor and heptachlor epoxide
Animal feed, processed
Artichokes
Asparagus
Beans, except snap beans
Citrus fruits
Cucumbers
Eggs
Eggplant
Figs
Fish and shellfish (edible portion)
Hay
Leafy vegetables
Melons
Okra
Pears
Pimentoes
Pumpkins
Quinces
Rice
Small fruits
Stone fruits
Squash (summer or winter)

Lead
Pottery (ceramics): flatware
Pottery (ceramics): small hollowware
Pottery (ceramics): large hollowware
Pottery (ceramics): silver-plated hollowware for children
Pottery (ceramics): silver-plated hollowware for adults

Lindane
Animal feed, processed
Artichokes
Barley
Beans
Citrus fruit
Cocoa beans, whole raw bean
Corn, fresh sweet
Corn
Eggs
Endive
Figs
Hay
Milk and dairy products (fat basis)
Oats
Peas
Rice
Root vegetables
Rye
Small fruits,
Sorghum (milo)
Turnip greens
Wheat

Mercury
Fish, shellfish, crustaceans, other aquatic animals; fresh, frozen, or processed (methyl mercury) in edible portion
Wheat (pink kernels only)

Methyl alcohol
Imported brandy (% alcohol)

Mirex
Fish and shellfish (edible portion)

*N*-Nitrosamines
Rubber baby bottle nipples

**Table 9.5.1 (continued**

| | |
|---|---|
| Paralytic shellfish toxin | Figs |
| Clams, mussels, oysters | Fish and shellfish (edible portion) |
| (fresh, frozen, canned) | Melons |
| Polychlorinated biphenyls (PCB) | Mustard greens |
| Red meat (fat basis) | Plums |
| Toxaphene | Pumpkins |
| Animal feed, processed | Root vegetables, |
| Artichokes | Small fruits |
| Asparagus | Squash (summer or winter) |
| Cherries | Turnip greens |

## 9.6. Uses and Limitations of Selected Legal Food Chemicals

**Table 9.6.1. Color Additives Exempt from Certification and Used in Foods**

| | | |
|---|---|---|
| Annatto extract | Dried algae meal | *Tagetes* (Azetic |
| β-Apo-8′-carotenal | Ferrous gluconate | marigold) meal |
| Canthaxanthin | Fruit juice | and extract |
| Caramel | Grape color extract | Titanium dioxide |
| β-Carotene | Grape skin extract | Toasted, partially |
| Carrot oil | (enocianina) | defatted, cooked |
| Cochineal extract; | Paprika | cottonseed flour |
| carmine | Paprika oleoresin | Turmeric |
| Corn endosperm oil | Riboflavin | Turmeric oleoresin |
| Dehydrated beets | Saffron | Ultramarine blue |
| (beet powder) | Synthetic iron oxide | Vegetable juice |

Source: 21 CFR 73, 1991.

**Table 9.6.2. Color Additives Subject to Certification and Used in Foods**

| | | |
|---|---|---|
| FD&C Blue No. 1 | Citrus Red No. 2 | FD&C Yellow No. 6 |
| FD&C Blue No. 2 | FD&C Red No. 3 | FD&C Yellow No. 10 |
| FD&C Green No. 3 | FD&C Red No. 40 | |
| Orange B | FD&C Yellow No. 5 | |

Source: 21 CFR 74, 1991.

**Table 9.6.3. Legal Uses and Limitations of Food Preservatives[a]**

| Food | Limitation (ppm) | Uses and Remarks |
|---|---|---|
| Anoxomer | | |
| All | 5000 | Limitation based on fat and oil content of foods |
| BHA | | |
| Dehydrated potato shreds | 50 | ppm = BHA + BHT |
| Active dry yeast | 1000 | ppm = BHA |
| Beverages and desserts prepared from dry mixes | 2 | ppm = BHA |
| Dry breakfast cereals | 50 | ppm = BHA + BHT |
| Dry diced glazed fruit | 32 | ppm = BHA |
| Dry mixes for beverages and desserts | 90 | ppm = BHA |
| Emulsion stabilizers for shortenings | 200 | ppm = BHA + BHT |
| Potato flakes | 50 | ppm = BHA + BHT |
| Potato granules | 10 | ppm = BHA + BHT |
| Sweetpotato flakes | 50 | ppm = BHA + BHT |
| BHT | | |
| Dehydrated potato shreds | 50 | ppm = BHA + BHT |
| Dry breakfast cereals | 50 | ppm = BHA + BHT |
| Emulsion stabilizers for shortenings | 200 | ppm = BHA + BHT |
| Potato flakes | 50 | ppm = BHA + BHT |
| Potato granules | 10 | ppm = BHA + BHT |
| Sweetpotato flakes | 50 | ppm = BHA + BHT |
| Calcium Disodium EDTA [All calculations based on the anhydrous compound.] | | |
| Cabbage, pickled and texture | 220 | Promote color, flavor, and texture retention |
| Canned carbonated soft drinks | 33 | Promote flavor retention |
| Canned white potatoes | 110 | Promote color retention |
| Clams (Cooked canned) | 340 | Promote color retention |
| Crabmeat (cooked canned) | 275 | Retard struvite formation; promote color retention |
| Cucumbers pickled | 220 | Promote color, flavor, and |

**Table 9.6.3 (continued)**

| Food | Limitation (ppm) | Uses and Remarks |
|---|---|---|
| | | texture retention |
| Distilled alcoholic beverages | 25 | Promote stability of color, flavor, and/or product clarity |
| Dressings, nonstandardized | 75 | Preservative |
| Dried lima beans (cooked canned) | 310 | Promote color retention |
| Egg product that is hard-cooked and consists, in a cylindrical shape, of egg white with an inner core of egg yolk | 200 | Preservative, by weight of egg yolk portion |
| Fermented malt beverages | 25 | Antigushing agent |
| French dressing | 75 | Preservative |
| Mayonnaise | 75 | Preservative |
| Mushrooms (cooked canned) | 200 | Promote color retention |
| Oleomargarine | 75 | Preservative |
| Pecan pie filling | 100 | Promote color retention |
| Potato salad | 100 | Preservative |
| Processed dry pinto beans | 800 | Promote color retention |
| Salad dressing | 75 | Preservative |
| Sandwich spread | 100 | Preservative |
| Sauces | 75 | Preservative. |
| Shrimp (cooked canned) | 250 | Retard struvite formation; promote color retention |
| Spice extractives in soluble carriers | 60 | Promote color and flavor retention |
| Spreads, artificially colored and lemon flavored or orange flavored | 100 | Promote color retention |
| Calcium Disodium EDTA + Disodium EDTA | | |
| Dressings, nonstandardized | 75 | Preservative |
| French dressing | 75 | Preservative |
| Mayonnaise | 75 | Preservative |
| Salad dressing | 75 | Preservative |
| Sandwich spread | 100 | Preservative |
| Sauces | 75 | Preservative |

| | | |
|---|---|---|
| <u>Dehydroacetic acid and/or Its Sodium Salt</u> | | |
| Squash | 65 | For cut or peeled squash 65 ppm expressed as dehyhydroacetic acid is the highest level permitted in or on the prepared squash |
| <u>Dimethyl Dicarbonate</u> | | |
| Wine | 500 | Intended for use as an inhibitor of yeast in wine under normal circums tances of bottling where the viable yeast count has been reduced to 500 mL or less by current GMP such as flash pasteurization or filtration. |
| <u>Disodium EDTA</u> | | |
| Aqueous multivitamin preparations | 150 | With iron salts as a stabilizer for vitamin $B_{12}$ in liquid multivitamin preparations |
| Canned black-eyed peas | 145 | Promote color retention |
| Canned cooked chickpeas | 165 | Promote color retention |
| Canned kidney beans | 165 | Preservative |
| Canned strawberry pie filling | 500 | Promote color retention |
| Cooked sausage | 36 | As a cure accelerator with sodium ascorbate or ascorbic acid |
| Dressings, nonstandardized | 75 | Preservative |
| French dressing | 75 | Preservative |
| Frozen white potatoes including cut potatoes | 100 | Promote color retention |
| Gefilte fish balls or patties in packing medium | 50 | Inhibit discoloration; based on total weight of finished product including packing medium |
| Mayonnaise | 75 | Preservative |

**Table 9.6.3 (continued)**

| Food | Limitation (ppm) | Uses and Remarks |
|---|---|---|
| Nonnutritive sweeteners | 1000 | Calculated as anhydrous calcium disodium EDTA |
| Ready-to-eat cereal products containing dried bananas | 315 | Promote color retention In dried banana component of cereal product |
| Salad dressing | 75 | Preservative |
| Sandwich spread | 100 | Preservative |
| Sauces | 75 | Preservative |
| <u>Disodium EDTA + Calcium Disodium EDTA</u> | | |
| Dressings, nonstandardized | 75 | Preservative. |
| French dressing | 75 | Preservative |
| Mayonnaise | 75 | Preservative |
| Salad dressing | 75 | Preservative |
| Sandwich spread | 100 | Preservative |
| Sauces | 75 | Preservative |
| <u>Ethoxyquin</u> | | |
| Chili powder, paprika, ground chili | 100 | As an antioxidant for preserv ation of color |
| Uncooked fat of meat from animals except poultry | 5 | In or on product |
| Uncooked liver and fat of poultry | 3 | In or on product |
| Uncooked muscle meat of animals | 0.5 | In or on product |
| Poultry eggs | 0.5 | In product |
| Milk | 0 | In product |
| <u>Heptylparaben</u> | | |
| Fermented malt beverages | 12 | To inhibit microbiological spoilage |
| Noncarbonated soft drinks and fruit-based beverages | 20 | Unless a standard of identity prohibits its use |
| <u>4-Hydroxymethyl-2,6-di-*tert*-butylphenol</u> | | |
| All | 20 | An antioxidant; calculation based on the oil or fat |

| | | |
|---|---|---|
| | | content of food including the essential (volatile) oil content of food |
| <u>Natamycin (Pimaricin)</u> | | |
| Surface of cuts and slices of cheese | | Inhibit mold spoilage. May be applied by dipping or by spraying, using an aqueous solution containing 200-300 ppm of the additive. Used only if standard of identity permits. |
| <u>Potassium Nitrate</u> | | |
| Cod roe processing | 200 | A curing agent |
| <u>Quaternary Ammonium Chloride Combination</u> | | |
| Sugar cane juice (additive added prior to clarification when further processing must be delayed) | 0.25-1.0 | *n*-Dodecyl dimethylbenzyl-ammonium chloride |
| | 3.4-13.5 | *n*-Dodecyl dimethyl ethyl-benzylammonium chloride |
| | 1.5-6.0 | *n*-Hexadecyl dimethylbenzyl-ammonium chloride |
| | 0.25-1.0 | *n*-Octadecyl dimethylbenzyl-ammonium chloride |
| | 3.0-12.0 | *n*-Tetradecyl dimethylbenzyl-ammonium chloride |
| | 1.6-6.5 | *n*-Tetradecyl dimethyl ethyl-benzylammonium chloride |
| <u>Sodium Nitrate</u> | | |
| Smoked, cured sable fish; smoked, cured salmon; and smoked, cured shad | | As a preservative and color fixative |
| | 500 | Sodium nitrate |
| | 200 | Sodium nitrite |
| Meat-curing preparations for the home curing of meat and meat products (including poultry and wild game) | | As a preservative and color fixative |
| | 500 | Sodium nitrate |
| | 200 | Sodium nitrite |

**Table 9.6.3 (continued)**

| Food | Limitation (ppm) | Uses and Remarks |
|---|---|---|
| Sodium Nitrite | | |
| Smoked cured tunafish products | | As a color fixative |
| | 10 | Sodium nitrite |
| Smoked, cured sable fish; smoked, cured salmon; and smoked, cured shad | | As a preservative and color fixative |
| | 200 | Sodium nitrite |
| | 500 | Sodium nitrate |
| Meat-curing preparations for the home curing of meat and meat products (including poultry and wild game) | | As a preservative and color fixative |
| | 200 | Sodium nitrite |
| | 500 | Sodium nitrate |
| Sodium Nitrite[a] + Sodium Chloride | | |
| Smoked chub | 100-200 | Sodium nitrite |
| | 3500 | Sodium chloride |
| Stannous Chloride | | |
| Asparagus packed in glass | 20 ppm | Color retention; calculated as tin |
| TBHQ [2-(1,1-dimethylethyl)-1,4-benzenediol] | | |
| All | 20 | An antioxidant used alone or in combination with BHA and/or BHT; calculation based on oil or content of food, including the essential (volatile) oil content of food. |
| THBP (2,4,5-trihydroxybutyrophenone) | | |
| All | 20 | An antioxidant used alone or in combination with other permitted antioxidants; calculation based on oil or content of food, including |

| | | the essential (volatile) oil content of food. |
|---|---|---|

Source 21 CFR 172, 1991.

[a]The food additive sodium nitrite may be safely used in combination with salt (NaCl) to aid in inhibiting the outgrowth and toxin formation from *Clostridium botulinum* type E in chub in accordance with the following prescribed conditions:

1. All fish in smoking establishments shall be clean and wholesome and shall be expeditiously processed, packed, and stored under adequate sanitary conditions in accordance with good manufacturing practice.
2. The brining procedure is controlled in such a manner that the water-phase portion of the edible portion of the finished smoked product has a salt (NaCl) content of not less than 3.5%, as measured in the loin muscle, and the sodium nitrite content of the edible portion of the finished smoked product is 100-200 ppm, as measured in the loin muscle.
3. Smoked chub shall be heated by a controlled heat process that provides a monitoring system positioned in as many strategic locations in the smokehouse as necessary to assure a continuous temperature throughout each fish of at least 160°F for a minimum of 30 min.
4. The finished product shall be cooled to a temperature of 50°F or below within 3 hr after smoking and further cooled to a temperature of 38°F or below shall be maintained during all subsequent storage and distribution. All shipping containers, retail packages, and shipping records shall indicate with appropriate notice the perishable nature of the product and specify that the product shall be held under refrigeration (38°F or below) until consumed.

**Table 9.6.4. Synthetic Flavoring Substances and Adjuvants Permitted in Foods**

Acetal; acetaldehyde diethyl acetal
Acetaldehyde phenethyl propyl acetal
Acetanisole; 4'-methoxyacetophenone
Acetophenone; methyl phenyl ketone
Allyl anthranilate
Allyl butyrate
Allyl cinnamate
Allyl cyclohexaneacetate
Allyl cyclohexanebutyrate
Allyl cyclohexanehexanoate
Allyl cyclohexaneproprionate
Allyl cyclohexanevalerate
Allyl disulfide
Allyl 2-ethylbutyrate
Allyl hexanoate; allyl caproate
Allyl $\alpha$-inone; 1-(2,6,6-trimethyl-2-cyclohexene-1-yl)-1,6-heptadiene-3-one
Allyl isothiocyanate; mustard oil
Allyl isovalerate
Allyl mercaptan; 2-propene-1-thiol
Allyl nonanoate
Allyl octanoate
Allyl phenoxyacetate
Allyl phenylacetate
Allyl propionate
Allyl sorbate; allyl 2,4-hexadienoate
Allyl sulfide
Allyl tiglate; allyl *trans*-2-methyl-2-butenoate
Allyl 10-undecenoate
Ammonium isovalerate
Ammonium sulfide
Amyl alcohol; pentyl alcohol
Amyl butyrate
$\alpha$-Amylcinnamaldehyde
$\alpha$-Amylcinnamaldehyde dimethyl acetal
$\alpha$-Amylcinnamyl acetate
$\alpha$-Amylcinnamyl alcohol
$\alpha$-Amylcinnamyl formate
$\alpha$-Amylcinnamyl isovalerate
Amyl formate
Amyl heptanoate
Amyl hexanoate
Amyl octanoate
Anisole; methoxybenzene
Anisyl acetate
Anisyl alcohol; *p*-methoxybenzyl alcohol
Anisyl butyrate
Anisyl formate
Anisyl phenylacetate
Anisyl propionate
Beechwood creosote
Benzaldehyde glyceryl acetal; 2-phenyl-*m*-dioxan-5-ol
Benzaldehyde propylene glycol acetal; 4-methyl-2-phenyl-*m*-dioxolane
Benzenethiol; thiophenol
Benzoin; 2-hydroxy-2-phenylacetophenone
Benzophenone; diphenylketone
Benzyl acetate
Benzyl acetoacetate
Benzyl alcohol
Benzyl benzoate
Benzyl butyl ether
Benzyl butyrate
Benzyl cinnamate
Benzyl 2,3-dimethylcrotonate; benzyl methyl tiglate
Benzyl disulfide; dibenzyl disulfide

Benzyl ethyl ether
Benzyl formate
3-Benzyl-4-heptanone; benzyl dipropyl ketone
Benzyl isobutyrate
Benzyl isovalerate
Benzyl mercaptan; $\alpha$-toluenethiol
Benzyl methoxyethyl acetal; acetaldehyde benzyl $\beta$-methoxyethyl acetal
Benzyl phenylacetate
Benzyl propionate
Benzyl salicylate
Birch tar oil
Borneol; *d*-camphanol
Bornyl acetate
Bornyl formate
Bornyl isovalerate
Bornyl valerate
$\beta$-Bourbonene; 1,2,3,3a,3b$\beta$,4,5,6,6a$\beta$,6b$\alpha$-decahydro-l$\alpha$-isopropyl-3$a_a$-methyl-6-methylene-cyclobuta [1,2:3,4] dicyclopentene
2-Butanol
2-Butanone; methyl ethyl ketone
Butter acids
Butter esters
Butyl acetate
Butyl acetoacetate
Butyl alcohol; 1-butanol
Butyl anthranilate
Butyl butyrate
Butyl butyryllactate; lactic acid, butyl ester, butyrate
$\alpha$-Butyl cinnamate
Butyl 2-decenoate
Butyl ethyl malonate
Butyl formate
Butyl heptanoate
Butyl hexanoate
Butyl *p*-hydroxybenzoate
Butyl isobutyrate
Butyl isovalerate
Butyl lactate
Butyl laurate
Butyl levulinate
Butyl phenylacetate
Butyl propionate
Butyl stearate
Butyl sulfide
Butyl 10-undecenoate
Butyl valerate
Butyraldehyde
Cadinene
Camphene; 2,2-dimethyl-3-methylenenor-bornane
*d*-Camphor
Carvacrol; 2-*p*-cymenol
Carvacryl ethyl ether; 2-ethoxy-*p*-cymene
Carveol; *p*-mentha-6,8-dien-2-ol
4-Carvomenthenol; 1-*p*-menthen-4-ol; 4-terpinenol
*cis* Carvone oxide; 1,6-epoxy-*p*-menth-8-en-2-one
Carvyl acetate
Carvyl propionate
$\beta$-Caryophyllene
Caryophyllene alcohol
Caryophyllene alcohol acetate
$\beta$-Caryophyllene oxide; 4-12,12-trimethyl-9-methylene-5-oxatricylo[8.2.0.$0^{4,6}$]dodecane
Cedarwood oil alcohols
Cedarwood oil terpenes
1,4-Cinole
Cinnamaldehyde ethylene glycol acetal
Cinnamic acid
Cinnamyl acetate
Cinnamyl alcohol; 3-phenyl-2-propen-1-ol

**Table 9.6.4 (continued)**

Cinnamyl benzoate
Cinnamyl butyrate
Cinnamyl cinnamate
Cinnamyl formate
Cinnamyl isobutyrate
Cinnamyl isovalerate
Cinnamyl phenylacetate
Cinnamyl propionate
Citral diethyl acetal; 3,7-dimethyl-2,6-octadienal diethyl acetal
Citral dimethyl acetal; 3,7-dimethyl-2,6-octadienal dimethyl acetal
Citral propylene glycol acetal
Citronellal; 3,7-dimethyl-6-octenal; rhodinal
Citronellol; 3,7-dimethyl-6-octen-1-ol; *d*-citronellol
Citronelloxyacetaldehyde
Citronellyl acetate
Citronellyl butyrate
Citronellyl formate
Citronellyl isobutyrate
Citronellyl phenylacetate
Citronellyl propionate
Citronellyl valerate
*p*-Cresol
Cuminaldehyde; cuminal; *p*-isopropyl benzaldehyde
Cyclohexaneacetic acid
Cyclohexaneethyl acetate
Cyclohexyl acetate.
Cyclohexyl anthranilate
Cyclohexyl cinnamate
Cyclohexyl formate
Cyclohexyl isovalerate
Cyclohexyl proprionate
*p*-Cymene
γ-Decalactone; 4-hydroxy-decanoic acid, γ-lactone
γ-Decalactone; 5-hydroxy-decanoic acid, δ-lactone
Decanal dimethyl acetal
1-Decanol; decylic alcohol
2-Decenal
3-Decen-2-one; heptylidene acetone
Decyl actate
Decyl butyrate
Decyl propionate
Dibenzyl ether
4,4-Dibutyl-γ-butyrolactone; 4,4-dibutyl-4-hydroxy-butyric acid, γ-lactone
Dibutyl sebacate
Diethyl malate
Diethyl malonate; ethyl malonate
Diethyl sebacate
Diethyl succinate
Diethyl tartrate
2,5-Diethyltetrahydrofuran
Dihydrocarveol; 8-*p*-menthen-2-ol; 6-methyl-3-isopropenylcyclohexanol
Dihydrocarvone
Dihydrocarvyl acetate
*m*-Dimethoxybenzene
*p*-Dimethoxybenzene; dimethyl hydroquin-one
2,4-Dimethylacetophenone
α, α-Dimethylbenzyl isobutyrate; phenyldimethylcarbinyl isobutyrate
2,6-Dimethyl-5-heptenal
2,6-Dimethyl octanal; isodecylaldehyde
3,7-Dimethyl-1-octanol; tetrahydrogeraniol
α,α-Dimethylphenethyl acetate; benzylpropyl acetate; benzyldimethylcarbinyl acetate
α,α-Dimethylphenethyl alcohol; dimethylbenzyl carbinol
α,α-Dimethylphenethyl butyrate; benzyldimethylcarbinyl butyrate

$\alpha,\alpha$-Dimethylphenethyl formate; benzyldimethylcarbinyl formate
Dimethyl succinate
1,3-Diphenyl-2-propanone; dibenzyl ketone
$\delta$-Dodecalactone; 5-hydroxy-dodecanoic acid, $\delta$-lactone
$\gamma$-Dodecalactone; 4-hydroxy-dodecanoic acid $\gamma$-lactone
2-Dodecenal
Estragole
*p*-Ethoxybenzaldehyde
Ethyl acetoacetate
Ethyl 2-acetyl-3-phenylpropionate; ethyl-benzyl acetoacitate
Ethylaconitate, mixed esters
Ethyl acrylate
Ethyl *p*-anisate
Ethyl anthranilate
Ethyl benzoate
Ethyl benzoylacetate
$\alpha$-Ethylbenzyl butyrate; $\alpha$-phenylpropyl butyrate
Ethyl brassylate; tridecanedioic acid cyclic ethylene glycol diester; cyclo 1,13-ethylenedioxytridecan-1,13-dione
2-Ethylbutyl acetate
2-Ethylbutyraldehyde
2-Ethylbutyric acid
Ethyl cinnamate
Ethyl crotonate; *trans*-2-butenoic acid ethylester
Ethyl cyclohexanepropionate
Ethyl decanoate
2-Ethylfuran
Ethyl 2-furanpropionate
4-Ethylguaiacol; 4-ethyl-2-methoxyphenol
Ethyl heptanoate
2-Ethyl-2-heptenal; 2-ethyl-3-butylacrolein
Ethyl hexanoate
Ethyl isobutyrate
Ethyl isovalerate
Ethyl lactate
Ethyl laurate
Ethyl levulinate
Ethyl maltol; 2-ethyl-3-hydroxy-4H-pyran-4-one
Ethyl 2-methylbutyrate
Ethyl myristate
Ethyl nitrite
Ethyl nonanoate
Ethyl 2-nonynoate; ethyl octyne carbonate
Ethyl octanoate
Ethyl oleate
Ethyl phenylacetate
Ethyl 4-phenylbutyrate
Ethyl 3-phenylglycidate
Ethyl 3-phenylpropionate; ethyl hydrocinnamate
Ethyl propionate
Ethyl pyruvate
Ethyl salicylate
Ethyl sorbate; ethyl 2,4-hexadienoate
Ethyl tiglate; ethyl *trans*-2-methyl-butenoate
Ethyl undecanoate
Ethyl 10-undecenoate
Ethyl valerate
Eucalyptol; 1,8-epoxy-*p*-methane; cineole
Eugenyl acetate
Eugenyl benzoate
Eugenyl formate
Eugenyl methyl ether; 4-allylveratrole; methyl eugenol

**Table 9.6.4 (continued)**

Farnesol; 3,7.11-trimethyl-2,6,10-dodecatrien-1-kol.d-Fenchone; d-1,3,3-trimethyl-2-norbornan-one
*d*-Fenchone; d-1,3,3-trimethyl-2-norbornanone
Fenchyl alcohol; 1,3,3-trimethyl-2-norbornanol
Formic acid
(2-Furyl)-2-propanone; furyl acetone
1-Furyl-2-propanone; furyl acetone
Fusel oil, refined (mixed amyl alcohols)
Geranyl acetoacetate; *trans*-3,7-dimethyl-2,6-octadien-1-yl acetoacetate
Geranyl acetone; 6,10-dimethyl-5,9-undecadien-2-one
Geranyl benzoate
Geranyl butyrate
Geranyl formate
Geranyl hexanoate
Geranyl isobutyrate
Geranyl isovalerate
Geranyl phenylacetate
Geranyl propionate
Glucose pentaacetate
Guaiacol; $\mu$-methoxyphenol
Guaiacyl acetate; $\mu$-methoxyphenyl acetate
Guaiacyl phenylacetate
Guaiene; 1,4-diemthyl-7-isopropenyl-$\Delta$9,10-octahydroazulene
Guaiol acetate; 1,4-dimethyl-7-($\alpha$-hydroxyisopropyl)-$\Delta$9,10-octahydroazulene acetate
$\gamma$-Heptalactone; 4-hydroxyheptanoic acid, $\gamma$-lactone
Heptanal; enanthaldehyde
Heptanal dimethyl acetal
Heptanal 1,2-glyceryl acetal
2,3-Heptanedione; acetyl valeryl
3-Heptanol
2-Heptanone; methyl amyl ketone
3-Heptanone; ethyl butyl ketone
4-Heptanone; dipropyl ketone
*cis*-4-Heptenal; *cis*-4-hepten-1-al
Heptyl acetate
Heptyl alcohol; enanthic alcohol
Heptyl butyrate
Heptyl cinnamate
Heptyl formate
Heptyl isobutyrate
Heptyl octanoate
1-Hexadecanol; cetyl alcohol
$\omega$-6-Hexadecenlactone; 16-hydroxy-6-hexadecenoic acid, $\omega$-lactone; ambrettolide
$\gamma$-Hexalactone; 4-hydroxyhexanoic acid, $\gamma$-lactone; tonkalide
Hexanal; caproic aldehyde
2,3-Hexanedione; acetyl butyryl
Hexanoic acid; caproic acid
2-Hexenal
2-Hexen-1-ol
3-Hexen-1-ol; leaf alcohol
2-Hexen-1-yl acetate
3-Hexenyl isovalerate
3-Hexenyl 2-methylbutyrate
3-Hexenyl phenylacetate; *cis*-3-hexenyl phenylacetate
Hexyl acetate
2-Hexyl-4-acetoxytetrahydrofuran
Hexyl alcohol
Hexytl butyrate
$\alpha$-Hexylcinnamaldehyde
Hexyl formate
Hexyl hexanoate
2-Hexylidene cyclopentanone
Hexyl isovalerate

Hexyl 2-methylbutyrate
Hexyl octanoate
Hexyl phenylacetate; *n*-hexyl phenylacetate
Hexyl propionate
Hydroxycitronellal; 3,7-dimethyl-7-hydroxyoctanal
Hydroxycitronellal diethyl acetal
Hydroxycitronellal dimethyl acetal
Hydroxycitronellal; 3,7-dimethyl-1,7-octanediol
*N*-(4-Hydroxy-3-methoxybenzyl)-nonanamide; pelargonyl vanillylamide
5-Hydroxy-4-octanone; butyroin
4-(*p*-Hydroxyphenyl)-2-butanone; *p*-hydroxybenzyl acetone
Indole
$\alpha$-Ionone; 4-(2,6,6-trimethyl-2-cyclohexen-1-yl)-3-buten-2-one
$\beta$-Ionone; 4-(2,6,6-trimethyl-1-cyclohexen-1-yl)-3-buten-2-one
$\alpha$-Irone; 4-(2,5,6,6-tetramethyl-2-cyclohexene-1-yl)-3-buten-2-one; 6-methylionone
Isoamyl acetate
Isoamyl acetoacetate
Isoamyl alcohol; isopentyl alcohol; 3-methyl-1-butanol.
Isoamyl benzoate
Isoamyl butyrate
Isoamyl cinnamate
Isoamyl formate
Isoamyl 2-furanbutyrate; $\alpha$-isoamyl furfurylpropionate
Isoamyl 2-furanpropionate; $\alpha$-isoamyl furfurylacetate
Isoamyl hexanoate
Isoamyl isobutyrate
Isoamyl isovalerate
Isoamyl laurate
Isoamyl-2-methylbutyrate; isopentyl-2-methylbutyrate
Isoamyl nonanoate
Isoamyl octanoate
Isoamyl phenylacetate
Isoamyl propionate
Isoamyl pyruvate
Isoamyl salicylate
Isoborneol
Isobornyl acetate
Isobornyl formate
Isobornyl isovalerate
Isobornyl propionate
Isobutyl acetate
Isobutyl acetoacetate
Isobutyl alcohol
Isobutyl angelate; isobutyl *cis*-2-methyl-2-butenoate
Isobutyl anthranilate
Isobutyl benzoate
Isobutyl benzoate
Isobutyl butyrate
Isobutyl cinnamate
Isobutyl formate
Isobutyl 2-furanpropionate
Isobutyl heptanoate
Isobutyl hexanoate
Isobutyl isobutyrate
$\alpha$-Isobutylphenethyl alcohol; isobutyl benzyl carbinol; 4-methyl-1-phenyl-2-pentanol
Isobutyl phenylacetate
Isobutyl propionate
Isobutyl salicylate
2-Isobutylthiazole
Isobutyraldehyde
Isobutyric acid
Isoeugenol; 2-methoxy-4-propenylphenol
Isoeugenyl acetate

**Table 9.6.4 (continued)**

| |
|---|
| Isoegenyl benzyl ether; benzyl isoeugenol |
| Isoeugenyl ethyl ether; 2-ethoxy-5-propenyl-anisole; ethyl isoeugenol |
| Isoeugenyl formate |
| Isoeugenyl methyl ether; 4-propenylveratrole; methyl isoeugenol |
| Isoeugenyl phenylacetate |
| Isojasmone; mixture of 2-hexylidenecyclopentanone and 2-hexyl-2-cyclopenten-1-one |
| $\alpha$-Isomethylionone; 4-(2,6,6-trimethyl-2-cyclohexen-1-yl)-3-methyl-3-buten-2-one; methyl $\gamma$-ionone |
| Isopropyl acetate |
| *p*-Isopropylacetophenone |
| Isoprpyl alcohol; isopropanol |
| Isopropyl benzoate |
| *p*-Isopropylbenzyl alcohol; cuminic alcohol; *p*-cymen-7-ol |
| Isopropyl butyrate |
| Isopropyl cinnamate |
| Isopropyl formate |
| Isopropyl hexanoate |
| Isopropyl isobutyrate |
| Isopropyl isovalerate |
| *p*-Isopropylphenylacetaldehyde; *p*-cymen-7-carboxaldehyde |
| Isopropyl phenylacetate |
| 3-(*p*-Isopropylphenyl)-propionaldehyde; *p*-isopropylhydrocinnamaldehyde; cuminyl acetaldehyde |
| Isopropyl propionate |
| Isopulegol; *p*-menth-8-en-3-ol |
| Isopulegone; *p*-menth-8-en-3-one |
| Isopulegyl acetate |
| Isoquinoline |
| Isovaleric acid |
| *cis*-Jasmone; 3-methyl-2-(2-pentenyl)-2-cyclopenten-1-one |
| Lauric aldehyde; dodecanal |
| Lauryl acetate |
| Lauryl alcohol; 1-dodecanol |
| Lepidine; 4-methylquinoline. |
| Levulinic acid |
| Linalool oxide; *cis*- and *trans*-2-vinyl-2-methyl-5-(1′-hydroxy-1′-methylethyl) tetrahydrofuran |
| Linalyl anthranilate; 3,7-dimethyl-1,6-octadien-3-yl anthranilate |
| Linalyl benzoate |
| Linalyl butyrate |
| Linalyl cinnamate |
| Linalyl formate |
| Linalyl hexanoate |
| Linalyl isobutyrate |
| Linalyl isovalerate |
| Linalyl octanoate |
| Linalyl propionate |
| Maltol; 3-hydroxy-2-methyl-4*H*-pyran-4-one |
| Menthadienol; *p*-mentha-1,8(10)-dien-9-ol |
| *p*-Mentha-1,8-dien-7-ol; perillyl alcohol |
| Menthadienyl acetate; *p*-mentha-1,8(10)-dien-9-yl acetate |
| *p*-Menth-3-en-1-ol |
| 1-*p*-Menthen-9-yl acetate; p-menth-1-en-9-yl acetate |
| Menthol; 2-isopropyl-5-methylcyclohexanol |
| Menthone; *p*-menthan-3-one |
| Menthyl acetate; *p*-menth-3-yl acetate |
| Menthyl isovalerate; *p*-menth-3-yl isovalerate |
| *o*-Methoxybenzaldehyde |

*p*-Methoxybenzaldehyde; *p*-anisaldehyde
*o*-Methoxycinnamaldehyde
2-Methoxy-4-methylphenol; 4-methylguaiacol; 2-methoxy-*p*-cresol
4-(*p*-Methoxyphenyl)-2-butanone; anisyl acetone
1-(4-Methoxyphenyl)-4-methyl-1-penten-3-one; methoxystyryl isopropyl ketone
1-(*p*-Methoxyphenyl)-1-penten-3-one; methyl-anisylidene acetone; ethone
1-(*p*-Methoxyphenyl)-2-propanone; anisylmethyl ketone; anisic ketone
2-Methoxy-4-vinylphenol; *p*-vinylguaiacol
Methyl acetate
4-Methylacetophenone; *p*-methylacetophenone; methyl *p*-tolyl ketone
2-Methylallyl butyrate; 2-methyl-2-propenyl butyrate
Methyl anisate
*o*-Methylanisole; *o*-cresyl methyl ether
*p*-Methylanisole; *p*-cresyl methyl ether; *p*-methoxytoluene
Methyl benzoate
Methylbenzyl acetate, mixed *o*-, *m*-, *p*-
α-Methylbenzyl acetate; styralyl acetate
α-Methylbenzyl alcohol; styralyl alcohol
α-Methylbenzyl isobutyrate; styralyl isobutyrate
α-Methylbenzyl formate; styralyl formate
α-Methylbenzyl propionate; styralyl propionate
2-Methyl-3-buten-2-ol
2-Methylbutyl isovalerate
Methyl *p-tert*-butylphenylacetate
2-Methylbutyraldehyde; methyl ethyl acetaldehyde
3-Methylbutyraldehyde; isovaleraldehyde
Methyl butyrate
2-Methylbutyric acid
α-Methylcinnamaldehyde
*p*-Methylcinnamaldehyde
Methyl cinnamate
2-Methyl-1,3-cyclohexadiene
Methylcyclopentenolone; 3-methylcyclopentane-1,2-dione
Methyl disulfide; dimethyl disulfide
Methyl ester of rosin, partially hydrogenated (21 CFR 172.615); methyl dihydroabietate
Methyl heptanoate
2-Methylheptanoic acid
6-Methyl-3,5-heptadien-2-one
Methyl-5-hepten-2-ol
6-Methyl-5-hepten-2-one
Methyl hexanoate
Methyl 2-hexanoate
Methyl *p*-hydroxybenzoate; methylparaben
Methyl α-ionone; 5-(2,6,6-trimethyl-2-cyclohexen-1-yl)-4-penten-3-one
Methyl β-ionone; 5-(2,6,6-trimethyl-1-cyclohexen-1-yl)-4-penten-3-one
Methyl Δ-ionone; 5-(2,6,6-trimethyl-3-cyclohexen-1-yl-)-4-penten-3-one
Methyl isobutyrate
2-Methyl-3-(*p*-isopropylphenyl)-propionaldehyde; α-methyl-*p*-isopropylhydro-cinnamal-dehyde; cyclamen aldehyde
Methyl isovalerate
Methyl laurate

**Table 9.6.4 (continued)**

Methyl mercaptan; methanethiol
Methyl *o*-methoxybenzoate
Methyl *N*-methylanthranilate; dimethyl anthranilate
Methyl 2-methylbutyrate
Methyl 2-methylthiopropionate
Methyl 4-methylvalerate
Methyl myristate
Methyl *β*-naphthyl ketone; 2′-acetonaphthone
Methyl nonanoate
Methyl 2-nonenoate
Methyl 2-nonenoate; methyloctyne carbonate
2-Methyloctanal; methyl hexyl acetaldehyde
Methyl octanoate
Methyl 2-octynoate; methyl heptine carbonate
4-Methyl-2,3-pentanedione; acetyl isobutyryl
4-Methyl-2-pentanone; methyl isobutyl ketone
*β*-Methylphenethyl alcohol; hydratropyl alcohol
Methyl phenylacetate
3-Methyl-4-phenyl-3-butene-2-one
2-Methyl-4-phenyl-2-butyl acetate; dimethylphenylethyl carbinyl acetate
2-Methyl-4-phenyl-2-butyl isobutyrate; dimethylphenyl ethylcarbinyl isobutyrate
3-Methyl-2-phenylbutyraldehyde; *α*-isopropyl phenylacetaldehyde
Methyl 4-phenylbutyrate
4-Methyl-1-phenyl-2-pentanone; benzyl isobutyl ketone
Methyl 3-phenylpropionate; methyl hydrocinnamate
Methyl propionate
3-Methyl-5-propyl-2-cyclohexen-1-one
Methyl sulfide
3-Methylthiopropionaldehyde; methional
2-Methyl-3-tolylpropionaldehyde, mixed *o*-, *m*-, *p*-
2-Methylundecanal; methyl nonyl acetaldehyde
Methyl 9-undecenoate
Methyl 2-undecynoate; methyl decyne carbonate
Methyl valerate
2-Methylvaleric acid
Myrcene; 7-methyl-3-methylene-1,6-octadiene
Myristaldehyde; tetradecanal
*d*-Neomenthol; 2-isopropyl-5-methylcyclohexanol
Nerol; *cis*-3,7-dimenthyl-2,6-octadien-1-ol
Nerolidol; 3,7,11-trimethyl-1,6,10-dodecatrien-3-ol
Neryl acetate
Neryl butyrate
Neryl formate
Neryl isobutyrate
Neryl isovalerate
Neryl propionate
2,6-Nonadien-1-ol
*γ*-Nonalactone; 4-hydroxynonanoic acid, *γ*-lactone; aldehyde C-18.
Nonanal; pelargonic aldehyde
1,3-Nonanediol acetate, mixed esters
Nonanoic acid; pelargonic acid
2-Nonanone; methylheptyl ketone
3-Nonanon-1-yl acetate; 1-hydroxy-3-nonanone acetate
Nonyl acetate
Nonyl alcohol; 1-nonanol

Nonyl octanoate
Nonyl isovalerate
Nootkatone; 5,6-dimethyl-8-isopropenyl-bicyclo[4.4.0]dec-1-en-3-one
Ocimene; *trans*-$\beta$-ocimene; 3,7-dimethyl-a,3,6-octatriene
$\gamma$-Octalactone; 4-hydroxyoctanoic acid, $\gamma$-lactone
Octanal; caprylaldehyde
Octanol; octyl alcohol
2-Octanol
3-Octanol
2-Octanone; methyl hexyl ketone
3-Octanone; ethyl amyl ketone
3-Octanon-1-ol
1-Octen-3-ol; amyl vinyl carbinol
1-Octen-3-yl acetate
Octyl acetate
3-Octyl acetate
Octyl butyrate
Octyl formate
Octyl heptanoate
Octyl isobutyrate
Octyl isovalerate
Octyl octanoate
Octyl phenylacetate
Octyl propionate
$\omega$-Pentadecalactone; 15-hydroxypentadecanoic acid, $\omega$-lactone; pentadecanolide; angelica lactone
2,3-Pentanedione; acetyl propionyl
2-Pentanone; methyl propyl ketone
4-Pentenoic acid
1-Penten-3-ol
Perillaldehyde; 4-isopropenyl-1-cyclohexene-1-carboxaldehyde; *p*-mentha-1,8-dien-7-al
Perillyl acetate; *p*-mentha-1,8-dien-7-yl acetate
$\alpha$-Phellandrene; *p*-mentha-1,5-diene
Phenethyl acetate
Phenethyl alcohol; $\beta$-phenylethyl alcohol
Phenethyl anthranilate
Phenethyl benzoate
Phenethyl butyrate
Phenethyl cinnamate
Phenethyl formate
Phenethyl isobutyrate
Phenethyl isovalerate
Phenethyl 2-methylbutyrate
Phenethyl phenylacetate
Phenethyl propionate
Phenethyl salicylate
Phenethyl senecioate; phenethyl 3,3-dimethylacrylate
Phenethyl tiglate
Phenoxyacetic acid
2-Phenoxyethyl isobutyrate
Phenylacetaldehyde; $\alpha$-toluic aldehyde
Phenylacetaldehyde 2,3-butylene glycol acetal
Phenylacetaldehyde dimethyl acetal
Phenylacetaldehyde glyceryl acetal
Phenylacetic acid; $\alpha$-toluic acid
4-Phenyl-2-butanol; phenylethyl methyl carbinol
4-Phenyl-3-buten-2-ol; methyl styryl carbinol
4-Phenyl-3-buten-2-one
4-Phenyl-2-butyl acetate; phenylethyl methyl carbinyl acetate
1-Phenyl-3-methyl-3-pentanol; phenylethyl methyl ethyl carbinol
1-Phenyl-1-propanol; phenylethyl carbinol
3-Phenyl-1-propanol; hydrocinnamyl alcohol
2-Phenylpropionaldehyde; hydratropaldehyde

**Table 9.6.4 (continued)**

3-Phenylpropionic acid; hydrocinnamic acid
3-Phenylpropyl acetate
2-Phenylpropyl butyrate
3-Phenylpropyl cinnamate
3-Phenylpropyl formate
3-Phenylpropyl hexanoate
2-Phenylpropyl isobutyrate
3-Phenylpropyl isobutyrate
3-Phenylpropyl isovalerate
3-Phenylpropyl propionate
2-(3-Phenylpropyl)-tetrahydrofuran
$\alpha$-Pinene; 2-pinene
$\beta$-Pinene; 2(10)-pinene
Pine tar oil
Pinocarveol; 2(10)-pinen-3-ol
Piperidine
Piperine
*d*-Piperitone; *p*-menth-1-en-3-one
Piperitenone; *p*-mentha-1,4(8)-dien-3-one
Piperitenone oxide; 1,2-epoxy-*p*-menth-4-(8)-en-3-one
Piperonyl acetate; heliotropyl acetate
Piperonyl isobutyrate
Polylimonene
Polysorbate 20; polyoxyethylene (20) sorbitan monolaurate
Polysorbate 60; polyoxyethylene (20) worbitan monostereate
Polysorbate 80; polyoxyethylene (20) sorbitan monooleate
Potassium acetate
Propenylguaethol; 6-ethoxy-*m*-anol
Propionaldehyde
Propyl acetate
Propyl alcohol; 1-propanol
*p*-Propyl anisole; dihydroanethole
Propyl benzoate
Propyl butyrate
Propyl cinnamate
Propyl disulfide
Propyl formate
Propyl 2-furanacrylate
Propyl heptanoate
Propyl hexanoate
Propyl *p*-hydroxybenzoate; propylparaben
3-Propylidenephthalide
Propyl isobutyrate
Propyl isovalerate
Propyl mercaptan
$\alpha$-Propylphenethyl alcohol
Propyl phenylacetate
Propyl propionate
Pulegone; *p*-menth-4(8)-en-3-one
Pyridine
Pyroligneous acid extract
Pyruvaldehyde
Pyruvic acid
Rhodinol; 3,7-dimethyl-7-octen-1-ol; *l*-citronellol
Rhodinyl acetate
Rhodinyl butyrate
Rhodinyl formate
Rhodinyl isobutyrate
Rhodinyl isovalerate
Rhodinyl phenylacetate
Rhodinyl propionate
Rum ether; ethyl oxyhydrate
Salicylaldehyde
Santalol, $\alpha$ and $\beta$
Santalyl acetate
Santalyl phenylacetate
Skatole
Sorbitan monostearate
Styrene
Sucrose octaacetate
$\alpha$-Terpinene
$\gamma$-Terpinene

$\alpha$-Terpineol; *p*-menth-1-en-8-ol
$\beta$-Terpineol
Terpinolene; *p*-menth-1,4(8)-diene
Terpinyl acetate
Terpinyl anthranilate
Terpinyl butyrate
Terpinyl cinnamate
Terpinyl formate
Terpinyl isobutyrate
Terpinyl isovalerate
Terpinyl propionate
Tetrahydrofurfuryl acetate
Tetrahydrofurfuryl alcohol
Tetrahydrofurfuryl butyrate
Tetrahydrofurfuryl propionate
Tetrahydro-pseudo-ionone; 6,10-dimethyl-9-hundecen-2-one
Tetrahydrolinalool; 3,7-dimethyloctan-3-ol
Tetramethyl ethylcyclohexenone; mixture of 5-ethyl-2-3-4-5-tetramethyl-2-cyclohexen-1-one and 5-ethyl-3,4,5,6-tetramethyl-2-cyclohexen-1-one
2-Thienyl mercaptan; 2-thienylthiol
Thymol
Tolualdehyde glyceryl acetal, mixed *o*-, *m*-, *p*-
Tolualdehydes, mixed *o*-, *m*-, *p*-
*p*-Tolylacetaldehyde
*o*-Tolyl acetate; *o*-cresyl acetate
*p*-Tolyl acetate; *p*-cresyl acetate
4-(*p*-Tolyl)-2-butanone; *p*-methylbenzylacetone
*p*-Tolyl isobutyrate
*p*-Tolyl laurate
*p*-Tolyl phenylacetate
2-(*p*-Tolyl)-propionaldehyde; *p*-methylhydratropic aldehyde
Tributyl acetylcitrate
2-Tridecenal
2,3-Undecadione; acetyl nonyryl
$\gamma$-Undecalactone; 4-hydroxyundecanoic acid $\gamma$-lactone; peach aldehyde; aldehyde C-14
Undecenal
2-Undecanone; methyl nonyl ketone
9-Undecenal; undecenoic aldehyde
10-Undecenal
Undecen-1-ol; undecylenic alcohol
10-Undecen-1-yl acetate
Undecyl alcohol
Valeraldehyde; pentanal
Valeric acid; pentanoic acid
Vanillin acetate; acetyl vanillin
Veratraldehyde
Verbenol; 2-pinen-4-ol
Zingerone; 4-(4-hydroxy-3-methoxyphenyl)-2-butanone

---

Source: 21 CFR 172.515, 1991.

[a]According to FDA regulations, synthetic flavoring substances and adjuvants may be safely used in food in accordance with the following conditions:

1. They are used in the minimum quantity required to produce their intended effect, and otherwise in accordance with all the principles of good manufacturing practice.
2. They consist of one or more of the chemicals listed above used alone or in combination with flavoring substances and adjuvants generally recognized as safe in food, prior, sanctioned for such use, as governed by the appropriate FDA regulations.

# CHAPTER 10
# FOOD LAWS AND REGULATIONS

It is probably not an exaggeration to state that this country has the safest food supply in the world. It is also true to state that the United States has the most stringent food laws and regulations in the world.

Food processors in this country and those who export to the United States have to spend an enormous amount of effort to become familiar with U.S. food laws and regulations so that their products will not violate government requirements. This chapter provides data on the following categories of laws and regulations enforced by the U.S. Food and Drug Administration (FDA):

1. Major responsibilities of the FDA.
2. Legal and administrative terms used by the FDA.
3. Food additives.
4. Food standards.
5. Tables of contents for the FDA Inspection Operations Manual (IOM).
6. Tables of contents for the FDA Regulatory Procedures Manual (RPM).

Selected data are presented for each category of information. The subject matters for items 1 to 3 are self-explanatory. However, items 4 to 7 deserve a brief discussion.

Under the Freedom of Information Act, the FDA makes it possible for interested parties to purchase two important manuals which detail some of the major workings of the agency.

The IOM describes in detail the inspection operations by FDA inspectors. This document contains thousands of pages and is continuously updated and increased in size. Ideally, every food processor should purchase a copy, but the status of this publication discourages this practice. First, it is too big; second, it is difficult to read because the letters are small; and third, keeping up with update pages for any form of technical literature is a big task.

The RPM is the master guide used by regulators and lawyers within the FDA. It contains legal and administrative procedures adopted by the FDA in enforcing its mandates. The reasons why many food processors do not own a copy of RPM are same as those for the IOM.

Obviously, it is not possible for this book to reproduce the two manuals. The alternative of including their tables of contents in this chapter will permit the reader an instant overview of the contents of these two important FDA documents.

The citation of legal document follows the standard format. For example, 21 CFR 133 refers to the U.S. Code of Federal Regulations, Volume 21, Part 133. The same applies to U.S. Federal Register (FR). Both legal documents and the U.S. Tariff Schedules are available form the Government Printing Office, Washington, D.C.

All documents published by the FDA may be available from the FDA (Washington, D.C.) or from public (private) agencies designated by the FDA to distribute some publications.

Other abbreviations used in this chapter include:

EDRO: Executive Director of Regional Operations.
EI: Establishment Inspection.
GMP: Good Manufacturing Practice.
GRAS: Generally Recognized As Safe.
HACCP: Hazards Analysis Critical Control Points.
IMS: Interstate Milk Shippers.
LACF: Low-acid canned foods.
PHS: Public Health Service.
USDA: United States Department of Agriculture.

## 10.1. Responsibilities of the U. S. Food and Drug Administration

The U.S. Food and Drug Administration has summarized the major requirements of laws and regulations it enforces in a governemnt document titled *Requirements Laws and Regulations Enforced by the U.S. Food and Drug Administration*, DHHS Publication No. (FDA) 89-1115, 1989. Several tables in this section have been adapted from this document. Where applicable, this document will be cited as "Legal Requirements."

**Table 10.1.1. Legal Responsibilities of the FDA**

Drugs and feeds for human food-producing animals: to regulate drugs and feeds for human food-producing animals

Enforcement actions: to initiate corrective action such as citations, injunctions, recalls, and seizures whenever a potential hazard is identified

Food additives: to make decisions on the safety of food additives and to regulate the conditions of their usage

Food colors: to determine the safety of food colors and to test batches of those subject to certification

Imports: to inspect imported food products to assure compliance with domestic federal laws

Industrial chemicals: to monitor food supply to detect possible contamination with industrial chemicals or other deleterious substances and to take proper regulatory actions

Inspections: to make periodic inspections of food establishments and to examine food product samples for adulteration or misbranding

Interagency cooperation: to maintain adequate communication with and to extend maximal cooperation to State authorities and other Federal agencies in urgent situations such as massive food contamination by floods, explosions, fires, and food poisoning

Labeling: to develop and enforce regulations for proper food labeling

Pesticide residues: to enforce safe limits on the amount of pesticide residues that may remain on food crops, if any

Prosecution: to initiate court proceedings against firms and individuals that have violated the law(s)

Public information: to disseminate information to consumers so as to promote and support consumer self-help activities

Research: to conduct research to identify, reduce, and control consumer risks

Standards: to establish standards of identity, quality, and fill of containers for food products

Travel and food: to enforce sanitation and safety regulations for food and water served to passengers traveling in regular interstate vehicles
Voluntary compliance: to assist the food industry in voluntary compliance with the law and to issue regulations designed to prevent violations
Warning and alert: to warn and alert the general public when a hazardous food product is identified

Source: Adapted from "Legal Requirements."

**Table 10.1.2. Statutes Administered by the FDA**

The Federal Food, Drug, and Cosmetic Act of 1938, as amended
The Fair Packaging and Labeling Act of 1966
The Radiation Control for Health and Safety Act of 1968
The Public Health Service Act of 1944
The Infant Formula Act of 1983
The Tea Importation Act of 1897, as amended
Import Milk Act of 1927
Federal Meat Inspection Act of 1967
Poultry Products Inspection Act of 1957, as amended
Egg Products Inspection Act of 1970

Source: Adapted from "Legal Requirements."

**Table 10.1.3. Health Safeguards of the U.S. Federal Food, Drug, and Cosmetic Act**

A food is illegal (adulterated) if it contains an added poisonous or deleterious (harmful) substance that *may* render it injurious to health [402(a)(1)].
A food is illegal if it contains a natural poisonous or deleterious substance that *ordinarily* renders it injurious to health [402(a)(1)]
Food additives [201(s) and (409)] must be determined to be safe by FDA before they may be used in a food or become a part of a food as a result of processing, packaging, transporting, or holding the food.
Raw agricultural products are illegal if they contain residues of pesticides not authorized by, or in excess of, tolerances established by regulations of the Environmental Protection Agency (408).
A food is illegal if it is prepared, packed, or held under insanitary conditions whereby it *may* have been rendered injurious to health [402(a)(4)].

**Table 10.1.3. (continued)**

Food containers must be free from any poisonous or deleterious substance that may cause the contents to be injurious to health [402(a)(6)]. Some packaging materials, for example plastic or vinyl containers, may be "food additives" subject to regulations (409).

Colors added to food must be only those established by the FDA as being safe [402(c) and 706]. Unless exempt, colors for use in food must be from batches tested and certified by the FDA [706(c)].

A food is illegal if any part of it is filthy, putrid, or decomposed [402(a)(3)].

A food is illegal if it is prepared, packed, or held under insanitary conditions whereby it may have become contaminated with filth [402(a)(4)].

A food is illegal if it is the product of a diseased animal or one that has died otherwise than by slaughter [402(a)(5)].

Source: Adapted from "Legal Requirements."
Note: Information in parentheses refers to the relevant sections of the Act.

**Table 10.1.4. Economic Safeguards of the U.S. Federal Food, Drug, and Cosmetic Act**

Damage or inferiority in a food must not be concealed in any manner [402(b)(3)]. Example: artificial coloring and/or flavoring added to a food to make it appear a better grade than it is, as in the case of yellow coloring used to make a food appear to contain more eggs than it actually contains.

Food labels or labeling (circulars, etc.) must not be false or misleading in any particular [403(a)]. Labeling is misleading not only if it contains false or misleading statements but also if it fails to reveal material facts [201(n)].

A food must not be sold under the name of another food [403(b)]. Example: canned bonito labeled as tunafish.

A substance recognized as being a valuable constituent of a food must not be omitted or abstracted in whole or in part, nor may any substance be substituted for the food in whole or in part [402(b)(1) and (2)]. Example: an article labeled "milk" or "whole milk" from which part of the butterfat has been skimmed.

Food containers must not be so made, formed, or filled as to be misleading [403(d)]. Example: a closed package filled to less than its capacity. A food for which a standard of fill of container has been prescribed (401) must comply with the full requirements, and if the fill falls below that specified its label must bear a statement that it falls below such standard [403(h)(2)].

Source: Adapted from "Legal Requirements."
Note: Information in parentheses refers to the relevant sections of the Act.

**Table 10.1.5. FDA Regulations for Sanitation and Safety**

| |
|---|
| Sanitation requirements and factory inspection |
| Current good manufacturing practice regulations |
| Regulations |
| Personnel |
| Exclusions |
| Buildings and facilities: plants and grounds |
| Buildings and facilities: sanitary facilities and control |
| Buildings and facilities: sanitary operations |
| Equipment and procedures |
| Production and process controls |
| Emergency permits, low-acid foods and acidified foods |
| Emergency permit |
| Thermally processed low-acid foods packaged in hermetically sealed containers |
| Acidified foods |
| Unavoidable contaminants |
| Establishment of regulations |
| Added poisonous and deleterious substances |
| Unavoidability |
| Exceptions |
| Polychlorinated biphenyls (PCBs) |
| Food action levels |

Source: Adapted from "Legal Requirements."

**Table 10.1.6. FDA Regulations for Food and Agricultural Chemicals, Food Service, and Drugs**

| | |
|---|---|
| Food and color additives | Basic responsibilities |
| Food additives | Regulations promulgated |
| Color additives | Animal feeds |
| Pesticides residues on raw agricultural products | Animal foods for special dietary uses |
| Food service sanitation | Food additives |
| Drugs and feeds for food-producing animals | Animal dosage form drugs and medicated feeds |

Source: Adapted from "Legal Requirements."

**Table 10.1.7. FDA Regulations and Product Categories**

- Baby food
- Bakery products
  - Standards
  - Inspection instruction
- Beverages and beverage materials
  - Alcoholic beverages
  - Nonalcoholic beverages
- Cacao products
- Cereal flours and related products
- Cheese and cheese products
- Eggs and egg products
- Fish and shellfish
  - Basic requirements
  - Shellfish sanitation program
  - Special seafood inspection program
  - Specific FDA inspection instructions
- Food dressings and flavorings
- Frozen desserts
- Fruit and fruit products
  - An overview
  - Fresh fruit juices
  - Canned fruits and fruit juices
  - Frozen fruits and juices
  - Dried fruits
  - Fruit jams (preserves), jellies, fruit butters, marmalades
- Grain and related products
- Macaroni and noodle products
- Margarine
- Meat, poultry, and related products
  - An overview
  - Jurisdiction of the FDA over meat food products
  - Detention of carcasses
  - Recall of food products with meat or poultry ingredients
  - Drug, pesticide, and industrial chemical residues in animal feeds, meat, and poultry
- Milk and milk products
  - Milk sanitation
  - Inspection procedures
  - Standards for milk and cream
- Nuts and nut products
- Spice, condiments, and related products
- Sweeteners, sirups, and related products
- Vegetables and vegetable products
  - Canned vegetables
  - Vegetable juices and frozen vegetables
  - Pickles
- Vegetable oil

Source: Adapted from "Legal Requirements."

**Table 10.1.8. FDA Enforcement Actions**

- Imminent hazard
- Publicity
- Recalls
- Notice of adverse findings and regulatory letters
- Seizures
- Citations
- Injunctions
- Criminal prosecution
- Compliance followup on insanitary conditions in food plants

Source: Adapted from "Legal Requirements."

## 10.2. Terms Used by the FDA

**Table 10.2.1. FDA's Administrative and Legal Terminology**

| Term: Definition |
|---|
| Abate: to put an end to; demolish; to do away with, to nullify, to make void. |
| Adjournment: a putting off or postponing of business or of a session until another time or place; the act of a court |
| Adjudicate (adjudication): to make a final judgment in a lawsuit; the judgment of the court |
| Admonition: a reprimand from a judge to a person accused, on being discharged, warning of the consequences of his conduct and should he be guilty of same fault, he will be punished with greater severity |
| Affidavit: a written statement made voluntarily under oath. An affidavit may be made on (a) personal knowledge or (b) information and belief. |
| Allegation: an assertion or statement, made in a pleading, setting forth what its maker intends to prove |
| *Amicus curiae*: a friend of the court; A person who, although not a party in a suit, is allowed to intervene to protect his interests |
| Analysis: An examination of a sample |
| Answer: After filing of a Claim of Ownership in a seizure, the claimant files an *Answer* in which he may deny any or all of the allegations of the *Complaint for Forfeiture*. |
| Appeal: the formal review by a higher court of a lower court's disposition of a lawsuit |
| Arraignment: the appearance of the defendant (in any criminal prosecution) before the court to answer that allegation made against him and to enter his plea (guilty or not guilty) |
| Avoirdupois: 16 oz/1 lb; used in weighing articles other than medicine, metals, and precious stones. |
| Bill of lading, B/L: the written order from a shipper to a carrier lading, to move goods from one place to another. When available, this is the best source of shipping dates, origin, and name of shipper. |
| Bill of particulars: used in criminal cases only; a motion requesting the specifics of a charge, so that defense against the charges may be properly prepared |
| Bona fide: literally "in good faith." A bona fide document is a genuine one. |
| Bond: a written promise to pay a certain sum of money in the event agreed upon conditions are not met. For example, Section 304(d) of the FD&C Act requires a bond as a condition of the release of the condemned article for salvaging. |

**Table 10.2.1. (continued)**

Burden of proof: the necessity or duty of proving a fact or facts in dispute on an issue. If the evidence produced by the party having the burden of proof is insufficient, the court will rule against that party.

Certiorari: Certiorari means "to be certified"; a writ commanding a court to certify records to a superior court. When the supreme court grants certiorari, that means it has agreed to hear the case. If it denies certiorari, it will not hear the case.

Charge sheet: the specification of the violation attached to a Notice of Hearing. It is the proposed statement of charges that, should prosecution ensue, would become the government's allegation in the criminal investigation.

Check analysis: a second examination of a sample which has been examined and found violative. Check analyses are to be conducted by other than the analyst performing the original analysis.

Citation: see "Hearing."

Citee: one who is cited. See "Hearing."

Civil suit: a lawsuit brought under the rules of civil procedure. For example, seizure, injunction, and civil contempt are actions initiated under civil rules.

Claim: a statement of interest in seized property, affirmed by sworn oath, entered by a person in response to seizure of a lot

Commingled shipment: two or more separate shipments of the same product, stored together and bearing no distinguishing marks that enable easy separation. Codes normally do not make a sufficient distinction, since there may be many codes in one shipment.

Complainant: person registering a complaint, as a consumer complaining about an article

Complaint for forfeiture: the document furnished to the U.S. Attorney for filing with the Clerk of the Court to initiate a seizure of a lot

Consent decree of condemnation: the document entered by the court in a seizure action based on the claimant's agreement that the article seized is in violation as alleged in the Complaint for Forfeiture and the article is subject to condemnation. It is also a declaration of Claimant's intent to provide a bond and to recondition the article under supervision of FDA and to pay costs.

Contempt of court: any act that is calculated to embarrass, hinder, or obstruct a court in the administration of justice, or that is calculated to lessen its authority or its dignity

Count: a statement of a violation of the Act in a Criminal Information or Indictment

Contest: to make defense to an adverse claim in a court of law; to oppose, resist, or dispute the case made by a plaintiff

Counterclaim: a claim presented by a defendant in opposition to the claim of the plaintiff, which claim could have been filed in a separate suit. If asserted in the same suit brought by plaintiff, it is called a counterclaim.

Contraband: against the law; prohibited; goods exported or imported into a country against its laws

Criminal suit: a lawsuit brought under the Rules of Criminal Procedure, whereby the defendant may be punished for having violated the law, or a court order. Actions initiated under the criminal rules are prosecutions and contempt.

Dealer: The term is agency jargon to identify the party from whom a sample was collected; this does not have to be a business concern and often is a private party. Under the Radiation Act, this term is defined slightly differently, as found in 21 CFR 1000.3(L).

Dealer statement (D/S): the written statement taken from the dealer at the time a sample is collected and in which the dealer asserts that the sample obtained was taken from a lot received by him from a certain source and that those documents showing the movement of the lot were furnished for copying

Dealer violation: a violation of Section 301(k) of the FD&C Act indicating the doing of some act, after shipment in interstate commerce, that has resulted in the article being adulterated or misbranded

Default decree: A Default Decree of Condemnation is a court order entered when lots under seizure are not claimed or defended. The order condemns the product as being in violation and provides for destruction, donation to charity, sale, or disposal of the court may elect to decree. When signed by the court, it signifies the adjudicatory step in a seizure action.

Defendant: the party or parties named in an Information, Indictment, or Complaint for Injunction, and against whom the government is proceeding. The article against which a seizure action is brought is also referred to as the defendant.

De minimus: not enough to be considered

Denature: The decharacterization of an article generally through the addition of some foreign substance in sufficient quantity so as to preclude its future use for its original purpose

Deposition: the sworn testimony of a witness obtained outside the courtroom before trial through examination and crossexamination by the attorneys for the parties

Detention: the act of preventing entry into the country of any article that appears to be in violation of the Act. The detention allows the owner to

**Table 10.2.1. (continued)**

come forth and define the lot, much in the same manner as a hearing prior to prosecution.

Discovery: the process of obtaining information concerning the other party's case. Methods of discovery are contained in Rule 16 of the Rules of Criminal Procedure, and Rules 26 through 37 of the Rules of Civil Procedure.

Dismissal: an order of judgment disposing of an action, suit, motion, etc., by sending it out of court. A dismissal would always be in the defendant's favor.

Docket: a chronological listing of all pleadings and orders filed in each case. It is usually maintained by the Clerk of Court in the form of a "Docket Sheet."

Documentation: the collection of documents to support a certain element of proof, i.e., the copying of a Bill of Lading to demonstrate that interstate commerce has been accomplished

Effectiveness check: the determination of the effectiveness or completeness of a recall

Efficacy: worth, merit, value

Enjoin: a restraint through formal court order; to cease a practice or to stop further violation of the Act

Entrapment: the act of agents of the government in inducing a person to commit a crime not contemplated by him, for the purpose of bringing criminal charges against him. Engaging in artifice and deceit to afford someone an opportunity to commit a crime is not entrapment.

Et al.: and others; for example, samples 021-77-198 et al.

Ex parte: on one side only (cases in which only one side is represented)

Execution of decree: the carrying out of the court's order, e.g., the destruction of goods under seizure by the Marshal in response to a Default Decree of Condemnation.

Fiat: a command; order directing some legal act to be done

Final order: The last order issued by the court in a lawsuit; it terminates the action.

Forfeiture of bond: literally, the loss of a bond, as a result of a failure to perform the conditions agreed upon when the bond was made

Freight Bill (F/B): the document stating the transportation charges incurred by the carrier (a bill for the freight charges on a shipment). Next to a B/L, the freight bill is the next most authentic document that supports movement of a shipment in interstate commerce.

General Council (GC): A group of attorneys employed by the Department of Health and Human Services (DHHS) and assigned to handle lawsuits brought by or against FDA. The chief lawyer is an Assistant to the General Counsel of DHHS.

Grand jury: a jury of inquiry composed of 23 citizens whose duty is to determine, in the case of proposed felony charges, if there is sufficient evidence to warrant a trial

Guaranty: a formal and signed agreement between buyers and sellers in which the latter verifies the goods he sells are not in violation of the FD&C Act when shipped

Hearing: the opportunity for a party to present views. In imports, the importer may explain why his goods should not be refused entry and returned to the shipper. A hearing prior to consideration of criminal proceedings is provided pursuant to Section 305 of the Act. Many types of Administration hearings are provided for in the various sections of the Code of Federal Regulations.

Indictment: the formal statement of the charges against an individual presented by a grand jury. The U.S. Attorney (with witnesses deemed neccessary) appears before a grand jury and attempts to convince this body that an offense has been committed. If the grand jury concurs, it approves, signs, and returns the True Bill of Indictment to the court and the matter is then entered on the court docket. This procedure is utilized in second offense cases to afford the court the latitude of imposing the stronger penalties available for the punishment of those convicted.

Information: the formal statement of the charges against an individual in a misdemeanor, issued by the U.S. Attorney, and not by a grand jury. It may always be used in misdemeanor prosections and may be used in felony prosections only if the defendant waives his right to be indicted.

Information letter: a letter to responsible officials of a firm warning of the existence of a violative practice

Injunction: an order issued by the court requiring a defendant to do or refrain from doing a specified act

In personam: against the person

In rem: against the thing

Invoice: a document showing the seller's intent to sell an article and showing the amount due. This document does not prove actual movement of the goods.

Interrogatories: written questions propounded by one party in a suit and served on the other party, who must serve written answers thereto under oath within a given period of time

Interstate (IS): between one State and any place outside of that state, or within the District of Columbia or any other U.S. territory

Intrastate: within a State.

**Table 10.2.1. (continued)**

Ipso facto: by the act itself

Judicial district: physical parameters, set by Congress, designating the counties under the jurisdiction of a U.S. District Court

Label/labeling: These words are defined in the FD&C Act.

Legal status: the legal constitution of a business enterprise as to whether it is a corporation, partnership, cooperative, or sole ownership

Litigation: the act or process of carrying on a lawsuit

Magistrate: formerly U.S. Commissioner; a part-time or full-time court official authorized to try misdemeanor cases without a jury, and to perform certain other duties of the court, such as issuing inspection warrants.

Mandamus: a command that a superior court issues to a lower court or a person (including a government agency), ordering it to do its duty

Manifest: an itemized listing of a vessel's cargo, with other particulars, for the facility of the customs officers

Mitigation: alleviation; abatement or diminution of a penalty or punishment imposed by law, as a result of the circumstances surrounding the crime, or the acts of the defendant (e.g., he gave the money back after robbing the bank).

Modus operandi: manner of working

Monition: the summons to appear and answer, issued on filing the complaint

Motion: an application to a court, either written or oral, for a rule or order, such as to permit postseizure sampling

Nolle prosequi (nol. pros.): Not willing to prosecute, either as to some of the counts, or some of the defendants, or both

Nolo contendere: a plea in a criminal case which is essentially the same as a guilty plea. The only difference is that a guilty plea may be used as evidence in a subsequent civil suit against the defendant, but a nolo plea may not be so used. A nolo plea counts as a first offense in second offense cases.

Nunc pro tunc: Now for then. A phrase applied to acts allowed to be done after the time when they should have been done, with a retroactive effect.

Official compendia: See FD&C Act Section 201(g)(1)(A).

Order: a formal written statement from a court, requiring action or simply stating a ruling

Pendente lite-pending: while litigation continues (or is pending)

Permanent injunction: A Decree of Permanent Injunction perpetually restrains the defendants from engaging in specified violative practices and remains in force until termination.

Plaintiff: the party who institutes the lawsuit; the opposite of the defendant. The government is the plaintiff in all the actions it initiates. When the government is sued, the suing party is the plaintiff in the action.

Pleadings: the papers filed in court to initiate and defend a suit, such as the Complaint, and answer, setting forth the allegations and defense in the case.

Preliminary injunction (or temporary injunction): the intermediate segment of the injunction process which is initiated through filing of a Motion of Preliminary Injunction. If granted and signed by the court, the defendant is preliminarily restrained from committing future violations of the Act; it is not a final process, and is an unadjudicated civil case as far as the court is concerned.

Presentence investigation: Many courts request an investigation of convicted parties prior to the imposition of sentence. These investigations are conducted by the U.S. Probation Officer and are governed by Rule 32(c) of the Federal Rules of Criminal Procedure.

Pretrial conference: a meeting between the court and the attorney for the parties in preparation for trial, e.g., to consider various ways to limit the issues for trial through admissions or agreements of counsel

Prima facie: evidence sufficient to support the allegations of a Complaint unless refuted by some evidence to the contrary

Procedure in Admiralty: Seizure actions taken pursuant to the FD&C Act are [per Section 304(b)] to be in conformity with the procedure in Admiralty, which means the government may initiate confiscation of articles it deems to be contraband without having to first show proof of the allegation or responsibility for the violation.

Pro forma: as a matter of form

Promulgation: to propose and publish as an official announcement; i.e., to promulgate new regulations pursuant to the FD&C Act

Prosecution: the institution of a criminal proceeding against an individual

Quash: to overthrow; to vacate, to make void

Rebuttal: an answer or response to a statement; to refute, oppose, or contradict that which has been stated

Recall: Removal of goods from the market (or repair in place) through voluntary action by shipper in lieu of seizure.

Recondition: the process of reworking a lot of goods under seizure pursuant to a Consent Decree of Condemnation, in an attempt to bring the goods into compliance with the Act

Regulatory letter: a letter to the responsible party of a firm, warning of violations of the Act and admonishing that, unless immediate remedy is

**Table 10.2.1. (continued)**

initiated, the violation may be the subject of regulatory action be taken by the government

Relabeling: the application of different labels to a product

Res: things; includes real and personal property

Res judicata: Matters that have been decided between the same parties need not be decided again. If a product has been seized for misbranding and adjudicated by order of a court, the same article with the same labeling containing the same misbranding may be seized subsequently and summary judgment obtained without a trial.

Respondent: one who answers; as in Notice of Hearing, the response is given by the party to whom notice is given; he then is the respondent.

Return date: After a lot has been seized by the U.S. Marshal, a period of time is permitted by the court so that a claimant may appear. The date when this interval, which varies according to the judicial district, will expire is known as the "return date." If no one appears as claimant before expiration of the return date in the proceedings, the U.S. Attorney will secure a Default Decree of Condemnation, forfeiting and disposing of the article.

Return of service: The person serving the process shall make proof of service to the court.

Reversal: the act of reversing a court order or verdict; i.e., an appellate court may reverse a decision rendered by a district court.

Revocation of probation: Any person convicted of a criminal charge may be granted probation for a specified period by the district court. If, during the period of probation, the person violates the law, his probation may be revoked and the original penalty imposed at time of sentencing can be placed in effect.

Sample jacket: the file containing the investigation, analytical, and compliance or regulatory recommendation and legal documents on any civil or criminal case

Sample summary (SS): Form FD465, prepared to record summarized information about a sample (description of sample, IS movement, value, manufacturing information, and district's recommendation)

Segregation: See Recondition.

Seizure: attachment of goods through court order by a U.S. Marshal pursuant to Section 304 of the FD&C Act

Sentence: the judgment formally pronounced by the court upon the defendant after his conviction in a criminal prosecution, awarding the punishment to be inflicted

Service: the delivery of writ, subpoena, notice, injunction, or other court order by an authorized person, to a person, thereby officially notifying him of some action or step that he is commanded to take or not to take

Shipper: the party whose name and address appears on documents (B/L, F/B) as the one who introduced an article into interstate commerce

Standards: the term applied to formal statements of definitions and standards for foods promulgated pursuant to Section 401, FD&C Act

Statute of limitation: This is the period in which any criminal action contemplated must be brought. In all FD&C matters, the period is five (5) years; if no prosecution is filed within that time, FDA may not proceed.

Subpoena: a document, issued by the Clerk of Court which, when delivered to the person named therein, requires him to appear in court, or at a deposition, or before a grand jury

Subpoena duces tecum: same as above, but which also requires the person to bring with him certain records or items named in the subpoena

Summary judgment: a judgment given to one party or the other without the necessity of a formal trial. This may be done when the record before the court shows there are no genuine issues of fact that are in dispute. It is literally "summarily deciding" the case.

Summary and recommendation (S & R): the written justification from the district in its request for institution of a criminal prosecution

Summons: a document that is served upon the defendant along with the complaint, advising him that a case has been filed against him, and requiring him to appear and answer the complaint

Temporary abeyance (TA): the holding of sample(s) in abeyance for possible future regulatory action

Temporary restraining order (TRO): The first segment of most injunction suits in which the government is requesting the imposition of immediate restraint of the defendants; it is, as its name applies, temporary. It lasts for 10 days during which, unless the defendants consent to a Decree of Preliminary or Permanent Injunction, a hearing may be held on the Preliminary Injunction.

Terminated: ended; adjudicated

Venue: place of trial

Verdict: the decision of the jury in a civil or criminal case

Warrant (for arrest): a writ, directed to the U.S. Marshal or other proper officer, requiring him to arrest the defendant or article named and bring him before the court or U.S. magistrate

Waybill: This record accompanies the shipment during transit

Witness: a person who will appear to present evidence as testimony in any proceeding in a lawsuit. An expert witness is one who is skilled in some

**Table 10.2.1. (continued)**

art, science, trade, profession, or other human activity and may give opinions while under oath. A fact witness may not express opinions while testifying.
Worksheet: a document on which the analyst records his analytical work and findings
Writ: an order of the court

Source: FDA Regulatory Procedures Manual, 1991.

## 10.3. Food Additives

**Table 10.3.1. Table of Contents for FDA Regulations on Food Additives**

| 21 CFR Part | Title |
|---|---|
| 170 | Food additives |
| 171 | Food additive petitions |
| 172 | Food additives permitted for direct addition to food for human consumption |
| 173 | Secondary direct food additives permitted in food for human consumption |
| 174 | Indirect food additives: general |
| 175 | Indirect food additives: adhesives and components of coatings |
| 176 | Indirect food additives: paper and paperboard components |
| 177 | Indirect food additives: polymers |
| 178 | Indirect food additives: adjuvants, production aids, and sanitizers |
| 179 | Irradiation in the production, processing, and handling of food |
| 180 | Food additives permitted in food on an interim basis or in contact with food pending additional study |
| 181 | Prior-sanctioned food ingredients |
| 182 | Substances generally recognized as safe |
| 184 | Direct food substances affirmed as generally recognized as safe |
| 186 | Indirect food substances affirmed as generally recognized as safe |
| 189 | Substances prohibited from use in human food |

Source: 21 CFR 170-189, 1991.

**Table 10.3.2. Classification of Foods by The FDA for the Purpose of Establishing Tolerances or Limitations for the Use of Direct Human Food Ingredients**

1. Baked goods and baking mixes, including all ready-to-eat and ready-to-bake products, flours, and mixes requiring preparation before serving
2. Beverages, alcoholic, including malt beverages, wines, distilled liquors, and cocktail mix
3. Beverages and beverage bases, nonalcoholic, including only special or spiced teas, soft drinks, coffee substitutes, and fruit- and vegetable-flavored gelatin drinks
4. Breakfast cereals, including ready-to-eat and instant and regular hot cereals
5. Cheeses, including curd and whey cheeses, cream, natural, grating, processed, spread, dip, and miscellaneous cheeses
6. Chewing gum, including all forms
7. Coffee and tea, including regular, decaffeinated, and instant types
8. Condiments and relishes, including plain seasoning sauces and spreads, olives, pickles, and relishes, but not spices or herbs
9. Confections and frostings, including candy and flavored frostings, marshmallows, baking chocolate, and brown, lump, rock, maple, powdered, and raw sugars
10. Dairy product analogs, including nondairy milk, frozen or liquid creamers, coffee whiteners, toppings, and other nondairy products
11. Egg products, including liquid, frozen, or dried eggs, and egg dishes made therefrom, i.e., egg roll, egg foo young, egg salad, and frozen multicourse egg meals, but not fresh eggs
12. Fats and oils, including margarine, dressings for salads, butter, salad oils, shortenings, and cooking oils
13. Fish products, including all prepared main dishes, salads, appetizers, frozen multicourse meals, and spreads containing fish, shellfish, and other aquatic animals, but not fresh fish
14. Fresh eggs, including cooked eggs and egg dishes made only from fresh shell eggs
15. Fresh fish, including only fresh and frozen fish, shellfish, and other aquatic animals
16. Fresh fruits and fruit juices, including only raw fruits, citrus, melons, and berries, and home-prepared "ades" and punches made therefrom
17. Fresh meats, including only fresh or home-frozen beef or veal, pork, lamb, or mutton and home-prepared fresh meat-containing dishes, salads, appetizers, or sandwich spreads made therefrom

**Table 10.3.2. (continued)**

18. Fresh poultry, including only fresh or home-frozen poultry and game birds and home-prepared fresh poultry-containing dishes, salads, appetizers, or sandwich spreads made therefrom
19. Fresh vegetables, tomatoes, and potatoes, including only fresh and home-prepared vegetables
20. Frozen dairy desserts and mixes, including ice cream, ice milks, sherbets, and other frozen dairy desserts and specialties
21. Fruit and water ices, including all frozen fruit and water ices
22. Gelatins, puddings, and fillings, including flavored gelatin desserts, puddings, custards, parfaits, pie fillings, and gelatin base salads
23. Grain products and pastas, including macaroni and noodle products, rice dishes, and frozen multicourse meals, without meat or vegetables
24. Gravies and sauces, including all meat sauces and gravies, and tomato, milk, buttery, and specialty sauces
25. Hard candy and cough drops, including all hard-type candies
26. Herbs, seeds, spices, seasonings, blends, extracts, and flavorings, including all natural and artificial spices, blends, and flavors
27. Jams and jellies, home-prepared, including only home-prepared jams, jellies, fruit butters, preserves, and sweet spreads
28. Jams and jellies, commercial, including only commercially processed jams, jellies, fruit butters, preserves, and sweet spreads
29. Meat products, including all meats and meat-containing dishes, salads, appetizers, frozen multicourse meat meals, and sandwich ingredients prepared by commercial processing or using commercially processed meats with home preparation
30. Milk, whole and skim, including only whole, low-fat, and skim fluid milks
31. Milk products, including flavored milks and milk drinks, dry milks, toppings, snack dips, spreads, weight-control milk beverages, and other milk origin products
32. Nuts and nut products, including whole or shelled tree nuts, peanuts, coconut, and nut and peanut spreads
33. Plant protein products, including the National Academy of Sciences/National Research Council "reconstituted vegetable protein" category, and meat, poultry, and fish substitutes, analogs, and extender products made from plant proteins
34. Poultry products, including all poultry and poultry-containing dishes, salads, appetizers, frozen multicourse poultry meals, and sandwich ingredients prepared by commercial processing or using commercially processed poultry with home preparation

35. Processed fruits and fruit juices, including all commercially processed fruits, citrus, berries, and mixtures; salads; juices and juice punches; concentrates; dilutions; "-ades"; and drink substitutes made therefrom
36. Processed vegetables and vegetable juices, including all commercially processed vegetables, vegetable dishes, frozen multicourse vegetable meals, and vegetable juices and blends
37. Snack foods, including chips, pretzels, and other novelty snacks
38. Soft candy, including candy bars, chocolates, fudge, mints, and other chewy or nougat candies
39. Soups, home-prepared, including meat, fish, poultry, vegetable, and combination home-prepared soups
40. Soups and soup mixes, including commercially prepared meat, fish, poultry, vegetable, and combination soups and soup mixes
41. Sugar, white, granulated, including only white granulated sugar
42. Sugar substitutes, including granulated, liquid, and tablet sugar substitutes
43. Sweet sauces, toppings, and syrups, including chocolate, berry, fruit, corn syrup, and maple sweet sauces and toppings

Source: 21 CFR 170.3, 1991.

**Table 10.3.3. Terminology for the Physical or Technical Functional Effects for Which Direct Human Food Ingredients May Be Added to Foods**

1. Anticaking agents and free-flow agents: substances added to finely powdered or crystalline food products to prevent caking, lumping, or agglomeration
2. Antimicrobial agents: substances used to preserve food by preventing growth of microorganisms and subsequent spoilage, including fungistats, mold and rope inhibitors, and the effects listed by the National Academy of Sciences/National Research Council under preservatives
3. Antioxidants: substances used to preserve food by retarding deterioration, rancidity, or discoloration due to oxidation
4. Colors and coloring adjuncts: substances used to impart, preserve, or enhance the color or shading of a food, including color stabilizers, color fixatives, color-retention agents, etc.
5. Curing and pickling agents: substances imparting a unique flavor and/or color to a food, usually producing an increase in shelf life stability
6. Dough strengtheners: substances used to modify starch and gluten, thereby producing a more stable dough, including the applicable effects listed by the National Academy of Sciences/National Research Council under dough conditioner

**Table 10.3.3. (continued)**

7. Drying agents: substances with moisture-absorbing ability, used to maintain an environment of low moisture
8. Emulsifiers and emulsifier salts: substances that modify surface tension in the component phase of an emulsion to establish a uniform dispersion or emulsion
9. Enzymes.: enzymes used to improve food processing and the quality of the finished food
10. Firming agents: substances added to precipitate residual pectin, thus strengthening the supporting tissue and preventing its collapse during processing
11. Flavor enhancers: substances added to supplement, enhance, or modify the original taste and/or aroma of a food, without imparting characteristic tastes or aromas of their own
12. Flavoring agents and adjuvants: substances added to impart or help impart a taste or aroma in food
13. Flour-treating agents: substances added to milled flour, at the mill, to improve its color and/or baking qualities, including bleaching and maturing agents
14. Formulation aids: substances used to promote or produce a desired physical state or texture in food, including carriers, binders, fillers, plasticizers, film-formers, and tableting aids, etc.
15. Fumigants: volatile substances used for controlling insects or pests
16. Humectants: hygroscopic substances incorporated in food to promote retention of moisture, including moisture-retention agents and antidusting agents
17. Leavening agents: substances used to produce or stimulate production of carbon dioxide in baked goods to impart a light texture, including yeast, yeast foods, and calcium salts listed by the National Academy of Sciences/National Research Council under dough conditioners
18. Lubricants and release agents: substances added to food contact surfaces to prevent ingredients and finished products from sticking to them
19. Nonnutritive sweeteners: substances having less than 2% of the caloric value of sucrose per equivalent unit of sweetening capacity
20. Nutrient supplements: substances that are necessary for the body's nutritional and metabolic processes
21. Nutritive sweeteners: substances having greater than 2% of the caloric value of sucrose per equivalent unit of sweetening capacity
22. Oxidizing and reducing agents: Substances that chemically oxidize or reduce another food ingredient, thereby producing a more stable product,

including the applicable effect listed by the National Academy of Sciences/National Research Council under dough conditioners

23. pH control agents: substances added to change or maintain active acidity or basicity, including buffers, acids, alkalies, and neutralizing agents
24. Processing aids: substances used as manufacturing aids to enhance the appeal or utility of a food or food component, including clarifying agents, clouding agents, catalysts, flocculents, filter aids, and crystallization inhibitors, etc.
25. Propellants, aerating agents, and gases: gases used to supply force to expel a product or used to reduce the amount of oxygen in contact with the food in packaging
26. Sequestrants: substances that combine with polyvalent metal ions to form a soluble metal complex, to improve the quality and stability of products
27. Solvents and vehicles: substances used to extract or dissolve another substance
28. Stabilizers and thickeners: substances used to produce viscous solutions or dispersions, to impart body, improve consistency, or stabilize emulsions, including suspending and bodying agents, setting agents, jellying agents, and bulking agents, etc.
29. Surface-active agents: substances used to modify surface properties of liquid food components for a variety of effects, other than emulsifiers, but including solubilizing agents, dispersants, detergents, wetting agents, rehydration enhancers, whipping agents, foaming agents, and defoaming agents, etc.
30. Surface-finishing agents: substances used to increase palatability, preserve gloss, and inhibit discoloration of foods, including glazes, polishes, waxes, and protective coatings
31. Synergists: substances used to act or react with another food ingredient to produce a total effect different or greater than the sum of the effects produced by the individual ingredients
32. Texturizers: substances which affect the appearance or feel of the food

Source: 21 CFR 170.3, 1991.

**Table 10.3.4. Food Additives Permitted for Direct Addition to Food for Human Consumption**

| 21 CFR 172 Section | Food Additives |
|---|---|
| | Food Preservatives |
| 105 | Anoxomer |
| 110 | BHA |
| 115 | BHT |
| 120 | Calcium disodium EDTA |
| 130 | Dehydroacetic acid |
| 135 | Disodium EDTA |
| 140 | Thoxyquin |
| 145 | Heptyl paraben |
| 150 | 4-Hydroxymethyl 2,6-di-*tert*-butylphenol |
| 155 | Natamycin (pimaricin) |
| 160 | Potassium nitrate |
| 165 | Quaternary ammonium chloride combination |
| 170 | Sodium nitrate |
| 175 | Sodium nitrite |
| 177 | Sodium nitrite used in processing smoked chub |
| 180 | Stannous chloride |
| 185 | TBHQ |
| 190 | THBP |
| | Coatings, Films, and Related Substances |
| 210 | Coatings on fresh citrus fruit |
| 215 | Coumarone-indene resin |
| 225 | Methyl esters of fatty acids produced from edible fats and oils |
| 230 | Microcapsules for flavoring substances |
| 235 | Morpholine |
| 250 | Petroleum naphtha |
| 255 | Polyacrylamide |
| 260 | Oxidized polyethylene |
| 275 | Synthetic paraffin and succinic derivatives |
| 280 | Terpene resin |

## Special Dietary and Nutritional Additives

| | |
|---|---|
| 310 | Aluminum nicotinate |
| 315 | Nicotinamide-ascorbic acid complex |
| 320 | Amino acids |
| 325 | Bakers yeast protein |
| 330 | Calcium pantothenate, calcium chloride double salt |
| 335 | D-Pantothenamide |
| 340 | Fish protein isolate |
| 245 | Folic acid (folacin) |
| 350 | Fumaric acid and salts of fumaric acids |
| 365 | Kelp |
| 370 | Iron-choline citrate complex |
| 372 | *N*-Acetyl L-methionine |
| 375 | Potassium iodide |
| 385 | Whole fish protein concentrate |
| 395 | Xylitol |
| 399 | Zinc methionine sulfate |

## Anticaking Agents

| | |
|---|---|
| 410 | Calcium silicate |
| 430 | Iron ammonium citrate |
| 480 | Silicon dioxide |
| 490 | Yellow prussiate of soda |

## Flavoring Agents and Related Substances

| | |
|---|---|
| 510 | Natural flavoring substances and natural substances used in conjunction with flavors |
| 515 | Synthetic flavoring substances and adjuvants |
| 520 | Cocoa with dioctyl sodium sulfosuccinate for manufacturing |
| 530 | Disodium guanylate |
| 535 | Disodium inosinate |
| 560 | Modified hop extract |
| 575 | Quinine |
| 580 | Safrole-free extract of sassafras |
| 585 | Sugar beet extract flavor base |
| 590 | Yeast-malt sprout extract |

**Table 10.3.4. (continued)**

| 21 CFR 172 Section | Food Additives |
|---|---|
| | Gums, Chewing Gum Bases, and Related Substances |
| 610 | Arabinogalactan |
| 615 | Chewing gum base |
| 620 | Carrageenan |
| 623 | Carrageenan with polysorbate 80 |
| 626 | Salts of carrageenan |
| 655 | Furcelleran |
| 660 | Salts of furcelleran |
| 695 | Xanthan gum |
| | Other Specific Usage Additives |
| 710 | Adjuvants for pesticide use dilutions |
| 712 | Dimethyl dialkyl ammonium chloride |
| 715 | Calcium lignosulfonate |
| 720 | Calcium lactobionate |
| 725 | Gibberellic acid and its potassium salt |
| 730 | Potassium bromate |
| 735 | Glycerol ester of wood rosin |
| 755 | Stearyl monoglyceridyl citrate |
| 765 | Succistearin (stearoyl propylene glycol hydrogen succinate) |
| 770 | Ethylene oxide polymer |
| 775 | Methacrylic acid-divinylbenzene copolymer |
| | Multipurpose Additives |
| 802 | Acetone peroxides |
| 804 | Aspartame |
| 806 | Azodicarbonamide |
| 808 | Copolymer condensates of ethylene oxide and propylene oxide |
| 810 | Dioctyl sodium sulfosuccinate |
| 812 | Glycine |
| 814 | Hydroxylated lecithin |
| 816 | Methyl glucoside-coconut oil ester |
| 818 | Oxystearin |
| 820 | Polyethylene glycol (mean molecular weight 200-9500) |
| 822 | Sodium lauryl sulfate |

| | |
|---|---|
| 824 | Sodium mono- and dimethyl naphthalene sulfonates |
| 826 | Sodium stearyl fumarate |
| 828 | Acetylated monoglycerides |
| 830 | Succinylated monoglycerides |
| 832 | Monoglyceride citrate |
| 834 | Ethoxylated mono- and diglycerides |
| 836 | Polysorbate 60 |
| 838 | Polysorbate 65 |
| 840 | Polysorbate 80 |
| 841 | Polydextrose |
| 842 | Sorbitan monostearate |
| 844 | Calcium stearoyl-2-lactylate |
| 846 | Sodium stearoyl-2-lactylate |
| 848 | Lactylic esters of fatty acids |
| 850 | Lactylated fatty acid esters of glycerol and propylene glycol |
| 852 | Glyceryl-lacto esters of fatty acids |
| 854 | Polyglycerol esters of fatty acids |
| 856 | Propylene glycol mono- and diesters of fats and fatty acids |
| 858 | Propylene glycol alginate |
| 859 | Sucrose fatty acid esters |
| 861 | Fatty acids |
| 862 | Oleic acid derived from tall oil fatty acids |
| 863 | Salts of fatty acids |
| 864 | Synthetic fatty alcohols |
| 866 | Synthetic glycerin produced by the hydrogenolysis of carbohydrates |
| 868 | Ethyl cellulose |
| 870 | Hydroxypropyl cellulose |
| 872 | Methyl ethyl cellulose |
| 874 | Hydroxypropyl methyl cellulose |
| 876 | Castor oil |
| 878 | White mineral oil |
| 880 | Petrolatum |
| 882 | Synthetic isoparaffinic petroleum hydrocarbons |
| 884 | Odorless light petroleum hydrocarbons |
| 886 | Petroleum wax |
| 888 | Synthetic petroleum wax |
| 890 | Rice bran wax |
| 892 | Food starch, modified |
| 894 | Modified cottonseed products intended for human consumption |
| 896 | Dried yeasts |

**Table 10.3.4. (continued)**

| 21 CFR 172 Section | Food Additives |
|---|---|
| 898 | Bakers yeast glycan |

Source: 21 CFR 172, 1991.

**Table 10.3.5. Secondary Direct Food Additives Permitted in Food for Human Consumption**

| 21 CFR 173 Section | Secondary Direct Food Additives |
|---|---|
| | Polymer Substances and Polymer Adjuvants for Food Treatment |
| 5 | Acrylate-acrylamide resins |
| 10 | Modified polyacrylamide resin |
| 20 | Ion-exchange membranes |
| 25 | Ion-exchange resins |
| 40 | Molecular sieve resins |
| 45 | Polymaleic acid and its sodium salt |
| 50 | Polyvinyl polypyrrolidone |
| 55 | Polyvinyl pyrrolidone |
| 60 | Dimethylamine-epichlorohydrin copolymer |
| 65 | Divinylbenzene copolymer |
| 70 | Chloromethylated aminated styrene-divinylbenzene resin |
| 75 | Sorbitan monooleate |
| | Enzyme Preparations and Microorganisms |
| 110 | Amyloglucosidase derived from *Rhizopus niveus* |
| 120 | Carbohydrase and cellulase derived from *Aspergillus niger* |
| 130 | Carbohydrase and cellulase derived from *Rhizopus oryzae* |
| 135 | Catalse derived from *Microccocus lysodeikticus* |
| 140 | Esterase-lipase derived from *Mucor miehei* |
| 145 | $\alpha$-Galactosidase derived from *Morteirella vinnaceae* var. *raffinoseutilizer* |
| 150 | Milk-clotting enzymes, microbial |
| 160 | *Candida guilliermondii* |
| 165 | *Candida lipolytica* |

Solvents, Lubricants, Release Agents, and Related Substances

| | |
|---|---|
| 210 | Acetone |
| 220 | 1,3-Butylene glycol |
| 228 | Ethyl acetate |
| 230 | Ethylene dichloride |
| 240 | Isopropyl alcohol |
| 250 | Methyl alcohol residues |
| 255 | Methylene chloride |
| 270 | Hexane |
| 275 | Hydrogenated sperm oil |
| 280 | Solvent extraction process for citric acid |
| 290 | Trichloroethylene |

Specific Usage Additives

| | |
|---|---|
| 310 | Boiler water additives |
| 315 | Chemicals used in washing or to assist in the lye peeling of fruits and vegetables |
| 320 | Chemicals for controlling microorganisms in cane-sugar and beet-sugar mills |
| 322 | Chemicals used in delinting cottonseed |
| 340 | Defoaming agents |
| 345 | Chloropentafluoroethane |
| 350 | Combustion product gas |
| 355 | Dichlorodifluoromethane |
| 357 | Materials used as fixing agents in the immobilization of enzyme preparations |
| 360 | Octafluorocyclobutane |
| 385 | Sodium methyl sulfate |
| 395 | Trifluoromethane sulfonic acid |

Source: 21 CFR 173, 1991.

**Table 10.3.6. Indirect Food Additives: Adhesives and Components of Coatings**

| 21 CFR 175 Section | Indirect Food Additives |
|---|---|
| | Substances for Use Only as Components of Adhesives |
| 105 | Adhesives |
| 125 | Pressure-sensitive adhesives |
| | Substances for Use as Components of Coatings |
| 210 | Acrylate ester copolymer coatings |
| 230 | Hot-melt strippable food coatings |
| 250 | Paraffin (synthetic) |
| 260 | Partial phosphoric acid esters of polyester resins |
| 270 | Poly(vinyl fluoride) resins |
| 300 | Resinous and polymeric coatings |
| 320 | Resinous and polymeric coatings for polyolefin films |
| 350 | Vinyl acetate/crotonic acid copolymer |
| 360 | Vinylidene chloride copolymer coatings for nylon film |
| 365 | Vinylidene chloride copolymer coatings for polycarbonate film |
| 380 | Xyylene-formaldehyde resins condensed with 4,4′-isopropylidene diphenol-epichlorohydrin epoxy resins |
| 390 | Zinc-silicon dioxide matrix coatings |

Source: 21 CFR 175, 1991.

**Table 10.3.7. Indirect Food Additives: Paper and Paperboard Components**

| 21 CFR 176 Section | Indirect Food Additives |
|---|---|
| | Substances for Use Only as Components of Paper and Paperboard |
| 110 | Acrylamide-acrylic acid resins |
| 120 | Alkyl ketene dimers |
| 130 | Anti-offset substances |
| 150 | Chelating agents used in the manufacture of paper and paperboard |

| | |
|---|---|
| 160 | Chromium (Cr III) complex of *N*-ethyl *N*-heptadecyl fluoro-ocatane sulfonyl glycine |
| 170 | Components of paper and paperboard in contact with aqueous and fatty foods |
| 180 | Components of paper and paperboard in contact with dry food |
| 200 | Defoaming agents used in coatings |
| 210 | Defoaming agents used in the manufacture of paper and paperboard |
| 230 | 3,5-Dimethyl 1,3,5,2*H*-tetrahydrothiadiazine-2-thione |
| 250 | Poly-1,4,7,10,13-pentaaza-15-hydroxyhexadecane |
| 260 | Pulp from reclaimed fiber |
| 300 | Slimicides |
| 320 | Sodium nitrate-urea complex |
| 350 | Tamarind seed kernel powder |

Source: 21 CFR 176, 1991.

**Table 10.3.8. Indirect Food Additives: Polymers**

| 21 CFR 177 Section | Indirect Food Additives |
|---|---|
| | Substances for Use as Basic Components of Single- and Repeated-Use Food Contact Surfaces |
| 1010 | Acrylic and modified acrylic plastics, semirigid and rigid |
| 1020 | Acrylonitrile-butadiene-styrene copolymer |
| 1030 | Acrylonitrile-butadiene-styrene-methyl methacrylate copolymer |
| 1040 | Acrylonitrile-styrene copolymer |
| 1050 | Acrylonitrile-styrene copolymer modified with butadiene-styrene elastomer |
| 1200 | Cellophane |
| 1210 | Closures with sealing gaskets for food containers |
| 1240 | 1,4-Cyclohexylene dimethylene terephthalate and 1,4-cyclohexylene dimethylene isophthalate copolymer |
| 1310 | Ethylene-acrylic acid copolymers |
| 1315 | Ethylene-1,4-cyclohexylene dimethylene terephthalate copolymers |
| 1320 | Ethylene-ethyl acrylate copolymers |
| 1330 | Ionomeric resins |
| 1340 | Ethylene-methyl acrylate copolymer resins |
| 1350 | Ethylene-vinyl acetate copolymers |

**Table 10.3.8. (continued)**

| 21 CFR 177 Section | Indirect Food Additives |
|---|---|
| 1360 | Ethylene-vinyl alcohol copolymers |
| 1380 | Fluorocarbon resins |
| 1390 | Laminate structures for use at temperatures of 250ºF and above |
| 1395 | Laminate structures for use at temperatures between 120ºF and 120ºF |
| 1400 | Hydroxyethyl cellulose film, water-insoluble |
| 1420 | Isobutylene polymers |
| 1430 | Isobutylene-butene copolymers |
| 1440 | 4,4′-Isopropylidenediphenol-epichlorohydrin resins, minimum molecular weight 10,000 |
| 1460 | Melamine-formaldehyde resins in molded articles |
| 1480 | Nitrile rubber-modified acrylonitrile-methyl acrylate copolymers |
| 1500 | Nylon resins |
| 1520 | Olefin polymers |
| 1550 | Perfluorocarbon resins |
| 1555 | Polyarylate resins |
| 1560 | Polyarylsulfone resins |
| 1570 | Poly-1-butene resins and butene-ethylene copolymers |
| 1580 | Polycarbonate resins |
| 1590 | Polyester elastomers |
| 1595 | Polyetherimide resin |
| 1600 | Polyethylene resins, carboxyl-modified |
| 1610 | Polyethylene, chlorinated |
| 1615 | Polyethylene, fluorinated |
| 1620 | Polyethylene, oxidized |
| 1630 | Polyethylene, phthalate polymers |
| 1635 | Poly(*p*-methyl styrene) and rubber-modified poly(*p*-methyl styrene) |
| 1640 | Polystyrene and rubber-modified polystyrene |
| 1650 | Polysulfide polymer-polyepoxy resins |
| 1655 | Polysulfone resins |
| 1660 | Poly(tetramethylene terephthalate) |
| 1670 | Polyvinyl alcohol film |
| 1680 | Polyurethane resins |
| 1810 | Styrene block polymers |
| 1820 | Styrene-maleic anhydride copolymers |
| 1830 | Styrene-methyl methacrylate copolymers |

| | |
|---|---|
| 1850 | Textryls |
| 1900 | Urea-formaldehyde resins in molded articles |
| 1950 | Vinyl chloride-ethylene copolymers |
| 1960 | Vinyl Chloride-hexene-1 copolymers |
| 1970 | Vinyl chloride-lauryl vinyl ether copolymers |
| 1980 | Vinyl chloride-propylene copolymers |
| 1990 | Vinylidene chloride-methyl acrylate copolymers |
| 2000 | Vinylidene chloride-methyl acrylate-methyl methacrylate polymers |
| | Substances for Use Only As Components of Articles Intended for Repeated Use |
| 2210 | Ethylene polymer, chlorosulfonated |
| 2250 | Filters, microporous polymeric |
| 2260 | Filters, resin-bonded |
| 2280 | 4,4′-Isopropylidene diphenol-epichlorohydrin thermosetting epoxy resins |
| 2355 | Mineral-reinforced nylon resins |
| 2400 | Perfluorocarbon cured elastomers |
| 2410 | Phenolic resins in molded articles |
| 2420 | Polyester resins, crosslinked |
| 2430 | Polyether resins, chlorinated |
| 2440 | Polyethersulfone resins |
| 2450 | Polyamide-imide resins |
| 2460 | Poly(2,6-dimethyl 1,4-phenylene) oxide resins |
| 2470 | Polyoxymethylene copolymer |
| 2480 | Polyoxymethylene homopolymer |
| 2490 | Polyphenylene sulfide resins |
| 2510 | Polyvinylidene fluoride resins |
| 2550 | Reverse osmosis membranes |
| 2600 | Rubber articles intended for repeated use |
| 2710 | Styrene-divinylbenzene resins, crosslinked |
| 2800 | Textiles and textile fibers |
| 2910 | Ultrafiltration membranes |

Source: 21 CFR 177, 1991

**Table 10.3.9. Indirect Food Additives: Adjuvants, Production Aids, and Sanitizers**

| 21 CFR 178 Section | Indirect Food Additives |
|---|---|
| | Substances Utilized To Control the Growth of Microorganisms |
| 1005 | Hydrogen peroxide solution |
| 1010 | Sanitizing solutions |
| | Antioxidants and Stabilizers |
| 2010 | Antioxidants and/or stabilizers for polymers |
| 2550 | 4-Hydroxymethyl 2,6-di-*tert*-butylphenol |
| 2650 | Organotin stabilizers in vinyl chloride plastics |
| | Certain Adjuvants and Production Aids |
| 3010 | Adjuvant substances used in the manufacture of foamed plastics |
| 3120 | Animal glue |
| 3125 | Anticorrosive agents |
| 3130 | Antistatic and/or antifogging agents in food-packaging materials |
| 3280 | Castor oil, hydrogenated |
| 3290 | Chromic chloride complexes |
| 3295 | Clarifying agents for polymers |
| 3297 | Colorants for polymers |
| 3300 | Corrosion inhibitors used for steel or template |
| 3400 | Emulsifiers and/or surface-active agents |
| 3450 | Esters of stearic and palmitic acids |
| 3480 | Fatty alcohols, synthetic |
| 3500 | Glycerin, synthetic |
| 3505 | Glyceryl tri(12-acetoxystearate) |
| 3520 | Industrial starch, modified |
| 3530 | Isoparaffinic petroleum hydrocarbons, synthetic |
| 3550 | Kaolin, modified |
| 3570 | Lubricants with incidental food contact |
| 3600 | Methyl glucoside-coconut oil ester |
| 3610 | $\alpha$-Methylstyrene-vinyltoluene resins, hydrogenated |
| 3620 | Mineral oil |
| 3650 | Odorless light petroleum hydrocarbons |
| 3690 | Pentaerythritol adipate-stearate |

| | |
|---|---|
| 3700 | Petrolatum |
| 3710 | Petroleum wax |
| 3720 | Petroleum wax, synthetic |
| 3730 | Piperonyl butoxide and pyrethrins as components of bags |
| 3740 | Plasticizers in polymeric substances |
| 3750 | Polyethylene glycol (mean molecular weight 200-9500) |
| 3760 | Polyethylene glycol (400) monolaurate |
| 3770 | Polyhydric alcohol diesters of oxidatively refined (Gersthoffen process) montan wax acids |
| 3780 | Polyhydric alcohol esters of long-chain monobasic acids |
| 3790 | Polymer modifiers in semirigid and rigid vinyl chloride plastics |
| 3800 | Preservatives for wood |
| 3850 | Reinforced wax |
| 3860 | Release agents |
| 3870 | Rosins and rosin derivatives |
| 3900 | Sodium pentachlorophenate |
| 3910 | Surface lubricants used in the manufacture of metallic articles |
| 3930 | Terpene resins |
| 3940 | Tetraethylene glycol di(2-ethyl hexoate) |
| 3950 | Tetrahydrofuran |
| 3970 | Ultramarine blue |

Source: 21 CFR 178, 1991.

**Table 10.3.10. Food Additives: Interim and Prior Sanctioned**

| 21 CFR Section | Subject |
|---|---|
| | Specific Requirements for Certain Interim Food Additives |
| 180.22 | Acrylonitrile copolymers |
| 180.25 | Mannitol |
| 180.30 | Brominated vegetable oil |
| 180.37 | Saccharin, ammonium saccharin, calcium saccharin, and sodium saccharin |
| | Specific Prior-Sanctioned Food Ingredients |
| 181.22 | Certain substances employed in the manufacture of food packaging materials |
| 181.23 | Antimycotics |

**Table 10.3.10. (continued)**

| 21 CFR Section | Subject |
|---|---|
| 181.24 | Antioxidants |
| 181.25 | Driers |
| 181.26 | Drying oils as components of finished resins |
| 181.27 | Plasticizers |
| 181.28 | Release agents |
| 181.29 | Stabilizers |
| 181.30 | Substances used in the manufacture of paper and paperboard products used in food packaging |
| 181.32 | Acrylonitrile polymers and resins |
| 181.33 | Sodium nitrate and potassium nitrate |
| 181.34 | Sodium nitrite and potassium nitrite |

Source: 21 CFR 180-181, 1991.

**Table 10.3.11. Food Additives: Substances Generally Recognized as Safe**

| 21 CFR 182 Section | Regulations or Specific Food Additive |
|---|---|
| General Provisions | |
| 1 | Substances that are generally recognized as safe (GRAS) |
| 10 | Spices and other natural seasonings and flavorings |
| 20 | Essential oils, oleoresins (solvent-free), and natural extractives (including distillates) |
| 40 | Natural extractives (solvent-free) used in conjunction with spices, seasonings, and flavorings |
| 50 | Certain other spices, seasonings, essential oils, oleoresins, and natural extracts |
| 60 | Synthetic flavoring substances and adjuvants |
| 70 | Substances migrating from cotton and cotton fabrics used in dry food packaging |
| 90 | Substances migrating to food from paper and paperboard products |
| 99 | Adjuvants for pesticide chemicals |
| | Multiple-Purpose GRAS Food Substances |

| | |
|---|---|
| 1033 | Citric acid |
| 1045 | Glutamic acid |
| 1047 | Glutamic acid hydrochloride |
| 1057 | Hydrochloric acid |
| 1073 | Phosphoric acid |
| 1087 | Sodium acid pyrophosphate |
| 1125 | Aluminum sulfate |
| 1127 | Aluminum ammonium sulfate |
| 1129 | Aluminum potassium sulfate |
| 1131 | Aluminum sodium sulfate |
| 1180 | Caffeine |
| 1195 | Calcium citrate |
| 1217 | Calcium phosphate |
| 1235 | Caramel |
| 1320 | Glycerin |
| 1324 | Glyceryl monostearate |
| 1480 | Methyl cellulose |
| 1500 | Monoammonium glutamate |
| 1516 | Monopotassium glutamate |
| 1625 | Potassium citrate |
| 1711 | Silica aerogel |
| 1745 | Sodium carboxymethyl cellulose |
| 1748 | Sodium caseinate |
| 1751 | Sodium citrate |
| 1778 | Sodium phosphate |
| 1781 | Sodium aluminum phosphate |
| 1810 | Sodium tripolyphosphate |
| 1866 | High-fructose corn syrup |
| 1901 | Triacetin |
| 1911 | Triethyl citrate |

Anticaking Agents

| | |
|---|---|
| 2122 | Aluminum calcium silicate |
| 2227 | Calcium silicate |
| 2437 | Magnesium silicate |
| 2727 | Sodium aluminosilicate |
| 2729 | Sodium calcium aluminosilicate, hydrated |
| 2906 | Tricalcium silicate |

**Table 10.3.11. (continued)**

| 21 CFR 182 Section | Subject |
|---|---|
| | Chemical Preservatives |
| 3013 | Ascorbic acid |
| 3041 | Erythorbic acid |
| 3089 | Sorbic acid |
| 3109 | Thiodipropionic acid |
| 3149 | Ascorbyl palmitate |
| 3169 | Butylated hydroxyanisole |
| 3173 | Butylated hydroxytoluene |
| 3189 | Calcium ascorbate |
| 3225 | Calcium sorbate |
| 3280 | Dilauryl thiodipropionate |
| 3616 | Potassium bisulfite |
| 3637 | Potassium metabisulfite |
| 3640 | Potassium sorbate |
| 3731 | Sodium ascorbate |
| 3739 | Sodium bisulfite |
| 3766 | Sodium metabisulfite |
| 3795 | Sodium sorbate |
| 3798 | Sodium sulfite |
| 3862 | Sulfur dioxide |
| 3890 | Tocopherols |
| | Emulsifying Agents |
| 4101 | Diacetyl tartaric acid esters of mono- and diglycerides of edible fats or oils, or edible fat-forming fatty acids |
| 4505 | Mono- and diglycerides of edible fats or oils, or edible fat-forming acids |
| 4521 | Monosodium phosphate derivatives of mono- and diglycerides of edible fats or oils, or edible fat-forming fatty acids |
| | Dietary Supplements |
| 5013 | Ascorbic acid |
| 5065 | Linoleic acid |
| 5159 | Biotin |
| 5191 | Calcium carbonate |

| | |
|---|---|
| 5195 | Calcium citrate |
| 5201 | Calcium glycerophosphate |
| 5210 | Calcium oxide |
| 5212 | Calcium pantothenate |
| 5217 | Calcium phosphate |
| 5223 | Calcium pyrophosphate |
| 5245 | Carotene |
| 5250 | Choline bitartrate |
| 5252 | Choline chloride |
| 5260 | Copper gluconate |
| 5301 | Ferric phosphate |
| 5304 | Ferric pyrophosphate |
| 5306 | Ferric sodium pyrophosphate |
| 5308 | Ferrous gluconate |
| 5311 | Ferrous lactate |
| 5315 | Ferrous sulfate |
| 5370 | Inositol |
| 5375 | Iron, reduced |
| 5431 | Magnesium oxide |
| 5434 | Magnesium phosphate |
| 5443 | Magnesium sulfate |
| 5446 | Manganese chloride |
| 5449 | Manganese citrate |
| 5452 | Manganese gluconate |
| 5455 | Manganese glycerophosphate |
| 5461 | Manganese sulfate |
| 5464 | Manganous oxide |
| 5530 | Niacin |
| 5535 | Niacinamide |
| 5580 | D-Pantothenyl alcohol |
| 5622 | Potassium chloride |
| 5628 | Potassium glycerophosphate |
| 5676 | Pyridoxine hydrochloride |
| 5695 | Riboflavin |
| 5697 | Riboflavin-5-phosphate |
| 5772 | Sodium pantothenate |
| 5778 | Sodium phosphate |
| 5875 | Thiamine hydrochloride |
| 5878 | Thiamine mononitrate |
| 5890 | Tocopherols |
| 5892 | $\alpha$-Tocopherol acetate |

**Table 10.3.11. (continued)**

| 21 CFR 182 Section | Subject |
|---|---|
| 5930 | Vitamin A |
| 5933 | Vitamin A acetate |
| 5936 | Vitamin A palmitate |
| 5945 | Vitamin $B_{12}$ |
| 5950 | Vitamin $D_2$ |
| 5953 | Vitamin $D_3$ |
| 5985 | Zinc chloride |
| 5988 | Zinc gluconate |
| 5991 | Zinc oxide |
| 5994 | Zinc stearate |
| 5997 | Zinc sulfate |
| | Sequestrants |
| 6033 | Citric acid |
| 6085 | Sodium acid phosphate |
| 6195 | Calcium citrate |
| 6197 | Calcium diacetate |
| 6203 | Calcium hexametaphosphate |
| 6215 | Monobasic calcium phosphate |
| 6285 | Dipotassium phosphate |
| 6290 | Disodium phosphate |
| 6386 | Isopropyl citrate |
| 6511 | Monoisopropyl citrate |
| 6625 | Potassium citrate |
| 6751 | Sodium citrate |
| 6757 | Sodium gluconate |
| 6760 | Sodium hexametaphosphate |
| 6769 | Sodium metaphosphate |
| 6778 | Sodium phosphate |
| 6787 | Sodium pyrophosphate |
| 6789 | Tetrasodium pyrophosphate |
| 6810 | Sodium tripolyphosphate |
| 6851 | Stearyl citrate |
| | Stabilizers |
| 7255 | *Chondrus* extract |

| | Nutrients |
|---|---|
| 8013 | Ascorbic acid |
| 8159 | Biotin |
| 8195 | Calcium citrate |
| 8201 | Calcium glycerophosphate |
| 8217 | Calcium phosphate |
| 8223 | Calcium pyrophosphate |
| 8250 | Choline bitartrate |
| 8252 | Choline chloride |
| 8301 | Ferric phosphate |
| 8304 | Ferric pyrophosphate |
| 8306 | Ferric sodium pyrophosphate |
| 8308 | Ferrous gluconate |
| 8311 | Ferrous lactate |
| 8315 | Ferrous sulfate |
| 8375 | Iron, reduced |
| 8455 | Manganese glycerophosphate |
| 8458 | Manganese hypophosphite |
| 8628 | Potassium glycerophosphate |
| 8778 | Sodium phosphate |
| 8890 | Tocopherols |
| 8892 | $\alpha$-Tocopherol acetate |
| 8985 | Zinc chloride |
| 8988 | Zinc gluconate |
| 8991 | Zinc oxide |
| 8994 | Zinc stearate |
| 8997 | Zinc sulfate |

Source: 21 CFR 182, 1991.

**Table 10.3.12. Food Additives: Direct Food Substances Affirmed as Generally Recognized as Safe**

| 21 CFR 184 Section | Food Additives |
|---|---|
| | General Provisions |
| 1 | Substances added directly to human food affirmed as generally recognized as safe (GRAS) |

**Table 10.3.12. (continued)**

| 21 CFR 184 Section | Subject |
|---|---|
| | Listing of Specific Substances Affirmed as GRAS |
| 1005 | Acetic acid |
| 1007 | Aconitic acid |
| 1009 | Adipic acid |
| 1011 | Alginic acid |
| 1021 | Benzoic acid |
| 1025 | Caprylic acid |
| 1027 | Mixed carbohydrase and protease enzyme product |
| 1061 | Lactic acid |
| 1065 | Linoleic acid |
| 1069 | Malic acid |
| 1077 | Potassium acid tartrate |
| 1081 | Propionic acid |
| 1090 | Stearic acid |
| 1091 | Succinic acid |
| 1095 | Sulfuric acid |
| 1097 | Tannic acid |
| 1099 | Tartaric acid |
| 1115 | Agar-agar |
| 1120 | Brown algae |
| 1121 | Red algae |
| 1133 | Ammonium alginate |
| 1135 | Ammonium bicarbonate |
| 1137 | Ammonium carbonate |
| 1138 | Ammonium chloride |
| 1139 | Ammonium hydroxide |
| 1141a | Ammonium phosphate, monobasic |
| 1141b | Ammonium phosphate, dibasic |
| 1143 | Ammonium sulfate |
| 1155 | Bentonite |
| 1157 | Benzoyl peroxide |
| 1165 | *n*-Butane and isobutane |
| 1185 | Calcium acetate |
| 1187 | Calcium alginate |
| 1191 | Calcium carbonate |
| 1193 | Calcium chloride |

| | |
|---|---|
| 1199 | Calcium gluconate |
| 1205 | Calcium hydroxide |
| 1206 | Calcium iodate |
| 1207 | Calcium lactate |
| 1210 | Calcium oxide |
| 1212 | Calcium pantothenate |
| 1221 | Calcium propionate |
| 1229 | Calcium stearate |
| 1230 | Calcium sulfate |
| 1240 | Carbon dioxide |
| 1245 | $\beta$-Carotene |
| 1257 | Clove and its derivatives |
| 1259 | Cocoa butter substitute primarily from palm oil |
| 1260 | Copper gluconate |
| 1261 | Copper sulfate |
| 1262 | Corn silk and corn silk extract |
| 1265 | Cuprous iodide |
| 1271 | L-Cysteine |
| 1272 | L-Cysteine monohydrochloride |
| 1277 | Dextrin |
| 1278 | Diacetyl |
| 1282 | Dill and its derivatives |
| 1287 | Enzyme-modified fats |
| 1293 | Ethyl alcohol |
| 1295 | Ethyl formate |
| 1317 | Garlic and its derivatives |
| 1318 | Glucono $\delta$-lactone |
| 1321 | Corn gluten |
| 1322 | Wheat gluten |
| 1328 | Glyceryl behenate |
| 1330 | Acacia (gum arabic) |
| 1333 | Gum ghatti |
| 1339 | Guar gum |
| 1343 | Locust (carob) bean gum |
| 1349 | Karaya gum (sterculia gum) |
| 1351 | Gum tragacanth |
| 1355 | Helium |
| 1366 | Hydrogen peroxide |
| 1370 | Inositol |
| 1372 | Insoluble glucose isomerase enzyme preparations |
| 1388 | Lactase enzyme preparation from *Kluyveromyces lactis* |

**Table 10.3.12. (continued)**

| 21 CFR 184 Section | Subject |
|---|---|
| 1400 | Lecithin |
| 1408 | Licorice and licorice derivatives |
| 1409 | Ground limestone |
| 1425 | Magnesium carbonate |
| 1426 | Magnesium chloride |
| 1428 | Magnesium hydroxide |
| 1431 | Magnesium oxide |
| 1434 | Magnesium phosphate |
| 1440 | Magnesium stearate |
| 1443 | Magnesium sulfate |
| 1444 | Maltodextrin |
| 1445 | Malt syrup (malt extract) |
| 1446 | Manganese chloride |
| 1449 | Manganese citrate |
| 1452 | Manganese gluconate |
| 1461 | Manganese sulfate |
| 1490 | Methyl paraben |
| 1530 | Niacin |
| 1535 | Niacinamide |
| 1537 | Nickel |
| 1540 | Nitrogen |
| 1545 | Nitrous oxide |
| 1553 | Peptones |
| 1555 | Rapeseed oil |
| 1560 | Ox bile extract |
| 1563 | Ozone |
| 1585 | Papain |
| 1588 | Pectins |
| 1610 | Potassium alginate |
| 1613 | Potassium bicarbonate |
| 1619 | Potassium carbonate |
| 1622 | Potassium chloride |
| 1631 | Potassium hydroxide |
| 1634 | Potassium iodide |
| 1635 | Potassium iodate |
| 1639 | Potassium lactate |
| 1643 | Potassium sulfate |
| 1655 | Propane |

| | |
|---|---|
| 1660 | Propyl gallate |
| 1666 | Propylene glycol |
| 1670 | Propyl paraben |
| 1676 | Pyridoxine hydrochloride |
| 1685 | Rennet (animal-derived) |
| 1695 | Riboflavin |
| 1697 | Riboflavin-5′-phosphate (sodium) |
| 1698 | Rue |
| 1699 | Oil of rue |
| 1721 | Sodium acetate |
| 1724 | Sodium alginate |
| 1733 | Sodium benzoate |
| 1736 | Sodium bicarbonate |
| 1742 | Sodium carbonate |
| 1754 | Sodium diacetate |
| 1763 | Sodium hydroxide |
| 1764 | Sodium hypophosphite |
| 1768 | Sodium lactate |
| 1769a | Sodium metasilicate |
| 1784 | Sodium propionate |
| 1792 | Sodium sesquicarbonate |
| 1801 | Sodium tartrate |
| 1804 | Sodium potassium tartrate |
| 1807 | Sodium thiosulfate |
| 1835 | Sorbitol |
| 1845 | Stannous chloride (anhydrous and dihydrated) |
| 1848 | Starter distillate |
| 1875 | Thiamine hydrochloride |
| 1878 | Thiamine mononitrate |
| 1890 | $\alpha$-Tocopherols |
| 1923 | Urea |
| 1930 | Vitamin A |
| 1945 | Vitamin $B_{12}$ |
| 1950 | Vitamin D |
| 1973 | Beeswax (yellow and white) |
| 1976 | *Candelilla* wax |
| 1978 | *Carnauba* wax |
| 1979 | Whey |
| 1979a | Reduced-lactose whey |
| 1979b | Reduced-minerals whey |
| 1933 | Bakers yeast extract |

**Table 10.3.12. (continued)**

| 21 CFR 184 Section | Subject |
|---|---|
| 1984 | Zein |

Source: 21 CFR 184, 1991.

**Table 10.3.13. Indirect Food Substances Affirmed as Generally Recognized as Safe**

| 21 CFR 186 Section | Indirect Food Additives |
|---|---|
| | General Provisions |
| 186.1 | Substances added indirectly to human food affirmed as generally recognized as safe (GRAS) |
| | Listing of Specific Substances Affirmed as GRAS |
| 1025 | Caprylic acid |
| 1093 | Sulfamic acid |
| 1256 | Clay (kaolin) |
| 1275 | Dextrans |
| 1316 | Formic acid |
| 1551 | Hydrogenated fish oil |
| 1557 | Tall oil |
| 1673 | Pulp |
| 1750 | Sodium chlorite |
| 1756 | Sodium formate |
| 1770 | Sodium oleate |
| 1771 | Sodium palmitate |
| 1797 | Sodium sulfate |
| 1839 | Sorbose |

Source: 21 CFR 186, 1991.

**Table 10.3.14. Substances Prohibited from Use in Human Food**

| 21 CFR 189 Section | Food Additives |
|---|---|
| | Substances Generally Prohibited From Direct Addition or Use as Human Food |
| 110 | *Calamus* and its derivatives |
| 113 | Cinnamyl anthranilate |
| 120 | Cobaltous salts and its derivatives |
| 130 | Coumarin |
| 135 | Cyclamate and its derivatives |
| 140 | Diethyl pyrocarbonate (DEPC) |
| 145 | Dulcin |
| 155 | Monochloroacetic acid |
| 165 | Nordihydroguaiaretic acid (NDGA) |
| 175 | P-4000 |
| 180 | Safrole |
| 190 | Thiourea |
| 191 | Chloroflurocarbon propellants |
| | Substances Prohibited From Indirect Addition to Human Food Through Food-Contact Surfaces |
| 220 | Flectol H |
| 250 | Mercaptoimidazoline and 2-mercaptoimidazoline |
| 280 | 4,4-Methylene bis(2-chloroanaline) |

Source: 21 CFR 189, 1991.

## 10.4. Food Standards

**Table 10.4.1. Standards Promulgated by the FDA**

| 21 CFR Part | Standards Promulgated |
|---|---|
| 131 | Milk and cream |
| 133 | Cheeses and related cheese products |
| 135 | Frozen desserts |
| 136 | Bakery products |
| 137 | Cereal flours and related products |
| 139 | Macaroni and noodle products |
| 145 | Canned fruits |
| 146 | Canned fruit juices |
| 150 | Fruit butters, jellies, preserves, and related products |
| 152 | Fruit pies |
| 155 | Canned vegetables |
| 156 | Vegetable juices |
| 158 | Frozen vegetables |
| 160 | Eggs and egg products |
| 161 | Fish and shellfish |
| 163 | Cacao products |
| 164 | Tree nut and peanut products |
| 165 | Nonalcoholic beverages |
| 166 | Margarine |
| 168 | Sweeteners and table sirups |
| 169 | Food dressings and flavorings |

Source: 21 CFR 131-169, 1991.

**Table 10.4.2. FDA Standards for Milk and Cream**

| 21 CFR 131 Section | Standards Promulgated |
|---|---|
| 110 | Milk |
| 111 | Acidified milk |
| 112 | Cultured milk |
| 115 | Concentrated milk |
| 120 | Sweetened condensed milk |
| 122 | Sweetened condensed skimmed milk |

| | |
|---|---|
| 123 | Low-fat dry milk |
| 125 | Nonfat dry milk |
| 127 | Nonfat dry milk fortified with vitamins A and D |
| 130 | Evaporated milk |
| 132 | Evaporated skimmed milk |
| 135 | Low-fat milk |
| 136 | Acidified low-fat milk |
| 138 | Cultured low-fat milk |
| 143 | Skim milk |
| 144 | Acidified skim milk |
| 146 | Cultured skim milk |
| 147 | Dry whole milk |
| 149 | Dry cream |
| 150 | Heavy cream |
| 155 | Light cream |
| 157 | Light whipping cream |
| 160 | Sour cream |
| 162 | Acidified sour cream |
| 170 | Eggnog |
| 180 | Half-and-half |
| 185 | Sour half-and-half |
| 187 | Acidified sour half-and-half |
| 200 | Yogurt |
| 203 | Low-fat yogurt |
| 206 | Nonfat yogurt |

Source: 21 CFR 131, 1991.

**Table 10.4.3. FDA Standards for Cheese and Related Products**

| 21 CFR 133 Section | Standards Promulgated |
|---|---|
| 102 | Asiago fresh and asiago soft cheese |
| 103 | Asiago medium cheese |
| 104 | Asiago old cheese |
| 106 | Blue cheese |
| 108 | Brick cheese |
| 109 | Brick cheese for manufacturing |
| 111 | Caciocavallo siciliano cheese |
| 113 | Cheddar cheese |
| 114 | Cheddar cheese for manufacturing |

**Table 10.4.3. (continued)**

| 21 CFR 133 Section | Standards Promulgated |
|---|---|
| 116 | Low-sodium cheddar cheese |
| 118 | Colby cheese |
| 119 | Colby cheese for manufacturing |
| 121 | Low-sodium colby cheese |
| 123 | Cold-pack and club cheese |
| 124 | Cold-pack cheese food |
| 125 | Cold-pack cheese food with fruits, vegetables, or meats |
| 127 | Cook cheese, koch kaese |
| 128 | Cottage cheese |
| 129 | Dry curd cottage cheese |
| 131 | Low-fat cottage cheese |
| 133 | Cream cheese |
| 134 | Cream cheese with other foods |
| 136 | Washed curd and soaked curd cheese |
| 137 | Washed curd cheese for manufacturing |
| 138 | Edam cheese |
| 140 | Gammelost cheese |
| 141 | Gorgonzola cheese |
| 142 | Gouda cheese |
| 144 | Granular and stirred curd cheese |
| 145 | Granular cheese for manufacturing |
| 146 | Grated cheeses |
| 147 | Grated American cheese food |
| 148 | Hard grating cheeses |
| 149 | Gruyere cheese |
| 150 | Hard cheeses |
| 152 | Limburger cheese |
| 153 | Monterey cheese and Monterey jack cheese |
| 154 | High-moisture jack cheese |
| 155 | Mozzarella cheese and scamorza cheese |
| 156 | Low-moisture mozzarella and scamorza cheese |
| 157 | Part-skim mozzarella and scamorza cheese |
| 158 | Low-moisture part-skim mozzarella and scamorza cheese |
| 160 | Muenster and munster cheese |
| 161 | Muenster and munster cheese for manufacturing |
| 162 | Neufchatel cheese |
| 164 | Nuworld cheese |
| 165 | Parmesan and reggiano cheese |

| | |
|---|---|
| 167 | Pasteurized blended cheese |
| 168 | Pasteurized blended cheese with fruits, vegetables, or meats |
| 169 | Pasteurized process cheese |
| 170 | Pasteurized process cheese with fruits, vegetables, or meats |
| 171 | Pasteurized process pimento cheese |
| 173 | Pasteurized process cheese food |
| 174 | Pasteurized process cheese food with fruits, vegetables, or meats |
| 175 | Pasteurized cheese spread |
| 176 | Pasteurized cheese spread with fruits, vegetables, or meats |
| 178 | Pasteurized neufchatel cheese spread with other foods |
| 179 | Pasteurized process cheese spread |
| 180 | Pasteurized process cheese spread with fruits, vegetables, or meats |
| 181 | Provolone cheese |
| 182 | Soft ripened cheeses |
| 183 | Romano cheese |
| 184 | Roquefort, sheep's milk blue-mold, and blue-mold cheese from sheep's milk |
| 185 | Samsoe cheese |
| 186 | Sap sago cheese |
| 187 | Semisoft cheeses |
| 188 | Semisoft part-skim cheeses |
| 189 | Skim milk cheese for manufacturing |
| 190 | Spiced cheeses |
| 191 | Part-skim spiced cheeses |
| 193 | Spiced, flavored standardized cheeses |
| 195 | Swiss and Emmentaler cheese |
| 196 | Swiss cheese for manufacturing |

Source: 21 CFR 133, 1991

**Table 10.4.4. FDA Standards for Frozen Desserts**

| 21 CFR 135 Section | Standards Promulgated |
|---|---|
| 110 | Ice cream and frozen custard. |
| 115 | Goat's milk ice cream |
| 120 | Ice milk |
| 125 | Goat's milk ice milk |
| 130 | Mellorine |

**Table 10.4.4. (continued)**

| 21 CFR 135 Section | Standards Promulgated |
|---|---|
| 140 | Sherbet |
| 160 | Water ices |

Source: 21 CFR 135, 1991.

**Table 10.4.5. FDA Standards for Bakery Products**

| 21 CFR 136 Section | Standards Promulgated |
|---|---|
| 110 | Bread, rolls, and buns |
| 115 | Enriched bread, rolls, and buns |
| 130 | Milk bread, rolls, and buns |
| 160 | Raisin bread, rolls, and buns |
| 180 | Whole wheat bread, rolls, and buns |

Source: 21 CFR 136, 1991.

**Table 10.4.6. FDA Standards for Cereal Flours and Related Products**

| 21 CFR 137 Section | Standards Promulgated |
|---|---|
| 105 | Flour |
| 155 | Bromated flour |
| 160 | Enriched bromated flour |
| 165 | Enriched flour |
| 170 | Instantized flours |
| 175 | Phosphated flour |
| 180 | Self-rising flour |
| 185 | Enriched self-rising flour |
| 190 | Cracked wheat |
| 195 | Crushed wheat |
| 200 | Whole wheat flour |
| 205 | Bromated whole wheat flour |
| 211 | White corn flour |
| 215 | Yellow corn flour |
| 220 | Durum flour |
| 225 | Whole durum flour |

| | |
|---|---|
| 230 | Corn grits |
| 235 | Enriched corn grits |
| 240 | Quick grits |
| 245 | Yellow grits |
| 250 | White corn meal |
| 255 | Bolted white corn meal |
| 260 | Enriched corn meals |
| 265 | Degerminated white corn meal |
| 270 | Self-rising white corn meal |
| 275 | Yellow corn meal |
| 280 | Bolted yellow corn meal |
| 285 | Degerminated yellow corn meal |
| 290 | Self-rising yellow corn meal |
| 300 | Farina |
| 305 | Enriched farina |
| 320 | Semolina |
| 350 | Enriched rice |

Source: 21 CFR 137, 1991.

**Table 10.4.7. FDA Standards for Macaroni and Noodle Products**

| 21 CFR 139 Section | Standards Promulgated |
|---|---|
| 110 | Macaroni products |
| 115 | Enriched macaroni products |
| 117 | Enriched macaroni products with fortified protein |
| 120 | Milk macaroni products |
| 121 | Nonfat milk macaroni products |
| 122 | Enriched nonfat milk macaroni products |
| 125 | Vegetable macaroni products |
| 135 | Enriched vegetable macaroni products |
| 138 | Whole wheat macaroni products |
| 140 | Wheat and soy macaroni products |
| 150 | Noodle products |
| 155 | Enriched noodle products |
| 160 | Vegetable noodle products |
| 165 | Enriched vegetable noodle products |
| 180 | Wheat and soy noodle products |

Source: 21 CFR 139, 1991.

**Table 10.4.8. FDA Standards for Canned Fruits**

| 21 CFR 145 Section | Standards Promulgated |
|---|---|
| 110 | Canned applesauce |
| 115 | Canned apricots |
| 116 | Artificially sweetened canned apricots |
| 120 | Berries |
| 125 | Canned cherries |
| 126 | Artificially sweetened canned cherries |
| 130 | Canned figs |
| 131 | Artificially sweetened canned figs |
| 134 | Canned preserved figs |
| 135 | Canned fruit cocktail |
| 136 | Artificially sweetened canned fruit cocktail |
| 140 | Canned seedless grapes |
| 145 | Canned grapefruit |
| 170 | Peaches |
| 171 | Artificially sweetened canned peaches |
| 175 | Canned pears |
| 176 | Artificially sweetened canned pears |
| 180 | Canned pineapple |
| 181 | Artificially sweetened canned pineapple |
| 185 | Canned plums |
| 190 | Canned prunes |

Source: 21 CFR 145, 1991.

**Table 10.4.9. FDA Standards for Canned Fruit Juices**

| 21 CFR 146 Section | Standards Promulgated |
|---|---|
| 113 | Canned fruit nectars |
| 114 | Lemon juice |
| 120 | Frozen concentrate for lemonade |
| 121 | Frozen concentrate for artificially sweetened lemonade |
| 126 | Frozen concentrate for colored lemonade |
| 132 | Grapefruit juice |
| 135 | Orange juice |
| 137 | Frozen orange juice |
| 140 | Pasteurized orange juice |

| | |
|---|---|
| 141 | Canned orange juice |
| 145 | Orange juice from concentrate |
| 146 | Frozen concentrated orange juice |
| 148 | Reduced-acid frozen concentrated orange juice |
| 150 | Canned concentrated orange juice |
| 151 | Orange juice for manufacturing |
| 152 | Orange juice with preservative |
| 153 | Concentrated orange juice for manufacturing |
| 154 | Concentrated orange juice with preservative |
| 185 | Pineapple juice |
| 187 | Canned prune juice |

Source: 21 CFR 146, 1991.

**Table 10.4.10. FDA Standards for Fruit Butters, Jellies, Preserves, and Related Products**

| 21 CFR 150 Section | Standards Promulgated |
|---|---|
| 110 | Fruit butter |
| 140 | Fruit jelly |
| 141 | Artificially sweetened fruit jelly |
| 160 | Fruit preserves and jams |
| 161 | Artificially sweetened fruit preserves and jams |

Source: 21 CFR 150, 1991.

**Table 10.4.11. FDA Standards for Vegetable Products**

| 21 CFR Section | Standards Promulgated |
|---|---|
| | Canned Vegetables |
| 155.130 | Canned green beans and canned wax beans |
| 155.131 | Canned field corn |
| 155.170 | Canned peas |
| 155.172 | Canned dry peas |
| 155.190 | Canned tomatoes |
| 155.191 | Tomato concentrates |
| 155.194 | Catsup |

**Table 10.4.11. (continued)**

| 21 CFR 133 Section | Standards Promulgated |
|---|---|
| 155.200 | Certain other canned vegetables |
| 155.210 | Canned mushrooms |
| | Canned Vegetable Juices |
| 156.145 | Canned tomato juice |
| | Frozen Vegetable |
| 158.170 | Frozen peas |

Source: 21 CFR 155, 156, 158 (1991).

**Table 10.4.12 FDA Standards for Eggs and Egg Products**

| 21 CFR 160 Section | Standards Promulgated |
|---|---|
| 100 | Eggs |
| 105 | Dried eggs |
| 110 | Frozen eggs |
| 115 | Liquid eggs |
| 140 | Egg whites |
| 145 | Dried egg whites |
| 150 | Frozen egg whites |
| 180 | Egg yolks |
| 185 | Dried egg yolks |
| 190 | Frozen egg yolks |

Source: 21 CFR 160, 1991.

**Table 10.4.13. FDA Standards for Fish and Shellfish**

| 21 CFR 161 Section | Standards Promulgated |
|---|---|
| 130 | Oysters |
| 131 | Extra large oysters |
| 132 | Large oysters |

| | |
|---|---|
| 133 | Medium oysters |
| 134 | Small oysters |
| 135 | Very small oysters |
| 136 | Olympia oysters |
| 137 | Large Pacific oysters |
| 138 | Medium Pacific oysters |
| 139 | Small Pacific oysters |
| 140 | Extra small Pacific oysters |
| 145 | Canned oysters |
| 170 | Canned Pacific slamon |
| 173 | Canned wet packed shrimp in transparent or nontransparent containers |
| 175 | Frozen raw breaded shrimp |
| 176 | Frozen raw lightly breaded shrimp |
| 190 | Canned tuna |

Source: 21 CFR 161, 1991.

**Table 10.4.14. FDA Standards for Cacao Products**

| 21 CFR 163 Section | Standards Promulgated |
|---|---|
| 110 | Cacao nibs |
| 111 | Chocolate liquor |
| 112 | Breakfast cocoa |
| 113 | Cocoa |
| 114 | Low-fat cocoa |
| 117 | Cocoa with dioctyl sodium sulfosuccinate for manufacturing |
| 123 | Sweet chocolate |
| 130 | Milk chocolate |
| 135 | Buttermilk chocolate |
| 140 | Skim milk chocolate |
| 145 | Mixed dairy product chocolates |
| 150 | Sweet cocoa and vegetable fat (other than cacao fat) coating |
| 153 | Sweet chocolate and vegetable fat (other than cacao fat) coating |
| 155 | Milk chocolate and vegetable fat (other than cacao fat) coating |

Source: 21 CFR 163, 1991.

**Table 10.4.15. FDA Standards for Tree Nut and Peanut Products**

| 21 CFR 164 Section | Standards Promulgated |
|---|---|
| 110 | Mixed nuts |
| 120 | Shelled nuts in rigid or semirigid containers |
| 150 | Peanut butter |

Source: 21 CFR 164, 1991.

**Table 10.4.16. FDA Standards for Sweeteners and Table Sirup**

| 21 CFR 168 Section | Standards Promulgated |
|---|---|
| 110 | Dextrose, anhydrous |
| 111 | Dextrose monohydrate |
| 120 | Glucose sirup |
| 121 | Dried glucose sirup |
| 122 | Lactose |
| 130 | Cane sirup |
| 140 | Maple sirup |
| 160 | Sorghum sirup |
| 180 | Table sirup |

Source: 21 CFR 168, 1991.

**Table 10.4.17. FDA Standards for Food Dressings and Flavorings**

| 21 CFR 169 Section | Standards Promulgated |
|---|---|
| 115 | French dressing |
| 140 | Mayonnaise |
| 150 | Salad dressing |
| 175 | Vanilla extract |
| 176 | Concentrated vanilla extract |
| 177 | Vanilla flavoring |
| 178 | Concentrated vanilla flavoring |
| 179 | Vanilla powder |
| 180 | Vanilla-vanillin extract |
| 181 | Vanilla-vanillin flavoring |

| | |
|---|---|
| 182 | Vanilla-vanillin powder |

Source: 21 CFR 169, 1991.

**Table 10.4.18. FDA Standards for Other Products**

| 21 CFR Section | Standards Promulgated |
|---|---|
| | Fruit Pies |
| 152.126 | Frozen cherry pie |
| | Nonalcoholic Beverages |
| 165.175 | Soda water |
| | Margarine |
| 166.110 | Margarine |

Source: 21 CFR 152, 165, 166 (1991).

# INDEX

The nature of the contents in this handbook requires an index that differs from the standard in one important aspect. A regular index lists routine terms that are considered significant. This index, in addition, has included all chemical names, food products, microorganisms, standards, definitions, and other similar categories that are already alphabetized in the text. This massive inclusion of terms accounts for a 100 page index. However, it is noted that the fish list is not indexed since Chapter 7 provides an exhaustive alphabetical listing of the scientific and market names of food fishes. To repeat them in this index will require an additional 50-70 pages.

B

F

I

J